INTERSTITIAL ALLOYS

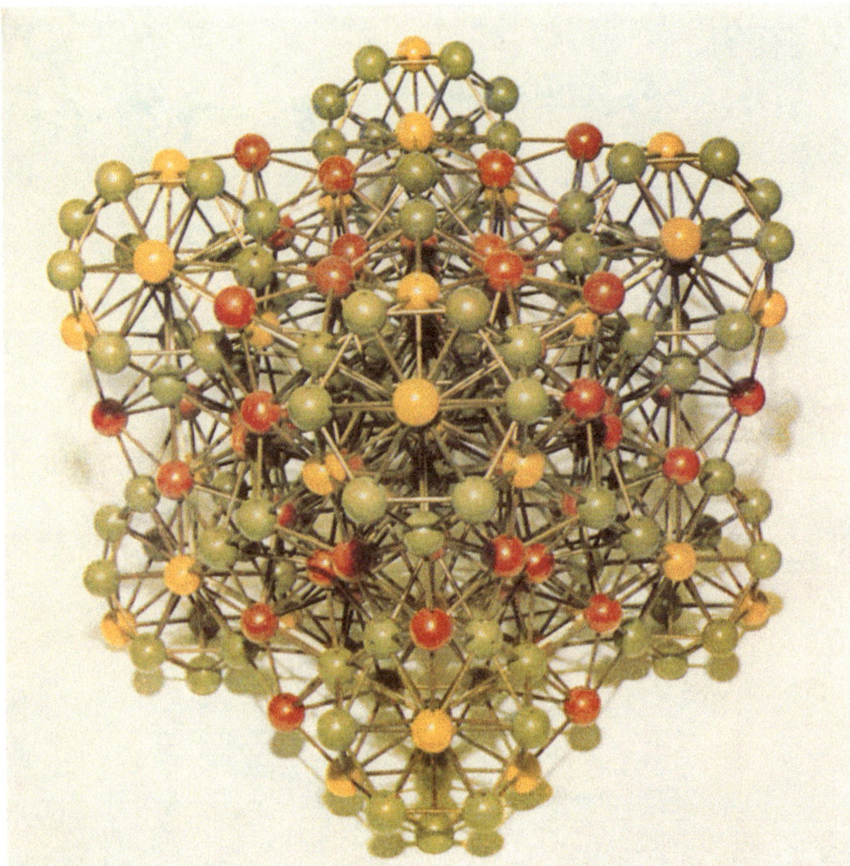

Frontispiece. Space model of the structure of the χ phase (α-Mn type)

INTERSTITIAL ALLOYS

H. J. GOLDSCHMIDT, D.Sc., F.Inst.P., F.I.M.
Head of Crystallographic Section,
The B.S.A. Group Research Centre,
Birmingham, England

Springer Science+Business Media, LLC

ISBN 978-1-4899-5882-2 ISBN 978-1-4899-5880-8 (eBook)
DOI 10.1007/978-1-4899-5880-8

Originally published by Butterworth & Co. (Publishers) Ltd. in 1967.

Softcover reprint of the hardcover 1st edition 1967

Suggested U.D.C. *No.* 669·018: 546·3
Library of Congress Catalog Card Number 67–31095

CONTENTS

PREFACE

There has been outstanding progress in the development and understanding of interstitial alloys during recent years. Interstitial alloys, especially those between transition metals and the smaller atoms of carbon, nitrogen, boron, silicon, oxygen and hydrogen, either taken singly or together, constitute a significant section of the field of modern metallurgy. Many of these advances took place empirically or in specific technological contexts. This book reviews the subject as a whole, combining a stock-taking of present knowledge with a systematic survey of the underlying principles and trends. This conspectus reveals unexpected parallels between behaviour patterns in seemingly remote fields such as alloy steels and refractory metals. An incidental result is to show up many gaps in our knowledge which still exist, and this should stimulate further research. A number of comprehensive diagrams and tables will facilitate the use of this book as a work of reference, and selected literature is appended to each Chapter.

Grateful acknowledgement is made to Professor G. K. T. Conn (University of Exeter) for his initial stimulation of this book and for kindly reading the first draft; to Professor R. E. Smallman (University of Birmingham) also for reading the manuscript, and to Professor A. D. McQuillan (at the same University) for doing so on Chapter 9 (Hydrides).

My cordial thanks are due to Mr. D. A. Oliver, C.B.E., Director of Research, the B.S.A. Group Research Centre, for his constant encouragement and support. The writer's colleagues, Dr. G. D. McAdam and Mr. E. A. Catherall, have made valued comments on some of the Chapters.

I wish to express my special gratitude to the staff of Messrs. Butterworths for their great help with all aspects of the book and also to Miss D. Burton who has been most helpful in typing the manuscript. Finally, the book could not have been completed without my wife's patient forbearance, untiring help and constant encouragement over the years.

H.J.G.

1

INTRODUCTION

Interstitial alloys have played their part in man's history ever since, about 4,000 years ago, he first chanced to intermingle small carbon atoms with large iron atoms, and thus to forge both sword and ploughshare (and so his destiny). However, only in recent decades has the scope of interstitial alloys multiplied so greatly and in so many novel directions, because only now has the real demand arisen in science and industry. Moreover, most of the transition metals upon which interstitial alloys depend, have only recently become more readily available. It would seem appropriate, therefore, in this book to try and take stock, to see how 'iron-age chance' has developed into the scientific design of today, and to review some of the main recent trends.

A few examples may give some idea of the variety of ways in which interstitial atoms take part in alloys and in related processes:

(1) Interstitial alloys are clearly the *steel industry's* chief product, both by intention, for hardening and carbide formation, and involuntarily, as impurity which must be minimized.

(2) The refractory *hard-metals* used in carbide tools and the like are substantially interstitial alloys.

(3) Solid state *diffusion* and *oxidation* processes depend on the, at least transitory, presence of interstitial atoms during flow.

(4) *Surface-hardening* processes such as nitriding involve their (generally unstable but persistent) entry into the crystal lattice.

(5) The occurrence of interstitial atoms in *microsegregates*, e.g. at dislocations, grain boundaries or in inclusions, frequently determines mechanical strength or fracture of alloys, for use in normal, and in high- and low-temperature service.

(6) 'Artificial interstitial' alloys can be created by *nuclear irradiation* of metals, and many new implications arise from this.

(7) The problem of *heat-resisting materials* able to stand higher and higher temperatures, e.g. in gas turbines, rocket components or furnace elements, requires exceptional refractory properties which only certain interstitial compounds can provide.

Thus it may well be said that, whilst the bond of carbon atoms in iron and its variability in different structures of the metal launched man on his early progress, the greater bonding strength of interstitials in other transition metals may yet launch him from his planet! It has indeed already helped in doing so.

A curious antithesis of research aims arises here in recent metallurgical development: (*a*) to *introduce* interstitial atoms into metals, and (*b*) to *eliminate* them, on the one hand, for the enhancement of creep, hardenability and refractory properties, on the other, upon the demand for super-high-purity alloys (to less than 1 impurity atom in 10^7 if possible) made in many new fields of science; for only thus can, in many cases, the source points for cracks and mechanical weakness be avoided. It is noteworthy that both these opposite routes have, to a large extent, a common goal, namely to promote high strength.

Meaning of 'Interstitial Alloy'

To define an interstitial alloy presents a logical difficulty. The term was first used in Hägg's[1] classical work on the structure of transition metal carbides, nitrides, hydrides and borides, and although it has the merit of recognizing the crystal lattice as the basic concept, it would imply the prior existence of a pure metal lattice, acting as host to further atoms (of smaller size) which fill the 'interstices', but capable as such of existing without them. However, in some of the most important cases this does not apply. In all carbides and other interstitial compounds of the transition metals, the non-metal atom forms an integral part, and without it the metal lattice would differ entirely. Only primary solid solutions are in this sense truly 'interstitial'.

A wider definition must lay emphasis not merely on the geometry but also on the manner of bonding. We may refer to an interstitial alloy as one in which the metal–metal atom bond remains the dominant one, and the non-metal atoms are sufficiently small to be accommodated within the metal lattice without, or with only a limited degree of, distortion from metal-type symmetry. The proviso of a 'flexibility condition', allowing for *some* distortion, is important to cover many cases of interstitial alloys*.

As a matter of principle, and from the aspect of certain applications, it might indeed be preferred formally to consider all metals as *a priori* interstitial, and the pure metal only as the special case of eventual lattice depletion. This is by no means trivial but, as will become clear later, is a helpful concept, since the vacant interstices themselves remain significant as atom-equivalent 'holes' or favoured places for electron or atom occupation without much change in lattice energy: in a sense, the vacancies remain 'furnished' for occupation by interstitial atoms, and they can diffuse and otherwise behave as though they were real atoms.

An essential feature of interstitial alloys are their metallic properties (conductivity, opacity, etc.) implied by the predominant metal–metal bond. Hägg[1] put the definition more restrictively by referring to the insertion of 'H, B, C or N atoms within simple metal lattices', which comprises only interstitial compounds of sufficiently small non-metal–metal atom radius ratio. In recent years, however, interstitial alloys have much outgrown this definition: oxygen, silicon and other elements in many cases also qualify as interstitials.

It is useful to take for a moment a synoptic view of crystal structures amongst all elemental metals. It is known that, in any arrangement of contacting equal spheres, the maximum numbers of nearest neighbours to any one (co-ordination number) is 12, and that this represents the densest

* Indeed, it seems to emerge more and more generally in modern science that, the more it advances, the less clear-cut any preconceived definitions become, and the greater the need for a flexibility clause! A kind of philosophical 'uncertainty principle'.

occupation of space. If now we survey the factually occurring types of crystal structure of metallic elements, the following statistics can be drawn up.

Of 109 pure metal phases known (shared by 75 metals and including different polymorphic forms to many of these), 55 (i.e. 50 per cent) have a co-ordination number of 12 or about 12, the rest one of 8 or less. This is to say that there is a strong tendency for the atoms to be close packed, as in the face-centred cubic and hexagonal close-packed structures, which represent the great majority of the 50 per cent, viz. FCC: 23 (= 21 per cent), HCP: 28 (= 26 per cent); and any deviation, notably the body-centred cube, is due to the superposition of other factors. A more detailed analysis is shown in *Figure 1.1.* However, whilst of the narrower group of 18 elemental transition metal forms of Groups IV, V, VI* in the Periodic Table only 4 (22 per

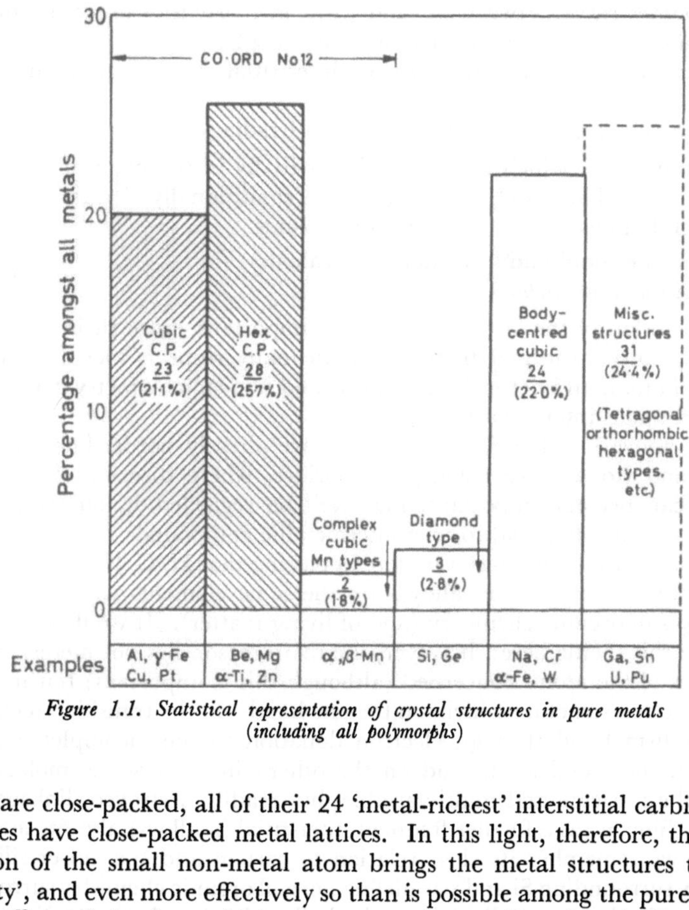

Figure 1.1. Statistical representation of crystal structures in pure metals (including all polymorphs)

cent) are close-packed, all of their 24 'metal-richest' interstitial carbides and nitrides have close-packed metal lattices. In this light, therefore, the introduction of the small non-metal atom brings the metal structures to 'conformity', and even more effectively so than is possible among the pure metals. This effect of the interstitial atoms must be attributed basically to their acting as electron donors to the metal lattice, as will be discussed later. We may regard this as amounting, electronically, to an increase in the atomic

* Those which form stable interstitial compounds, as discussed later.

number of the metal. The analogy comes to mind of the small atom acting as a medium to 'straighten out' the degenerated surrounding metal framework, by a kind of gentle inflation, the interstitial atom acting not so much through its mere volume as through its influence in filling the electron orbitals of the host metal atom.

Classification

We may subdivide interstitial alloys into the following classes (which will also serve as a guide in the rest of this book):

(*a*) interstitial *solid solutions*, e.g. C in austenite; O in Ti, Zr;

(*b*) interstitial *compounds*, e.g. steel carbides; WC, TiC, Fe_4N, etc., and transition metal hydrides, carbides, nitrides, borides generally; certain oxides and silicides;

(*c*) *transition stages* between (*a*) and (*b*), e.g. martensite in steels; metal-rich 'semi-carbides'; certain nitriding constituents;

(*d*) *transiently interstitial conditions* or 'interstitial flow', e.g. in diffusion, oxidation, etc.;

(*e*) *artificially created* interstitials: through irradiation.

The atom *species* which may occur interstitially in the metal lattices are:

(1) Non-metal atoms H, C, N, O, B, Si; (conditionally) P and S;

(2) metal atoms of their own or foreign kinds.

To these we should add (as atom-equivalents)

(3) *vacancies* and *electrons*.

The inclusion of the latter in this list is important, since their behaviour and compliance to formal treatment is in many respects like that of any interstitial atom, and defect lattice effects and non-stoichiometry can rightly be classed under interstitial phenomena.

In the above list, the Si and B atoms are already border-line, and the silicides and (to a lesser extent) the borides, where direct Si–Si or B–B linkages can become important, may well be regarded in our system as lead-overs from interstitial to true intermetallic compounds.

It seems of profound interest to note in passing that H, C, N and O, which are our chief interstitials in alloys, are the same elements which form the whole basis of organic chemistry (and of living matter). It would go too far to pursue this theme here; however, the main reason lies not merely in the small sizes of the atoms concerned (although this is important) but in their special electron structure which, on the one hand—in the transition metals—creates a firm bond through electron donation to the incompleted inner orbitals of the metal atom, and on the other—in the organic molecule—permits the intricate and multi-variable, but readily destroyed, linkages we know of. Silicon (as a further frequent interstitial in alloys) also fits into the theme if we remember its close relation to carbon and the newer silicone chemistry surrounding Si and paralleling normal organic chemistry surrounding C; so do the (borderline) interstitials P and S in relation to N and O.

Techniques of Detecting Interstitial Atoms

Methods of locating interstitial atoms have, until perhaps 15 years ago, been largely those of x-ray crystallography. However, in view of the low scattering power of the light atoms relative to those of the metal, it is usually not possible to locate interstitials from diffraction line intensity*, and mostly the position is derived from geometrical considerations involving metal lattice dimensions and known atomic sizes. The small atoms can reveal their presence also *indirectly* (1) through straining the host lattice, causing line broadening; (2) by distortion of the unit cell, e.g. from cubic to tetragonal symmetry; or, indeed, (3) by an entire change in metal structure. However, where x-ray diffraction fails to 'see' the interstitial atom, *neutron* diffraction (see e.g. Bacon[3]) succeeds, since neutrons are not scattered by electrons but through the magnetic moment of the nucleus; thus light atoms like carbon or hydrogen exert a relatively large influence on diffraction intensity. An early example[4] is that of NaH, where the hydrogen positions could be located by neutron diffraction as those of Cl in NaCl found by x-ray diffraction. Many more cases will arise, particularly in discussing the Hydrides (Chapter 9). (Different isotopes of the light atom are often used in conjunction, and for example the advantage of using deuterides rather than hydrides for structural investigations lies in the fact that the strong incoherent neutron scattering of hydrogen is greatly reduced in deuterium.) Another powerful field for neutron diffraction, as far as interstitial alloys are concerned, lies in the determination of magnetic structure, such as in the anti-ferromagnetism of MnO[5] and of other oxides. It proved particularly useful in the ferrites (see Chapter 8), e.g. for the siting of metal ions of different valency in spinels (Fe_3O_4, etc.).

Amongst newer physical techniques of studying interstitial alloy solid solutions, *internal friction* measurement has been very successful; briefly, it depends on the fact that the damping capacity of an alloy under externally imposed vibrations varies with the sites of the interstitial atom within the metal lattice and can give an abnormal rise and fall at certain temperatures, when the time of vibration is equal to the mean time between jumps from one interstitial site to another. The position of this 'internal friction peak' is characteristic of the type of interstitial element in the metal solution and the height and width of the peak of the interstitial amount and its distribution (for more detail, see Chapter 3).

Electrical resistivity measurement has in some cases also been found a useful, though limited, tool to study interstitial solubilities and signs of phase change, by observing temperature or heat-treatment dependence. Also, by measuring the resistivity at very *low* temperatures (near absolute zero) and its temperature dependence, a sensitive indication of interstitial lattice impurity content can be obtained; however, at present this does not identify the impurity and only shows its physical presence (and that of vacancies); the analysis would have to be a further step.

Orthodox *chemical analysis* and *metallography* have, of course, their well established place in the identification of interstitial compounds (for example, of inclusions and carbides in steels), whilst perhaps the most outstanding recent advances have been obtained by the *electron microscope* and associated techniques[6–10].

* Except in some favourable cases by extremely sensitive measurement: a notable example was Lipson and Petch's[2a] location of the carbon atom in cementite by three-dimensional Fourier analysis, and Petch's[2b] of carbon in austenite and martensite.

Physically, one of the frequent features of interstitial compounds is their dispersed nature in metallic matrices, which has presented the need for special extraction techniques including, more recently, that of extraction replicas in which the topology of detached particles can be maintained as in the matrix, and that of preferential etching.

This characteristic fine-particle dispersion of interstitial compounds has led to a new field of application, namely that of 'dispersed phase strengthening', in which finely divided hard particles are distributed within the metallic mass, and their insolubility and unreactivity are utilized to hinder slip and dislocation movement and to improve mechanical properties at elevated temperatures. *Figure 1.2* shows some typical examples of how interstitial phases can reveal themselves by such different techniques.

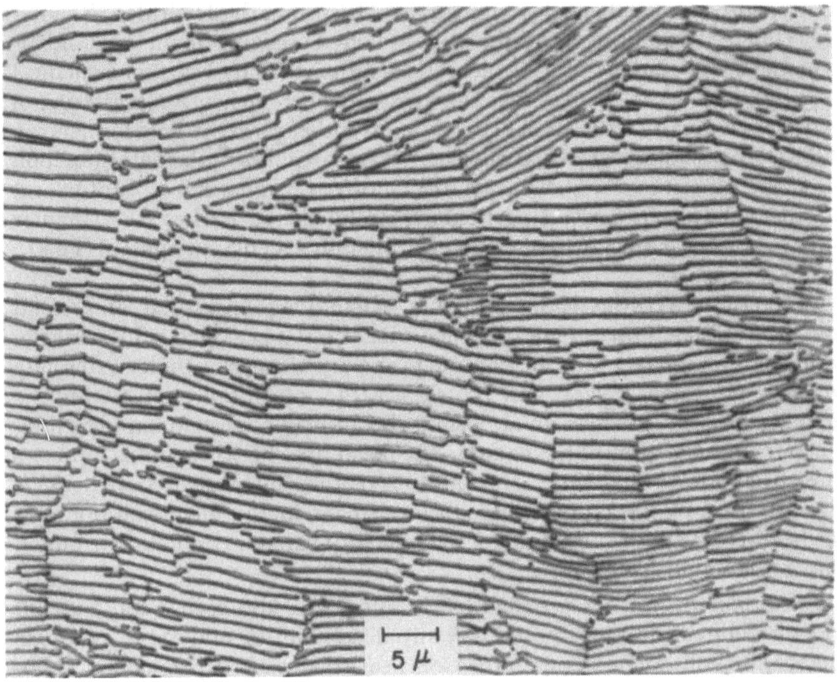

Figure 1.2. Examples of interstitial phases as shown by various techniques: (a) Photomicrograph of a carbide: cementite lamellae in α-iron (pearlite), × 1,700. (From Honeycombe[6a] and Darken and Fisher[8], by courtesy of Dr. R. M. Fisher)

By courtesy of the Iron and Steel Institute, London; Interscience Publishers, New York; and the Editor, *Acta Metallurgica*

Analysis by the x-ray and electron *microprobe* scanning techniques recently developed is a major new advance, in that segregates and inclusions can be analysed within areas down to $< 1\mu$ in diameter. As far as interstitials are concerned, the present drawback is still that lower atomic number elements like C, N, O, B are yet hard to detect by current techniques; but much active development has gone on in this direction, and advances such as, say, plotting a 'carbon-contour map' within microscopic areas of a steel, now possible already for the heavier elements, may be anticipated.

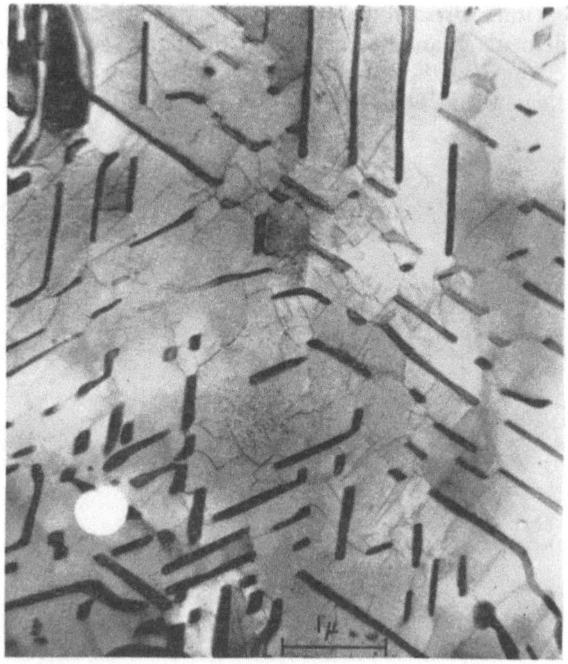

Figure 1.2b Electron micrographs: thin-film technique[6a] (I) Cementite lamellae in α-iron grown from γ-solution; coherent orientation on lattice planes; $\times$ 17,000

Photograph by courtesy of Dr. R. M. Fisher

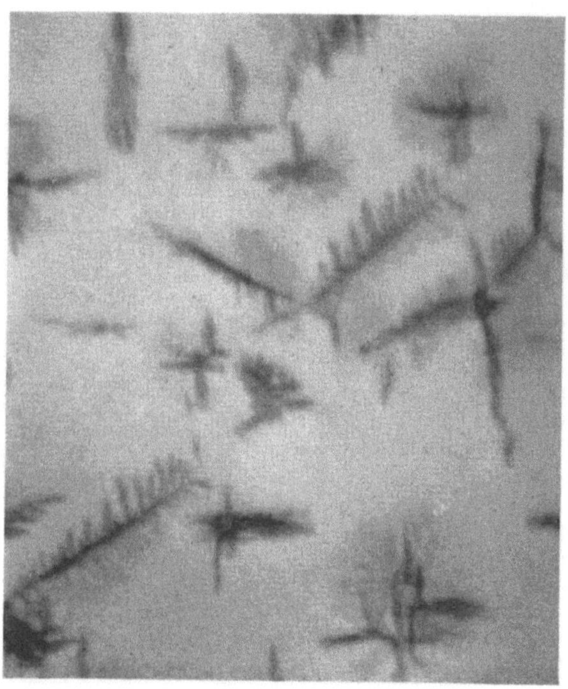

(II) Hexagonal ε-iron carbide precipitating in a Fe–C alloy, quenched 690°C and aged for 18 months at room temperature; $\times$ 33,000 (Hale and McLean[10])

Photograph by courtesy of Dr. D. McLean; *Crown Copyright reserved*

Other physical techniques to locate and follow the movement of interstitial elements in an alloy are those of using *radioactive tracers* and *autoradiography*.

Amongst the numerous new techniques that have become important in their special fields of solid-state physics, there are several which also have,

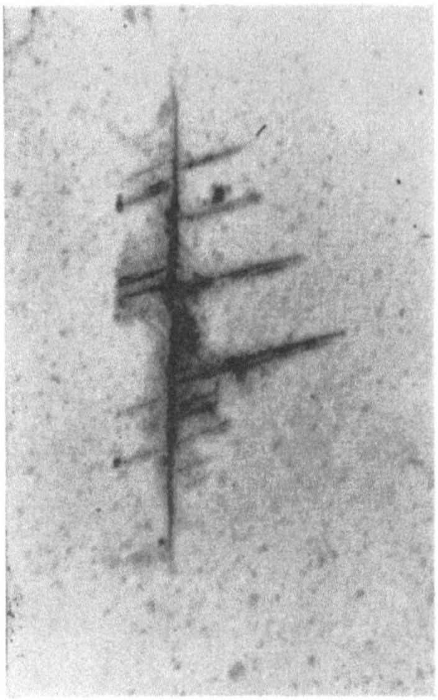

(I)

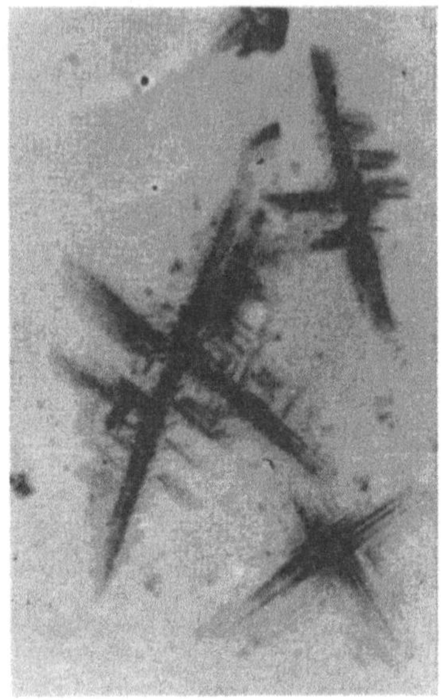

(II)

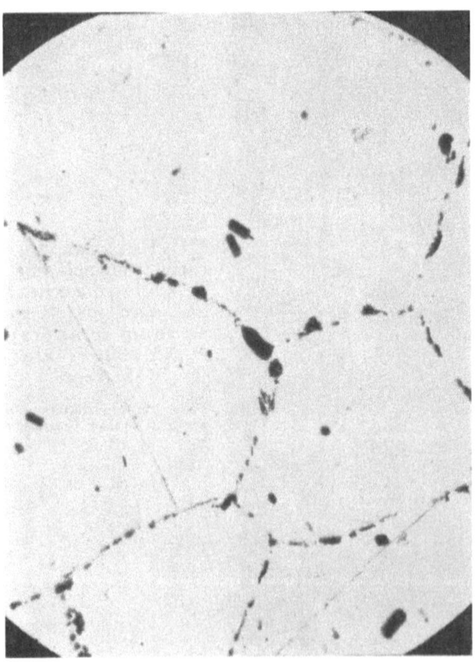

Figure 1.2c Electron micrographs: extraction replica technique. Extracted cementite in a low plain carbon steel, quenched 740°C, then aged at (I) 170 h at 150°C, (II) 3 min at 260°C. Fe_3C oriented along ferrite lattice planes; × 20,000, (III) $M_{23}C_6$ particles in a Ni–Cr-based alloy, precipitated at grain boundaries; × 6,000

From Honeycombe[6c] and Leslie *et al*[9] (III) (I and II), Nutting[6b] (III); photographs by courtesy of Prof. J. Nutting and Mr. Dewey

Figure 1.2c (*iii*)

more specifically, thrown light on the behaviour of interstitials. Any physical property which measures the amount and determines the nature and location of interstitials (e.g. random or ordered, whether displaced from normal sites, static or transient in position, their mobility, bond character, vibrational state, etc.) will thus be relevant. Examples, besides those already mentioned, are nuclear and electron magnetic resonance, magnetic susceptibility, the Hall effect and thermal EMF, as well as optical and infra-red, etc. absorption and photoconductivity. Similarly, behaviour trends (e.g. discontinuities)

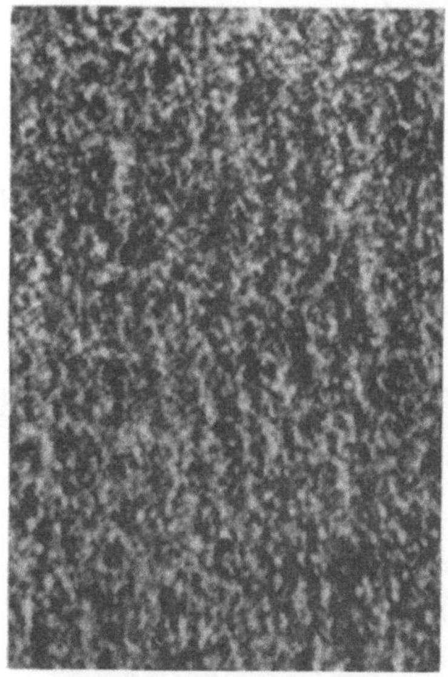

Figure 1.2d X-Ray microradiograph using a thin foil. η-carbide in a high-speed steel matrix; cobalt radiation: carbide white owing to heavier absorption; × 80

in mechanical properties, such as elastic constants, tensile strength, creep, etc. can often be regarded at least as 'indicators' manifesting changes in location and mobility in interstitials, holes and electrons.

Thermodynamic properties such as heat capacity, specific heats, heat of formation, dissociation pressures in dependence of temperature, time and composition clearly are similar useful indicators. The work of Richardson and his school[11] on the stability variations of interstitial phases with temperature is here important, and many valuable thermal data are now available (e.g. [12]), though still with wide gaps.

Again, the determination of *gases in metals* largely, though not exclusively, amounts to the determination of interstitial contents, be it in solution or as compounds (e.g. oxides). A considerable literature on these methods, such as vacuum fusion, exists; for reviews, see e.g. [13–18].

Applications

As indicated above, the areas of modern metallurgy in which interstitial alloys find practical application (or create practical difficulties) are numerous

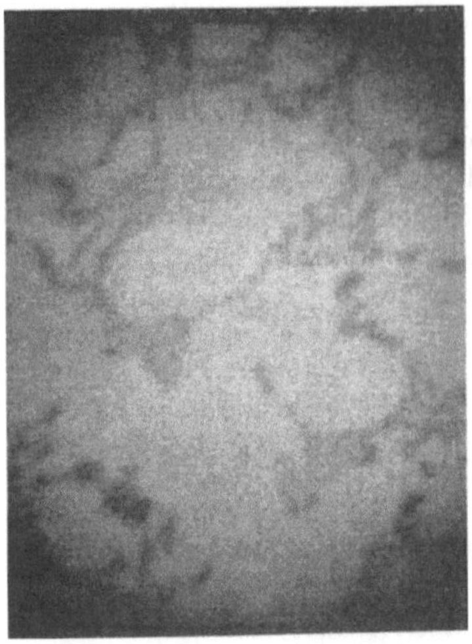

(I)

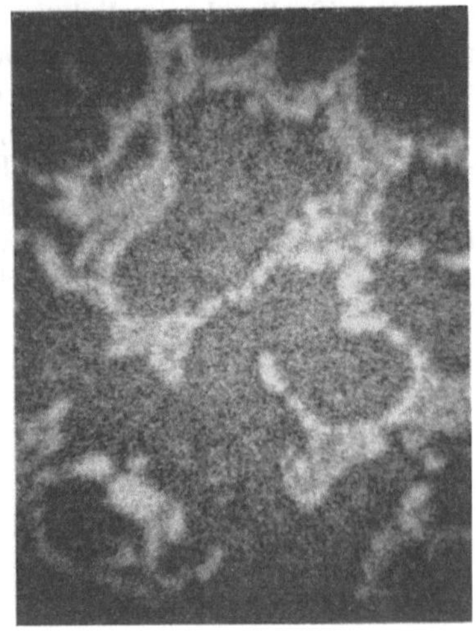

(II)

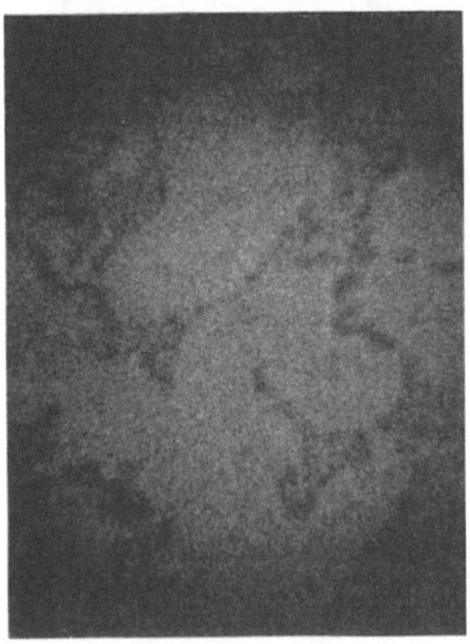

(III)

Figure 1.2e Microprobe analysis photographs of carbide segregates in a chromium steel showing (same field) (I) electron image; (II) x-ray CrK_α image; (III) x-ray FeK_α image: chromium enrichment within carbide (carbon not detectable at this stage) × 1,400

Photographs by Mr. E. A. Catherall, in author's laboratory

Figure 1.2f X-ray diffraction photographs of a high-speed steel containing (I) ferrite and η-carbide; (II) the extracted pure carbide

Figure 1.2f (ii)

and diverse. Physically, however, they stem, in the main, all from a central theme. These aspects will be dealt with in later Chapters.

SELECTED LITERATURE

X-ray Crystallography

Bragg, W. H. and W. L. *The Crystalline State I*; James, R. W. *The Crystalline State II: The optical principles of the diffraction of X-rays*, London (Bell) 1962 and 1963

Buerger, M. J. *Elementary Crystallography*, New York (Wiley) 1963

Lonsdale, K. *Crystals and X-rays*, London (Bell) 1948

Philips, F. C. *An Introduction to Crystallography*, London (Longmans) 1963

Bunn, C. W. *Chemical Crystallography*, Oxford University Press, 1961; *Crystals: Their Role in Nature and in Science*, New York and London (Academic Press) 1964

Azaroff, L. V. and Buerger, M. J. *The Powder Method in X-ray Crystallography*, New York (McGraw-Hill) 1958

D'Eye, R. W. M. and Wait, E. *X-ray Powder Photography*, London (Butterworths) 1960

Henry, N. F. M., Lipson, H. and Wooster, W. A. *The Interpretation of X-ray Diffraction Photographs*, London (Macmillan) 1960

Klug, H. P. and Alexander, L. E. *X-ray Diffraction Procedures for Polycrystalline and Amorphous Materials*, New York (Wiley) and London (Chapman & Hall) 1954

X-ray Diffraction by Polycrystalline Materials (Ed. Peiser, H. S., Rooksby, H. P. and Wilson, A. J. C.) London (Institute of Physics) 1960.

Barrett, C. S. and Massalski, T. B. *Structure of Metals*, New York (McGraw-Hill) 1966

Glocker, R. R. *Materialprüfung mit Röntgenstrahlung*, Berlin (Springer) 1958

Guinier, A. *X-ray Crystallographic Technology*, London (Hilger & Watts) 1952

Guinier, A. and Dexter, D. L. *X-ray Studies of Materials*, New York (Interscience) 1963

Taylor, A. *X-ray Metallography*, New York (Wiley) 1961

Advances in X-ray Analysis (Ed. Mueller, W. M.), New York (Plenum Press)
Annual Series of Denver Conference Proceedings from 1952 onwards

Cullity, B. D. *Elements of X-ray Diffraction*, Reading, Mass. (Addison-Wesley), 1959

Mirkin, L. I. *Handbook of X-ray Structure Analysis of Polycrystalline Materials*, New York (Plenum Press) 1964

Neutron Diffraction

Bacon, G. E.[3]; *Applications of Neutron Diffraction in Chemistry*, London (Pergamon) 1963; *Research* 7 (1954) 312

Electron Microscopy and Diffraction

Techniques of Electron Microscopy, Diffraction and Microprobe Analysis, STP 372, Philadelphia (American Society for Testing and Materials) 1965

Cosslett, V. E. *Practical Electron Microscopy*, London (Butterworths) 1951

Fisher, R. B. *Applied Electron Microscopy*, Indiana University Press, 1954

Haine, M. E. *The Electron Microscope. The Present State of the Art*, London (Spon) 1961

Hirsch, P. B. *et al.*, *Electron Microscopy of Thin Crystals*, London (Butterworths) 1965

Wyckoff, R. W. G. *The World of the Electron Microscope*, Yale University Press, 1958

Field Emission Microscopy

Brandon, D. G. *Modern Techniques in Metallography*, London (Butterworths) 1966

Muller, E. W. *J. phys. Soc. Japan* 18 Suppl. 11 (1963) 1

Cottrell, A. H. *J. Inst. Metals* 90 (1962) 449
May Lecture on 'Metallography at the Atomic Level'

Gomer, R. *Field Emission and Field Ionization*, Harvard University Press, 1961

Ehrlich, G. *Discovery* 26 (11) (1965) 25

Optical Microscopy (Numerous metallurgical texts, e.g.)

Desch, C. H. *Metallography*, London (Longmans) 1937

Nutting, J. and Baker, R. G. *The Microstructure of Metals*, London (Institute of Metals) 1966

Greaves, R. H. and Wrightson, H. *Practical Microscopical Metallography*, London (Chapman & Hall) 1957

Hartshorne, N. H. and Stuart, A. *Practical Optical Crystallography*, Amsterdam (Elsevier) 1964

Internal Friction (cf. Chapter 3)
Entwistle, K. M. *Metall. Rev.* 7 (1962) 175 [Review]

Microprobe Analysis
Symposium on X-ray and Electron Probe Analysis, STP 349, Philadelphia (American Society for Testing and Materials) 1964
Theisen, R. *Quantitative Electron Microprobe Analysis,* Berlin (Springer) 1965
Birks, L. S. *Electron Probe Microanalysis,* New York (Interscience) 1963
X-ray Optics and X-ray Microanalysis (Ed. Pattee, H. H., Cosslett, V. E. and Engstrom, A.) New York and London (Academic Press) 1963 [Stanford Symposium Report]

Nuclear and Electron Magnetic Resonance
Andrew, E. R. *Nuclear Magnetic Resonance,* Cambridge University Press, 1958
Electronic Magnetic Resonance and Solid Dielectrics (Ed. Servant, R. and Charru, A.), [12th Ampère Colloquium] Amsterdam (North Holland) and New York (Interscience) 1964

Radioactive Tracer Techniques and Autoradiography
Leymonie, C. *Radioactive Tracers in Physical Metallurgy,* London (Chapman & Hall) 1963
Cahn, R. W. *Trans. Inst. Metal Finish.* 35 (1958) 1
Review article on application of radioactive tracers to surface studies

Thermodynamic Techniques
Kubaschewski, O. and Catterall, J. A. *Thermochemical Data of Alloys,* London (Pergamon) 1956
Kubaschewski, O. and Evans, E. L. *Metallurgical Thermochemistry,* London (Pergamon) 1958
(and many standard thermodynamic texts)
Experimental Thermochemistry (Ed. Rossini, F. D.), New York (Interscience) 1956
Experimental Thermochemistry (2 Vols.) (Ed. Skinner, H. A.), New York (Interscience) 1956, 1962
Hume-Rothery, W., Christian, J. W. and Pearson, W. B. *Metallurgical Equilibrium Diagrams* London (Institute of Physics) 1951

Phase Extraction Methods
Andrews, K. W. and Hughes, H. *Iron Steel Lond.* 31 (1958) 43
Beeghly, H. F. *Analyt. Chem.* 21 (1949) 1513
Booker, G. R. and Norbury, J. *Brit. J. appl. Phys.* 8 (1957) 109, 158; 9 (1958) 361; *J. scient. Instrum.* 36 (1959) 368
Especially for small-sample amounts from surfaces, e.g. for x-ray examination
Klinger, P. and Koch, W. *Beiträge zur Metallkundlichen Analyse,* Düsseldorf (Stahleisen) 1949
The authors are perhaps the best known in this field for their development work
Pemberton, R. *Analyst, Lond.* 77 (1952) 285
Schrader, A. *Arch. EisenhüttWes.* 29 (1958) 793
Seal, A. K. and Honeycombe, R. W. K. *J. Iron Steel Inst.* 188 (1958) 9
Smith, E. and Nutting, J. *Brit. J. appl. Phys.* 7 (1956) 214; *J. Iron Steel Inst.* 187 (1957) 314
The last three papers are typical of methods of extraction *replica* preparation for electron microscopy
Hughes, H. *J. Iron Steel Inst.* 204 (1966) 804

REFERENCES

[1] Hägg, G. *Z. phys. Chem.* (B) 6 (1930) 221; 12 (1931) 33
[2a] Lipson, H. and Petch, N. J. *J. Iron Steel Inst.* II (1940) 95; [b] Petch, N. J. 145 (1942) 111
[3] Bacon, G. E. *Neutron Diffraction,* Oxford University Press, 1962
[4] Shull, C. G., Morton, G. A. and Davidson, W. L. *Phys. Rev.* 73 (1948) 842
[5] Shull, C. G., Strauser, W. A. and Wollan, E. O. *Phys. Rev.* 83 (1951) 333
[6a] Honeycombe, R. W. K. In *Metallography 1963, Spec. Rep. Iron Steel Inst.* 80 (1964) 245; [b] Nutting, J., *ibid.* 154
[7] Smallman, R. L. *Modern Metallography,* London (Pergamon) 1966
[8] Darken, L. S. and Fisher, R. M. In *Decomposition of Austenite by Diffusional Processes* (Ed. Zackay, V. F. and Aaronson, H. I.), New York (Interscience) 1962
[9] Leslie, W. C., Fisher, R. M. and Sen, H. *Acta metall.* 7 (1959) 632
[10] Hale, K. F. and McLean, D. *J. Iron Steel Inst.* 201 (1963) 337

REFERENCES

[11] Richardson, F. D. and Jeffes, J. H. E. *J. Iron Steel Inst.* 160 (1948) 261; Richardson, F. D. 175 (1953) 33

[12] Kubaschewski, O. and Evans, E. L. *Metallurgical Thermochemistry*, London (Pergamon) 1958

[13] Smithels, C. J. *Gases in Metals: An Introduction to the study of gas–metal equilibria*, London (Chapman & Hall) 1937; *J. Iron Steel Inst.* 198 (1961) 124

[14] Cupp, C. R. In *Progress in Metal Physics IV*, London (Pergamon), p. 105, 1953

[15] James, J. A. *Metall. Rev.* 9 (1964) 33, 93

[16] Promisel, N. E. In *The Science and Technology of W, Ta, Mo, Nb and their Alloys* (AGARD Cont. Proc.), London (Pergamon) 1964, p. 377

[17] Smith, D. P. *et al.*, *Gases in Metals*, Cleveland (American Society of Metals) 1953

[18] Turovtseva, Z. M. and Kunin, L. L. *Analysis of Gases in Metals*, New York (Consultants Bureau) 1961

2

METAL PHYSICS OF INTERSTITIAL ALLOYS

2.1 ATOMIC SIZES

Favourable atomic size ratios are a necessary, though not a sufficient, condition for the formation of interstitial alloys. In *Figure 2.1* we have represented the atomic diameter for all elements in the manner of concentric circles*; the periods of the Periodic Table are allotted different sectors. This shows the relative dimensions as well as the variations with atomic number, group and period. For our purpose, the diagram indicates the elemental combinations favouring the formation of interstitial alloys. As mentioned already, the radius ratio, r_x/r_m, of non-metal and metal atom played the basic part in Hägg's classical survey of interstitial compounds. Below a critical value (about 0·59), very simple structures arise in which the non-metal atoms may be considered as 'inserted' within the metal atom lattice. The latter itself is dominant and belongs to one of the structure types well known amongst the metals themselves—the cubic or hexagonal close-packed (with deviations in stacking sequence) and body-centred cubic. The co-ordination number of the metal atoms is 12 or 8 as in the pure metals. Above 0·59, more complex structures form. Hägg's rule obviously arises from *Figure 2.1*, where the potential interstitial atoms are indicated by shading. The atomic radii are C 0·77, N 0·71, B 0·91, Si 1·17, H 0·46 Å—figures, however, subject to strong reservations and variability, as will be discussed later. In general, the effective radius of the interstitial atom depends on the metal surround and increases slightly with the higher interstitial contents.

It is an essential feature of interstitial alloys that metallic properties (e.g. conductivity) are maintained. Consequently, compounds of ionic or homopolar bonding are excluded although the relative atomic sizes (cf. *Figure 2.1*) might *seem* to justify their inclusion. Nevertheless, there are borderline cases, and silicon, oxygen, sulphur and phosphorus can often act as interstitial atoms and give transition metal compounds of metallic bond and character (for example, TiO, ZrP, CeS and numerous silicides). Furthermore, the boundary between interstitial and intermetallic compounds is by no means sharp, and even metal atoms themselves are capable of entering interstitially, if sometimes only transiently†. Therefore, though generally

* To regard the atom as a rigid sphere is, of course, only a rough approximation; it can be fairly applied to interstitial alloys, although the size concept altogether becomes controversial when electron exchange between large and small atoms is considered. Indeed, deviations in apparent 'atomic size' in different solid-state environments will in themselves reflect differences in degree of ionization or covalency. The atomic sizes in the figure are those of V. M. Goldschmidt, in some cases as amended by later workers and from other recent sources, and adjusted to 12-co-ordination.

† There is, for example, the broad *twilight area* of semiconductors, etc. where the boundary of what are interstitial compounds is far from clear-cut, and where the evolution to ionic bonding becomes predominant. This applies to oxides in particular. Although there is no doubt about the interstitial nature of C, N, H and 'holes', the borderline cases of Si, B, O, S, and P have their various restrictive conditions. Similarly, the delineation towards true intermetallic compounds is a diffuse one.

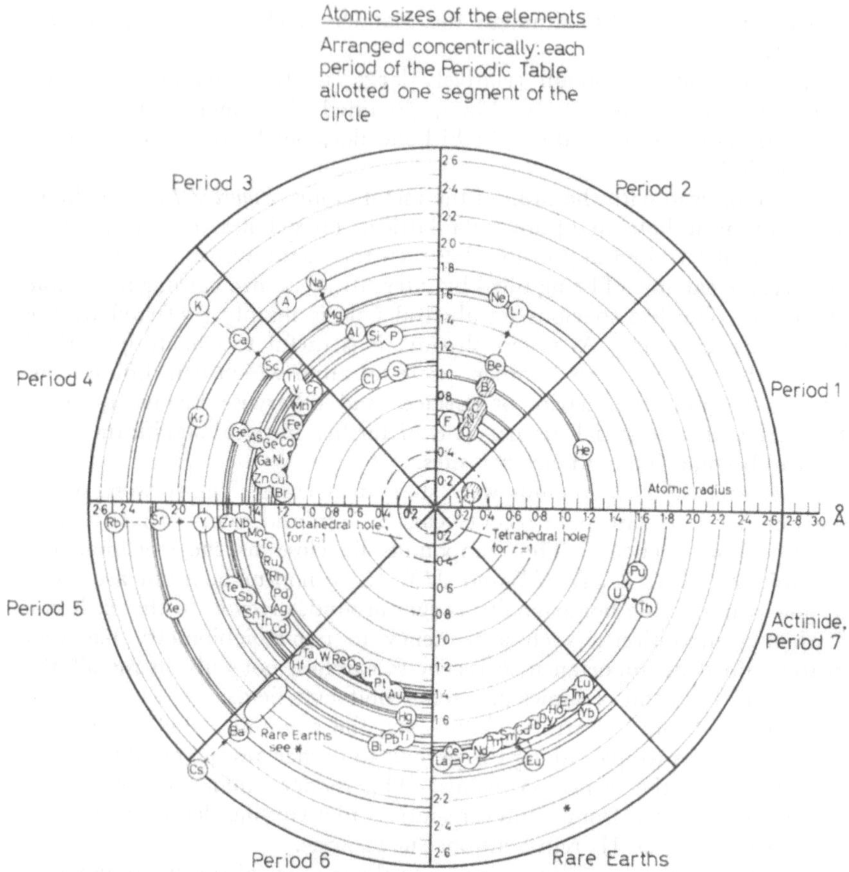

Figure 2.1. Atomic sizes of the elements, Å

Arranged concentrically, one segment to each period of the Periodic Table but Rare Earths separate; main interstitial atoms shaded

we are concerned with non-metal atoms as interstitials, this is not an inviolate condition, and indeed, the theory of metallic diffusion makes substantial use of the occurrence of interstitial metal atoms. A further facet is that some theories of interstitial structures[1,2] actually regard the non-metal atom, when built in, as though in effect *being* a metal atom: the 'expanded' metal lattice surrounding the interstitial atoms exerts such extreme internal pressure upon the latter that its valency electrons are made to contribute to the *d* orbital of the metal atom, and it has effectively *become* of metallic electron configuration.

However, a limit is set by the fact that strong electronegativity of the interstitial atom will prevent metallic alloy formation; this excludes for instance the halogens, which combine to form ionic salts. (Nevertheless, even the case of ionic bonding may be viewed organically as a kind of evolution or 'consolidation' from interstitial alloys, as far as internal bonding is concerned, making it possible for NaCl ⟷ TiC and CsCl ⟷ FeAl still to be structural isomorphs, although the manner of electron sharing differs so widely.)

The silicon atom, though relatively large, is still capable of forming interstitial silicides akin to carbides, provided the metal host atom is sufficiently large, and the direct Si–Si bond does not become dominant (as in disilicides).

For comparison with the radii of the metal atoms, *Figure 2.1* also includes that of a typical 'hole' in a pure metal lattice. These holes are interspaces in octahedral or tetrahedral positions between the metal atoms, and it is these sites that are capable of being filled by interstititial atoms, causing a variable expansion. The less spacious tetrahedral hole is mainly preferred by the small hydrogen atom. As noted already, it is useful to classify interstitial holes as though they were another species of atom which can diffuse, concentrate at grain boundaries or surfaces and otherwise behave quite in character with interstitial atoms; it is of further physical significance that they can become favoured electron sites.

Many interstitial compounds were studied in Hägg's[3] early work which has been continued in more recent years by him and many of his students of the Uppsala school. The compounds lie close to the stoichiometric compositions M_4X, M_2X, MX and MX_2 (X = non-metal), but are liable to show considerable homogeneity ranges around, or even to the exclusion of, these compositions; such a tendency to non-stoichiometry has very greatly gained in importance in recent researches, not only in metallurgy but also in the metal ceramics and semi-conductor fields.

Incompleted inner electron orbitals, viz. the *s*, *p* and *d* shells, and their gradual filling as we pass along the periods to higher atomic numbers, are a characteristic of the transition metals. This renders them receptive to the donation of electrons from whatever source: to a varying degree, interstitial atoms (e.g. C, N, O, H) represent one such source.

There has been much controversy as to the manner of this bonding which depends on a process referred to by Pauling as 'hybridization' of the *s*, *p* and *d* orbitals. Only a relatively small difference of energy exists between these electron levels, and their hybridization is made possible through a ready change-over of electrons from one to the other. The 'spread' into bands of the energy levels of the 3*d* and 4*s* orbitals in the transition metals, for instance, is intensified with decreasing interatomic distance in the lattice, more so for 4*s*, so that an overlap and eventual cross-over of the levels occurs. This distribution is also highly temperature- and pressure-sensitive. Furthermore, it is manifested in the liability of the transition metals to undergo polymorphic phase changes, which tends to be considerably greater than for the non-transition metals. Basically, electron supply from (and, possibly, exchange with) the non-metal atom can enhance this hybridization and facilitate the metal–non-metal bond. The process does not advance to ionization of the non-metal atom, however, and the essence of metallic bonding is maintained, i.e. the interstitial atom remains *substantially* in the atomic state. Thus, for example, it can be shown that in tungsten carbide, WC, the molecular volume (12·94 $Å^3$) results additively from the atomic volumes known in the elements tungsten (= 9·53 $Å^3$) and carbon (= 3·37 $Å^3$); the latter is significantly the value for carbon in diamond and not that in graphite. This atomic bonding differs greatly from that in compounds of metals *other* than transition metals with the same non-metal; thus, for example, aluminium or calcium carbides have non-metallic properties and more nearly covalent bonding; the crystal lattice is governed mainly by the non-metal atom rather than the metal atom.

Hydrogen occupies a special position; with its small atomic size (see *Figure 2.1* again), it is commensurate to a normal vacancy. This atom is especially capable of entering into the lattice interstices, either in solution or to form hydrides. Easy diffusion is possible, utilized e.g. in the well known case of palladium to instil the gas into otherwise sealed vacuum enclosures. Hydrides are physically much more akin to the parent metals than are carbides, nitrides, borides and oxides; they are (despite 'hydrogen embrittlement', which has a different basis) relatively soft and plastic and of moderate melting point or decomposition temperatures, if, as frequently, they are unstable. Furthermore, hydrogen is the only interstitial atom which can be widely absorbed by the body-centred cubic transition metal lattice, e.g. by Ti or Nb, without causing a transformation, though sometimes with major and interesting lattice distortions (to be discussed in Chapter 9). The next largest solubility in the BCC, that of oxygen, is much more limited. In the two close-packed metal structures (cubic and hexagonal), the hydrogen atom prefers the tetrahedral to the more spacious octahedral hole where, as it is sometimes expressed, it would 'rattle'; similarly, it seeks the smaller interstices in the body-centred cubic (BCC) structure. The compositions of hydrides are generally ill-defined; they form variable and structure-sensitive solution ranges which defy stoichiometry.

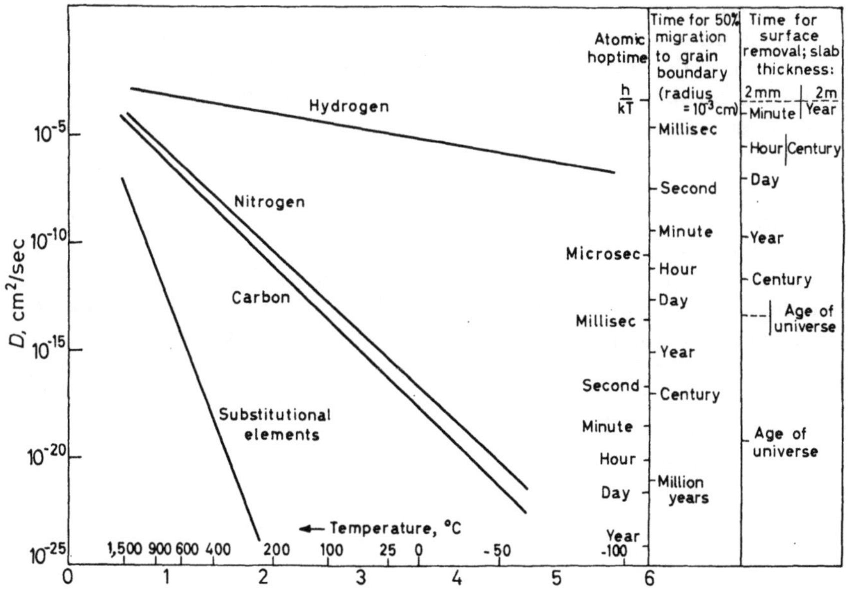

Figure 2.2. Diffusivity and time of migration in ferrite (After L. S. Darken[4])

By courtesy of H.M. Stationery Office

A consequence of the small size of the interstitial atoms (C, N, O and particularly H) is that they can possess high mobility in the solid state, even at relatively low temperatures. *Figure 2.2* shows the diffusivity of hydrogen in ferrite (α-Fe) and also of nitrogen, oxygen and of a typical substitutional element, in descending order. The times required for diffusion through a given distance, say to a grain boundary or a surface on degassing, can vary from microseconds to infinity, depending on temperature and the atoms concerned*.

2.2. SYSTEMATIC SURVEY OF INTERSTITIAL COMPOUNDS

A systematic survey of interstitial compounds is given in *Figures 2.3* and *2.4*, which should be read in conjunction: they are intended not merely to record the data but also to bring out the pattern obeyed. Many things of interest arise; only the more important points can be discussed here. In *Figure 2.3* the structures of the transition metals themselves, of their hydrides, nitrides, carbides, borides and oxides, are shown conjointly according to the Periodic Table, with the areas of isostructural types of metal atom lattice shaded in. Also indicated are the existence of two or more polymorphous forms, and (where this applies) of several compounds occurring in any one system. *Figure 2.4* shows the *melting points* of the metals and their interstitial compounds, in relation to Group and Period.

It may be seen that, from left to right along each period, the progressive transition from body-centred cubic (BCC) to face-centred cubic (FCC) or close-packed hexagonal (HCP) structures, which occurs for the pure metals, becomes in the interstitial phase drastically displaced to the left. With the hydrides, some adherence to the BCC structure is still recognizable, though with major lattice distortions. For many of the hydrides, however, and for the carbides and nitrides, the switch to close packing is immediate; the process continues with increasing atomic number to the structurally more complex and decreasingly stable compounds of Groups VII and VIII.

It is interesting to relate this to carbide formation in alloy steels and to note the borderline position of iron, manganese and (still to some extent) chromium in being liable to form *metastable* and *readily transformable* (low-alloy) steel carbides, as against the very *stable* refractory carbides of titanium, vanadium and chromium and the *unstable* ones of cobalt and nickel. This critical position of iron, together with the difference in carbon solubility in its γ and α forms, makes the metal so unique and is, fundamentally, the key to all ferrous metallurgy.

Significantly, however, *Figure 2.3* also brings out that, while for the 2nd and 3rd Long Period the simple interstitial structures cease with Group VI, in the next (Actinide) period they prevail beyond this, to the transuranic metals. Uranium once more obeys the theme of the BCC structure (γ–U)

---

* Further aspects of the concept of atomic size have recently been discussed with reference to semi-conductors[5] and to an effective *change* in size during phase transformations[6] BCC ⟷ FCC or HCP (see also 'Selected Literature', p. 55).

changing to the FCC metal lattice with carbon, nitrogen and oxygen addition; similarly neptunium and plutonium*.

Amongst the oxides, the sodium chloride-type structure systematically recurs right up to Group VIII, and throughout it might be considered as forming by direct interstitition into the FCC lattice of the parent metal (see γ-Fe $\rightarrow$ FeO, Co $\rightarrow$ CoO, etc.). This mode of interstition can take an alternative form in the CaF_2 (fluorspar) rather than the NaCl-type structure, but these two do only differ in non-metal occupancy and not in the metal atom lattice itself (see *Figure 2.7* below, p. 35). The process is illustrated by the oxide successions in transuranic elements: $M \rightarrow MO$ (NaCl type) $\rightarrow MO_2$ (CaF_2 type), where $M =$ U, Np, Pu or Am; here structure and stoichiometry are very simply related to the lattice sites available. The rutile form, MO_2, preferred by many oxides (see *Figure 2.3*), presents a related but rather more complex variant.

Another interesting variation to this theme occurs in the hydrides, e.g. Ti $\rightarrow$ TiH $\rightarrow$ TiH_2, where the final CaF_2 type of occupancy is reached by way of the ZnS (zincblende) structure, a diamond isomorph.

The close association between the FCC structures in the metal and its oxide has been demonstrated e.g. by observations on nickel and nickel oxide using electron diffraction[9]. Here thin nickel films grown on sodium

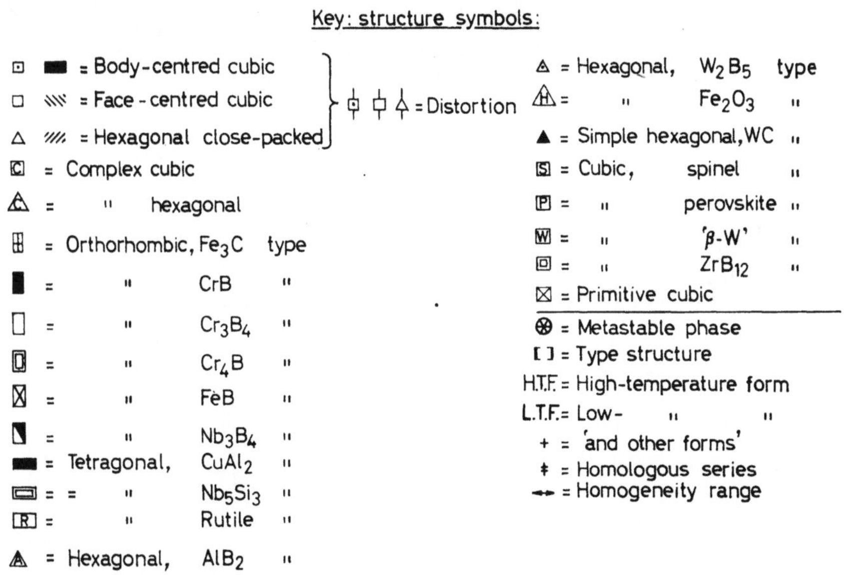

Figure 2.3. Occurrence and structure of interstitial transition metal phases (From Pearson[7], Hansen[8], etc. and original papers)

* This relation is graphically reflected in practice by the fact that frequent impurities (inclusions) found in metallic uranium are UC, UN, UO or UO_2: thus the interstitial addition results in the simple and immediate switch from BCC (γ–U) to the close-packed metal atom constellation which, under the different set of influences referred to, also occurs for the 1st, 2nd and 3rd Long Period elements. (The phase change γ–U $\rightarrow$ β–U in *pure* uranium is one from the BCC to the σ-type structure, as has indeed also a parallel occurrence, upon decreasing temperature, in certain of the pure metal alloys of the earlier periods.)

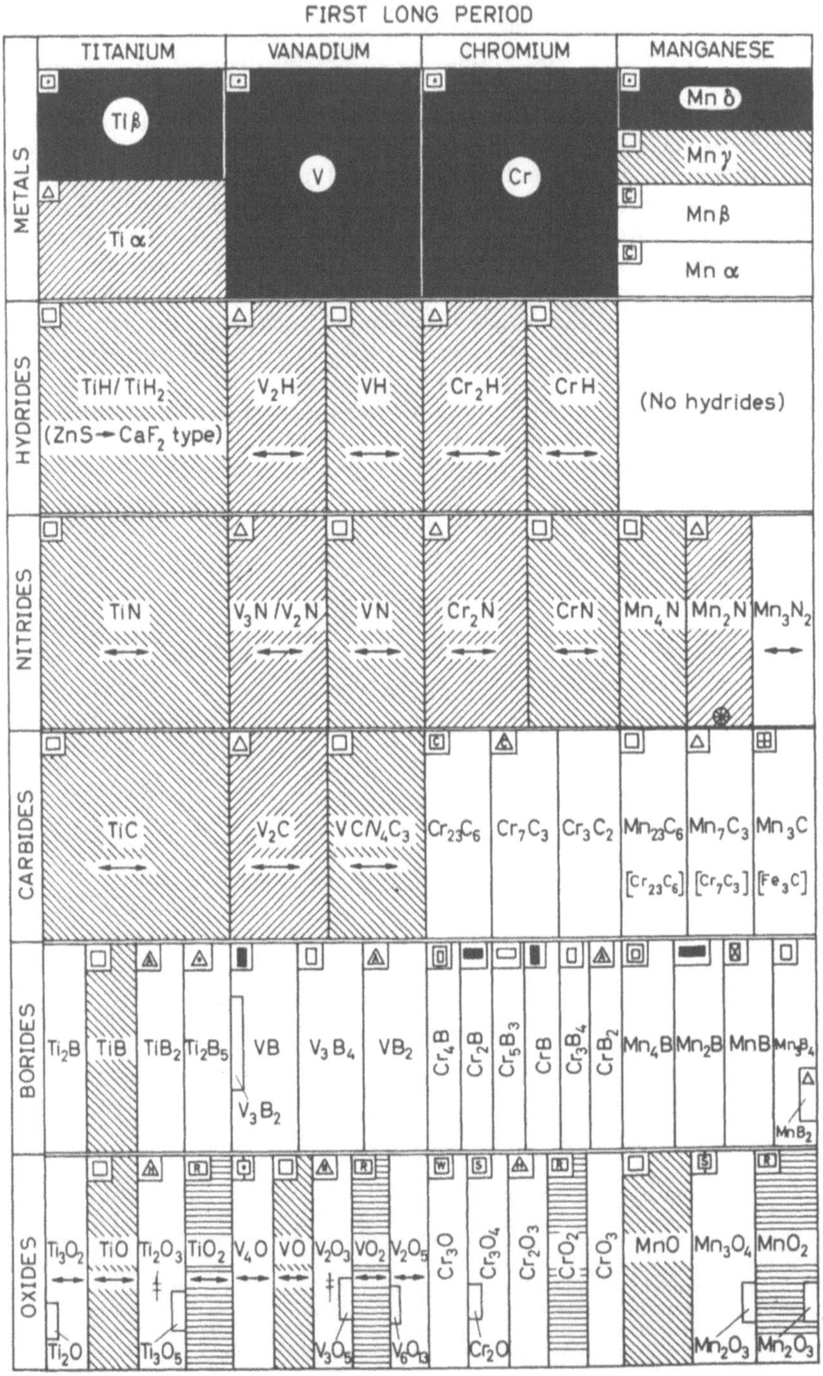
FIRST LONG PERIOD
TITANIUM
VANADIUM
CHROMIUM
MANGANESE
METALS
HYDRIDES
NITRIDES
CARBIDES
BORIDES
OXIDES
Ti β
Ti α
V
Cr
Mn δ
Mn γ
Mn β
Mn α
TiH/TiH2
(ZnS → CaF2 type)
V2H
VH
Cr2H
CrH
(No hydrides)
TiN
V3N/V2N
VN
Cr2N
CrN
Mn4N
Mn2N
Mn3N2
TiC
V2C
VC/V4C3
Cr23C6
Cr7C3
Cr3C2
Mn23C6
Mn7C3
Mn3C
[Cr23C6]
[Cr7C3]
[Fe3C]
Ti2B
TiB
TiB2
Ti2B5
VB
V3B2
V3B4
VB2
Cr4B
Cr2B
Cr5B3
CrB
Cr3B4
CrB2
Mn4B
Mn2B
MnB
Mn3B4
MnB2
Ti3O2
Ti2O
TiO
Ti2O3
Ti3O5
TiO2
V4O
VO
V2O3
V3O5
VO2
V6O13
V2O5
Cr3O
Cr3O4
Cr2O
Cr2O3
CrO2
CrO3
MnO
Mn3O4
Mn2O3
MnO2
Mn2O3

Figure (2.3a) I

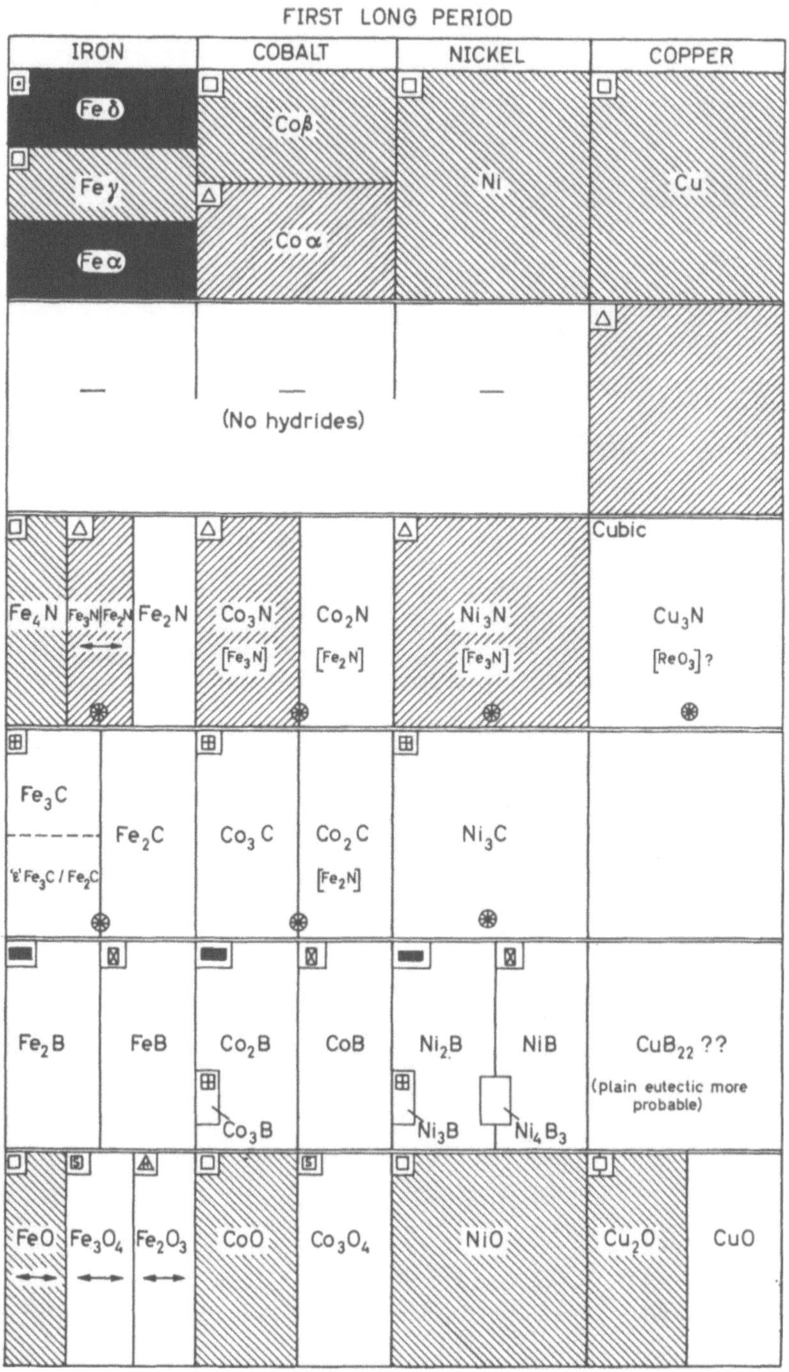

Figure (2.3a) II

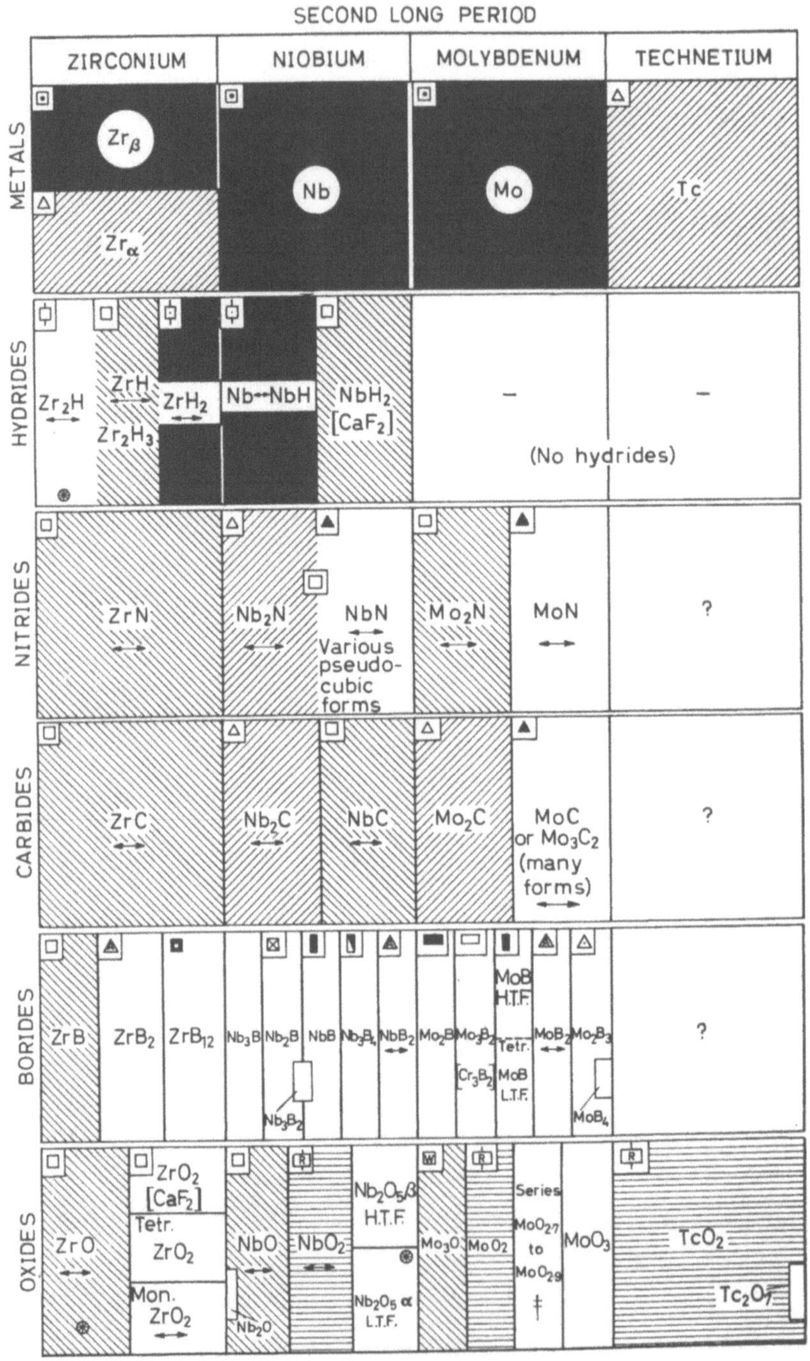
SECOND LONG PERIOD
ZIRCONIUM
NIOBIUM
MOLYBDENUM
TECHNETIUM
METALS
HYDRIDES
NITRIDES
CARBIDES
BORIDES
OXIDES
Zrβ
Zrα
Nb
Mo
Tc
Zr2H
ZrH
Zr2H3
ZrH2
Nb↔NbH
NbH2
[CaF2]
(No hydrides)
ZrN
Nb2N
NbN
Various pseudo-cubic forms
Mo2N
MoN
?
ZrC
Nb2C
NbC
Mo2C
MoC or Mo3C2 (many forms)
ZrB
ZrB2
ZrB12
Nb3B
Nb2B
NbB
Nb3B4
NbB2
Nb3B2
Mo2B
Mo3B2
[Cr3B2]
MoB H.T.F.
Tetr. MoB L.T.F.
MoB2
Mo2B5
MoB4
ZrO
ZrO2 [CaF2]
Tetr. ZrO2
Mon. ZrO2
NbO
Nb2O
NbO2
Nb2O5 β H.T.F.
Nb2O5 α L.T.F.
Mo3O
MoO2
Series MoO2.7 to MoO2.9
MoO3
TcO2
Tc2O7

Figure (2.3b) I

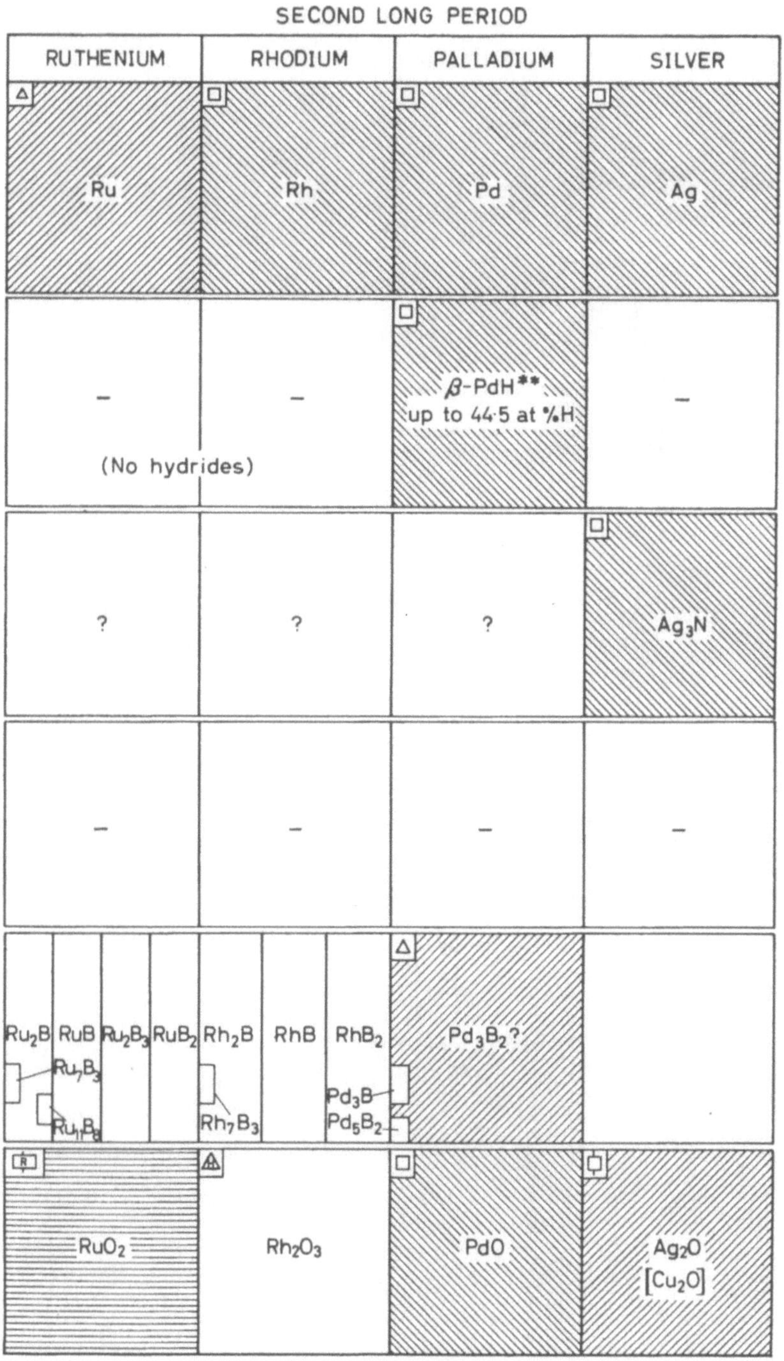

Figure (2.3b) II

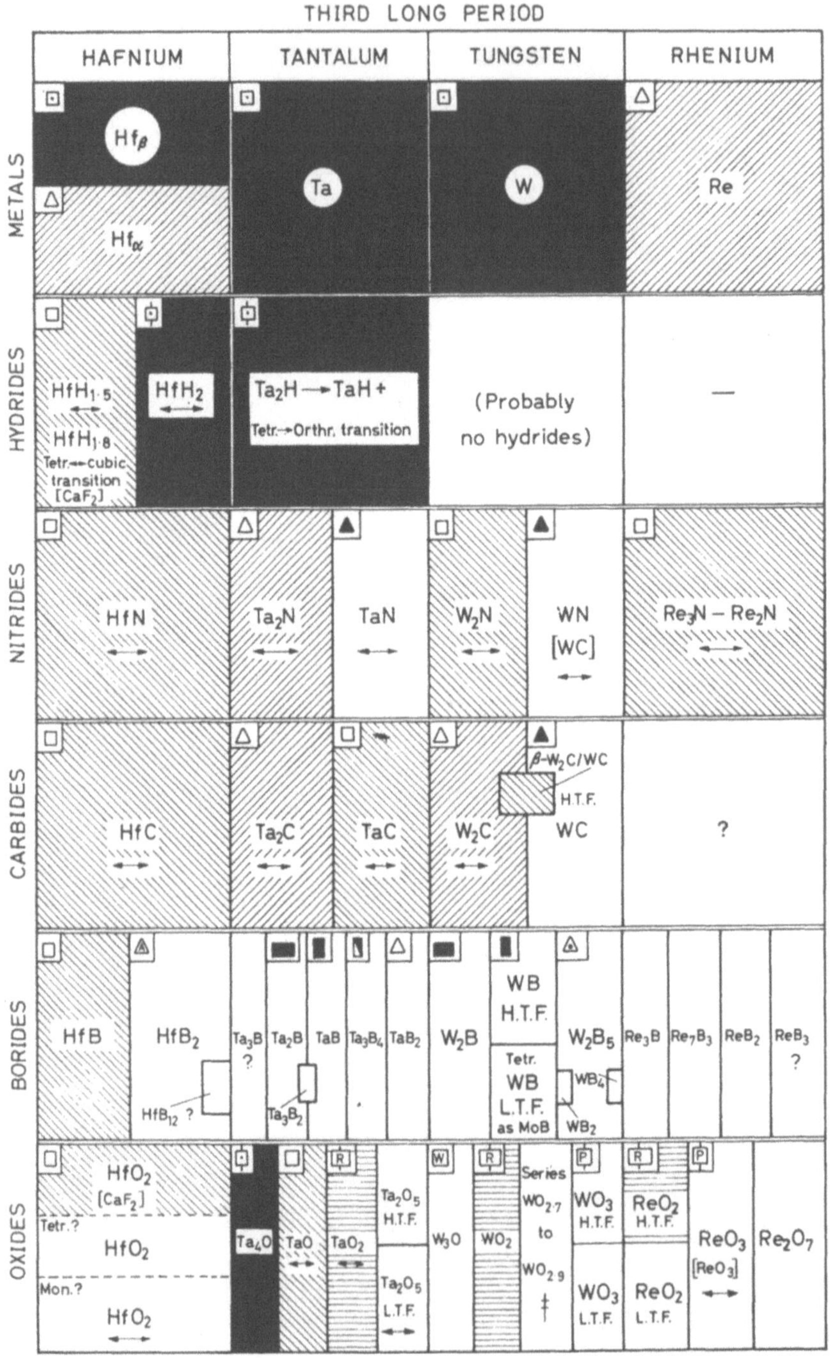
THIRD LONG PERIOD
HAFNIUM
TANTALUM
TUNGSTEN
RHENIUM
METALS
Hfβ
Hfα
Ta
W
Re
HYDRIDES
HfH1·5
HfH1·8
Tetr.→cubic transition
[CaF2]
HfH2
Ta2H → TaH +
Tetr.→Orthr. transition
(Probably no hydrides)
—
NITRIDES
HfN
Ta2N
TaN
W2N
WN
[WC]
Re3N – Re2N
CARBIDES
HfC
Ta2C
TaC
W2C
β-W2C/WC
H.T.F.
WC
?
BORIDES
HfB
HfB2
HfB12 ?
Ta3B
?
Ta2B
TaB
Ta3B4
TaB2
Ta3B2
W2B
WB
H.T.F.
Tetr.
WB
L.T.F.
as MoB
W2B5
WB4
WB2
Re3B
Re7B3
ReB2
ReB3
?
OXIDES
HfO2
[CaF2]
Tetr.?
HfO2
Mon.?
HfO2
Ta4O
TaO
TaO2
Ta2O5
H.T.F.
Ta2O5
L.T.F.
W3O
WO2
Series
WO2·7
to
WO2·9
WO3
H.T.F.
WO3
L.T.F.
ReO2
H.T.F.
ReO2
L.T.F.
ReO3
[ReO3]
Re2O7

Figure (2.3c) I

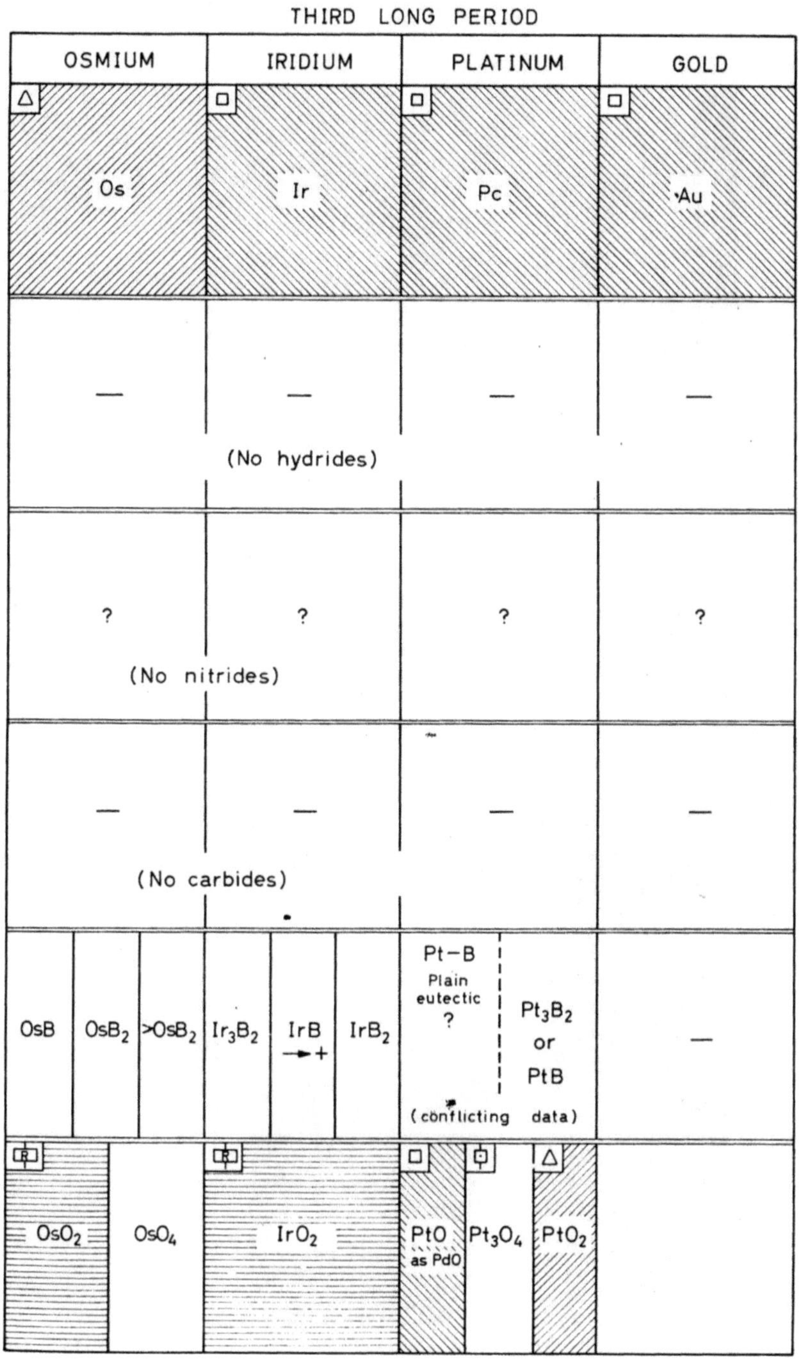

Figure (2.3c) II

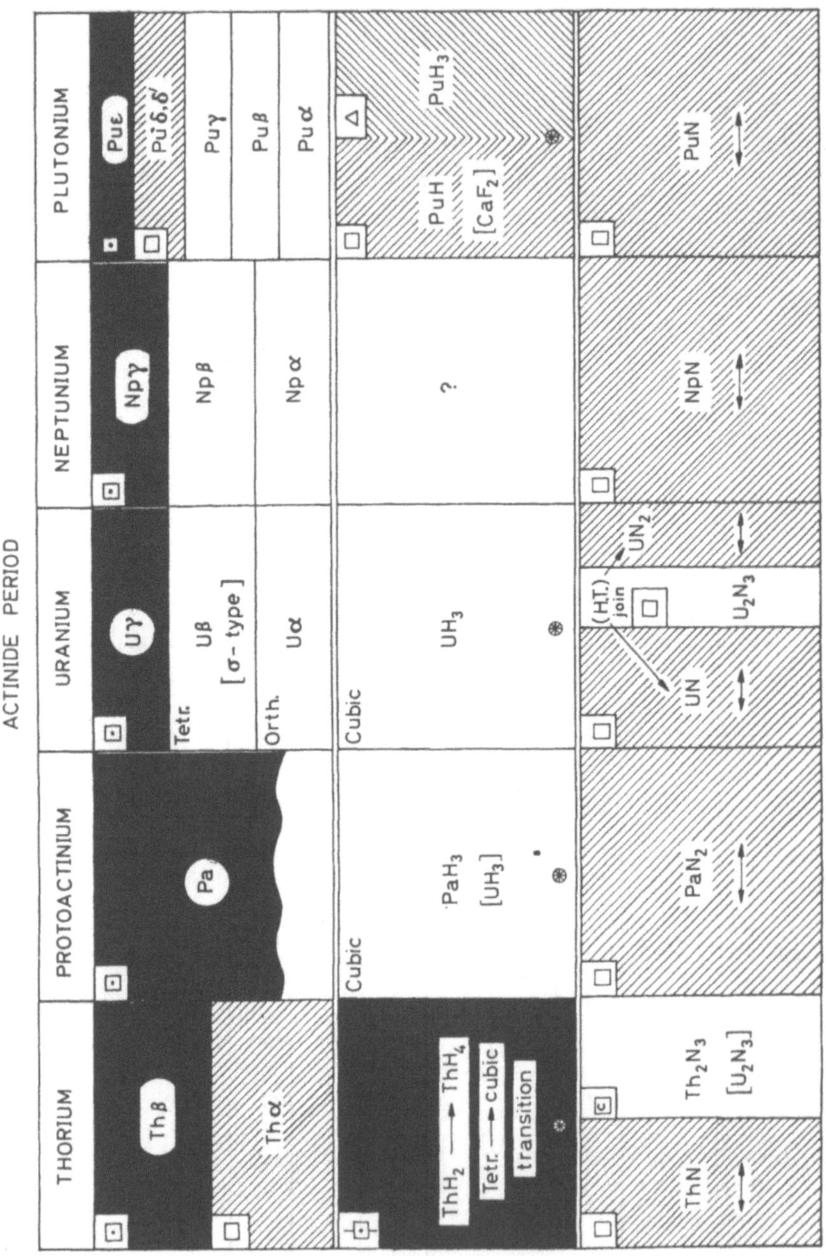
ACTINIDE PERIOD
THORIUM
PROTOACTINIUM
URANIUM
NEPTUNIUM
PLUTONIUM
METALS
HYDRIDES
NITRIDES
Thβ
Thα
Pa
Uγ
Tetr.
Uβ
[σ-type]
Orth.
Uα
Npγ
Npβ
Npα
Puε
Puδ,δ'
Puγ
Puβ
Puα
ThH2 → ThH4
Tetr. → cubic
transition
Cubic
PaH3
[UH3]
UH3
?
PuH
[CaF2]
PuH3
ThN
Th2N3
[U2N3]
PaN2
UN
(H.T.) join
U2N3
UN2
NpN
PuN

Figure (2.3d) I

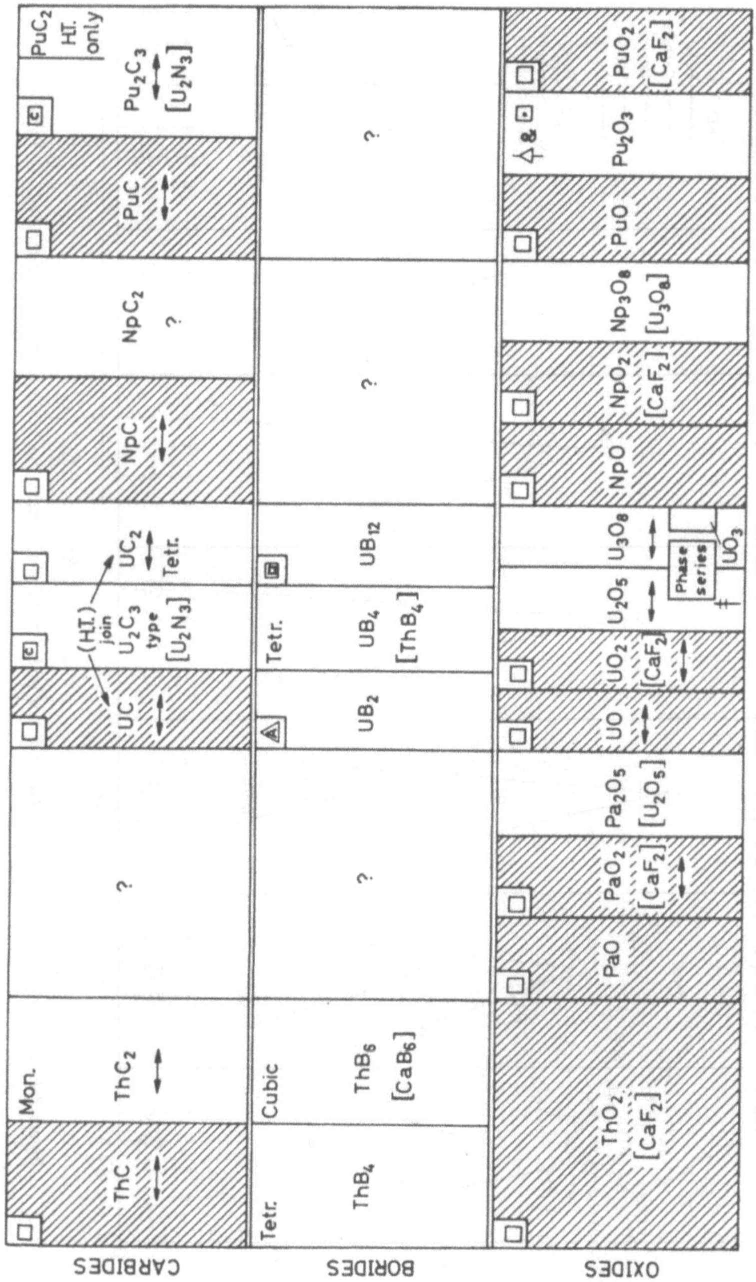

Figure (2.3d) II

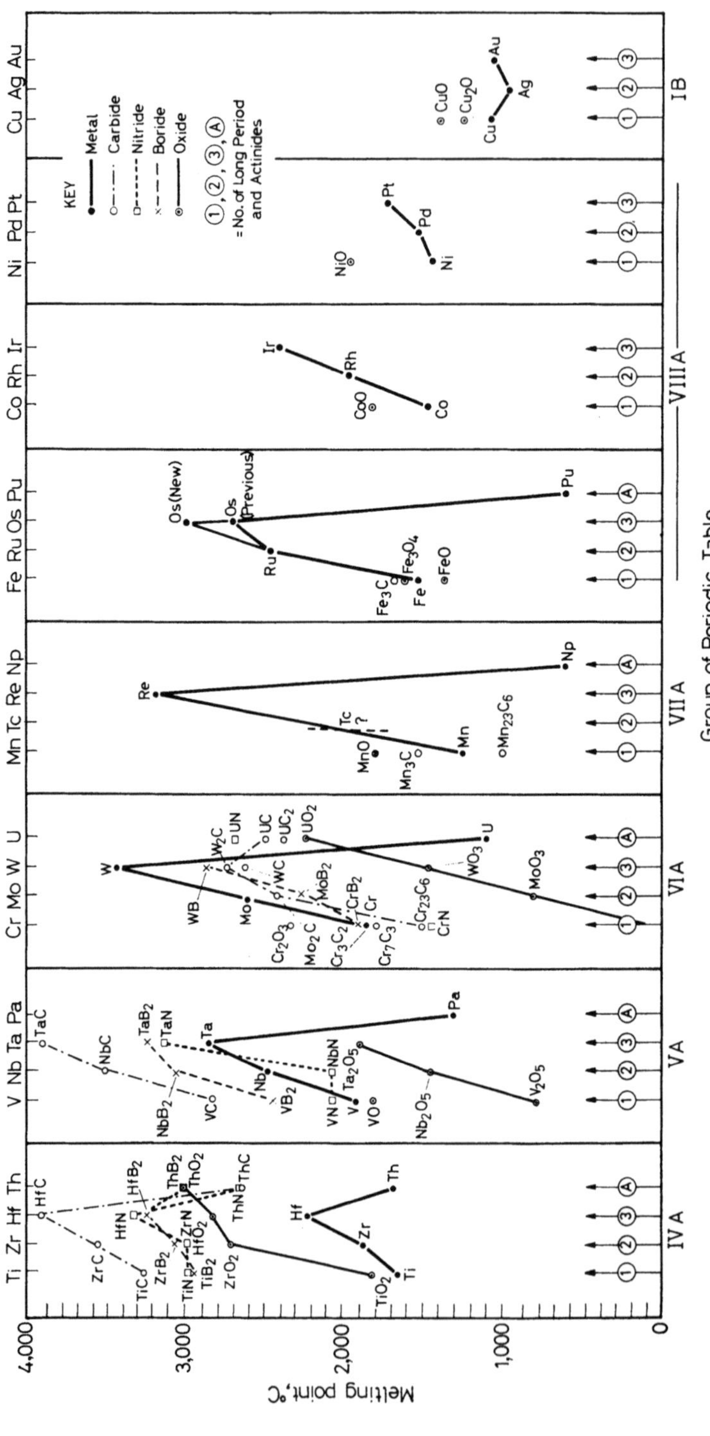

Figure 2.4. Melting points of transition metals and their carbides, nitrides, borides and oxides, as function of the Periodic Table

chloride (100) faces showed 'forbidden reflections' upon annealing: initially coexisting NiO disappeared and a *superlattice* appeared in nickel through oxygen diffusion into the octahedral sites.

That 'interstitialcy' is not limited to alloys but extends to solid matter generally and ionic crystals in particular, is obvious and important. It is, of course, the foundation of semi-conductor science. As just one example, pure silicon can dissolve Li^{3+} ions with a lattice expansion[10], which can be attributed essentially to electrostatic repulsion between Si^{4+} and Li^{3+} ions. Further cases will occur later, e.g. in Chapters 8 and 9.

The *melting points* of the metals and interstitial compounds shown in *Figure 2.4* reveal variations of great interest. We may regard this property broadly (with some qualification) as a measure of affinity between metal and non-metal and of strength of bonding in the interstitial compounds. Heats of formation or compressibility would be a stricter yardstick than melting point: however, such data are as yet insufficient, and melting points may be used as a fair substitute.

Some of the features arising from *Figure 2.4* are:

(1) Within each group, passing from 1st to 3rd Period, the melting points both of the metals and their interstitial phases increase steadily and in parallel; they drop sharply on passing to the next Period (Actinides), with the exception of the oxides for which the increase continues.

(2) The NaCl-type interstitial compounds have by far the highest melting points, θ, greatly exceeding those of their parent metals*. The *difference* $\Delta\theta = (\theta_{\text{interstitial}}) - (\theta_{\text{metal}})$, however, decreases on passing from Group IV to V to VI. In VI it becomes nearly zero or negative (except for uranium), i.e. the metal surpasses its compound at the same time as the NaCl-type structures are abandoned. On continuation to Groups VII and VIII, this trend develops into the instability and disappearance of carbides and nitrides and, at a later stage, oxides. The transition proceeds hand in hand with that from BCC to FCC structures in the pure metals. In other words, the particular structure which had previously been possessed by the metal lattice in the high-melting interstitial phases of the earlier Groups, is now taken over by the pure metals themselves.

If we plot the ratio $\Delta\theta = (\theta_{\text{carbide}} - \theta_{\text{metal}})/\theta_{\text{metal}}$ (θ = °abs), as in *Figure 2.5*, a further new feature comes to light: the 2nd Long Period (Zr, Nb, Mo) always shows a singularity as against the 1st and 3rd, namely a minimum inverting progressively towards a maximum.

Also we may see that the carbide formers can be divided into three groups, according to whether $\Delta\theta$ is (*A*) positive, (*B*) about zero or (*C*) negative. It is characteristic that this division is paralleled by the carbide crystal structures adopted. To Class (*A*) belong all close-packed cubic or hexagonal carbides of Ti, Zr, Hf, V, Nb, Ta; to Class (*C*) the simple and close-packed hexagonal one of W; to Class (*B*) the complex and multiple carbide structures for Cr, Mo, Mn, Fe (with Mo in a borderline position). For nitrides, with due modifications, the position is similar.

* Indeed, they include the highest-melting solids known, the absolute maximum having been reported at an HfC—TaC solution.

(3) The fact that the melting points of the nitrides are lower than those of the carbides, and—although with exceptions—those of the oxides lower still, can be explained by the predominance of the metal–non-metal bond in these phases: the C, N and O atoms contribute, in this order, a decreasing number of electrons to each of their metal neighbours, so that the main bonds are progressively weakened and the melting points decreased. The borides, on the other hand, are essentially different, as the new non-metal–non-metal bond now contributes to cohesion. The latter is shown by a recovery in the melting points, despite the weakening of the metal bonds to metal and non-metal.

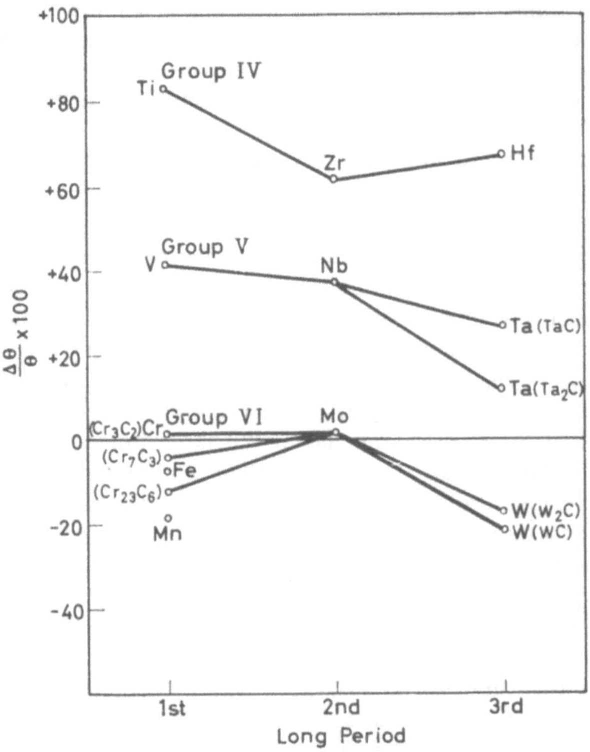

Figure 2.5. Melting-point ratios (θ carbide–θ metal) / θ metal, $\times$ 100, *in relation to the Periodic Table* (θ = temperature, °abs)

(4) Of other regularities which arise from *Figure 2.4*, one is remarkable: the almost perfect *linearity* of increase in melting point of the pure metals along each Group from 1st to 2nd to 3rd Long Period. This applies from Ti → Zr → Hf through to Co → Rh → Ir (the value for Tc is not known, but about 2,100°C would follow by interpolation). The linearity has been emphasized by the recent case of osmium. With the earlier reported melting point for Os of 2,700°C, the sequence Fe → Ru → Os (see *Figure 2.4*) did seem to show a discontinuity, but a recent redetermination gives a new value

of 3,000°C, which falls well on to the straight line extrapolated from Fe and Ru.

Furthermore, the *gradients* of this family of straight lines follow a definite law in that they systematically increase, attain a maximum, then decrease. This is illustrated in *Table 2.1* where the gradients are expressed as mean difference $\Delta\theta_{3-1}$ between melting points in 3rd and 1st Long Period.

Table 2.1 Gradient of melting point increases $\Delta\theta_{3-1}$ (°C) within the Groups

Group	*Metals*	*Carbides*	*Nitrides*	*Oxides*
IV	Hf–Ti 570	HfC–Tic 640	HfN–TiN 360	HfO_2–TiO_2 1,020
V	Ta–V 940	TaC–VC 1,060	TaN–VN, 1,100	Ta_2O_5–V_2O_5 1,210
VI	W–Cr 1,560	WC–Cr_3C_2 750	WN–CrN	WO_3–CrO_3 1,220
VII	Re–Mn 1,910	(unstable)	(unstable)	
VIII	Os–Fe 1,470 Ir–Co 960 Pt–Ni 320			
IB	Au–Cu* –20			

*Minimum at Ag

The trend is repeated, though less strongly, for the interstitial compounds. For carbides and other interstitial phases, *Figure 2.4* also shows melting point rises which are broadly similar but with distinct anomalies (these may partly be due to imprecise data or may be genuine). A near-linearity seems, however, certainly to apply to the carbides, and at least to the Group V and VI oxides; the gradients, as far as can be determined, are also included in *Table 2.1*, and a maximum similar to that of the metals occurs certainly for the carbides; compared to the metals, this is displaced to the left—from the iron to the vanadium Group—in line with the structural characteristics discussed.

Again referring to *Figure 2.4*, the linearity in rise of the melting point of the metals begins to fault at Group Ni–Pd–Pt with a central deflection, and breaks down in the next Group (Cu–Ag–Au), where the difference in values vanishes and the prior deflection at Pd develops into a full minimum at Ag. This, however, occurs beyond the region in which interstitial compounds are stable.

(5) There is an evident correlation between atomic *radii* of metals and $\Delta\theta$ (for the carbides): their strength of bonding and stability, within each period, are favoured by the larger transition metal atoms; with decreasing atomic size of the metal, the metallic bond strength becomes sufficient to militate against, and then to prevent, interstitial compound formation.

An interesting sidelight is also provided by the *compressibilities* of the metal. The inverse of this property is a direct measure of strength of bonding, that is, of the resistance of the atoms to being pressed together and therefore to being pulled apart; and so is the interatomic distance (other factors being equal). In *Figure 2.6* the metallic compressibilities, χ, taken from Hume-Rothery and Coles[11], are shown as a function of atomic radius. Within each period, χ increases with atomic size: both the radius, r_M, and χ show joint minima, and whilst the broad fact of interdependence of r_M and χ holds good irrespective of crystal structure, the slopes of each branch *are* structure-sensitive. Significantly, the reversal on traversing each period begins at the point of passing from the BCC to the close-packed (i.e. FCC or HCP) structures of the metals, the gradients being larger for the close-packed than for the BCC structures. The minimum also marks the boundary between those metals forming and those not forming interstitial compounds, the carbide- and nitride-formers being restricted to the *descending* branch.

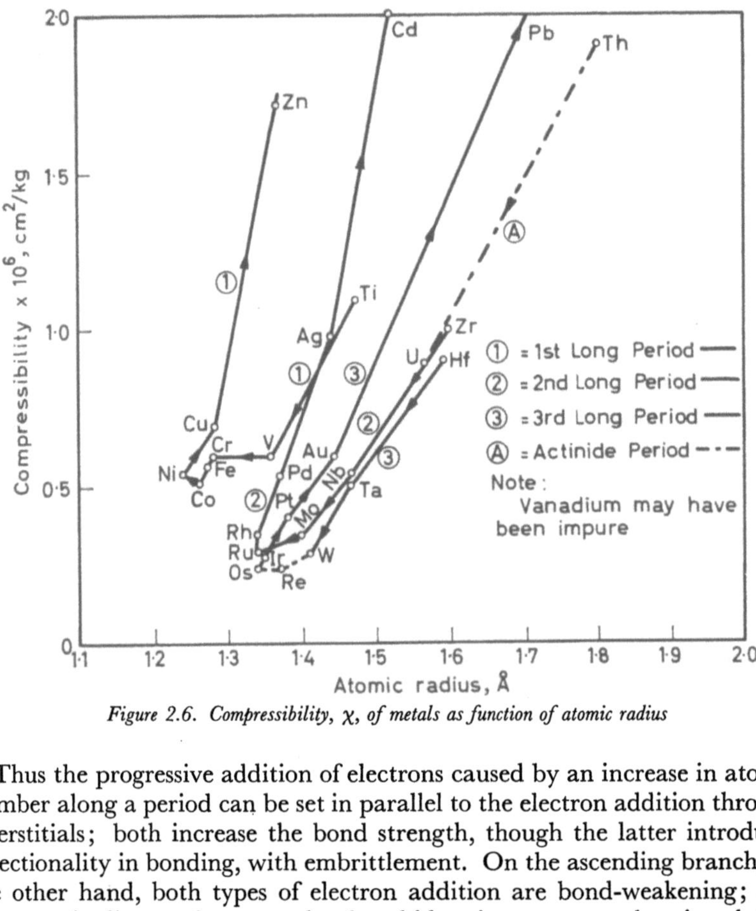

Figure 2.6. Compressibility, χ, of metals as function of atomic radius

Thus the progressive addition of electrons caused by an increase in atomic number along a period can be set in parallel to the electron addition through interstitials; both increase the bond strength, though the latter introduces directionality in bonding, with embrittlement. On the ascending branch, on the other hand, both types of electron addition are bond-weakening; the interatomic distance increases, hard carbides give way to carbon in solution and graphite, and nitrides to nitrogen in solution and the free gas. The negligible primary solubility for C in the descending-branch metals (Groups IV to VI) falls consistently into this pattern, and interstitial addition gives rise to the immediate close-packed configuration of the carbides for the *descending*-branch metals and the generally larger primary solubility for the (already close-packed) *ascending*-branch metals.

It is revealing to consider the *thermal expansion coefficients* in this light. For the hexagonal interstitial compounds (as for hexagonal metals) they are larger in a direction parallel to the *c*-axis than in the basal plane. Thus the addition of carbon to tungsten creates the hexagonal close-packed structure

(W_2C), whilst that of an electron to the tungsten atom 'creates' the metal rhenium (neighbouring tungsten to the right in the same period), which is also hexagonal close-packed. The coefficients of expansion are

	$\times 10^6$/°C	
Pure W (isotropic)	4·5	(0–100°C)
Pure Re	4·7 in basal plane	(0–100°C)
	12·4 ‖ *c* axis	
W_2C	1·2 in basal plane	(0–2,400°C)*
	11·4 ‖ *c* axis	
WC	5·2 in basal plane	(0–2,400°C)*
	7·3 ‖ *c* axis	

Considering that high thermal expansion coefficients reflect greater freedom to increase the amplitude of thermal vibration, the greater bond strength within the basal plane follows naturally. Thus, both the metals (where the tungsten value is similar to that for rhenium in the basal plane), and the carbides (where in W_2C the basal plane bonding is greatly strengthened and the axial bonding weakened, whereas in WC the basal bonding becomes weakened, the axial one strengthened), tend to become more nearly alike.

Reverting to *Figure 2.6*, the compressibility of the interstitial *compounds* is considerably lower compared with their parent metals, and this again reflects increased strength of bonding through interstitial insertion. For instance, for cubic TiC, $\chi = 0{\cdot}47 \times 10^{-6}$ cm²/kg, against 1·1 for Ti.

Mixed Compounds

The *binary* interstitial compounds, only considered so far, are of course capable of forming manifold mutual solid solutions, both by metal and by non-metal replacement, e.g. ZrC ⟷ NbC or ZrC ⟷ ZrN. The degree of this intersolubility varies widely, depending not only on isomorphy of the end members but on relative atomic size and electronegativity; the governing rules are similar to those formulated by Hume-Rothery for pure metals, though with important distinctions. Thus titanium forms a 'carbonitride', Ti(C,N), and an 'oxycarbonitride', Ti(C,N,O), which are really solid solutions and not separate compounds, as sometimes thought. One implication is that, for instance, TiC and WC, of different structure, are intersoluble only partly and to an extent dependent on temperature. This relation is important in cemented hard metals, as it determines their properties and the effect of heat treatment, as well as introducing varied precipitation effects; in these the bond metal (usually cobalt) can participate, as it, too, possesses variable solubilities.

In addition to these solution series, there exist, however, also distinct ternary interstitial compounds of new structure, such as the so-called 'Eta-carbides', Fe_4W_2C/Fe_3W_3C, found in high-speed steels, which only represent a prototype for a large family of compounds with both metal and non-metal substitutions. Another recent new type of compound, with two interstitials, is a molybdenum silicocarbide, Mo(Si,C), ('Nowotny phase'), in which

* This larger temperature range would tend to bias these values somewhat on the high side, but the point at issue is still brought out.

again various replacements are possible. (We shall have to say more about this in later Chapters.)

Progressive Instability

In connection with our Periodic Table survey of carbides and nitrides (*Figure 2.3*), it is interesting to remember that all iron carbides and nitrides are in fact metastable and that, just as the true equilibrium is between iron and graphite and not iron and cementite, so it is ideally between iron and nitrogen *gas*. The realistic Fe–N system—with Fe_4N and all the associated phase-change phenomena found by Jack[12], and utilized e.g. in nitriding—is metastable and possible only because time is insufficient for the trapped lattice nitrogen to migrate and be released. Not only one but a whole sequence of unstable but persistent states can thus be obtained, to be described later. This is yet another manifestation of the borderline position of iron within the pattern outlined in *Figure 2.3*, where the nitrides and carbides of Mn and Cr are truly stable, whereas those of Co and Ni are entirely fleeting.

A further implication of the relative instability of iron nitrides is that small additions to Fe–N alloys of elements which are themselves stronger nitride formers, such as aluminium (to give AlN) or silicon (to give Si_3N_4) readily stabilize these compounds and, in so doing, act as grain refiners in steels. Moreover, AlN does not only precipitate as nuclei of the actual compound but is already incipient in the Fe base matrix containing N and Al atoms in solution, N atoms tending to aggregate around Al atoms within the iron lattice (Darken). This is quite analogous to the action of carbide-former elements upon carbon atoms with which they may, at higher temperatures, be in joint austenite solution.

Layer Lattices

One important aspect of interstitial compounds is that, similar to close-packed metals, they involve layer lattices in which the well known varying types of stacking sequence repeat in the manner *A B A B A B* . . . (e.g. Nb_2C, W_2C) and *A B C A B C* . . . (e.g. NbC, TiC). The non-metal atoms are inserted between the metal layers, and their bonds can indeed be envisaged as preventing easy glide by 'locking' of the metal atom layers, leading to hardness and brittleness. Non-stoichiometry, say carbon deficiency or oxygen excess, has here a special significance in creating local distortions of the metal atom layers and irregular 'puckering', with further enhanced resistance to glide. It can also have the curious side-effect of promoting ease of diffusion by the creation of holes and larger interspaces: some interstitials act, as it were, as a 'lock' to help others to move. This can also occur through heterogeneous combinations of interstitial atom species, such as mixtures of C and O, or B and C.

Tungsten monocarbide, WC, with its isomorphs of MoC, MoN and others, is here of special interest. Its simple hexagonal structure (see *Figure 2.7f*) implies a metal-stacking type *A A A A* . . . which does not occur in pure metals, being made possible only by the carbon layers interlacing at single repeat. It is worth considering these carbon sheets in tungsten carbide as similar already to those in pure carbon, viz. graphite, so that the WC structure would literally contain alternate hexagonal tungsten and graphite-like layers;

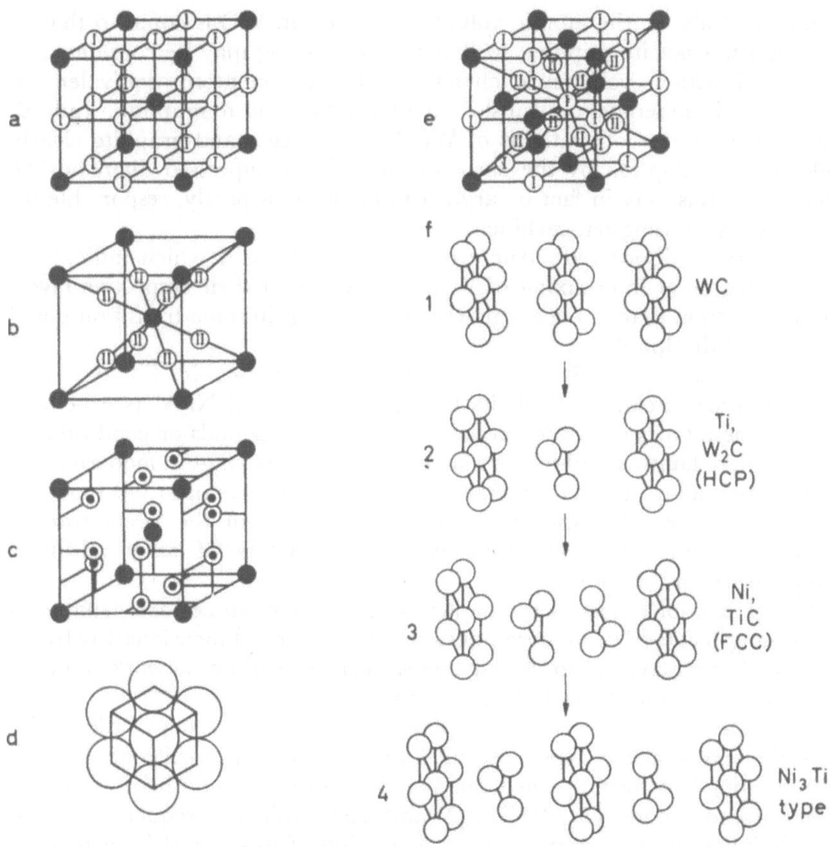

Figure 2.7. Some basic interstitial structures

(*a*) Favoured interstitial positions in BCC (Type I); (*b*) less favoured but possible positions in BCC (Type II); (*c*) 'β-tungsten' (= W_3O)-type structure with different interstitial positions; (*d*) view along BCC cube diagonal showing pseudo-hexagonal 'puckered' layers into which interstitials can fit, distorted by their entry; (*e*) interstitial positions in FCC: Types I and II, both favoured in varying cases (also basis for rock salt- and fluorspar-type compounds evolving to diamond type); (*f*) hexagonal layering sequences in interstitial structures, with examples

the analogy fails, of course, insofar as the covalent C—C bond has broken down and is replaced by the W—C one. However, it may well be said that a decomposition or synthesis

$$WC \rightleftharpoons W_2C + \text{carbon (graphite)} \qquad (1)$$

consists basically in a rearrangement of hexagonal layers which remain intact as such. The loss of interstitial carbon on passing from WC to W_2C permits a metal atom switch from the simple to the close-packed hexagonal type sequence: i.e. reaction (1) can in terms of layers be written

$$\begin{array}{lcl} WC & \rightleftharpoons & W_2C \qquad\qquad + \text{ graphite} \\ \ldots A\,\mathit{C}\,A\,\mathit{C}\,A\,\mathit{C}\ldots & \rightleftharpoons & \ldots A\,\tfrac{1}{2}\mathit{C}\,B\,\tfrac{1}{2}\mathit{C}\,A\,\tfrac{1}{2}\mathit{C}\,B\,\tfrac{1}{2}\mathit{C}\ldots \\ & & \qquad + \ldots \mathrm{C\,C\,C\,C}\ldots \end{array} \qquad (2)$$

where C = the interstitial carbon
C = free carbon

As noted above, the atomic volume of carbon in WC is equal to that in diamond but not in graphite, so that this re-layering may be considered as connected with a volumetric change of the carbon atom, equivalent to that which it undergoes on transforming from diamond to graphite. Indeed, the reverse process of synthesis of WC from tungsten and graphite clearly enforces the adoption by the carbon atom of its compressed diamond-like character; this may in fact be argued to be, at least partly, responsible for the hardness of tungsten carbide.

A 'partially disordered' structure of WC is known[13], which indicates a certain degree of intermixing of W and C atoms in their respective layers (a phenomenon that can occur more generally, e.g. in some transition metal mono- and disulphides).

To complete *Figure 2.7f*, we have added the cases of cubic, TiC-type, and of Ni_3Ti-type stacking. With its hexagonal variant of Ni_3Al (see below), each of these types stands for definite observed compounds or conditions of intermediate faulting within the matrix. Each is important in modern high-temperature alloys and hard metals, and in all these materials these modes of layering can be said basically to form the essence of the 'orchestration' directed by the thermodynamic controlling factors of alloy addition, temperature and stress.

In the case of *borides* (see Chapter 6), the layer lattices are capable of containing bonded boron–boron networks (1-, 2- or 3-dimensional) between the metal atom layers; no similar bonded carbon–carbon network can exist in the carbides, but it can in free graphite.

In *nitrides*, too, this curious interplay of hexagonal packing of the simple and the close-packed types prevails, and the tantalum–nitrogen system is interesting in affording a new variant to the same theme: here we have a series of compounds $Ta_{2\cdot5}N$, $Ta_{1\cdot2}N$ and TaN which, in sequence, possess the close-packed, the simple, and a novel kind of mixture of both types of packing. A whole variety of new mixed nitrides has been discovered in recent years by Hägg's school, notably by Kiessling and Schoenberg, in which, by varying the metal combinations and proportions, the changes can be rung almost at will, either in isomorphous replacement or in stacking sequence, degree of atomic ordering or of lattice distortion. The same applies to various new carbides and double oxides, as well as to compounds with mixed interstitials of (N, C, O, B, 'hole') combination; thus, for instance, there is no niobium nitride of the strict composition 'NbN' with the cubic close-packed structure, but it can be created by substituting a little nitrogen by oxygen to give '$NbN_{0\cdot9}O_{0\cdot1}$'.

Some *sulphides* and *phosphides* of high metal content are, in this light, structurally akin as compounds, the chief difference being much increased lattice dimensions caused by the larger sulphur and phosphorus atoms (see *Figure 2.1*). To get a measure of scale: compared with the pure metals, the metal–metal atom distances in carbides and nitrides are 4–10 per cent larger, those in sulphides and phosphides 14–21 per cent larger. Again we have the WC-type lattice in MoP, TiS or NbS, whilst ZrP shows the *A B C A B C . . .* (close-packed cubic) type sequence in one of its forms, but the uncommon *A A B B A A B B . . .* type in another. Such a double-layer sequence is made possible because the larger phosphorus atom is barely able to enter the metal lattice as a true interstitial; but it can still do so by creating spacious intermetal gaps, so that S and P atoms can remain isolated within the host metal lattice. The metal–metal bond, however, is weakened compared with carbides and nitrides, the metal–non-metal bond strengthened; this is well reflected by the physical properties.

Interstitials as Structure-Controlling Agents

The basic theme of structure sequences throughout may be summarized schematically as shown in *Figure 2.8.* It manifests itself in various forms, like the elemental progression within the Periodic System, or temperature change within any one transition metal, or the sequence upon interstitial addition in a given binary system. The theme is clearly reflected in *Figure 2.3*; note, for instance, the sequences $V \rightarrow V_2C \rightarrow VC$, $Nb \rightarrow Nb_2C \rightarrow NbC$; within a given metal (e.g. $\beta \rightarrow \alpha$-Ti, etc.) and with increasing atomic number (e.g. for the Pt-Group metals: Ru → Rh, Os → Ir, etc.).

A notable hybridization of the processes occurs, for example, for titanium where there exists no 'Ti_2C' but the change Ti → TiC (BCC → FCC) occurs directly, i.e. the HCP stage is transposed to the metal itself and the primary α-(Ti,C) solid solution which evolves into a pseudo-compound (see Chapter 4, *Figure 4.1* and discussion). Structurally we have here really a link-up between interstitial compounds and primary solid solutions. *Thus the function of the interstitial atom is to control these structure transformations, but its presence is not an intrinsic condition to their occurence.* An instructive example of this aspect has recently been reported[14]: in single crystals of vanadium, a metal BCC → HCP transformation can occur through the presence of a trace of nitrogen; this is not simply V_2N but a structural change in the matrix lattice which is stimulated by only a few interstitial atoms. Seeing that titanium, the neighbour of vanadium in Group IV, also shows this transformation, but already in the pure metal, it may well be asked whether even in 'pure' Ti (and Zr) the phase change is not actually due to a minute interstitial content and would be suppressed in ideally pure Ti*. Very probably, however, such a question must remain academic, as the very least traces of interstitial atoms (e.g. C, N, O or H) may fundamentally be irremovable from the Ti lattice. This raises the subtle point whether in some metals the affinity to interstitial atoms is not so extreme that the presence of at least some should be considered as an integral part of the metal lattice which has become electronically necessary. In practical terms, the gettering action of Ti and Zr (or, yet more extremely, the use of the Group III metals, such as Sc, Y, rare earths, as scavengers) is well known and, on approaching complete purity, resistance to last-trace removal may well become impenetrable. How important this is is shown by the present-day drive towards highest-purity metals which has as a main aim that of interstitial removal, with a profound effect upon the properties of the metal (e.g. the brittle-to-ductile transition). A fundamentally new class of metals is thus emerging by

* A related instance of the theme of a given (otherwise typically 'pure' metal) structure requiring some interstitial atoms to become stabilized, occurs in manganese. Here the high-temperature β-Mn form, which for the pure metal is unquenchable, can occur stably as a carbide, namely the π phase in Fe–Cr–W–C steels[15] requiring carbon as an integral atom (to be considered later); also γ-Mn (FCC), the next higher form, becomes stabilized through carbon to much lower temperatures, as does of course austenite.

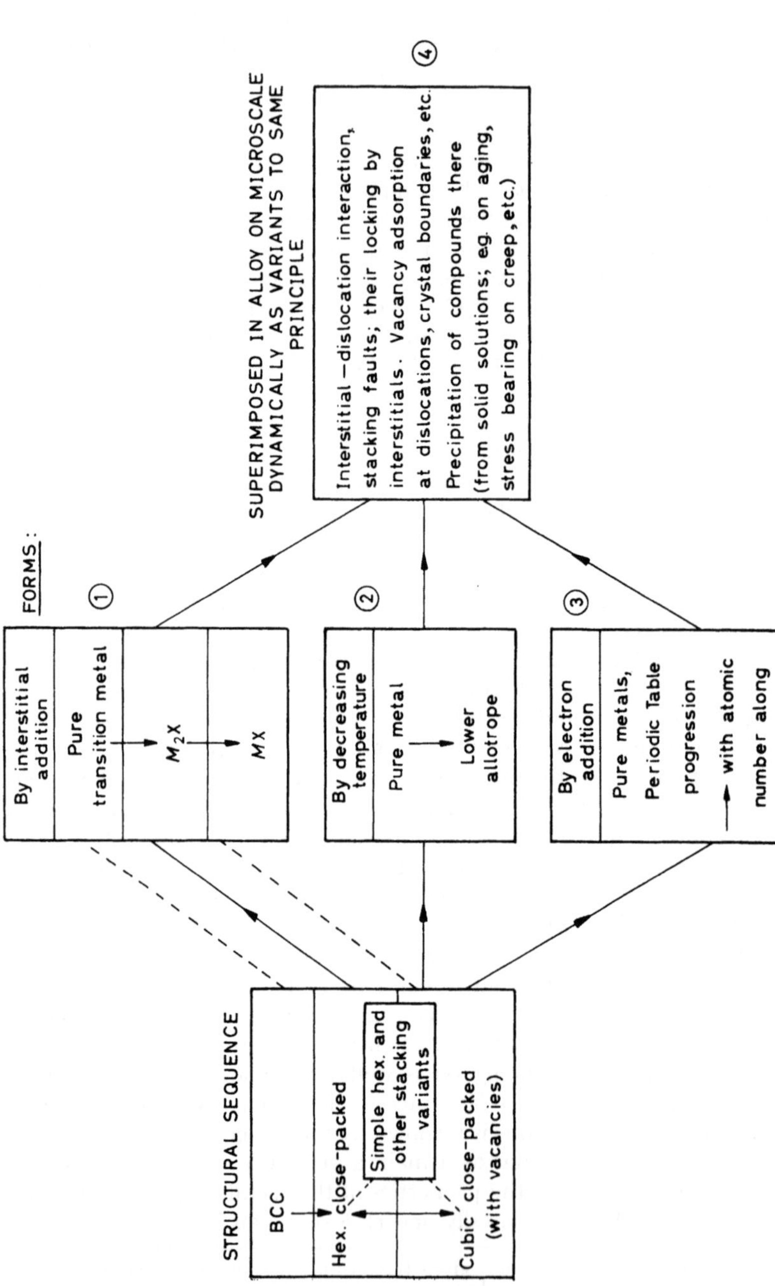

Figure 2.8. Evolutions of interstitial effects

the use of vacuum metallurgy and allied techniques. This affects, of course, not only interstitial atoms in lattice solution but also those in segregation at grain boundaries, dislocations and other imperfections.

We must in principle expect that the capacity to absorb interstitial atoms at crystal interfaces may in some cases become greater than the solubility within a single crystal lattice. This can well be envisaged, like grain-boundary diffusion, as due to a 'loosening-up' of the structure and the consequent increase in the number of vacancies at and near these points; dislocations similarly attract an aggregation of interstitial atoms which tend to impede their movement and cause hardening (see later). Similarly, cubic close-packed titanium hydride is capable of nucleating from Ti metal: it does so coherently in platelet formation, such that the close-packed (111) planes of the hydride lie parallel to the $(10\bar{1}0)$ slip planes of parent Ti, a phenomenon which could be observed directly under the electron microscope. Through their effect on slip and the interference of such platelets (whether nascent or advanced) with dislocation movement, H, C, O or N atoms in solution in Ti can to a large extent govern properties such as yield point, hardness and work-hardenability.

The effect of ductile $\longleftrightarrow$ brittle transition in metals, notably Cr, Mo and W, must be considered in this light; the poor ductility is generally attributed to the presence of very small amounts of interstitial atoms. The detailed mechanism is still very much the subject of current investigation, but basically this embrittlement can happen, for example, through formation of thin hydride or nitride platelets coherent with the matrix, which would interfere, under stress conditions, with the twinning process along certain lattice planes and the mobility of dislocations. Upon impact, fracture would then naturally also be initiated at such platelets (or, put more generally, at any interstitial microsegregates, from nucleation to the precipitation stage). The success in producing e.g. ductile chromium, or molybdenum with much lowered transition point, was thus largely bound up with the development of vacuum metallurgy and of extreme purification processes, such as zone refining. Besides their effect on dislocation density and mobility, interstitial segregates have a direct 'frictional' effect upon flow stress: the latter is highly temperature-dependent, the more so the higher the impurity content, as shown by O_2- and N_2-loaded titanium.

Whilst brittleness is most pronounced in 'pure' Cr, Mo and W (the Group VI metals), owing to the mere trace amounts of interstitials needed to make this a seemingly 'inherent' property, a ductile $\rightarrow$ brittle transition temperature can be acquired also, for example, by niobium, if the interstitial (O, N or H) content is sufficiently high. This situation applies in a varying degree to all Group IV, V, VI metals. The increase in hardness involved is in fact a well known practical measure of interstitial content, e.g. in Ti, Zr or Mo.

The appearance of defect lattices in interstitial alloys frequently precedes the change-over in stacking sequence. Thus, in a system such as NbC-Nb_2C, the non-stoichiometry, i.e. loss of carbon from NbC towards $NbC_{0.75}$, which occurs in niobium-bearing, creep-resisting steels, may be regarded as a first stage in a process which culminates in a stacking sequence switch towards hexagonal. That heat treatment can also cause such changes may be illustrated e.g. on vanadium nitride, 'VN', where it was found that, with increasing temperatures, nitrogen is first expelled from the cubic lattice down to $VN_{0.71}$; this is followed first by a transformation to hexagonal '$VN_{0.5}$,' then by N deficiency, to a range $VN_{0.43}$–$VN_{0.37}$. In 'NbN', too, the non-stoichiometry is marked, about 5 per cent of the nitrogen sites being vacant.

The system Ti–N is instructive in this connection. Here, a relatively large solid solution of nitrogen in Ti entails, first, the stabilization of the Ti (hexagonal) form, followed by FCC 'TiN' with, initially, a high-defect lattice before stoichiometry is attained: i.e. the additive solution of interstitial nitrogen in (hexagonal close-packed) titanium is matched by the subtractive one of interstitial vacancies in (cubic close-packed) TiN. Similar relations apply to Zr–N.

These are instances of metal excess through non-metal deficiency, but vacancies can in other cases go in the direction of metal deficiency: thus in the W–N–O system occurs a defect *ternary* 'WX' compound of NaCl-type structure[16] which in fact is $W_{0.62}X$, i.e. with a high proportion of metal vacancies. Similarly, ferrous oxide, FeO, has a number of vacant iron sites. A further important point is that vacancies, like atoms, can form '*super-lattices*', i.e. there can be ordered or random vacancies and varying degrees of order. Such a case is known in the Fe–N system where the random solid solution of nitrogen in γ-iron is followed by Fe_4N (γ' phase), which really is isomorphous with it but contains N atoms and holes in ordered arrangement, as against the random one in γ; in other words, ordering of the interstitials has become necessary, if more of their kind are to be accommodated—as happens indeed in other ordering processes in nature.

A further example of order–disorder transformations in interstitial alloys occurs in Ta–H. Here the extensive (random) primary solid solution of lattice hydrogen divides at very low temperatures into two, namely almost pure Ta and the ordered but otherwise isomorphous Ta_2H.

Yet another type of close-packed stacking sequence can occur, namely *A B A C A B A C* . . . (*Figure 2.7*), known for instance in the intermetallic compound Ni_3Ti but also amongst interstitial compounds. This type (hexagonal) has its cubic counterpart in Ni_3Al. These Ni_3Ti and Ni_3Al phases, together with their interstitial analogues, present different modes of precipitation prevalent in different composition fields of alloy steels (and, especially, nickel-base alloys of the 'Nimonic' kind), and add two further variants to our fundamental theme of phase changes by changes in layering.

This variability in stacking sequence is only one of the many common factors tending to blur the boundary between the concepts 'element', 'solid solution', 'intermetallic' and 'interstitial' compound, and whilst the *ideal* case of each is clear enough, the fundamental fluidity must be stressed. Continuous transitions are possible, which have direct practical consequence on phase transitions in alloy steels and heat-resisting alloys of various bases. It would thus appear that the more metallurgical research advances, and the more precisely detailed knowledge is accumulated, the *less* precise the previously conceived distinctions become (which, incidentally, is the case not only in metallurgy but also in many other fields of science and between sciences).

Amongst intermetallic compounds, a very frequent and flexible structure is that of the nickel arsenide, NiAs, type, which comprises pyrrhotite, FeS, and other transition-metal sulphides. These compounds give an interesting lead-over from intermetallic to interstitial phases, for the NiAs structure is geometrically the same simple hexagonal one as in WC or MoN. Thus we can have, in equivalent interstices, on the one hand the very large, readily polarized atoms of As, Se, Sn, S, Te or Bi and, on the other, the very small C or N atoms. The common denominators are the transition-metal base lattice and the 1:1 proportion. It might have been expected that the axial ratio c/a were the smaller for the small-type interstitials (C, N); in fact, however, it is the larger, because a closer mutual approach of the transition-metal atoms within the basal layers is allowed, whilst the larger interstitials (As, S, etc.) protrude into these layers, causing greater separation and weakening of the bond (also manifested by the much lower melting points). The NiAs phases are also characterized by non-stoichiometry which, in different cases, can proceed to both excess or deficiency as from 1:1. Thus 'FeSb' can exist only with an excess of iron and, rather remarkably, is attained by *insertion* of additional Fe atoms into the FeSb lattice interstices. We meet here an 'interstitial compound' of a novel type, made possible because the Fe atom is in this case relatively *small* enough to fit in (a case of an inverse Hägg's rule). The crucial condition for stabilizing this type of interstitial compound is the existence of large size ratios between atomic species, whether the transition metal atom is the larger (WC) or the smaller (NiAs). A case of *non-metal* excess, as against that of metal in FeSb, can occur in FeS, where the mechanism is the more orthodox one of subtraction of iron from some of its sites, the sulphur lattice layers remaining unchanged.

It is clear from the foregoing, and indeed from a glance at *Figures 2.3* and *2.4*, that the bond strength between metal and non-metal atom decreases with increasing atomic number within each period, and this is also reflected in the partition of metal atoms between matrix and carbide in ternary systems $M(1)$–$M(2)$–C, and more generally in equilibria wherever two or more metal species compete for one non-metal. By implication, this relation dominates of course alloy-steel equilibria profoundly, as discussed below and in Chapter 4. As an illustration, it has been shown[17] that when the First Long Period Metals are added in sequence to iron–carbon alloys, the partition as between ferrite and cementite runs as follows:

M	Cr	Mn	Fe	Co and Ni	Cu
Partition ratio (M in Fe_3C)/(M in ferrite)	28	11	0·9	≈ 0·2 – 0·3	0

(The elements Ti and V, preceding Cr, form of course their own stable carbides but are still slightly partitioned with the matrix and certain other carbides.)

2.3 STEEL: CONSTITUTIONAL ASPECTS

The structural system of carbides, their melting points and other properties have, of course, a specific bearing on the formation of carbides in alloy steels. The author had reviewed this in earlier work[15,18], and numerous investigations have been presented in recent Symposia (e.g. by the Iron and Steel Institute[19,20]) and by Nutting's, Quarrell's, Honeycombe's, Andrew's and Owen's schools. As pointed out already, the critical position of iron, manganese and chromium carbides in the Periodic Table, relative to those on the left (stable) and those on the right (unstable), represents one of the key features underlying the properties of alloy steels and is closely linked to the primary solubilities of carbon in austenite and ferrite (and, unstably, in martensite). Austenite we may in fact regard as a continuation, structurally, of the cubic close-packed carbides of Groups IV and V, derived by carbon abstraction, or as a highly defect 'carbide' lattice, to give an occupation by C of only up to 1 in 12 sites. Also the four polymorphous forms of metallic manganese (α-, β-, γ- and δ-Mn) are capable of being reproduced structurally as alloy-steel constituents, even without the presence of Mn: thus they occupy a borderline position, playing the double role of 'matrix phase' and of 'carbides'[15].

This position makes it possible to bring about precipitation of various stable or unstable carbides, not only from the α- and γ-matrix but also from another carbide. Under decarburizing influences, this may evolve to that of intermetallic compounds like the σ- or χ-phase. Additionally, certain *ternary* carbides such as the η-carbide, M_6C (Fe_4W_2C, prominent in high-speed steels), form under certain Group [VII–VIII] with Group [V–VI] combinations, whilst known binary carbides, such as $M_{23}C_6$ (basic: $Cr_{23}C_6$), conversely can form some ternary isomorphous carbides, and notably carbo-borides. The same applies naturally to other interstitial compounds, e.g. nitrides, and to intermetallics, where for instance σ- or Laves phases can occur in binary or ternary forms. These matters will be discussed in later Chapters but, seen methodically, it may be said that alloy steels and their properties are founded on an interplay of phases which, whether matrix or secondary, form one structural family, and in which interstitial elements and their 'go-between' character form an essential link. They provide a means of steering the lattice movements by directing or partitioning the transition metals according to electron structure and atomic size: heat treatment, mechanical working and alloy additions are only the external means to effect this and to control the degree of solution, dislocation movement, nucleation, coherence in growth and lattice strain.

2.4. 'INVERSE' INTERSTITIALS AND FURTHER SYSTEMATIC CONSIDERATIONS

Size factor as a structure-governing one leads to another class of 'interstitial alloy' which might be termed 'inverse interstitials'. For here the non-metal atoms form a primary lattice in which the *metal* atoms enter interstitially. An example are phases[21] between Group IIB atoms (S, Se, Te) and Group IB atoms (Cu, Ag). The former occupy a BCC lattice, if $R_{metal}/R_{non\text{-}metal} > 0{\cdot}60$, and a FCC one if $< 0{\cdot}60$; the latter occupy the holes which can occur either randomly or ordered. We might liken this again to a 'Hägg's rule in reverse', insofar as the simple, cubic close-packed, structure is this time favoured by the *non-metal* atom, when the radius ratio is small enough.

Another borderline example of an inverse interstitial compound is molybdenum disulphide, MoS_2, where we have a *hexagonal* close-packed layer structure of large S atoms, with small Mo atoms inserted at two-fold repeat. However, here the bonding is non-metallic and extremely weak between the hexagonal sheets, as in graphite, making for easy glide and the good lubricating properties of this material. Although many of the present interstitial compounds are also layer structures, there can be no such easy glide owing to the strong interlayer bonds, and we may contrast the great hardness of metallic interstitial compounds ($R_M > R_X$) with the softness and easy glide of the inverted kinds ($R_X > R_M$).

Referring to the sizes of the interstitial atoms C, N, O, B, H (see *Figure 2.1*), these are inherently about as precise as the 'size of a ball of cotton wool' when subjected to varying external forces. The radius for nitrogen was generally thought smaller than that for carbon, and this is indeed the case, for example in TiN and TiC. However, Jack's[12] work on the Fe–N and Fe–C systems has shown that the sizes are virtually the same, not only in the compounds but also in the nitrogen and carbon austenites and martensites, where the Fe atom is irregularly displaced from its lattice positions (see Chapters 4 and 5). It appears that this near-equality in size of the C and N atoms is associated with the weaker types of bond to the Group VIII transition metal, whilst the strong bonds for the hard metal compounds in Groups IV–VI are required to bring about the effective size difference.

Borides occupy a special position amongst interstitial compounds because here the B–B atom bond becomes more prominent relative to the *M*–B bond; this gives rise to the characteristic sequence of structures in which (1) isolated bonds, (2) chain bonds, (3) double-chain bonds, and (4) network bonds of B–B type can prevail (see Chapter 6). This non-metal–non-metal bond cannot yet occur for carbides and nitrides but evolves further in silicides—a trend which is clearly size-dependent (approximate radius for O = 0·66; N = 0·70; C = 0·77; B = 0·86; Si = 1·17 Å), although the effect of electron structure obviously superimposes.

This theme develops in an interesting way if we consider briefly also the borderline cases of *phosphides* and *sulphides* within the frame of the series (truly interstitial elements printed in bold, elements in hybrid position italicized).

$$\begin{array}{ccccccccc}
\mathbf{B} & \longrightarrow & \mathit{Al} & \longrightarrow & \mathrm{Ga} & \longrightarrow & \mathrm{In} & \longrightarrow & \mathrm{Tl} \\
\downarrow & & \downarrow & & \downarrow & & \downarrow & & \downarrow \\
\mathbf{C} & \longrightarrow & \mathit{Si} & \longrightarrow & \mathrm{Ge} & \longrightarrow & \mathrm{Sn} & \longrightarrow & \mathrm{Pb} \\
\downarrow & & \downarrow & & \downarrow & & \downarrow & & \downarrow \\
\mathbf{N} & \longrightarrow & \mathit{P} & \longrightarrow & \mathit{As} & \longrightarrow & \mathrm{Sb} & \longrightarrow & \mathrm{Bi} \\
\downarrow & & \downarrow & & \downarrow & & \downarrow & & \downarrow \\
\mathbf{O} & \longrightarrow & \mathit{S} & \longrightarrow & \mathit{Se} & \longrightarrow & \mathrm{Te} & \longrightarrow & \mathrm{Po}
\end{array}$$

as 'interstitial' atoms to partner the transition metals. The passage from N to P, as that from C to Si, results in an increased metal–metal distance so that, in the phosphides, the metal–metal bond is greatly weakened; however, with further increase in size of the 'interstitial' atom, strength is regained through the increasingly metallic bond, evolving into true intermetallic compounds (arsenides, antimonides and bismuthides).

In the sulphides, the tendency is one to the 'inverted interstitials', as discussed already, with layer lattice formation and easy glide (reaching an extreme in MoS_2); in the selenides and tellurides, this still applies but degenerates to greater hardness through the gain in metallic bond strength.

In this sense, the new classes of semi-conducting materials developed, such as the 'III–V' compounds (Al, Ga, In antimonides, etc.), although clearly not interstitial but intermetallic phases, can at least be recognized as structures evolving from the interstitial ones. The silicides occupy a transitional position (cf. Chapter 7).

Thus the science of new materials such as *semi-conductors* and *intermetallic compounds* clearly is closely linked to the concept of interstitial alloys, and the vast exploitation of their properties in recent years, in transistors and other electronic devices, needs no emphasis.

The 'N'- and 'P'-type defects effectually behave in many ways like interstitial particles. Thus solid silicon and germanium can act as a solvent, comparable, in analogy with liquids, to the solution of OH ($\approx$'N') and H ($\approx$'P') ions in the case of water. If Group IIIB atoms are added to Si and Ge (Group IVB), the analogy is to a base; if Group VB atoms, it is to an acid, and the behaviour as electron donor and acceptor is quite comparable (cf. Selected Literature at the end of this Chapter).

2.5 BONDING IN INTERSTITIAL COMPOUNDS

There has been considerable controversy on the theory of interstitial compounds. Rundle[22], following Pauling's basic concepts, considers the 'MX'-type compounds as electron-deficient structures, in which the non-metal forms more bonds than it has bond orbitals, and suggests that the non-metal–metal bonds are aided by hybridization of *sp* and *p* bonding orbitals. The idea of 'half-bonds' is introduced, which means that one orbital with one electron pair is used to form two bonds. It involves an internal rearrangement of electron shells of the metal atom which would enable two of the six bonds of the non-metal atom to become electron-pair bonds, and by *resonance* all six would become equivalent and be statistically distributed amongst all atoms, resulting in Pauling's fractional bonds[23]. These interstitial compounds can then be regarded as largely covalently bonded, and their brittleness and hardness be explained by *directionality* of the metal–non-metal bonds; this bond is octahedral, which would also explain the NaCl-type crystal structure.

Rundle's concepts were, however, partly challenged by Hume-Rothery[24], particularly as octahedral occupation would equally apply to close-packed hexagonal structures, which are not adopted by MX compounds, and on the grounds of known atomic distance and melting-point data; the much simpler picture is put forward that the FCC metal lattice is adopted because this permits the metal atom to have mutually perpendicular bonds to six neighbours, a condition satisfying the non-metal (C,N) bonds in the lattice. The covalency of bond can still be construed as valid but, as explained earlier, the viewpoint of essentially metallic bond is a justified and in some ways preferable alternative. Also, comparing carbides and nitrides, Hume-Rothery concludes that the latter must have a lower melting point, because the bonding in the nitride will be weaker, involving fewer electrons per atom, as is borne out by fact (see *Figure 2.4*); the metal electrons added on passing from Group IV to VA tend to weaken the bond and thus also to lower the melting points.

The theory has not so far been extended from the cubic to the hexagonal close-packed structures of M_2X type but these are in principle likely to be similar to the cubic case.

The bond energies have been extensively studied from the thermodynamic viewpoint. Thus, for example, Brewer and Krikorian[25] plotted the energies for various MX compounds and compared these curves with those for the heats of sublimation of the pure metals. These bonding-energy curves (see *Figure 8.1*, p. 350) reveal a close similarity and also the fact that the type of bonding in all these interstitial compounds still resembles that in metals. On the other hand, the curves for the dioxides and fluorides of the transition metals, which have *ionic* bonding, are very different from those for the metals or for the MX compounds.

It is important to note how, amongst interstitial compounds, for instance of a given transition metal, the nature of the bond can vary radically. Thus, for instance, for *chromium* as base, Cr-*carbides* tend to be homopolar, Cr-*nitrides* metallic and Cr-*oxides* ionic in character. However, the distinction is by no means so clear-cut: a *degree* of the other two types of bonding applies to all three cases; the bond for each compound can indeed vary with temperature, pressure and composition. Chromium with its borderline position provides a good example of the fluid nature of electron distributions and of the gradualness of change. But it is valid for the other transition metals, and the presence, in a given case, of an undertone of all types of bonding is one of the characteristic features of interstitial compounds. The same applies to the bond-type *variation* also within the transition-metal block in the Periodic Table, for a *given* interstitial atom, as will become evident later on. For example, when plotting the x-ray emission spectra of niobium metal, of Nb_2C, NbC and Nb_2O_5, it could, from the shape of these curves, be shown that the bonding in Nb and Nb_2C is similar (metallic), but that in NbC it was more akin to Nb_2O_5, indicative of partial covalent bonding[26].

It is outside our scope to deal with the electron theories of the transition metals and alloys (for example, the influence of interaction of incomplete *d* orbitals, and the idea of bond hybrids and their effect upon crystal structure, magnetic properties and alloying behaviour); and it must suffice to refer for instance to the work of Mott and Jones, Pauling, Zener, Griffith and Orgel, as well as to that of Hume-Rothery's school, Raynor, Smallman (see Selected Literature), and to a valuable brief review by Kiessling[27] on the manner of bonding in interstitial compounds. The subject, even as the pure metals are concerned, is still highly controversial and incompletely explored, and this applies even more to the effect of interstitial addition.

The number of bonding electrons per atom when related to the co-ordination number of the transition metal atom can explain many of the properties of their carbides, borides and silicides (see e.g. the theory of Robins[28]).

Furthermore, the bonding energies in these and the oxides and halogenides have been studied e.g. by Brewer's school in the United States[29] and by Dempsey[30] who gave a useful survey: proving in the Group IV–VI hard metals an electron donation from the non-metal to the metal, which involves a change in the light of modes of splitting of *d* bands. Much theoretical work done on the *pure* transtition metals also is inherently relevant to interstitial alloys; for example, theories of the hybridization of bond orbitals in the pure metal structures[31], which underlie transformations of BCC→FCC/HCP-type, contain the seed for those of interstitial alloys in effecting similar electron and structure changes. In this sense, a generalization could well be proposed which regards most atomic bond theories developed for the pure transition metals as also implicitly referring to interstitial alloys—with increasing emphasis towards the earlier Groups.

2.6 A NOTE ON THE INTERSTITIAL ELEMENTS WHEN FREE

It may be noted that those atoms which are potential interstitial atoms in transition metal compounds, C, N, B, Si, O, are also those which, when alone in *elemental* form, or in certain mutual combinations, provide some of the *hardest* substances existing. For example:

Carbon: diamond;

Nitrogen: not as element, but in BN ('Borazon') and Si_3N_4;

Boron: not as element, but in B_4C and BN;

Silicon: not as element, though this has diamond structure, but in SiC (carburundum), Si_3N_4 and SiO_2 (quartz and its polymorphs);

Oxygen: not as element, but in SiO_2 and minerals where SiO_2 frameworks predominate—Considering the close electronic Si-Al relations, we might also include Al_2O_3 in its gemstone forms and in minerals where Al can substitute Si in SiO_2 tetrahedra.

In the light of the Periodic Table, this may be represented as shown in *Figure 2.9.*

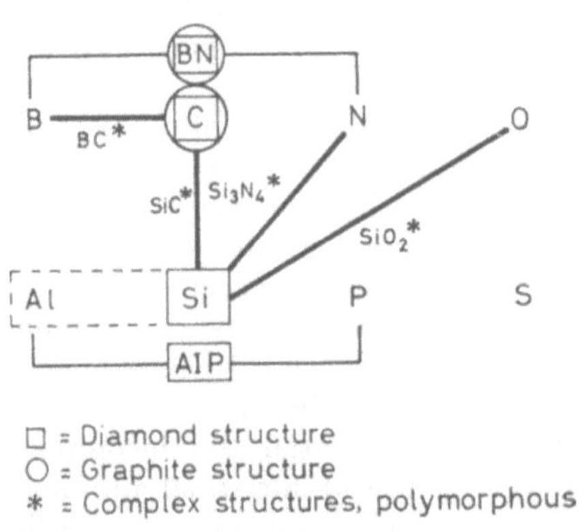

Figure 2.9. Compounds of some 'free interstitial elements' when combined

A further common factor is that these potentially interstitial atoms *can*, of course (and thermodynamically more readily), adopt the softer layer-type hexagonal forms, like the graphitic ones in carbon and in boron nitride, and that it requires extreme pressures to produce the hard forms (just as the pressures, it might be argued, to which the interstitial atoms are subjected in the hard carbide lattice itself, e.g. tungsten carbide, through the surrounding metal framework).

'Borazon', the diamond-type form to the normally graphitic boron nitride, BN, clearly represents a position symmetrical to carbon (see *Figure 2.9*) and is a close analogue to it. It seems very possible (though this is not known) that the transformation from graphite type to diamond type could be aided by the presence of seeds of silicon or of aluminium phosphide, AlP, both of which already have the diamond structure (but less hard, in accordance with the larger atoms concerned).

The synthesis of cubic from hexagonal boron nitride can be achieved by very high pressures at temperatures of about 1,500°C, but it can be catalysed to take effect at lower pressures and temperatures by the presence of alkalis and alkaline earths and their nitrides. The extensive work on synthetic diamonds, by Wentorff's Group[32] at the General Electric Co.*, by Milledge[34] on the graphitization mechanism in diamond, Grenville-Wells[35] and by Lipschutz and Anders[36] on Fe–Ni-based meteorites is here particularly interesting. A metastable cementite-type compound, Fe_3C–Ni_3C ('cohenite'), can be made to act as a catalytic seed from which the diamond originates (rather than graphite), consistent with what will follow later on this subject (cf. Chapter 4). The direct conversion from graphite to diamond can also be achieved by using high static pressures and a 'flash-heating' technique; a new pressure–temperature transformation diagram for carbon has been derived by Bundy[37]. Similarly, the synthesis of *boron carbide* under high pressure–temperature conditions has recently proved to be possible.

The similarity of the cubic ⟷ hexagonal transition obtained in such cases to that amongst the interstitial compounds discussed is striking, and a detailed comparison, based on *Figures 2.3* and *2.9*, would be fruitful, though going too far in this account.

Hydrogen, the further potential interstitial atom omitted from the above list, does not in itself form hard compounds with other interstitials; it should be recalled, however, that the various crystalline forms of ice, H_2O, are isostructural with those of silica, O_2Si (cristobalite, tridymite, quartz; even vitreous ice exists, like silica), and we might well enquire from bond relations whether, at *very low temperatures*, these forms of ice would not attain exceptionally great hardness.

Reverting to interstitial alloys, we should emphasize one further point. *Pressure*, as a thermodynamic variable, has so far tacitly been assumed constant (at atmospheric). However, high pressures are certain to have a radical effect upon interstitial alloy formation and, in particular, would be likely to stabilize many interstitial compounds and structures which so far are either unstable or non-existent. In this sense, *Figure 2.3* should really be regarded as an isobaric section, and with increased stability relations, many of the structures involved would generally tend to shift to the right in the Periodic Table, implying new compounds and phase transformations.

* This is well described in '*Progress in Very High Pressure Research*'[33], which also contains various other papers of present relevance, e.g. by Lazarus on the mobility of lattice vacancies and interstitials at high pressures.

This, and at the same time Ubbelohde's idea of regarding the interstitial atoms (particular hydrogen) as of metallic state, can be linked to the concept of 'metallic' diamond. By extreme pressure, the Group IVB elements (C, Si, Ge) can all be rendered metallic (as the next one, Sn, is normally already); so can compounds of symmetrical Group [III/VB] and [II/VIB] combinations[38]. The metallic diamond is of greater density than the normal one. This may be viewed as a three-dimensional analogue to *graphite* which indeed can, two-dimensionally, already be taken as metallic.

We might here note that the FCC → BCC transition amongst the metals, and in the interstitial compounds' progression of the Periodic Table as discussed above, can be directly 'copied' by high-pressure techniques. The same is true in ionic crystals. Thus sodium chloride (FCC) can be converted to the cesium chloride (ordered BCC)-type structure at high pressures (shown e.g. by using a high-pressure x-ray technique)[39].

It is quite generally the case in the solid state that increasing pressure tends towards increased metallic bonding, and the view is sometimes expressed that, at sufficiently high pressures, *all* elements will become 'metallic' as the electron orbitals begin to merge. To obtain metallic hydrogen (as present already e.g. in Pd), a pressure of the order of 10^6 atm would be necessary which is not at present attainable.

2.7 VACANCIES AND ELECTRONS AS INTERSTITIALS. NON-STOICHIOMETRY

The point has been made in our initial classification that vacant interstitial lattice sites can, in their formal behaviour, in many ways be considered as though being interstitial atoms. This leads to the whole subject of non-stoichiometry, and whilst it is not intended to enlarge on this too much, some additional observations are necessary here.

The similarity in physical behaviour of vacancies to interstitial atoms is emphasized by three further effects:

In the first place, holes in metals can be '*quenched in*' from temperatures near melting point, and subsequently annealed out to produce strain-hardening effects just as carbon or nitrogen can. Thus a very pure metal, e.g. copper, when quenched very rapidly, remains soft at first, as the high temperature vacancies are retained in random dispersion, but when it is then warmed ('tempered'), hardening can occur, through diffusion of vacancies to dislocations, or through their lattice concentration and local distortion, quite similarly to interstitials.

Secondly, lattice holes in materials such as uranium, graphite or ceramic oxides, created through *nuclear irradiation*, can be healed out again in time, and more rapidly so at elevated temperatures, as is the case with (non-equilibrium) interstitial atoms, say those of occluded hydrogen or of carbon in metastable martensite, which would coagulate to a carbide or to graphite on tempering, as vacancies could to voids, at dislocations and grain-boundaries. The effect of *irradiation* is to store energy in the lattice which increases as atoms are knocked into interstitial positions and vacancies form; quite considerable heat can be evolved on healing out this damage (for heavy irradiation, spontaneous annealing can in fact occur, and extreme temperatures be produced).

The third property of vacancies which they share with interstitial atoms is the tendency to *aggregate* at dislocations, grain boundaries and other lattice imperfections; we shall return to this presently.

In many compounds, non-stoichiometry is, of course, a stable condition (e.g. in UO_{2+x}, FeO_{1+x}, Nb_4C_3), and these we may compare more properly with permanent primary solid solutions of interstitial holes in the ideally occupied lattice; in these, supersaturation can also occur to cause precipitation, so that we must distinguish between stable and unstable vacancies, as for interstitial atoms. The example of uranium dioxide is here illustrative; non-stoichiometric UO_{2+x} can slowly be decomposed into stoichiometric UO_2.

In their practical consequences, such vacancy defects influence metal properties like creep, fatigue and hardenability greatly and, with due qualifications*, quite on the same plane as atomic interstitials. Furthermore, creep and fatigue straining in metals, like irradiation, are potential agents to create vacancies in large amounts, and this can constitute a major source of hardening and failure. Another source of vacancies is provided by volume changes within a solid metal through transformation or phase precipitation. Superlattices and varying degrees of order–disorder transitions of vacancies, as of interstitial atoms, can occur as a further common feature. Lastly, vacancies can also be produced by the Kirkendall effect, i.e. by interdiffusion of a metal couple at differential rates across the boundary, which can leave lattice (or larger-scale) voids within that metal, possessing the higher diffusion rate.

The tendency to non-stoichiometry amongst carbides and nitrides has already been referred to, and is also exemplified in *Figure 2.3.* To V_4C_3/VC and Nb_4C_3/NbC we may add a deviation[4,26] of 'TaC' to $TaC_{0\cdot 96}$, typifying a frequent inability to attain perfect stoichiometry.

How interstitials tend to segregate and precipitate at stacking faults has now been demonstrated by much beautiful electron microscope work. *Figure 2.10* shows an example[40] on a stainless steel (18 per cent Cr, 10 per cent Ni, 1 per cent Nb, 0·1 per cent C). The steel is single-phase austenitic as solution-treated but, on tempering, dislocation-initiated stacking faults appear, with eventual finely dispersed NbC forming.

The diffusion of vacancies is naturally also subject to similar laws as that of interstitial atoms, as will be further discussed later. It should be noted that, when a migrating lattice defect meets a dislocation, it tends to be caught; the dislocation thus becomes a sink for defects and a nucleus for precipitation of second phases. Thus dislocations tend not only to attract interstitials and vacancies into themselves but also, as Cottrell expresses it, to be surrounded by a definite 'atmosphere' of this kind, diminishing in density with distance. In their movement, dislocations can carry these atmospheres with them or leave them partially behind, trail-like, to dissipate in time or to remain in

* One obvious difference is that vacancies can be abundantly 'created' within a solid by external means, unlike normal elemental interstititals (but like some of the latter through fission); however, given this source difference, their behaviour, once created, is comparable.

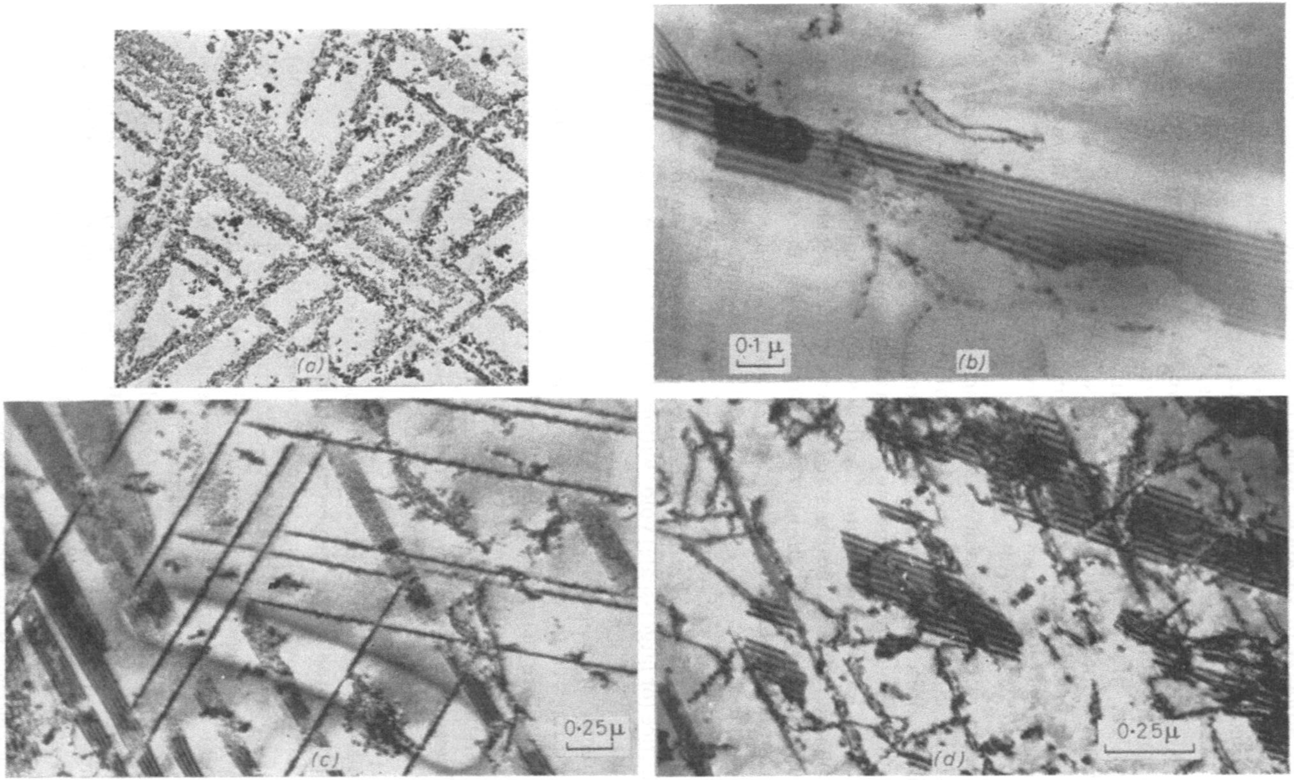

Figure 2.10. Electron micrographs from an austenitic steel (18 Ni/10 Cr/1 Nb) *showing lattice precipitates of* NbC *carbide in various states.* (From Van Aswegen, Honeycombe and Warrington[40])

(*a*) **Banded** precipitate formed on tempering at 700°C, 24 h, after solution treatment: coarser precipitates between the bands (carbon extraction replica × 16,000) (*b*) **Stacking** faults grown on dislocations formed on tempering at 700°C, 5 h, after solution treatment (thin film, × 80,000); (*c*) Further development of stacking faults and precipitates, after 350 h at 700°C; number of faults here $1{\cdot}5 \times 10^{12}/cm^3$ (thin film, × 32,000); (*d*) Slow precipitation, after tensile deformation and holding at 600°C for 2500 h; many more stacking faults appear (here $24 \times 10/cm^3$), increasing with time: deformation multiplies the stacking faults and points of carbide precipitation (thin film, × 64,000) By courtesy of the Editor, *Acta Metallurgica*; photographs by courtesy of Prof. R. W. K. Honeycombe clusters. This has a profound bearing on the *yield point* in that the latter occurs only when there is a break-away of dislocations from the interstitial atmosphere (observed e.g. for hydrogen in titanium), and not at temperatures at which the dislocations move with their atmospheres.

Furthermore, the ductile-to-brittle transition temperature notably in chromium (but common to all transition metals) is not a constant but subject to rate of cooling from the high temperature, a fast rate tending towards brittleness, slower rates towards ductility. This, too, illustrates the point, as it depends sensitively on the relation between dislocation and interstitial mobility.

A major effect of interstitials is not only their tendency to aggregate at dislocations but, conversely, that dislocation movements can be stopped through interstitials, to entail hardening. At a more advanced stage, the same can be expressed by saying that actual carbide nuclei precipitate at dislocations, whose movement is stopped by such nuclei already existing. This is the case in creep-resisting steels and alloys, and explains for instance the action of niobium carbide particles in niobium-bearing austenitic steels. There are means, such as 'warm work', to increase the density of dislocations, and thus that of interstitial nuclei and precipitates, to give grain refinement as well as enhanced creep strength. Carbides, borides and other interstitial compounds can thus have the further important virtue of acting as grain refiners whilst in fine dispersion and *in statu nascendi* (i.e. essentially in matrix solution, but in aggregates producing marked local lattice distortions). Dislocation 'pile-ups' occur at such interstitial-caused obstructions, a phenomenon now much studied by the thin-film techniques under the electron microscope.

Dislocations are, however, not only a sink but also a potential *source* of vacancies, an effect which makes them important in our understanding of them as the origin of cracks and fracture during fatigue or creep.

The number of dislocations (which can be calculated from crystallographic data), and thus the number of sinks for interstitials and vacancies, can vary between $10^8/cm^3$ for normal annealed metals and $10^{12}/cm^3$ for heavily cold-worked materials; thus cold work would, on the one hand, create a storehouse for interstitials lodged at dislocations and, on the other hand, insofar as interstitials will obstruct the dislocation movement, cause mechanical hardening. Such obstruction could cause a pile-up and, under more force, a merging of dislocations which, as Stroh[41] has so beautifully shown, is the germ of a crack and of eventual fracture. In FCC metals it has been suggested that interstitial atoms can occur as 'interstitialcy pairs' which, dumbbell-like, can migrate by a rotary motion (Seeger *et al.*[42]), 'swinging forward' as it were through the lattice.

One particular aspect of the effect of interstitials, as regards mechanical properties, is that *hydrogen* in steels can induce 'delayed' brittle fracture, implying an incubation period in order to initiate and develop the crack; this, in turn, is determined by the time needed for hydrogen to concentrate by diffusion into a region of high tri-axial stress such as may be created by notches, inclusions or similar stress-raising points, as well as by dislocation arrays. There has recently been increasing evidence of a significance of this behaviour much wider than merely in the hydrogen-in-steel case. Such delayed brittle fracture and associated phenomena can be induced more generally in metals and alloys by interstitials 'pacing' the dislocation movement, so that diffusion rates and their temperature dependence become highly determinative factors. This applies equally to other mechanical properties, and basically it may be said that strength or failure of a material is to a major extent determined by the nature, amount and distribution of interstitials present.

Vacancies can be produced by quenching strains, and the more effectively so if there is deformation at higher than at lower temperatures. Thus, by mechanical work, loci for subsequent fine interstitial nucleation can be artificially created.

Furthermore, dislocations created by working at high temperatures can be 'inherited' during phase transformations to lower temperatures; they can also serve as point sources of *new* ones formed during the transformation: again increasing dislocation density and inviting denser interstitial segregation.

Amongst interstitial vacancies, we must differentiate between the Frenkel- and Schottky-type defects and between N- and P-type conductivity based on the electronic and positive-hole kinds of defects. Thus non-stoichiometry can in principle have four causes, as set out in *Table 2.2.*

Table 2.2

Types of non-stoichiometry caused by interstitial vacancies (in compounds M_aX_b)

Type	Examples
(1) Excess of X, due to cation vacancies	Cu_2O, NiO, FeO, CoO, Cr_2O_3, Bi_2O_3, SnS
(2) Excess of metal, due to anion vacancies	V_4C_3, Nb_4C_3, TiC_{1-x}; also Nb_2O_5 (among ionic salts, KCl with excess K, and many others)
(3) Excess of interstitial anions	$CaF_2 + YFe_3$
(4) Excess of interstitial cations	ZnO, CdO, TiO, ZrO_2 + MgO

Throughout, an evident condition is that the electronic balance must be maintained. Thus cases (1) and (4), for example, would imply N-type semi-conduction, i.e. vacancies in which electrons can be trapped ('F' centres). It is then a basic property of vacancies that they can attract electrons (as well as sufficiently small interstitial negative ions). Silicon and germanium, with controlled impurities like boron, now so well established as semi-conducting materials, generally act by positive hole migration. (Semi-conductors represent, as mentioned already, a natural evolution from interstitial alloys.)

One consequence may well be illustrated by the oxidation of nickel. Addition to nickel oxide, NiO, of lithium oxide, Li_2O, owing to interstitial electron supply and decrease in the number of vacancies, increases its conductivity and its catalytic action; it also retards the oxidation rate. On the other hand, InO_2 addition has the opposite effect, both atomically and physically.

A further interesting trend, which joins up with more fundamental considerations on interstitial character (see *Appendix*, p. 602), is shown by some of the oxides. For instance, MgO and ThO_2, when heated to 500°C or more, are capable of exchanging some of their lattice oxygen for interstitial gaseous oxygen molecules (O_2) which can be adsorbed at grain boundaries, dislocations or at the surface*. In practice, this exchange is a prime cause of porosity and mechanical weakness, and no doubt also occurs in other ceramic compounds, as in oxide scales on metals where it will profoundly affect their behaviour, e.g. during the oxidation process. A similar hybrid interstitial state between atoms in lattice positions and in gaseous form is shown by the phenomenon that carbon monoxide and nitrous oxide reaction upon N-type oxides like MnO_2, Al_2O_3, Cu_2O or ZnO can create *metal nucleation* within the oxide lattice and an 'incipient metal ceramic'. Carbon monoxide, as a strongly reducing gas, is capable of being adsorbed at relatively low temperatures, the effect being to remove an interstitial oxygen atom from the lattice (to form CO_2 which remains adsorbed). In Cr_2O_3 it has even been discovered that a *carbonate* (CO_3) ion can replace oxygen in isolated places, leaving a surrounding oxygen-impoverished defect lattice field. For the oxides of the strong carbide formers of Groups IV, V, VI, it may be suggested as likely that this process can continue towards the nucleation of a true carbide lattice configuration, e.g. that of TiC out of TiO_2 by way of TiO and Ti (O,C,CO_3). Analogous effects can be produced in ceramic oxides by *hydrogen* reduction or, indeed, in some cases by vacuum treatment which is capable of introducing a degree of plasticity into the brittle oxide and an improvement in thermal shock resistance. The fundamental mechanism is probably similar, in that absorbed hydrogen (or holes) produce (*a*) vacant interstitial oxygen sites, (*b*) OH ion groups replacing some oxygen and (*c*) incipient nucleation of metal (or, at lower temperatures, a hydride) in the lattice or reduction towards a lower oxide. That a *vacuum* treatment tends in the same direction is shown, for instance, by the fact that on heating pure Al_2O_3 at high temperatures, the spinel structure and eventually aluminium metal can be liberated.

The existence, production and controlled re-occupation of interstitial vacant lattice sites in both metals and ceramics thus has important implications in manifold ways, a major one being that of the oxidation mechanism on metals.

The wide variation of non-stoichiometry, involving both metal and non-metal sites, is well illustrated by titanium monoxide, 'TiO'. This has been

---

* This has its parallel in alloys. Thus it is of interest to note that *hydrogen* atoms in iron and steel can (*a*) occur in lattice solution, causing brittleness, and can (*b*) aggregate in elemental form in voids and dislocations within the matrix, thus being rendered more harmless, except for pressure build-ups, when this can be a serious source of stress and cracking.

A characteristic of hydrogen as an interstitial atom in iron or steel, in contrast to carbon or nitrogen, is that it does not form distinct compounds, so that its only alternative in coming out of lattice solution is to build up (potentially very large) gas pressures concentrating at imperfections and voids; this is facilitated the more through the high mobility of hydrogen atoms in interstitial solution. Hydrogen tends to cluster particularly at dislocations, an effect that can become tremendous through cold work. Again, Stroh-type cracks can be the direct development of hydrogen-loaded dislocations, and thus of mechanical failure, e.g. under fatigue stress.

shown[43] on a range of densities as set out in *Table 2.3* (the same basic NaCl-type structure being maintained).

Table 2.3. Homogeneity range in 'TiO' (Ehrlich[43])

Towards β-(Ti,O) *solution* (hexagonal)	*Vacant* Ti *sites* per cent	*Vacant* O *sites* per cent
↑		
$TiO_{0.89}$	4	34
$TiO_{1.00}$	15	15
$TiO_{1.12}$	19	9
$TiO_{1.25}$	23	4
$TiO_{1.33}$	26	2
⋮ ↓		
Series of structures towards TiO_2		

Thus, even a true stoichiometric composition does by no means imply perfect lattice occupation, and the importance of such a condition in facilitating flow of atoms in various states of ionization and flow of electrons (e.g. as in oxidation and semi-conduction) is evident. Also, such an open lattice clearly invites to be 'stuffed' by foreign atoms of whatever source, both metal and interstitial, provided they are of favourable size and electron structure.

The defect lattices in UO_2, to give UO_{2+x}, and in Nb_2O_5 form other such cases. In the latter, for instance, it is known that, during niobium oxidation, Nb_2O_5 can lose both Nb and O atoms, with significant effect on oxidation resistance[44]; in the former (UO_{2+x} can range to about $U_{2.16}$), the non-stoichiometry is known to improve the high-temperature plasticity of UO_2, with valuable consequences in its use in nuclear reactors.

We might regard a surface layer of gaseous ions adhering to the solid by chemisorption as though being *already* a (non-stoichiometric) interstitial phase but two-dimentional in extent and a source or a sink for atoms, electrons or vacancies to the substrate crystal. The rate and type of diffusion would be governed by the particular compositions of the latter, as of course by the type of chemisorbed atom.

Another mechanism of producing vacancies (in turn attracting interstitials) is through the Kirkendall effect, i.e. by micro-voids created as a result of differential rates of diffusion in a diffusion couple.

Chemically, lattice vacancies at surfaces have a vital influence on catalytic behaviour of metals, on sintering, oxidation and tarnishing reactions as well as on photographic processes. One significant factor is their interchange with potentially interstitial ions and atoms and their ability to act as 'anchor' and as sinks to adsorbed molecules of many kinds. In chemisorption, such externally absorbed gas or carbon atoms will become an integral part of the main lattice. These wide further topics, which link up with the existence of surface interstitials (whether as atom or hole), can only be mentioned here.

Electrons, as occupants of interstitial vacancies in a statistically random or ordered manner, play a special part in pure transition metals, for example iron. Schubert[45] and others here have developed the concept of a site correlation ('Ortskorrelation') of the outer electrons to definite points in lattice space, to explain the degree of solid solubility and of added alloy elements in α -and γ-iron and their effect in opening or closing the γ-loop. The concept means that effectively an '*electron lattice*' is assumed to be present within the BCC or FCC atom lattice at a prescribed set of positions which, however, is variable; the number of available electrons to be accommodated (which may either be only the valency electrons or all those outside the nearest completed inert gas orbital, depending on the place of the metal in the Periodic Table) in turn determines the atom structure adopted. This electron lattice can, by translatory movements and extra insertions, take various configurations relative to the crystal lattice, so that, for instance in α- and β-Ti, in α- and γ-Fe and in α- and β-Co, electron lattices of different kinds and number of electrons per atom would occur. Metals such as nickel, when added substitutionally to iron, increase the number of electrons per unit cell and therefore stabilize austenite which can accommodate them more readily; the higher electron density can alternatively be produced by adding interstitial *carbon* to γ-Fe, which brings along 4 additional electrons, similar even to the extent of the closely analogous tendency to martensite formation in the Fe–C and Fe–Ni systems. Although this picture is certainly oversimplified and debatable, it is a helpful one in this context.

The following selection of literature, with some brief annotations, gives additional material on subjects referred to in the text.

SELECTED LITERATURE

Metal Physics and Solid State Physics Aspects

Chalmers, B. *Physical Metallurgy*, New York (Wiley) 1959

Hume-Rothery, W. and Raynor, G. V. *The Structure of Metals and Alloys*, London (Institute of Metals) 1955

Hume-Rothery, W. *Atomic Theory for Students of Metallurgy*, London (Institute of Metals) 1960

Smallman, R. E. *Modern Physical Metallurgy*, London (Butterworths) 1964

Reed-Hill, R. E. *Physical Metallurgy Principles*, Princeton (Van Norstrand) 1964

Salli, I. V. *Structure Formation in Alloys*, New York (Consultants Bureau) 1964

Dekker, A. J. *Solid State Physics*, London (McMillan) 1958

Kittel, C. *Introduction to Solid State Physics*, New York (Wiley) 1961

Sachs, M. *Solid State Theory*, New York (McGraw-Hill) 1963

Solid State Physics, Advances in Research and Applications (Ed. Seitz, F. and Turnbull, D.), New York (Academic Press) from 1955 onwards

Wannier, G. H. *Elements of Solid State Physics*, Cambridge (University Press) 1959

Weiss, R. J. *Solid State Physics for Metallurgists*, London (Pergamon) 1963

Christian, J. W. *The Theory of Transformations in Metals and Alloys*, London (Pergamon) 1960

Phase Transformations in Solids (Ed. Smoluchowski, R., Mayer, J. E. and Weyl, W. A.), New York (Wiley) and London (Chapman & Hall) 1957

Cottrell, A. H. *Theoretical Structural Metallurgy*, London (Arnold) 1957

Honeycombe, R. W. K. In *Progress in Materials Science*, Vol. 9, London (Pergamon) 1961
Effect of temperature and alloy additions, including interstitials, on deformation

Mott, N. F. *Atomic Structure and the Strength of Metals*, London (Pergamon) 1960

Mott, N. F. and Jones, H. *The Theory and the Properties of Metals and Alloys*, Oxford University Press, 1962

Seitz, F. *Trans. metall. Soc. A.I.M.E.* 215 (1959) 354
Irradiation effects; discussed by Bleakney, H. H. *Can. J. Phys.* 39 (1961) 947

Atomic sizes

Goldschmidt, V. M. *Trans. Faraday Soc.* 25 (1929) 253
One of the historic papers

Donohue, J.[6]
Critical discussion of 'change of atomic size' during structural transitions

Pauling, L. *J. Am. chem. Soc.* 69 (1947) 542

Pearson, W. B.[5]
Theories specific to relative atomic radii in semi-conductors

Samsonov, G. V. and Umanskii, Ya. S. *Hard Compounds of Refractory Metals, N.A.S.A. Technical Translation F–102*, Washington (National Aeronautics and Space Administration) 1962
Schwarzkopf, P. and Kieffer, R. *Refractory Hard Metals*, New York (Mcmillan) 1953
'AGARD' Advances in Materials Research in the N.A.T.O. Nations (Ed. Brooks, H. *et al.*), London (Pergamon) 1963
N.A.T.O. Symposium, containing several basic papers

The Electronic Structure and Alloy Chemistry of the Transition Elements (Ed. Beck, P. A.), New York (Interscience) 1963

Metallurgical Achievements (Ed. Alexander, W. O.), London (Pergamon) 1965
Collection of review papers, Birmingham Metallurgical Society Diamond Jubilee Session
Chambers, R. C. *Br. J. appl. Phys.* 15 (1964) 341
Bristol Symposium Report on Solid State Physics

The Physical Chemistry of Metallic Solutions and Intermetallic Compounds; The Relations between the Structure and Mechanical Properties of Metals, 2 vols. each, London (H.M. Stationery Office) 1959 and 1963
Two Symposia at the National Physical Laboratory, London

Materials Science Research (Ed. Stadelmaier, H. H. and Austin, W. W.), New York (Plenum Press) 1964
Russian Conference on structure and properties

Structural Chemistry

Wells, A. F. *Structural Inorganic Chemistry*, Oxford University Press, 1962
Emeleus, H. J. and Anderson, J. S. *Modern Aspects of Inorganic Chemistry*, London (Routledge) 1960
Evans, R. C. *An Introduction to Crystal Chemistry*, Cambridge University Press, 1964
Kristallphysik, 2 Vols. (Ed. Flügge, S.) In *Handbuch der Physik*, Berlin (Springer) 1955, 1958
Garner, W. E. *Chemistry of the Solid State*, London (Butterworths) 1955
Ketelaar, J. A. A. *Chemical Constitution*, Amsterdam (Elsevier) 1958
Schubert, K. *Kristallstrukturen Zweikomponentiger Phasen*, Berlin (Springer) 1964
Orgel, L. E. *An Introduction to the Chemistry of Transition Metals*, London (Methuen) 1961
Makarov, E. S. *Crystal Chemistry of Simple Compounds* (Actinides), New York (Consultants Bureau) and London (Chapman & Hall) 1959
Rice, F. O. and Teller, E. *The Structure of Matter*, New York (Science Edition) 1961
Van Arkel, A. E. *Molecules and Crystals in Inorganic Chemistry*, London (Butterworths) 1956

Bond Theory, Electron Structure, Valency, etc.

Pauling, L. *Nature of the Chemical Bond*, Cornell University Press, 1960
Cottrell, T. L. *Strength of Chemical Bonds*, London (Butterworths) 1958
Seel, F. *Atomic Structure and Chemical Bonding*, London (Methuen) 1963
Raynor, G. V. 'The Next Stages in the Electron Theory of Metals'. In *Metallurgical Achievements* (Ed. W. O. Alexander), London (Pergamon) 1965
Coulsen, C. A. *Valence*, Oxford University Press, 1952
Sidgwick, N. V. *Electronic Theory of Valence*, Oxford University Press, 1946
Jones, H. *The Theory of Brillouin Zones and Electronic Holes in Crystals*, Amsterdam (North Holland) 1964
Pilcher, G. and Skinner, H. A. *J. inorg. nucl. Chem.* 24 (1962) 937
Theory of valence of B, C, N and O
Zener, C. *Phys. Rev.* 81 (1951) 440
Orbital interaction in transition metals
Griffith, J. S. and Orgel, L. E. *Nature, Lond.* 191 (1958) 170
Localized electrons in BCC metals

Defect Lattice, Dislocation, etc.

Note—The numerous pioneering papers, from the 1950s onwards, on dislocations, their movement, interstitial interaction by Hirsch, Wheelan, Cottrell, Smallman, Nutting, Honeycombe and Seeger, and at many other centres, need only be referred to and are available in many journals, notably *Phil. Mag.* and *Acta metall.*

The Defect Solid State (Ed. Gray, T.) New York (Interscience) 1957
Kröger, F. A. *Chemistry of Imperfect Crystals*, Amsterdam (North Holland) 1964
Proc. int. Conf. Crystal Lattice Defects, 3 vols., Tokyo (Physical Society of Japan) 1963
Faraday Society Symposium, *Dislocations in Solids*, London (Butterworths) 1964
Rees, A. L. G. *Chemistry of the Defect Solid State*, London (Methuen) 1954
Roberts, J. P. *Jnl Br. ceram. Soc.* 1 (1964) 409
Point defects, especially in ceramic materials
British Ceramic Society Symposium on *Point Defects, Proc. Br. ceram. Soc.* 1 (1964)
Though mainly on ionic-type materials, several basic papers on interstitial bonding

SELECTED LITERATURE

Imperfections in Nearly Perfect Crystals (Ed. Shockley, W.), New York (Wiley) and London (Chapman & Hall) 1952

Van Bueren, H. G. *Philips Res. Rep.* 12 (1957) 190
Introduction to point defects

Vacancies and other Point Defects in Metals and Alloys [Harwell Symposium] London (Institute of Metals) 1958

Ascoli, A. and Damask A. C. *J. Phys. Chem. Solids* 21 (1961) 124
Vacancy theory

Damask, A. C. and Dienes, G. J. (e.g.) *Phys. Rev.* 125 (1962) 128, 444, 447, 542; *Acta metall.* 12 (1964) 797; *Point Defects in Metals*, New York (Gordon & Breach) 1963

Nonstoichiometric Compounds [Symposium] (Ed. Ward, R.) *Adv. Chem.* 39, 1963
An excellent review by J. S. Anderson is representative of his extensive work in this field

Kinetics of High Temperature Processes [Symposium] (Ed. Kingery, W. D.), New York (M.I.T., Wiley) and London (Chapman & Hall) 1958
Papers on lattice imperfections and interstitial diffusion

Corless, G. K. and March, N. H. *Phil. Mag.* 6 (1961) 1285
Defects

Frenkel, J. *Z. Phys.* 35 (1926) 652; *Phys. Z. Sowj. Un.* 8 (1935) 185. Schottky, W. *Z. phys. Chem.* 11B (1930) 163; 29B (1935) 335. Wagner, C. *ibid.* 22B (1933) 181
Some early basic papers on lattice defects, leading e.g. to semi-conductivity

Hauffe, K. *Oxidation of Metals*, New York (Plenum Press) 1965
Introduction to vacancy theory, etc.

Rhodes, R. G. *Imperfections and Active Centres in Semiconductors*, London (Pergamon) 1963

Moss, T. S. *Optical Properties of Semiconductors*, London (Butterworths) 1959

Schulman, J. H. and Compton, W. D. *Color Centers in Solids*, London (Pergamon) 1963

Friedel, J. *Dislocations*, Paris (Gauthier-Villars) and London (Pergamon) 1963

De Wit, R. *Growth of Helical Dislocations, Nat. Bur. Stand. Monogr.* 59 (1963) 13
Introduction to theory of dislocations

Thermal Chemistry, etc.

Kubaschewski, O. and Evans, E. L. *Metallurgical Thermochemistry*, London (Pergamon) 1958

Darken, L. S. and Gurrey, R. W. *Physical Chemistry of Metals*, New York (McGraw-Hill) 1953

Guggenheim, E. A. *Thermodynamics*, Amsterdam (North Holland) 1959

Ubbelohde, A. R. and Lewis, F. A. *Graphite and its Crystal Compounds*, Oxford University Press, 1960

Data Works

Hansen, M.[8]; continued by Elliott, R. P. *Constitution of Binary Alloys*, 1st Supplement, New York (McGraw-Hill) 1965

Pearson, W. B.[7]

Smithells, C. J. *Metals Reference Book*, 4th ed., London (Butterworths) 1967

A.S.M. Metals Handbook, 8th Ed., New York and London (Chapman & Hall) 1964

Goldsmith, A. *et al.* *Handbook of Thermophysical Properties of Solid Materials*, 5 Vols., London (Pergamon) London 1961, 1963

Hultgreen, R. *et al.*, *Selected Values of Thermodynamic Properties of Metals and Alloys*, New York (Wiley) 1963

Handbook of Chemistry and Physics, New York (Chem. Rubber Co.) 1959

Structure Reports and *Strukturbericht*, Vols. 1–7, Leipzig (Akad. Verlagsges.) 1913–1939; Vols. 8–21, International Union of Crystallography, Utrecht (Oosthoek) 1940–1957; continuing

Wyckoff, R. W. G. *Crystal Structures*, New York (Interscience) 1958
Comprehensive collection

Taylor, A. and Kagle, B. J. *Crystallographic Data on Metal and Alloy Structures*, New York (Dover Publ.) 1963

*Applications**

Reactive Metals [Met. Soc. AIME Conf.] (Ed. Clough, W. R.) New York (Interscience) 1959
Many papers relevant to interstitial alloys

Lozinskii, M. G. *High Temperature Metallurgy*, London (Pergamon) 1961
Translated from Russian

Handbooks of High Temperature Materials, 4 vols., New York (Plenum Press) 1964 and *High-Temperature Technology* (Ed. Campbell, I. E.), New York (Wiley) and London (Chapman & Hall) 1956

Specific Refractory Metals

The Science and Technology of Tungsten, Tantalum, Molybdenum and Niobium and their Alloys (Ed. Promisel, N. E.), [AGARD Conf. Proc. Oslo, NATO], London (Pergamon) 1964

* Almost by definition much of the comprehensive literature on *vacuum* technology is relevant to interstitial alloys, but not here included.

Niobium, Tantalum, Molybdenum and Tungsten (Ed. Quarrell, A. G.) [Proc. Sheffield Conf.] Amsterdam (Elsevier) 1961
Refractory Metals and Alloys [AIMME Conf. Proc.] (Ed. Semchyshen, M. and Perlmutter, I.) New York (Interscience) 1963
[Ti] McQuillan, A. D. and M. K. *Titanium*, London (Butterworths) 1956
[Zr] Miller, G. *Zirconium*, London (Butterworths) 1957
[Nb, Ta] Miller G. *Tantalum and Niobium*, London (Butterworths) 1959
[Cr] Sully, A. H. *Chromium*, London (Butterworths) 1954
[Mo] Northcott, L. *Molybdenum*, London (Butterworths) 1956
[W] Smithels, C. J. *Tungsten*, London (Chapman & Hall) 1952
[U] Gittus, J. H. *Uranium*, London (Butterworths) 1963
Peaceful Uses of Atomic Energy, Vol. VI: Basic Metallurgy and Fabrication of Fuels, Geneva (United Nations) 1958
Interstitial alloys in atomic energy
Wilkinson, W. D. and Murphy, W. F. *Nuclear Reactor Metallurgy*, Amsterdam (Van Nor-strand) 1958
Sections on uranium interstitial compounds
Advances in Semiconductor Science (Ed. Brooks, H.) [Rochester Symposium], London(Pergamon) 1959
Typical of numerous symposia and books on semi-conductors

Note—The vast basic literature on carbon (and related) steels refers, by implication, also to interstitial alloys.

REFERENCES

1 Ubbelohde, A. R. *Trans. Faraday Soc.* 28 (1931) 284
2 Umanski, S. Ya. *Dokl. Akad. Nauk SSSR* 26 (1943) 127
3 Hägg, G. *Z. phys. Chem.* B6 (1930) 221; 12 (1931) 33
4 Darken, L. S. In *The Physical Chemistry of Metallic Solutions and Intermetallic Compounds*, Vol. 2, London (H.M. Stationery Office) 1959
5 Pearson, W. B. *J. Phys. Chem. Solids* 23 (1962) 103
6 Donohue, J. *J. Am. chem. Soc.* 85 (1963) 1238
7 Pearson, W. B. *A Handbook of Lattice Spacings and Structures of Metals and Alloys*, London (Pergamon) 1958
8 Hansen, M. *The Constitution of Binary Alloys*, Supplement by Elliott, J. P., New York (McGraw-Hill) 1958, 1965
9 Alessandrini, E. I. and Freedman, J. F. *Acta crystallogr.* 16 (1963) 54
10 Priener, J. S. and Williams, F. E. *J. chem. Phys.* 35 (1961) 1803
11 Hume-Rothery, W. and Coles, B. R. *Phil. Mag., Suppl.* 3 (1954) 149
12 Jack, K. H. *Acta crystallogr.* 3 (1950) 392; 5 (1952) 404; *Proc. R. Soc.* 195 (1948) 34; 208 (1951) 216
13 Schonberg, N. *Acta metall.* 2 (1954) 427
14 Sandor, E. and Wooster, W. A. *Acta crystallogr.* 13 (1960) 339
15 Goldschmidt, H. J. *Metallurgia* 56 (1957) 17
16 Kiessling, R. and Peterson, L. *Acta metall.* 2 (1954) 675
17 Hultgren, A. *Jernkont. Annlr* 135 (1951) 403
18 Goldschmidt, H. J. *J. Iron Steel Inst.* 166 (1948) 345; 170 (1952) 185; and 186 (1957) 68; *Nature, Lond.* 162 (1948) 837
19 *Symposium*: *Precipitation Processes in Steels* (Ed. Quarrel, A. G.), *Spec. Rep. Iron Steel Inst.* 64, 1959
20 *Metallography* 1963 [Sorby Centenary Symposium], *Spec. Rep. Iron Steel Inst.* 80, 1964
21 Rahlphs, P. *Z. phys. Chem.* B31 (1935) 157
22 Rundle, R. E. *Acta crystallogr.* 1 (1948) 180
23 Pauling, L. *Phys. Rev.* 54 (1938) 899
24 Hume-Rothery, W. *Phil. Mag.* 44 (1953) 1154
25 Brewer, L. and Krikorian, O. *Rep. Univ. Calif. Rad. Lab.* UCRL–2888 (1955), 3352 (1956)
26 Renner, H., Brauer, G. and Faessler, A. *Z. Naturf.* A10 (1954) 17
27 Kiessling, R. 'Bonding in Metals', *Metall. Rev.* 2 (1957) 77
28 Robins, D. A. *Powder Metall.* 1/2 (1958) 172
29 (e.g.[25]); Brewer, L. and Krikorian, O. *J. electrochem. Soc.* 103 (1956) 38; Brewer, L. and Haraldsen, H. 102 (1955) 399
30 Dempsey, E. *Phil. Mag.* 8 (1963) 285
31 Altmann, S. L., Coulson, C. A. and Hume-Rothery, W. *Proc. R. Soc.* A240 (1957) 145
32 Wentorff, R. H. *J. chem. Phys.* 26 (1957) 956; 34 (1961) 809
33 Hibbard, W. R. and Strong, H. M. *Progress in Very High Pressure Research* (Ed. Bundy, F. P.), New York (Wiley) 1960

REFERENCES

[34] Milledge, H. J. *Nature, Lond.* 190 (1961) 1181

[35] Grenville-Wells, H. J. *Mineralog. Mag.* 24 (1952) 803

[36] Lipschutz, M. E. and Anders, E. *Geochim. cosmochim. Acta* 24 (1961) 83; *Science, N.Y.* 134 (1961) 2095

[37] Bundy, F. P. *J. chem. Phys.* 38 (1963) 631

[38] Libby, W. F. *Phys. Rev.* 130 (1963) 548

[39] Evdokimova, V. V. and Vereschchagin, L. F. *Soviet Abstr. Phys.* 16 (1963) 855

[40] Van Aswegen, J. S. T., Honeycombe, R. W. K. and Warrington, D. H. *Acta metall.* 12 (1964) 1

[41] Stroh, A. N. *Proc. R. Soc.* A. 223 (1954) 402; 232 (1955) 548

[42] Seeger, A., Schiller, P. and Kronmüller, H. *Phil. Mag.* 5 (1960) 853

[43] Ehrlich, P. *Z. anorg. Chem.* 247 (1941) 53

[44] Argent, B. B. and Phelps, B. *J. Inst. Metals* 88 (1959) 301

[45] Schubert, K. *Z. Metallk.* 46 (1955) 43; *Z. Naturf.* 14a (1959) 650

3

INTERSTITIAL SOLID SOLUTIONS

The discussion of some general features of primary solid solutions of interstitial atoms in the transition metals in Chapter 2 will now be amplified.

3.1 SOLUBILITIES

In *Figure 3.1* an attempt has been made to co-ordinate data, where known, on the solubilities of carbon, nitrogen, boron, oxygen and hydrogen in relation to atomic number and crystal structure of the solvent metals*. It was necessary to assemble these data from widely scattered sources, and many are still uncertain, particularly those on boron. Nevertheless, a number of trends clearly arise.

(1) There is the evident structure dependence which, for C and N, takes the form of a generally greater solubility in the two close-packed than in the BCC structures—a 'rule' which, however, fails notably in some cases of unfavourable atomic size ratio and electronegativity, e.g. for copper.

(2) For the same structures, there is a steady decrease in solubility along the periods with rising atomic number, and close similarity exists in this respect between the C and N behaviour. The oxygen solubility tends to be higher, with exceptions (see *Figure 3.1*). The small hydrogen atom, on the other hand, occupying basically different lattice sites (to be discussed in Chapter 9) has a much larger BCC solubility than have O, N, and C for given pressures. This manifests itself, for example, in Ti and Zr, where the β-forms have much the higher solubilities for H than the HCP α-forms, the reverse to those for O, N and B. (Curiously, the carbon case seems in the same direction as for H, though the difference between α and β in regard to carbon is only small and not too certain.)

The process of hydrogen solution may be considered as corresponding essentially to one of addition of electrons to the metal atom, i.e. it is more akin to a 'change in condition' of the latter than to the alloying of a separate new element. The extensive solution of hydrogen in the BCC metals of Groups IV and V, contrasting with the restricted solution in the HCP forms (Ti, Zr, Hf), may in fact be regarded as 'continued', in shape of a change in stacking sequence, namely to the hydrides of FCC structure (see *Figure 2.3*), which are basically Ti or Zr solution phases stabilized by hydrogen. The Group VI, VII and VIII metal atoms, on the other hand, are already so electron-saturated that hydrogen solutions remain small and endothermic in formation, and are not followed by any hydrides.

The case of the palladium–hydrogen solution is of special interest, in that at lower temperatures the remarkably high H solution divides into two (H-poorer and H-richer) isomorphs, an effect which has been studied in

---

* *Maximum* solubilities, whatever the temperature: normally at the eutectic (for atmospheric pressure).

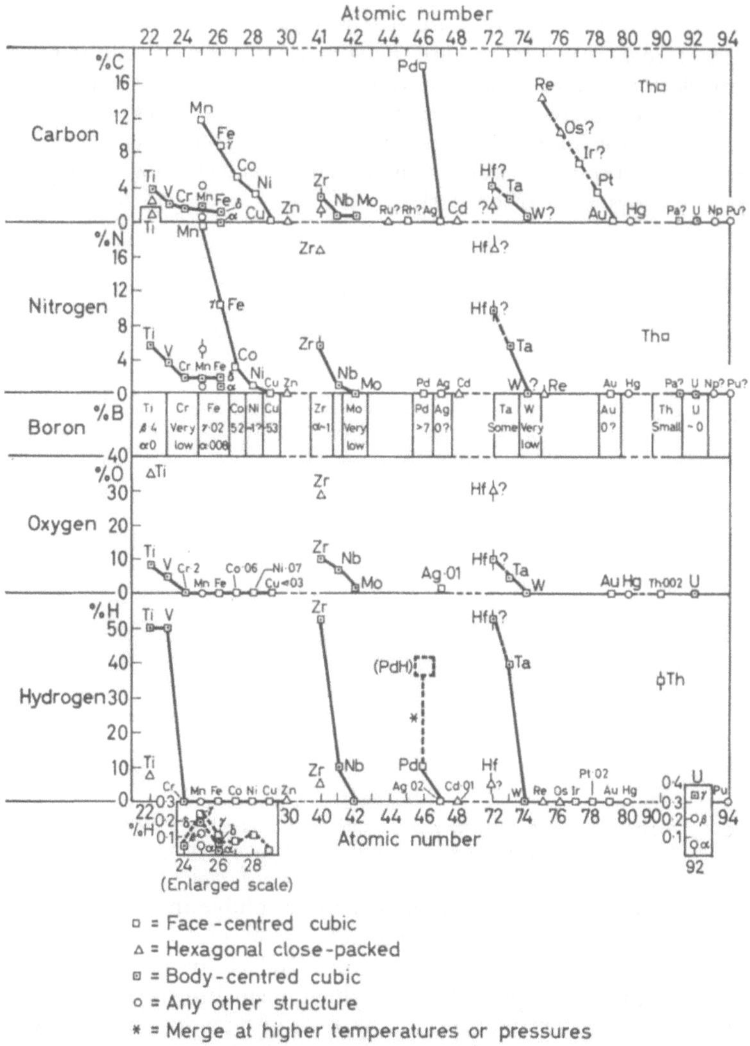

Figure 3.1. Maximum solid solubilities of interstitial elements in transition metals (at. per cent) *as function of atomic number and crystal structure*

much detail, e.g. by lattice-parameter[1] and physical property changes (see Chapter 9).

The practical bearing of these differential solubilities is manifold. For example, that of hydrogen in the steel matrix phases (austenite and ferrite) may become the cause of embrittlement and, according to some authors, of hair-line cracking; more rigorously, we should perhaps express this as the capacity to supersaturation of this solution, which on decomposition gives coagulation of hydrogen atoms near dislocations and grain boundaries.

Generally, depending on structure, the interstitial solubility changes discontinuously when a pure metal undergoes a polymorphic change; a sequence like this is strikingly illustrated in *Figure 3.2* showing the changes for hydrogen in manganese[2], where the stepped curve well reflects the Mn phase transformations in a manner which, conversely, almost could suggest this as one indirect method of determining the latter. A further point arising from *Figure 3.2* is the increasing phase hysteresis with decreasing temperature, an effect which, of course, is at the root of many vital technological properties of interstitial alloys, e.g. carbon steels.

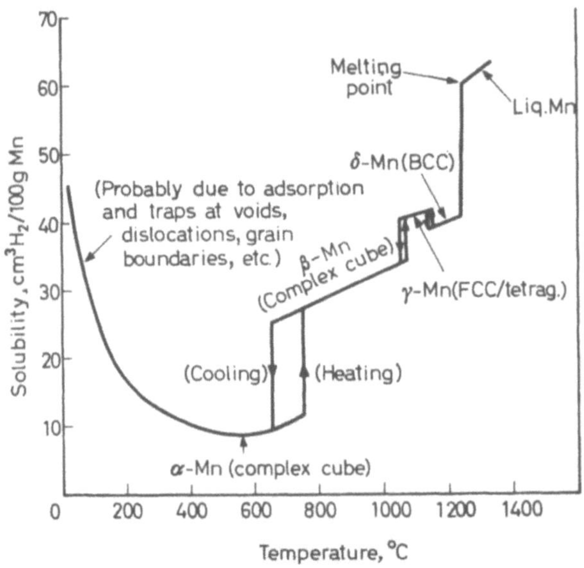

Figure 3.2. Solubility of hydrogen in manganese, at atmospheric pressure
(After Sieverts and Moritz[2], with annotations)

Since gases (and carbon) are generally more soluble in the *liquid* than in the solid metal, freezing involves a large decrease in solubility, to the amount the interstitial lattice capacity would permit. Indeed, we might say that the breakdown to almost complete disorder which the melting process represents, provides in *solid-state* terminology, a multitude of 'pseudo grain boundaries' on an atomic cluster scale, each with an interstitial accommodating power similar to that known of normal grain boundaries. (This difference in liquid and solid solubility is, incidentally, utilized for instance in the *zone-refining* processes, in which the metal is purified by moving a liquefied or hot solid zone along a bar sweeping the interstitials to one end.)

The solubility of *oxygen* in the primary BCC lattices of Ti, Zr, Hf, V, Nb, Ta (refractory metals of Groups IV and V) is particularly prominent and currently important because, during the high-temperature oxidation of these metals, before there is any external sign of 'oxidation' in the sense of oxide formation, this considerable matrix solution occurs, to a degree which is strongly pressure- and temperature-dependent. Thus, HCP (α)-titanium is capable of absorbing the remarkable amount of 34 atomic per cent oxygen; in this process, the *c*-axis is expanded, the *a*-axis practically unaffected, in accord with interstitial oxygen entering the ($\frac{2}{3}$, $\frac{1}{3}$, $\frac{1}{4}$) and ($\frac{2}{3}$, $\frac{1}{3}$, $\frac{3}{4}$) positions randomly between the basal planes. The behaviour is analogous to that of the anistropic thermal expansion of α-titanium and e.g. of rhenium and tungsten carbide discussed, and to that of preferred electron sites. When niobium (Group V) oxidizes, it does so, as a first stage, by interstitial solution expanding the lattice, which then evolves into a tetragonal distortion and an ordering transformation before the first *actual* oxide occurs. This process of primary solution is accompanied by hardening and embrittlement of the metallic mass itself; it presents, along with the actual oxide formation (see Chapter 8), one of the major difficulties in the, otherwise mechanically attractive, high-temperature use of Nb and Nb alloys, which the design of an effective protective coating or suitable alloy addition must meet. In the Group VI elements (Cr, Mo, W), in which the primary oxygen solubility becomes much less, the effect nevertheless continues, but in a modified form: a new structure arises, such as that of 'β-tungsten', which in fact is already a lower oxide W_3O (see *Figure 2.7c*); this is still based on the parent body-centred cube, but oxygens are squeezed in at additional sites, causing the tungsten atoms to be deformed to oblate shapes. Continuing further right in the Periodic Table (Fe Group, etc.), the oxygen solubility ceases almost completely, whatever the structures.

In titanium taking oxygen into solid solution (and in other instances), it has proved possible, by internal friction measurement, to make a fine distinction between (1) the oxygen content required to saturate the dislocations and (2) that entering true interstitial sites (at 800°C, (1) = 1·5 atomic per cent O, (2) = about 34 per cent O). This had a direct practical use because, by increasing the dislocation absorption capacity, the ductile→brittle transition, i.e. the beginning of oxygen embrittlement, could be deferred to higher oxygen contents.

A major consequence of the insertion of interstitials into the hexagonal metal lattices is their effect on ductility and on work-hardening. Whilst in the *pure* metals, slip in single crystals depends largely on the tensile axis and elongation on the operative plane*, this can be drastically altered by the presence of interstitial impurities (as would the consequent stress to fracture), and indeed, reported differences amongst 'pure' hexagonal close-packed (HCP) metals (e.g. Ti, Zr, Rh, Be, Mg, Zn) may, in *some* cases at least, be due to trace impurities.

The interstitial solubility in the transition metals itself is greatly dependent on secondary alloy additions, and their effect upon the electron/atom ratio of the matrix; thus it has been shown e.g. for niobium[3] that an E/A increase through alloy metals reduces the oxygen solubility and can do so to zero at a certain ratio (here E/A = 5·7). A similar relation occurs for hydrogen solubilities in the transition metals (see Chapter 9).

It has been discovered[4,5] that *rhenium* solid solutions in molybdenum and tungsten were able to increase the ductility of the alloys to a remarkable extent, and this effect has again partly been attributed to a decrease in interstitial solubility. From this arises an interesting generalization in that the interstitial effects disappear again at about the same E/A ratio, ≈5·8,

* This can vary from one HPC metal to another, but in rhenium, for example, slip occurs in the <11$\bar{2}$0> direction, and elongation in the (0001) plane, being the ideal on geometric, but frequently upset on physical grounds.

throughout. One example is shown[6] in *Figure 3.3.* It demonstrates that, in Ta–Re alloys[6], any differences in hardness through thermal treatment, which *within* the solution region had been considerable, tend to disappear on approaching the phase boundary; this is shown to conform well with Pauling's theory of the resonating valency bond and its extensions (Hume-Rothery *et al.*[7] and Robins[8]).

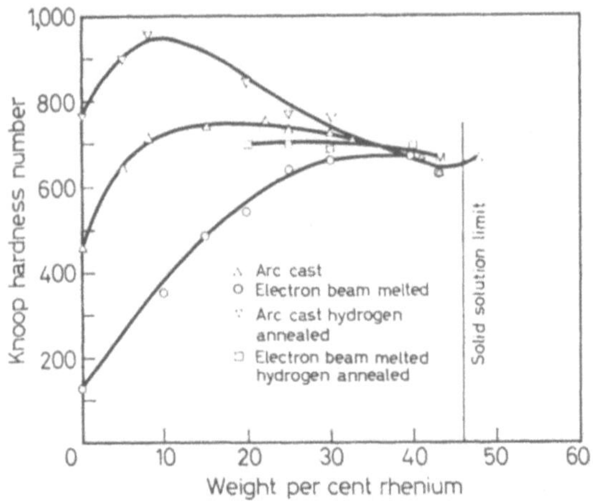

Figure 3.3. Hardness divergence within the Ta → Re *primary solid solution region* (After Brophy, Rose and Wulf[6])

The divergence, due to variations in interstitial contents associated with different methods of preparation, disappears on approaching the solution boundary

By courtesy of the Editor, *Journal of Less Common Metals*

3.2 FURTHER SOLUBILITY TRENDS

Reverting to *Figure 3.1,* it may be seen that, for comparable structures and solvents, the solubility increases in the order B→C→N→O→H, i.e. with decreasing size of the solute atom; similary, the *rate* of decrease with atomic number decreases. Size factor alone is here not determinative, however, and the trend is accentuated by the increase in electronegativity of the non-metal in this same order; it means that, considering the solutes as electron donors to the parent metal, we have also a weakening, in this order, of the donating capacity and a delay in the structure change to the first compound.

Again, in *Figure 3.1,* the very high solid solubility of carbon in γ-Mn and in Re, i.e. the cubic and hexagonal close-packed metals of Group VII, marks the beginning of new descending sequences: Mn→Cu, Re→Au, again in accordance with the increasing electronegativity, as precedingly in the BCC series of Group IV–VI metals. These constant solute descents in solubility with varying solvent are a counterpart to the constant solvent descents with varying solute (as discussed above, i.e. in sequence H→O→N→C→B), both amounting to increasing electron addition.

The solubility of different interstitial species in a given metal is competitive insofar as, for instance, that of H in zirconium is decreased if O, N or C are also present[9], and quantitatively so in this order, i.e. that of rising donor capacity. An immediate implication of this competitive character is that, e.g. in nitriding processes, the rate of diffusion of nitrogen into the metal can be adversely affected by the effect of oxygen and other impurities present.

The question *which* interstitial position in the BCC, octahedral or tetrahedral, will be preferred, depends on the strain energy produced for interstitial atoms of varying size (e.g. [10]).

From strain energy and thermodynamic considerations it can be shown[11] that the solubility of interstitials must be smaller in BCC than in FCC and HCP host lattices, such as α,γ-Fe-(C and N) and α,β-Ti-(C, N and O) solid solutions. Furthermore, it has been possible to calculate the effective atomic sizes of the interstitial solute atoms in binary solid solutions[12] on lattice-geometrical grounds, and to derive as radii, for C in γ-Fe, 0·68 Å (i.e. less than the 0·76 radius, such as in the cubic *carbides* TiC, ZrC, NbC, etc.); for N in γ-Fe, 0·73 Å (i.e. unusually large and slightly greater than the 0·70 radius of the corresponding nitrides); N in α-Ti, 0·70 Å which is the same as in the nitride TiN; O in α-Ti and in α-Zr, 0·68 Å, i.e. similar to the value for oxygen in the low oxides except when ionized (cf. Chapter 2).

The stability of interstitial atoms in FCC metals has been the subject of much controversy. For example, Seeger *et al.*[13] conclude that, in equilibrium, interstitial atoms cannot strictly exist in FCC solution and must eventually anneal out, in contrast to their stable state in BCC solution; within this 'metastable' state, however, various alternative interstitial configurations are considered possible. This seems strange in view of the large interstitial solubilities known, e.g. in austenite, unless even this is considered strictly non-equilibrium though persistent for indefinite times.

Superlattice formation by interstitial atoms when entering into dilute solution is another issue arising. Depending on temperature (and, presumably, on time for the interstitials to find their stable positions), such entry can be random or ordered, to varying degrees. For instance, in *tantalum*, atomic ordering of nitrogen atoms was observed[14], with an indication that the ordered N atoms occupy the tetrahedral rather than the octahedral sites. Ordering occurs already for as little as 0·1 atomic per cent N, so that there is a giant unit cell, of 27 BCC unit cells of BCC tantalum, containing two N atoms*.—Ordered *oxygen* solutions in niobium and tantalum are to be discussed in Chapter 8.

We may note, incidentally, that generally nitrogen and carbon are only soluble in those BCC transition metals, which also are capable of forming nitrides and carbides, with some important exceptions, however (cf. *Figures 3.1* and *2.3*).

3.3 CARBON IN FERRITE

The large solubility of carbon in γ-iron, as against the small one in α, is obviously the principal basis of ferrous metallurgy. However, the solubility of carbon in α-iron, even though so minute (about 0·01 per cent C maximum), is still of great importance in steels and has been the subject of extensive research because carbide nucleation from α gives rise to hardening effects at low temperatures. The solution of carbon in ferrite is capable of a considerable degree of supersaturation (from the viewpoint of lattice

* The paper by Seraphim *et al.*[14] is also of interest for techniques and for references in this field.

mechanics, this is akin to martensite, although the latter really represents the opposite extreme of high carbon content inherited from austenite); the mode of precipitation of cementite platelets upon quench aging of ferrite, along the lattice planes and at dislocation lines, has been studied with special success by electron microscopy.

As mentioned in Chapter 1, one of the most powerful tools in determining interstitial solubility is *internal friction.* It is helpful to envisage its operation by considering that the interstitials at cube edge positions I (see *Figure 2.7a*) 'strain the metal atoms apart', and randomly so if the crystal is externally unstressed. However, if a stress is applied, tensile or compressive, in the cube edge direction, either more or less space will be afforded in these positions so that for alternating stress (as in damping experiments), the interstitial atoms are caused to jump; this in turn enhances damping of the imposed vibrations in bulk, say the wire or bar sample used. The magnitude of damping depends on the diffusion coefficient, the lattice spacing and frequency, so that, with changing temperature, an internal friction peak arises; its position is characteristic of the given system and diffusion rate, its height of the amount of interstitial dissolved. This makes possible the measurement both of solubility and of diffusion coefficient, as well as of the kinetics of precipitation out of solution.

In *ferrite*[15], the carbon solubility falls from 0·018 per cent C at 720°C (eutectoid) to 0·0006 per cent C at 300°C and to (an estimated) 2×10^{-7} per cent C at room temperature. Atomically this means that assuming random distribution, *one* carbon atom is to be found in every 600th α-iron unit cell at 720°C, in every 17,000th at 300°C, and in every 100 millionth at 20°C.

In contrast, in γ-iron (FCC) 1 carbon atom can exist randomly in up to every *3rd* unit cell at 1,147°C, the eutectic composition*. In chromium, for comparison with α-iron, carbon can fill up to every 40th unit cell at 1,500°C, decreasing to every 5,000th at 900°C. The age-hardening capacities of the two BCC host metals, although the absolute solubility may differ, are thus on a more comparable scale, and the same seems to be generally true for the other BCC transition metals, bearing in mind their different absolute melting points.

We must, however, look at these hardening reactions kinetically. Whilst the (equilibrium) solubility may fall with temperature, the *rate* of diffusion and of carbide nucleation does so, too. Thus, in ferrite for example, when quenched from the eutectoid (720°C), precipitation can occur within seconds, even during rapid cooling through the upper temperature regions; as *Table 3.1* shows, at 300°C full precipitation occurs in minutes, at 20°C within months. The resulting morphology and especially the particle size and dispersion of the carbide—as well as the physical properties obtained—depend therefore sensitively upon the precise path of heat treatment.

The large difference in solution carbon between γ- and α-Fe which underlies the whole of steel metallurgy, is in many respects paralleled by the metallurgy of Ti, Zr and Hf, where differential solubilities of interstitials, e.g. of O or H, in the α- and β-forms have similar implications, though with

---

* This wide range of interstitial carbon solutions encompasses the whole sequence of martensitic phenomena, in which the diffusionless $\gamma \rightarrow \alpha$ change and the residual ferrite precipitations form the opposite termini.

differences resulting from the *low*- instead of the *high*-temperature form having the greater solubility.

Table 3.1

Fraction of available carbon precipitated from α-iron solution in time and at temperature stated[15]

Time	Temperature 20°C	100°C	200°C	300°C
	per cent			
1 sec	—	—	—	0·55
1 min	—	—	2·8	76
1 h	—	7	99·99	> 99·99
1 day	0·25	99·93		
1 month	68	> 99·99		
1 year	> 99·99			

The literature regarding specific types of interstitial solid solutions in α-iron is, as might be expected, voluminous. A few typical examples will be given later; for the solubility of carbon and nitrogen in iron many good compendia can of course be referred to; to that of hydrogen in iron, with its associated embrittling effects, etc. we shall return in Chapter 9. *Boron* occupies a unique position vis-à-vis iron, because its atomic size is too large to permit interstitial entry without considerable lattice strain, yet too small (and electronically unsuited) for substitutional solution. The same is true in γ-Fe. Very small boron additions are known to be able to exert a beneficial effect on properties of ferrous alloys (e.g. [16,17]) which, when joined with that of carbon, has given interesting new and strengthened materials. Economically this has the attraction of using boron, either on its own or in concert with other interstitials, as a cheaper substitute for more expensive alloy elements, though this has its limits.

Basically the function of solution boron may be visualized as follows. Because of its intermediate size, the boron atom tends to aggregate at any lattice defects and dislocations, which indeed it helps to create and multiply, and to 'anchor'. Grain boundaries thus become particularly enriched in boron, and crystal growth inhibited. The boron atom can also act as a impediment to diffusion of the main substitutional metal atoms (e.g. Ni, Cr), so that effects like grain-boundary denudation become reduced. This, in turn, promotes creep resistance because, without boron, creep is often associated with preferential lattice glide near such composition-weakened grain-boundary zones (which, once started, generates cracks). We might then look upon the boron atoms as (1) effective crystal boundary strengtheners and stabilizers; (2) dislocation anchors; and (3) inhibitors of undesirable diffusion effects by maintaining homogeneity in alloy content and preventing precipitations at grain boundaries other than boron. By filling up vacancies normally enriched at these boundaries (and sub-domain boundaries), boron atoms effectively help to stiffen them (e.g. [18], typical of much other work).

It is of interest to note that an internal friction peak through the presence of *boron* atoms in α-Fe solution could be detected[19,20], indicating that they enter *interstitially*. This does not, however, exclude substitutional entry as well, as other evidence appeared to support[21], and indeed, their preference for lattice defects, dislocations and grain boundaries. Another great interest lies in the distribution of the boron atom, along with carbon, in γ-Fe grains. Adsorption occurs principally at grain boundaries[21], increasingly so with increasing temperatures; effectively, boron can act both as grain-growth inhibitor and as retarder of austenite decomposition.

The solubility of boron in α- and γ-Fe according to Nicholson[17] has the following limits: α-Fe: 0·002 wt. per cent B_{max} at 915°C, γ-Fe from 0·001 per cent at 915°C (eutectoid), to 0·020 per cent B at 1,165°C (eutectic). Thus the α- and γ-solubilities for boron are much more commensurate than those, for instance, for carbon—again a manifestation of the borderline role of the boron atom.

3.4 ROLE OF INTERNAL FRICTION

A valuable point of detail which internal friction measurements can clarify is that the diffusion of two different types of interstitial atom in the same matrix can be distinguished. Thus, for example, tantalum loaded with oxygen shows two peaks superimposing: one, at 137°C, due to the diffusion of separate oxygen atoms, another, at 162°C, to that of two coupled oxygen atoms, mutually interacting. A similar twin peak obtains for nitrogen, and in the case of *both* O and N being present, a successive broadening is obtained, representing a series of peaks for varying O/N ratios. Such an example is shown in *Figure 3.4*, the main peak being due to O alone, the broadening to diffusion of interacting (O–N) pairs. Moreover, data such as these permit the evaluation of activation energies for the diffusion of each species.

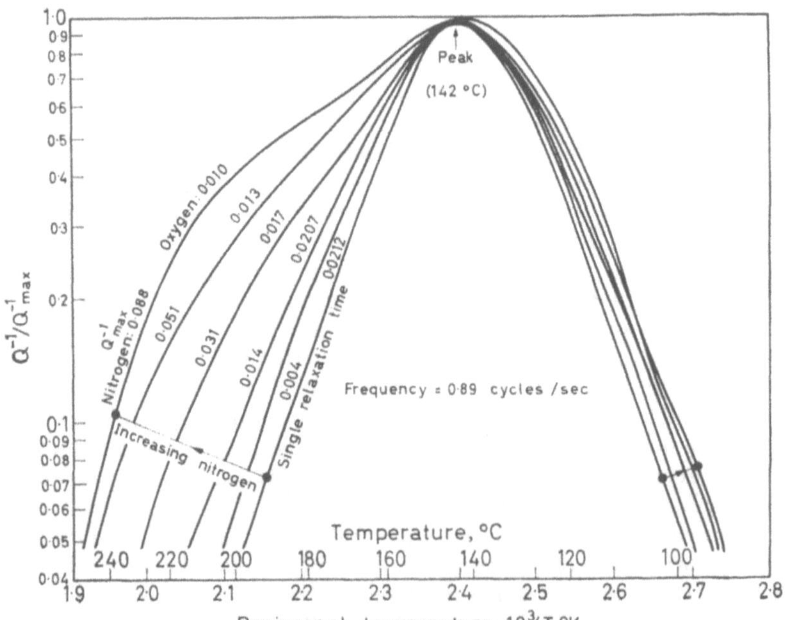

Figure 3.4. Broadening of internal friction peak for oxygen in tantalum by nitrogen additions
Typifying internal friction effects due to different interstitial atoms in solution

Similarly, it can occur that, in a substitutional type of solution, such as of Si or V in α-Fe, a major proportion of, say, nitrogen also present is linked with one solvent atom species rather than another (e.g. Si or V). Thus, in a silicon iron, the solubility of N showed[22] a predisposition for N atoms, already whilst in solution, to cluster preferentially at sites around silicon in the lattice. This is, in fact, preparatory to the formation of silicon nitride, Si_3N_4, a compound actually detected by later researches in a more advanced state of precipitation, in nitrogen-bearing silicon irons.

The presence of grain boundaries in a solution matrix is a further important factor, and the damping relaxation curves are strongly grain-size-dependent; the effect of interstitial impurities in this has also been much studied. We may view this effect as one of sliding friction of grain boundaries over each other, and if the alloy contains more than one phase, different peaks may arise for different types of grain boundaries. For instance, in an iron–chromium alloy containing α- and γ-phases side by side, the α/α, α/γ or γ/γ grain boundaries give rise to separate internal friction peaks; this is again greatly dependent on whether a given solute is interstitial or substitutional, and can be a criterion to decide this. In the same sense, the solid solution of hydrogen in iron and steel (both the interstitial and the dislocation kind) has been studied by internal friction, such as H absorbed through electroplating or pickling, with results usefully relating to embrittlement. (The Bibliography at the end of the Chapter gives further literature on internal friction.)

3.5 ELECTRODEPOSITED METALS

Electrolytic deposits frequently provide good examples of *metastable* interstitial solid solutions, possessing a large facility for supersaturation. In chromium, for instance, such solutions are linked with x-ray line broadening, which can be interpreted as due to heterogeneous absorption of hydrogen within the crystallites. The internal gradients thus set up entail considerable lattice strains, which are at the basis of the great hardness and wear resistance of so-called 'hard chromium' deposits[23]. A major effect is here a metastable hydrogen solution occurring in the iron substrate, which permits keying of the boundary layer of the chromium lattice onto that of the iron (substrate); a large stress thus persists in the cold, but is removed upon heating, so that degassing by annealing causes progressive softening and strain relief. This may be illustrated by x-ray photographs and spacing changes shown in *Figure 3.5.*

The literature concerning electrodeposition and the interstitial gas occlusions thereby attained is of course voluminous, and it is not intended to deal with this here in detail.

However, it is basically interesting to note one property of interstitial solid solutions: namely that, under the influence of an outer *electric field*, the interstitial non-metal atom can behave as though ionized. Thus, in zirconium, the solution oxygen has been shown[24] to be mobile, diffusing, when a potential is applied along an oxygen-loaded Zr wire, in the manner of a negative (O^{2-}) ion; it is almost, as in electrolysis, a direct current, thus causing the oxygen to migrate to the positive pole, leaving the negative end of the wire practically the pure zirconium. This does not mean that the

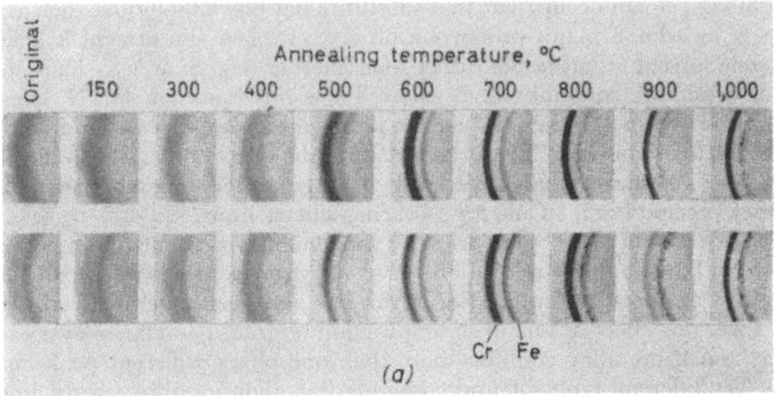

Figure 3.5. Chromium plating on a steel containing hydrogen in solution (Goldschmidt[23])

(*a*) x-ray diffraction line appearances: lattice strains through interstitial H atoms are relieved on their ejection by annealing; (*b*) associated changes in lattice dimensions and line width; (*upper graph*) originally hard sample—extreme strains; (*lower graph*) originally soft sample—low strains (*P*, *Q*, *R*, *S* = maxima due to various causes)

By courtesy of the Editor, *Metallurgia*

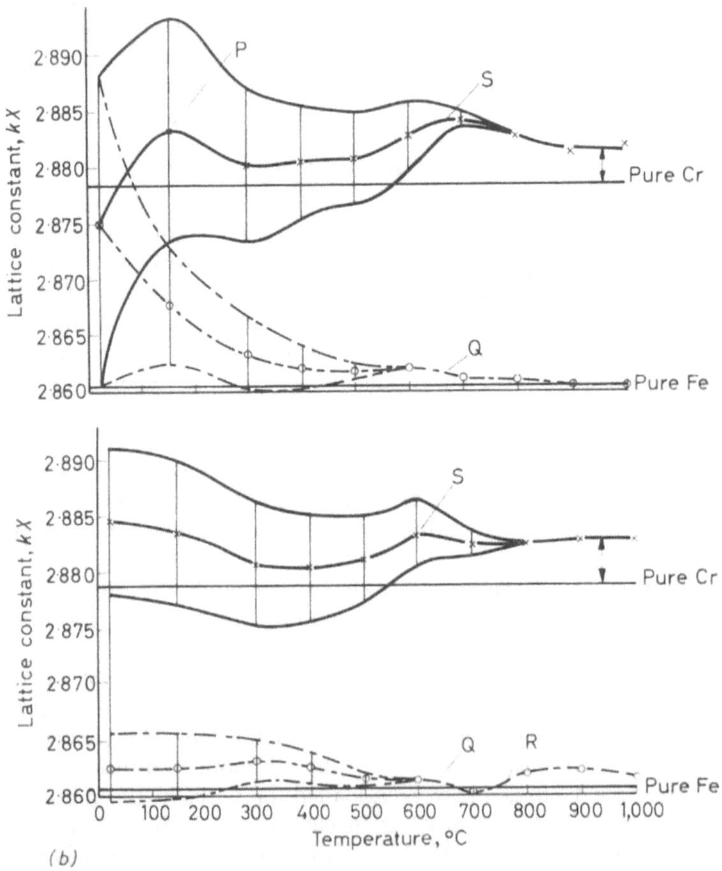

Figure 3.5b

normal (no-field) solid solution of oxygen in zirconium should be considered an ionic lattice (there the bonding is more probably pseudo-metallic, as discussed in Chapter 2); it does, however, lend support to the contention that, already without the field, a part-way electron exchange between the non-metal and the metal atom has occurred, which only requires the outer potential to transform the interstitial into a *complete* ion.

We might look upon this group of phenomena as 'dry electrolysis'.

It is similarly possible for interstitial carbon in transition metals to be transported in an applied electric field like a charged electrolyte[25].

3.6 EFFECT ON PROPERTIES

The manifold implications of interstitial solutes on the magnetic and electrical properties of the host metals cannot possibly all be discussed here, and brief reference must suffice. The increase in resistivity, for instance, has in many cases been found a valuable measure of solubility, and its magnitude and dependence upon temperature an index of mode of solution.

The *mechanical* properties too, are, as mentioned already, vitally dependent on interstitial solutes*, and some amplification, by way of examples, seems here appropriate. Their function can generally be reduced to that of dislocation locking, but the detailed manner of the process and outward consequence, as far as the strength of the material is concerned, varies widely. Thus the yield-point phenomenon in BCC metals such as iron or tungsten is due, first, to locking, and, upon stress application, to the tearing away of dislocations from the interstitial atmospheres which surround them. It is important to differentiate two types of interstitial locking operating when stress is applied, namely what might be called the 'Snoek effect' (S) and the 'Cottrell effect' (C); (S) is that already familiar and due to the interstitials occupying preferentially, instead of randomly, one of the three cube edges through directional strain (upon alternating strain, atomic jumps then cause the sites to fluctuate). Effect (C) is due to the presence of interstitial atmospheres around the dislocations and their subsequent break-away. (C) requires diffusion over fair distances, (S) only displacements on the short-range scale of the unit cell, thus proceeding much more rapidly. The basic situation as a function of stress and temperature can be well represented for iron[28] in *Figure 3.6.* At lower (sub-zero) temperatures, the flow stress is strongly temperature-dependent, as effect (C) is mainly operative; this is followed (in the range of room temperature to 200°C) by a less temperature-dependent region where (S) becomes dominant, followed again, at higher temperatures, by a region of serrated (jerky) yield. We might consider

* For instance, in super-high-purity metals only a few parts per million of interstitial impurities can lower the *recrystallization temperature* appreciably and raise the ductile–brittle transition temperature, whilst in the newer refractory alloys fatigue behaviour can be adversely affected. For chromium, the embrittlement through minutest amount of nitrogen is especially notorious, and for a long time 'pure chromium' has been thought of as an inherently brittle metal. However, this has now been shown otherwise[26], and ductile Cr can be produced, behaving like other normal metals; for instance, within the ductile range, strain aging and work-hardening effects can be obtained[27]. Also the ductile–brittle transition temperature can be depressed, e.g. from 350° to 150°C, by reducing the nitrogen content from 0·02 to 0·0005 per cent; it can then be reduced still further. One artifice can be (see later) the use of the scavenger action of the earlier transition metals such as those of Group III and the Rare Earths.

the *first* as the region where dislocations are largely frozen in by their surrounding interstitial Cottrell atmospheres; the *second* as that where their movement becomes freed from the latter through sufficient thermal energy and triggered off by relatively small stresses, but is still impeded by the

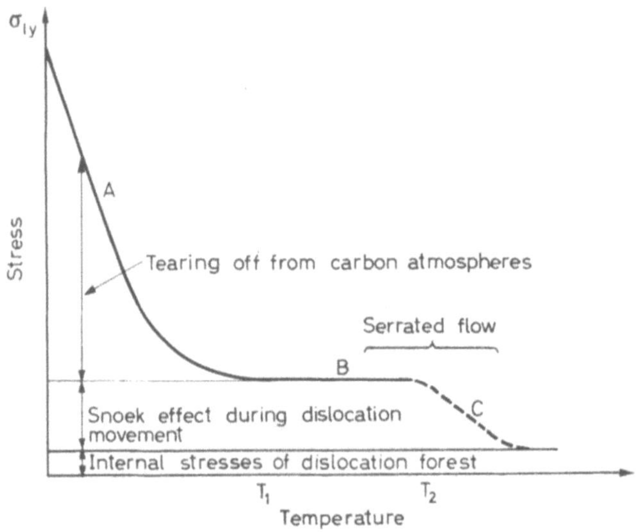

Figure 3.6. Temperature dependence of flow stress in α-iron containing C or N in solution (After Schoeck and Seeger[28])
By courtesy of the Editor, *Acta Metallurgica*

Snoek atmospheres; the *third* as that where these also begin to be surmounted. Even at high temperatures, however, both Cottrell and Snoek atmospheres remain operative, but in a more dynamic and subtle fashion. The resistance a dislocation meets to its movement is itself speed-dependent, because if too slow, interstitial (e.g. C or N) atoms have time to diffuse towards it from greater distances, thus automatically increasing the resistance (both through drag by the Cottrell cloud and through greater interstitial population to give enhanced Snoek friction). Thus, any slow-down is self-reinforcing, and jerky flow can arise through hindrances and their distribution in the metallic mass becoming erratic and non-statistical. At highest temperatures, the chief remaining dislocation-impeding factor is not so much that of interstitial atoms (which have enough mobility to travel with the dislocations in any case) but the mutual interaction between members of the dislocation forest itself and the stress field surrounding each.

A considerable 'drag' can also be exerted on the movement of a grain boundary by an impurity atmosphere, depending on the diffusivity of the interstitials. For instance, it was found[29] that, at *high* grain boundary velocities, the *faster* diffusing impurities have the greater drag, at *low* velocities the *slower* diffusing ones. The whole kinetics of this movement, therefore, can be quite erratic, and can, in its manner of progress, critically manifest the types of interstitial atmospheres present and their local variations.

By studying the interaction between nitrogen atoms and dislocations in iron after aging and cold working, Leak[30] found that the stress field of dislocations has no effect on the relaxation time of internal friction, which itself would be associated with the movement of the nitrogen atoms in the BCC lattice; he concludes that only a small number of atomic displacements can induce the change from random solution to interstitial locking in dislocation atmospheres.

In quenched metals, the clustering of vacancies can be greatly affected by impurities. For instance, a new kind of defect has been found[31], denoted 'black-spot defect', consisting of a spherical cluster of a limited number (some hundreds) of vacancies; it only occurs on quenching from *below* a certain critical temperature, and when the ratio of impurities/quenched-in vacancies exceeds a certain value. The clustering becomes reduced for quenching from high temperatures, the vacancy dispersion becoming more uniform.

The above picture of dislocation–interstitial interaction has in some cases already succeeded in explaining observed mechanical properties, notably yield stress, embrittlement and fracture, but generally the interfering factors are at present still much too complex and various to permit complete answers or predictions. The problem is complex enough for one pure metal plus one interstitial (e.g. Fe–C) and would appear more formidable still in the multi-metal/multi-non-metal systems which characterize, for instance, the modern commercial creep-resisting steels and alloys. However, after some no doubt initially empirical approach in these alloy developments, good progress has been made also on the fundamental side, to explain their behaviour in terms of composition, heat treatment and consequent lattice structure and dislocation effects. We must here remember that in these cases the complexity is increased still further in that, within the micro-aggregate, solid solution superimposes on compound formation. Yet both aspects form really part of the same overall picture of what might be called 'Interstitial Science' which we are attempting here to present. Completed carbide crystallites, particularly those of finest particle size, can also be essentially effective to pin down dislocations, in a sense similar to that of the more nascent state represented by the Cottrell atmospheres. Indeed, the borderline between an advanced Cottrell concentration of interstitials around dislocations in the matrix and an actually nucleated particle, say of a carbide, is by no means a sharp one, and we might envisage intermediate, coherently forming semi-carbide stages (like the finely dispersed HCP ε-phase in steels) as originating from interstitial-induced stacking-fault areas within the primary solution lattice, the stacking faults in turn becoming effective sources of new dislocations.

Any strains, whether induced by external applied stress or by internal stress, tend to enhance such precipitation, and much recent work has demonstrated this. For example, in studies of high-temperature ductility of austenitic stainless steels containing niobium (18 per cent Ni, 12 per cent Cr, 1 per cent Nb), it has been shown[32] how NbC, initially in solid solution in the austenite, can, through plastic strain, be precipitated in varying grain sizes and degrees of dispersion, depending on stress, time and temperature of loading, with directly demonstrable influence on creep, strength, rupture and hardness. Here again, the niobium carbide nucleates at dislocations, preceded by the

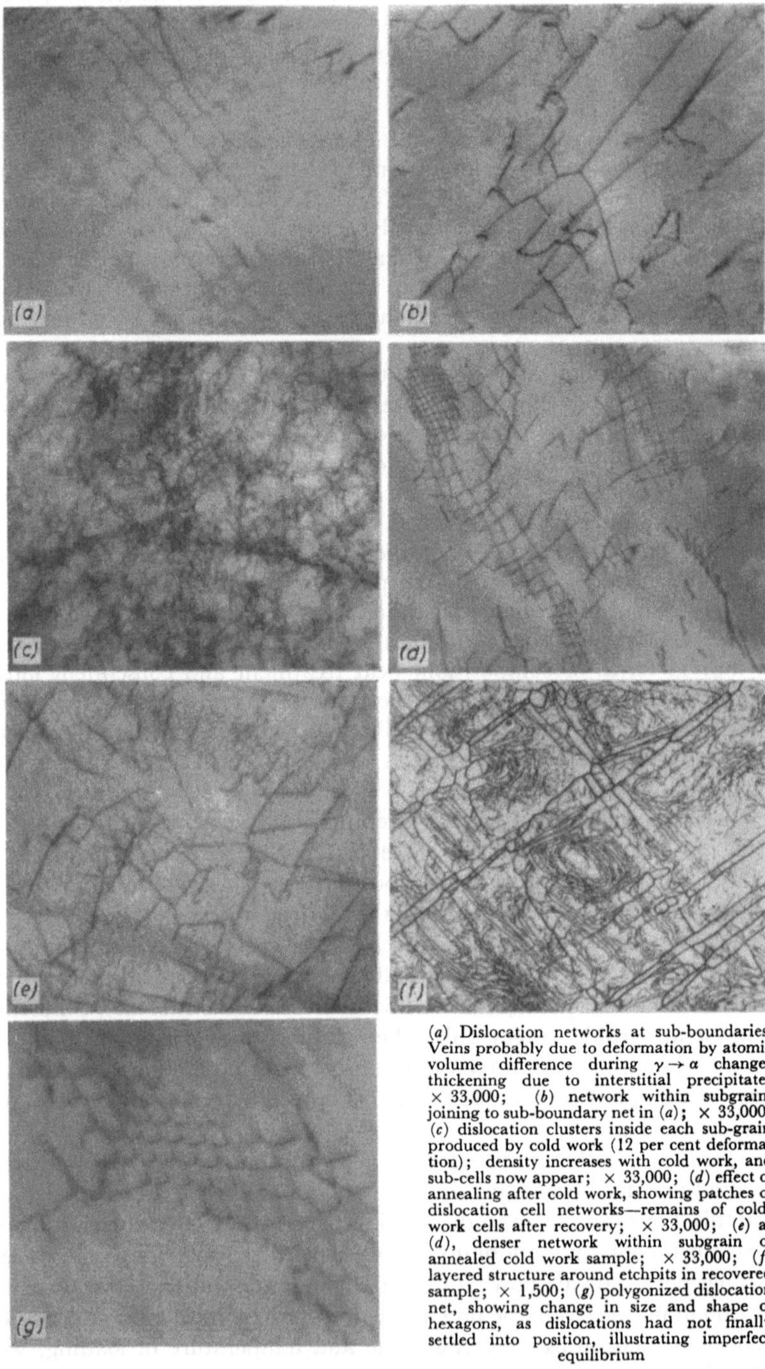

(*a*) Dislocation networks at sub-boundaries. Veins probably due to deformation by atomic volume difference during $\gamma \rightarrow \alpha$ change; thickening due to interstitial precipitate; × 33,000; (*b*) network within subgrain, joining to sub-boundary net in (*a*); × 33,000; (*c*) dislocation clusters inside each sub-grain produced by cold work (12 per cent deformation); density increases with cold work, and sub-cells now appear; × 33,000; (*d*) effect of annealing after cold work, showing patches of dislocation cell networks—remains of cold-work cells after recovery; × 33,000; (*e*) as (*d*), denser network within subgrain of annealed cold work sample; × 33,000; (*f*) layered structure around etchpits in recovered sample; × 1,500; (*g*) polygonized dislocation net, showing change in size and shape of hexagons, as dislocations had not finally settled into position, illustrating imperfect equilibrium

Figure 3.7. Thin-film electron micrographs of dislocations in iron, as loci for interaction with interstitials
(From Carrington, Hale and McLean[34], by courtesy of The Royal Society; photographs by courtesy of Dr. D. McLean; *Crown Copyright reserved*)

aggregation of carbon atmospheres in the FCC matrix. In a related study[33], the electron microscope showed upon aging fine *threads* of NbC forming along dislocation lines, but also, around *existing* NbC particles, the generation of new dislocations at which then *more* NbC can nucleate. Examples of such effects[34] are shown in *Figure 3.7.*

Stresses, both external or 'implanted', are then amongst the most effective dislocation generators. Near dispersed particles, dislocation loops can be produced through stresses set up in the solution matrix surrounding them. In creep-resisting (austenitic) high-temperature steels containing niobium[35], the undissolved NbC particles produce, on cooling, contractive thermal stresses and, thus, dislocations. These in turn can absorb the excess vacancies, and so inhibit precipitation hardening. 'Warm working' (i.e. mechanical work at elevated temperatures within the range where precipitation is critical) can increase the dislocation density greatly, providing loci for new very finely and stably dispersed NbC and other carbide particles, which are initially only some 10 or 20 Å in size. One aim is then to minimize their growth both during heat treatment and later service at temperature, so as to render them effective as dispersion strengtheners.

An interesting related phenomenon, though in a sense the opposite, is that of 'superplasticity'. This is a high degree of mechanical deformability occurring in the material *during* a phase transformation under stress, and it can be attained in primary solid solutions at temperatures near the solubility limit. These conditions can produce a transient breakdown of atomic bonding under stress during the incipient stage of precipitation of the new structure, and an abnormal degree of plasticity (for a useful review, see Underwood[36]). Although reported mainly for substitutional solid solutions, this is likely to be also inducible in interstitial ones or in mutual combinations.

The recent advances in transmission electron microscopy provide a powerful means of observing the creation, movement and clustering of dislocations and their interaction with interstitials; some further examples are shown in *Figure 3.7.*

A useful approach in measuring the interstitial–dislocation interaction is to follow the *temperature dependence* of tensile properties, such as yield stress as well as the grain size dependence. This work has so far mainly concentrated on the (high-melting-point) BCC transition metals, because of their immediate practical interest for use in gas turbine components and aerospace applications. Pioneering work was here that of Heslop and Petch[37] who showed that the lower yield (or shear) stress, σ_y, of plain carbon steels can be related to the mean grain diameter by an equation

$$\sigma_y = \sigma_i + \frac{k_y}{\sqrt{d}}$$

where σ_i and k_y are parameters significant and derivable in terms of the dislocation model[38,39]; σ_i is the lattice friction stress acting on a dislocation and opposing the applied stress, and is strongly temperature-dependent; k_y, which increases with interstitial content, is an index of the strength of dislocation locking (or of the force σ_D required to 'un-pin' a dislocation), and of the mean distance, l, from a grain boundary to the nearest dislocation source, so that it is sensitive to interstitial impurity content. A relation $k_y = \sigma_D/\sqrt{l}$ is found to hold good. In the (low-temperature) brittle range, dislocation pile-ups produce high local stress concentrations, which initiate a Stroh crack (see above) and lead to fracture. In this case, Petch[40] has shown an equation

$$\sigma_f = \sigma_i + \frac{k_f}{\sqrt{\mathrm{d}}}$$

to apply for the fracture stress, σ_f, i.e. similar to the above applying in the ductile range. Yield propagation of dislocations is here hindered by locking; however, in basic principle failure still arises through dislocation pile-ups, which are largely interstitial-induced, and it is qualitatively the same process, only with a change of emphasis as between governing factors in the ductile and the brittle ranges. In several cases, the above coefficients have now been successfully worked out by carefully tracing the mechanical properties as a function of temperature, crystal size and strain rate, as well as of pre-treatment of the sample and of interstitial content in solution. For niobium and molybdenum the temperature dependence of deformation and of fracture has been studied in this sense[41a,42], and it seems now to become possible gradually to explain even quantitative differences in ductile/brittle transition between the various BCC metals.

These effects have assumed fundamental importance in explaining the aging properties of metals, both strain-aging and temper-aging and their superposition. The dislocation pinning by carbon and nitrogen has at low temperatures its counterpart in that by *hydrogen* atmospheres, because at higher temperatures the H atom becomes the less effective in this owing to its greater mobility. However, in niobium for example, hydrogen has a marked effect on dislocation locking, as recently shown by strain-aging and yield-stress experiments[43]; Wilcox and Huggins[41b] extended this to Nb, Ta and a mild steel. The activation energy worked out from the temperature dependence of strain-aging agreed well with that obtained from diffusion of H in Nb. It is interesting in this connection that BCC metals can, by strain-aging, be 'over-aged', which means that, through the presence of lattice strain, *more* interstitial atoms can coagulate near a dislocation than can in true solid solution: thus a fine precipitate of, say, carbide, nitride or hydride can form, which in its turn gives rise to increased dislocation locking.

On molybdenum Cottrell's and Petch's theories have been extensively tested, e.g. by critical deformation experiments[44]. Altogether, the ductile/brittle transition in BCC metals, which so greatly depends on the presence of interstitials, forms in itself the subject of intensive current researches where, of course, the alloy development and the fundamental aspects go hand in hand. Conversely, the transition temperature is often taken as a measure of interstitial content: not always rightly. The effect of hydrogen on the yield properties is of closely related interest (for Mo, see[45]).

Fundamental advantage can be taken of the fact of the primary solubility increasing with decreasing Group number in the transition metals, because this lies at the basis of their scavenger action, and of their addition to the later Group metals. Thus interstitials can be scavenged by yttrium, scandium and the rare earths in Cr, Mo, W, V, Nb and Ta, and (less strongly so) by Zr and Ti.

Direct indications of an interstitial atmosphere around dislocations are given by the many fascinating electron micrographs of dislocation network now available (cf. *Figure 3.7*), which frequently include a visible thickening of some of the dislocations; this is, in fact, a 'decorating effect' due to the aggregation of interstitials. It is, of course, on a much finer scale than grain-boundary segregation, but the two kinds of segregation (the first observable optically, the second in the electron microscope) form the beginning of an instructive progression in resolution: to complete the series, a third (necessarily hypothetical!) picture might show the actual cloud of interstitial atoms adjoining defects within the lattice. Field-ion microscopy may well achieve this third step.

Upon cold work, the numbcr and density of dislocations over the matrix becomes greatly increased, and consequently that of the places of interstitial nucleation. The degree of cold work vitally affccts their distribution, and essentially, the heavier it is, the more a grouping of dislocations into clusters is promoted. Eventually they evolve into a cellular structure, of the order of microns, which becomes the finer the heavier the strain. This progressive subdivision then culminates in the formation of actual sub-boundaries and crystallites—an evolution which beautifully links up with and explains the long known empirical effect of cold work to cause particle refinement, each particle being the outcome of a dislocation-surrounded cell of this kind. In iron, Carrington *et al.*[34] (see *Figure 3.7*) have performed this link-up quantitatively and quite precisely by thin-film electron microscopy; they could similarly demonstrate the effect of creep, work-hardening and recovery on the evolving dislocation network. The implication of this, as far as interstitials are concerned, is that these loci predetermine the distribution, first, of interstitial atmospheres and, second, of compound particles. 'Warm working', a process in which mechanical work is applied at elevated temperatures (about 700°C)[46], greatly improving the creep resistance of modern niobium-bearing austenitic steels, similarly increases the dislocation density and thus, the number of points where niobium carbide (and other compounds) can precipitate, yet remain for long times of small particle size and stable even at high temperatures of service.

Moreover, the interaction is a converse one in that particles, e.g. of NbC, *once formed* in the matrix, surround themselves by a stress field and consequent fresh dislocation loops: these are ring-shaped in association with their centre of origin. Similarly, differential thermal expansion or contraction between, say, an existing carbide crystal and the embedding matrix—a situation which commonly arises during heating or cooling in heat treatment or during service—would, as mentioned earlier, create local (compressive or tensile) stresses around the particles and thus further sources of dislocations which, in turn, will attract solution interstitials and vacancies. On cooling, these contraction stresses will again produce loop-shaped dislocation rings 'imaging' the particles which generate them and thus new types of nucleation sources. These loops, once created, again determine the further pattern of sinks for attraction of interstitial atoms and of second phases within the metallic mass. Indirectly, then, this interaction process provides a sensitive means of manipulating, whether by design or fortuitously, the mechanical properties of these alloys and their temperature dependence: the term 'dislocation engineering' might not be amiss, the interstitials acting sometimes as tools, sometimes as constructional units.

One very interesting observation[47] should be recorded. In determining the solid solubility of carbon in Fe–Ni alloys, a *reduction* in the solubility value was found at Ni_3Fe, i.e. where the superlattice occurs. The wider theme can be raised, therefore, that atomic *ordering* in parent alloys can reduce interstitial solid solubility. That interstitial solubility should, in principle, be the higher for the disordered state, seems consistent again with the above.

A further fascinating aspect of interstitial atom behaviour in solution is that, in magnetic materials, they can be used to stabilize definite *domain walls*, e.g. in iron–silicon crystals[48]. The interstitial atoms can in effect be 'trained' to segregate at regular domain boundaries to reinforce and stabilize the latter, as they do in the normal, less orderly fashion at crystal and grain boundaries. The thickness of such an '*interstitial-stiffened*' cell wall was in the instance cited of the order of 160 Å for 180° and 90° walls, but can of course be made to vary. The possibilities, in wider principle, of positive utilization of training interstitial segregation into regular textural sub-regions seem to be great and merit much further study.

In a similar context, it is interesting to note the influence of interstitial solutes on processes such as magnetic annealing. For example, one research finding[49] is that if Ni–Fe and Ni–Co–Fe alloys—'Permalloy' type—are heat-treated in a magnetic field, and their oxygen content is very low (< 14 p.p.m.), magnetic polarization of domains *fails* to occur. But it *does* occur for oxygen higher than this because these oxygen atoms are required to segregate into definite crystallographic planes at temperature, so that the domain walls can be stabilized. The magnetic orientation only then becomes possible on heat treatment in the magnetic field and anisotropy will be maintained: another example of 'training' of solute interstitials for a positive physical purpose. In this case, the upper limit of effective oxygen content is set by the solubility limit (still very low) above which level free oxide formed would act detrimentally.

The scope of such 'training' of interstitials by special heat treatment (though not magnetic), for creation of special properties, extends even more substantially to alloys of the earlier Groups (e.g. IV, V, VI) where the solubilities are the higher and the more temperature-dependent.

In a less ordered, but nevertheless related manner we have the process of grain-boundary hardening induced by oxygen and other interstitials in metallic matrices. The latter can be both primary solid solutions and primary intermetallic compounds. The process can be produced, for instance, by oxygen-diffusion treatment, and large increases in hardness at grain boundaries (above that in the interior) and at free surfaces can occur. This is due not to actual oxide precipitation—although in some cases it can go so far—but to interstitial oxygen preventing lattice slip and causing local dislocation locking. Considerable research now proceeds on this subject of 'internal oxidation' (a term which, however, also embraces oxide formation and dispersion effects and their utilization).

3.7 SOME SPECIFIC ASPECTS OF INTERSTITIAL SOLID SOLUTIONS

It is not the purpose of this Section to give a comprehensive collection of data on primary solubilities of interstitials in the transition metals; but a few additional examples from the rapidly increasing literature may typify some of the aspects.

In the refractory alloys based on the early transition-metal Groups, the solubilities of interstitials become considerable not only in absolute terms (C, O, N and H are on the 'crescendo' from Groups VI to V to IV to III*), but their differentials with temperature and pressure also are large. The effect of this on properties can be considerable. The strengthening and precipitation mechanisms in the refractory alloys are often dominated by interstitial presence. Hardness increase is probably the best known effect, but it is only one, and much intense research is now proceeding to determine these solubilities and their effects. The effect of interstitials through strain aging, for instance, upon the fatigue limit can be marked (for titanium, see e.g. [50]), and the interstitial strengthening in binary and ternary *alloys* is receiving increasing attention (e.g. on systems such as Nb–Zr–W or Mo–Ti–Zr[51] with interstitial addition). The apparent presence or absence of miscibility gaps in systems *between* transition metals can, in the author's experience, be greatly interstitial-affected. Likewise can the temperature dependence for interstitial solubilities be strong. For example, carbon solubility in *tantalum*[52] may be as high as 7·5 atomic per cent at ≈ 2,900°C, falling to 2·5 at ≈ 2,000°C, to 0·3 at ≈ 1,500°C. A similar abrupt decrease was observed for *niobium*[53] and for tungsten–carbon[54], as shown in *Figure 3.8*; W_2C nuclei are here the precipitating phase. The implication of age-hardenability (by tempering alloys pre-quenched from the higher solution region) is clear, and this has indeed been demonstrated on W–C (author, unpublished). *Figure 3.8* also gives a diagram of carbon solubilities in various transition metals (as per 1962; data are not necessarily final). To typify variations between interstitials for constant parent metal, the C, N and O solubility limits in *tantalum* decrease with temperature[55]:

Solute	1,500°C	1,000°C	500°C
C	0·7	< 0·07	—
N	3·70	2·75	1·8
O	3·65	2·97	2·5

with Ta_2C, 'Ta_XN' and Ta_2O_5 precipitating on excess. The eutectics lie here, however, at temperatures higher than 1,500°C, so that this does not represent the maximum.

For alternative solubilities of carbon in tantalum, conflicting results have been reported[56,57], reflecting the very sensitive dependence on detailed experimental conditions and on precautions, e.g. to avoid tramp interstitials. Strong effects of C and N in niobium on hardness, strength, workability and

* Utilized in the successive gettering and scavenging actions, increasing in this sequence.

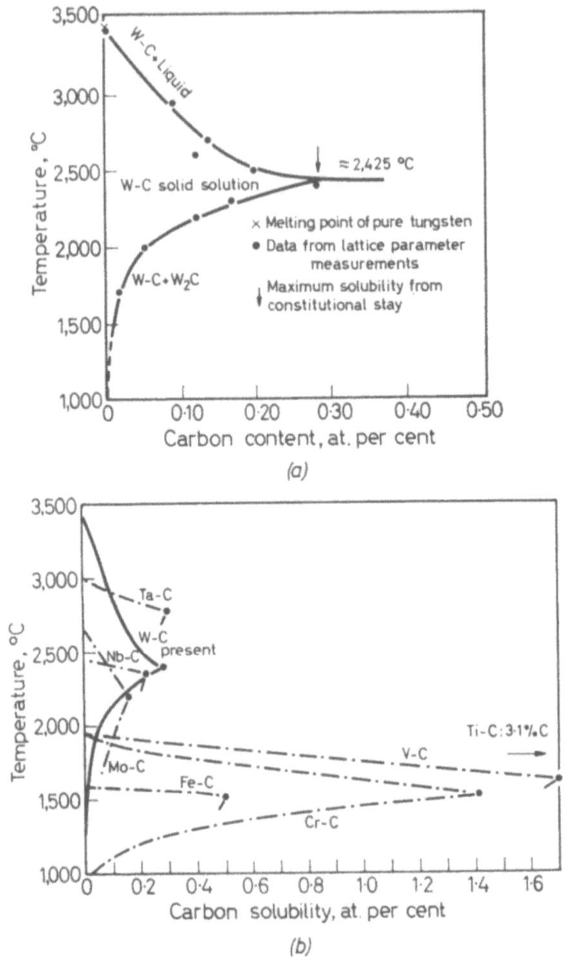

Figure 3.8. Solid solubilities of carbon in metals (From Goldschmidt and Brand[54]): (*a*) Carbon in tungsten; (*b*) comparison for various transition metals, as function of temperature

By courtesy of the Editor, *Journal of Less-common Metals*

recrystallization have been observed[58], and likewise of O and H on the mechanical properties of niobium and tantalum[59] (cf. Bibliography).

Of the greatest significance is obviously the interstitial solubility of oxygen in regard to the oxidation process of the metals, in its first stage, and here the transition to the (still FCC-related) lowest oxides has proved of special interest (phases Nb_xO, Ta_yO, etc., including interstitial ordering, as shown for instance in the work by Kofstad's school[60]; cf. Chapter 8). Altogether, however, we should stress, as a point of principle, the essential *lack of demarcation* here between solid solutions and interstitial compounds, and the continuity in transition between the two.

Closely associated with the equilibrium solubility is the kinetic question, i.e. of *diffusion* of oxygen in the refractory metals exposed either to the gas isothermally or to a *thermal gradient*; the latter has recently been discussed in some detail[61], also in connection with the rate of evaporation of oxygen which must proceed simultaneously.

The solubilities in *iron* (α-, γ-, δ- and liquid) of C, H, N and O have been very widely studied; as just one instance, *Figure 3.9* shows the case of nitrogen in α-Fe[62]. The receding phase boundary should, in fact, be viewed in

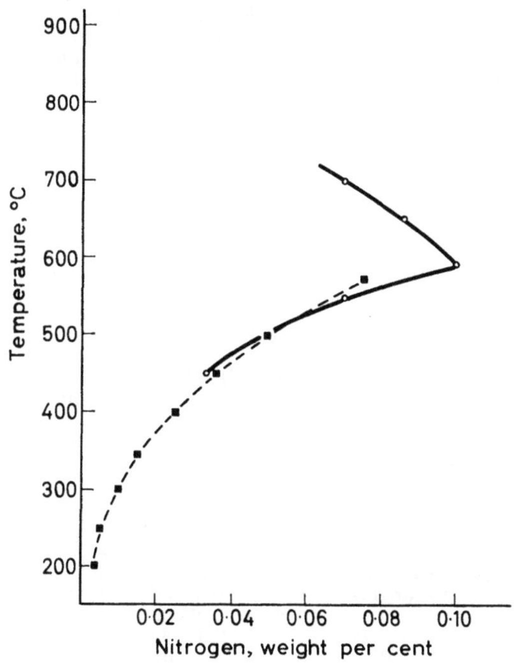

Figure 3.9. Solid solubility of nitrogen in α-iron as function of temperature (From Paranjpe, Cohen, Bever and Floe[62])

By courtesy of the Metallurgical Society of AIME

conjunction with the iron nitrides then precipitating (Fe_4N, etc.) to be discussed in Chapter 5: phases which can form coherently with the primary solution, and in varying degrees of dispersion and lattice morphology. This is very critically temperature- (and preheat-treatment-) dependent, and so are the resulting properties. Qualitatively, the same type of recession occurs for α-Fe–C*, α-Fe–B, α-Fe–O and α-Fe–H, which will be subjects of later Chapters.

* With all the important quench-aging (e.g. pearlite-forming) effects known and implied in ferrous metallography. This should be kept distinctive from the martensite, bainite, etc. effects which are inherited from *austenite*; although the two sets of effects largely overlap, depending on thermal history (see e.g. Chapter 4). In practical terms, they closely interweave for instance in the modern development of high-strength steels.

The following Selected Literature on solubilities may typify and amplify some of the aspects of interstitial solid solution.

SELECTED LITERATURE

A. *General*

International Symposium on Metallic Solid Solutions [Orsay, 1962] (Ed. Friedel, J. and Guinier, A.), Paris (Benjamin) 1963
Metal physics of solid solutions, electronic and atomic structure; solid solutions generally, e.g. substitutional, but papers relevant to interstitials

Seeger, A., Mann, E. and Jan, R. von *Physics Chem. Solids* 23 (1962) 639
Interstitial atoms in FCC crystals, in particular Cu; location and movements; interpretation and verification of various crystalline models; cf.[13]

Krasko, G. L. and Lyubov, B. Ya. *Dokl. Akad. Nauk SSSR* 147 (1962) 562
Continuum theory of elastic interaction between atoms in interstitial BCC solid solutions

Gulbransen, E. A. and Andrew, K. F. *Trans. Inst. Min. Engrs* 185 (1949) 513, 741; 188 (1950) 586
Kinetics of Zr, Ti, Nb, Ta solution for O, N and H

Stringer, J. and Rosenfield, A. R. *Nature, Lond.* 119 (1962) 337; Jacob, A. J. *ibid.* 200 (1963) 1310
Nature of solutions in BCC metals

Hume-Rothery, W., Irving, H. M. and Williams, J. R. P.[7]
Theoretical aspects; electron donation through later Group metals

Robins, D. A.[8]
Theory; relates and partly predicts interstitial solubilities and mechanical properties

Thermodynamics of interstitial solutions

Khachaturyan, A. G. *Fiz. tverd. Tela* 5 (1963) 15
Thermodynamic theory

Fastov, N. S. *Soviet Phys. Metals Metallogr.* 11 (1961) 35
Thermodynamics of interstitial solid solutions in BCC metals; theory of elastically deformed solutions; free energies under equilibrium and non-equilibrium conditions

Rees, A. L. G. *Trans. Faraday Soc.* 50 (1954) 335

Speiser, R. and Spretnak[11] [Statistical mechanics]
Dealing more generally with thermodynamics of interstitial solutions

Zener, C. *Trans. metall. Soc. A.I.M.E.* 167 (1946) 513; *Elasticity and Anelasticity of Metals*, Chicago Univ. Press 1948. Darken, L. S. and Smith, R. P. *J. Am. chem. Soc.* 68 (1946) 1163; Kaufman, L. and Bernstein, O. *U.S.A.F. Rep. AD 285984* (1962)
Last of these more specifically on Ti $\rightarrow$ O solid solution series, TiO, etc.

De Keyser, W. *4th Int. Symp. Reactivity of Solids 1960*, p. 376, Amsterdam, 1961
Review of lattice defects related to reactivity in solutions

Dienes, G. J. and Damask, A. C. *Phys. Rev.* 125 (1962) 444, 447; 128 (1962) 2542
Kinetics of vacancy-interstitial annihilation: (I) Impurity-interstitial trapping; (II) Di-interstitial formation; (III) Interstitial migration to sinks

Bullough, R. and Newman, R. C. *J. phys. Soc. Japan* 18 (1963) Suppl. III, 27
Kyoto Conference on crystal lattice defects, 1962: theoretical discussion of vacancy and impurity atom distribution around edge dislocations

Saada, G. V. *J. phys. Soc. Japan* 18 (1963) 41
Theoretical discussion on stability of quenched-in interstitial and vacancy loops in FCC metals

Danilenko, V. M. *Fizika Metall.* 9 (1960) 810, 940
Extension of theory of Cottrell atmospheres in dilute interstitial solutions

Hirsch, P. B. *J. phys. Soc. Japan* 17 (1962) 45 [Kyoto 1961] Conference 'Magnetism & Crystallography'
Defects in crystals: electron microscope studies; review of applications, theory and techniques. Hirsch's classical work in this field, though perhaps not specific to interstitial solutions, is of basic relevance

Weertman, J. *Phil. Mag.* 8 (1963) 967
Interstitial-producing extended 'jogs' in FCC lattices; comprehensive theory, relation of interstitials and vacancies to various types of stacking fault

Pashley, D. W. and Presland, A. E. B. *Phil. Mag.* 7 (1962) 1407
Dislocation movement in thin films under electron microscope: effect on it of interstitial contamination

Jackson, K. A. *Phil. Mag.* 7 (1962) 1117
Nucleation of dislocation loops from vacancies; theory of vacancy loop configurations and formation

Grinberg, B. A. *Soviet Phys. solid St.* 4 (1963) 1901
Theory of conditions for pinning of dislocations in clusters, by impurtiy atoms

Internal friction as bearing on interstitial solutions
Internal Friction Conference, Cornell, 1961. [Report] *Acta metall.* 10 (1962) 267
Papers on dislocations, point defects, interstitial pinning, etc.
A further Symposium on Internal Friction was organised by The Institute of Physics and The Physical Society, Manchester, Sept. 1965
Entwistle, K. M. *Metall. Rev.* 7 (1962) 175 [Review]
Novick, A. S. 'Internal Friction in Metals', in *Progress in Metal Physics*, Vol. 4, London (Pergamon) 1953
Snoek, J. L. *Physica, s'Grav.* 8 (1941) 711
Kê, T. S. *Phys. Rev.* 70 (1946) 105; 71 (1947) 142, 533; *Trans. metall. Soc. A.I.M.E.* 176 (1948) 448; *J. appl. Phys.* 20 (1949) 274
Wert, C. A. *J. appl. Phys.* 20 (1949) 943
Typical of numerous papers from Wert's school in this field
Dijkstra, L. J. *Philips Res. Rep.* 2 (1947) 357, 399, etc.
Schoeck, G. *Acta metall.* 11 (1963) 617
Internal friction in interstitial solute atom/dislocation interactions in BCC metals
Fast, J. D. *Metaux Corros. Inds* 36 (1961) 383, 431
General review of internal friction of metals; interstitial action
Johnson, A. A. *Phil. Mag.* 7 (1962) 177
Ductile–brittle transition in BCC transition metals; discussion of Petch relation and parameters; comprehensive criteria of brittle fracture
Chaudron, G. *Am. Soc. Metals Seminar on Ultra High Purity Metals*, 1961, p. 149
Metallurgical problems presented by high-purity metals; effect of trace interstitials on recrystallization temperature of Zr, Fe, Cu, Al; its decrease on ultra-purification; grain-boundary effects and dislocation/vacancy interaction; solubility of oxygen in iron
Cottrell, A. H. *Fracture* [Symposium on Reactor Materials, Berkeley Castle, 1961]
General theory of fracture mechanism; ductile → brittle behaviour and crack initiation
Symposia on Refractory Metals containing papers on solubility of interstitials in Nb, Ta, Mo and W (particularly oxidation and mechanical properties):
Refractory Metals and Alloys I and II (Ed. Semchyshen, M., Harwood, J. J. and Perlmutter, I.), A.I.M.E. Symposia, 1960, 1962, New York (Interscience) 1961, 1963
Columbium Metallurgy (Ed. Douglass, D. L. and Kunz, F. W.) [A.I.M.E. Symposium, 1960], New York (Interscience) 1961
Niobium, Tantalum, Molybdenum and Tungsten (Ed. Quarrell, A. G.) [Conference at Sheffield 1960] Amsterdam (Elsevier) 1961
The Science and Technology of W, Ta, Mo, Nb and their alloys (Ed. Promisel, N. E.) [Proc. AGARD Conference 1963], London (Pergamon) 1964

B. *Specific Systems* (typical papers only)
Notes—(1) $M \rightarrow X$ means solid solution of interstitial X in the metal M
(2) On solubilities of C, N, B, O and H in Ti, Zr, Nb, Ta, Cr, Mo, W and U, see also monographs on these in Bibliography to Chapter 2

Ti Gupta, K. P., Cheng, C. H. and Beck, P. A. *J. Metals* 14 (1962) 90
α-Ti → O, bond character and effect on electrical resistivity and specific heat; also on the large primary solid solution ranges of oxygen in Ti, Zr and Hf
Palty, A. E. *et al.*, *Trans. Am. Soc. Metals* 46 (1954) 312 [α,β-Ti → N]
Cadoff, I. and Nielsen, J. P. *J. Inst. Metals* 5 (1953) 248; Van Thyne, R. J. and Kessler, H. D. *Trans. Am. Inst. Min. metall. Engrs* 200 (1954) 193 (α,β-Ti → C]
McQuillan, A. D. *Proc. R. Soc.* A204 (1950) 307; *J. Inst. Metals* 79 (1951) 371; 80 (1952) 363 [Ti → H]

Zr Holmberg, B. and Dagerham, T. *Acta chem. scand.* 15 (1961) 915
α-Zr → O; solubility, studied by x-rays, found to be independent of temperature at $\approx ZrO_{0.4}$; effect of oxygen on *a* and *c* axes shows ordering and other structure changes
Domegala, R. F. *et al.*, *Trans. Am. Inst. Min. metall. Engrs* 206 (1956) 98 [Zr → N]
Mallett, M. W. *et al.*, *J. electrochem. Soc.* 101 (1954) 1 [Zr → N]
Gebhardt, E. *et al.*, *J. nucl. Mat.* 4 (1961) 254, 255, 269
Zr → O solubilities; electrical resistance, constitution, reaction kinetics, recrystallization

Hf Dagerham, T. *Acta chem. scand.* 15 (1961) 214
α-Hf → O; solubility, studied by x-rays, shows spacing expansions of *a* and *c*; random distribution of O atoms throughout, in contrast to Zr–O
Sara, R. V. *Trans. metall. Soc. A.I.M.E.* 233 (1965) 1683 [Hf → C]

Hf—*cont.* Hf can take up to 10·5 at. per cent C into solution at 2,208°C (melting point)
Benesovsky, F. and Rudy, E. *Planseeber. PulvMetall.* 8 (1960) 66 [Hf → C]
Rudy, E. and Benesovsky, F. *Mh. Chem.* 92 (1961) 415 [Hf → N]

Th Peterson, D. T. *Trans. Am. Soc. Metals* 53 (1961) 765 [Th → C]
Smith, M. D. and Honeycombe, R. W. K. *J. nucl. Mat.* 1 (1959) 345 [Th → C]
Gerds, A. F. and Mallett, M. W. *J. electrochem. Soc.* 101 (1954) 175 [Th → N]

V Gurevich, M. A. and Ormont, B. F. *Russ. J. inorg. Chem.* 2 (1957) 183 [V → C]

Nb, Ta Pemsler, J. P. *Publ. U.S.A.F. Rep.* AD 262060, 60–655, 1961
Comprehensive report of Nb and Ta → O and N solutions, at temperatures near melting point; thermodynamics of atom interactions

Nb Boone, D. H. and Wert, C. A. *J. phys. Soc. Japan* 18 (1963) 141 [Symposium on Lattice Defects, Tokyo 1962]
Nb → O and N; interstitial–dislocation interaction by cold work and internal friction; interstitial clustering, etc.
De Batist, R. *Physics Status Solidi* 3 (1963) 1475
Deformation-induced internal friction peaks in Nb
Stiegler, J. O. *et al., Acta metall.* 11 (1963) 851
Details of dislocations in deformed and annealed Nb single crystals, acting as sinks for interstitials

Nb, Ta, Mo Peacock, D. E. and Johnson, A. A. *Nature, Lond.* 195 (1962) 169
Point defects in Nb, Ta, Mo; effect of irradiation and deformation on internal friction peaks

Nb, Ta Ogden, H. R. *et al., Can. metall. Q.* 1 (1962) 13
Nb and Ta → O and H; effect of interstitials on tensile flow, strength and fracture

Gebhardt, E. *et al., Z. Metallk.* 55 (1964) 432
Nb and Ta → O and N: processes during *de*gassification, property changes

Nb Schofield, T. H. *J. Inst. Metals* 85 (1957) 372
Rawson, J. D. W. and Argent, B. B. *J. Inst. Metals* 91 (1962) 223 [Nb → C]
Nb → O; effect on yield strength
Hobson, D. O. *U.S. At. Energy Comm., Publ. Rep. ORNL* 03245, 1962
Aging phenomena in Nb-base alloys; effect of interstitial impurities on mechanical properties of a Nb –1 per cent Zr alloy; study of strengthening and ductility decrease
Evans, P. R. V. *J. less-common Metals* 4 (1962) 78
Nb → N; effect on yield strength, N acting as solution hardener; theoretical analysis in light of Petch equation
Gregory, D. P. *Acta metall.* 11 (1963) 455
Effect of impurities on vacancy mobility, dislocation interactions and flow stress variations with temperature, in Nb and other BCC metals
Cost, J. R. and Wert, C. A. *Acta metall.* 11 (1963) 4, 231
Nb → N; and metal–gas equilibrium: comprehensive thermodynamic study, Sievert's law obeyed; discussion on nitrides formed
Elliott, R. T. and Komjathy, S. In *Columbium Metallurgy*, New York (Interscience) 1961
Carver, M. D. *et al., U.S. Bur. Mines, Res. Invest.* 5872, 1961
Nb → C and N; effect on mechanical properties at *low* temperatures; increase by interstitials of ductile–brittle transition temperature
Gebhardt, E. and Rothenbacher, R. *Z. Metalllk.* 54 (1963) 443, 623; Gebhardt, E. *et al., ibid.* 55 (1964) 423; *Metall* 18 (1964) 691
Nb → O and Nb → N solubilities; effect on properties

Ta Pemsler, J. P. *J. electrochem. Soc.* 108 (1961) 744 [Ta → N]
Gebhardt, E. *et al. Z. Metallk.* 46 (1955) 560; 48 (1957)
Ta → O solubility; effect on properties
Fromm, E. and Roy, U. *J. less-common Metals* 8 (1965) 73 [Ta → C solubility]
Note a good summary of work on (Nb and Ta) → interstitial of Gebhardt and Rothenbacher, AGARD Symposium [see p. 83], p. 157
Schoeck, G., Mondino, M. *J. Phys. Soc. Japan* 18 (1963) Suppl. I, 149 [Symposium on Lattice Defects, Tokyo 1962] Ta → O; internal friction peak due to interstitial–dislocation interaction and as function of cold work

Nb, Ta Vaughan, D. A. *et al., Trans. metall. Soc. A.I.M.E.* 221 (1961) 937; 224 (1962) 620 (discussion)
Ta → C, N and O; solid solubility limits and first precipitates determined
N.B.—For Nb and Ta oxidation kinetics, see Chapter 8

Pochon, M. L. *et al.*[56] [Nb and Ta → C]

Cr Smith, W. H. *Trans. metall. Soc. A.I.M.E.* 209 (1957) 47 [Cr → C]
Seybolt, A. U. and Oriani, R. A. *J. Inst. Metals* 8 (1956) 556 [Cr → N]

Mo Speiser, R. *et al., J. Inst. Metals* 4 (1952) 275 [Mo → C]
Few, W. E. and Manning, K. *Trans. metall. Soc. A.I.M.E.* 194 (1952) 271 [Mo → C]
Ham, J. L. and Herzig, A. J. *Climax Molybdenum Co. Report* No. 2 (1951) [Mo → B]
Norton, F. J. and Marshall, A. L. *Trans. Am. Inst. Min. metall. Engrs* 156 (1949) 351 [Mo → N]

W Samsonov, G. V. *Proc. Acad. Sci. USSR* 113 (1957) 1 [W → B]
Allen, B. C., Maykuth, D. J. and Jaffee, R. I. *J. Inst. Metals* 90 (1961) 120 [W → C]
Goldschmidt, H. J. and Brand, J. A.[54] [W → C]
Dushman, S. In *Scientific Foundations of Vacuum Technique*, New York (Wiley) 1949, p.599

U Brailsford, A. D. *J. nucl. Mater.* 8 (1963) 248
α-U → *interstitials*; electrical resistivity as affected by interstitial and vacancy contributions
Blumenthal, B. *J. nucl. Mater.* 2 (1960) 197 [U → C]

Fe Bass, J. and Lazarus, D. *Phys. Chem. Solids* 23 (1962) 1820
α-Fe → C; no effect of pressure on the mobility of C atoms found; anelastic relaxation technique used up to 7,000 atm
Bosman, A. J. *et al., Physica, s'Grav.* 26 (1960) 533
α-Fe → C; again effect of pressure, on the mean time of stay of C in α-Fe, from decrease in permeability; little effect found
Swartz, J. C. *Acta metall.* 10 (1962) 406
α-Fe → C; internal friction and modulus changes in cold-worked iron; dislocation interactions with carbon
Petrova, E. F. *et al., Metallov. term. Obrab. Metall.* 4 (1960) 22
α-Fe → C; effect of alloying elements (Co, Mn, Cr, Si) on solubility and thermodynamic activity
Crussard, C. *Acta metall.* 10 (1962) 749
α-Fe → C; on stacking faults in α-iron, suggesting that these can be due to carbon
Smith, R. P. *Trans. metall. Soc. A.I.M.E.* 224 (1962) 105
Diffusivity and solubility of C in α-Fe determined from single-phase boundary motion in aggregate
Gendrel, C. R., Crussard, C. and Jacque, L. *C.r. hebd. Séanc. Acad. Sci., Paris* 254 (1961) 276
γ-Fe → C; maximum solubility of C in austenite studied by x-rays; also effect on this by oxygen
Wriedt, H. A. and Zwell, L. *Trans. metall. Soc. A.I.M.E.* 224 (1962) 1242
α-Fe → N; linear lattice expansion through N; relation to nitrogen martensite; changes during low-temperature aging
Carswell, W. S. *Acta metall.* 9 (1961) 670
α-Fe → N; internal friction study of strain aging of quenched α-Fe–N solutions; interpretation in terms of Cottrell mechanism and absorption of N by dislocations

Fe and alloys Verner, V. D. *et al., Fizika tverd. Tela* 3 (1961) 3363; *Fizika Metall.* 14 (1962) 880
γ-Fe → N; internal friction study on Fe–Ni, Fe–Mn and Fe–Ni–Mn alloys saturated with nitrogen: peak showing beginning of diffusion; fuller investigation in second paper
Sugeno, T. *et al., J. phys. Soc. Japan* 18 (1963) 154
α-Fe → N; interstitial–dislocation interaction in cold-worked Fe; by internal friction
Kamber, K. *et al. Acta metall.* 9 (1961) 403
α-Fe → N and C; internal friction study on interactions of interstitials with dislocations in Fe, discussed in terms of Cottrell atmospheres, etc.
Keefer, D. and Wert, C. A. *J. phys. Soc. Japan* 18 (1963) 110
α-Fe → N and C; clustering of N and C in solution; derived lattice arrangement of interstitials from anelastic measurements: at low concentrations mostly singly dissolved in lattice, but some already in pairs or triplets; binding energies calculated; differences shown between C pairs in BCC Fe and FCC Ni
Barrand, P., Leak, G. M. and Rosenfield, A. R. *Nature, Lond.* 198 (1963) 279
α-Fe → C, N, O; discussion of a 250°C internal friction peak in terms of possible interstitial presence

Fe and alloys *—cont.*
Ward, R. G. *The Physical Chemistry of Iron and Steel Making*, London (Arnold) 1962
Contains curves for Fe → C, N, O and H)
Conrad, H. and Schoeck, G. *Acta metall.* 8 (1960) 791
Fe → *interstitials* generally: Cottrell locking and flow stress; effect of temperature and grain size on yield and flow stress; detailed theoretical study in light of Cottrell atmospheres and Petch parameters
Leslie, W. C. and Keh, A. S. *J. Iron Steel Inst.* 200 (1962) 722
Fe→*interstitials* generally: strain aging; electron transmission study
Warrington, D. H. *J. Iron Steel Inst.* 201 (1963) 610
Fe: flow stress–subgrain size relationships; dependence on interstitials

Fe-Ni Wriedt, H. A. and Gonzalez, O. D. *Trans. metall. Soc. A.I.M.E.* 221 (1961) 532
Fe–Ni alloys → C and N; solubilities determined

Fe-Ti Köster, W. and Haber, J. *Arch. EisenhüttWes.* 33 (1962) 23
Fe–Ti → C; solubility determined by damping measurements

Fe-Mn Fast, J. D. *et al.*, *Metaux Corros. Inds* 36 (1961) 112; *Colloquium on Internal Friction, St. Germain 1963* 145
Fe–Mn alloys → N; internal friction study; effect of Mn on N mobility; 'capture' of N atoms by Mn

Fe-Cr Rawlings, R. and Koshal, T. B. *Publ. U.S.A.F. Rep.* AD 275094, 1962
Fe–Cr alloys → N; internal friction and allied studies

Fe alloys Schenck, H. *et al.*, *Stahl Eisen* 83 (1963) 93
Fe alloys → N; solubilities of nitrogen in Fe–Cr, Fe–Mn, Fe–Mo and Fe–Co alloys; relation to pressure; systematic considerations; effect of C, Si, P, Co, Ni
Enrietto, J. E. *Trans. metall. Soc. A.I.M.E.* 224 (1962) 43
Fe alloys → N; internal friction, effect of low Mn on solubility and nitride precipitation
Floridis, T. P. and Chilsott, W. R. *Trans. Am. Soc. Metals Qly* (March, 1964) 360
Fe alloys → N; solubility variations at high temperatures
Smith, R. P. *J. Am. chem. Soc.* 68 (1946) 1163; 70 (1948) 2724; *Trans. metall. Soc. A.I.M.E.* 218 (1960) 62
Binary and ternary *austenites* → C and other interstitials; thermodynamic activities of these in solution
Kirkaldy, J. S. and Purdy, G. R. *Can. J. Phys.* 40 (1962) 202
Binary and ternary austenite (γ-Fe) → C solutions; thermodynamic theory, also in presence of substitutional solute atoms: γ-Fe–C, γ-Fe–Mn–C, γ-Fe–Ni–C, γ-Fe–Si–C

Co Dilewuns, J. *Cobalt* 11 (1961) 28
Co → C; solubility, supersaturation and nodular graphite formation; effect of further elements

Co-Ni Rao, K. K. and Nicholson, M. E. *Trans. metall. Soc. A.I.M.E.* 227 (1963) 1029
Co–Ni alloys → C; solubility of C for whole Co–Ni range

Ni Flinn, P. A. *J. Inst. Metals* 14 (1962) 75
Ni → C; mechanical deformation, yield stress; after solution treatment, hardening effects considerable at low temperatures

Ni-Cr Tszya-Lun, S. *et al.*, *Izv. Akad. Nauk SSSR, Otdel. tẹkh. nauk* 5 (1963) 58
Ni–Cr alloys → O; solubility of oxygen in *liquid* Ni–Cr alloys at 1,525°C, lowering of O solution in Ni by *small* Cr, but rapid increase above 2 per cent

Cu Bennemann, K. H. *Z. Phys.* 165 (1961) 445
Cu → interstitials; their lattice positions in Cu; theoretical discussion on possible equilibrium configurations

(*N.B.*—For hydrogen solubilities in the metals, see also Chapter 9.)

REFERENCES

[1] Owen, E. A. and Jones, R. I. *Proc. phys. Soc. Lond.* 49 (1937) 587, 603
[2] Sieverts, A. and Moritz, H. *Z. anorg. Chem.* 24 (1941) 124
[3] Bryant, R. T. and Evans, E. W. *Phil. Mag.* 6 (1961) 805; Bryant, R. T. *J. less-common Metals* 4 (1962) 62
[4] Geach, G. A. and Hughes, J. R. *Plansee Proc.* (1955) 245
[5] Jaffẹe, R. I. *et al.*, *Plansee Proc.* (1958) 380
[6] Brophy *et al.*, *J. less-common Metals* 5 (1963) 90
[7] Hume-Rothery, W. *et al.*, *Proc. R. Soc.* A208 (1951) 431
[8] Robins, D. A. *J. less-common Metals* 1 (1959) 396
[9] McQuillan, A. D. and M. K. '*Zirconium*', London (Butterworths) 1957

REFERENCES

[10] Beshers, D. N. *U.S. At. Energy Comm. Rep. U.C.R.L.* 7184, 1963
[11] Speiser, R. and Spretnak, J. W. *Trans. Am. Soc. Metals* 47 (1955) 493
[12] Speiser, R., Spretnak, J. W. and Taylor, W. J. *Trans. Am. Soc. Metals* 46 (1954) 1168
[13] Seeger, A., Schiller, P. and Kronmüller, H. *Phil. Mag.* 5 (1960) 853
[14] Seraphim, D. P. *et al.*, *J. appl. Phys.* 32 (1962) 136
[15] Allen, N. P. *J. Iron Steel Inst.* (Jan. 1959) 1
[16] Leak, G. M. *Metal Treat. Drop Forg.* 23 (1956) 21
[17] Nicholson, M. E. *J. Inst. Metals* 4 (1952) 148; 6 (1954) 185; 9 (1957) 1
[18] Bungardt, K. and Lennartz, G. *Arch. EisenhüttWes.* 34 (1963) 531
[19] Thomas, W. R. and Leak, G. M. *Nature, Lond.* 176 (1955) 29
[20] Hasigati, R. R. and Kamoshita, O. G. *J. phys. Soc. Japan* 9 (1954) 646
[21] Goldhoff, R. M. and Spretnak, J. W. *J. Inst. Metals* 6 (1954) 1278
[22] Leak, G. M. *et al.*, *Acta metall.* 3 (1955) 501
[23] Goldschmidt, H. J. *Metallurgia* 36 (1947) 297
[24] de Boer, J. H. and Fast, J. D. *Recl Trav. chim. Pays-Bas Belg.* 55 (1936) 459
[25] Kovenskii, I. I. *Fizika tverd. Tela* 5 (1963) 1423
[26] Sully, A. H. *Chromium*, London (Butterworths) 1954; Burr, A. A. *et al.*, *Ductile Chromium*, Cleveland (American Society of Metals) 1957
[27] Weaver, C. W. *J. Inst. Metals* 89 (1961) 385
[28] Schoeck, G. and Seeger, A. *Acta metall.* 7 (1959) 469
[29] Cahn, J. W. *Acta metall.* 10 (1962) 789
[30] Leak, G. M. *Metaux Corros. Inds* 36 (1961) 232
[31] Cotterill, R. M. J. and Segall, R. L. *5th Int. Conf. Electron Microsc.* I, J13, New York (Academic Press) 1963
[32] Irvine, K. J. *et al.*, *J. Iron Steel Inst.* 195 (1960) 336; 196 (1960) 166; Irvine, K. J. and Pickering, F. B. 194 (1960) 137; 188 (1958) 101
[33] Nutting, J. and Arrowsmith, J. M. *Spec. Rep. Iron Steel Inst.* 70 (1961) 147
[34] Carrington, W., Hale, K. F. and McLean, D. *Proc. R. Soc.* A259 (1960) 209
[35] Oliver, D. A. and Harris, G. T. *Jl W. Scott. Iron Steel Inst.* 54 (1946) 97
[36] Underwood, E. E. *J. Inst. Metals* 14 (1962) 914
[37] Heslop, J. and Petch, N. J. *Phil. Mag.* 1 (1956) 866; 2 (1957) 649; 3 (1958) 1128
[38] Cottrell, A. H. and Bilby, B. A. *Proc. phys. Soc. Lond.* A62 (1949) 49
[39] Eshelby, J. D., Frank, F. C. and Nabarro, F. R. N. *Phil. Mag.* 42 (1951) 351
[40] Petch, N. J. *J. Iron Steel Inst.* 174 (1953) 25
[41] *Niobium, Tantalum, Molybdenum and Tungsten* (Ed. Quarrell, A. G.), Amsterdam (Elsevier) 1961; (*a*) Johnson, A. A., p. 277; (*b*) Wilcox, B. A. and Huggins, R. A., p. 289
[42] Adams, M. A. *et al.*, *Acta metall.* 8 (1960) 328; Churchman, A. T. *J. Inst. Metals* 88 (1960) 221
[43] Wilcox, B. A. and Huggins, R. A. *U.S.A.F. Rep. ASD. TR 61–39*, 1962
[44] Wronski, A. S. and Johnson, A. A. *Phil. Mag.* 7 (1962) 213
[45] Lawley, A. W. *et al.*, *Acta metall.* 9 (1961) 841
[46] Harris, G. T. and Bailey, W. H. *Spec. Rep. Iron Steel Inst.* 43 (1951) 60
[47] Petrova, E. F. and Shvartsman, L. A. *Dokl. Akad. Nauk SSSR* 146 (1962) 646
[48] Bindels, J., Bijvoet, J. and Rathenau, G. W. *Physica, s'Grav.* 26 (1960) 163
[49] *Bell Labs Rec.* 38 (1960) 192
[50] Lipsitt, H. A. and Wang, D. V. *Trans. metall. Soc. A.I.M.E.* 221 (1961) 918
[51] Hoch, M. and Desjardins, M. *Trans. metall. Soc. A.I.M.E.* 224 (1962) 821
[52] Fromm, E. and Roy, U. *J. less-common Metals* 8 (1965) 73
[53] Kimura, H. and Sasaki, Y. *Nippon Kink. Gakk.* 2 (1961) 98
[54] Goldschmidt, H. J. and Brand, J. A. *J. less-common Metals* 5 (1963) 181
[55] Vaughan, D. A. *et al.*, *Trans. metall. Soc. A.I.M.E.* 221 (1961) 937
[56] Pochon, M. I. *et al.*, *Reactive Metals* (Ed. Clough), New York (Interscience) 1959
[57] Smirnova, V. I. and Ormont, B. F. *Doklady Akad. Nauk SSSR* 96 (1954) 557
[58] Erben, E. and Spencer, F. *Metall.* 17 (1963) 694
[59] Ingram, A. G., Bartlett, E. S. and Ogden, H. R. *Trans. metall. Soc. A.I.M.E.* 227 (1963) 131
[60] Kofstad, P., e.g. *J. Inst. Metals* 91 (1962) 209
[61] Rieck, G. D. and Vogel, D. L. *5th Plansee Seminar*, Paper No. 25, 1964
[62] Paranjpe, V. G. *et al.*, *Trans. Am. Inst. Min. metall. Engrs* 188 (1950) 261

4

CARBIDES

INTRODUCTORY NOTE ON INTERSTITIAL COMPOUNDS

It is proposed to consider in the following Chapters the true interstitial compounds, as distinct from primary solid solutions or stages intermediate to both, in some more detail than was possible within the framework of Chapter 2. We shall, in sequence, deal with carbides (Chapter 4); nitrides Chapter (5); borides (Chapter 6); silicides (Chapter 7) and oxides, in as far as these are interstitial (Chapter 8); then with hydrides (Chapter 9). Mixed interstitials, e.g. carbonitrides, carbosilicides or -oxides, which in some respects have assumed a new and independent importance, will follow in Chapter 10.

The natural further cross-divisions of (*a*) structural and fundamental aspects, (*b*) systems between, or incorporating, the compounds, and (*c*) technological aspects and applications will, as far as possible, also be observed within each Section. However, this approach is not rigid, and sometimes a blending of these topics becomes more appropriate.

4.1 BASIC SURVEY

The assembly of interstitial carbides and their structural and property trends have been introduced in Chapter 2. A continuous and systematic change occurs within the Periodic System, which shows itself in many facets.

Substantially, we have the carbide families of cubic and hexagonal close-packed metal atoms, centred in Groups IV and V, which are also the most refractory. These give way, with decreasing stability, to the Group VI carbides, of which those of Mo and W still form variants upon the simple structures, but the Cr carbides show large and more complex unit cells. Thence, after an overlap for Mn in Group VII, the pattern evolves to the cementite-type structures of the iron Group VIII, which fade in stability in order $Mn_3C \rightarrow Fe_3C \rightarrow Co_3C \rightarrow Ni_3C$. Group VII is also unique in its borderline position, in that stable carbides persist only in the first Long period (Mn), whilst an abrupt carbide breakdown occurs for the higher periods (beginning of Pt metals), except possibly under high pressures.

Another important variant is that only single carbides exist in Group IV (TiC, ZrC, HfC), with a considerable degree of non-stoichiometry, whilst in Group V (VC, NbC, TaC) the lower M_2C carbides (HCP structure) appear in addition. The homogeneity ranges, though still large, tend to contract. (The complementary high primary solubility of C in the Group IV metals (Chapter 3) may be recalled, which decreases in parallel order of Groups IV $\rightarrow$ V $\rightarrow$ VI.) The multiplicity and decrease in homogeneity ranges continues for the Group VI carbides. The metal atoms become now relatively so small that the metal lattice undergoes either a major distortion

(Cr) or a change in stacking sequence (Mo,W), though only in the mono-carbides, not yet in the hemicarbides). In the carbon-richest Cr carbide (Cr_3C_2), there are indeed already indications of direct C–C bonds, with the

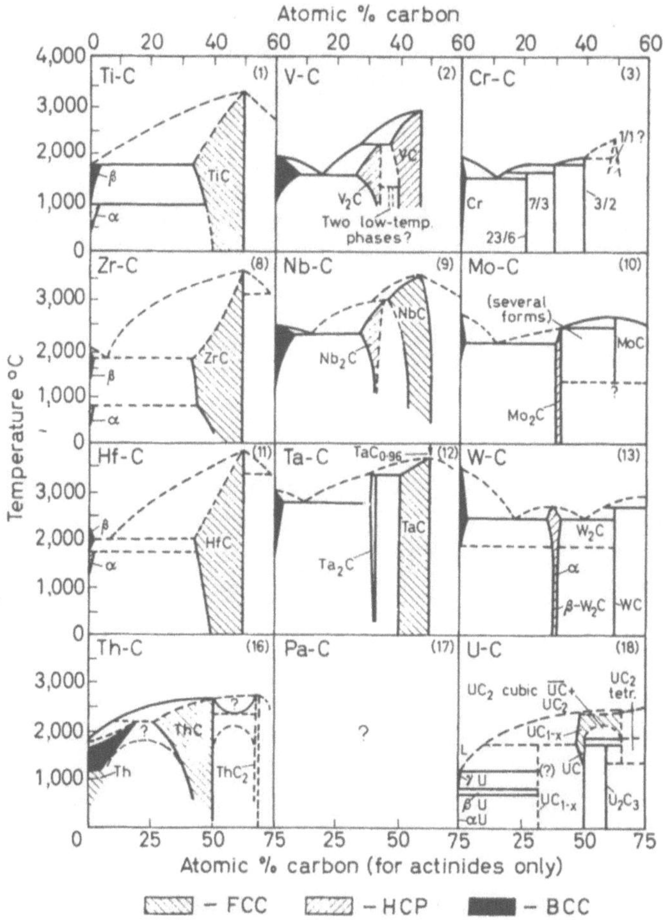

Figure 4.1a. The pattern of binary metal–carbon systems

non-metal atoms forming chains of a type greatly strengthened in the borides and silicides. There is also an initiation of C–C bonds in the ambiguous 'MoC' compound (see later) and in the simple hexagonal layer lattice of WC (see *Figure 2.7*). In Group VII they strengthen to such an extent that, for rhenium, a carbide cannot exist and graphite becomes immediately stable*. In Group VIII, for Fe, Co and Ni, the decreasing *M*–C and increasing C–C bond strength again evolves into *graphite* being the true equilibrium phase, but by way of a sequence of metastable states (cementite, Jack-type structures and, in a more incipient form, in martensite).

In *Figure 4.1* we have systematically collated all the transition metal–carbon systems as far as known to date. They are arranged according to

* Nothing seems known of a hypothetical technetium carbide, between Mn and Re.

Groups and Rows of the Periodic Table, which brings out many basic points. *Table 4.1* gives data of structural properties to serve for reference in discussion. [*Figure 4.1* should also be read in conjunction with the corresponding

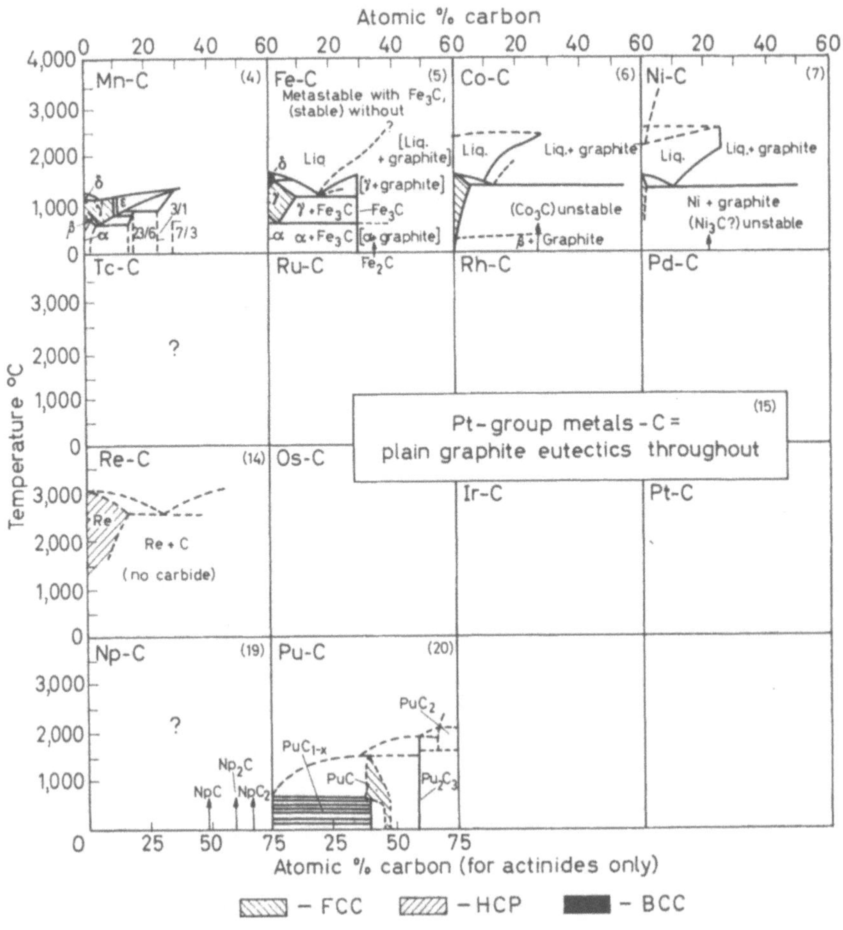

Figure 4.1b

Figures 5.1 (nitrides), *6.1* (borides), *7.1* (silicides), *8.2* (oxides) and *9.3* (hydrides).]

In most respects, *Figure 4.1* and *Table 4.1* are self-explanatory. However, additional comments are needed on individual carbides and literature cited.

The iso-structural areas of carbides within the transition metal block may be represented as in *Figure 4.2* which also indicates the inter-carbide solid solutions as far as known. The atomic size factor is here largely the determinative parameter. Amongst the NaCl-type carbides, e.g. TiC, NbC, etc., Hume-Rothery's often misused '15 per cent rule' that there should be about 15 per cent maximum difference in atomic size for favourable solid-solution formation, still proves to hold well. Indeed, although originally formulated

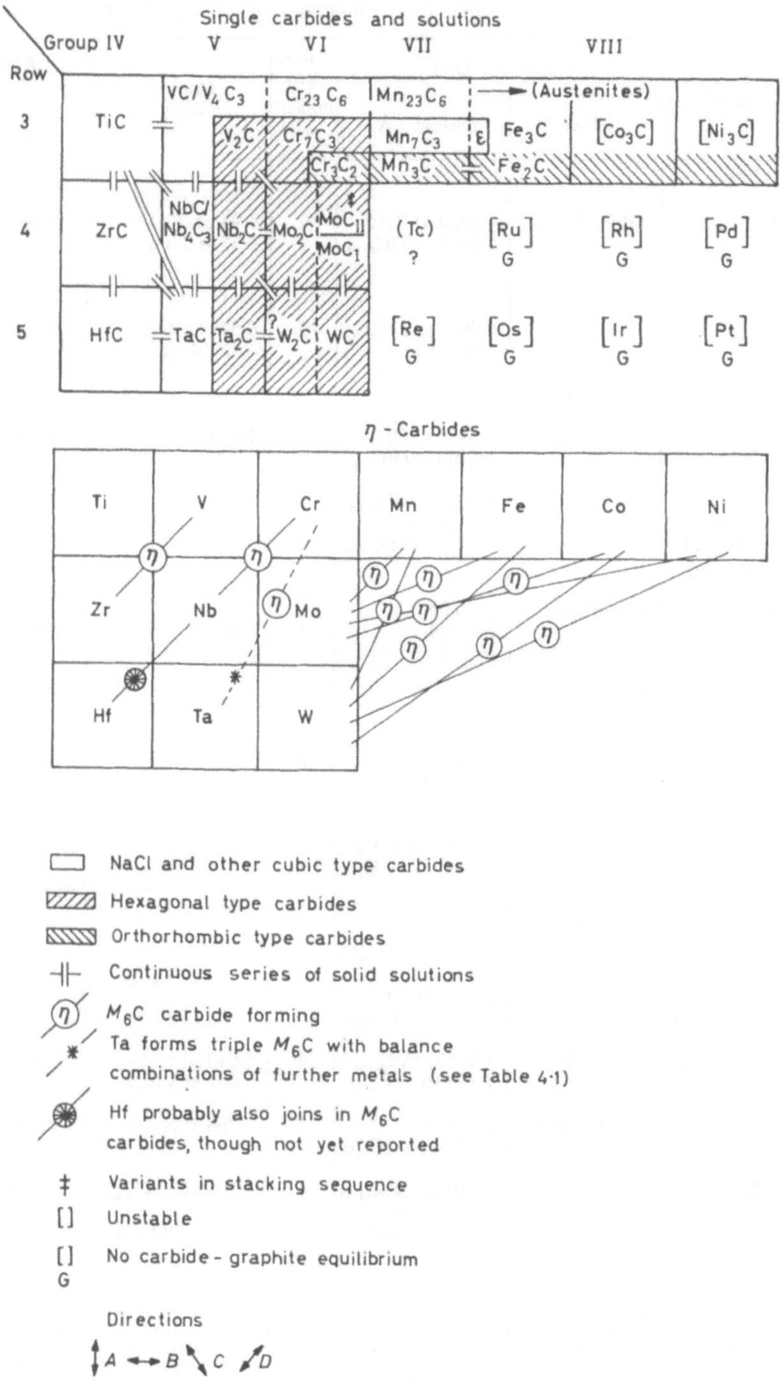

Figure 4.2. Carbides: structure relations

Table 4.1. Summary of binary transition metal carbides,

Base metal	Carbide*	Crystal structure†		Molecules per unit cell	Density (x-ray)	Melting point (approximate)	
		Type	*Spacings, Å*			°C	ΔT‡
Group IV							
Ti	$TiC_{\leftrightarrow}$	cubic, B1	4·3280 at TiC— 4·3127 at $TiC_{0 \cdot 61}$	4	4·94 at TiC— 4·57 at $TiC_{0 \cdot 6}$	3,140	+1,340
Zr	$ZrC_{\leftrightarrow}$	B1	4·6983 at $ZrC_{0 \cdot 95}$ —4·691 at $ZrC_{0 \cdot 55}$	4	6·572 at $ZrC_{0 \cdot 98}$ —6·294 at $ZrC_{0 \cdot 55}$	3,550	+1,680
Hf	$HfC_{\leftrightarrow}$	B1	4·6430 at HfC— 4·6190 at $HfC_{0 \cdot 51}$	4	12·67	3,890	+1,660
Th	$ThC_{\leftrightarrow}$ (joining to cubic ThC_2 at high temps.)	B1	5·346 at $ThC_{0 \cdot 99}$ —5·305 at $ThC_{0 \cdot 80}$	4	10·67	2,625	+925
	$ThC_{2\leftrightarrow}$	monoclinic	a=6·53 b=4·24 β=10·4	4	9·6	2,655	+955
		alternatively					
		orthorhombic	a=8·233 b=10·555 c=4·201	8			
		cubic (>1,415°C)	5·808 (for 1,500° C)				
Group V							
V	$V_2C_{\leftrightarrow}$	hex. C.P.L'3	a=2·902 c=4·577 (at V_2C) —to a=2·8855 c=4·5705 at $VC_{0 \cdot 47}$	1	5·665 at V_2C	2,200	+300
	$VC/V_4C_{3\leftrightarrow}$	B1	4·1655 at $VC_{0 \cdot 87}$— 4·1310 at $VC_{0 \cdot 73}$	4	5·649 at VC— 5·607 at $VC_{0 \cdot 78}$	2,830	+1,030
Nb	$Nb_2C_{\leftrightarrow}$	hex. C.P.L'3	a=3·127 c=4·972 at Nb_2C	1	7·796 at Nb_2C	3,100	+630
	$NbC/Nb_4C_{3\leftrightarrow}$	B1	4·4707 at $NbC_{0 \cdot 99}$ —4·4318 at $NbC_{0 \cdot 77}$	4	7·788 at $NbC_{0 \cdot 99}$ —7·716 at $NbC_{0 \cdot 785}$	3,480	+1,030

Note: Spacing, density, etc. data quoted apply to stoichiometric composition, or nearest, unless otherwise stated. Many of the data on properties of carbides are subject to divergencies in the literature, particularly for those (TiC, UC, etc.) liable to dissolve oxygen or nitrogen, or to have non-stoichiometric ranges. It has been endeavoured to choose the most recent and reliable data, but some are still subject to special experimental conditions.

*$_{\leftrightarrow}$ = homogeneity region appreciable

† *Strukturbericht* nomenclature

‡ΔT = Difference of melting points of carbide and metal

their structures and properties (excluding double carbides)

Heat of formation, kcal/g-atom at R.T., 1 atm	*Hardness*, Moh (microhardness, kg/mm² at 50–200 g load)	*Other properties*§	*Main applications; remarks*
−57	8–9 (3,200)	ρ=68, χ=0·12×10^{-6} E=32·200 β=7·8×10^{-6} (10–100) SPCD <1·2°K	cemented hard metals; high-temperature materials (potentially); dispersion strengtheners
−48	8–9 (2,600)	ρ=63 χ=0·22×10^{-6} SPCD <3°K β=6·6×10^{-6} (10–500)	
−81	9–10 (2,900) (850)	ρ=37 SPCD <1·3°K β=6·6×10^{-6} (0–600)	
−7	9–10 (850)	ρ=25, χ=0·069 SPCD <9°K	atomic energy
−25	(600)	(decomposed by H_2O) ρ=30, χ=0·057	
—	8–9 (2,140)	—	constituents in alloy steels and high-temperature materials
−24·5	9–10 (2,850–3,000)	ρ=60 E=27,600	
−45 (?)	—	—	in creep-resisting and refractory alloys and steels; dispersion strengtheners; possibly in hard metals
−33	8–9 (2,400)	ρ=35 (or 60?) E=3,4500 χ=0·034 SPCD <10°K β=6×10^{-6} (0–500)	

§ ρ = electrical resistivity, μ Ωcm at room temperature (there are wide discrepancies according to data of different authors)

χ = thermal conductivity, cal/cm °C

E = Young's modulus

β = thermal expansion coefficient/°C (linear, mean for temperature range stated in brackets, °C)

SPCD = superconducting below this temperature (°K)

Base metal	Carbide*	Crystal structure† Type	Crystal structure† Spacings, Å	Molecules per unit cell	Density (x-ray)	Melting point °C (approximate)	Melting point ΔT‡ (approximate)
Group V cont. Ta	$Ta_2C_{\longleftrightarrow}$	hex. C.P.	$a=3{\cdot}106$ $c=4{\cdot}945$ at Ta_2C— $a=3{\cdot}101$ $c=4{\cdot}933$ at $TaC_{0\cdot41}$	1	15·22	3,400 (dec.)	+400
	$TaC_{\longleftrightarrow}$	B1	4·4555 at $TaC_{0\cdot99}$ —4·4100 at $TaC_{0\cdot70}$	4	14·48 at $TaC_{0\cdot99}$	3,880, or 3,540?	+850
Pa	not known, but PaC, PaC_2 presumed, similar to Th—C and U—C analogues						
Group VI Cr	$Cr_{23}C_6$	complex cubic $D8_4$	10·60	4 (i.e. 116 atoms)	6·97	1,580 (1,520)	−300
	Cr_7C_3	complex hex. P31C	$a=4{\cdot}53$ $c=14{\cdot}01$	8 (i.e. 80 atoms)	6·92	1,665 (1,780)	−200
	Cr_3C_2	orthorhombic D^{16}_{2h}	$a=2{\cdot}82$ $b=5{\cdot}53$ $c=11{\cdot}47$	4 (i.e. 20 atoms)	6·68	1,895	0
Mo	Mo_2C	hex. C.P. L′3	$a=3{\cdot}0028$ $c=4{\cdot}7288$	1	9·06	2,410	+110
	$MoC_{\longleftrightarrow}$ (alternatively reported as Mo_3C_2)	(1) γ-MoC= simple hex. (WC type)	$a=2{\cdot}898$ $c=2{\cdot}809$	1	9·46	2,690 (2,575 for 'Mo_3C_2')	+120
		(2) γ′-MoC: hex. mod	$a=2{\cdot}932$ $c=10{\cdot}97$	4	—	—	—
		(3) ζ-MoC: cubic B1 (at high temps.), etc.	4·28	4			
		(2) and (3) metastable for pure MoC, but each stabilized by different additions					
W	$W_2C_{\longleftrightarrow}$	hex. C.P. L′3	$a=2{\cdot}992$ $c=4{\cdot}722$	1	17·20	2,800	−530
	β-W_2C/WC (at high temps. only)	cubic B1	4·16 at ≈$WC_{0\cdot5}$ 4·215 at ≈$WC_{0\cdot82}$	4			

Heat of formation, kcal/g-atom at R.T., 1 atm	*Hardness*, Moh (microhardness, kg/mm² at 50–200 g load)	*Other properties*§	*Main applications; remarks*
−46	(947)	—	cemented hard metals; refractory alloys; in filaments and other heating elements; dispersion strengtheners, etc.
−38	9–10 (1,800–2,400) 840 Brinell	$\rho=25$ (or 42 and higher?) $E=29{,}500$ $\chi=0{\cdot}045$ SPCD $<9{\cdot}5°K$ $\beta=6{\cdot}2\times10^{-6}$ (0–100)	
−16·4	9 (1,000)	$\rho=127$	prominent in chromium-bearing alloy steels
−42·5	(1,600)	$\rho=109$	
−21	(1,300)	$\rho=75$	cemented carbides; a phase 'CrC' sometimes reported but doubtful
−11 (?)	7–9 (1,479)	$\beta=4{\cdot}9//a$ $8{\cdot}2//c$ axis, $\times10^{-6}$ (0–800) $\rho=71$ SPCD $<2{\cdot}9°K$	in alloy steels; hard metals?
−2·5 (?)	7–8 (1,800)	$\rho=49$ for Mo_3C_2 $\rho=19$ for MoC (?) SPCD $<7{\cdot}9°K$	refractory alloys
—	—	$E=23{,}000$ SPCD $<3°K$	Mo_2C recently reported only pseudo-hexagonal, viz. *orthorhombic*, with $a=4{\cdot}724$, $b=6{\cdot}004$, $c=5{\cdot}199$ Å
−6·3	9–10 (1,450)	$\rho=80$ $E=42{,}800$ SPCD $<2{\cdot}74°K$	cubic H.T. form confirmed above 2,450°, ranging from WC to $\approx WC_{0\cdot66}$, decomposing into W_2C+WC

Base metal	*Carbide**	*Crystal structure†*		*Molecules per unit cell*	*Density* (x-ray)	*Melting point*	
		Type	*Spacings,* Å			°C (*approximate*)	ΔT‡
Group VI cont.							
	WC	simple hex., D'_{3h}	$a=2{\cdot}9063$ $c=2{\cdot}8368$	1	15·77	dec. >2,755	−540
U	$UC_{\leftrightarrow}$	Bl	4·9605	4	13·6	2,560	+1,300
	U_2C_3	complex BC cubic, $D5_c$	8·088	8	12·9	dec. >1,800 to UC/UC_2	—
	$UC_{2\leftrightarrow}$	(1) BC tetragonal C11a (CaC_2 type)	$a=3{\cdot}5241$ $c=5{\cdot}9962$	2	11·68	2,350—2,475 (melting point)	+1,200
		(2) cubic (CaF_2 type) above 1,820°C	$a=5{\cdot}475$ at 1,900°C				
		but diffusionless transformation from (1) to (2) i.e. tetragonal to cubic UC_2 occurs (martensite-like). Also solubility 'UC–UC_2' qualified by slight U-atom displacements by C interstition. U_2C_3 forms similarly by alternative mechanism at lower temperatures					
Group VII Mn	$Mn_{23}C_6$	complex cubic, $D8_4$ (as $Cr_{23}C_6$)	10·586	4	—	1,010	−250
	Mn_3C	orthorhombic DO_{11} (as Fe_3C)	$a=5{\cdot}080$ $b=6{\cdot}772$ $c=4{\cdot}530$	4	6·9	1,520	—
	Mn_5C_2	monoclinic C2/c	$a=5{\cdot}806$ $b=4{\cdot}573$ $c=11{\cdot}66$ $\beta=97{\cdot}75°$	4	—	—	—
	Mn_7C_3	complex hex. P31C (as Cr_7C_3)	$a=13{\cdot}838$ $c=4{\cdot}539$	8	—	1,340	+80
Tc	not known but Mn–C and Mo–C analogues possible, less stable than these						
Re	no rhenium carbide occurs						
Np	NpC	Bl	5·004	4	—	—	—

Heat of formation, kcal/g-atom at R.T., 1 atm	*Hardness,* Moh (microhardness, kg/mm² at 50–200 g load)	*Other properties*§	*Main applications; remarks*
−8·4	9+ (2,100)	β=5·0//*a*4·2//*c* axis, × 10^{-6} (0–800) ρ=25 *E*=72,200 χ=0·05 SPCD <2°K	most important phase in cemented carbides; crucibles, dies, furnace parts, hard linings, etc.
−21	(600–800)	ρ=35–40 χ=0·06 β=11·4× 10^{-6} (0–950)	atomic reactors; possible in rarer alloy steels and refractory alloys
−49	(1,080 or 1,500?)	ρ=200 β=10·5× 10^{-6} (20–1,000)	
−21	(500)	120	$UC \leftrightarrow UC_2$ joining and miscible at high temperatures C–C atom bonds begin to be established in U_2C_3 and UC_2
—	—	—	in alloy steels, generally with Fe_3C, $Cr_{23}C_6$ and Cr_7C_3 as primary solvents, a phase 'Mn_7C_2' (?) also reported at high temperatures, 850–1,000°C only. Mn_3C has been reported in two modifications (α, β), with transformation≈1,040°C
−23	—	—	
—	—	—	
—	—	—	
—	—	—	

Base metal	*Carbide**	*Crystal structure*†		*Molecules per unit cell*	*Density* (x-ray)	*Melting point*	
		Type	*Spacings,* Å			°C (*approximate*)	ΔT‡
Group VIII							
Fe	Fe_3C (*cementite,* metastable)	orthorhombic DO_{11}	$a=5{\cdot}088$ $b=6{\cdot}744$ $c=4{\cdot}524$	4	7·2	1,650	+100
	'ε' carbide χ-Fe_3 C(unstable)	hex.	$a=2{\cdot}73$ $c=4{\cdot}33$	1	—	—	—
	$Fe_2C_{\longleftrightarrow}$ (or $Fe_{20}C_9$)	orthorhombic	$a=9{\cdot}061$ $b=15{\cdot}695$ $c=7{\cdot}936$ (but possibly hex. form)	4	—	—	—
Ru, Os	no carbides: graphitic systems						
Pu	Pu_3C_2 (below 575°C only)	(unknown)	—	—	—	—	—
	$PuC_{\longleftrightarrow}$	cubic, Bl	4·9730 (at $PuC_{0{\cdot}94}$)—4·953 (at $PuC_{0{\cdot}74}$)	4	14·0	dec. >1,600	—
	$Pu_2C_{3\longleftrightarrow}$	complex cubic, $D5_6$ (as U_2C_3)	8·1258—8·1317	8	12·7	2,050	—
	PuC_2 (stable above 1,750°C only)					dec. >2,250	
Co	Co_3C (unstable)	orthorhombic (as Fe_3C)	$a=5{\cdot}103$ $b=6{\cdot}244$ $c=4{\cdot}432$	4	—	—	—
	Co_2C (unstable)	orthorhombic (as Co_2N and Fe_2C)	$a=4{\cdot}4465$ $b=4{\cdot}3707$ $c=2{\cdot}8969$	2	—	—	—
Ni	Ni_3C (unstable)	orthorhombic (as Fe_3C)	?	4	—	—	—
Pt, etc.	no Pt–metal group carbides occur, but fleeting carbide-like configurations possible at surfaces, etc.						

Heat of formation, kcal/g-atom at R.T., 1 atm	*Hardness,* Moh (microhardness, kg/mm² at 50–200 g load)	*Other properties§*	*Main applications; remarks*
+2·5	840 Brinell	$\beta = +10^{-6}$ (0–500)	in plain-carbon and low-alloy steels
—	—	—	intermediate stages towards cementite or graphitization
—	—	—	
—	—	—	atomic energy uses
—	(750)	—	
—	(900)	—	peritectic to Pu_2C_3 at 2,050°C
			peritectic to PuC_2 at 2,250°C
—	—	—	e.g. in catalytical reactions
—	—	—	a 'hexagonal Ni-carbide' observed is probably a metastable Ni→C solution with stacking sequence changes similar to α→β-Co
—	—	—	

for certain pure metal systems only, the similarity in application of this rule also to carbides is quite striking. For example, in the two ternary diagrams Nb–Ti–Zr and NbC–TiC–ZrC, the single-phase solid-solution fields and miscibility bays, allowing for different temperature levels, are so similar that the presence or absence of the interstitial atom becomes almost incidental—a fair reflection of the essentially metallic bonding also in the carbides. We may here refer in advance to *Figure 4.28* (p. 161) and compare it to the small added Nb–Cr–Mo diagram (determined by the writer) to illustrate this point. Also, for example, cementite, Fe_3C, possesses a considerable solid solubility for Cr (~ 17 per cent Cr; ratio of atomic radii, $r_{Cr}/r_{Fe} = 1{\cdot}01$) but a very restricted one for the larger yet equivalent Mo atom ($\sim$ 2 per cent; $r_{Mo}/r_{Fe} = 1{\cdot}10$). These conditions are also reflected in the Fe–Cr and the Fe–Mo systems. These factors have an obvious and immediate bearing on alloy steel behaviour, and (in this case) on precipitations in Cr–Mo steels.

Again, Fe is hardly soluble in the cubic carbides, TiC, ZrC, NbC, TaC, etc. which exist in virtually pure form in alloy steels; however, in VC (also

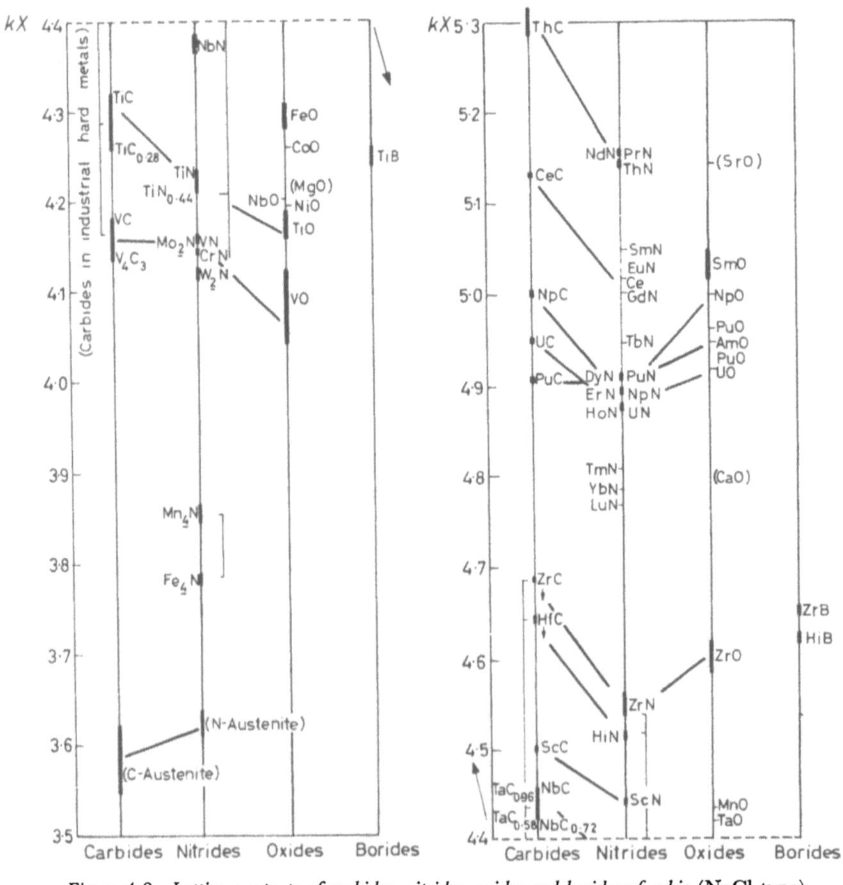

Figure 4.3. Lattice constants of carbides, nitrides, oxides and borides of cubic (NaCl-type) *structure; intersolubilities and homogeneity ranges shown*

cubic), Fe dissolves much more already, as the smaller V atom approaches Fe in size, a trend greatly advanced for neighbouring Cr and Mn. This has its implications, for instance, in high-speed steels, where it is critical whether or not the cubic *M*C carbide will be stabilized by varying alloying or heat treatment.

Wide solid solutions are also formed in non-isomorphous cases like TiC $\longleftrightarrow$ WC, when the metal atoms are sufficiently compatible in size (see p. 15). The intersolubility, however, there becomes highly temperature-dependent—a point of basic importance in hard metals*.

It is useful to get an idea of scale by plotting the lattice dimensions of the known cubic *M*C-type carbides, as shown in *Figure 4.3.* This also includes the isomorphous nitrides, oxides and borides (for later comparison). The compounds forming mutual solid solutions are joined, and for comparison of overall scale the spacings of the isomorphous C- and N-bearing austenites have also been inserted, as well as of some non-metallic ceramic oxides of the same structure, CaO, MgO.

4.2 DOUBLE CARBIDES

The above carbides are all based on single transition metals, and we shall return to them later. However, it is important first to introduce the 'double carbides', i.e. those requiring the integration of two metals to be stable. Probably the chief one is the so-called η-carbide, M_6C, but there are others, and later it will be one of our objects to see whether a pattern exists in their condition of formation.

η-carbide—The original η-carbide was Westgren's and Phragmén's[1a,b] high-speed steel carbide, Fe_4W_2C/Fe_3W_3C cubic, spacing 11·04 *kX* but variable; 112 atoms per unit cell, i.e. 96 metals, 16 carbons. However, many isomorphs to it are now known between the Iron Group and Groups V–VI metals, with conditional extensions to other metals and interstitials. A complete list is given in *Table 4.2.* In the lower part of *Figure 4.2* their placings in the Periodic Table have been indicated. It is characteristic that the metal pairs combining in η-carbides always follow the reciprocal diagonal relation shown.

The basic reason for a new structure forming under such conditions might be envisaged on the high-speed steel carbide, Fe_4W_2C, itself. The ternary diagram Fe–W–C (*Figure 4.4a*) may help here, together with the quaternary system Fe–Cr–W–C. Consider the addition of tungsten to cementite, such as in a plain carbon steel. In trace amounts, the large W atom can substitute Fe in the Fe_3C lattice (as Cr and Mo, also in Group VI, can more extensively). However, this is associated with considerable lattice strains and with a tendency of the W atom to attract an atmosphere of C atoms around itself.

* It is instructive to observe how two apparently separate technological subjects such as (*a*) steel metallurgy and (*b*) hard metals metallurgy find their natural link-up in the present carbide system: (*a*) operates at the right, (*b*) at the left of the Periodic Table representation, but with a large intermediate field of overlap. Both technologies can in fact glean from each other, which is not always appreciated in practice. High-speed steels occupy already a position inclined towards (*b*), as will arise later, and in modern tool materials research there may well be much room for the development of new alloys of special properties filling this gap. A similar 'bridging position' applies to high-temperature alloys.

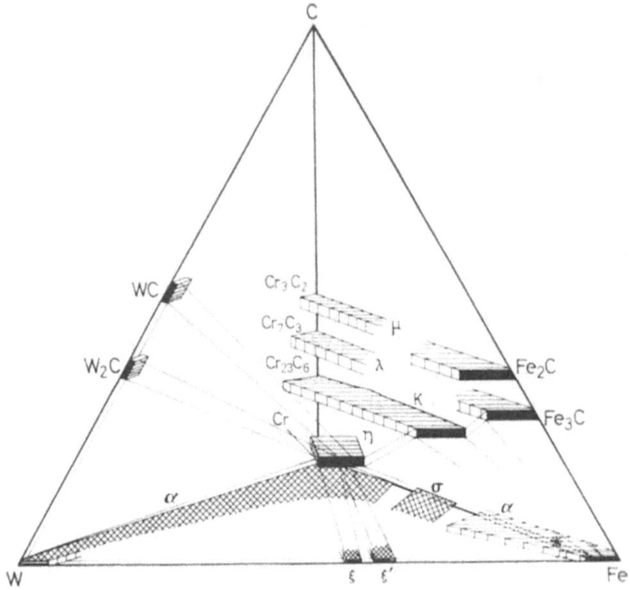

(a)

Figure 4.4a. Fe–Cr–W–C *diagram (schematic)*, quaternary model (After Goldschmidt[2a])

* Indicates approximate position of high-speed steels

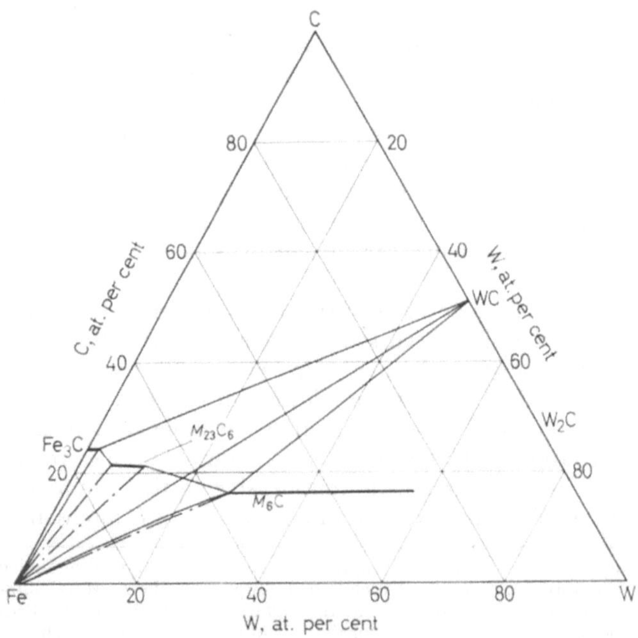

(b)

Figure 4.4b. Equilibrium diagram of Fe–W–C *system at* 700°C (After Kuo[3a])

—— = stable, – · – = metastable

By courtesy of the Iron and Steel Institute, London

A direct phase transition to W_2C might be expected as the next stage, i.e. for W exceeding saturation in Fe_3C. However, for the hexagonal tungsten carbide to form from Fe_3C would require a major degree of diffusion (and a drastic lattice change); for this, the W atom has insufficient mobility, at any rate at moderate temperatures and for limited time periods. Thus intermediate structures are adopted as a compromise, more readily derived from Fe_3C by shear-type movements and by short-range displacement. As from Fe_3C, these new forms are, first, that of the $M_{3\cdot8}C$ carbide (viz. $M_{23}C_6 = Fe_{21}W_2C_6$), and then that of the M_6C ($= Fe_4W_2C$) carbide. The three structures, Fe_3C, $M_{23}C_6$ and M_6C, are shown in *Figures 4.5–4.7.*

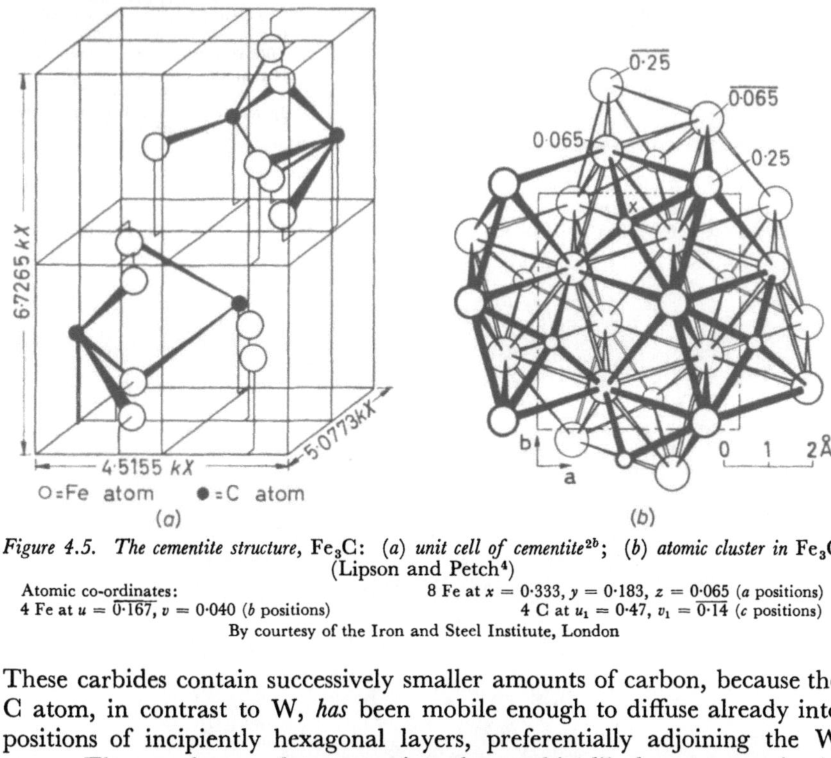

Figure 4.5. The cementite structure, Fe_3C: *(a) unit cell of cementite*[2b]; *(b) atomic cluster in* Fe_3C (Lipson and Petch[4])

Atomic co-ordinates: 8 Fe at $x = 0{\cdot}333, y = 0{\cdot}183, z = \overline{0{\cdot}065}$ (*a* positions)
4 Fe at $u = \overline{0{\cdot}167}, v = 0{\cdot}040$ (*b* positions) 4 C at $u_1 = 0{\cdot}47, v_1 = \overline{0{\cdot}14}$ (*c* positions)

By courtesy of the Iron and Steel Institute, London

These carbides contain successively smaller amounts of carbon, because the C atom, in contrast to W, *has* been mobile enough to diffuse already into positions of incipiently hexagonal layers, preferentially adjoining the W atoms. These evolve, at a later stage, into the graphite-like layers appearing in the layer lattice of tungsten carbide itself (in the sense of *Figures 2.7* and discussion, p. 34). The resulting phase equilibria are expressed in *Figure 4.4b.*

In this rather schematic picture*, Fe and W are of course only representative of the other η-carbide combinations (*Table 4.2* and *Figure 4.2*). More

* Directly reflected, however, in practice in the results of modern alloy steel research which show a high degree of heat-treatment dependence of η-carbide formation. Thus, for instance, Fe–Mo–C alloys quenched from high (austenitic) temperatures precipitate, upon tempering in a moderate temperature range, first the Mo_2C carbide but subsequently the η-carbide. In ferritic alloy steels this carbide can, but need not be, the ultimate product of a transformation series; this depends critically on temperature and on further alloy additions.

The manner of this type of reaction has been demonstrated in work at Sheffield University[5], showing by transmission electron microscopy how precipitation occurred, preferably at dislocations. The $M_7C_3 \rightarrow M_{23}C_6$ transformation has been similarly studied[6] recently.

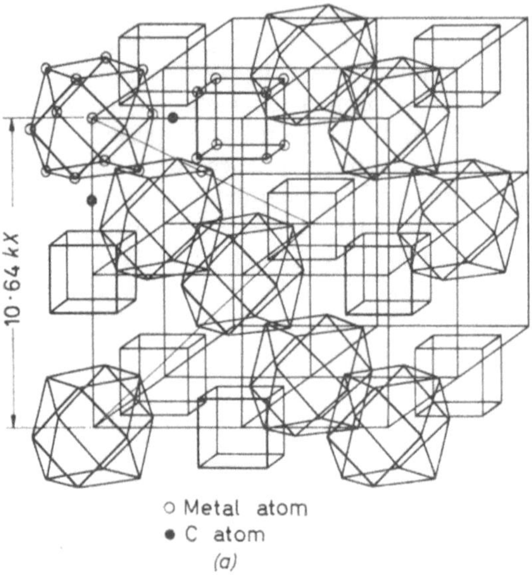

Figure 4.6a. *The* $M_{23}C_6$ *structure: (a) unit cell for* (Cr, Fe, W, Mo)$_{23}$C$_6$ (Goldschmidt[2b]); *(b) space model of this structure*

Atomic co-ordinates in $M_{23}C_6$ ($Cr_{23}C_6 \leftrightarrow Fe_{21}W_2C_6$):

4 Cr at 000 (*a* positions)

32 Cr at uuu, $u\bar{u}\bar{u}$, $\bar{u}u\bar{u}$, $\bar{u}\bar{u}u$, $\bar{u}\bar{u}\bar{u}$, $\bar{u}uu$, $u\bar{u}u$, $uu\bar{u}$; $u = 0{\cdot}385$, $v = 0{\cdot}165$, $w = 0{\cdot}275$ (*f* positions)

8 Cr at $\frac{1}{4}\frac{1}{4}\frac{1}{4}$, $\frac{3}{4}\frac{3}{4}\frac{3}{4}$ (*c* positions)

48 Cr at $v\bar{v}0$, $0v\bar{v}$, $\bar{v}0v$, $vv0$, $0vv$, $v0v$, $\bar{v}\bar{v}0$, $0\bar{v}\bar{v}$, $\bar{v}0\bar{v}$, $\bar{v}v0$, $0\bar{v}v$, $v0\bar{v}$ (*h* positions)

24 C at $w00$, $\bar{w}00$, $0w0$, $0\bar{w}0$, $00w$, $00\bar{w}$ (*e* positions)

By courtesy of the Iron and Steel Institute, London

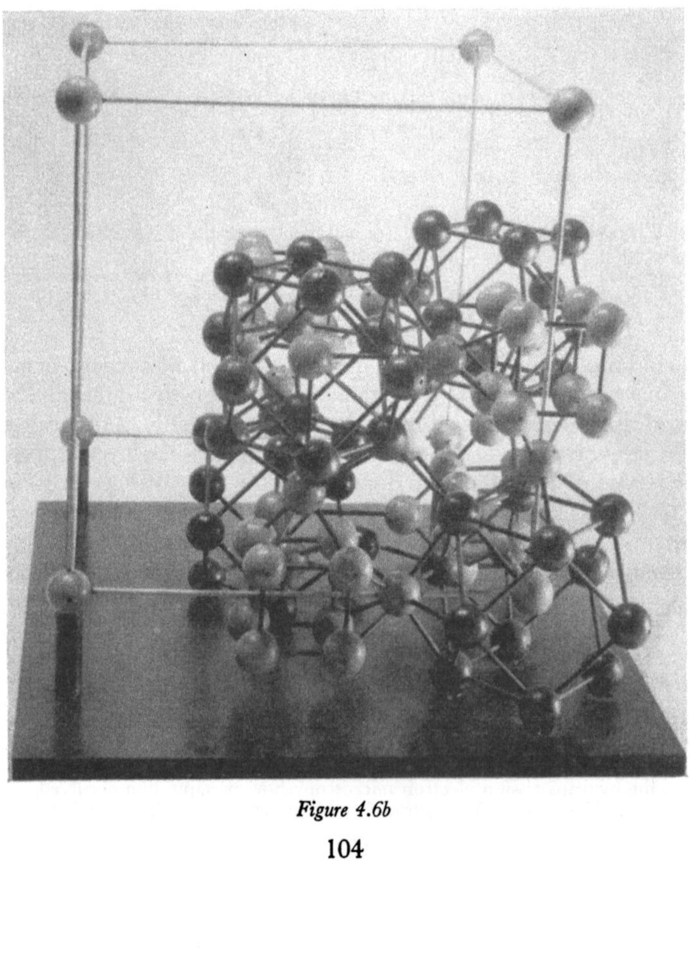

Figure 4.6b

Figure 4.7a. The M_6C (η-carbide) structure

(*a*) Unit cell for $Fe_4(W,Mo)_2C/Fe_3(W,Mo)_3C$ (Goldschmidt[2b])

(*b*) Structure model of η-carbide ($\frac{1}{4}$ unit cell); atoms: A = W,Mo or Fe, random replacement at octahedra; B = Fe (at tetrahedra); C = W or Mo; D = carbon

Atomic co-ordinates:

16 Fe at $\frac{1}{8}\,\frac{3}{8}\,\frac{7}{8}$, $\frac{1}{8}\,\frac{5}{8}\,\frac{1}{8}$, $\frac{7}{8}\,\frac{3}{8}\,\frac{1}{8}$, $\frac{7}{8}\,\frac{5}{8}\,\frac{7}{8}$ (*d* positions)

32 Fe at uuu, $\frac{1}{4}-u\ \frac{1}{4}-u\ \frac{1}{4}-u$, $u\bar{u}\bar{u}$, $\frac{1}{4}-u\ \frac{1}{4}-u\ u+\frac{1}{4}$, $\bar{u}u\bar{u}$, $u+\frac{1}{4}\ \frac{1}{4}-u\ u+\frac{1}{4}$, $\bar{u}\bar{u}u$, $u+\frac{1}{4}\ u+\frac{1}{4}\ \frac{1}{4}-u$; $u = -0{\cdot}175$, $v = 0{\cdot}195$ (*e* positions)

32 W and 16 Fe randomly distr. at $v00$, $\frac{1}{4}-v\ \frac{1}{4}\ \frac{1}{4}$, $\bar{v}00$, $v+\frac{1}{4}\ \frac{1}{4}\ \frac{1}{4}$, $0v0$, $\frac{1}{4}\ \frac{1}{4}-v\ \frac{1}{4}$, $0\bar{v}0$, $\frac{1}{4}\ v+\frac{1}{4}\ \frac{1}{4}$, $00v$, $\frac{1}{4}\ \frac{1}{4}\ \frac{1}{4}-v$, $00\bar{v}$, $\frac{1}{4}\ \frac{1}{4}\ v+\frac{1}{4}$ (*f* positions)

16 C at $\frac{1}{8}\,\frac{1}{8}\,\frac{1}{8}$, $\frac{1}{8}\,\frac{3}{8}\,\frac{3}{8}$, $\frac{3}{8}\,\frac{1}{8}\,\frac{3}{8}$, $\frac{3}{8}\,\frac{3}{8}\,\frac{1}{8}$ (*c* positions)

By courtesy of the Iron and Steel Institute, London

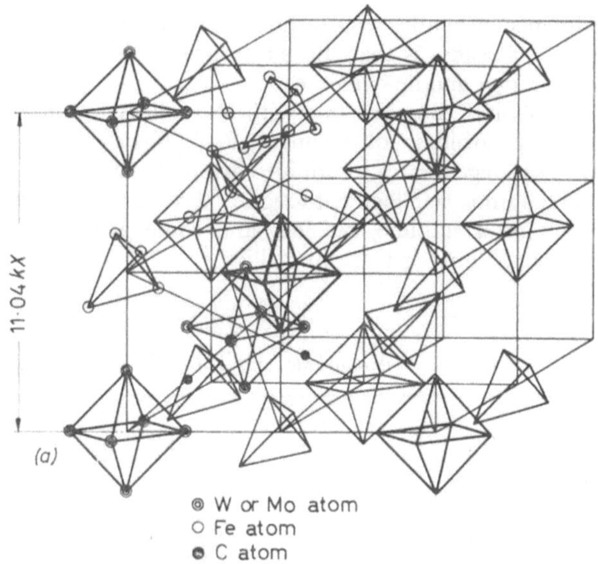

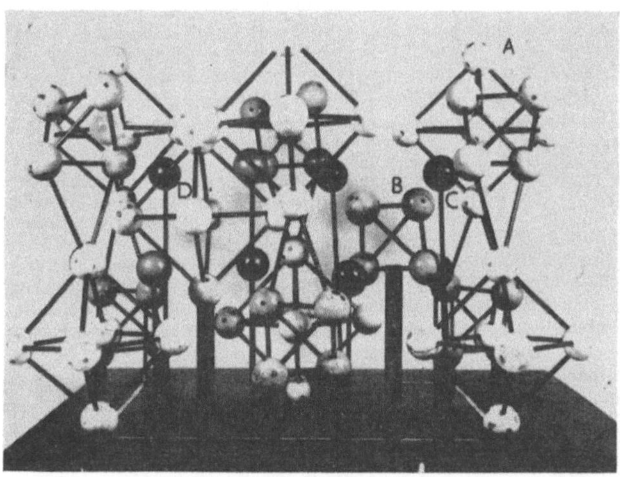

Figure 4.7b

By courtesy of the Iron and Steel Institute, London

generally, these vary in stability and tend to form wherever a pair of a smaller and a larger transition metal atoms, though electronically compatible for isomorphous replacement (as e.g. in Fe ⟷ Cr, Fe ⟷ V), are prevented from doing so by unfavourable atomic sizes and by low activation energy for diffusion. The diagonal relation (*Figure 4.2*) also clearly expresses the condition that a double carbide is favoured when one of the two metals is itself a stronger, the other a weaker carbide former. This is to say that one atom should have a stronger, the other a weaker bond to C, although the difference in bond strength should not be too extreme. A further condition, implicit already, is that the more strongly bonding metal atom should also be the larger and heavier one.

We should compare this with the solid-solution-forming trend amongst the cubic carbides: directions A, B, C (*Figure 4.2*) favour these, but in direction D it degenerates to double carbide formation.

The atomic size factor, though a necessary, is clearly not a sufficient condition for a stable η-carbide, and some aspects recall those of electron compounds amongst intermetallics.

Table 4.2

η-Carbide, M_6C, family of compounds

Carbides		
Fe_3W_3C–Fe_4W_2C	Fe_2W_4C	η_2
Fe_3Mo_3C–Fe_4Mo_2C	Fe_2Mo_4C	η_2
Co_3W_3C–Co_4W_2C	Co_2W_4C	η_2
Co_3Mo_3C	Co_2Mo_4C	η_2
Ni_3W_3C–Ni_4W_2C	Ni_2W_4C	η_2
Ni_3Mo_3C	Ni_2Mo_4C	η_2
V_3Zr_3C	Also η-carbide isomorphs bearing one non-transition metal:	
Cr_3Nb_3C		
Mn_3Mo_3C	Fi_2ZnC_x ($x < 1$)	
Mn_3W_3C	Zr_2ZnC_x	
	Hf_2ZnC_x	
	Nb_2ZnC_x	
Triple carbides		
$(Cr, Fe)_3Ta_3C$	$Ni_2(Ti,Ta)_4C$	η_2
$(Cr, Co)_3Ta_3C$	$Co_2(Ti,Ta)_4C$	η_2
$(Cr, Ni)_3Ta_3C$	$Ni_2(Ti,Nb)_4C$	η_2
$(Cr, Cu)_3Ta_3C$	plus carbides of varying composition in alloy steels	
$(V, Fe)_3Ta_3C$	*Recent additions:* an aluminium-bearing group	
$(V, Co)_3Ta_3C$	Nb_3MAl_2C and Ta_3MAl_2C where M=V, Cr,	
$(V, Ni)_3Ta_3C$	Mn, Fe, Co, Ni	
$(V, Ni)_3Nb_3C$	(no doubt others)	
Nitrides		
Mn_3Ti_3N	Ti_2ZnN_x	
Fe_3Ti_3N	Zr_2ZnN_x	
(Ni, Co, Si) (Mo, Cr, W) (C, N)	Hf_2ZnN_x	
(extract from Ni-base alloy)	(no doubt others)	
Oxides		
Mn_3Ti_3O	Cu_3Ti_3O	
Fe_3Ti_3O	V_3Zr_3O	
Co_3Ti_3O	Cr_3Zr_3O	
Ni_3Ti_3O	Mn_3Zr_3O	
Interstitial vacated isomorphs		
Sc_2Pd HfMn Hf_2Rh		
Ti_2Co Hf_2Fe Hf_2Ir	Nb_3Fe_2 (others Group V-based?)	
Ti_2Ni Hf_2Co Hf_2Pt		

It is of interest that *tantalum* does not seem to participate in any of the true double carbides so far reported, but it can do so in 'triple carbides', i.e. for suitable combinations of two elements of the First Long Period. Thus Ti, V or Cr are required in order *also* to admit Ta. The occurrence of further,

Nb-, Zr- and Hf-bearing η-carbides seems highly probable, though not reported yet.

Table 4.2 also shows the η-*nitrides* and *oxides* known so far—a list which no doubt requires extending. Two closely related structural sub-types exist, denoted η_1 and η_2. These differ in the relative proportions of smaller and larger metal atoms: $\eta_1 = A_4B_2C/A_3B_3C$, $\eta_2 = A_2B_4C$, where A atoms belong to the First Long Period (Ni, Co, Fe, with Mn, Cr and even V, Ti possible) and B atoms to the Second and Third Long Periods (Mo, W, Nb, Zr, with Ta, Hf possible). The lattice dimensions of η_2 are thus always larger than those of η_1.

The crystal structure of the η-carbide (η_1), with a table of atomic co-ordinates, is shown in the diagram and model of *Figure 4.7.* There are 112 atoms per unit cell, viz. 96 M and 16 C atoms. The structure consists of octahedral and tetrahedral units. The former are randomly occupied by A and B atoms, the latter by A atoms only, whilst a further class of A positions exists in the interspaces between these units. Carbon atoms are located centrally between any two octahedra, and this produces, for Fe_3W_3C, a six-fold carbon environment chiefly of W atoms (see also *Figure 4.9* below). Significantly, this is a configuration similar to that known in W_2C, even as regards interatomic distances. There is, however, a slight contraction in η_1, consistent with the partial W $\rightarrow$ Fe substitution in the octahedra: a flexibility which also accounts for the known homogeneity ranges in η (such as in $Fe_3W_3C \longleftrightarrow Fe_4W_2C$).

In the η_2-carbides, the atomic positions are similar[7,8] to those in η_1 except that the A atoms fill some of the 32 (*e*) positions in *Figure 4.7*, the B atoms, some of the 48 (*f*) and 16 (*d*) positions.

The possibility of mutual substitution (partly or fully) in the η-carbides of the transition metals, as indicated in *Table 4.2*, clearly exists; similarly, that of C by N or O* for corresponding variation of metal ratios. This wide compositional tolerance of the η-carbide structure is of considerable practical consequence. It allows the carbide to act as solvent when present as micro-constituent in steels and refractory alloys; it also makes the compound adaptable to a variety of practical applications. A further point of interest is that, as found by the author, *silicon* can enter as a substitutional or interstitial element in η-carbide. Very recently, its range has also been extended to some replacements in which *aluminium* is an integral metal, besides the transition metals[13-4]. Such metal replaceability also applies to the η-oxides, e.g. from about M_3Ti_3O to M_2Ti_4O, where M can be Mn, Fe, Co or Cu in varying proportions[15]. The C $\longleftrightarrow$ O substitution in η is of special interest in permitting a continuous carbide $\longleftrightarrow$ oxide transition from the interior to the surface of a material, and lattice 'anchoring' of an oxide.

Regarding physical and technological properties, η-carbide is of great hardness and has a number of attractive features in common with the more refractory carbides of Groups IV, V, VI in its own right, quite apart from

---

* To this we should add 'vacancies', inasfar as the stoichiometry 'M_6C' may not necessarily be rigid but move towards higher metal contents; i.e. η *intermetallic compounds* are able to exist free or low in interstitial, and are in fact known[2c,9-11] (cf. *Table 4.2*).

The ubiquity of the η-carbide structure as a host compound for wide replacements is exemplified by a phase of composition[12] $(Ni_{0\cdot58}\,Co_{0\cdot30}\,Si_{0\cdot12})_3\,(Mo_{0\cdot41}\,W_{0\cdot07}\,Cr_{0\cdot44})_3$.

that as microconstituent. Thus it has an unusually high thermal expansion (commensurate with that of iron), good oxidation and acid resistance and, in contrast to the more refractory carbides, the property of 'self-binding' and sintering at relatively low temperatures[16].

The $M_{23}C_6$ Carbide—An important double carbide, often associated with η on the one hand and with cementite on the other, is $M_{23}C_6$. This *can* in fact be a single-element carbide (in contrast to η), namely one of chromium, $Cr_{23}C_6$; but $Fe_{21}W_2C_6$ is an example of its double carbide role. This carbide occupies a key position not only in the carbide system but also in the alloy-steel metallurgy bound up with it. Thus it occurs in stainless and in many other types of Cr-bearing steel.

In *Figure 4.2*, chromium (and, similarly, still manganese) has a singular position, in that each of its three carbides has a complex structure related to the symmetry of one of the adjoining types, but not identical with it:

$Cr_{23}C_6$, cubic, as the VC and austenite class;
Cr_7C_3, hexagonal, as the Mo_2C class;
Cr_3C_2, orthorhombic, as cementite.

The cubic unit cell[16] of $M_{23}C_6$ is shown in *Figure 4.6*, together with a list of co-ordinates. It is face-centred, with 92 metal and 34 carbon atoms, and of cube edge $a = 10{\cdot}638$ *kX* (for $Cr_{23}C_6$). We may divide the unit cell into 8 octants (one only is shown in the model) in each of which the corners are alternately surrounded by cubo-octahedra (48 '*h*' positions [Westgren's nomenclature]) and by simple cubes (32 *f* positions). An additional metal atom occupies the centres of the cubo-octahedra (4 *a* positions: unit cell corners in this representation), and one the more spacious centre of each octant (8 *c* positions).

This carbide may well serve as illustration of how crystal chemistry can govern the actual composition ranges of carbides. Thus, starting from pure chromium carbide, $Cr_{23}C_6$, Fe can replace Cr extensively, such that the cubo-octahedra (*h*) remain occupied preferentially by Cr, but the cubic (*f*) positions are preferred by the smaller Fe atom. Such Cr → Fe substitution can proceed up to about 30 atomic per cent Cr, when the available space is occupied; thence onward the presence of *some* tungsten (or Mo) is necessary in the lattice, in order to enable the carbide to dissolve more Fe. This is because, upon excess substitution of Cr by Fe, the octant centres (*c*-positions) become too spacious to continue taking either Fe or Cr only, and a larger atom, W or Mo, is required to fill them.

In terms of phase diagrams, this is reflected in the Fe–Cr–W–C quaternary system (see *Figure 4.4*) by the fact of the single-phase field of $Cr_{23}C_6$ base detaching itself from the ternary Cr–Fe–C side at an Fe/Cr proportion dictated by the *f* sites available (about 30/70). This gives rise to a $M_{23}C_6$ 'tongue' continuing right across the quaternary model to the Cr-free composition $Fe_{21}W_2C_6$. Here again, the final W or Mo capacity is clearly defined by the proportion of *c* positions available (8 out of 92 metal atoms). Any excess W would contribute to stabilizing the η-carbide Fe_4W_2C, as the next higher one.

In all these isomorphous replacements of *M* atoms, the interstitial *carbon* remains constant both in proportion and position, but excess Fe changes it to that in cementite. These relations may also be expressed in a wider context, as in *Figure 4.8*.

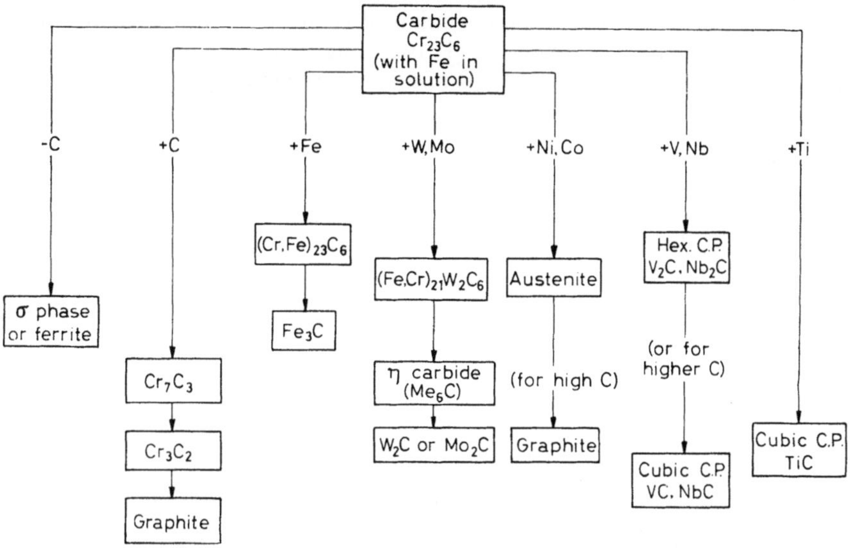

Figure 4.8. Phase evolutions from the $Cr_{23}C_6$ *(high-chromium steel) carbide*

Table 4.3

Atomic co-ordinates in Cr_7C_3 *and* Cr_3C_2 (From Westgren[16b] and Hellborn and Westgren[17]); *for* $Cr_{23}C_6$, see *Figure 4.6*

Cr_7C_3: (56 Cr, 24 C per unit cell); trigonal

2Cr at $\frac{1}{3}\ \frac{2}{3}\ z;\ \frac{2}{3}, \frac{1}{3}, \frac{1}{2} + z;\ z=\frac{3}{4}$

30Cr at $x\ \bar{x}\ z;\ 2\bar{x}\ xz;\ x\ 2xz;\ \bar{x}, x, \frac{1}{2} + z;\ 2x, x, \frac{1}{2} + z;\ \bar{x}, \frac{1}{2} + z,$
where (five-fold) $x_1 = 0{\cdot}06,\ z_1 = \frac{1}{2};\ x_2 = \frac{1}{6},\ z_2 = \frac{3}{4};\ x_3 = 0{\cdot}23,$
$z_3 = \frac{1}{4},\ x_4 = 0{\cdot}27,\ z_4 = \frac{3}{4};\ x_5 = 0{\cdot}56,\ z_5 = 0$

24Cr at $xyz;\ y\text{–}x, \bar{x}, z;\ \bar{y}, x\text{–}y, z;\ \bar{x}, \bar{y}, \frac{1}{2} + z;\ x\text{–}y, x, \frac{1}{2} + z;$
$y, y\text{–}x, \frac{1}{2} + z;\ \bar{x}, y\text{–}x, \frac{1}{2} + z;\ y, x, \frac{1}{2} + z;\ x\text{–}y, \bar{y}, \frac{1}{2} + z;$
$x, x\text{–}y, z;\ \bar{y}, \bar{x}, z;\ y\text{–}x, y, z$
where (two-fold): $x\mathrm{Cr}_5 = 0{\cdot}27,\ y\mathrm{Cr}_5 = 0{\cdot}04,\ z\mathrm{Cr}_5 = \frac{1}{4}$
$x\mathrm{Cr}_6 = 0{\cdot}44,\ y\mathrm{Cr}_6 = 0{\cdot}06,\ z\mathrm{Cr}_6 = \frac{1}{2}$

12C (two-fold) at $x\bar{x}z$, etc. where $x_{C_1} = 0{\cdot}10,\ z_{C_1} = 0,\ x_{C_2} = 0{\cdot}40,\ z_{C_2} = 0$

12C at xyz, etc. where $x_{C_3} = 0{\cdot}40,\ y_{C_3} = 0{\cdot}10,\ z_{C_3} = \frac{1}{2}$

Cr_3C_2: (12Cr, 8C per unit cell); orthorhombic

4CrI	$xy\frac{1}{4};\ \overline{xy}\frac{3}{4};\ \frac{1}{2}-x, \frac{1}{2}+y, \frac{1}{4};\ \frac{1}{2}+x, \frac{1}{2}-y, \frac{3}{4}$
	where $x_{\mathrm{CrI}} = 0{\cdot}406,\ y_{\mathrm{CrI}} = 0{\cdot}03$
4CrII at (*c*)	where $x_{\mathrm{CrII}} = -\ 0{\cdot}230,\ y_{\mathrm{CrII}} = 0{\cdot}175$
4CrIII	where $x_{\mathrm{CrIII}} = -\ 0{\cdot}070,\ y_{\mathrm{CrIII}} = -\ 0{\cdot}150$
4CI	where $x_{\mathrm{CI}} = 0{\cdot}109,\ y_{\mathrm{CI}} = -\ 0{\cdot}100$
4CII	where $x_{\mathrm{CII}} = -\ 0{\cdot}057,\ y_{\mathrm{CII}} = 0{\cdot}217$

It is helpful at this stage to consider the *carbon environment*, not only in $M_{23}C_6$ but in its broader relation to that in the allied carbides, M_6C, M_3C (cementite) and others. These environments are shown in *Figure 4.9*. In $M_{23}C_6$, each carbon atom has 8 metal neighbours (*Figure 4.6*), whereas in all other carbides there is 6-co-ordination of carbon. However, there exist two types of the latter: namely, of (1) two congruent triangles and (2) two inverted triangles (*Figure 4.9*). Fe_3C and WC represent type (1), η-carbide,

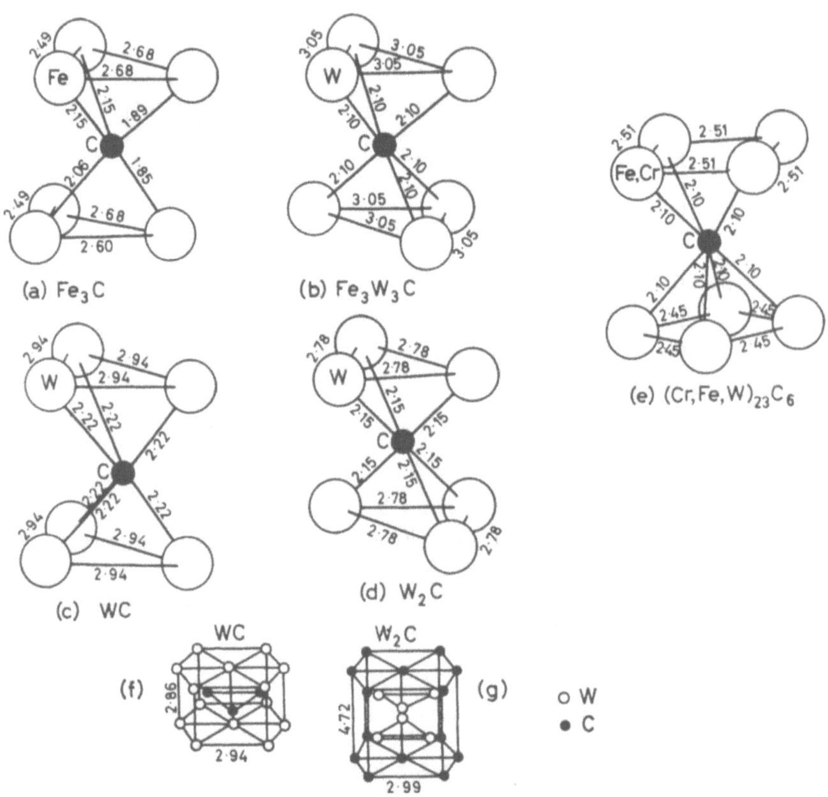

Figure 4.9 (a)–(e): Carbon environment in carbides of iron, chromium and tungsten; (f) and *(g): unit cells of* WC *and* W_2C (Goldschmidt[2b])
By courtesy of the Iron and Steel Institute, London

and W_2C, type (2), as is clear from the structures shown. This fact is noteworthy because it shows that, in Fe_4W_2C, the carbon atoms prefer already the same configuration and surround themselves entirely by W, not Fe, consistent with this being a stage transitional to forming tungsten carbide.

For the two pure tungsten carbides, WC → W_2C, the (1) → (2) inversion of C environment applies again. The *cubic* close-packed structure, e.g. in VC, NbC, etc., again represents the congruent type (1), and it may be noted that the (1) → (2) change is generally also implicit in the cubic → hexagonal close-packed transition, as for instance in NbC → Nb_2C. The 8-fold environment in $M_{23}C_6$ may be considered as a hybrid state between (1) [$=Fe_3C$] and (2) [$=M_6C$], as indeed it falls into line compositionally.

Again, the reason why the total carbon content should diminish in this carbide sequence ($M_3C \rightarrow M_{3\cdot 8}C \rightarrow M_6C$) derives naturally from an inspection of the crystal structures: it is directly bound up with the decreasing number of lattice sites available for an interstitial atom. All these crystallographic relations are clearly reflected in the resulting phase diagram (see *Figure 4.4*).

Furthermore, the $M_{23}C_6$ carbide is structurally related to austenite, in a manner permitting the ready precipitation or re-solution of the carbide in an austenitic matrix, as in fact commonly occurs in practice. Structurally the cubo-octahedra in $M_{23}C_6$ (*h* positions) may be considered a 'hybrid configuration' between a face-centred and a body-centred cube, namely a FCC in which the insertion of a further atom in the body centre has forced those at the face centres 'outwards', such that all become equidistant from the central one. Thus, by the manipulation of inserting or removing one metal atom, a ferrite-like or austenite-like cluster can be produced in the carbide proper. The simple cubic *f* positions, which in this carbide alternate with the cubo-octahedra, can be made into a body-centred cube by the obvious device of displacing the remaining 8 *c* atoms, previously at the octant centres, into those of sub-cube (*f*). These 'operations' account for the total number of available atoms per unit cell*. The *c* positions are those occupied by the large W or Mo atoms, for which there would be insufficient space inside the *f* cube. Upon slight decarburization of the carbide, the transformation begins either to the BCC lattice (Cr, Fe, W solution) or to σ-phase.

Implications in steel reactions—These reactions have a direct bearing upon the relation of the $M_{23}C_6$ carbide with the α- or the γ-matrix. During the decomposition, for example of austenites in some stainless steels, one may in fact consider $M_{23}C_6$ as a stable alternative to martensite. Thus in many austenitic or semi-austenitic steels, the $M_{23}C_6$ carbide is that nucleating at high temperatures, evolving at lower temperatures to reactions of type

$$\begin{array}{ccc} \text{austenite} & \rightarrow & \text{martensite} \\ \searrow & \swarrow & \searrow \\ M_{23}C_6 & & \text{ferrite} \end{array}$$

These can occur coherently and to varying degree in different lattice regions, without necessarily involving major atomic diffusion (except of carbon).

Similar considerations can readily be extended to other carbides, $M_{23}C_6$ being here discussed in some more detail by way of illustration. Thus, for instance, the 3 carbides $M_{23}C_6$, M_3C and M_7C_3 [$=$ $(Cr,Fe)_{23}C_6$, $(Fe,Cr)_3C$, $(Cr,Fe)_7C_3$] are all frequently alternative crystalline precipitates for higher-chromium steels (e.g. 12 per cent Cr, 1 per cent C) depending on heat treatment. In ferritic lower-Cr steels, [cementite + martensite] form upon quenching from the austenitic region, [$M_{23}C_6$ + ferrite], upon annealing. In this case, the cementite lattice, with its 12-co-ordination and direct

* These movements are not, of course, necessarily the real ones occurring during the process of formation of the carbide, but at least help to bring out formally the relationship with the α and γ matrix and the other carbides.

kinship with austenite, can occur as a spontaneous transition stage formed relatively easily, before the more complex movement towards $M_{23}C_6$ can materialize. M_7C_3 is in this context a further variant that can be produced by critical lower-temperature heat treatments, either stably or metastably, depending upon composition (e.g. Cr and C content). *Which* transformation applies in a given case is sensitively guided by the alloy metals present, their atomic sizes, Group number, proportion (e.g. Fe/Cr/Mo/W) and, of course, M/C ratio.

This strong dependence on heat treatment of the type of carbide formed in a given steel may be illustrated by an example[18] in which the diffusion dependence of the type of carbide formed is particularly marked. Upon tempering martensitic steels in the range 4–6 per cent Cr, 0·35–0·5 per cent C, we obtain: Fe_2C at 316°C; Fe_3C at 427°C; ($Fe_3C + Cr_2C_3$) at 538°C; Cr_7C_3 at 583°C and higher.

Many similar sequences could be cited. However, the essential point is that, although the chromium-base carbide is the true equilibrium one and that forming at the higher tempering temperatures, the less stable iron-base carbides are formed at the lower temperatures, since the carbon derived from the decomposing martensite tends to associate with the ubiquitous iron atoms forming their immediate environment; it requires a much higher degree of diffusion of chromium atoms before the more stable chromium carbide can form. Similar effects dominate the more highly alloyed (e.g. Mo, W, V) steels, the central theme being that the more stable carbide can be temporarily inhibited in favour of the less stable one in which, despite their stronger carbon affinity, these larger atoms can be trapped owing to insufficient diffusion energy. This delicate balance between diffusion rates and carbide stability is a chief cause for the sensitive property dependence upon heat treatment in so many alloy steels. In the case, for example, of higher W or Mo steels, e.g. high-speed steels, similar considerations apply to the formation of the M_6C carbide, with $M_{23}C_6$ as a critical alternative.,

The formation of pearlite, bainite and similar microconstituents is, of course, closely interlinked with all these effects; this will be further discussed later.

Even within a given carbide grain embedded in a steel matrix, non-homogeneous compositions can occur. Thus, for example, a steady (micro) gradient in W content can occur from core to edge of an M_6C particle, shown by varying spacing within it and conforming with a range like Fe_4W_2C/Fe_3W_3C. Also, upon decarburization, a *carbon* gradient can be developed; this can take the extreme form of a transition towards an intermediate compound or matrix phase, e.g. (1) $M_6C \rightarrow Fe_3W_2$ (μ) phase or (2) $M_{23}C_6 \rightarrow$ FeCr (σ) phase or (3) $Fe_3C \rightarrow$ ferrite. The actual course is clearly determined by structure relations and by diffusion rates*.

Some other double carbides—Although M_6C and $M_{23}C_6$ are probably the most important double carbide types in practice, a number of others do exist: some of initially undefined composition, such as the 'M_aC_b' carbide, a

* Adelskjöld, Sundelin and Westgren[19], in their original work discovering the η-carbide in high-speed steel, report that this in fact decomposes on heating to very high temperatures, Fe_3W_3C forming WC, Fe_3Mo_3C–Mo_2C. With some modifications (see later), this still conforms with more recent experience in alloy steel constitutions.

combination of M–Fe–W* found by Kuo[3b]. A further double carbide, $Co_3W_9C_4$, was found[22] in work on the Co–W–C system (*Figure 4.10*). It is here clearly associated with the two η-carbides. This, incidentally, again conforms with the observation of the diagonal relationship (*Figure 4.2*)

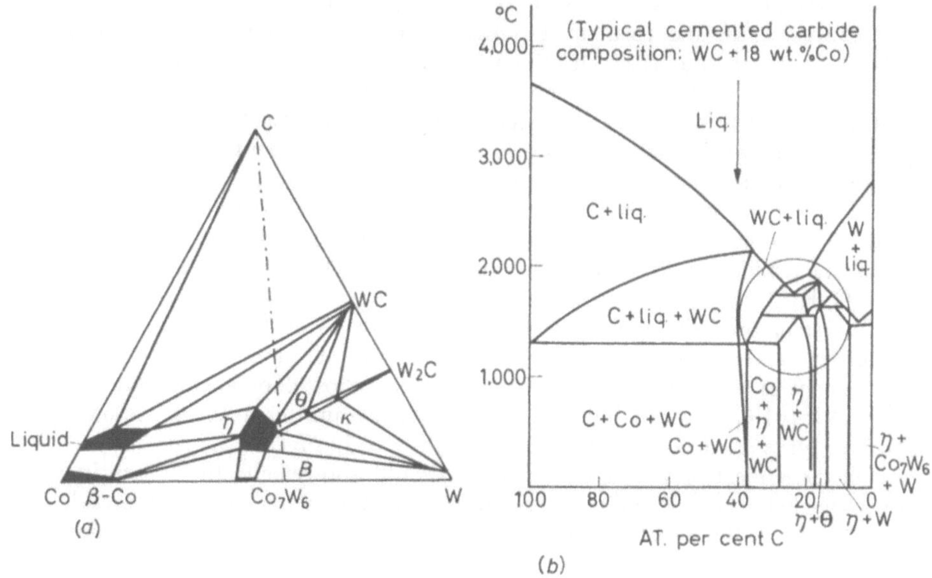

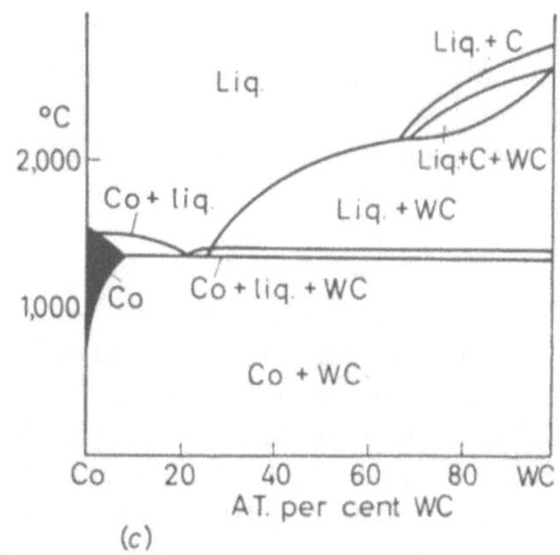

Figure 4.10. Co–W–C *system, basic to hard metals* (Rautala and Norton[22])

(*a*) Co–W–C equilibrium diagram at 1,400°C $\eta = Co_3W_3C$; $\theta = Co_3W_6C_2$; $\kappa = Co_3W_{10}C_4$

(*b*) Temperature section along line *B–C* in (*a*) (the ringed area represents multiple fields of liquid, WC, η, κ, θ, Co, Co_7W_6 and C)

(*c*) Temperature section along Co–WC in (*a*)

By courtesy of the Metallurgical Society of AIME

---

* The structure of Kuo's M_aC_b carbide and its composition have recently been solved by Dyson and Andrews[20]: the formula is Fe_2MoC, i.e. stoichiometrically M_3C, like cementite. Indeed, it is structurally derived from cementite, and of an enlarged orthorhombic unit cell with $a = 16{\cdot}27_6$, $b = 10{\cdot}03_4$, $c = 11{\cdot}32_3$ Å. The phase represents an intermediate stage between Fe_3C and the M_6C η-carbide, and clearly must fit, as Andrews points out, significantly into the reaction sequence of Mo-bearing alloy steels. It seems likely to have a tungsten analogue and to be only the first prototype to a wider isomorphous carbide (also nitride?) family, though as yet unknown.

A further study of composition and interplanar spacings of M_aC_b is due to Cadek *et al.*[21].

applying to the η-carbides, and suggests that this rule governs the occurrence of double carbides more generally; the $M_{23}C_6$ (viz. $Fe_{21}W_2C_6$, etc.) carbide is another case in point. All double carbides so far known appear to apply in directions D, none in directions A, B, C. This is clearly in line with the requirement, mentioned above, of one type transition metal atom with a strong, the other with a weak M–C bond energy. Also should we remember the existence, in the case of the $M_{23}C_6$ double carbide, of $Cr_{23}C_6$ and $Mn_{23}C_6$, i.e. singular cases in which it 'degenerates' into a single-element carbide. This is of particular interest because we may well consider that varying *types* of Cr and Mn atoms occur here, of different electronic configuration. These are, in fact, well known for Mn as an element (β-Mn and π-phase), and Cr could also adopt a similar atomic duplicity. This raises the point whether such a singular case of a single-element carbide could not possibly also occur for the η-carbide. It is not so far known, but hypothetically such a carbide could conceivably occur for technetium or ruthenium (Tc_6C, Ru_6C?).

The structure of $Co_3W_9C_4$ is hexagonal with 24 metal and 8 carbon atoms[23a]. It is of interest to note that, out of the 8C atoms, 2 are located at the centres of trigonal prisms of 6W atoms, 6 at those of octahedra containing (5W + 1 Co) atoms. The net result is that within this structure we have the *two* types of carbon environment, A and B, as in *Figure 4.9*.

On the subject of double carbides, we should also refer to the U-bearing ones (see below, p. 156,) to those in which *one* metal is a non-transition metal, e.g. Al, and to the so-called 'Nowotny phases' which, however, were revealed as carbosilicides and will be discussed in Chapter 10.

4.3 CEMENTITE

Cementite occupies a special position amongst the carbides in that, although persistent, it is unstable at any rate for iron; the stable system is Fe–graphite. In the isomorphous sequence $Mn_3C \rightarrow Fe_3C \rightarrow Co_3C \rightarrow Ni_3C$, cobalt and nickel increasingly favour graphitization but manganese and, when in solution, chromium stabilize the cementite structure, until their excess causes $M_{23}C_6$ or M_7C_3 formation. The Group V and IV metals Ti, Zr, Ta, Nb, can no longer replace Fe in Fe_3C, producing an immediate change-over to the very stable NaCl-type carbide. The intervening smaller vanadium atom does so less readily, still having a small solubility in Fe_3C. Mo and W, as already explained, break down the cementite in favour of the double carbide. Again the theme appears here of the FCC structure of austenite being placed in the transitional position of a semi-carbide already isomorphous with the cubic carbide family (TiC, etc.). The relationship is also well brought out by the eutectoid decomposition in the iron–carbon diagram itself, shown on the carbide chart (*Figure 4.1*). The γ-Fe solution nucleates the cementite and martensite simultaneously.

These relations between austenite, cementite and martensite are immediately manifested in the microstructure of steels, such as in pearlite or bainite. With varying heat treatment and diffusivity of C in Fe at given temperatures, we can produce an intergrowth of the three structures in different stages of progress. This is represented, after quenching out of the austenitic region, by a variety of microassemblies of Fe_3C, ferrite, martensite and residual austenite, and by varying factors such as C content, rate of cooling, holding time at different temperature levels, etc. Each ensuing structure, basic in ferrous metallurgy, is distinctive in the physical properties associated with it. It is instructive to note here that, after quenching from γ, we have

(1) *pearlite*, being formed at higher sub-eutectoid temperatures and for slower cooling rates;

(2) *bainite*, at intermediate temperatures upon isothermal annealing; and

(3) *martensitic* temper-carbide types at low temperatures.

All these (and other) microconstituents are basically assemblages of matrix/cementite fine structure at various stages of evolution and of relative physical size, from near-atomic to the order of microns. Once again, this leads up to our theme of essential unity between carbide and matrix phases, which will, of course, find vital application in steels.

It also gives special interest to the cementite structure itself, now to be considered, and to the problem of its relation to austenite, ferrite and to the other carbides. The structure of cementite has been solved by Lipson and Petch[4], although the unit cell and the Fe atom positions had been found previously[1c,24]. The C positions had long been in question but could be ascertained finally by careful Fourier synthesis. *Figure 4.5* gives a picture of the network of Fe and C atoms found. The carbon atoms, with their 6 iron neighbours, are placed at the centre of slightly distorted Fe prisms (cf. *Figure 4.9*). The orthorhombic unit cell is indicated by the dotted rectangle in *Figure 4.5b*; the co-ordinates show 16 atoms = 12 Fe, 4C, comprising 2 unit cells of austenite*. The structure may be described as *near* close-packing of iron atoms, with carbon in the interstices, and in this respect there is a close kinship to austenite, the parent phase upon ($\gamma \rightarrow \alpha + Fe_3C$) breakdown. The detailed transformation mechanism is not easy to visualize and by no means finally solved; however, by the insertion of a carbon atom and a shear and slight puckering of the hexagonal (III) planes in γ, we can at least formally derive the cementite network from that of austenite. The fact that the Fe atom structure in Fe_3C is more closely related to that in γ- than in α-Fe is itself important in accounting for the much greater solubility of Fe_3C in γ. It is closely bound up with martensite as an intermediate stage and the lattice strain developed through the suppressed nucleation of actual Fe_3C.

One further point is of interest here: considering the iron–carbon phase diagram (*Figure 4.1*), it may be seen that, *above the melting point*, the solubility for carbon in iron is 25 atomic per cent, i.e. Fe_3C, indicating the existence even in the liquid of molecular *aggregates* of Fe_3C; the liquid solubility further increases with temperature to a maximum of 33 atomic per cent C at 2,200°C, i.e. attaining Fe_2C. In other words, solid graphite is in equilibrium with a carbon-saturated liquid iron which quite possibly contains carbide-like configurations already. A similar situation occurs in the Ni–C and Co–C systems (*Figure 4.1*). The existence of *colloidal solutions* of iron carbide has

* The interplanar spacings of Fe_3C have been given accurately and fully[25] as $a = 4{\cdot}515$, $b = 5{\cdot}078$, $c = 6{\cdot}728$ *kX* (cf. *Table 4.1*).

been assumed in the liquid metal and, at higher C contents, of graphite[26]. However, this is a subject of some controversy, in which the presence of discontinuities in liquid solubility at the values Fe_3C and Fe_2C has been denied[27]. For the carbon atom configuration in liquid iron, finely dispersed 'micro-heterogeneities' or miscelle-like carbon-enriched molecular groupings have been stipulated within the liquid.

In view of the metastability of Fe_3C, we might well consider its structure kinetically as a transitional state towards the stable Fe $\longleftrightarrow$ graphite system. The carbon diffusion towards graphite needed to achieve full segregation of the γ- or α-matrix cannot occur sufficiently fast, in view of the metal–carbon atom bond being still fairly strong for Fe. For Co and Ni, it becomes so weak that immediate graphitization occurs, although the very unstable Co_3C and Ni_3C do exist.

The iron percarbide, Fe_2C, first described by Hägg[28], later studied by Jack[29a], is also orthorhombic and closely related to Fe_3C; it is likewise metastable. The axes of the Fe_3C and Fe_2C unit cells are simply related:

(*a*) $a_{Fe_2C} \approx 2 \times aFe_3C$

(*b*) $b_{Fe_2C} \approx 3 \times bFe_3C$

(*c*) $c_{Fe_2C} \approx cFe_3C$

Thus 6 Fe_3C cells stacked together form a volume similar to, though slightly smaller than, one unit cell of Fe_2C. In Fe_2C, this volume contains 8 more iron and 12 more carbon atoms than in Fe_3C, an excess well accounted for by the difference in equivalent volumes:

in Fe_3C, 96 atoms (= 72 Fe, 24 C) occupy 924 kX^3

in Fe_2C, 96 atoms (= 64 Fe, 32 Fe) occupy 934 kX^3

Therefore, no major expansion or contraction is involved on passing from Fe_3C to Fe_2C. In contrast, the equivalent γ-Fe volume of 24 unit cells* is 1,164 kX^3. We shall return to this subject below (Section 4.4).

An interesting sidelight on cementite stability may be obtained, by way of example, from the following experiment. It concerned a plain carbon steel containing Fe_3C in which the effects of small, increasing amounts of chromium and molybdenum were studied; there were two series of steels, one without Cr, the other with 0·2 per cent Cr, and within each series the Mo content increased from nil to 0·8 per cent Mo. Upon x-ray analysis of the extracted carbides, they were all found to be Fe_3C only. However, characteristic changes occurred in two respects, (1) in lattice dimensions, (2) in the state of stress of the cementite lattice. *Figure 4.11* shows the *a*, *b* and *c* axes plotted against Mo contents. It may be seen that the Cr atom, substituting Fe, consistently expands the unit cell. With increasing Mo contents, however, lattice strains are developed, manifested by line broadening, and whilst the *a* axis remains fairly constant, the *c* axis rises and attains a maximum, the *b* axis remaining constant initially, then attaining a minimum. This is repeated at both Cr levels. The essential point here is that the Cr atom 'comfortably' dissolves in cementite, but not so the larger Mo atom: this does enter unstably in cementite, but with increasing distortion of the

* At 910°C, i.e. above A_3; but a (hypothetical) room-temperature value would be much lower and approach the above.

lattice. True equilibrium would correspond to another carbide higher in Mo (the $M_{23}C_6$ carbide, here $(Fe_2,Cr)_{21}Mo_{26}$), so that the tendency is to precipitate this. However, in the circumstances, this occurs only sluggishly and remains incipient, the Mo atoms being retained in the Fe_3C lattice under a high state of stress, not only through its size but also by producing fluctuations in carbon content from one unit cell to the other, rather as occurs in

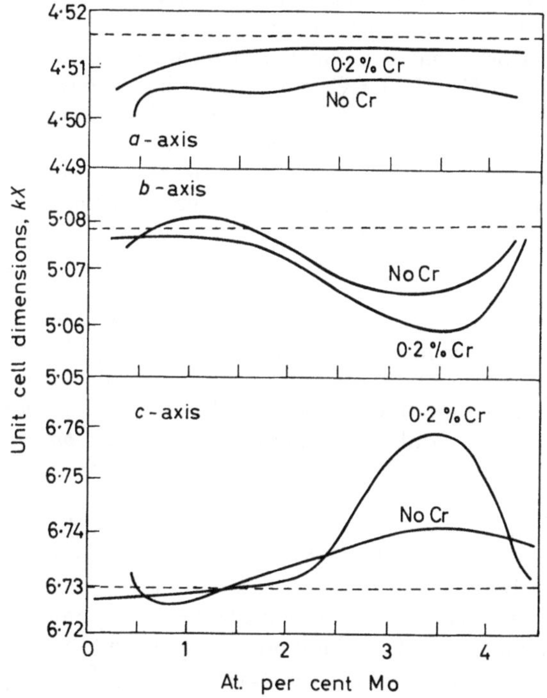

Figure 4.11. Axes of cementite unit cell, in dependence on Cr *and* Mo *content of a steel* (Goldschmidt, unpublished)

Dashed line = values for Fe_3C in plain carbon steel, unalloyed

martensite. The Mo atom probably tends already to form around itself a carbon environment, as in $M_{23}C_6$, which is that of 8 nearest neighbours to the carbon (cf. *Figure 4.9* and earlier discussion) instead of 6, with a consequent displacement of Fe atoms and high lattice strains.

The main point of principle is that both these Group VI metals, Cr and Mo, being carbide stabilizers, act distinctively upon cementite: the former rendering cementite stable (from its inherent metastability for pure Fe), the latter less stable, favouring another carbide which, once formed, also accommodates more chromium. All this is reflected in the lattice structure of this next, $M_{23}C_6$, carbide, as explained earlier. It also illustrates a type of occurrence frequent and important amongst pairs of interstitial compounds. Manganese plays here an intermediate part as cementite stabilizer insofar as it 'goes with

* As demonstrated for higher Cr, Mo alloy steels.

the majority': it favours the Mn_3C structure for Fe excess, the $Mn_{23}C_6$ or Mn_7C_3 structure for Cr excess. This critical position of Mn has also other consequences, as will be amplified below.

In connection with cementite, it is of interest also to refer to its Curie point. This lies at 210°C (for Fe, at 768°C) but is strongly composition-dependent. It has been shown[30] that the Curie temperature of Fe_3C decreases with increasing manganese content in solid solution, and does so linearly. Physical property changes such as these often serve generally as an auxiliary analytical guide for carbide composition.

More recent work[31] has been able to determine the carbon positions in Fe_3C (as well as in $Cr_{23}C_6$, Cr_3C_2 and WC) using *neutron diffraction.* In an electron diffraction study on cementite extracts[32], the validity of carbon positions reported previously had been questioned. The substitutional distribution of solvent metal atoms in cementite is equally important, such as that of chromium, manganese and nickel critically poised around iron in the Periodic Table (for Cr, see [33]).

An interesting issue is whether cementite when embodied in a steel can be subject to plastic deformation—the more refractory carbides certainly cannot. This has been shown to be the case in electron-microscope work[34]: a defect structure appears in cementite, with a density increasing with plastic deformation. There are defects of dislocation and stacking fault type, quite as in the primary solution (α,γ-matrix) phases: in further support for the kinship to the latter referred to above. Another facet is that *pressure* is indeed a parameter that can affect the isothermal austenite $\rightarrow$ ferrite + Fe_3C transformations in Fe–C alloys[35,36]. In this sense, the Fe–C diagram is of course only an *isobaric* and, strictly, an *isochronic* section towards Fe/graphite.

Electron microscope studies[37,38] on the crystallographic relationship between γ-matrix, martensite and precipitating cementite identified the intermediate stages, such as a transient platelet precipitate from austenite; this is the hexagonal ε-Fe_3C carbide preceding cementite. Again we might look at this kinetically as a carbon-induced stacking-fault region from the FCC γ-phase.

One unusual aspect of cementite, briefly mentioned already (Chapter 2), is its occurrence in iron meteorites, where in the form of 'cohenite' (= Fe_3C) the phase is able to act as intermediary nuclei to *diamonds*, formed under shock-heating conditions, as against graphite[39]. The effect, along with that of Ni-cementite and related substances, enters also into the laboratory synthesis of diamonds (G.E. process).

Figure 4.12 shows a pressure/temperature diagram for carbon in the presence of iron, according to Lipschutz and Anders[39]*. It was derived from free-energy calculations in connection with explaining the conditions of formation of diamonds in iron-based meteorites. Using this in the manner

---

* The authors point out that the temperature scale may be considerably in error, owing to an uncertainty in the enthalpy of cementite formation; it would indeed conflict with the known metastability of Fe_3C at normal pressures. However, the diagram should be read as one of principle, and is most useful as such. This stimulating work is worth referring to in the original, both for the fundamental issues involved and for further references.

Clearly, there is again a mutually valuable link-up between seemingly unrelated scientific disciplines such as those of natural meteorites, 'diamond-making' and ferrous metallurgy.

of a phase diagram, it shows that high pressures would increasingly *stabilize cementite*, then acting as a catalysing agent and intermediary stage to diamond ($Fe_3C \rightarrow 3Fe + C$) as it does to *graphite* (in steels and cast irons) at lower pressures. The same applies to Ni_3C in nickel base and to $(Fe,Ni)_3C$ in meteorites.

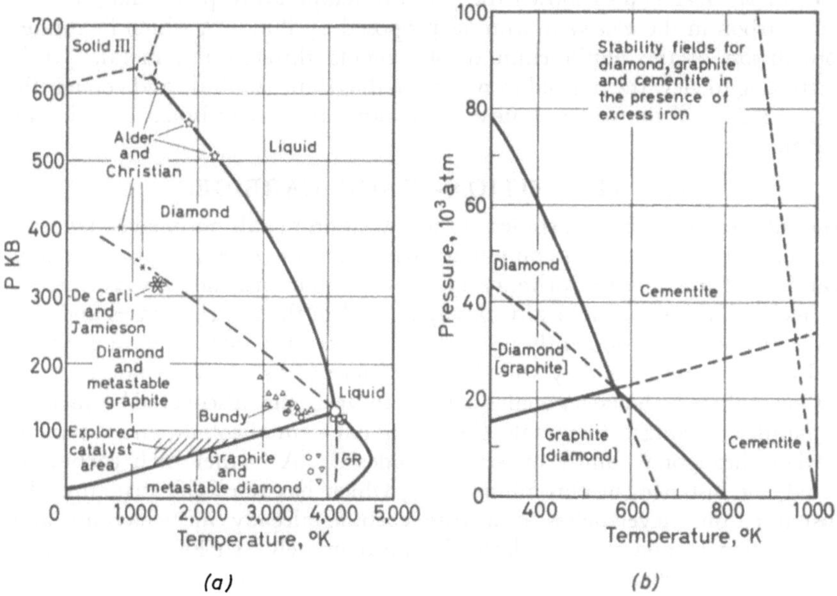

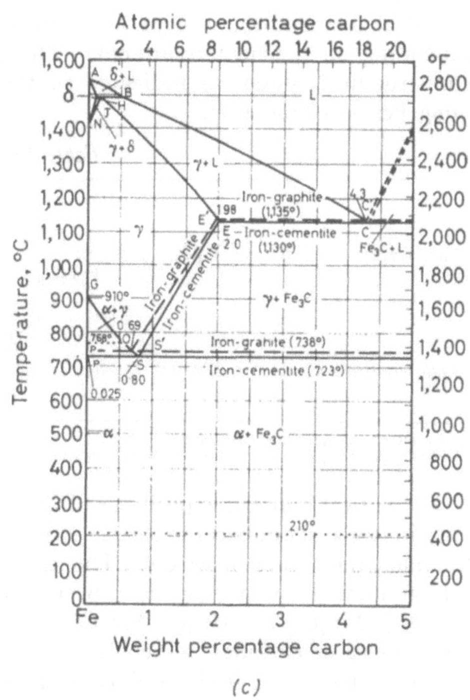

Figure 4.12. Aspects of iron–carbon equilibria

(*a*) Carbon (pressure, temperature) equilibrium diagram, without iron present[40–42]; (*b*) As (*a*), with iron present in excess (Lipschutz and Anders[39]); (*c*) The normal Fe–C phase diagram (stable graphite, metastable cementite), i.e. isobaric section at 1 atm; to be related to (*a*) and (*b*) (A.S.M. Metals Handbook[43])

By courtesy of the American Institute of Physics; the Editor, *Geochimica et Cosmochimica Acta;* and the American Society for Metals

Nevertheless, this is a subject of much controversy (see also Wentorff's work) and, in fact, Lipschutz and Anders themselves show evidence tending to disprove the carbide intermediary theory of diamond formation previously held, in favour of one of direct shock-heating conversion from graphite.

In *Figure 4.12* is also shown the pressure/temperature phase diagram for pure carbon in the *absence* of iron, as proposed by Bundy[40], which facilitates comparison of the equilibrium displacements through the metal. It is interesting to view the normal iron–carbon diagram, graphitic and cementitic (*Figure 4.12c*), in the wider context of the other two, as an isobaric section at 1 atm.

4.4 TRANSITION-STAGE CARBIDES

The substantially stable carbides of the transition metals outlined so far lead on to a class of interstitial compounds which are in certain respects already transitional towards the primary metal solid solutions, notably α- and γ-, in steels. They are structurally closely related to the matrix phases and can nucleate and grow coherently upon their lattices; however, in most physical and chemical respects they still show the characteristics of carbides.

One of these is the hexagonal ε-carbide in steels, first discovered by Jack[29b]. It parallels, amongst the metals, the hexagonal form of cobalt, and it typifies a whole hexagonal family of 'semi-carbides'[44]. A further such carbide is found, for instance, in chromium steels, which has the FCC structure like austenite, but nevertheless is a true carbide already in behaviour and composition (ψ-phase). The latter is approximately $(Cr,Fe)_7C_3$, similar to the established chromium carbide of more complex structure. In fact, the whole phase sequence, namely (*a*) the ψ-phase, (*b*) M_7C_3, (*c*) $M_{23}C_6$ and (*d*) M_3C (cementite), is capable of forming in a given chromium steel, depending on the heat treatment. The phase again fits organically into our structural system as indicated in the following series along the first Long Period (*Table 4.4*).

Table 4.4

Group	IV	V	VI	VII	VIII	
Cubic carbides	TiC	V_4C_3	ψ-phase ≡ $(Cr, Fe)_7C_3$	⟶		Austenite $(Fe, Mn, Co, Ni)_nC$
				$(Cr, Mn, Fe)_3C$	$(Cr, Mn, Fe)_{23}C_6$	
Ratio metal/carbon	≥ 1·0*	1·3	2·3	3·0	3·8	> 12 → ∞ for pure FCC metal†

* But already of wide C-deficient range
† For C-saturated γ-Fe; lower with Mn and Cr in solution

The critical borderline positions of Cr and Mn, which are of such importance in steels, again become clear. It is of interest in this connection to note the manganese–carbon system. The diagram is included in *Figure 4.1*, but a large number of carbides have been reported (although not all verified): $Mn_{23}C_6$, Mn_7C_2(?), Mn_3C, Mn_5C_2 and Mn_7C_3. The very fact of this multiplicity of carbides, which is greater than in any other single metal/carbon system, is in itself significant of their unstable character and ready transformability. The existence of the cementite analogue, Mn_3C, had been claimed in early work, was later denied but more recently re-asserted. $Mn_{23}C_6$ and Mn_7C_3 are isomorphous with their Cr analogues. In the Mn–C system, the wide interstitial C solution in γ-Mn is followed by a high-temperature carbide, also of wide homogeneity range around 25 atomic per cent. Whatever its structure, this is clearly the phase that (*a*) can be considered as the high-temperature form common to $Mn_{23}C_6$ and to Mn_3C, and (*b*) possesses a large mutual solubility with γ-Mn. It may correspond to a derivative of the cubic close-packed structure, conceivably our ψ-phase analogue; this is unconfirmed, however. This key system should actually be much better known. An x-ray high-temperature study would be rewarding: the existence of a metastable hexagonal form, 'ε'-Mn_3C, transitional to cementite seems quite likely. Furthermore, the above considerations imply that a carbon-saturated austenite may well be considered in terms already of 'a carbide' of the cubic group, with a highly defect lattice, and several of its physical property traits indeed confirm this*.

Two examples of practical implications may be given.

(*a*) In principle, it will be possible for a cubic carbide, for instance V_4C_3, to form large or continuous solid solutions at high temperatures with austenite, within the restriction set by atomic sizes. The manner of precipitation would then essentially be one of an initial carbon atom settling in the parent lattice, primarily at a dislocation or grain boundary associated with diffusion of the Cr, Mo (and other metal atoms of high carbon affinity), followed by coherent precipitation. The formation of cubic $M_{23}C_6$ in austenitic stainless steel is a case in point. At high temperatures, the carbide dissolves completely in the austenitic matrix but segregates at lower ones with a modified (but still related) lattice.

(*b*) During the formation from austenite of cementite, which still possesses a near-close-packed configuration of metal atoms like γ, there is reason to suppose that a transient hexagonal state exists. Support comes from many other cases in which the cubic lattice tends to change transitorily into one of hexagonal symmetry. The 'ε-carbide'[29b] is an example, and perhaps the most important case of this transitional type. It occurs as an intermediate stage during the decomposition of martensite and is isomorphous with Fe_3N and the carbonitride series $Fe_3(N,C)$, of which it forms the metastable end member. A characteristic is that it forms coherently from parent martensite

* The intensive recent developments of *high-strength steels* largely depend upon these and related concepts, namely the judicious blending, by thermal treatment and, in some cases, mechanical work at elevated temperatures, to direct the progress of the γ- $\rightarrow$ α- (= martensite/ferrite) $\rightarrow$ carbide transformation in such a way as to produce extreme strengths (to > 100 tons/in.^{2}) without, however, creating embrittlement. This includes processes now known as '*ausforming*' and '*mar-aging*' (a good review[45]). Basically, the essential aim is to adjust the interactions of dislocations (amongst themselves, with the interstitial solute atoms, with finely dispersed phases, stacking faults and grain and sub-grain boundaries) and to balance the various hardening mechanisms so as to yield maximum flow stress.

Whilst in *iron*, with its special phase changes, the potentialities are especially favourable, this *principle* of creating high-strength alloys is of course a general one and applies also to other metals (e.g. Ti) exhibiting phase changes with differential interstitial solubilities and lattice precipitates.

upon low-temperature tempering. The composition, although also near Fe_3C, is not rigidly bound to this (in contrast to cementite), and broad heterogeneity ranges as well as high internal stresses are manifested by pronounced x-ray line broadening. The lattice dimensions of ε-Fe_3C are $a = 2{\cdot}73$, $c = 4{\cdot}33\,kX$, similar to Fe_3N (cf. Chapter 5). Also the coherency is associated with a marked crystalline directionality of the phase on precipitation from martensite. Thus the (101) plane of ε-Fe_3C tends to be parallel to, and of almost identical spacing as, the (101) plane of the body-centred tetragonal martensite from which it forms. Only during a later stage of steel tempering (or, for the same time, at increased temperatures) is this lattice continuity and two-dimensional nature broken down. The ε-carbide then transforms to finely dispersed true cementite, at first strained and distorted; but with time, growth, re-crystallization and strain relief occurs and coherency ceases. It should be noted that the configuration of iron atoms in hexagonal ε-Fe_3C and in cementite Fe_3C is quite similar, so that only a small Fe atom movement is required for this phase change, but a larger one of C atoms. In ε-Fe_3C, as in cementite, the 12-co-ordination of the face-centred cube of *austenite* is very nearly maintained. We might regard the formation of both ε-Fe_3C and cementite as two different modes of re-establishment of the close-packed metal structures which had been destroyed during the original $\gamma \rightarrow$ martensite transformation.

Another consequence of the two-dimensional continuity between martensite and ε-Fe_3C is the setting up of considerable lattice strains at the interface, with consequent age-hardening effects, accompanied by exceptional x-ray line broadening. The composition of the ε-carbide is not rigidly tied to Fe_3C, and there exists an initial tendency of a homogeneity range towards higher carbon contents; 'Fe_2C' is a possible theoretical limit. This again is consistent with its incipient formation from martensite, no doubt preceded by carbon migration to dislocations and crystal boundaries at which it forms.

The fact of isomorphy of ε-Fe_3C with several other hexagonal metal carbides could well suggest its possible stabilization by the presence of alloying elements such as Cr, V and Nb, though only within the limits set by atomic size and electron concentration. Manganese, however, may very well stabilize it.

The *iron percarbide*, 'Fe_2C', had been reported already in Hägg's early work[28], and can be considered either as orthorhombic ($a = 9{\cdot}04$, $b = 15{\cdot}66$, $c = 7{\cdot}93\ kX$; 80 atoms per unit cell) or as hexagonal ($a' = 2a$, $c' = b$). The formula has been re-defined by Jack[29b] as $Fe_{20}C_9$. This compound, like cementite, is metastable only, decomposing upon heating into α-Fe and graphite; the percarbide, however, does so via a cementite stage. Whilst the latter can be produced as a free compound (e.g. by carburization of the nitride), it is important to note that this is impossible for the ε-carbide which, as far as is known, can only exist within its martensitic soil.

The unstable *cobalt percarbide*, Co_2C, first synthesized as a new compound by Hofer and Peebles[46], has also been prepared and examined for its crystal structure by Clarke and Jack[47]. It is isomorphous with Co_2N. The arrangement of metal atoms is here again hexagonal close-packed but distorted as from Fe_2(C,N), the basic difference being one of different interstitial positions. An interesting and close relationship between the Co_2C and TiO_2 structure has been pointed out by these authors (*Figure 4.13*), and forms a notable example of interstitial atom inversion. On the other hand, the compound Co_3C is truly isomorphous to the ε-Fe carbide and nitride.

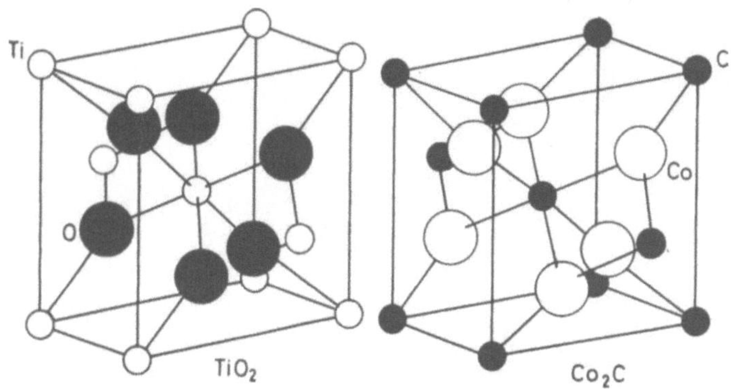

Figure 4.13. Similarity of rutile and Co_2C *structures* (After Clarke and Jack[47])

By courtesy of the Editor, *Chemistry & Industry*

In this connection, it has been shown by electron diffraction work on single crystals of α-iron[48] that, during carburization at elevated temperatures in carbon monoxide and hydrogen, a definite orientation relationship develops. The (001) plane of α-Fe lies parallel to (111) of the evolving Fe_3C plane, and the (110) of α-Fe to $(1\bar{2}1)$ of Fe_3C. Dimensionally these planes are in fact those of easiest fit. The relation may be represented as in *Figure 4.14*. However, the orientation breaks down when the film attains a certain thickness, upon carburization for extended times or for too high a temperature. This is mentioned here because it falls well into line with the idea of generation of a carbide from the body-centred cubic α-matrix. The known $\gamma \longleftrightarrow \alpha$ orientation relationship and that between the near close-packing of Fe atoms in Fe_3C, as in γ, would indicate the close crystallographic interdependence of these phases. Jack's separate evidence of the transient ε-Fe_3C, with eventual Fe_3C formation within a carbon steel, may in fact be regarded as complementary to the experiment of synthetic carburization, and it is indeed not excluded that a hexagonal state might occur also in the latter case.

Electron diffraction could further reveal definite differences in the manner of formation and in carbon atom location of these metastable carbides[49] ('carbidic states' may be a better term), namely ε-$Fe_{2-3}C$, χ-$Fe_{2-3}C$, Fe_3C, Co_2C, Co_3C, Ni_3C. Incidentally, such carbidic states can also be produced experimentally in thin metal films vaporized in carbon monoxide, quite apart from their occurrence at various stages of precipitation in bulk alloys. No doubt the presence, in traces, of oxygen and interstitials other than carbon will have a powerful controlling influence.

From another viewpoint, Fe_3C can be considered more as a substitutional 'solid solution' of γ-Fe than as a compound[50]. This seems somewhat controversial, but the C atom partial lattice in cementite is looked upon as a near relation to the C atom lattice in the saturated face-centred cube of austenite as is the metal atom lattice in Fe_3C to that in austenite, though greatly distorted. It is even postulated[50] that C atoms can *substitute* Fe in lattice positions and shown that then the interatomic distances and symmetries closely approach during the phase change.

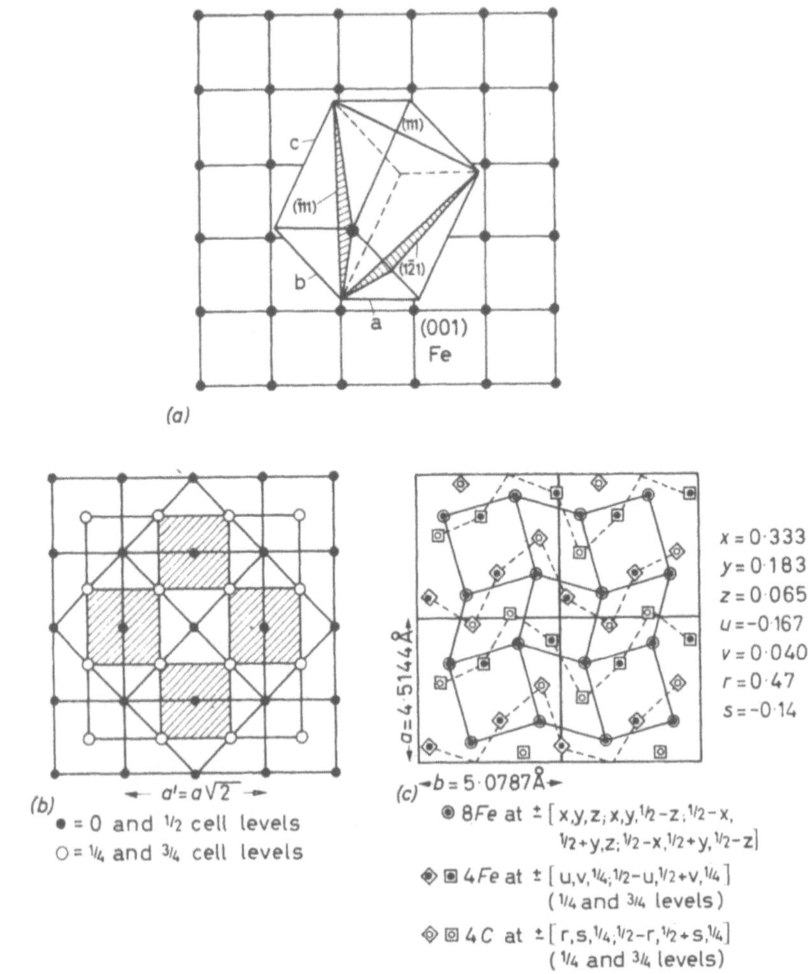

*Figure 4.14. Cementite/α-*Fe *and γ-*Fe *lattice relationships*

(*a*) Orientation of Fe_3C in α-Fe base (Trillat and Oketani[48])

(*b*) Projection of γ-Fe lattice positions onto (001) and tetragonal cell derived, with $a' = a\sqrt{2}$ and $c' = 2a$; shaded areas become Fe–C octahedral chains in Fe_3C (Lõhberg[49])

(*c*) Projection of Fe_3C lattice positions onto the (001) plane; octahedral chains: squares Fe–Fe, diamonds Fe–C (Löhberg[49])

By courtsey of the Editors, *Acta Crystallographica* and *Archiv für das Eisenhüttenwesen*

An important and comprehensive recent investigation on the structure of cementite and on its orientation relation to ferrite and ε-Fe_3C and austenite is due to Andrews[25b]. He traces the crystallographic change from BCC ferrite to cementite to a fairly simple mechanism of atom movements of

Fe atoms (with limited C diffusion), in which close-packed rows of Fe atoms in parent ferrite become 'puckered' in zig-zag fashion upon transformation, parallel to the {112} planes of ferrite. The detailed mechanism, also in relation to the observed twinning and the intermediate ε-stage, has been elaborated in a further paper[25c].

In cobalt, the structures of Co_3C and Co_2C have been confirmed[51], also in their lattice relationship with parent cobalt. They can form and decompose in the 450°–500°C region, leaving graphite in a cobalt matrix rich in stacking faults. We may see here a direct illustration of interstitial carbon rendering the close-packed cubic → hexagonal transformation in Co entirely labile, depending on lattice zones of fluctuating C content and on regions of lower or higher C content; the latter being inherited from points of decomposing carbide. Again, the coherent relationship between the cementite-type and the FCC lattice arises, as in Fe_3C and austenite, with the graphite phase produced with considerably greater readiness than in Fe–C. This may again be viewed as the retention of the ordered hexagonal carbon configuration previously innate in the supersaturated hexagonal cobalt lattice (cf. *Figure 2.7f*). The process of cobalt graphitization[52], also in connection with spheroidization (nodular graphite), likewise has a bearing on this topic. By very rapid quenching it has been possible to retain highly supersaturated solid solutions of C in cobalt and nickel, up to 1·65 and 1·85 weight per cent, respectively, and both graphite and carbide can later precipitate in metastable equilibria, in a eutectic system.

In connection with unstable iron carbides, Johnston *et al.*[53] were able to obtain a fine powder of a close-packed *hexagonal* form of Fe_2C (as distinct from the more common 'Hägg form'), by passing a $H_2 + CO$ mixture over pure iron at 240°C; it requires the exclusion of Fe_3O_4, using a hydrocarbon, otherwise the normal orthorhombic Fe_2C is formed. This finely powdered hexagonal Fe_2C has a considerable coercive force, about 800 Oersted. Such forms of iron, cobalt and nickel carbides are of special interest in catalysis[54] and in much applied industrial work. The particular structure formed depends sensitively on temperature and detailed chemical conditions during synthesis (CO or H_2 as carrier, etc.).

Another, more recently found[55] cementite-type of carbide, of variable composition, is the so-called 'χ-Fe_xC' where $x < 3$. Its formation is influenced by plastic cold deformation of the steel in which it is contained, and the composition change is also reflected by a lowering in Curie point. It is of much interest to see, as in *Figure 4.15*, how this Fe/C ratio and the Curie point vary with degree of deformation as well as with particle size of the carbide after given heat treatments; the variation can also be related to dislocation density[55].

Further investigation[56] of the nature of this iron carbide has shown that the *same* carbide can be produced by tempering and by plastic deformation. It has a lower density than Fe_3C (6·73 as against 7·68), in line with higher carbon content; as to composition, figures of x can vary from 1·7 to 2·0, depending on thermal treatment (the paper[56] also includes some electron diffraction and magnetic data for this phase). Further crystallographic relations between '$Fe_{2\cdot5}C$' and Fe_3C were given by Seneteur and Fruchart[57].

There is a useful method, using electron bombardment[58], whereby such iron carbides of varying composition can be prepared at will in the form of thin films for electron diffraction studies, by evaporating carbon to a given extent.

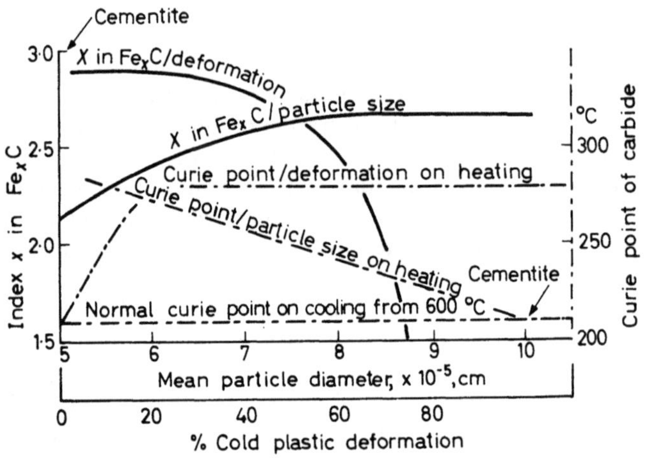

Figure 4.15. Changes in composition and Curie point of the metastable carbide 'Fe_xC' in steel as function of particle size and cold deformation (Belorus and Cherepin[55])

By courtesy of Pergamon Press

The position of *nickel* vis-à-vis carbon shows several points of interest in this connection.

(1) Although it is a strong graphitizer, that transient Ni carbide formation is still possible as has been shown.

(2) In diamond synthesis, nickel assists the graphite → diamond transformation catalytically. Here an interfacial nickel carbide is likely to play a part*, but the theory has also been put forward that the FCC Ni lattice matches the lattice planes of diamond and forms a basis for its epitaxial growth.

(3) Ni and C in iron behave similarly in many ways, being not merely austenite stabilizers but also promoting martensite and sluggish transformations.

4.5 CARBON-DISSOLVING INTERMETALLIC COMPOUNDS: α- and β-Mn STRUCTURES

There exists a further carbide variant which can play the two-fold role of either an intermetallic compound or of a carbide: the so-called χ-phase†. This has the elemental crystal structure of α-manganese, but it occurs in a number of transition metal systems not bearing Mn, e.g. Fe–Cr–Mo, Fe–Cr–W, Nb–Re, Ta–Re, Mo–Re, W–Re, Hf–Mo, etc. The compound is

* Probably in the form of cementite-like layers. It is interesting in this connection that in meteorites Ni, Fe and C are closely associated, and a constituent 'cohenite' of $(Fe,Ni)_3C$ has been reported (e.g.[39]).

† Different from the χ-Fe_xC above—a clash in nomenclature.

generally associated with the occurrence of the σ-phase, of which FeCr is a prototype. The position of Mn between Fe and Cr is significant in this connection. Different 'types' of Mn atom occur in fact in the manganese structure, which can be regarded as electronically equivalent to Fe and Cr. The χ-phase, as σ, also is known in steels[2e,3d,25d], such as those based on the Fe–Cr–Mo and Fe–Cr–W systems[25,59].

A characteristic of χ—and this is the salient point—is that *carbon* can dissolve in this structure and stabilize it. This is quite in contrast to the e-phase which does *not* dissolve appreciable C but, upon carbon addition, transforms into a proper carbide, $M_{23}C_6$. Typical compositions of pure χ are $Fe_{28}Cr_{23}W_7$ and $Fe_{36}Cr_{12}Mo_{10}$ (calculated to the 58 atoms per unit cell). However, the phase possesses, within each system, an appreciable composition area around this, i.e. a tolerance for metal interchange. It is of interest that, upon carbon addition, the metal–atom proportion within χ is shifted towards (W, Mo), at the cost of (Fe, Cr), i.e. towards the stronger carbide formers. Typical phase assemblages surrounding the χ-phase are shown in *Figure 4.16*.

The unit cubic cell of the χ-phase is shown in the Frontispiece*. There are four types of metal atom sites, distinguished already in Bradley and Preston's original α-Mn structure and preferred in the compound by the different species of atoms. A thorough structure investigation of χ-(Fe–Cr–Mo) is due to Kasper[60]. The main point here is that in the χ-lattice sufficiently large holes exist to accommodate limited amounts of carbon and other interstitials, so as to convert it to a carbide. Stoichiometrically, the χ carbide may be written as '$M_{18}C$' ($\approx Fe_8Cr_{6\cdot3}W_{3\cdot7}C$). The close association (cf. *Figure 4.16*) of the χ- and *η-carbide* (M_6C) phases should be noted, partly in mutual equilibrium, partly as alternatives during various heat treatments. This provides a further link with the η-carbide structure.

Pure manganese, as we know, has the four α, β, γ, δ allotropes stable at temperatures increasing in this order (see e.g. the Mn diagram in *Figure 4.1*). The γ and δ forms are the FCC and BCC isomorphs to austenite and ferrite, respectively, with which they form continuous solid solutions. The question arose whether *β-Mn* also has its analogues amongst either carbides or intermetallics. This is indeed the case. The β-Mn-type structure occurs for compositions, e.g. in the Fe–Cr–W–C system, adjacent to and overlapping that of α-Mn type (χ) but quenched from a higher temperature, quite in analogy with Mn metal. This phase may be considered as a high-temperature form of χ and, like the latter, shows carbide characteristics; however, the composition ranges and solution capacities for given elements differ between the two structures. We may well consider this 'π'-phase (= β-Mn type) carbide as another borderline case; it has been found that the presence of some carbon is in fact necessary to stabilize the phase and to render it retainable at room temperature. In the carbon-free systems (Fe–Cr–W and Fe–Cr–Mo), the π-phase could not so far be confirmed by quenching,

* *Atomic co-ordinates* for α-Mn—similar for χ phase (after Öhman[61]):
Cubic, 58 metal atoms per unit cell (C atoms interstitial but undetermined)
2 Mn I (at c = (0,0,0), ($\frac{1}{2},\frac{1}{2},\frac{1}{2}$) 8 Mn II (at c = (xxx; $x\bar{x}\bar{x}$, etc.); $x_{II} = 0{\cdot}317$
24 Mn III (at g = (xxz; $\bar{x}\bar{x}z$; $\bar{x}x\bar{z}$; $x\bar{x}z$, etc.) $x_{III} = 0{\cdot}356$; $z_{III} = 0{\cdot}042$
24 Mn IV (at g: $x_{IV} = 0{\cdot}089$; $z_{IV} = 0{\cdot}278$)

whilst it could upon the addition of carbon. Typical compositions for pure π range from about $M_{21}C$ to $M_{20}C$, i.e. they are less rich in carbon than χ, and in many instances a decomposition reaction $\pi \rightarrow \chi +$ ferrite could be observed. *Figure 4.16* also includes characteristic π compositions.

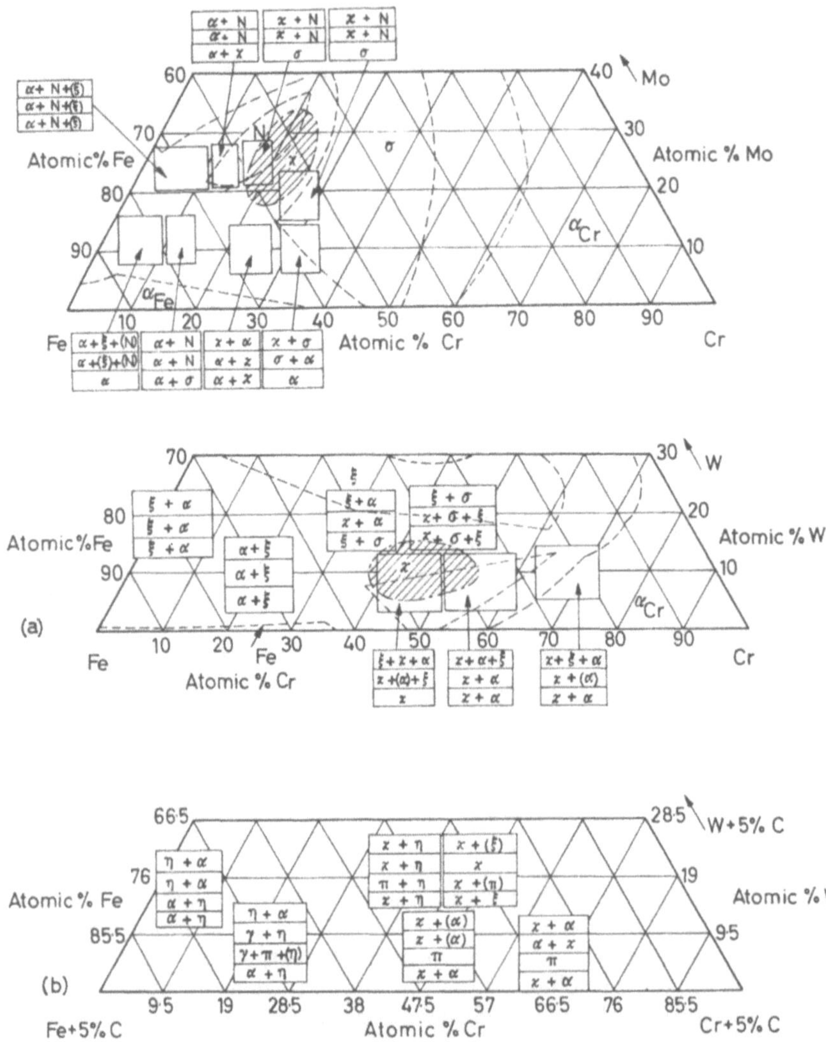

Figure 4.16. Some detailed equilibria involving the χ phase (Goldschmidt[26])

(*a*) Phase constitutions of Fe–Cr–Mo and Fe–Cr–W alloys as quenched from high temperatures

Phases in framed blocks correspond, reading from top to bottom, to 1,000, 1,200 and 1,400°C quenches, superimposed on thin outline of a 600°C diagram derived separately. The approximate position of the χ-phase field is shown shaded

(*b*) Typical constitutions of Fe–Cr–W–C alloys at 5 at. per cent C level

Phases in framed blocks correspond to temperature as in (*a*) and to 600°C anneals

By courtesy of the Editor, *Metallurgia*

The unit cell of π (β-Mn), also cubic, with 20 atoms, is shown in *Figure 4.17*, with co-ordinates. The spacing of this phase can vary from 6·275 to 6·39 *kX*, depending on the Fe/Cr/W/C ratio and, in any one alloy, on heat treatments. It is known from early structure work[1d,62,63] that two distinct

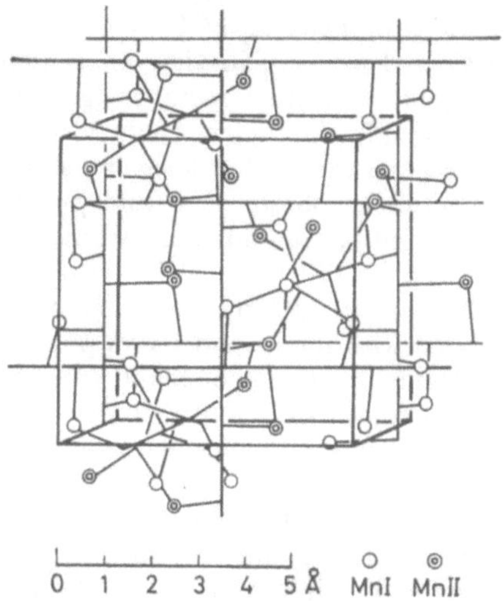

*Figure 4.17. The π phase structure, β-*Mn *type showing cubic unit cell and screw axes with the two types of metal atom* (After Preston[62])

Atomic co-ordinates (here for β-Mn; for π phase similar though expanded: Mn_I and Mn_{II} positions now occupied by different species):
Cubic, 20 metal atoms per unit cell; C interstitial but undetermined

8 Mn I at (c) = (xxx; $\frac{1}{2} + x, \frac{1}{2} - x, x$; $\frac{3}{4} - x, \frac{3}{4} - x, \frac{3}{4} - x$; $\frac{1}{4} - x, \frac{3}{4} + x, \frac{1}{4} + x$); $x_I = 0{\cdot}061$

12 Mn II at (d) = ($\frac{3}{8}, x, \frac{3}{4} + x$; $\frac{7}{8}, \frac{1}{2} + x$; $\frac{1}{4} - x$; $\frac{5}{8}$; $\frac{1}{2} - x, \frac{3}{4} - x$; $\frac{1}{8}, x, \frac{1}{4} + x$) $x_{II} = 0{\cdot}206$

By courtsey of the Editor, *Philosophical Magazine*

Mn atom sites exist in β-Mn, 12 and 8 per unit cell, occupied by two 'kinds' of Mn atom, differing in their electronic states and bond distances. In the π phase these will provide sites predisposed for filling by different species of atoms, e.g. the more spacious sites would preferentially take W, the others Fe, whilst Cr atoms could be shared but would substitute Fe in preference. The carbon positions in π, as those in χ, have not yet been ascertained, and indeed the detailed structure requires determining, preferably by neutron diffraction in conjunction with x-ray diffraction. The prevalence of 12 and 8 Mn atom sites in β-Mn is directly reflected in the empirical compositions found.

It may be recalled that the Hume-Rothery electron–atom ratio of 3/2 is characteristic of intermetallic compounds of the β-Mn type of structure (as well as of the BCC structure). The π-phase might therefore possibly be considered as an electron compound; it is, however, difficult to ascribe an effective electron valency to each atom (Fe, Cr, W, C); this is variable for the transition metal atoms, although an approximate total number of 30 electrons per unit cell might be assumed, as in β-Mn itself.

Constitutionally, the close link between π and η-carbide should again be noted, and the fact that π can be parent to precipitating either η or χ (see *Figure 4.16*). There are particular implications concerning deviations in high-speed steels and other tool steels. The principal point here is that both the χ- and π-phases find their place as true carbides, though borderline, within assemblages in which austenite, ferrite, η-carbide, σ-phase and Laves phase are closely associated, and that they can play the intermediary role both of a carbide and of an intermetallic compound capable of dissolving carbon. Indeed, the link to elemental manganese itself is here significant.

It implies that fundamentally we have a C solid solution in an elemental structure*. It is also reflected in the similar solution behaviour in Mn itself (see Mn–C diagram, *Figure 4.1*). It may be recalled that the η-carbide, too, can take the form of an interstitial-free intermetallic compound. In principle, therefore, we have a whole new class of phases 'bridging the gap' to primary solutions of FCC or BCC types. The hexagonal ε-carbide may also be considered as such, in view of its isomorphy and solution with the many hexagonal transition metals, e.g. Co and Re; the latter is particularly interesting in the present context, as it is homologous with Mn, can dissolve appreciable carbon but forms no stable carbide.

The above phases of α- and β-Mn structure are not only of academic interest but play an integral part in alloy steel transformations. In particular, they can be instrumental in creating high-strength materials, though also embrittlement (β-Mn in burnt W steels, by quenching[3d]) and are closely associated with the σ- and Laves phase problem, e.g. χ in stainless steels[64,65]. Analogues are also to be found among nitrides (Chapter 5).

Such considerations of carbide → intermetallic compound transition likewise enter into problems like *decarburization*, again a borderline subject to be discussed below.

A further example for the transitional aspect between carbides and intermetallics is the structure relationship between the WC and NiAs phases and their various isomorphs, as discussed earlier.

4.6 FURTHER DATA ON SINGLE METAL–CARBON SYSTEMS AND CARBIDES

The single-transition metal–carbon systems require, within this general survey, little supplementation to what has already been said and what is clear from *Figure 4.1* and *Table 4.1*. However, some features in individual systems will be discussed in the following notes (see also Bibliography).

Ti–C[66–80]

(*a*) Fundamentally interesting is the demonstration[80] *by thin-film electron microscopy*, of dislocation loops in TiC, and their development under deformation at high temperatures, quite in line with the still essentially metallic nature of FCC TiC.

(*b*) A still controversial report[81] of a titanium *di*carbide, TiC_2 (of simple cubic structure with $a = 3{\cdot}13$ Å), would, if confirmed, affect our systematic view of carbides and suggest other intermediary stages in Group IV carbides (ZrC_2, HfC_2?), between *M*C and graphite: in possible linkage to the Group III/Rare Earth carbides (e.g. YC_2), and indeed to ThC_2. A homopolar bonding would then be probable. However, the reality of such 'TiC_2' is questioned.

Zr–C[74,82–90]

The occurrence of a phase transformation in ZrC (and TaC) at high temperatures has been suggested from emission and resistance studies[86] above 2,000°C and exhibiting a strong hysteresis effect. The structure of this new H.T. form, if real, is not yet known; again, it has the wider implication that the characteristics, especially polymorphism, are by no means completely established even in the simpler carbides. (An analogous high-temperature phase change seems possible in TiC, and a search for this worth-while.)

Hf–C[83a,91–96]

(*a*) HfC is by no means as similar in properties to ZrC as might be expected. It has proved an interesting constituent in hard metals, is less brittle (i.e. tougher) than ZrC and

* Unless we prefer to look upon α- and β-Mn as basically already intermetallic compounds, namely structures between two species of Mn atom.

TiC, and in this respect already resembles TaC, its Group V neighbour. It also gains interest as a dispersion-strengthening agent for refractory alloys, along with other cubic carbides.

(*b*) Recent work[83a] on the Hf–C system reaffirmed HfC as the only intermediate phase, with a defect lattice range HfC_{1-x} from 48 to 34 at. per cent at 2,200°C, and 49·3–36 at 3,150°C. The melting point maximum was 3,830°C at 47·5 per cent C. Continuous lattice expansion occurs with carbon site filling, from 4·6190 to 4·6430 Å in the extreme limits.

V–C[23b,97–103]

(*a*) The lattice spacing changes within the VC homogeneity region are as shown in *Figure 4.21* (p. 145) together with those in TiC, ZrC, HfC, NbC and TaC.

(*b*) A V–VC diagram[99], incorporated in *Figure 4.1*, contains these principal points: a primary solubility to $VC_{0.09}$ at the eutectic; a eutectic with V_2C at 1,630°C; a peritectic at 2,165°C between $VC_{0.56}$ and $VC_{0.60}$; a further peritectic of 2,650°C between $VC_{0.85}$ and at least $VC_{1.2}$; a 'V_2C' homogeneity range from $VC_{0.47}$ to $VC_{0.50}$ at 1,100°C; a 'VC' homogeneity range from $VC_{0.73}$ to $VC_{0.87}$ at 1,100°C; and an indication of a new lower-temperature phase (below 1,344°C) at $VC_{0.50}/VC_{0.74}$. Furthermore[99], the *melting point* of vanadium *metal* was redetermined as 1,888°C and a preferential evaporation of V reported from all V–C compositions.

The hexagonal L'3 type structure for V_2C has recently[103] been reaffirmed by neutron diffraction.

Nb–C[104–12]

A more recent phase diagram Nb–C[107] suggests that the Nb_2C carbide has an extremely wide homogeneity range: from 4 to 50 at per cent C, below 1,400°C, and the following invariant reactions:

	at at. per cent C and	°C
Eutectic Liq. → Nb + Nb_2C	13·5	2,430
Peritectic Liq. + Nb_2C	33·3	3,080
Eutectic Liq. → NbC + graphite	60·5	3,300

This surprising result would conflict with previous ideas and requires confirmation.

Ta–C[69,86,110,113–20]

The *microhardness* of carbides Ta_2C and TaC (as of TiC) within their homogeneity ranges[91,121] shows that when the linear increase of microhardness with carbon is extrapolated backwards to the pure *metal*, the value for the latter is attained. This is taken to confirm, as a basic point, that the type of bond (metallic) is similar to that in the metal. It contrasts sharply to the case of *nitrides* where a fair degree of *ionic bond* appears already and the gradient in microhardness with nitrogen content becomes steeper.

Very recent neutron diffraction work[103] on Ta_2C has shown this to have a hexagonal structure of CdI_2 (anti-type), i.e. deviating slightly from that previously known.

Cr–C[1e,16b,100,122–9]

In addition to the carbides $Cr_{23}C_6$, Cr_7C_3 and Cr_3C_2 already referred to, a compound 'CrC' had been suggested in early work. However, this could not be confirmed in an extensive special investigation[124] (see diagram included in *Figure 4.1*). Nevertheless, the existence of a chromium monocarbide at high temperatures, but unquenchable, is still possible; for thermodynamic properties of Cr carbides, see also [125].

Mo–C[130–42]

(*a*) For the 'MoC' problem, see p. 140.

(*b*) Of particular interest is the ability[139] to retain the *cubic* MoC form under high pressures (40 to 70 kilobars), on cooling from 1,800–2,500°C.

(*c*) The composition 'Mo_3C_2' rather than 'MoC' has recently been confirmed[140].

W–C[1f,88,113,144–50]

The papers given here are a few representatives of numerous publications.

A cubic high-temperature form[151] of W_2C having a FCC (NaCl-type) structure and a finite homogeneity range has recently been affirmed. This was confirmed[147] with the modification that the phase is carbon-richer (≈ 37·5 per cent C) and decomposes eutectoidally to W_2C + WC below 2,525°C. Several other new high-temperature* features in the W–C diagram have also been introduced[147].

* In the light of the Mo–C system and the 'MoC' problem, as discussed, this W–C behaviour is of special interest, as it represents a further, though more stabilized, stage in phase relations and similar stacking sequences,

Mn–C[16,61b,152–6]

Numerous contradictory results reported are consistent with highly metastable conditions in this system.

Re–C[157,158]

The question as to whether a stable rhenium carbide exists is still controversial.

Fe–C

A few papers on the more basic and structural side[25,26,28–30,32,35–7,54,57,58,61c,159–74] are representative of innumerable further ones; an appraisal has been given recently[175].

Co–C[46,47,176–81]

The cementite structure of Co_3C has been confirmed[180], with $a = 4{\cdot}483$, $b = 5{\cdot}033$, $c = 6{\cdot}731$ Å; in Co_2C, as Fe_2C, $a = 2{\cdot}910$, $b = 4{\cdot}469$, $c = 4{\cdot}426$ Å. Both decompose above 450/480°C → Co + C.

Ni–C[153,182–93]

A metastable nickel carbide eutectic[193] produced at high temperatures, proved to be readily decomposed into Ni + graphite on annealing.

Platinum metals–C

Pt[194–5], Ru[196], Ru,Os[197], Os, Ir[158], Pt, Rh, Pd, Os, Ir, Ru[198], Pt, Pd, Rh, Ru[199]; cf. p. 140 and *Figure 4.1*. (The catalytic effects between the Pt metals and carbon have naturally also been the subject of extensive investigations in themselves, indirectly bearing on phase relationships.)

Some useful correlatory data have been given on the solidus and other melting temperatures in Pt metal and other single metal–carbon systems by Nadler and Kempter[198]. The following *eutectic minima* (°C) were obtained (rearranged to accord with the Periodic Table):

Nb–C* 2,328	Mo–C 2,210	Tc–C ?	Ru–C 1,942	Rh–C 1,694	Pd–C 1,504
Ta–C† 2,902	W–C 2,732	Re–C 2,486	Os–C 2,732	Ir–C 2,296	Pt–C 1,736

* Also Nb_2C peritectic at 3,080°C, NbC–C eutectic at 3,220°C
† Also Ta_2C peritectic at 3,500°C, TaC–C eutectic at 3,710°C

This shows a clear and continuous trend along Group and Period, and it is of particular interest to compare the data with those of the corresponding melting points of the metals and carbides (*Figure 4.1* and *Table 4.1* where some newer, revised[198] melting point values for NbC, Mo_2C, WC have been inserted).

For the Pt group metals this trend of melting-point depressions by carbon is remarkable. The following are the eutectic temperatures related to the melting points (°C) of the parent metals[198]:

$$\text{2nd}\quad\left\{\begin{array}{l}\text{Ru: } 2{,}310 \rightarrow 1{,}942, \Delta = -368\\ \text{Rh: } 1{,}960 \rightarrow 1{,}694, \Delta = -266\\ \text{Pd: } 1{,}552 \rightarrow 1{,}504, \Delta = -48\end{array}\right.$$

Long Period

$$\text{3rd}\quad\left\{\begin{array}{l}\text{Os: } 3{,}050 \rightarrow 2{,}732, \Delta = -318\\ \text{Ir: } 2{,}443 \rightarrow 2{,}296, \Delta = -147\\ \text{Pt: } 1{,}769 \rightarrow 1{,}736, \Delta = -33\end{array}\right.$$

This decrease along the periods may be looked upon as an expression, extraneous to the area of carbide stability, of the continued decrease in metal–carbon bond strength. The same occurs, as we know, for Fe → Co → Ni in the 1st Long Period of the same Group (VIII), yet there still with the metastable but weakening carbide appearance.

Incidentally, controversy has arisen concerning the existence or otherwise of carbides 'RuC' and 'OsC': Kempter and Nadler[200] affirming it and, after Jeantet and Knapton's[197] negation, reaffirming the phase (Kempter[196]), at least metastably, as of WC type structure.

Actinides–C

Th–C[201–9]

The structure of ThC_2, including carbon location, has also been studied[210], using neutron and x-ray diffraction. A monoclinic → cubic transformation in ThC_2 at 1,415°C has been demonstrated, the cubic form being unquenchable.

U–C

The examples[208,210–33] are representative of the very extensive literature.

Np–C[208,234], Pu–C[235–40]
(A recent appraisal is given in[175].)

(Sc,Y and Rare Earth)–C
Sc–C[241–4] Y–C[243,245,246] (R.E.)–C[243,245,247–51]

4.7 CARBIDES BORDERING THE TRANSITION-METAL BLOCK

Referring once more to the systematic listing of interstitial carbides, as shown e.g. in *Figure 4.2*, most of these are now well understood and require little further comment. Some of the 'outer-fringe' cases delineating the block of transition-metal carbides may, however, be briefly considered.

On three sides of the block in the Periodic Table we have: (1) the Group III and Rare Earth carbides; (2) the Actinide carbides, e.g. of uranium; (3) the Group VI–VII carbides, verging on the platinum metals. The fourth side, towards the lighter elements of the 2nd Short Period, can be excluded, for here the border is abrupt*: aluminium carbide, Al_4C_3, and silicon carbide, SiC, are in no sense interstitial and in most respects non-metallic. We might, in fact, regard this boundary as that towards the 'inter-interstitial' compounds in the sense discussed in Chapter 2. SiC and B_4C, of course, stand out by their high bond energy, hardness and refractoriness; some basic connection between these non-metallic carbides and TiC, and the other transition metal carbides has to be emphasized (e.g. the link TiC $\longleftrightarrow$ Al_3C_4). This is also reflected in their applications, e.g. in ceramics.

We shall deal with the fringe carbides in the order (1) Group III and Rare Earths; (2) Actinides; (3) Groups VI–VII.

4.7.1. Group III and Rare Earth Carbides

The Group III carbides, of scandium, yttrium and the Rare Earths, are incompletely known. Scandium carbide, ScC, and a number of lanthanide carbides, e.g. LaC_2, are established, though in the latter the extent of metallic bonding is not certain. In scandium carbide, which is still remarkably similar to its neighbour TiC, metallic bonding is beyond doubt, however.

Scandium carbide, ScC, has been synthesized[244] from the oxide, Sc_2O_3. It also has the NaCl-type structure. Recalling the position of Sc as the 'first transition metal' preceding Ti, this falls well into our structural system. ScC has the spacing 4·50 *kX* against 4·32 *kX* for TiC, with which it forms a (probably continuous) series of solid solutions. Much of the carbon can also be replaced by oxygen and/or nitrogen to form a Sc (C, O) or a Sc (C, N) series of solid solutions. The kinship to titanium carbonitrides and Ti (C, N, O) oxides is thus strong. The transition from ScC to CaO (alkaline earths adjoining Group III, on the other side), which of course is isomorphous, and to the broader series Ti(C,N,O) $\rightarrow$ Sc(C,N,O) $\rightarrow$ CaO, SrO, BaO would seem of great interest, both fundamentally, to observe the transition in bonding, and from technological aspects.

An analogous yttrium carbide, 'YC', is not yet known though it is possible. However, the known dicarbide, YC_2, forms an isomorphous family with

* In interesting and strong contrast to the other three sides, where the change is a much more gradual one.

all rare-earth carbides, CeC_2, LaC_2, PrC_2, etc. The structure is face-centred tetragonal (Stackelberg[248]) and has been shown, using neutron and x-ray diffraction, as being of the *calcium* carbide, CaC_2, type[252]. The structures can also be interpreted as body-centred tetragonal, of sides (for LaC_2) $a = 3{\cdot}934$, $c = 6{\cdot}572$ Å (2 mol. per unit cell), whilst La_2C_3 (8 mol. per unit cell) is body-centred cubic, $a = 8{\cdot}81$ Å, like Pu_2C_3 and related to U_2C_3. Characteristic throughout is the dipole grouping of 2 C atoms, typical of the calcium carbide structure. Both the di- and sesquicarbides are closely related, and a partial or complete homogeneity range between the above two structures is feasible. It is of interest to note that LaC_2 and the other rare-earth carbides are metallic and conductive—quite in contrast to the non-conducting and ionic but isomorphous CaC_2—which should be attributed io the additional valency electron.

Amongst the metal-richer rare-earth carbides, CeC is established as being of NaCl structure[247] and cube side 5·130 Å. Cerium also forms Ce_2C_3 and CeC_2 of Stackelberg's structure. New rare-earth carbides of a type M_3C are claimed[251,253].

The rare-earth carbides are high melting (m.p. > 2,000°C; for a useful survey, see[241]). A *hexagonal* form of ScC (as against the cubic one) could not be confirmed in Nowotny's investigations which established as beyond doubt the cubic NaCl structure (linking ScC $\longleftrightarrow$ TiC). It would, however, accord with a hexagonal cerium carbide reported[249]. An allotropic change seems quite possible but has not been proven.

Surveying the Group III carbides, it is of interest to observe the following series:

$B_4C \rightarrow Al_4C_3 \rightarrow ScC \rightarrow YC_2 \rightarrow LaC_2$ (and other Rare Earth carbides, MC_2)

This progression is accompanied by a decrease in radius ratio r_C/r_M in this order and by a systematic structural change. *Table 4·5* summarizes details available on the Group III and Rare Earth carbides.

Special points of interest are the simple and systematic changes in types and structures of the carbides and their distinctive link-up with other carbide groups already referred to. The cubic close-packed (NaCl-type) rare-earth carbides of composition M_3C retain their isomorphy with ZrC, HfC and the actinide carbides, but with a high degree of lattice defection of carbon; M_2C_3 and MC_2 with the actinide carbides on the one side, and the alkaline-earth carbides on the other. The lattice spacings of the metal-rich (Rare Earth)$_3$C series (which begins only at samarium, probably owing to too large a metal atom size in La, Ce, Pr and Nd), show a perfectly linear decrease with atomic number within the Rare Earth Group. The spacing changes in the (R.E.)$_2C_3$ and (R.E.)C_2 series also follow a linear course, but with some anomalies connected with valency changes. References on some further physical, etc. data of the rare-earth carbides are given in the Bibliography.

4.7.2. Actinon Carbides

The carbides of actinium, thorium, protactinium, uranium and the transuranic elements all show metallic bonding. It is worth devoting a little more space to the uranium carbides, first because of their practical

Table 4.5

Group III and Rare Earth carbides (Chiefly from Spedding and Daane[251] and Gschneidner[250])

System	*Compound*	*Structure, spacings* (Å), *properties*	*Remarks*
Sc–C	Sc_2C, Sc_4C_3	reported but unconfirmed	
	ScC	hexagonal, a = 5·46, c = 10·24; D = (density) 3·59); cubic, B1-type, a = 4·50 now established	
Y–C	Y_3C	cubic, Bl, a = 5·102; D = 3·41	
	Y_2C_3	not determined but different from 'Pu_2C_3 type'	
	YC_2	BC tetragonal, CaC_2-type; a = 3·664, c = 6·169; D = 4·528	
La–C	La_2C_3	H.R. (= homogeneity range) extensive; BC cubic; a = 8·8185–8·8034; decomp. peritectoidally at 1,415°C; H.R.	
	LaC_2	extensive; two forms: below 1,750°C, CaC_2-type (BC tetrag.), a = 3·934, c = 6·572; above 1,750°C, FC cubic (FeS_2-type), a = 6·0; D = 5·32; melts at 2,356°C	La–C phase diagram[245]
Ce–C	CeC	cubic, Bl; a = 5·130	*Note:* Structure recurs from Sm onwards at M_3C instead of MC (NaCl-type lattice with 3:1 C vacancies)
	Ce_2C_3	BCC, Pu_2C_3-type; a = 8·4476	
	CeC_2	CaC_2-type (BC tetrag.); a = 3·88, c = 649; D = 5·59	
Pr–C	Pr_2C_3	as Ce_2C_3; a = 8·5731–8·6072; H.R.; D = 6·621	
	PrC_2	as CeC_2; a = 3·855, c = 6·434; D = 5·728	
Nd–C	Nd_2C_3	as Ce_2C_3; a = 8·5207–8·5478; H.R.; D = 6·902	
	NdC_2	as CeC_2; a = 3·823, c = 6·405; D = 5·970	
Sm–C	Sm_3C	cubic, Bl; a = 5·172	*Note:* Beginning of M_3C-type series
	Sm_2C_3	as Ce_2C_3; a = 8·3989–8·4257; H.R.; D = 7·477	
	SmC_2	as CeC_2; a = 3·76, c = 6·31; D = 6·48	
Eu–C	—	—	
Gd–C	Gd_3C	as Sm_3C; a = 5·126	
	Gd_2C_3	as Ce_2C_3; a = 8·3221–8·3407; H.R.; D = 8·024	
	GdC_2	as CeC_2; a = 3·72, c = 6·27; D = 6·93	
Tb–C	Tb_3C	as Sm_3C; a = 5·107	
	Tb_2C_3	as Ce_2C_3; a = 8·2434–8·2617; H.R.; D = 8·335	
	TbC_2	as CeC_2; a = 3·690, c = 6·217; D = 7·176	
Dy–C	Dy_3C	as Sm_3C; a = 5·079	
	Dy_2C_3	as Ce_2C_3; a = 8·198	
	DyC_2	as CeC_2; a = 3·669, c = 6·176; D = 7·45	*Note:* Ceasing of H.R. in M_2C_3
Ho–C	Ho_3C	as Sm_3C; a = 5·061	
	Ho_2C_3	as Ce_2C_3; a = 8·176	
	HoC_2	as CeC_2; a = 3·643, c = 6·139; D = 7·701	
Er–C	Er_3C	as Sm_3C; a = 5·034	
	ErC_2	as CeC_2; a = 3·620, c = 6·094; D = 7·954	*Note:* Ceasing of M_2C_3-type series
Tm–C	Tm_3C	as Sm_3C; a = 5·016	
	TmC_2	as CeC_2; a = 3·600, c = 6·047; D = 8·175	
Yb–C	Yb_3C	as Sm_3C; a = 4·993	
	YbC_2	as CeC_2; a = 3·637, c = 6·109; D = 8·097	
Lu–C	Lu_3C	as Sm_3C; a = 4·965	
	LuC_2	as CeC_2; a = 3·563, c = 5·964; D = 8·728	

importance in atomic reactors, second because they well illustrate several points of principle (cf. the U–C system, *Figure 4.1*).

The uranium carbides form a simple structural sequence which, historically, was first identified and studied almost entirely by x-ray analysis, within the frame of the 'Manhattan Project' (e.g. Rundle *et al.*[217]). Simultaneously they were discovered by Litz *et al.*[219] and, incidentally, in unpublished work by the author (1946). The carbides now established are UC, of sodium chloride-type structure, but with some homogeneity range in the sense of carbon deficiency *and* excess; and UC_2, of a body-centred tetragonal structure (CaC_2-type). Between these two, a lower-temperature form, 'U_2C_3', exists, stable below about 2,000°C. There is, above this, the remarkable effect of an apparently continuous solid solution range between UC and UC_2 despite their non-isomorphy, with a continuous structural distortion[216] through carbon insertion. (A similar situation arises in the Th–C system.) The U_2C_3 phase then evolves peritectoidally. U_2C_3 was found[218] a BC cubic structure of enlarged unit cell, linking with that of UC_2: the latter can indeed be interpreted as having the geometry of a FC or BC tetragonal unit cell (*Figure 4.18*) elongated through carbon entry, and similar to that met in the early rare-earth and Group II and III dicarbides.

The adjustments in lattice geometry to perform this transformation consist essentially of a simple shear displacement of U atoms[221,228]. It is *martensitic* in type, and the analogy to Fe–C (steel) martensites is emphasized by the occurrence of the BCC (γ-U) and the carbon-stabilized FCC form of uranium.

It is further of interest that, at 1,700°C, UC can dissolve up to 27 mol. per cent U_2C_3 and, conversely, UC_2 up to 32 mol. per cent U_2C_3, both with marked spacing changes[218]. However, no solubility in U_2C_3 was found of either C or U.

Uranium carbides possess special practical importance in their uses as nuclear reactor fuel. Pure U metal undergoes two allotropic changes (α- $\rightarrow$ β- $\rightarrow$ γ-*U*) and, owing to the anisotropy of α- and β-, uranium shows anomalous thermal expansion, which causes difficulties. In contrast, uranium monocarbide offers the advantage of isotropy inherent in its cubic structure and of undergoing no allotropic change (except for carbon excess as discussed). This advantage is shared with the oxide, UO_2, which is similarly used, but with the added attraction for the carbide of the higher concentration of active uranium. Furthermore, the carbide is highly refractory (m.p. 2,600°C); a difficulty, however, lies in its brittleness and lack of thermal shock resistance, and this has been the subject of much specialized research, e.g. a Harwell Symposium[208]. Another advantage implicit in the cubic (NaCl-type) structure is that the carbide is intersoluble with several of its carbide isomorphs of Groups IV–VI which are capable of rendering it more stable. This applies, for instance, to ZrC, HfC and TaC as well as, within limits, to the hexagonal tungsten and molybdenum carbides. Similarly, plutonium and uranium carbides are intersoluble in their respective isomorphous pairs (cf. *Figure 4.3*); relevant changes in lattice parameter will be shown below. Of particular interest in this series is again the structural transition to the hexagonal state, with the liability to stacking-fault formation and hardening.

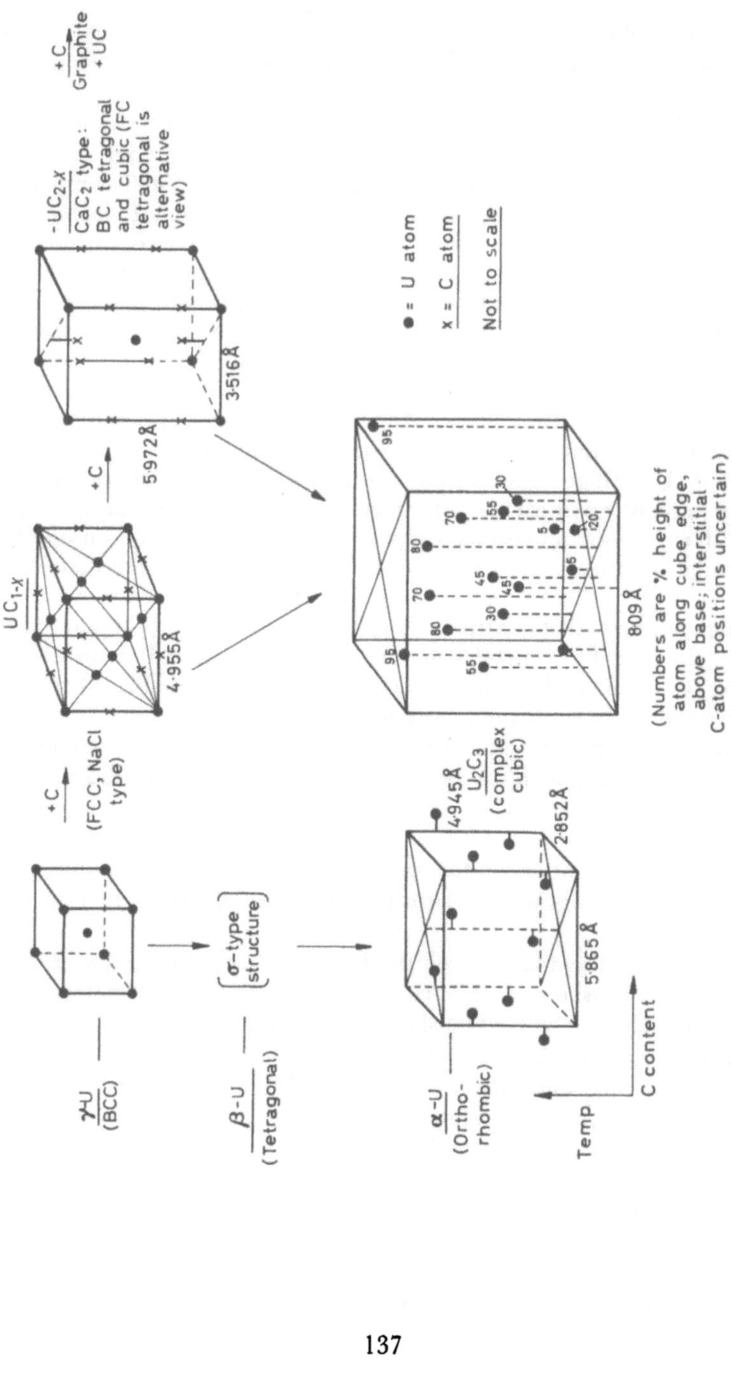

Figure 4.18. Structural progressions in the U–C *system, showing transitions and non-stoichiometry*

The non-stoichiometry of UC, in the direction UC_{1-x} $(0 < x < 1)$, has been found in early work by the writer, later confirmed[218]. Our observed lattice parameters ranged from 4·93–4·95 *kX*, depending on carbon content. There is also a considerable nitrogen and oxygen solubility in UC. The more general feature of the other carbides (Cr, Mo and W) of Group VI to tend towards carbon deficiency as from the 50/50 atomic composition, either in form of non-stoichiometry or of adopting a new structure, thus also still applies to UC.

The uranium dicarbide, UC_2, is of interest because of its isomorphy with calcium carbide, and indeed its chemical and physical behaviour shows some similarities. This is perhaps at the borderline of 'interstitial compounds', but the characteristic insertion into the body- or face-centred cubic metal structure of a carbon dipole would still justify this term. Indeed, the transition from the NaCl structure of UC, bearing single carbon atoms (*Figure 4.18*) to that of the dipole seems of interest; the process appears to be continued[217] by the high-temperature solubility for some further carbon in UC_2, to decompose into UC + graphite at lower temperatures. The changes are illustrated in *Figure 4.18.* This figure also includes, for completeness, the 3 polymorphs of U metal (α, β, γ). Although the lower-temperature β- and α-U seem less relevant, they should be noted on two scores: (*a*) we might regard the U-carbide transformations discussed as the descendants (or 'remanifestations' upon carbon addition) of the $\gamma \rightarrow \beta \rightarrow \alpha$ transformations in U, with a common underlying cause in electron structure; (*b*) noting that β-U has the structure of the σ-phase, so familiar amongst intermetallics in the earlier periods (FeCr), etc., we see that there it *also* transforms from a higher-temperature BCC lattice, as here γ-U; and also, upon carbon addition, that the FCC metal structure becomes stabilized, as in the First Long Period. This makes the re-emergence of martensite-like phenomena in this context, as in steel, even more interesting.

The wide homogeneity range of UC_2 is accompanied by an appreciable variation in axial ratio with carbon content, in the sense of a substraction-type lattice UC_{2-x}; this, however, was reported[220] to extend to compositions beyond UC and seems to conflict with other data, unless indeed there can be a metastable retention of the CaC_2-type structure to such metal-rich regions. These findings do not seem to have been confirmed by later work though.

The (cubic $\rightarrow$ tetragonal) transformation in UC_2, of diffusionless (martensitic) type[228], allows for some variation in carbon content in terms of a defect lattice, $UC_2 \rightarrow UC$. The shear transformation of the U atom base lattice necessary to perform the change from the UC_2 (BC tetragonal) to the U_2C_3 (BC cubic) unit cells has been shown[221], in a simple manner, to be truly diffusionless, but the C atom content of the unit cell (the actual stoichiometry) requires diffusion and is controlled by the space available in the U framework for interstitial insertion. Conversely, we might say that the U metal shear transformation is diffusion-controlled by the C atoms.

Variations in the lattice spacing of cubic, non-stoichiometric UC with excess C content[254] occur within the limits 4·9520–4·9600 Å. At the *low*-C end of the U–C system, the (α-, β-, γ-)U $\longleftrightarrow$ UC two-phase equilibria are of particular interest[231]. The eutectic point is at 1,117°C and 0·98 atomic per cent C; the eutectoids ($\gamma \rightarrow \beta$-U)—UC at 772°C, ($\beta \rightarrow \alpha$-U)—UC at 666°C, both temperatures being slightly raised from the change points in the pure metal. The C solubility in γ-U falls from ~ 0·30–0·09 atomic per cent between the eutectic and eutectoid temperatures; that in β- and α-U is negligible. (These data have been allowed for, as near as possible, in the U–C diagram of *Figure 4.1*). The non-stoichiometry of UC implies the feasibility of *quenching vacancies*, and these have in fact been manifested by an increase in electrical resistivity with quenching rate[255]. These vacancies can then be annealed out again at lower temperatures. A point of special interest arose from a high-temperature x-ray investigation[229] on the UC–UC_2 system, directly proving the transformation in UC_2 from tetragonal to cubic, at 1,820°C. This is shown to be from the CaC_2- to the CaF_2-type structure, the latter becoming increasingly intersoluble with UC. The adoption of the fluorspar structure is also significant from the aspect of oxides (Chapter 8) in that UO_2, of course, possesses this, and full lattice interchange C $\longleftrightarrow$ O seems likely, with subsequent decomposition and CO evolution at very high temperatures. The carbon loss would also result in a reversal to UC.

A summary of data of the uranium carbides is included in *Table 4.1*.

As to the *thorium*-carbon system, *Figure 4.1* shows a comparison with U–C. ThC and ThC_2 are isostructural with UC and UC_2 [201–2]. A common feature is also the apparent complete ThC–ThC_2 intersolubility at high temperatures, with immiscibility at lower; no U_2C_3 analogue appears. The ThC spacing varies from 5·34 to 5·29 Å, added interstitial carbon expanding the lattice. However, later work[256] contradicts the CaC_2-type structure of ThC_2 and reports a monoclinic unit cell (a = 6·53, b = 4·24, c = 6·56 Å; β = 104°; 4 mol./cell). Also it should be said that the Th–C diagram[201] in *Figure 4.1* is in some doubt, especially as Th and ThC are not isomorphous.

Neptunium and plutonium also, like U, form the monocarbides NpC and PuC of NaCl-structure, whilst Pu_2C_3 is isomorphous with U_2C_3 (for lattice dimensions, etc., see *Table 4.1*).

It is worth while again to note the continuity of the actinon-type carbides with (1) the transition-metal carbides of Groups IV–VI (largely expressed by isomorphy, continuous solid solution and defect lattice formation) but also with (2) the Group III and rare-earth carbides, as manifested by the dicarbide structures and properties. It is in the latter that the direct C–C bond makes its appearance, although as yet in the dipole form of the CaC_2 type in which the metallic character is maintained. In the silicides and borides (Chapters 6 and 7) this evolves much further into direct bonds between the non-metal atoms, to give chains and networks.

Some of the Group relations:

Rare Earths	Hf	Ta	W	Re	Os
↑	↑	↑	↑	↑	↑
(Ac)	Th	(Pa)	U	Np	Pu

may just be mentioned in this connection. The instability of any carbides of the platinum group metals, from rhenium onwards, contrasts strongly with the relative stability of the transuranic carbides beneath; in this sense the

1st and 4th Long Periods are similar and differ from the two intervening periods.

The NaCl-type structures of PuC, NpC, UC, PaC and ThC directly link with TaC and HfC, whilst in WC the change in stacking sequence occurs (cf. Chapter 2). UC and ThC are thus favourably placed for solid-solution formation with TaC, etc., as proven by the UC-base intercarbide system now widely investigated (see below).

Hafnium occupies an unique place in this context, not only through its position adjoining Th and the rare earths, but also in being the most refractory of any compounds known (melting point of HfC = 3,900°C, see *Table 4.1*). In contrast to HfC, the melting point of metallic Hf is low, namely 1,650°C *below* that of its carbide. The rare-earth carbides show the same feature: very high-melting carbide, low-melting parent metals, and so do the actinide carbides (though less extreme). This feature disappears along Groups IV → V → VI, and the sharp contrast, e.g. with tungsten, where the metal exceeds the carbide by about 500°C, is recalled (cf. p. 29 and *Figure 2.5*).

Quite generally (cf. Chapter 2), we can classify the carbides according to this difference into three clear-cut types: (*a*) those solidifying *after* their metals; (*b*) those in which the presence or absence of carbon makes little difference; and (*c*) those solidifying *before* their metals (viz. mainly the Group IV carbides). The correlation with structure should be noted, type (*c*) including all those of NaCl type, types (*a*) and (*b*) those of lower symmetry. Type (*c*) also extends to the actinide carbides and the Group III rare-earth carbides including the more complex dicarbides where the C—C bond introduces a new factor.

Some further physical and technological data on hafnium carbide have been given by Cotter and Kohn[94]; a HfC–'HfO' series of solid solutions, with continuous changing lattice dimensions, is of interest, as is also known in TiC, ZrC, NbC and TaC and, we should now add, in the Group III rare-earth and actinide carbides.

4.7.3. Boundary towards Platinum Group Metals. The MoC Problem

Continuing the consideration of 'outlying' carbides, platinum group carbides (counting from Group VII) are highly unstable or non-existent. Some have been suspected, but their existence was later disproved. For example, osmium and ruthenium carbides have been reported but could not be confirmed[197]. Similarly, the rhenium–carbon system has proved to be a simple graphitic one[157,257], in sharp contrast to the neighbouring tungsten–carbon. The transient formation of such carbides in thin surface layers is, however, distinctly possible and of importance in catalytic reactions. Under high pressures, carbide formation is also not excluded.

Although the Group VI (Cr, Mo, W) carbides are, of course, stable and well established, a complication arises here already, reflected in the elusive molybdenum monocarbide, MoC, as well as in the simple hexagonal nature of WC, singular for a monocarbide.

The conflicting position of the Mo–C system is remarkable (cf. *Figure 4.1*). The hexagonal Mo_2C of *A B A B A B*-type packing is established and straightforward, but a cubic modification has also been reported[107]. However, at least five structures of MoC are now known:

(1) γ-MoC: simple hexagonal[131], like WC ($a = 2{\cdot}898$, $c = 2{\cdot}809$ Å);

(2) face-centred cubic, with a complex and unexplained substructure[130];

(3) another hexagonal form[132], with $a = 3{\cdot}00$, $c = 14{\cdot}58$ kX (which, however, may be an alternative interpretation to (2);

(4) γ'-MoC (metastable): hexagonal[131], $a = 2{\cdot}932$, $c = 10{\cdot}97$ Å;

(5) a further form, 'η-MoC', in which a complex hexagonal close-packing was attained[139] employing high-pressure techniques. All these forms are greatly dependent on the manner of preparation and heat treatment, in a way by no means clarified. The earlier view (see Hansen) which had denied the very existence of any MoC is still significant, although its existence in *some* form seems now certainly assured. This also shows the uncertain energy relations and the possible contingency that graphite can directly form in coexistence with the Mo_2C—a first kinship to the platinum group behaviour.

(6) In more recent work[139], it proved possible also to produce a true *face-centred cubic* MoC by high-*pressure* treatment. This form, denoted α-MoC, prepared at 1,800°/2,500°C and 40/70 kilobars, could be retained to atmospheric pressure; the cube side was $a = 4{\cdot}27$ Å. The same cubic 'high-pressure' MoC could later be produced[258] in thin-film form by vaporization reaction between Mo and C already at 1,200°C.

Thus we have here a perfect link-up on the one hand, for high pressures, with the cubic group of NaCl-type carbides such as NbC, and on the other hand, for variable lower-pressure conditions, probably affected by impurities, varying and rather unstable MoC structures and critical stacking sequences, with a tendency to the graphitization characteristics of the Pt group metals.

The stacking sequence of *metal* atoms in all these forms can be summarized as follows:

Mo_2C: A B A B A B . . .
MoC: A A A A A A . . . (as WC)
or A B C A C B A B C A C B . . .
(approaching the NaCl-type A B C A B C . . . which is not attained)
γ'-MoC: A A B B A A B B . . .
η-MoC: A B A B A B . . .
α-MoC: A B C A B C A B C . . .

We may then say that, structurally and energetically, molybdenum in the 2nd Long Period quite clearly occupies a highly critical position as regards carbide formation. This is what happened at chromium in the 1st Long Period but with the difference that, whereas Cr merely marked the boundary between the simple stable and the more complex metastable carbide groups, and graphitization was deferred to iron/cobalt, Mo marks the abrupt breakdown towards entire carbide instability.

It may be noted that the γ' form of MoC was produced metastably by Kuo and Hägg[131] through carburizing the carbide Mo_2C at a low temperature, whilst at higher temperatures it transforms rapidly to γ-MoC.

Therefore, the γ'-type of stacking sequence appears also dynamically as a stage intermediate between that in Mo_2C and the later γ-MoC. It is imposed upon the Mo_2C lattice by the added interstitial carbon atoms, but only transiently so, the AABBAABB stacking then rearranging itself to the two alternative and more stable γ-MoC modes. Even here the possibility is still slightly open that all MoC forms are truly metastable and pass eventually towards a graphite equilibrium.

The fluid position of MoC is also well illustrated by the ease with which the stacking sequence can be 'steered' by added elements. Thus only very minor additions to MoC of NbC stabilized the cubic NaCl-type structure of NbC–MoC solid solution, of WC the simple hexagonal one of WC. It seems probable that Cr_3C_2 additions would enhance the more complex types, whilst the Pt metals, including Ru, Os and Re, would promote graphitization. On the other hand, Fe, Co and Ni additions, as we know, produce the double carbides.

In more recent work, the cubic form has also been obtained[135], by an extreme quenching rate from near-solid temperatures, with spacing of 4·2810 Å. However, the formula ascribed to it is Mo_3C_2, and it is very likely that such varying degrees of non-stoichiometric tolerance must be allowed in all 'MoC' allotropes. The kinship to 'Cr_3C_2' should also be recalled, and it might be investigated whether the latter carbide could not possibly possess a cubic form at highest temperatures.

The position has been complicated further by recent findings[142] on the structure of Mo_2C, using neutron diffraction; this showed an orthorhombic unit cell, with $a = 4{\cdot}72_4$, $b = 6{\cdot}00_4$ and $c = 5{\cdot}19$ Å ($M = 4$): of a new structure type with ordered carbon atom positions. The HCP *metal* lattice is still valid, however.

Viewing the hemicarbides as a whole, a trend is already detectable in Group V which precedes such carbon ordering in Mo_2C. Very recent neutron diffraction work by Bowman's team at Los Alamos[103] showed that, whilst V_2C has the normal hexagonal (L'3) structure, Ta_2C has a modified one of CdI_2 (C–6) anti-type, already containing a carbon *superlattice*. Thus in the M_2C carbides of Group V and VI, whilst all based on the hexagonal close-packed metal lattice, there occur already finer distinctions in carbon atom positioning, amounting to progressive ordering stages. In V_2C, carbon occupies half the octahedral holes randomly, in Ta_2C and Mo_2C ordered to increasing degrees. (For Nb_2C, the situation is not yet clear, but some partial ordering intermediate to V_2C and Ta_2C can probably be assumed.)

For W_2C, the hexagonal close-packed structure has been confirmed using electron diffraction[150], but a neutron diffraction study would be of special interest, to determine the degree of carbon ordering before the lattice switch to WC (simple hexagonal) or to the FCC high-temperature phase takes place.

The tungsten–carbon system also can by no means be regarded as finalized, and a redetermination[147] suggested quite a complex set of equilibria. A new face-centred cubic carbide, 'W_5C_3', is reported stable at very high temperatures (above 2,450°C) with a large non-stoichiometric range (see below), decomposing eutectoidally into $W_2C + WC$. Again it amounts to a transformation FCC → HCP + simple hexagonal, which in the later Groups, rhenium and Pt metals, is continued in its modified forms described (primary C solutions in rhenium and the controversial occurrence of the WC structure for OsC).

4.8 NON-STOICHIOMETRY IN CARBIDES

It is a characteristic of many carbides, as of interstitial compounds more generally, that they are liable to varying departures from stoichiometry. This is particularly so within the cubic (NaCl-type) group. In *Table 4.6* the data on degree of non-stoichiometry have been collated from the rather incomplete information available.

Table 4.6

Non-stoichiometric ranges in carbides

System	*Carbide* *Monocarbides range*	*Hemicarbides range*
Ti–C	TiC TiC to $Tic_{0.28}$[67] TiC to $TiC_{0.5}$ at 1,645°C max[66]	— —
Zr–C	ZrC ZrC to $ZrC_{0.28}$[82] ZrC to $ZrC_{0.54}$ (1,400°C)*[83] = $ZrC_{0.50}$ max)	— —
Hf–C	HfC HfC to $HfC_{0.56}$[91] HfC to $HfC_{0.6}$ (1,550°C)*[83] (= $HfC_{0.54}$ max)	—
Th–C	ThC ThC to $ThC_{0.61}$ (at 1,500°C)[259] (through-solid solution with ThC_2 at high temperatures)	—
V–C	VC $VC_{0.96}$ to $VC_{0.75}$[236] $VC_{0.9}$ to $VC_{0.6}$ max at 2,160°C (note non-occurrence at '$VC_{1.0}$')	V_2C $VC_{0.50}$ to $VC_{0.42}$[236,98] to $VC_{0.36}$ max at 1,630°C[99]
Nb–C	NbC NbC to $NbC_{0.7}$[109] to $NbC_{0.7}$ at 3,100°C[104]	Nb_2C $NbC_{0.50}$ to $NbC_{0.35}$[104–5] to $NbC_{0.39}$ max at 2,320°C[104]
Ta–C	TaC $TaC_{0.91}$ to $TaC_{0.58}$ at 3,500°C[115] (note non-occurrence of 'TaC'$_{1.0}$[69])	Ta_2C $TaC_{0.50}$ to $TaC_{0.38}$ at 2,900°C[115]
Cr–C	complex carbides: Cr_3C_2, Cr_7C_3, $Cr_{23}C_6$	—
Mo–C	MoC multiple stacking sequences; non-stoichiometry possibly extending to Mo_3C_2 and lower[138]; 'Mo_3C_2' also cited as alternative composition	Mo_2C $MoC_{0.5}$ to $MoC_{0.44}$ at 1,400°C[138]
W–C	WC very small range; 'β–WC' (H.T. form)[147] from WC to $WC_{0.6}$ between 2,785 and 2,526°C	W_2C small but finite[151] $WC_{0.52}$ at 2,540°C to $WC_{0.36}$ at 2,710°C[147]
U–C	UC small but finite[216,260] at lower but complete at high temperatures UC–UC_2 solution at high temperatures	— —
Pu–C	PuC $PuC_{0.92}$ to $PuC_{0.77}$ at 570°C[239] Pu_2C_3 $PuC_{1.50}$ to $PuC_{1.41}$[261]	— —

* Preferred values; later work

The values of the Table are also shown graphically in *Figure 4.19.*

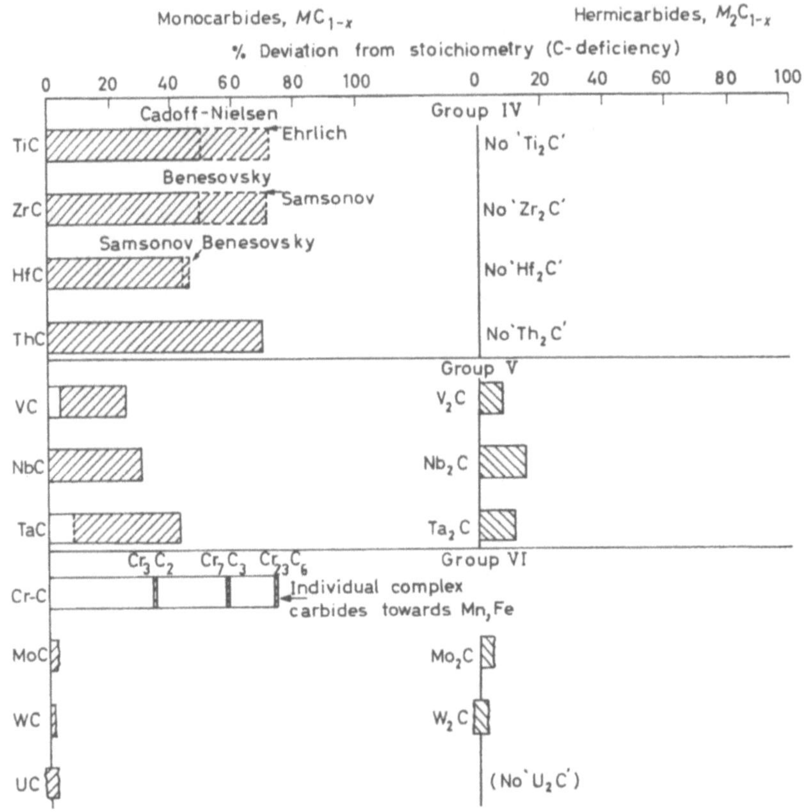

Figure 4.19. Non-stoichiometry in carbides

It is clear that, within the cubic group of monocarbides, the tendency towards C deficiency is greatest in Group IV and decreases to Group V (with a slightly indicated trend even to exclude the strict 50/50 composition from the range). It also emerges that down each column the range expands

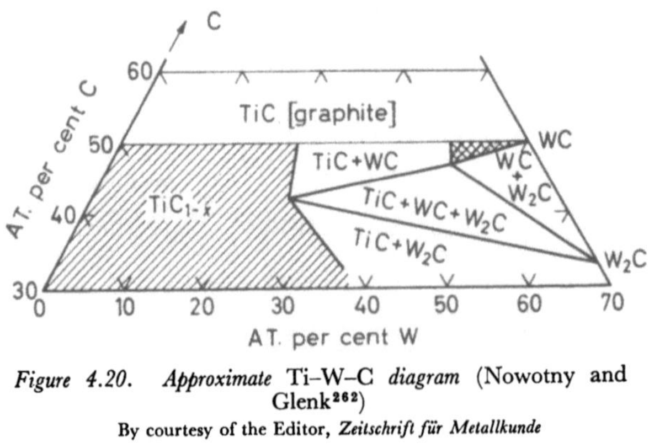

Figure 4.20. Approximate Ti–W–C *diagram* (Nowotny and Glenk[262])

By courtesy of the Editor, *Zeitschrift für Metallkunde*

with increasing atomic weight. In Group VI, we might consider that for chromium any hypothetical solution range has been inhibited and replaced by the collapse into the three distinct carbides of complex structure associated with the smaller metal atom. In MoC and WC, where we have still some semblance to the cubic structure but with changes in stacking sequence, the extent of non-stoichiometry appears still significant, though small (data insufficient), whilst for the cubic UC we have the strong tendency to *reverse* stoichiometry at high temperatures, i.e. to carbon excess (see UC $\longleftrightarrow$ UC_2 series above).

For the hexagonal carbides of M_2C type, the non-stoichiometry is considerably smaller than for the cubic ones: in fact, we may well consider the collapse to hexagonal stacking as the eventual outcome, upon C abstraction, of excessive vacancy/carbon ratio in the monocarbide lattice. It is significant that the non-occurrence of a M_2C carbide in Group IV (e.g. Ti_2C) goes together with the most extensive lattice defect in *M*C; in Group V, this is greatly reduced but supplemented by M_2C. Furthermore, as shown in the Ti–C diagram (*Figure 4.1*), the peritectoid reaction β-Ti + TiC $\rightarrow$ α-Ti corresponds to a transformation to a hexagonal phase, though there no longer expressed in the form of a carbide but of the primary solid solution.

The degree of non-stoichiometry possible may be illustrated (*Figure 4.20*) by the Ti–W–C system[262]. It is clear that in TiC the defect lattice

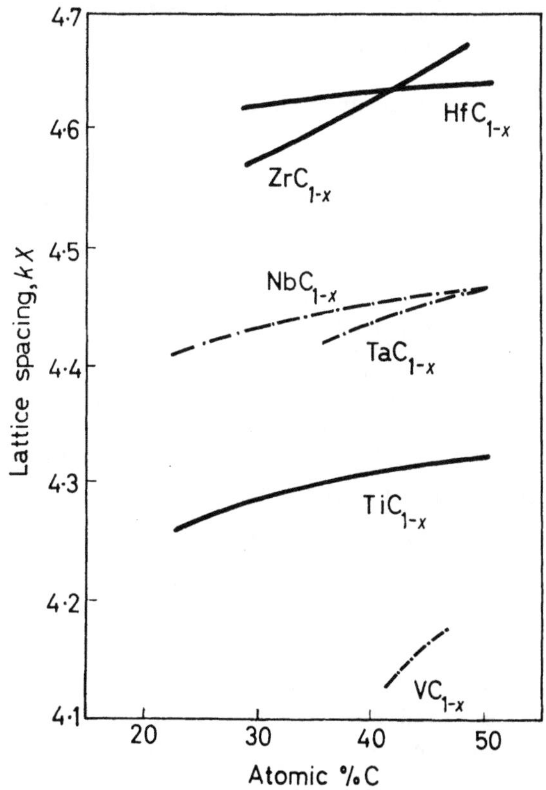

Figure 4.21. Lattice spacing variations within non-stoichiometric monocarbide ranges, MC_{1-x}, for Groups IV *(solid lines) and* V

range can span the whole range represented for tungsten by the two-carbide ($WC + W_2C$) field.

The importance of these deviations from stoichiometry arises in many fields, such as in controlling diffusion rates, oxidation or decarburization reactions, phase transformations or behaviour under irradiation, to name only a few. Also, the vacancies available clearly imply an enhanced capacity to dissolve third ('impurity') atoms such as oxygen.

Lattice-spacing variations within the TiC, ZrC, HfC, VC, NbC, TaC and UC ranges are shown in *Figure 4.21.* The limited accuracies of data (also of purities) from different sources make it doubtful whether the differences in slope between the ZrC–HfC and the NbC–TaC curves are in fact significant. However, the similarity in slopes throughout is probably the more noteworthy feature, independent of the absolute differences of parameters. The mean slopes (in terms of Å contraction/per cent C vacancy) are: TiC, $0{\cdot}002_2$; ZrC, $0{\cdot}004_8$; HfC, $0{\cdot}001_2$; VC, $0{\cdot}006_4$; NbC, $0{\cdot}002_2$; TaC, $0{\cdot}001_8$.

4.9 METHODS OF PREPARATION OF CARBIDES

The extensive subject of preparation of carbides, which falls outside our scope, has been well described in standard works[91,263a]. However, some points of principle will be briefly considered.

There are two broad methods of synthesis: (*a*) from the elements or (*b*) from their conveyor compounds. Additionally, the method of extraction from carbon-bearing melted alloys exists, for example in steels.

Each of the reaction types shown in *Table 4.7* can, in principle, be taken to represent a definite and established technological process. Thus, for instance,

Table 4.7

Basic reactions underlying methods of carbide preparation

$C + M \rightleftharpoons MC$*	(1)
$2C + MO \rightleftharpoons MC\ (+\ CO\nearrow)$	(2)
$C + MO + M'$ (auxiliary metal of higher oxygen affinity) $\rightarrow MC\ (+\ M'O)$	(3)
$2\ CO + M \rightarrow MC\ (+\ CO_2)$ $CO + MO \rightarrow MC\ (+\ CO_2)$ $CO_2 + M \rightarrow MC\ (+\ O_2)$	(4)
Hydrocarbon $A + H_2 + M \rightarrow MC\ (+\ H_2 +$ hydrocarbon $B)$	(5)
$CaC_2 + 2\ MO \rightarrow MC\ (+\ CaO)$	(6)
Hydrocarbon $+ H_2 +$ gaseous M salt (e.g. halide) $\rightarrow MC +$ acid (e.g. HCl)	(7)
Metal carbonyl $+ H_2 \rightarrow MC\ (+$ hydrocarbon$\nearrow)$	(8)
Metallo-organic compound $\xrightarrow[\text{Heat}]{} MC\ (+\ CO_2\nearrow$, etc.)	(9)
(Electrolysis) $M + M'$ carbonate (fused salt) $+ M \rightarrow MC\ (+\ CO_2 + M'$ salt)	(10)
'Steel Making', i.e. extraction of carbide from steels (electrolytic or acid)	(11)

* Varying stoichiometry and amounts have been omitted for simplification.

the direct reactions (1) and (2) have been used in Moissan's classical early syntheses of refractory carbides in the 1880s, based on fusion of the metal with carbon at extreme temperatures (~ 2,000°C) in the liquid, using a carbon electrode furnace.

Perhaps the modern heir to Moissan's method is that of synthesizing the carbide from the elements or the oxides and carbon, by fusion in vacuum or inert gas arc furnaces or in carbon resistance furnaces. Nevertheless, fusion methods have a drawback. They tend to suffer 'from the double-arrow' in the Table, i.e. the critical reversibility in direction, depending on a complex of variables like temperature, pressure, atmosphere, concentration. They also depend on the relative stabilities and melting ranges of the reactants at the high temperature concerned during synthesis, at which some carbides in fact decompose rather than form.

Carburization in the solid form, for instance of mixed metal (or metal oxide) and carbon powders, has the attraction of not only requiring lower temperatures but also of being more readily controllable and more certain to yield the desired product; furthermore, the method yields the smaller grain size, often a point of practical importance. The reactions of type

$$\text{Nb or } Nb_2O_5 + \text{carbon black} \rightarrow NbC$$

which become effective at about 1,300°C, typify most of the more conventional transition metal carbide syntheses, i.e. for V, Nb, Ta, Cr, Mo and W. However, in Group IV, the Ti, Zr or Hf *hydrides* preferably replace the pure metal as starting material. Hydrocarbons can act as efficient accelerators.

Carburization of the metal (or metal oxide) with *carbon-bearing gases* has been used in the form of exposing heated metal powders or filaments, e.g. W or Ta, to the passage of hydrocarbon gas. The latter can be methane, benzene vapours or CO; an H_2 or H_2–N_2 admixture is frequently employed.

An important method is that using the deposition of carbides from a *gaseous metal salt* phase: a gaseous mixture of (generally) a metal halide with CO, a hydrocarbon or H_2 is caused to decompose on the hot surface of a filament[263] (reaction type 7 in *Table 4.7*). The filament metal may but need not be that of the metal of which the carbide is required; it can be one of the Pt group metals as well as W, Ta, etc. The method has been successfully used for the preparation of smaller amounts of high-quality research samples rather than for large-scale carbide production. Many refinements of the technique have been developed, one interesting variant[120] being the use of *carbon* (not metal) filaments and build-up of the carbide in form of fine tubes. The gaseous decomposition method has been used for making many refractory carbides of the Group IV–VI metals; the decomposition of a carbonyl, e.g. of tungsten, molybdenum or chromium (cf. Chapter 10), has also been used, in which case quite low temperatures only (~500°C) and low pressures are required. The basic reaction then is (e.g. for molybdenum; for the carbonyls cf. Chapter 10)

$$\text{carbonyl } Mo(CO)_6 + H_2 \rightarrow Mo_2C + \text{hydrocarbon} \uparrow$$

A further method of carbide production which, however, works only under closely defined conditions, is that of 'Fused-Salt Electrolysis', the Andrieux–Weiss Process[264] in which carbonates are decomposed to form the refractory carbide in the presence of the metal. The carbonates concerned mainly are those of sodium or alkaline earths, with other salt admixtures.

For fundamental studies on carbide systems, the solid-state diffusion method is most commonly used rather than melting because, though slower, it allows better compositional control. The sinter process has now also come into more general industrial use, e.g. for producing hard carbides (e.g.[263b]; see also Bibliography).

The extraction of carbides from steels, cast irons or other naturally carbide-bearing alloys as a method of preparation amounts to equating 'steel making', etc. with 'carbide making'. In fact, though uneconomic, it is, for several carbides, the most satisfactory and efficient manner of procuring at least limited research amounts of carbides. Cementite, $Cr_{23}C_6$, Cr_7C_3, η-carbide and the χ-phase are examples of such carbides, which are very difficult to synthesize by other chemical means. It is indeed in the nature of these less stable carbides that their generation and retention depends on their direct co-existence with the primary metal-solution matrix from which in many cases they emanate by precipitation. Careful control of heat treatment is one good means of controlling the carbide desired, both in type and in physical state, e.g. particle size. We might say that, given the task of synthesizing such carbides, one would design, as an effective process, that of melting with excess metal, such as in steel making, with subsequent chemical isolation (although, if this were the prime purpose, one would no doubt choose compositions differently to those dictated by other considerations in industrial steels).

Many procedures of electrolytic or acid extraction of carbides in steels have been devised (cf. p. 12). Similarly, the more refractory carbides (TiC, NbC, TaC, etc.) can often be electrolytically isolated from ferro-alloys or higher alloyed carbon steels containing the respective metals. Fundamentally, we might put it that the refractory carbide, *M*C (e.g. NbC), will readily form in a molten second metal, M^1, of lower melting temperature and lesser carbon affinity (higher Group number) into which *M* and C have been introduced separately. Steels are only one example; others are the alloys favouring the formation of TaC from Ta + C in molten aluminium[265] or of solid-solution TiC–TaC in a similar Al melt[266]. Historically we might mention here again Moisson's ingenuity in having produced already molybdenum monocarbide from a Mo + C + Al melt[267].

Hot pressing of powders, generally in graphite dies, and using induction heating is a frequent method of preparing refractory carbides now in industrial use.

Pyrolytic methods of synthesizing carbides (as, indeed, of other interstitial compounds, e.g. borides, and of graphite) are important recent developments in connection with high-refractory applications (materials for aerospace uses).

4.10 INTER-CARBIDE SYSTEMS

4.10.1. Binary carbide systems

The various single-metal carbides introduced in the last Sections form systems with one another, which generally follow well the rules predictable from relative atomic sizes and electron concentration. Thus all the cubic monocarbides of Groups IV and V form complete series of solid solutions,

Table 4.8

Classification of transition-metal carbide systems by mutual intersolubility

Completely soluble		
TiC–ZrC	VC–NbC	UC–TaC
TiC–HfC	VC–TaC	UC–ThC
ZrC–HfC	VC–MoC*	UC–PuC
TiC–VC	NbC–TaC	U_2C_3–Pu_2C_3
TiC–NbC	NbC–MoC*	Also: V_2C–Nb_2C–Ta_2C
TiC–TaC	TaC–MoC	Mo_2C–W_2C (though in part doubtful)
ZrC–NbC	MoC†–WC	Metal radius ratios, $R_{M_1}/R_{M_2} = 1{\cdot}0 \pm 0{\cdot}1$
ZrC–TaC	$Cr_{23}C_6$–$Mn_{23}C_6$	
HfC–NbC	Cr_7C_3–Mn_7C_3	
HfC–TaC	UC–ZrC	
	UC–HfC	
	UC–NbC	
Partially or conditionally soluble		
ZrC–VC	NbC–Cr_3C_2 ($Cr_{23}C_6$)	$\left.\begin{matrix}Cr_{23}C_6\\Cr_7C_3\end{matrix}\right\}$–$\left\{\begin{matrix}Fe_3C\\(Co, Ni)–C\end{matrix}\right.$
HfC–VC	NbC–Mo_2C	Cr_3C_2, etc.–MoC(Mo_2C)
TiC–Cr_3C_2	NbC–WC	Cr_3C_2, etc.–WC(W_2C)
TiC–MoC (Mo_2C)	TaC–Cr_3C_2 ($Cr_{23}C_6$)[?]	UC–VC
TiC–WC (W_2C)	TaC–Mo_2C	UC–TiC
ZrC–MoC (Mo_2C)	TaC–WC (W_2C)	$R_{M_1}/R_{M_2} = 1{\cdot}0 \pm 0{\cdot}15$
ZrC–WC (W_2C)	VC–Mn_aC_b‡	(most size ratios still similar to those above, but greater electron differences)
HfC–MoC (Mo_2C)	VC–Fe_3C	
HfC–WC (W_2C)	VC–[Co, Ni–C]	
VC–Cr_3C_2 ($Cr_{23}C_6$)	Cr_3C_2–Mn_aC_b‡	
VC–MoC (other than cubic form)	Cr_3C_2–Fe_3C	
VC–WC (W_2C)	Cr_3C_2–[Co, Ni-C]	
Not (or very slightly) soluble		
TiC–Mn_aC_b	HfC–Fe_3C	MoC(Mo_2C)–(Co, Ni–C)**
TiC–Fe_3C	HfC–(Co, Ni–C)	WC(W_2C)–Mn_aC_b**
TiC–[Co, Ni–C]	NbC–Mn_aC_b‡	WC(W_2C)–Fe_3C**
ZrC–Cr_3C_2, etc.	NbC–Fe_3C	WC(W_2C)–(Co, Ni–C)**
ZrC–Mn_aC_b	NbC–(Co, Ni–C)	ThC‖–HfC
ZrC–Fe_3C	TaC–Mn_aC_b‡	UC–Cr_3C_2, etc.**
ZrC–(Co, Ni–C)	TaC–Fe_3C	UC–MoC(Mo_2C)**
HfC–Cr_3C_2, etc.	TaC–(Co, Ni–C)	UC–WC(W_2C)**
HfC–Mn_aC_b	MoC(Mo_2C)–Mn_aC_b‡**	UC–Mn_aC_b**
	MoC(Mo_2C)–Fe_3C**	UC–Fe_3C**
		UC–(Co, Ni–C)**
		$R_{M_1}/R_{M_2} = 1{\cdot}0 \pm 0{\cdot}30$

* Only the cubic form of MoC
† Only the simple hexagonal form of MoC
‡ Short for any of the various manganese carbides described
‖ Little known otherwise on ThC solutions, but solubilities probably limited on general grounds
** Double carbide forming

Note: Scandium carbides, yttrium carbides and the Rare Earth carbides are also likely to form extensive solutions with their Group IV and actinon isomorphs, but only limited data are available.

with the exception of the VC–ZrC and VC–HfC pairs which are beyond the borderline of favourable size.

Table 4.8 shows the inter-carbide systems which have experimentally been found to possess complete, partial and no or very small intersolubilities, along with the size ratios of the participating metal atoms. The data have been collected from a variety of sources[268–9, 91], brought up to date by more recent research. Much useful information on the simpler systems could also be gleaned from work on the more complex carbide equilibria in alloy steels and in refractory 'super-alloys' which, as a by-product to their development, has considerably furthered our knowledge of fundamental carbide relations.

It is evident from the Table that, for solution formation, isomorphy of the end members is an essential condition and that, within this limit, the size factor alone is almost sufficient. The conditions of similarity of bond and electron structure are in these cases also automatically fulfilled. UC, ThC and PuC, however, form a singular condition and will require special comment.

On the other hand, the insolubility between Groups IV, V, VI and the iron Group carbides again highlights the borderline position of the Cr and Mn carbides: although not completely soluble with any of the others, they are at least partially soluble with all of them*.

We propose first to consider the Group IV, V, VI inter-carbide systems, then advancing to Group VII$\rightarrow$VIII carbides.

Inspection of *Table 4.8* shows the general validity of the rule that metals forming continuous solutions amongst themselves also form the same amongst their carbides—with the qualification 'at any rate in some of their forms' allowing for cases like Mn_3C–Fe_3C, where the continuous series is restricted to γ- and δ-(FeMn).

The experimental data also confirm that the '15 per cent rule' regarding relative atomic sizes is here very satisfactorily fulfilled.

The lattice parameters, as a function of composition within the binary carbide solution systems of cubic NaCl type, are shown in *Figure 4.22*. As will be seen, the spacing–composition curves in the systems of cubic carbides, TiC–ZrC, TiC–VC, TiC–NbC, TiC–TaC, ZrC–NbC, ZrC–TaC, VC–NbC and NbC–TaC, are almost linear. Also brought out is the lack of solubility in ZrC–VC and HfC–VC. The figure also shows curves for non-isomorphous, but largely soluble, carbide end members, such as WC and MoC(Mo_2C) with TiC (where available). For the complete solution series Vegard's law is generally well obeyed. Only in some cases, e.g. TiC–NbC, a slight convexity towards the larger-size metal atom occurs. Assuming that this is not due to a residual lack of equilibration in experiment (as can easily be the case), it would imply that the addition of the larger metal atom—e.g.

* In apparent contrast, however, it was found[163] that, in alloy steels, (Nb, Ta)C did not dissolve any chromium carbide, whilst VC did. However, this may have been connected with the particular equilibrium conditions prevailing in a steel as against a pure carbide system. The chromium would tend to remain in α-matrix solution rather than enter the carbide. The substantial purity of NbC existing in Cr-bearing alloy steels is well in line with the writer's observations in this field. As a principle, caution should be used in applying pure carbide relations to those prevailing in metal co-existence, i.e. the question whether, in given cases, equilibrium has been attained or not must be considered critically.

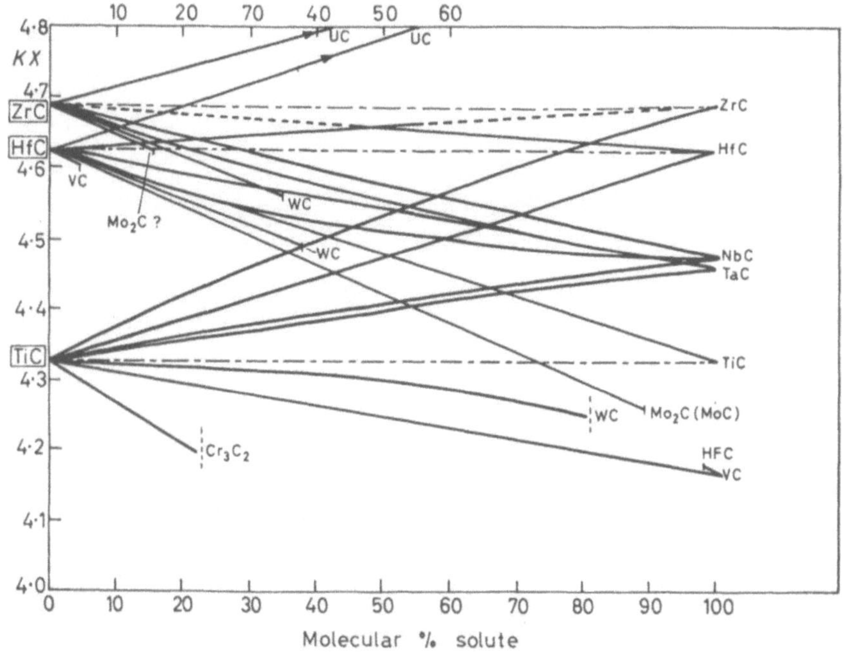

Figure 4.22. Lattice spacing changes in cubic carbides (NaCl *type*) *:* (*a*) *of Group IV*, upon addition of carbides of Groups V and VI; (*b*) *of Group V*, upon addition of carbides of Groups IV and VI; (*c*) *of Groups VI* (i.e. UC only), upon addition of carbides of Groups IV and V

ZrC–VC and HfC—VC (in *a*) are practically immiscible
I— denotes incomplete solid solutions

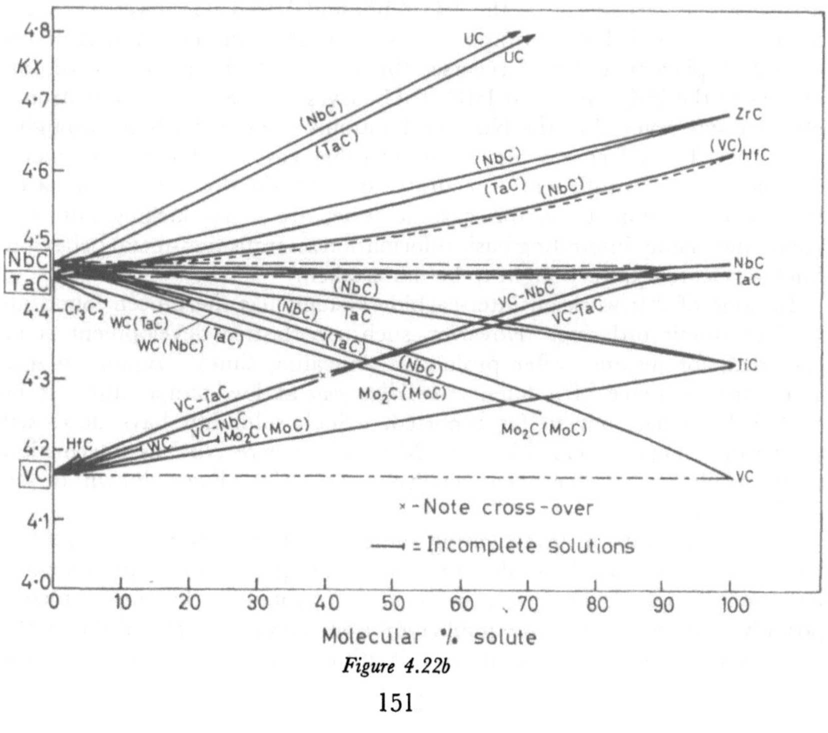

Figure 4.22b

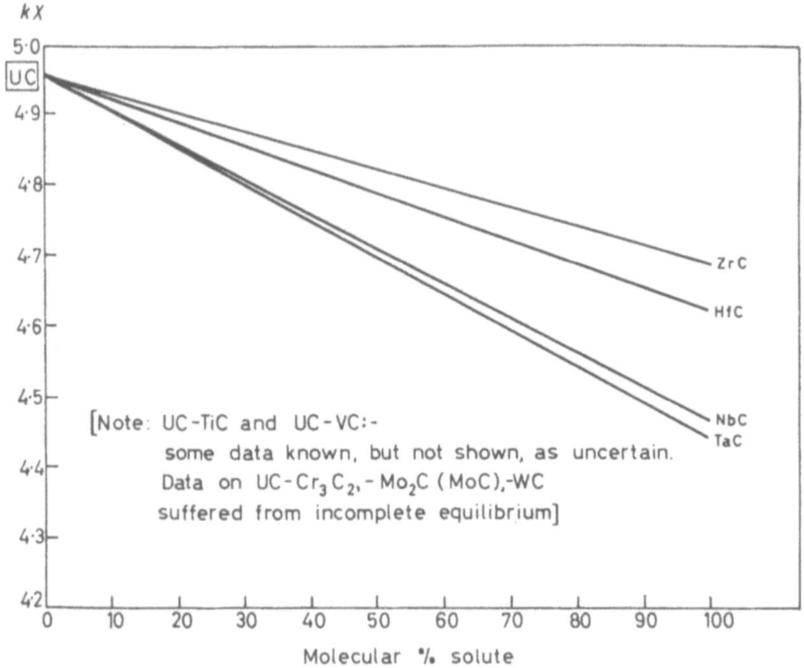

Figure 4.22c

Nb—is associated with a relative excess of electrons. An interesting point arising is the 'cross-over' of the VC–NbC and VC–TaC curves, i.e. whilst in pure NbC and TaC, Nb appears as the slightly larger atom, their progressive replacement by V reverses this order. A similar cross-over also applies to the HfC–NbC and HfC–TaC curves, indicating an even stronger effective size reversal of the Nb and Ta atoms through Hf than through V addition. The effect *may* also occur in other cases, but the experimental data are as yet insufficiently accurate to ascertain this. Precision lattice spacing measurements on this carbide series are so far lacking but could yield some highly interesting basic information on transition-metal behaviour. Such a research project ought to be undertaken.

In none of this work on inter-carbide systems has there been found any kind of atomic ordering. However, such superlattice development is not excluded, for instance after prolonged annealing times. Again, in some cases, the existence of a latent *miscibility gap* at low temperatures is not excluded, though not so far reported. Such solubility bays necessarily evolve in ternary systems like ZrC–NbC–VC, where NbC as intermediate solute would remove the heterogeneity in ZrC–VC. Examples will be seen in *Figure 4.23*.

A basic difficulty in investigating these refractory systems is the possible lack of attainment of equilibrium after the particular homogenization treatments applied, implying, in terms of structure, the appearance of possibly two or more non-equilibrium isomorphs. Much of the earlier experimental work suffered from this difficulty, and complete solubility could actually sometimes later be attained by gradual approach of spacings with time or increased temperature. Also, in many industrial processes such as alloy steel making or the production of hard metals, the attainment of equilibrium is by no means assured nor even necessarily desired. Therefore, phases like NbC and TiC, VC and $Cr_{23}C_6$, TiC and WC can appear

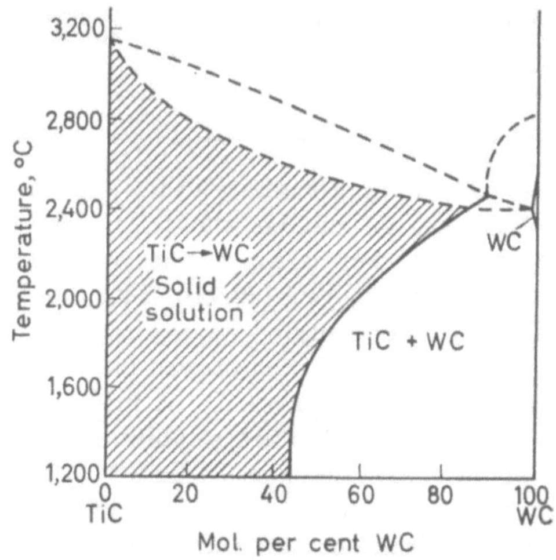

Figure 4.23. Tic–WC *pseudobinary section* (Metcalfe[272])

By courtesy of the Institute of Metals, London

separately through impeded diffusion or actual physical separation of potentially reacting particles, although true equilibrium may correspond either to a single phase or to two phases nearly in mutual solution.

Systems containing *hafnium carbide* as one component have been studied only comparatively recently, owing to its previous scarcity and the difficulty of separating Hf from Zr. However, it has been found[269] that these systems follow the predictions and the '15 per cent rule' very consistently. Complete solid solutions are formed of HfC with TiC, ZrC, NbC and TaC but (as for ZrC) there is only slight mutual solubility with VC and Cr_3C_2. On the other hand, HfC can dissolve up to at least at 90 per cent MoC, and about 40 per cent WC (at 2,000°C, at higher temperatures probably more). The changes in lattice spacing in HfC-base systems are incorporated in *Figure 4.22.* It is of interest to note that, for HfC–MoC, the solubility may indeed be completely in the cubic form of MoC[270] which HfC stabilizes.

Regarding the non-isomorphous pairs of the Group IV and V carbides with WC, it is noted that the WC solubility in the *cubic* carbides is large, but the converse one in WC is small or negligible. The relations with molybdenum carbide are similar, and here the metastable cubic form of MoC has been shown to be stabilized by and intersoluble with NbC—as no doubt with other cubic carbides. We may set this high solubility in parallel with the complete intersolubility of the parent metals, i.e. of Mo with Nb, V, Ta, Ti, etc. Early work[271] had already stated some of these relations with remarkable correctness (*Table 4.9*).

Table 4.9 (From Molkov and Vicker[271])

Dissolution					
by	*of*	*to wt. per cent*	*by*	*of*	*to wt. per cent*
TiC	WC	82	WC	TiC	only slightly
	Mo_2C	85	Mo_2C	TiC	only slightly
	Cr_3C_2	40	Cr_3C_2	TiC	10
Mo_2C	WC	53	VC	TiC	completely
VC	WC or Mo_2C	76			

These authors also gave an approximate outline of a ternary and quaternary system by observing, on molten alloys, homogeneous cubic solid solutions in defined regions. For example, these were bounded by: TaC (21–40 per cent), Mo_2C (0–40), WC (39–60), TiC (50–58), Mo_2C (6–9), WC (22–28) and VC (5–22 per cent).

The system TiC–WC, which has been of such basic importance in hard-metal development, is shown in *Figure 4.23.* A main feature is the strong temperature dependence of the mutual solubilities, and the necessarily sensitive influence of third carbides such as TaC[262,272–4].

The system Mo–Nb–C is shown in more detail[275] in *Figure 4.24.* In modification of previous ideas, they alter the γ'-MoC to a new 'Mo_3C_2' (i.e. defect lattice and now formal analogue to Cr_3C_2, though structurally the hexagonal cell of γ' is confirmed). It is of interest that the 'MoC'–NbC solution is now partial only and that, remarkably, Mo_2C–Nb_2C does not seem to form a continuous solution series but a very restricted one, broken by a wide two-phase field (Mo, Nb) metal + (Mo, Nb)C solid solution. This is a new and quite unexpected result, implying that the hemicarbides, stable for the single metals, become unstable in favour of the monocarbide series when alloyed. (If it is confirmed, this decomposition $M_2C \rightarrow MC + M$ for Group V–VI combinations would mean an appreciable revision of our ideas; however, the likelihood of a metastable equilibrium having persisted seems a strong one and, it is felt, needs checking.) The position would also bear on the closely related systems Ta–Mo–C, Nb–W–C and Ta–W–C and the stability of their mono- and hemi-carbide solutions.

Considering *tungsten carbide* as solute, the solid solubility of both Group IV and Group V carbides for WC decreases consistently with increasing size difference from the tungsten atom: namely in sequence TiC→ZrC→HfC and VC→NbC→TaC (see also *Figure 4.22*). On the other hand, whilst WC itself does not dissolve the cubic Group IV carbides, it does dissolve the Group V carbides: extensively so for VC, slightly for NbC and TaC as solutes.

These intersolubilities with tungsten and molybdenum carbides as one partner are all highly temperature-dependent, implying a sensitive controllability of the transformation hexagonal ⟷ cubic and of properties. In practice, this bears critically on hard metal relations as exemplified in the system WC–TiC. (References to specific inter-carbide systems will be found in the Bibliography.)

The Zr–Mo–C system, studied[138] for the 2,100°C section, is interesting and consistent with our systematic considerations in that it shows a large $ZrC \rightarrow Mo_3C_2$ solution region, extending right up to $Zr_{0\cdot02}Mo_{2\cdot9}C_2$, i.e. only the last abandonment of Zr causes a structure switch. The 'Mo_3C_2' phase (actually 38 at. per cent C; spacings: $a = 3{\cdot}010$, $c = 14{\cdot}62$ Å) may be regarded as a highly carbon-deficient derivative of MoC, showing MoC in yet another version; clarification is badly needed.

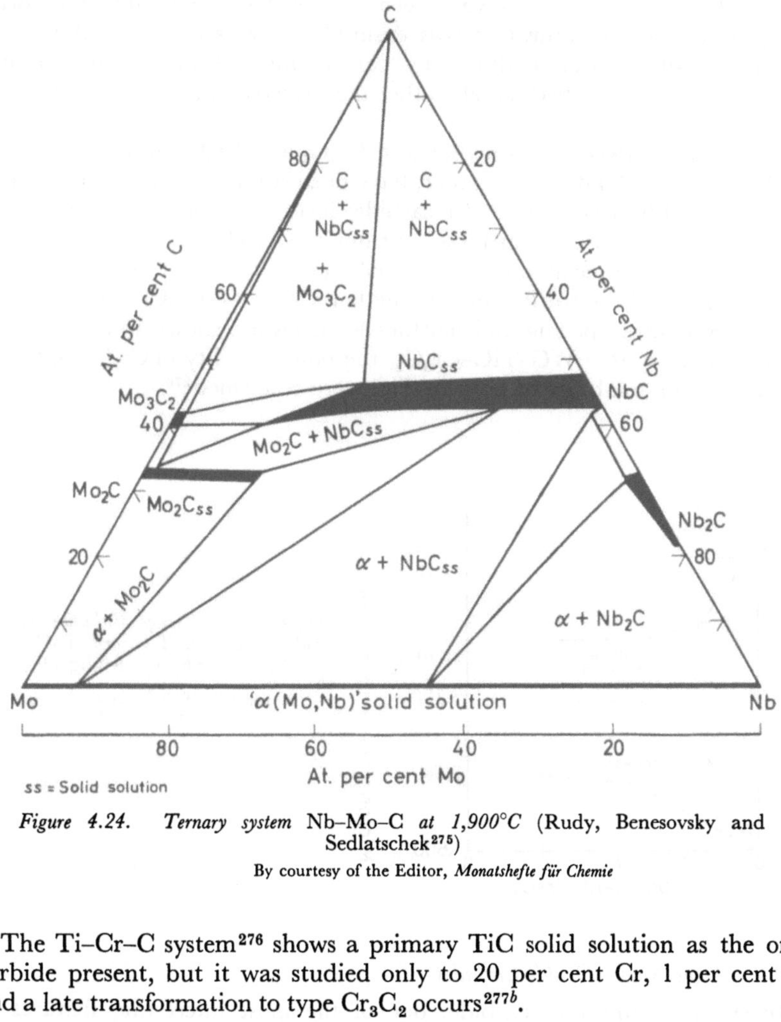

Figure 4.24. Ternary system Nb–Mo–C *at 1,900°C* (Rudy, Benesovsky and Sedlatschek[275])

By courtesy of the Editor, *Monatshefte für Chemie*

The Ti–Cr–C system[276] shows a primary TiC solid solution as the only carbide present, but it was studied only to 20 per cent Cr, 1 per cent C, and a late transformation to type Cr_3C_2 occurs[277b].

Uranium monocarbide, UC, also forms extensive series of solid solutions with its cubic isotypes. As shown in *Figure 4.22*, quite in accordance with size relations, this occurs with ZrC and HfC but doubtfully so with NbC and TaC. Only partial solutions are reported for TiC and VC, but here too the data are uncertain. The UC–ZrC, NbC and TaC relations have been shown by Brownlee *et al.*[278].

UC and PuC have proved[279–80] to be completely intersoluble (studied below 635°C only, but probably throughout), with small deviations in stoichiometric ranges.

A complication as regards carbide intersolubilities is liable to arise through the occurrence of *defect lattices* in some of the carbides, as discussed. This non-stoichiometry, through vacancies and through possible residual interstitial impurities such as O or N, would tend to decrease the observed lattice dimensions, and perhaps affect the solubility boundaries to some extent. This effect has not always been allowed for in experimental work, though often carbon saturation was ensured as far as possible. Also, the preparation of research carbides by carburizing the *hydrides* which are prepared from the purified metals rather than the oxides, has greatly helped in this direction.

The systems between *chromium carbides*, Cr_7C_3 and Cr_3C_2 on the one hand and WC, TaC and TiC on the other, have been studied[277] by measurements of spacing and hardness. Both Cr carbides form solid solutions in TiC and TaC (with an increase in hardness) but not conversely. Furthermore, WC is soluble to an appreciable extent in Cr_7C_3 and Cr_3C_2, but not conversely. The solubilities do not differ much whether Cr_7C_3 or Cr_3C_2 are chosen*. The corresponding spacing and hardness changes are shown in *Figure 4.25*. In the ternary system WC–TiC–Cr_3C_2, the non-solubility of Cr_3C_2 and TiC in WC and the solubility of both in TiC were confirmed[276].

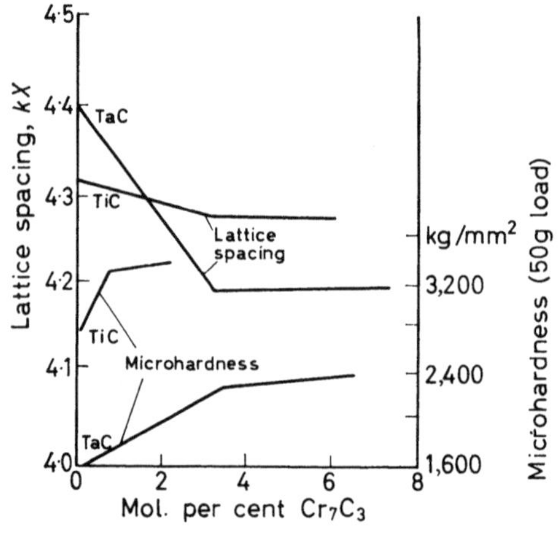

Figure 4.25. Change of lattice spacing and hardness in TiC *and* TaC *upon* Cr_7C_3 *addition* (Rüdiger[277])

By courtesy of the Editor, *Technische Mitteilungen Krupp*

4.10.2. Some Newer Double Carbides

Reverting to uranium carbide, one of the most interesting features is the occurrence, in the (UC—Group VI) carbide systems, of *ternary* carbides[281]. The three carbides in the systems U–Cr–C, U–Mo–C and U–W–C are isostructural and of composition $UCrC_2$, $UMoC_2$ and UWC_2. It was

* $Cr_{23}C_6$ is not considered, but its relation to WC is likely to be similar. The complete solution series $Cr_{23}C_6$–$Fe_{21}W_2C$ should here also be recalled. The actual pseudo-binary will be between Cr_3C_2 and WC.

possible, from single-crystal work, to determine their structure as of orthorhombic symmetry, with 16 atoms (i.e. 4 formula units) per unit cell. The lattice dimensions and other details are shown in *Table 4.10.* A structural relation with Cr_3C_2 was noted, involving the replacement of two Cr atoms by one of U, to give a formal exchange $Cr(Cr_2)C_2 \longleftrightarrow Cr_2UC_2$. The structure can also be considered as pseudo-hexagonal, and the carbon atoms could

Table 4.10. Structure of $UCrC_2$, $UMoC_2$ *and* UWC_2

(orthorhombic, 4 formula units per unit cell)

	Lattice spacings, kX			
	a	*b*	*c*	
$UCrC_2$	$5{\cdot}42_2$	$3{\cdot}22_5$	$10{\cdot}61_5$	(Nowotny *et al.*[281])
$UMoC_2$	$5{\cdot}61_2$	$3{\cdot}22_5$	$10{\cdot}9_5$	(Nowotny *et al.*[281])
	$5{\cdot}62_5$	$3{\cdot}24_9$	10·980	(Cromer *et al.*[282])
UWC_2	5·62	$3{\cdot}24_5$	$10{\cdot}9_5$	(Nowonty *et al.*[281])
For comparison				
Cr_3C_2	5·52	2·821	11·46	

Other isomorphs reported recently[208b]: $UMnC_2$, $UTcC_2$, $UReC_2$, i.e. extending to Group VII metals; and, related but with deviant stoichiometry, U_2RhC_2, U_2IrC_2; also $UCrC_x$ and $UTcC_x$ as above but carbon-deficient; probably many more occur.

Atomic parameters in $UCrC_2$	U *atom*	Cr *atom*	C *atoms* I	C *atoms* II
	$x = 0{\cdot}083$	0·350	≈0·5	0·72
	$y = \frac{1}{4}$	$\frac{1}{4}$	$\frac{1}{4}$	$\frac{1}{4}$
	$z = 0{\cdot}143$	0·860	0·04	≈0·0

be located in the tetrahedral interspaces between two layers of metal atoms; the exchange $U \longleftrightarrow (Cr_2)$ is accompanied by the formation of C_2 dipoles. A beginning link-up with boride and silicide structures may here already be detected (see later). The structure of $UCrC_2$ is shown in *Figure 4.26.*

These double carbides form clearly intermediate stages between UC on the one hand and WC, MoC and $CrC_{0{\cdot}67}$* on the other; they can still be considered as 'monocarbides' formed amongst Group VI metals for an appropriate ratio. The natural comparison is between the carbides that can formally be (in terms of the unit cell of the mixed carbide) written as

$$\underset{(1)}{U_8C_8}-\underset{(2)}{U_4(W,Mo,Cr)_4C_8}-\underset{(3)}{(W_8C_8,\ Mo_8C_8 \text{ or } Cr_8C_{5{\cdot}33})}$$

Basically, this sequence again represents a modulation of the stacking sequence. The cubic close-packed structure of (1) passes, upon the addition of the smaller but equivalent metal atom, to the pseudo-hexagonal close-packing of (2) and the ultimate split into the various types of structure amongst (3)—simple hexagonal in WC; the family of forms in MoC and orthorhombic in Cr_3C_2. This sequence is most interesting, as it again

* In lieu of an ideal but non-existent 'CrC' which has degenerated by carbon deficiency into the more complex Cr_3C_2.

typifies in a nutshell the principles of continuity in structure change we have been trying to emphasize.

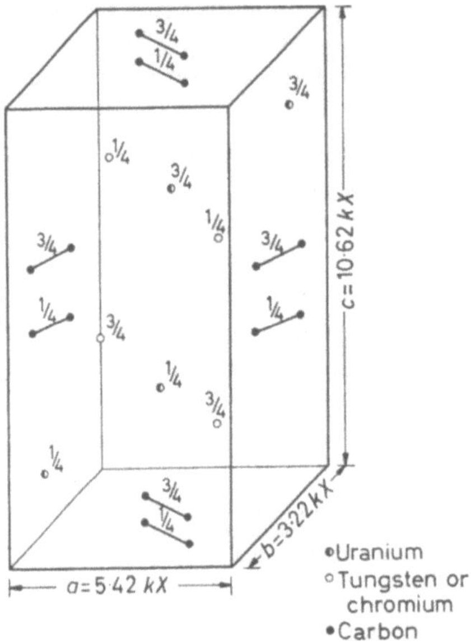

Figure 4.26. Unit cell of $UCrC_2$ *or* UWC_2 (Nowotny *et al.*[281])

It should also be noted that the development of carbon atom pairs in UMC_2 is not unique; it evolves further in the carbon-richer UC_2 structure towards which, it may be recalled, UC already inclines somewhat by its capacity for dissolving excess carbon. On a more general plane, we have here, for the first time, double carbides *amongst* metals of one Group, viz. Group VI, though we may recall the binary-based double carbides $(Cr,W)_{23}C_6$ and $(Cr,Mo)_{23}C_6$ of $Cr_{23}C_6$ structure and the η-carbides. This new one supplements the system of double carbides which may now be represented as in *Figure 4.27*. It would indicate a succession of double carbides with decreasing carbon content: (*a*) MC, (*b*) $M_{23}C_6$ ($\approx M_4C$) and (*c*) M_6C, where M

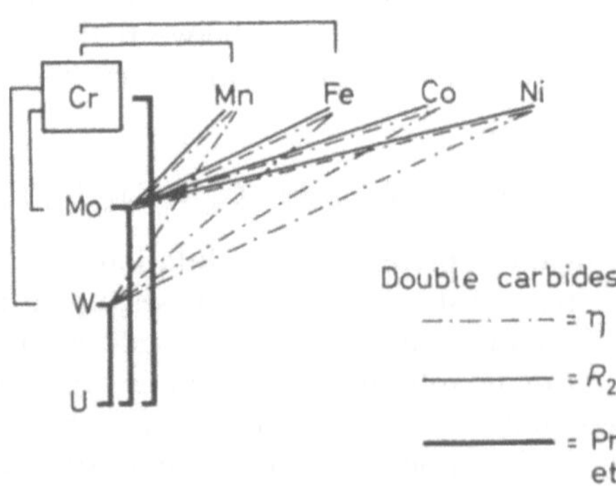

Figure 4.27. Overall diagram of double carbide formation, based on Group VI metals

is U and W,Mo or Cr in (*a*); Cr and W,Mo,Cr,Mn,Fe,Co or Ni in (*b*); (W or Mo) and Mn,Fe,Co or Ni in (*c*). A rule seems to emerge, according to which double carbides can represent both a diagonal (cf. *Figure 4.2*) and a vertical combination in the Periodic Table. The latter, however, only applies to Group VI and not, as far as known, to Group V or IV; but it would be of interest to investigate whether such a possibility might not indeed exist, for example, for lower temperatures. For instance, a possible more complex (V,Pa), (TiTh), (Zr,Th) or (Hf,Th) double carbide, perhaps by peritectic reaction from the immiscible end members, might be investigated.

4.10.3. Ternary and Higher Inter-Carbide Systems

From the two-carbide systems described, the ternary diagrams based on them can, in essence, be fairly well predicted, and the limited experimental work carried out generally confirms such 'ideal' diagrams. Thus, all the combinations of binary systems in set I of *Table 4.9* form complete ternary solid solutions whilst, for instance, the miscibility gaps in ZrC–VC and HfC–VC are closed by third monocarbides of intermediate metal atom size and form complete solutions with the two others. *Figure 4.28* summarizes the ternary systems so far substantiated; many of the diagrams are still hypothetical but safely predictable. A survey has also been made by Benesovsky and Rudy[83c].

The presence of miscibility bays in these systems, and their critical temperature dependence, is of obvious practical importance, both in affecting the structures developed and, by implication, in hard metal technology. Hardenability and precipitation effects become possible. In further development, properties such as hardness, toughness, etc. can be modified by fourth carbide additions, by borides, nitrides, silicides and, though often detrimentally, by oxides in solution, such as TiO (see also Section 4.13). The intersolubility with metallic matrices which are used as bonds, such as cobalt, also varies critically. All these reactions integrate to form the basis of a whole technology, viz. of hard metals, and now various newer refractory alloys.

Many systematic regularities are already apparent from *Figure 4.28*, though the picture is still incomplete; for example, in the series (NbC$\rightarrow$TaC$\rightarrow$TiC)–ZrC–VC, a progressive increase in the extent of the miscibility bay occurs. Lattice dimensional changes, i.e. contours within the single-phase fields and tie lines in two-phase fields, have mostly been omitted; they follow naturally from those of the binary sides. Typical examples for TiC–HfC–VC, TaC–HfC–VC and NbC–HfC–VC[283] and TiC–TaC–WC[284] are included, however.

The TiC–ZrC–WC system, derived by Tombriel[285] and shown in *Figure 4.29*, typifies the *kind* of phase relations, receding phase boundaries and spacing contours to be expected where the third carbide, WC, is incongruous. Of particular interest should here be the area where the incipient cubic $\rightarrow$hexagonal transformation will already induce stacking faults in the parent lattice.

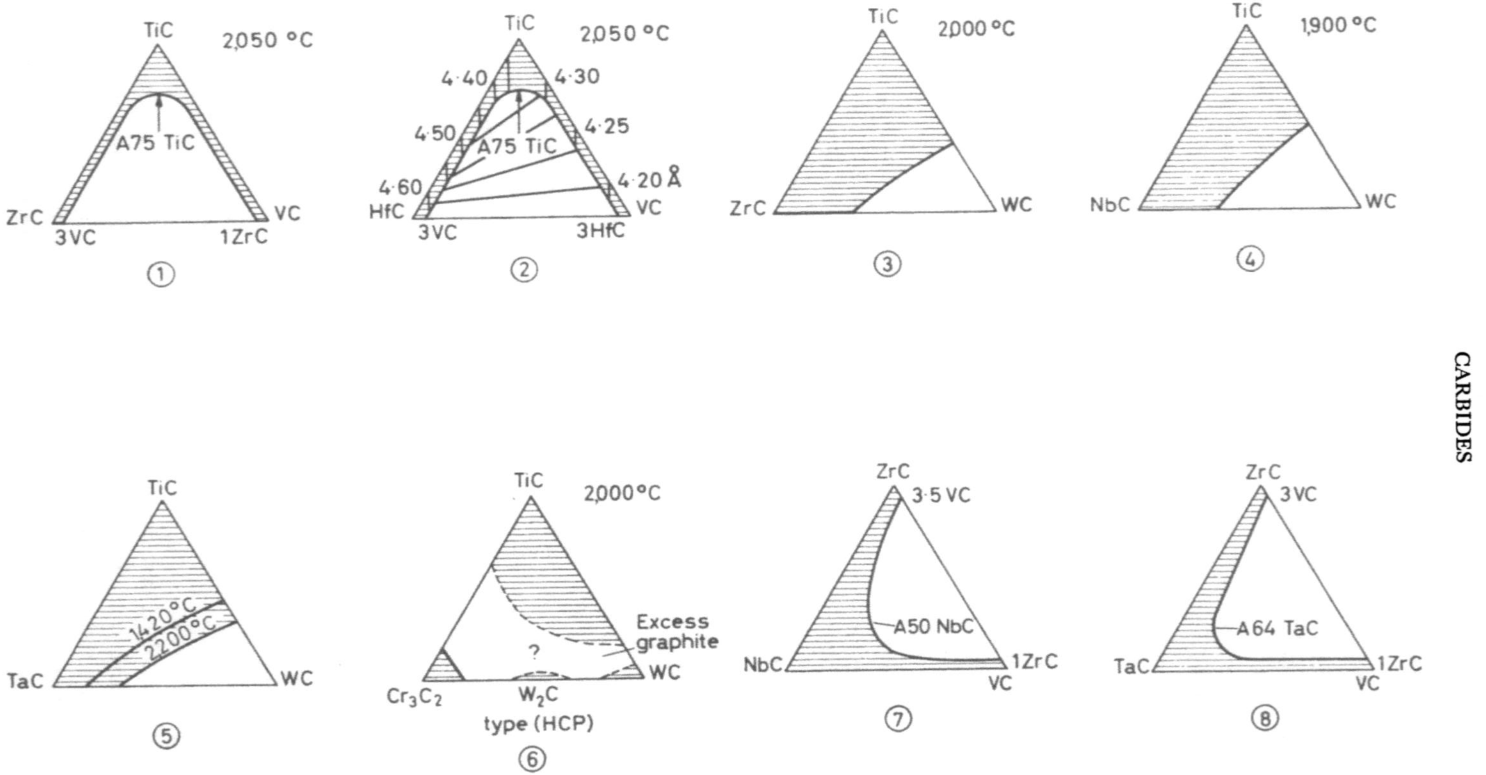

Figure 4.28. Ternary carbide systems: Group IV–V–VI carbide combinations
Nb–Mo–Cr (metal) diagram included for comparison (Goldschmidt and Brand[151])

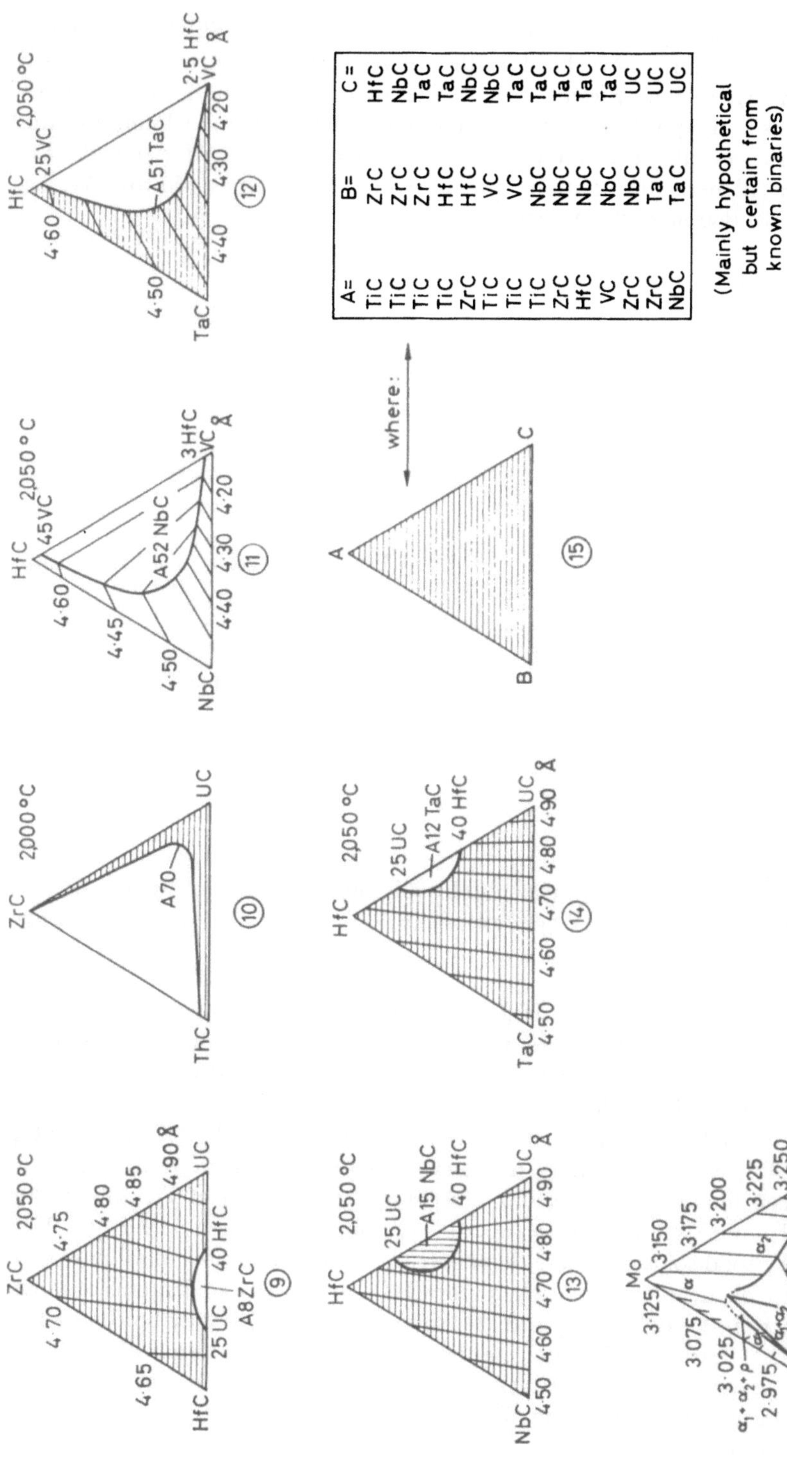

Figure 4.28 cont.

The ternary systems uranium–M–C where $M = $ Zr,Hf,Nb and Ta (see also Bibliography) have been further investigated[83b], particularly on the carbon-richer side (*Figure 4.30*). The figure shows both the solubility relations and equilibria, as well as the non-stoichiometry, inclining towards carbon excess at the UC and to carbon deficiency at the M–rich end. We may recall here the position evolving when M becomes Mo and W, i.e. the above double carbides emerge.

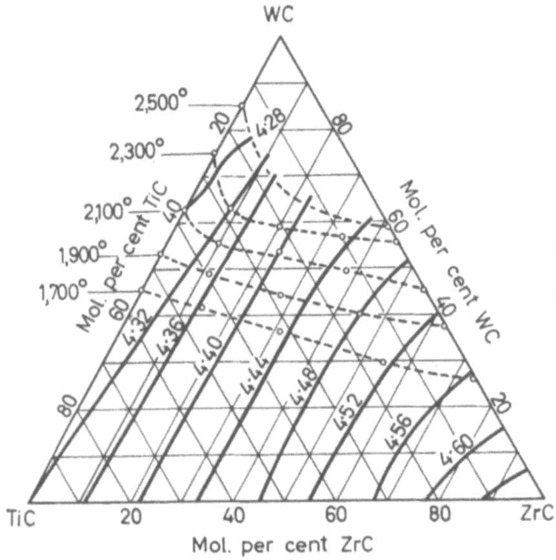

Figure 4.29. The TiC–ZrC–WC *diagram* (Tombriel[285])

By courtesy of the Editor, *Plansee Proceedings*

Chubb and Keller[208c] have substantially added to this in a paper on the constitution of the U–M–C systems, where $M = $ Nb, Mo, W, Re, Y, presenting new ternary diagrams with data on intersolubilities (e.g. UC–NbC, etc.), non-stoichiometry and ternary compounds formed (e.g. UWC_2 equilibrating with UC, U_2C_3, UC_2, W, W_2C and WC; much can here be gleaned to add to our systematic pattern). The U–Y–C system links the actinide carbides with the Rare Earth and Group III ones: UC–YC and UC_2–YC_2 are completely intersoluble.

As an example of interest, it may be mentioned that, in the TiC–WC–Cr_3C_2 system[277b], an extraordinary microhardness peak of above 4,300 kg/mm^2 has been found at a molecular percentage composition of TiC, about 80, WC, 10 and Cr_3C_2, 10 (*Figure 4.31*). It occurs in an area within the TiC primary solid solution where, however, Cr_3C_2 and WC precipitation would already be incipient and where the TiC lattice has undergone a large contraction through Cr_3C_2 and WC solid solution. This may be contrasted with the hardness diagram on the TiC–WC–TaC system, also given, which shows a uniform hardness decrease, concurrent with a lattice expansion within the dominating TiC solid-solution field (cf. also the phase diagrams TiC–WC–Cr_3C_2 and TiC–WC–TaC in *Figure 4.28*).

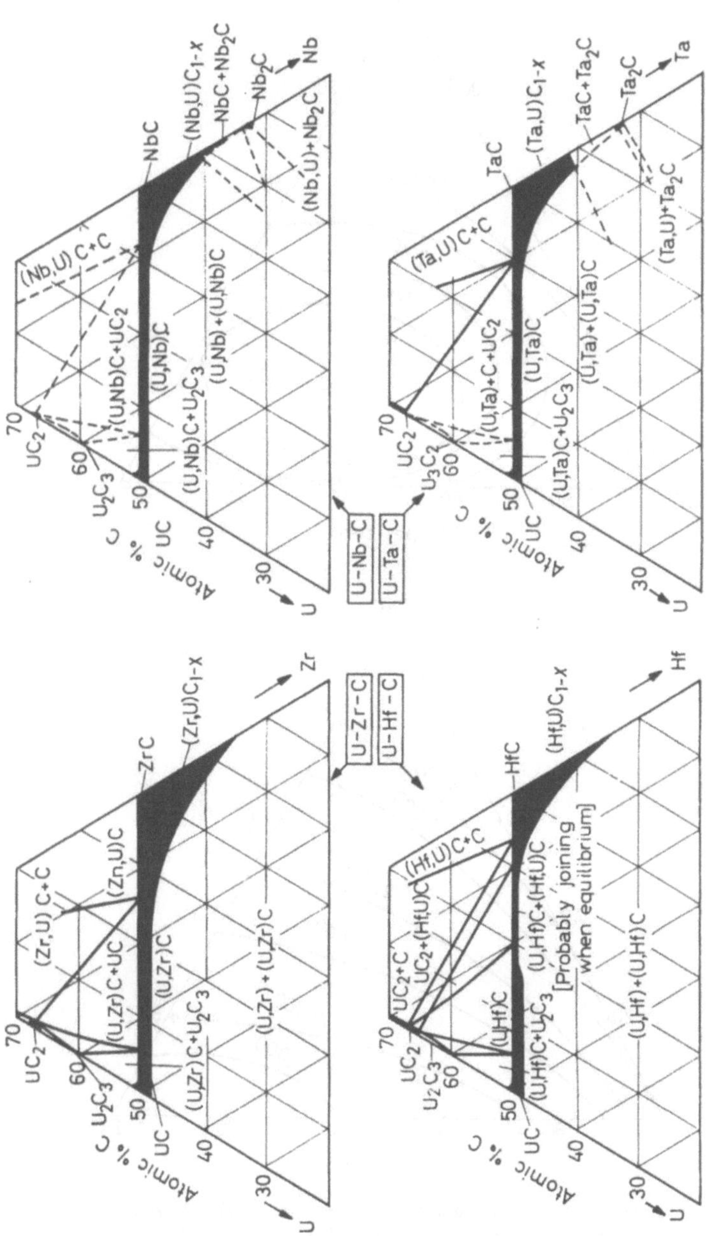

Figure 4.30. ***Phase fields in*** **U–*M*–C** ***diagrams, adjoining the*** **UC–*M*C** ***section*** **(*M* — Zr, Hf, Nb, Ta); 1,700°C sections (From Benesovsky and Rudy[83b])**

By courtesy of the Editor, *Planseeberichte für Pulvermetallurgie*

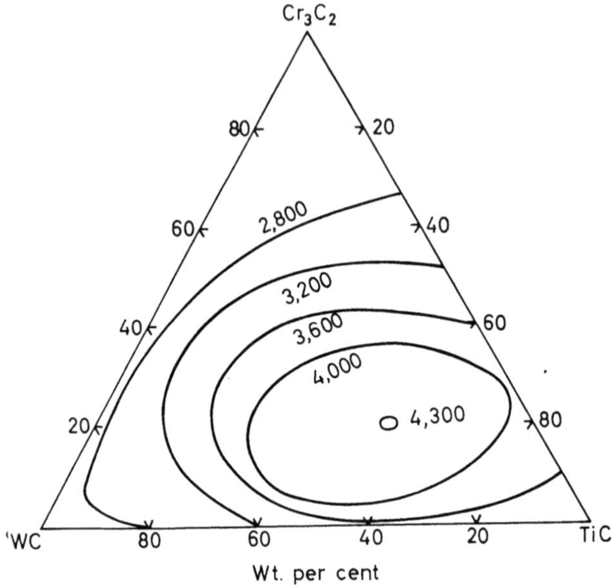

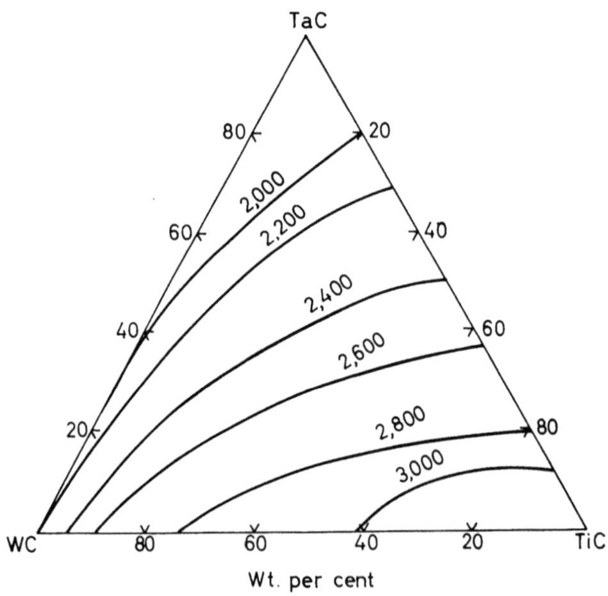

Figure 4.31. Microhardness (kg/mm^2) in the systems WC–TiC–Cr_3C_2 *and* WC–TiC–TaC (Rüdiger[277b])

By courtesy of the Editor, *Metall*

4.10 INTER-CARBIDE SYSTEMS

Of the systems Cr–Mo–C, Cr–W–C and Mo–W–C, i.e. 'inter-Group VI' carbide systems, the first two again contain the $M_{23}C_6$ carbides $(Cr,Mo)_{23}C_6$ and $(Cr,W)_{23}C_6$, with considerable Mo and W solubilities, though restricted, as discussed in connection with this structure. No complete investigation appears to have been carried out*. The occurrence of a true double carbide is improbable but possible, preceding the appearance of stable Mo_2C, W_2C and MoC (with unknown Cr solutions). In Mo–W–C, the continuous Mo–W solution will, upon carbon addition, certainly give way to the simple $(Mo,W)_2C$ and (Mo,W)C series, although again systematic work seems here desirable. (For Mo–W–C and Mo–Ti–C, see also Albert and Norton[287].)

The systems U–W–C and Th–W–C have been studied more widely[259] and provide an instructive contrast (*Figure 4.32*). The first clearly locates the above ternary compound, UWC_2, which equilibrates (in sequence) UC, U_2C_3, UC_2, W, W_2C, WC and C, with insignificant solubilities. The second contains no ternary compound, and the direct binary inter-carbide equilibria are as shown. Both systems clearly divide into two pseudo-ternaries, U(–Th)–W–UC(ThC) and UC(ThC)–W–C. It is highly probable that the U(Th)–Mo–C systems are similar, but this requires investigation.

The systems V–W–C and V–Mo–C have also been investigated[259], confirming that V_2C–W_2C and V_2C–Mo_2C form continuous solid solutions. On the other hand, VC–WC and VC–MoC, whilst forming pseudo-binary sections parallel to the former, only show partial solutions, namely in VC up to 43 mol. per cent WC and 70 per cent MoC† (at 1,800°C), in WC virtually no VC, but in MoC up to 15 per cent VC. However, there are indications at very high temperatures for *complete* solid solutions VC–Mo_3C_2, the latter now in its cubic (B1-type) form which, incidentally, can be stabilized by very small uranium and thorium additions, again well in line with our overall picture. The corresponding lattice spacing curves are included in *Figure 4.22.* The authors even envisage in the section VC–WC an almost complete B1 solid solution at extreme temperatures (2,500°C) and a possible cubic form of WC just below the solidus, though still unverified (a W_3C_2, similar to Mo_3C_2 and Cr_3C_2).

* A recent paper on the Cr–W–C system has[286] corrected this position. It shows (for a 1,350°C section) no solubility of Cr in WC, but a very extensive one in W_2C, to give a $(W,Cr)_2C$ series up to nearly 90 at. per cent Cr,10 W. Further chromium forms first a ternary compound of W_2C-related structure (orthorhombic Mo_2C type, with $a = 4{\cdot}47_5$, $b = 5{\cdot}680$ and $c = 4{\cdot}95_0 kX$), before limited solid solutions of W in $Cr_{23}C_6$, Cr_7C_3 and Cr_3C_2 appear. The point of principle is that the complex chromium carbide structures can thus be eliminated by only minor W (and, no doubt, Mo) additions, i.e. chromium can easily be rendered a 'normal' Group VI metal, effectively similar in its reaction with carbon to molybdenum and tungsten, its anomaly due to an excessive Hägg radius ratio being eliminated. A compound 'Cr_2C' thus *almost* exists. The interposition of a ternary phase of Mo_2C structure as an intermediate stage between $(W,Cr)_2C$ and the chromium carbide is also consistent. From the viewpoint of alloy-steel metallography all this is highly relevant because it confirms how the common occurrence of $M_{23}C_6$ and M_7C_3 in chromium steels can be replaced by that of M_2C through W or Mo additions. However, *iron* will counter-stabilize the former (cf. the Fe–Cr–W–C system) and, indeed, stabilize the η-carbide structure which is absent in Cr–W–C. For higher carbon, WC and Cr_3C_2 equilibrate directly with *graphite*, as does, notably, the Cr-saturated W_2C solid solution.

† Referred to by the authors as 'Mo_3C_2' (see previous discussion).

Systems of UC with Cr, Fe, Co and Ni and with UFe_2 have all been found[288] to contain eutectic sections (*Table 4.11*).

Table 4.11

	Eutectic temperature, °C	*Composition*, wt. per cent
UC–Cr	1,435	34 Cr
{UC–Fe	1,170	51·5 Fe
{UC–UFe_2	1,040	92 UFe_2*
UC–Ni	(more complex:) up to 30 Ni–UC + VC_2 + UNi_5*	
	40–50 Ni – UC_2 + C (graphite) + UNi_5	
	60–90 Ni – C + Ni + UNi_5.	
	low ternary eutectic probable	

* Related work[289] suggested the ternary compound $UFeC_2$, on the UC_2–Fe join. This compound may be regarded as an evolution, to Group VIII, of the $UCrC_2$ phase discussed earlier, and probably typifies other such phases UMC_2. Also, when Ni is added to UC, the formation of a phase $UNiC_2$ has been proven[290], tetragonal with $a = 4{\cdot}961$, $c = 7{\cdot}345$ Å

This is undoubtedly consistent with a decreasing carbide-forming tendency and the ascendence, in this order, of the formation of stable intermetallic compounds. It can probably be safely assumed that Mn and Co interpose

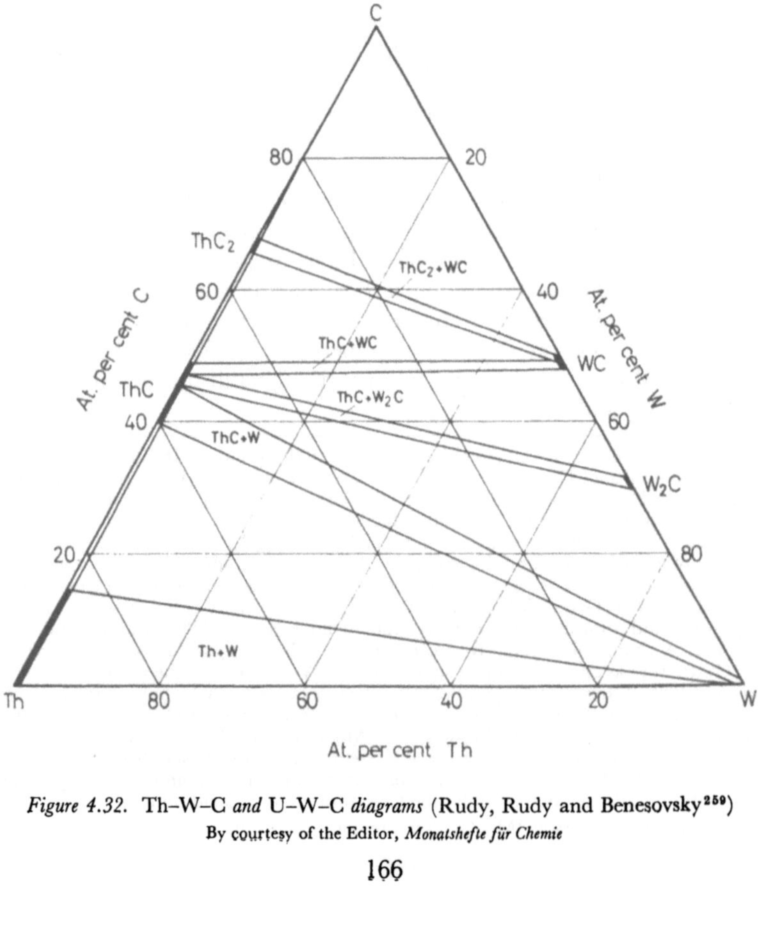

Figure 4.32. Th–W–C *and* U–W–C *diagrams* (Rudy, Rudy and Benesovsky[259])
By courtesy of the Editor, *Monatshefte für Chemie*

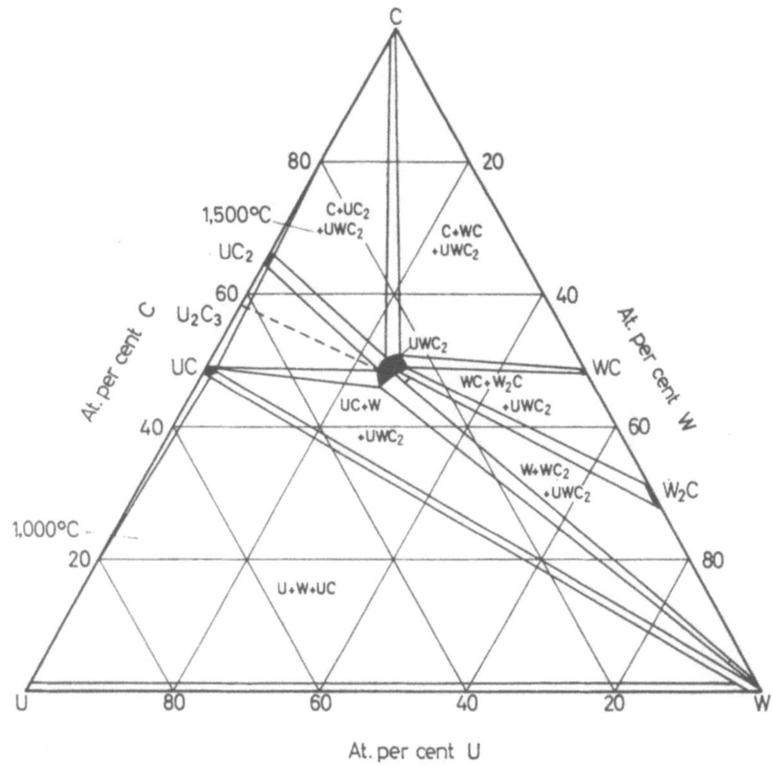

Figure 4.32b

consistently in the sequence. (For these and associated phase diagrams of U–*M*–C type, see also Bibliography.)

Of particular interest for our purpose of pointing out points of principle are the Re–W–C system and reactions therein, because of the isomorphous linkage of a later-Group metal with an earlier-Group carbide. It is found[291] that considerable solid solutions of WC in Re are formed, namely up to about 33 vol. per cent WC, with a simultaneous expansion of the hexagonal axes, from $a = 2{\cdot}760$, $c = 4{\cdot}458$ for pure Re to $a = 2{\cdot}882$, $c = 4{\cdot}564$ Å at saturation. The point, however, is that of the feasibility of isomorphous replacement of the hexagonal pure metal *by* the carbide. Thus, although Re is itself no carbide former, its high primary C solid solution (see Chapter 3) evolves into the true carbide upon electron abstraction (implied by Re → W replacement). The strict isomorphy, however, is between Re and W_2C, both with their close-packed hexagonal metal lattices. It would be worth enquiring whether in fact a complete solid solution is formed between these; this is supported by recent Russian work[292]. A very low melting eutectic occurs in this system.

4.11 CARBIDE SYSTEMS BASED ON THE Mn and Fe, Co, Ni GROUP METALS

So far, we have only considered ternary systems M_1–M_2–C, where M_1 and M_2 are Group IV, V and VI metals. We now pass on to the systems where

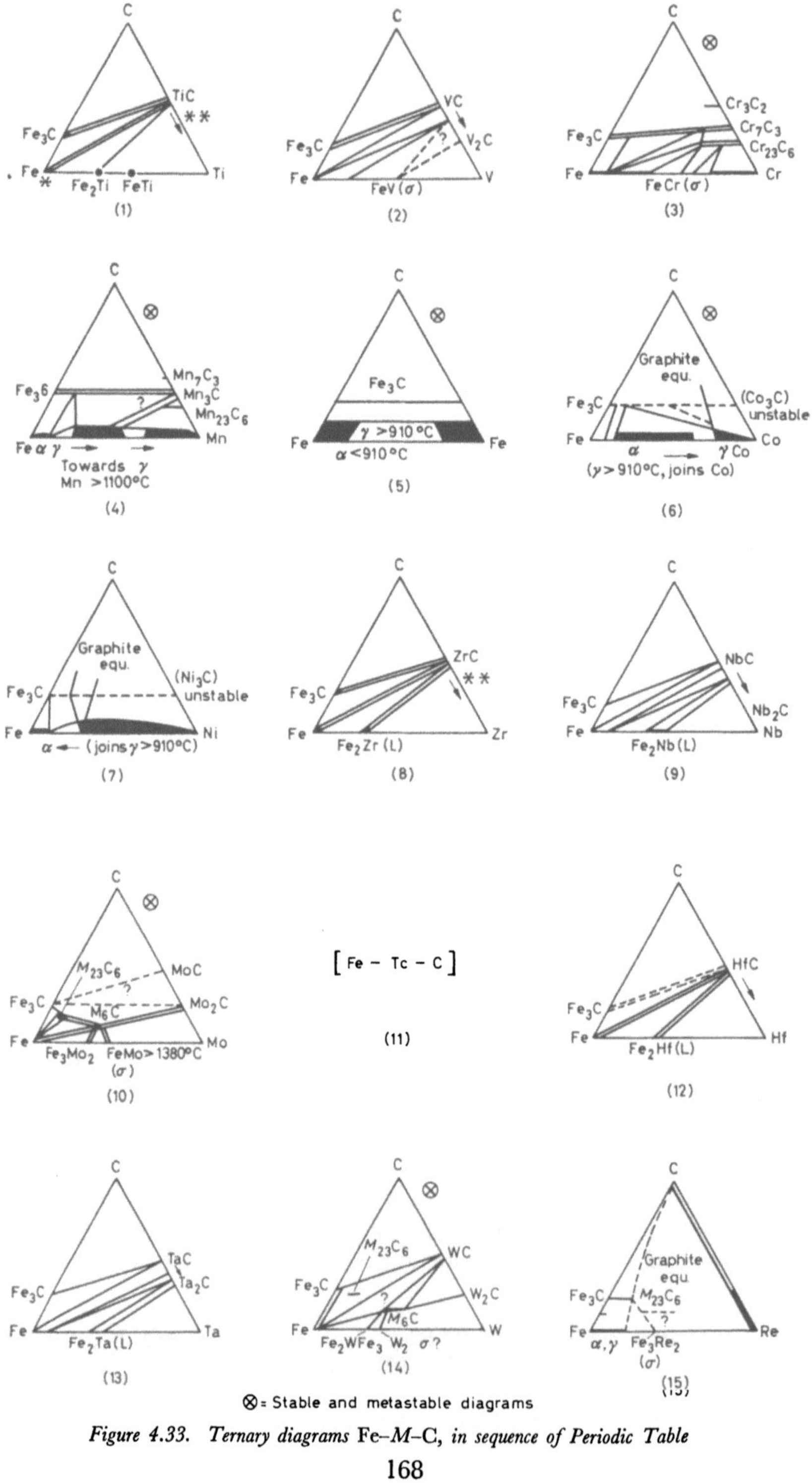

Figure 4.33. Ternary diagrams Fe–*M*–C, *in sequence of Periodic Table*

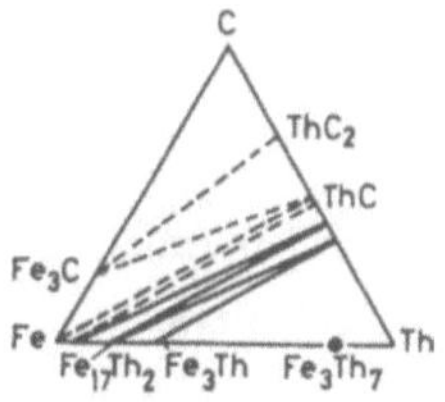

(16)

[Fe – Pa – C]

(17)

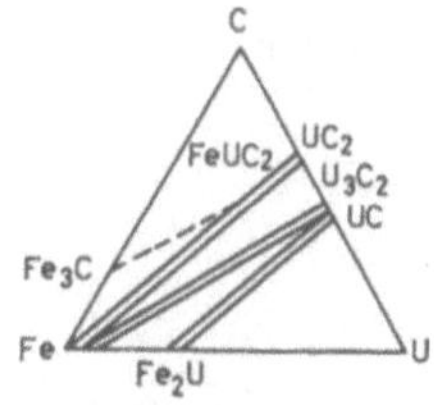

(18)

[Fe—Np—C, Pu—C, etc.—→]
probably similar to Fe-U-C

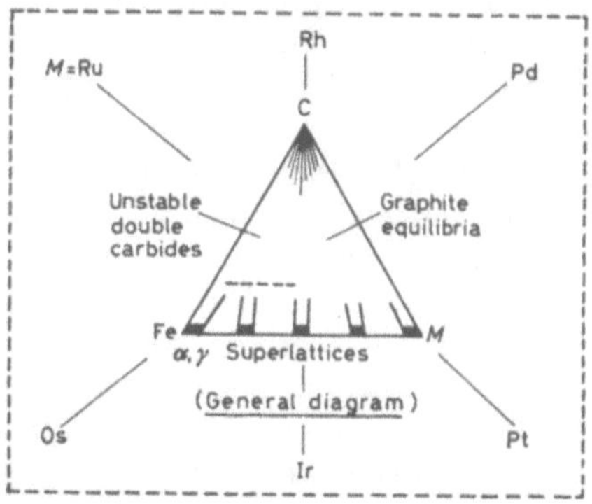

Figure 4.33b

M_1 lies in Group VII and VIII (Mn,Fe,Co,Ni), with a look at the platinum metals, and M_2 still lies in Group IV, V, VI; and to those in which both M_1 and M_2 fall in Groups VII and VIII.

In view of their basic importance in ferrous metallurgy, it is perhaps surprising how very considerable gaps in systematic investigation still exist in this field, and how information has grown largely in an ad hoc manner. Below, a summary attempt is made to co-ordinate the main constitutional features of the systems, as far as available. Much fuller information often exists on the quarternary and multi-element alloys than on the basic ternaries or even binaries. However, this is perhaps a natural consequence of the fact that, originally, the complex alloy steels were evolved in industry largely empirically on an optimum property basis, often with little regard to fundamental knowledge. Research on carbide constitutions, the part of dislocations, etc. then followed, proceeding from the chemically more complex to the simpler systems. This in no way detracts from the value of the more empirical pioneering data, and there are now strong signs of the process becoming reversed, i.e. newer materials being developed on the basis of the fundamental constitutional knowledge gradually built up. A similar evolution, from the more complex to the simpler systems, now again broadly reversed, has occurred in glass technology, mineralogy and other fields.

Apart from the commentary below, it is useful to show (*Figure 4.33*) an outline of the ternary carbide diagrams of *iron base*. Although this Chapter is mainly concerned with the carbides, not the α- and γ-primary solutions, the latter obviously enter as integral parts in this diagram.

The order of presentation will be

	where M =
Mn–M–C	Group IV → V → VI metals
Fe–M–C	Group IV → V → VI → VII → VIII metals
(Co and Ni)–M–C	Group IV → V → VI → VII → VIII metals

(1) *Mn—(Ti,Zr,Hf,Th)—C—*

None of these systems appears to have been adequately determined.

The surrounding *binaries* are, of course, well established: for Mn–C and (Ti,Zr,Hf,Th)–C, see preceding Section*.

The systems Mn–(Ti,Zr,Hf,Th) are characterized by the intermetallic compounds Mn_3Ti, Mn_2Ti and a σ-phase MnTi forming peritectoidally, by the Laves phases Mn_2Zr, Mn_2Hf and Mn_2Th and a further possible compound $Mn_{12}Th$. At the primary solutions, Mn stabilizes β-Ti to a limited extent (to a eutectoid), whilst Ti stabilizes δ-Mn, i.e. the BCC phases in each case. Little is known in this respect of Mn–Zr, Hf and Th.

The ternary systems *might* involve the occurrence, at least metastably, of a double carbide of η-carbide type, such as a hypothetical 'Mn_3Zr_3C' or 'Mn_3Hf_3C', but this has not been reported (a subject worth investigating). In Mn–Ti–C, it may be noted that although a ternary η-carbide is as yet unknown, an η-oxide, Mn_2Ti_4O, does exist[11], as does Mn_2V_4O. The dominant stability of TiC, ZrC, HfC and ThC over large parts of the ternaries seems in little doubt, probably also equilibrating with primary Mn and the intermetallic phases.

Craighead *et al.*[296], in their work on titanium-based alloys, include the Ti–Mn–C system, as well as (Ti–C) + Cu,Si,V,Cr,Fe and Co; (Ti–N) + Cr; (Ti–Cr) + V,Mo,W,Co and Ni, and (Ti–Mn) + Si,Cr,W and Fe†. Their viewpoint being largely that of mechanical properties, heat-treatment response and aging, constitutional information was limited to the extent of the α-/β-Ti transformation range. However, in Ti–Mn–C and Ti–Cr–C alloys an insoluble carbide phase was observed, namely TiC, which, interestingly, appears less soluble at temperatures where β-Ti is the stable phase than at lower temperatures where α and β coexist. This is again consistent with our earlier topic of the close-packed phase occurring upon C addition, be it in solution or as the carbide. In the system Ti–Mn, as well as in Ti–Fe, Ti–V and Ti–Cu, carbon addition tended to limit single-phase β to higher temperatures. (The paper[296] gives details on age hardening and mechanical properties, for which these alloys are notable, e.g. strengthening with loss of ductility; susceptibility to thermo-mechanical treatment; generally beneficial effect of limited carbon, etc.)

* For all adjoining M–C binary systems used here and subsequently, see earlier part of this Chapter. The metal–metal solutions and intermetallics are only mentioned but are outside the scope of this book. For full details, e.g. Hansen's[293,175] and Pearson's[294] books should be consulted. Whilst we are not so much concerned with the liquid solutions, work very relevant to carbide formation in this context is by Schenck *et al.*[295] who found the effect on the C solubilities in liquid Mn, Fe, Co and Ni by numerous third elements.

† This valuable earlier series of papers included binary, ternary and quaternary alloys of titanium base.

(2) *Mn—(V,Nb,Ta,Pa)—C*— None of these systems has been systematically determined. Electronically, a double carbide could well occur, but no η-carbides such as 'Mn_3Nb_3C' or 'Mn_3Ta_3C' are known.

On the Mn–*M* sides, we have a σ-phase $\approx$ Mn_3V evolving, for the heavier atoms, into the Laves-phases Mn_2Nb and Mn_2Ta. It is highly likely that all primary Mn solid solutions ($\alpha,\beta,\gamma,\delta$-Mn), with their varying carbon solubilities (see *Figure 4.1*), equilibriate directly with VC, NbC and TaC, as may also the σ and Laves phases. However, the hemicarbides V_2C, Nb_2C and Ta_2C *may* become restricted by double-carbide formation intervening between them and the intermetallics, particularly for the two heavier metals.

On the carbon side of the pseudo-binary section Mn–*M*C, it is again probable that the various Mn carbides soon become suppressed upon V, etc. addition by the monocarbide forming. However, all this must remain hypothetical, subject to experiment.

Nothing is known on Mn–Pa–C.

(3) *Mn—(Cr,Mo,W,U)—C*— The η-carbides Mn_3Mo_3C and Mn_3W_3C are well established, but this is about the only positive feature known on the ternaries, except for a report[3d] on a χ-phase existing in the Mn–W–C system. It would be of interest to enquire into the question how this phase would link up with α-Mn itself (its isomorph) and whether indeed a solid solution might occur.

Concerning Mn–Cr–C, it is surprising that so little detail is known; but $Mn_{23}C_6$–$Cr_{23}C_6$ and Mn_7C_3–Cr_7C_3 are certain to form complete series of solutions, whilst Mn_3C will only dissolve limited amounts of Mn before stabilization (probably) of $(Cr,Mn)_7C_3$. The Mn–Cr binary side is dominated by an extensive Cr $\rightarrow$ Mn solution, followed by a σ-phase Mn_3Cr, before the solid solutions $\alpha,\beta,\gamma,\delta$-(Mn $\rightarrow$ Cr) are stabilized. Further complex phase changes are involved, however, such as one of σ to a modified lower-temperature form σ' and an α'-phase structurally akin to α-Mn. The very complexity of transformations may here be recognized as the outstanding feature in the metallic Cr–Mn series; in the presence of carbon, this is projected into a complexity of carbide structures which is far from clarified.

Mn–Mo also contains a σ-phase at about Mn_2Mo, giving way, upon carbon addition, to the η-carbide and probably to a carbide $(Mn,Mo)_{23}C_6$. A similar position applies to Mn–W–C. The direct equilibrium MoC,Mo_2C, WC,W_2C with Mn is probably complicated by the interposition of these double carbides. Insufficient data are known.

In Mn–U–C, again, double-carbide formation is probable.

(4) *Fe—Ti—C* (*Figure 4.33*). The iron corner of this system[297–8] shows the following principal features. The bounding binaries contain Fe_3C, TiC, Fe_2Ti (Laves phase) and FeTi (BCC with possible superlattice), besides a doubtful compound $FeTi_2$; Ti rapidly closes the γ-loop in Fe, stabilizing α-, whilst at the Ti end, Fe stabilizes β-Ti.

In the ternary system, no new carbide occurs, and the following direct equilibrium sequence is met: (α-Fe + Fe_3C), (α + Fe_3C + TiC), (α + Tic), (α + TiC + Fe_2Ti), (α + Fe_2Ti) below A_3; (γ-Fe + Fe_3C), (γ +

Fe_3C + TiC), (γ + TiC), ($\gamma + \alpha$ + TiC), (α + TiC), (α + TiC + Fe_2Ti), (α + Fe_2Ti) above A_3.

The closed γ-field in Fe–Ti is initially widened by carbon addition, but beyond a certain carbon value ($\approx$ 0·35 wt. per cent) contracts again. The α/γ transformation is, with Ti addition, displaced to higher temperatures and carbon contents, so that the carbon-poorer alloys are rendered ferritic to melting point. In this connection, the diagram in *Figure 4.34a* is instructive in showing the progressive shrinkage and disappearance of the γ field with Ti content. It is here paralleled by the similar sequences in the Fe–V–C and Fe–Cr–C systems (the latter in *Figure 4.34b*). On the other hand, the disappearance of γ is accompanied by the appearance and increase of TiC,

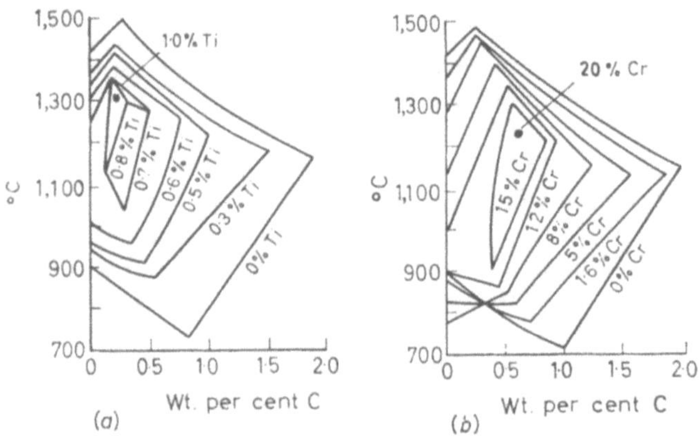

Figure 4.34. Projection of FCC *γ-field boundaries from ternary sections with varying* Ti *and* Cr *contents onto the* Fe–C *plane*

Note shrinkage and disappearance of γ but structural ' take-over ' by FCC *MC* carbides.—System Fe–C–V: intermediate position between (*a*) and (*b*)

By courtesy of the Editor, *Archiv für das Eisenhüttenwesen*

i.e. both the FCC metal structures again replace each other. Thus γ-Fe–TiC becomes the simple quasi-binary section shown: a cardinal theme extended below. (As to the Ti corner of this system, cf. [296].)

(5) *Fe—Zr—C* (*Figure 4.33*)

This system is closely akin to Fe–Ti–C. The pseudo-binary sections[299] are α- or γ-Fe–ZrC, Fe_3C–ZrC and Fe_3Zr_2–ZrC—not as might have been, Fe_3C–ZrC. Basically, therefore, we observe again throughout these system types that it is the FC cubic carbide which directly forms equilibria with the primary Group VIII metal, with cementite and the intermetallic compound.

The Fe–Zr binary contains only the compound Fe_2Zr of cubic Laves-phase structure ($MgCu_2$-type). Zr is practically insoluble in α- and γ-Fe, though rather more soluble in δ-Fe; Fe does not dissolve in α-Zr but slightly so in β-Zr. However, as carbon is added, a significant C solubility in γ-Fe arises already as the pseudo-binary Fe–ZrC is reached (to 0·8 per cent C, 6 per cent Zr at the eutectic of 1,460°C).

The ternary system is divided into two similar parts by the Fe–Zr–C quasi-binary, viz. Fe–ZrC–Fe_3C and Fe–ZrC–Fe_2Zr (for details, see [2b]).

(6) *Fe—Hf—C (Figure 4.33)*

No information is available regarding this ternary system*. However, there is little doubt that it is similar to Fe–Zr–C, the known Laves-phase Fe_2Hf and HfC entering analogously to Fe_2Zr and ZrC.

(7) *Fe—Th—C*

No information is available. Four intermetallics are known: Fe_7Th_2, Fe_5Th, Fe_3Th and Fe_3Th_7, and it is an open question how these and Fe_3C enter into equilibrium with ThC and ThC_2.

(8) *Fe—V—C (Figure 4.33)*

The Fe corner of this system has been comprehensively studied[301–3], with a useful summary by Desch[304]. Again, the pseudo-binary sections Fe–V_4C_3 and Fe_3C–V_4C_3 occur, but this time the new factor of complete solubility of parent metals (Fe–V) enters, though, according to more recent knowledge, broken by the σ-phase. It implies that, up to considerable V contents, it is this solid solution which equilibrates with V_4C_3. The occurrence of the σ-phase FeV was not at the time appreciated, and the appearance of V_2C, for higher V contents, would introduce a further new factor. The V corner is, in fact, still mainly unknown, and it may well be queried whether the two events of σ and of the lower-carbide formation would not be closely associated, suggesting a relation (so far hypothetical), as in *Figure 4.33*.

Since V rapidly closes the γ-loop in pure iron, the addition of vanadium to Fe–C alloys has the effect of progressively isolating the ternary γ field (cf. *Figure 4.34*). Thus we have again the progressive evolution of the γ-cubic carbide assembly, which in the later sequence V $\rightarrow$ Cr $\rightarrow$ Mn $\rightarrow$ Fe will develop into full stabilization, i.e. merging of the two separate FCC metal lattices into one. A γ-solution field thus extends, through the ternary, from γ-Fe–C to γ-Fe–V, with a γ minimum at a certain ratio, and its $\gamma \rightarrow \alpha$ and $\gamma \rightarrow$ carbide decomposition kinetics clearly dictates the microstructure and behaviour of vanadium steels at varying V levels.

The problem whether, in vanadium steels, one or both of the carbides 'VC' or 'V_4C_3' occur has been studied[303], but the result was not conclusive; in more modern understanding this should clearly be regarded as part of one homogeneity range with a defect lattice and is now so accepted. Which particular composition will prevail may well vary critically with composition and heat-treatment history of the steel concerned, and indeed a heterogeneous band of cubic carbide isomorphs may occur over a range of spacings, as indicated by some of the author's measurements. Furthermore, the occurrence or not of hexagonal V_2C will vary, and this links, for lower V contents, with the formation of metastable ε-carbide.

(9) *Fe—Nb—C (Figure 4.33)*

The Fe–Nb system, as distinct from Fe–V, does not form complete solid solutions but the intermetallic compounds Fe_2Nb (Laves phase, with a considerable homogeneity range), $\approx$ FeNb (σ) and Fe_2Nb_3 (η-carbide-type structure). Thus the iron corner of the Fe–Nb–C diagram[305] is again

* A Russian paper[300] on Fe–Hf–C gives detailed phase diagrams up to 20 wt. per cent Hf, 2·5 C. The general resemblance to Fe–Zr–C is confirmed, for instance the dominance and carbon stabilization of the (γ + HfC) field and its eutectoid decomposition, although there are quantitative differences.

one on which a pseudo-binary Fe–Nb_4C_3 section is flanked by (Fe + Nb_4C_3 + Fe_2Nb) and (Fe + Nb_4C_3 + Fe_3C) equilibria. The austenite (γ) region of Fe + Fe_3C is also constricted by Nb addition, associated with the appearance of Nb_4C_3. A decomposition of type $\gamma \rightarrow \alpha + Nb_4C_3 + Fe_3C$ occurs below 705°C; on the other side, a transformation $\gamma + Fe_2Nb \rightarrow \alpha + Nb_4C_3$ takes place at 920°C. However, since Eggers and Peter's work[305], the σ phase and the wide Laves-phase homogeneity range have been found on the Fe–Nb side[2g], which places the ternary Fe–Nb–C in a position intermediate to both Fe–Zr–C and Fe–V–C. Moreover, the occurrence of Nb_2C and the Nb_4C_3/NbC vacancy range must be allowed for, so that areas other than the Fe corner must be at present speculative.

It seems very likely, though unconfirmed, that the C-richer and C-poorer ends of the defect lattice range are successively in equilibrium with primary $\alpha(\gamma)$-Fe and with Laves phase. It would imply, on a more general plane, that carburization of the transition metals *Laves* phases produces the defect

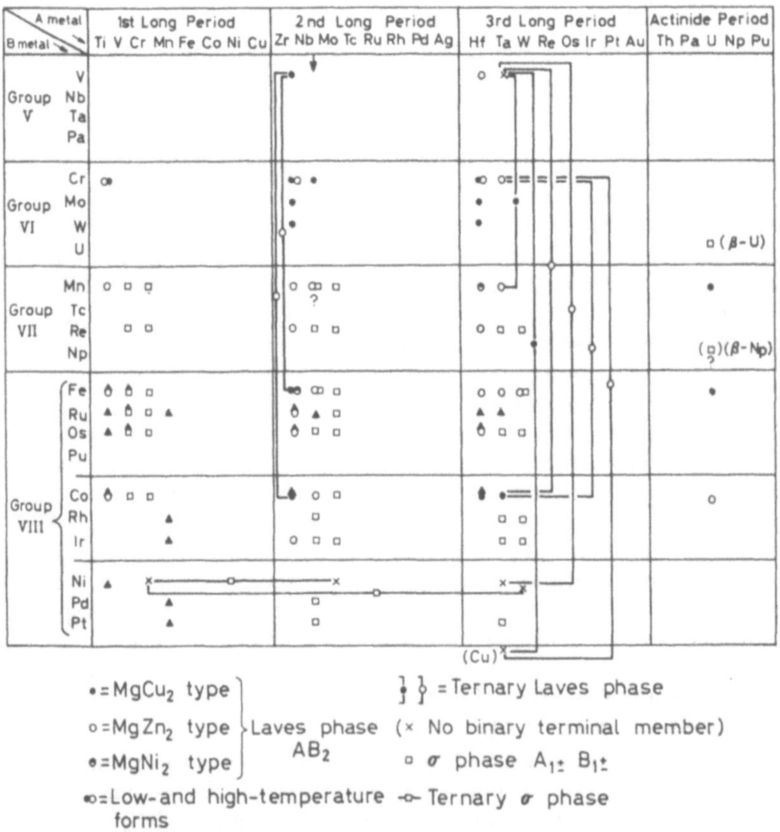

Figure 4.35. Binary and ternary Laves, σ- and CsCl-*type phases amongst transition metals* (After Goldschmidt[2h])

(To be correlated with the carbides formed by the same metal combinations)
By courtesy of the Editor, *Journal of less-common Metals*

NaCl-*type* carbide, that of iron-(or cobalt- or nickel-)richer alloys the more completely filled carbide. Similarly, the σ-phases, FeCr, FeMo, FeV, etc. tend, on carburization, to $M_{23}C_6$ carbide formation. It is instructive in this context to show (*Figure 4.35*) the general incidence of Laves-, σ- and CsCl-type phases amongst intermetallic systems[2h], and to compare this to the type of carbide formation, as presented here.

(10) *Fe—Ta—C* (*Figure 4.33*)

Any systematic knowledge of this ternary system appears to be lacking. However, from the binary similarities Fe–Nb/Fe–Ta (Laves phases, etc.) and Nb–C/Ta–C, a general analogy might be presumed.

(11) *Fe—Cr—C* (*Figure 4.33*)

This important system is dictated largely by the three known chromium carbides, $Cr_{23}C_6$, Cr_7C_3 and Cr_3C_2, and their considerable solid solubilities for Fe and, conversely, that of Cr in Fe_3C. Fe_3C–Cr_7C_3 represents a pseudo-binary system; the general phase assemblies are illustrated in *Figure 4.36*,

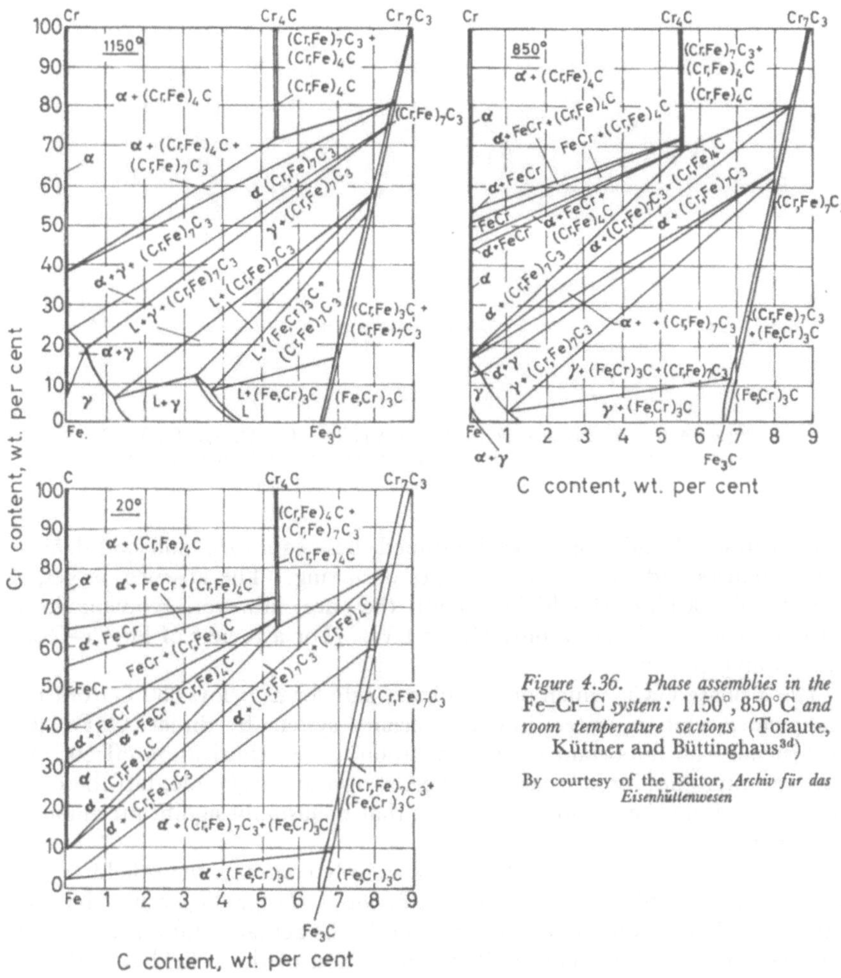

Figure 4.36. Phase assemblies in the Fe–Cr–C *system:* 1150°, 850°C *and room temperature sections* (Tofaute, Küttner and Büttinghaus[3d])

By courtesy of the Editor, *Archiv für das Eisenhüttenwesen* showing the equilibrium assemblies at 1,150°C, 850°C and room temperature[306]. The authors (including Tofaute's preceding work) have carried out perhaps one of the most comprehensive investigations of its kind. A fuller, slightly simplified, equilibrium diagram[307] is shown in *Figure 4.37.* The early x-ray crystallographic work on Fe–Cr–C by Westgren, Phragmén

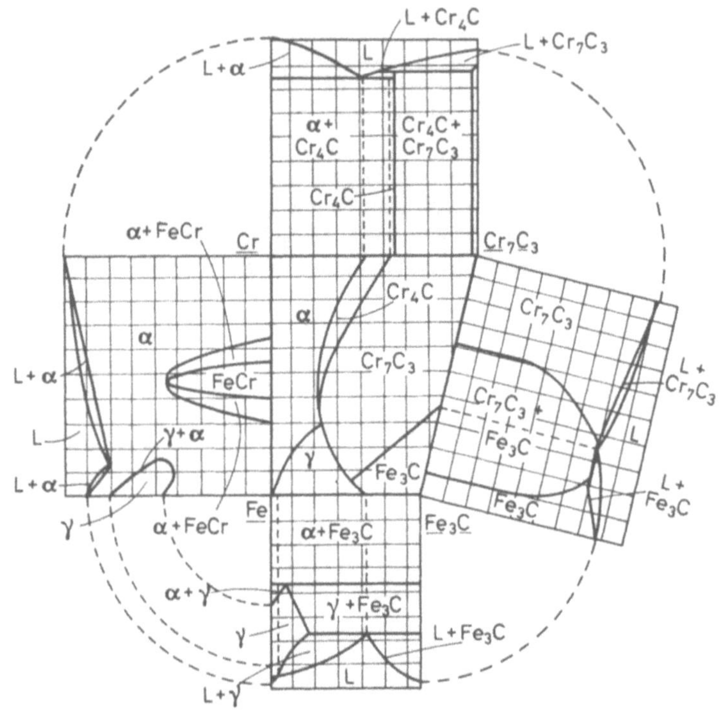

Figure 4.37. Solidification in the Fe–Cr–Cr_4C(= $Cr_{23}C_6$)–Cr_7C_3–Fe_3C *system* (After Tofaute, Küttner and Büttinghaus[306] and Monypenny[307])

By courtesy of Messrs. Chapman & Hall, London

and Negresco[308] had already well outlined these equilibria and had defined the structures and the spacing changes occurring. The absence, as yet, of any double carbides should be noted: they are, however, evolving down Group VI for the larger atoms Mo and W. (For a review of Fe–Cr–C see also Monypenny[307].)

The σ-phase FeCr, which on the binary side forms below 830°C from the complete α-solution, enters into the ternary system by direct $(Cr,Fe)_{23}C_6$ coexistence—an important point in relation to decarburization effects, etc. The carbon solubility in σ- as in α-(Fe,Cr) appears negligible, but there are indications from the author's own work that metastably significant amounts of carbon can dissolve in σ.

The addition of Cr to Fe alloys progressively contracts the γ field (*Figure 4.34b*); this should be compared with *Figure 4.34a* for Ti, noting the now rather less drastic contraction culminating in the actual γ-stabilization when we reach manganese. The limited constriction of the γ field is under 20 wt. per cent Cr,0·6 C. The complexity of reactions in this system is one of its outstanding features (cf. [307], where the implications on stainless steels are discussed), which also have a bearing on the modern developments of high-strength steels.

Upon decomposing, the γ-phase forms the carbides $(Cr,Fe)_7C_3$, $(Fe,Cr)_3C$ and $(Cr,Fe)_{23}C_6$, eventually to terminate in the α equilibria shown for these three carbides. It must be emphasized, however, that these equilibria are rarely achieved in practical Cr steels, and many of their properties are based on metastable conditions part-way towards them involving, however, these and apparently no new phases.

The equilibria of Cr_3C_2 (the refractory carbide) with Fe–Cr do not seem to have been systematically investigated. In connection with this system, an x-ray high-temperature analysis of Fe–Cr–C alloys[309] is of interest, showing how the $Cr_{23}C_6$ phase dissolves in the austenite lattice and the kinetics of the solution involved. The Fe–Cr–C system was recently further studied[310–12]; the latter paper, relating to the kinetics of carbide formation in chromium steels, is typical of the considerable literature bearing on the technology of Cr steels.

(12) *Fe—Mo—C and Fe—W—C* (*Figure 4.33*)

The Fe–Mo–C and Fe–W–C systems are characterized by double-carbide formation, namely (1) η-$Fe_4(Mo,W)_2C/Fe_3(Mo,W)_3C$, and (2) $Fe_{21}(Mo,W)_2C_6$, the $Cr_{23}C_6$ isomorph. A second M_6C-carbide, $(\eta_2)Fe_2Mo_4C$, has also been reported[3e]; only Fe_3Mo_3C, not η_2, was found when iron was in excess (as in high-speed steel).

For data on the ternary equilibria in the Fe corners of these systems, see [3a,163,313–4]. The low-temperature equilibria had for long been accepted as of the following sequence (starting from the Fe–C side):

$$\alpha + Fe_3C/\alpha + Fe_3C + M_{23}C_6/\alpha + M_{23}C_6/\alpha + M_{23}C_6 + M_6C/\alpha + M_6C/\alpha + M_6C + Fe_7Mo_3 \text{ (or } Fe_3W_2)/\alpha + Fe_7Mo_3 \text{ (or } Fe_3W)$$

However, this was amended[314] so as to include, after prolonged anneals, an *additional* ternary carbide phase, denoted ξ, and for higher Mo, a phase 'Mo_2C' which seems to be a considerable solid solution in Mo_2C. The latter diagram is shown in *Figure 4.38*.

The Fe–Mo and Fe–W binary sides contain the intermetallics Fe_2Mo and Fe_2W (Laves phase), Fe_7Mo_6–$Fe_3W_2(Fe_7W_6)$ of a rhombohedral structure and high-temperature σ-phase FeMo and FeW. A hexagonal Laves-phase Fe_2Mo–Fe_2W was also claimed. The solubilities of Mo and W in Fe are appreciable, and the γ loop is rapidly closed. Upon carbon addition, the $M_{23}C_6$ and M_6C carbides form instantly, i.e. without C solubility in the compounds. However, a χ (α-Mn type) and a π (β-Mn type) phase have also been found[2e,3d] in these systems. In investigations on carbides in Mo and W steels, 'metastable' and 'stable' Fe–W–C and Fe–Mo–C diagrams have been stipulated[3b]. The two Fe–W–C systems are shown superimposed in *Figure 4.4* and Kuo's conclusions[3b]—the work also included Fe–Cr–C—may be quoted:

(1) In the system Fe–Cr–C, the carbides present in equilibrium at 700°C are Fe_3C, Cr_7C_3, or $Cr_{23}C_6$, in order of increasing chromium content of the alloy. Carbide Cr_3C_2 does not exist in chromium steels.

(2) The stable carbides in the Fe–W–C system at 700°C are Fe_3C and WC, if the atomic ratio W:C in the steel is smaller than unity; otherwise, M_6C and WC coexist as stable carbides. Carbides $M_{23}C_6$ and W_2C are metastable. When carbides are formed directly from austenite at 700°C by isothermal transformation, they appear in the order Fe_3C, $M_{23}C_6$ and M_6C as the atomic ratio W:C in the steel increases.

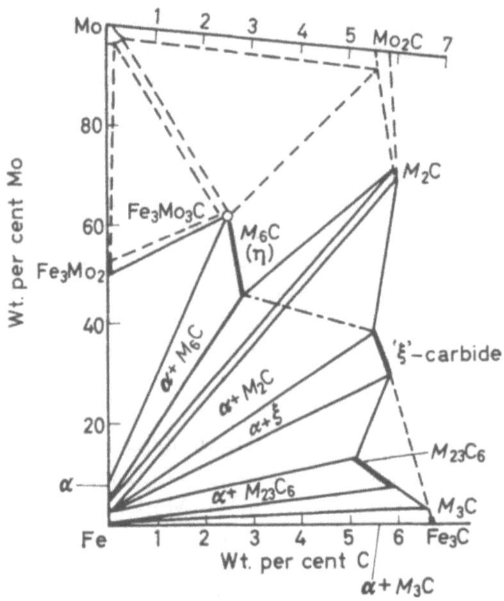

Figure 4.38. Phase diagram for Fe–Mo–C *alloys at* 700°C (Sato *et al.*[314])

— — — = final stable system

By courtesy of the Editor, *Journal of the Japan Institute of Metals*

(3) Stable and metastable carbides analogous to those described above are also found in the Fe–Mo–C system. In addition, a new iron–molybdenum double carbide, M_aC_b, with unknown structure, has been found in steels containing 1–5 per cent of molybdenum; its composition is probably between those of $Fe_{21}Mo_2C_6$ and Fe_4Mo_2C. This carbide occurs only after long annealing at 700°C and is metastable. Carbide Mo_2C also occurs in isothermally transformed molybdenum steels of very low molybdenum content (0·4 per cent).

(4) On tempering, Fe_3C with inherited metal composition from the matrix forms first (after Jack's ε-iron carbide) in all steels. On prolonged tempering, however, other alloy carbides formed at the expense of Fe_3C can dissolve as much as 18 per cent of chromium but less than 1–2 per cent of molybdenum or tungsten.

(5) The general carbide reaction in chromium steels is Fe_3C, Cr_7C_3,$Cr_{23}C_6$. The carbide reaction $Fe_3C \rightarrow Cr_7C_3$ is found to take place in two stages: the concentration of chromium in Fe_3C to its maximum solubility, and the transformation from Fe_3C into Cr_7C_3. The carbide reaction $Cr_7C_3 \rightarrow Cr_{23}C_6$ may follow the same stages. At high temperatures, $Cr_{23}C_6$ is found to be less stable than Cr_7C_3, the former dissolving first and completely in austenite.

(6) In tungsten steels, the tungsten-containing carbides formed after Fe_3C appear in the order of increasing atomic ratio W:C; i.e. $M_{23}C_6$ ($Fe_{21}W_2C_6$), WC, M_6C (Fe_4W_2C) and W_2C. On prolonged tempering, the carbide formed after W_2C is M_6C and after M_6C is $M_{23}C_6$. The metastable carbides W_2C and $M_{23}C_6$ later disappear completely. The carbide Fe_3C first redissolves in ferrite to supply carbon for the further formation of tungsten-containing carbides, and if the WC ratio is smaller than unity, reappears later when all the available tungsten is bound in WC.

(7) The carbide reactions in molybdenum steels are similar to those found in tungsten steels, with the following two exceptions: (*a*) on tempering, the carbides appearing after Fe_3C in steels with increasing atomic ratio Mo:C, $M_{23}C_6$, MoC, and Mo_2C; M_6C appears always after Mo_2C; (*b*) the carbide appearing after M_6C is M_aC_b, not $M_{23}C_6$.

(8) In explaining the observed carbide reactions, the chemical compositions of the carbides and matrix are considered instead of the crystal structure of the carbides. It is

tentatively suggested that Cr_7C_3 is the reaction product of Fe_3C and the matrix, whereas the molybdenum- and tungsten-containing carbides are formed directly from the matrix.

(9) The rate of carbide reaction is found to be higher in the quenched and tempered steel specimens. This is probably due to the internal stress induced by quenching. The reaction rate is high in chromium steels and extremely low in molybdenum steels.

It must be said that some of these conceptions are rather controversial and await much further work. However, the quote perhaps summarizes most suitably the nature of these complex equilibria of Group VI (Fe–C) compounds. Later work on steel constitutions[315] has helped to shed much light on the kinetics of carbide formation, but the need for a comprehensive investigation of equilibria and their progress of attainment in Fe–Cr–C, Fe–Mo–C and Fe–W–C is obvious.

In the Fe–Mo–C system, the 'M_aC_b' carbide[3b] is again relevant, now solved[20] (see p. 113) as Fe_2MoC, i.e. a Mo-bearing variant to Fe_3C, part-way towards the η-carbide Fe_3Mo_3C.

An important low-C carbide in the Fe–W–C system has been reported[316], of composition Fe_6W_6C. It is structurally closely related to Fe_3W_3C (the η-carbide) and confirms the idea of the M_6C composition being not invariant but containing potentially defect carbon sites, towards the intermetallic phase Fe_7W_6. Basically this is very liable to occur, e.g. in tool steels upon decarburization, etc.—The carbon positions in Fe_6W_6C have been determined by neutron diffraction (Leciejewicz[149a]).

It is of interest to record the atomic parameters in this '$M_{12}C$' carbide, which might be denoted $\bar{\eta}$, because derived by carbon substraction from the η lattice (cf. *Figure 4.7*).

Space group: Fd 3m

16 Fe atoms at $\frac{1}{2},\frac{1}{2},\frac{1}{2};\frac{1}{4},\frac{1}{2},\frac{1}{4};\frac{1}{2},\frac{1}{4},\frac{1}{4};\frac{1}{4},\frac{1}{4},\frac{1}{2}$; + FCT (*d* positions)
FCT = face-centring translations

32 Fe atoms at $\pm$ ($u,u,u;\ u,\frac{1}{4}\text{-}u,\ \frac{1}{4}\text{-}u;\ \frac{1}{4}\text{-}u,u,\frac{1}{4}\text{-}u;\ \frac{1}{2}\text{-}u,\frac{1}{4}\text{-}u,u$) + FCT
where $u = 0{\cdot}705$ (*e* positions)

48 W atoms at $\pm$ ($x,\frac{1}{8},\frac{1}{8};\frac{1}{8},x,\frac{1}{8};\frac{1}{8},\frac{1}{8},x;\ \frac{1}{4}\text{-}x,\frac{1}{8},\frac{1}{8};\frac{1}{8},\frac{1}{4}\text{-}x,\frac{1}{8};\frac{1}{8};\frac{1}{8},\frac{1}{4}\text{-}x$) + FCT,
where $x = 0{\cdot}070$ (*f* positions)

8 carbon atoms at $\pm$ ($\frac{1}{8},\frac{1}{8},\frac{1}{8}$) + FCT

There are thus 96 metal atoms and 8 carbon atoms per unit cell in $\bar{\eta}$ as against 96 and 16 in η_1 and η_2 (i.e. $\bar{\eta} = M_{96}C_8$, $\eta = M_{96}C_{16}$), with the same metal atom positions but a change in interstitial occupancy (only presumed, because the neutron-diffraction work of Leceijewicz on $M_{12}C$ does not yet seem to have been published for M_6C).

It is highly probable that a molybdenum analogue Fe_6Mo_6C would also occur and, more widely, that these are only first prototypes of a new family of $\bar{\eta}$ phases associated with the η family discussed earlier; that is, in $M_{12}C$, more generally M = [Group IV–VIA plus Group VII–VIII] metals and carbon replaceable by nitrogen, oxygen or, indeed, vacancies. Regarding the latter, the existence of the interstitial-free intermetallic compounds of η type (e.g. Ti_2Ni) and that of the $M_{12}X$ phases would simply represent stages of a continuous interstitial depletion process, starting at a maximum occupancy of M_6X* (see also *Table 4.2*).

* Not even this is certain, and carbon supersaturation is at least possible.

Reverting to the broad Fe–Mo–C and Fe–W–C phase diagrams, it is by no means clear yet how the newer η, $\bar{\eta}$ and M_aC_b (Fe_2MoC) phases will equilibrate and link to the Fe–Mo Laves phases. This suggests itself as a fruitful field for further research, important also from the practical aspect of alloy tool steels (decarburization, etc.).

(13) *Fe—U—C, Fe—Np—C—and Fe—Pu—C*

Little is known on the Fe–U–C system, except for a general statement[317] that 'U additions to carbon steels give two new constituents', which are interpreted as uranium carbides, and the recent information given in *Table 4.11*, in particular the double carbide $FeUC_2$[289].

No information is available on the Fe–Np–C and Fe–Pu–C systems.

(14) *Fe—Mn—C (Figure 4.33)*

The complex relations in the Mn–C system have been referred to already (p. 117; see also *Figure 4.1*). The effect of Fe is naturally to favour the cementite formation at the cost of $Mn_{23}C_6$ and Mn_7C_3. The stabilization of the $(Fe,Mn)_3C$ structure by Mn had been observed early[318].

The binary system Fe–Mn is characterized by the strong γ-stabilizing effect of Mn, in many ways similar to Ni, to which carbon addition will only contribute, and at high temperatures there is a complete solid solution with FCC γ-Mn, though tetragonally distorted upon quenching. No stable intermetallic compounds occur; however, the metastable and martensite-like $\gamma \rightarrow \varepsilon$ and $\gamma \rightarrow \alpha$ transformations in the region 5–15 per cent Mn have to be noted. A Fe–Mn σ-phase has been reported: a very likely contingency, but not yet confirmed.

The metastable ε-(FeMn) phase is hexagonal close-packed, and an interesting, though so far unresolved, question is that of its solution link-up with the isomorphous hexagonal carbides and, in particular, with Jack's ε-iron carbide. Newer Russian work had indicated that C addition to Fe–Mn in fact *hampers* ε formation.

The primary α-Mn and β-Mn structures dissolve up to about 30 per cent Fe.

Extensive lattice spacing work within the α-Fe, γ-(Fe,Mn), α-(Mn,Fe) and β-(Mn,Fe) solution series has been done, but apparently little upon carbon addition. Clearly an α-Mn–γ-phase join should occur, as will a β-(Mn,Fe)–π-phase join, again with significant carbon solubilities—another question awaiting an answer.

The ternary system Fe–Mn–C has been much studied[319–22], but only Vogel and Döring[154] present at least an approach to an equilibrium diagram, as shown in *Figure 4.39*, though with reservations: in particular $M_{23}C_6$ and M_7C_3 do not here find their place. Most of the other work quoted deals more with the metastable reactions in this system, such as ε formation and hardening effects through it, and the phase diagram is closely bound up with the intermediary and unstable states involved. Much valuable work on the Fe–Mn–C system is thus implicitly contained in general research on steels, hardenability and martensites (typical[318,320]).

The existence of a Fe–Mn *ternary carbide* has been claimed[61b], of monoclinic structure and composition 68 per cent Mn, 25 per cent Fe, 7 per cent C, but no later confirmation seems to have been forthcoming. A summary of work on Fe–Mn–C had been given by the author[2b] in 1948, and not much newer work seems to have been done.

For lower Mn (high Fe) contents, the equilibria are plainly those of ferrite, austenite and $(FeMn)_3C$.

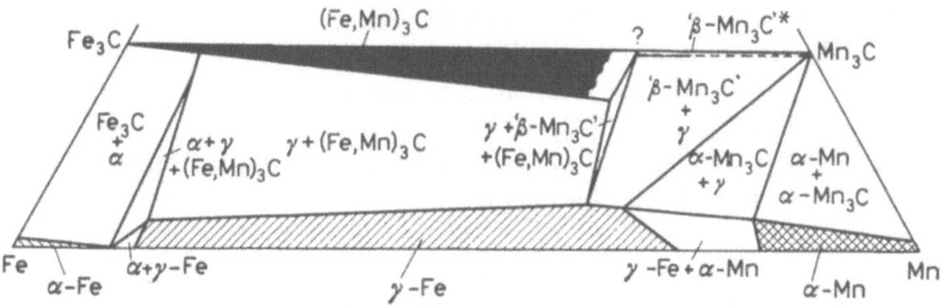

Figure 4.39. 'Ideal' Fe–Mn–C *diagram* (Vogel and Döring[154])

* This phase probably superseded by a, yet uncertain, $Mn_{23}C_6/Mn_7C_3/Mn_3C$ (assembly)

By courtesy of the Editor, *Archiv für das Eisenhüttenwesen*

However, by determination of thermodynamic activities in Fe–Mn–C alloys[323], the existence of Mn_7C_3 and extensive $(Fe,Mn)_7C_3$ solid solutions, in equilibrium with graphite could be re-affirmed, but also an *ideal* substitution of Mn by Fe: the system Mn_7C_3–'Fe_7C_3' is the strictly correct pseudo-binary, although the latter is hypothetical and, in fact, transformation to Fe_3C occurs on decreasing Mn. This would indicate a significant difference from Fe–Cr–C, where the pseudo-binary is an inclined one, $Cr_7C_3 \longleftrightarrow Fe_3C$. In Mn–Fe–C, the pseudo-binaries then would be $Mn_3C \longleftrightarrow Fe_3C$, and the above, $Mn_7C_3 \longleftrightarrow$ 'Fe_7C_3'. The question arises whether a finite homogeneity width in C content could occur [i.e. $(Fe,Mn)_3C$ and $(Fe,Mn)_7C_3$], but this is not known. Some non-stoichiometry in the two structures appears very possible.

The place of $Mn_{23}C_6$ in this system is not clear, but it is likely to equilibrate with both γ-FCC solution, from which it could precipitate (as $M_{23}C_6$ is known to do more generally, e.g. in austenitic Cr steel), and with σ and χ phase.

(15) *Fe—(Re and Pt-Group Metals)—C*

Nothing whatever appears to be known regarding the ternary systems Fe–Re–C and Fe–(Ru,Os; Rh,Ir; Pd,Pt)–C although the Fe–*M* binaries are of course well or fairly well established. Considering the increasing graphitizing effect of these metals upon carbon addition, new carbides will almost certainly not be formed in the ternaries. Also all the second metals concerned stabilize the γ-form of Fe and form either complete or partial solid solutions with it: an effect consonant with carbon addition and almost certainly promoted by this towards lower *M* contents.

In Fe–Re we still have the stable intermetallic Fe_3Re_2 of σ-type structure, and judging purely from analogy with Mn (above Re in Group VII) and our foregoing Fe–(Group VI)–C systems, it is well possible that ternary $M_{23}C_6$ and M_7C_3 type carbides exist. That is, an $(Fe,Re)_{23}C_6$ and

$(Fe,Re)_7C_3$ could be in equilibrium with σ-(Fe_3Re_2), as is the case elsewhere. This would be worth investigating.

One important characteristic of the pure metal binaries is *superlattice formation*, such as Fe_3Pt, FePt and $FePt_3$ of Cu_3Au,CuAu and (again) Cu_3Au-type, respectively. Much work has been done on this effect and the associated strong effects on magnetic properties.

In detail, the evolutionary trend is as follows:

Fe–Re: appreciable Re solution in γ and α; still stable intermetallic Fe_3Re_2 (σ)

Fe–Ru: Ru stabilizes γ; no compound or superlattice reported but incipiently possible

↓

Fe–Os: Os stabilizes γ; no compound or superlattice reported but possible

Fe–Rh: Rh stabilizes γ—complete solid solution; superlattice possible but information insufficient

↓

Fe–Ir: Ir stabilizes γ—complete solid solutions; *possible* ordering in FeIr reported

Fe–Pd: Pd stabilizes γ—complete solid solutions, no compound; superlattice known:
FePd (CuAl-type, FC tetragonal) and $FePd_3$ (Cu_3Au-type); Fe_3Pd possible but unconfirmed

↓

Fe–Pt: Pt stabilizes γ—complete solid solutions, no compound; ordering pronounced, extending to high temperatures within γ-(Fe–Pt): Fe_3Pt, FePt, $FePt_3$.

Interstitial carbon addition will certainly affect the degree and character of the order/disorder change in these systems, but the manner of it doing so is again unknown and should be investigated.

(16) *Fe—Co—C* (*Figure 4.33*)

The Fe–Co binary system is one of continuous FCC solid solutions γ-Fe–β-Co, and BCC α-Fe solutions extending up to 78 per cent Co; superlattices Fe_3Co and FeCo occur in the latter.

As discussed before (p. 114), cementite dissolves Co but is rendered progressively less stable, Co_3C being an entirely unstable carbide. With increasing cobalt there is also an increasing tendency for transition from $(Fe,Co)_3C$ to ferrite. No new carbide occurs in the ternary system which is dominated by an austenitic (Fe,Co,C) solid solution, whilst that of C in (Fe,Co) ferrite is negligible. The solution of Co in γ-(Fe,C) greatly stabilizes it, and renders it the more retainable, on tempering. The system has been well established[324]; *Figure 4.40* shows a sequence of sections through the ternary system for varying Fe:Co ratios.

The changes in γ-solution and eutectoid reaction are relatively slow, but progressively we have the replacement of cementitic by graphitic equilibria, reflected by the superimposition of eutectics containing $(Fe,Co)_3C$ and graphite. In this transitional region, only small changes in rate of cooling or pressure can sensitively determine in which form the carbon occurs, effects which greatly affect the metallographic structure, hardenability, etc. (see p. 119).

The Fe $\rightarrow$ Co replacement has been found[179] to increase the Curie point of Fe_3C; it was noted that, contrary to expectation, Co is a rather stronger graphitizing agent for cementite than Ni. Some sections in the *quarternary* system Fe–Co–Cr–C were studied by Coutsouradis and Habraken[325].

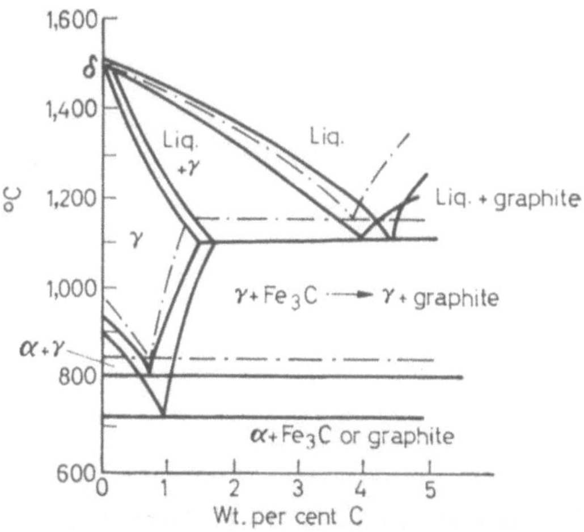

Figure 4.40. Fe–Co–C *phase relations* (Vogel and Sundermann[324])

Sections radiating to carbon corner, with the following Fe : Co ratios: —— 49 : 1, ▬ 2·3 : 1, —·—· 1 : 1

By courtesy of the Editor, *Archiv für das Eisenhüttenwesen*

(17) *Fe—Ni—C (Figure 4.33)*

Again the chief feature in this system is the continuous FCC solution of γ-(Fe,C) with the (Fe,Ni) solid solution and, by implication, the link-up and similarity of martensite-type reactions in (Fe,C) with the transitional martensite-like stages in (Fe,Ni). The very sluggish reactions, which make the $\gamma \longleftrightarrow \alpha$ transformation in the Fe–Ni binary system still problematic, continue in the ternary and make delineation of phase boundaries somewhat uncertain. However, in principle, the low-temperature phase fields [α solid soln., $\alpha + \gamma$, γ solid soln., $\alpha + \gamma +$ (Fe,Ni)$_3$C, $\alpha +$ (Fe,Ni)$_3$C, $\gamma +$ graphite] are beyond doubt[326–30].

Nickel, like cobalt, progressively renders cementite less stable, promoting graphitization. Ni raises the Fe_3C Curie point whilst still in solution, although the graphitizing influence for higher Ni contents is counteracted by the increasing stability of austenite, i.e. the reactions (Fe, Ni)$_3$C $\rightarrow \alpha +$ graphite and α-(Fe,Ni,C) $\rightarrow \gamma$-(Fe,Ni,C) are to a large extent self-compensating (for a full description of the reaction kinetics see e.g.[307]). The wide implications in steels and cast irons need only be mentioned. *Figure 4.41* shows a phase diagram for Fe–Ni–C alloys[320] containing 1 per cent C, 0·3 per cent Mn which, however, disregards graphitization.

(18) *Co—(Ti,Zr,Hf,Th)—C and Ni—(Ti,Zr,Hf,Th)—C*

Again, very little systematic data are available on these systems, except for the more technological study by Craighead *et al.*[296] on Ti-rich alloys (see p. 170). As to compound formation, the only information[3b] is the existence of the η-(M_6C) carbides, $Co_2(Ti,Ta)_4C$ and $Ni_2(Ti,Nb,Ta)_4C$.

However, from the nature of the surrounding binaries, i.e. (*a*) appreciable solubility of Ti in Co and in Ni; (*b*) Laves phases Co_2Ti and Ni_2Ti; (*c*) CoTi and NiTi of BCC structure; (*d*) dititanides $CoTi_2$ and $NiTi_2$; (*e*) stabilization of β-Ti to a eutectoid by both Co and Ni, ternary diagrams

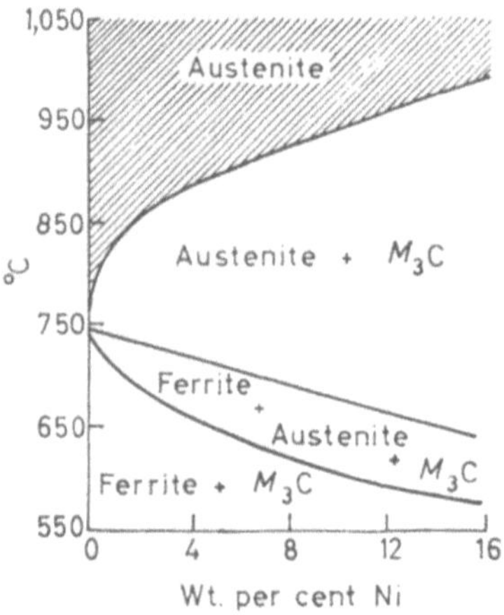

Figure 4.41. Fe–Ni–C *diagram for* 1 wt. per cent C *and* 0·3 wt. per cent Mn (Samuel, Finch and Rait[330])

By courtesy of the Editor, *Nature, London*

of the type shown for Fe–Ti–C (*Figure 4.33a*) seem probable. They must be amended, however, by the destabilization of cementite in favour of direct Co(Ni) solid solution/graphite equilibria. It is likely that pseudo-binary systems radiate from TiC in the sequence

$$\begin{aligned} &\text{TiC} \rightarrow \text{CoTi (NiTi)} \\ &\text{TiC} \rightarrow \text{Laves phases} \\ &\text{TiC} \rightarrow \text{FCC (Co,Ni) solution*} \end{aligned}$$

Ni–Ti–C alloys have also been discussed[331], and so have Co–Ti–C alloys[332]. In the former system, it has been shown[333] (at 870°C) that the Ni–Ti intermetallics form two-phase equilibria with TiC and that Ni_2Ti (the Laves phase) has a significant solubility for C. TiC and graphite have appreciable solubilities in primary nickel (for many valuable details see the original paper).

For *zirconium* we have the Laves-phase Co_2Zr and the further intermetallics Co_4Zr and Ni_4Zr; Ni_3Zr, NiZr and $NiZr_2$. For *hafnium*, a Laves-phase Co_2Hf also exists.

In the systems (Co,Ni)–Zr–C there is little doubt that the graphite equilibria of Co and Ni are rapidly replaced by carbide. However, before the stable (TiC, ZrC, HfC and ThC) are formed, one or more double carbide may first occur, of type Co_xZr_yC, Ni_xZr_yC (not necessarily of η type). This again suggests an interesting line of further investigation.

* Structurally, this sequence again well reflects the unifying effect of carbon addition: the three intermetallic compounds of diverse structures are all transformed by C to the one carbide of NaCl type.

(19) *Co—(V,Nb,Ta,Pa)—C and Ni—(V,Nb,Ta,Pa)—C*

Again, little information on these ternary systems is available, except for the knowledge of the existence of multiple η-carbides, in which Co–Nb and V–Nb–Ta enter in combination (*Table 4.2*).

The metal–metal binaries here involve: appreciable primary solutions; the σ-phases CoV and Ni_2V_3, and the compounds V_3Co, V_3Ni reported to be of β-W-type structure. Compounds $Co_3V(Co_2V?)$, and $Ni_3V(Ni_2V)$, possibly related to Laves-type structure, also exist. Laves-phases Co_2Nb and Co_2Ta again exist, whilst in Ni–Nb and Ni–Ta compounds Ni_3Nb (Ni_3Ta) of ordered HCP structure occur, possibly also analogous NiNb and 'Ni_2Ta_3'.

The ternary systems, though hypothetical, may once again be conjectured to be similar to those for Fe (*Figure 4.33*) with modifications.

(20) *Co—Cr—C*

Little is known on this ternary system, except for some limited work[334–5]. The Co–Cr binary side is characterized by the σ-phases as intermetallic and by considerable primary solid solutions, including stabilization of the hexagonal form of Co by Cr addition. Upon carbon addition, a wide solid solution in primary Co–Cr is certain, with eventual formation of $(Cr,Co)_{23}C_6$, $(Cr,Co)_7C_3$ and $(Cr,Co)_3C_2$: the first replacing σ, the second and last being the phases taking over from the graphite equilibrium and unstable Co_3C, upon Cr addition. (The paper by Coutsaradis and Habraken[325] deals more with the technological side of alloys on this base.)

(21) *Ni—Cr—C*

Surprisingly little systematic knowledge is available on this important system, though it is probable that the wider primary FCC (Ni,Cr) and BCC (Cr,Ni) solid solutions—Ni–Cr form a simple duplex system—directly equilibrate with the three chromium carbides until, for lower Cr, the Ni–graphite system prevails.

(22) *Co—(Mo,W,U)—C and Ni—(Mo,W,U)—C*

The Co–(Mo,W)–C and Ni–(Mo,W)–C systems are characterized by η-carbide formation, both of η_1 and η_2 type (see *Table 4.2*). The original work on these was that of Westgren *et al.*[19] and of Takeda[177]. The $M_{23}C$ compounds are also liable to enter. The binary inter-metal systems contain intermetallic phases, namely (for Co) the compounds Co_3Mo, Co_7Mo_6 (μ-phase, with Fe isomorph) and Co_3W, Co_7W_3, possibly a σ-phase; (for Ni) Ni_4Mo, Ni_3Mo, NiMo; and Ni_4W.

The Co–W–C system is the most thoroughly investigated, owing to its importance in hard metals; *Figure 4.10* shows Rautala and Norton's[22] phase diagram containing 3 double carbides, Co_3W_3C (η_1), $Co_3W_6C_2$ (Co_2W_4C [$=\eta_2$]) and $Co_3W_{10}C_4$. The latter has been redetermined[23] as hexagonal, the formula being amended to $Co_3W_9C_4$. The section Co–WC which is most relevant to hard-metals, is included in *Figure 4.10*. The fair, but receding, solubility of WC in Co is significant in the part played by Co as binder. Clearly the double carbides enter only upon carbon deficiency. Owing to sluggish reaction kinetics, e.g. during sintering, it is important, however, to differentiate stable from metastable equilibria (cf. [91, 263, 336–9]).

The stability of the tungsten-bearing η-carbides decreases in the order $(Fe \rightarrow Co \rightarrow Ni)_3W_3C$, as observed already in Westgren's early work.

No equivalent work on the Ni- and Mo-bearing systems appears to be known, although a close relationship to the above might be conjectured, but with significant differences. It is known[338] that the WC solubility in Ni is up to about 15 per cent WC (12 per cent in Co), and that free carbon is precipitated for Ni saturation by WC. This is one reason why nickel is less suitable than Co as a binder metal, and links up, as far as phase equilibria are concerned, with the general theme of the increased trend to graphitization as we proceed in the sequence $Fe \rightarrow Co \rightarrow Ni$.

No information whatever seems available on the Co–U–C and Ni–U–C systems, and some double-carbide formation can only be presumed. Co–U and Ni–U contain several stable intermetallic compounds which *may* equilibrate with UC, though complicated by stable or metastable mixed carbides intervening.

(23) *Co—Re—C; Ni—Re—C; Co—(Pt metals)—C; Ni—(Pt metals)—C*

These ternary systems are again undertermined, as far as the writer could find, although on the binary inter-metal systems there exists, of course, substantial knowledge. As for the Fe-base systems, the liability to superlattice formation becomes increasingly dominant. Also, it is virtually certain that no stable carbides occur. In passing, it may be pointed out that this transition also has a physical parallel in that from actual carbide formation and relatively strong though unstable bonding to merely adsorptive bonding of interstitials and catalytic reactions.

A note regarding rhenium may be added owing to its border line position in regard to carbide formation. It is known, for example, that Re forms stable intermetallic compounds with the Group IV, V, VI transition metals, such as Re_2Zr and Re_2Hf, which are of hexagonal Laves-phase type, stable and of high melting point[340] but, in the presence of carbon, immediately decompose to form ZrC or HfC. In principle, a similar situation will probably apply to Os and the other Pt metals, on the one hand, and Ti,V, Nb,Ta,Cr,Mo, W on the other, although with variations.

(24) *Mn—Co—C and Mn—Ni—C*

Again little systematic work has been done on these systems. The Mn–Co and Mn–Ni systems, as already Mn–Fe, are characterized by straight equilibria between primary FCC solid solutions (Mn,Co), (Mn,Ni); these are extensive and, at high temperatures, complete, linking with γ-Mn and the δ-Mn solid solutions, which are stabilized by Co and Ni. The Mn–Ni system is, however, complicated by a compound MnNi forming from FCCγ at lower temperatures; on quenching, this phase is of FCC tetragonal CuAu type, but at high temperatures it is, in fact, BCC. A superlattice occurs at Mn_3Ni, and in several aspects this system has much in common with the copper–gold system.

Upon carbon addition, the γ solubility of C will be extensive but decreasing towards Co and Ni. Manganese addition is likely to stabilize the unstable cementite structure of Co_3C and Ni_3C whilst, conversely, Co and Ni will rapidly eliminate $Mn_{23}C_6$ and Mn_7C_3. The effect of *carbon* on the effects surrounding MnNi is not known but, again, calls for investigation.

Note: In connection with some of the above systems, the solid solubilities of some Group IV, V and VI carbides in metallic Fe, Co and Ni have been investigated by Edwards and Raine[341] in relation to their hard metals applications. The data are shown in *Table 4.12.* The specimens had been melted and annealed at 1,250°C for 24 h. The use of this Table is, of course, limited to the specific data points and the heat treatment stated, but some significant trends with Group and Period are self-explanatory.

Table 4.12

Solid solubilities at* 1,250°C *of* TiC, V_4C_3, NbC, TaC, Cr_3C_2, Mo_2C, WC *and* (50/50 TiC–WC) *mixtures in* Fe, Co *and* Ni (Edwards and Raine[341])

Carbide	Solubilities in					
	γ-Fe	Co	Ni	Fe	Co	Ni
	molecular per cent			weight per cent		
TiC	<0.9	1·9	9·3	<0·5	1	5
V_4C_3	1·8	9·9	11·4	3	6	7
NbC	1·1	5·6	3·3	1	5	3
TaC	0·3	2·5	3·1	0·5	3	5
Cr_3C_2	11·9	18·2	18·2	8	12	12
Mo_2C	4·1	11·5	7·0	5	13	8
WC	4·1	14·5	7·5	7	22	12
TiC–WC (50/50)	0·4	1·8	4·6	0·5	2	5

* Converted to mol. per cent and rearranged

4.12 CARBIDES BETWEEN TRANSITION AND NON-TRANSITION METALS

There exist a number of carbides of type A_xB_yC which, though essentially transition-metal-(A)-based, also incorporate other metals (B) such as Al, Mg, Sn or Zn. These double carbides have their distinctive crystal structures and properties; η-carbides of this kind have been mentioned above (see *Table 4.2*). A most interesting class is that of the cubic ***perovskite*** structure which more usually is associated with double oxides such as barium titanate, $BaTiO_3$. Examples of these ternary carbides are Fe_3AlC and Mn_3ZnC, where the metal atoms occupy a FCC lattice, stabilized by the carbon atom. Fe_3AlC represents the ideal composition[342–4] but occupies an appreciable ternary homogeneity range in the Fe–Al–C system. This is shown in ***Figure 4.42.*** It should here be noted that, whilst in the Fe–Al system no FCC compound occurs, carbon addition is capable of producing this from the ordered BCC solid solution of Fe_3Al. Also, evidently the phase extends in direct continuation of the peninsula of FCC austenite (γ).

The perovskite of Fe_3AlC is shown in *Figure 4.43.* It represents a simple development from the plain FCC of γ-iron, where the metal atoms are randomly arranged. The Fe atoms become ordered into the face centres ($\frac{1}{2}\,\frac{1}{2}\,0$), the Al atoms into the cube corners (000), with the carbon atom occupying the octahedral interstices, i.e. the cube centre ($\frac{1}{2}\,\frac{1}{2}\,\frac{1}{2}$). This

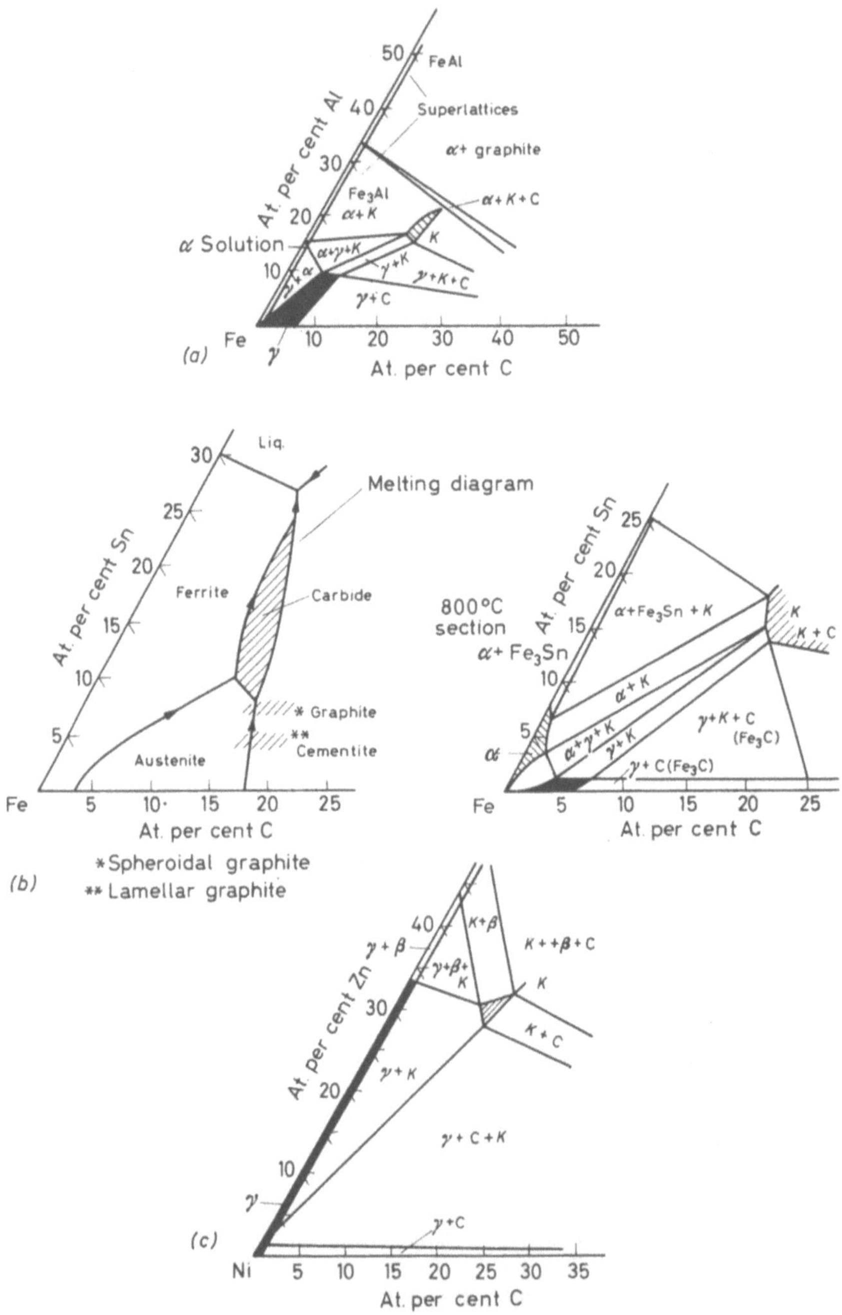

Figure 4.42. Phase equilibria involving the double carbide A_3BC

(*a*) Fe–Al–C diagram, for 1,000°C section (Morral[342]) $K \approx Fe_3AlC$
(*b*) Fe–Zn–C diagram (Stadelmaier and Waller[356]) $K \approx Fe_3ZnC$
(*c*) Ni–Zn–C diagram, for 800°C section (Stadelmaier and Hammad[354]) γ = primary FCC Nis olution; $K \approx Ni_3ZnC$; β = NiZn (AuCu type)

By courtesy of the Editor, *Metall*

ordered FCC structure may also be considered as a simple evolution from the Fe_3Al type of BCC superlattice forming within the Fe–Al ferrite solution, and thus as a 'carbon-stabilized product' of the $\alpha \rightarrow \gamma$ transformation, under ordered conditions as distinct from random conditions, e.g. in the NaCl-type carbides. In broader terms, this is now a *third* mode of carbide in which metal atoms occupy the face-centred cube, the other two being the Bl (rock salt) type, as in TiC, and the rarer Cl (fluorspar) type.

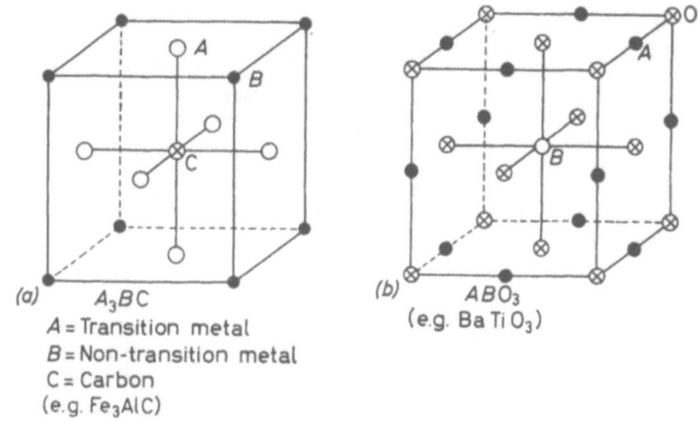

Figure 4.43. Unit cell of (a) carbides, (b) oxides of perovskite-type structures

Ideally cubic; non-stoichiometry of kinds A_3BC_x and ABO_{3-x} can alter interstitial content and sites, and cause anisotropy, e.g. cubic $\rightarrow$ tetragonal distortions

It might be enquired how this *carbidic* perovskite structure of A_3BC is related to the better known *oxidic* one ABO_3 (see also Chapter 8); the latter is, for comparison, included in *Figure 4.43.* It is partially an inversion of the first, in that oxygen now occupies the FCC metal lattice points, with metal atoms at the cube edges (similar to the rock-salt structure); the second metal atom now fills the cube centre. However, this is the 'ideal' structure, and variations can occur through non-stoichiometry and preferred occupancies, implying tetragonal distortion. The latter leads to the well known ferroelectric properties and transformations with temperature of perovskites, e.g. of the $BaTiO_3$ and $NaNbO_3$ types. Analogous effects also occur for the carbides. Thus, significant carbon deficiencies can occur (see below), implying 'vacant' unit cells. The body centre positions for C are only the more *probable* ones[345], the alternative of cube edge sites (similar to the oxides) not being included: indeed, it is suggested that both could coexist, depending on temperature and composition. Anisotropic distortions occur in the carbides, as in the oxides, and the possibility of mixed oxy-carbides of perovskite type is an obvious one, though, to our knowledge, not reported.

Representatives of the perovskite type of double carbides are shown in *Table 4.13.* Most of this work is due to Stadelmaier's and Huetter's school, as shown. They all have the basic $A:B = 3:1$ ratio in common, but with large variations in interstitial C content, mostly in the sense of a defect

Table 4.13

Ternary carbides of transition metals, A = Mn, Fe, Co, Ni *with non-transition metals, B, of A_3BC type**

Base structure: perovskite type (1 molecule per unit cell; (cf. *Figure 4.43*))

Carbide	*x* = (approx.)	*Lattice spacing, kX*	*Remarks*	*References*
with aluminium				
Mn_3AlC_x	1–1·20	3·86	ferromagnetic	344–7
Fe_3AlC_x	≈0·66	3·72–3·78	ferromagnetic	342–4, 348
Co_3AlC_x	0·59	3·69	ferromagnetic	349
Ni_3AlC_x	0·29	3·61	non-magnetic	344
with magnesium				
Co_3MgC_x	?	3·81	?	344
Ni_3MgC_x	0·5–1·25	3·72–3·81	non-magnetic	350
(Fe–Mg–C no double carbide found)				
with zinc				
Mn_3ZnC_x	1	3·92		345, 347, 351
Fe_3ZnC_x	—	3·80		351
Co_3ZnC_x	—	3·72		352–3
Ni_3ZnC_x	0·7	3·65		345, 354
(later modified to $Ni_2ZnC_{0.3}$)				
with gallium, germanium, indium and tin				
Mn_3GaC_x	≈1	3·86–3·87		351
Fe_3GaC_x	≈0·5	3·70–3·76		351
Co_3GaC_x	≈0·5	3·64		351
Ni_3GaC_x	0·1	3·59		351
Mn_3GeC_x	1	3·86–3·87		344
Fe_3GeC_x	0·45	3·65		344
Co_3GeC_x	0·25	3·59–3·62		344
Ni_3GeC_x	0·15	3·57		344
Ti_3InC_x	≈1	4·199		346
Mn_3InC_x	—	3·90–3·94		355
Fe_3InC_x	—	3·87		344
Co_3InC_x	0·75	3·81–3·89		344
Ni_3InC_x	0·5	3·77–3·79		349
Mn_3SnC_x	0·8	3·90–3·98		346
Fe_3SnC_x	1	3·85		351, 356
Co_3SnC_x	0·7	3·77		351, 357

* Further members of this family[358] are Ti_3AlC, Ti_3InC, Ce_3InC_X, Y_3TiC_X, Ce_3TiC_X, Ce_3SnC_X, Ce_3PdC_X and some nitride analogues

lattice (i.e. vacant unit cell, see *Figure 4.43*). The tendency to form at least partial solutions with FCC pure metal phases (e.g. austenite) therefore arises; in the case of Ni_3Ga (ordered FCC) → Ni_3GaC it actually evolves to a complete series of solutions. A lattice expansion through carbon occurs throughout, clearly in line with its interstitial entry.

Also, the A_3BC-type carbide is consistently in equilibrium with the *pure-metal* FCC (e.g. γ-Fe;Ni) on the one side, and with *graphite* on the other. We can therefore well interpret the double carbide as an intermediate structural step in the FCC metal → graphite equilibrium: this is a direct one for Ni–carbon, Co–carbon and (qualified) for γ-Fe–carbon but becomes impeded in the presence of the *B* metals (Al, etc.) to form this ordered intermediate lattice.

It is evident from *Table 4.13* that, for a given *B* metal, the carbon content decreases along the first Long Period (from Mn to Ni), i.e. in parallel with the electron filling of the *3d* orbital. It is interesting how, for nickel, the electron concentration then becomes sufficient for an ordered FCC isomorph to form already *without* C, e.g. in Ni_3Al which dissolves carbon up to Ni_3AlC. A further feature arising from *Table 4.13* is that, for a given transition metal, the lattice spacings follow in consistent sequence the atomic sizes of the non-transition metal (compare with *Figure 2.1*), and so does '*x*', the degree of carbon non-stoichiometry.

A carbide 'Cr_3AlC', of unknown structure, had been reported[124] which might have preceded our series in *Table 4.13* (starting with Mn–Al–C), though not confirmed. Stadelmaier did not find a double carbide around Cr_3AlC; instead, it is noteworthy that in the region of a hypothetical 'Cr_3AlC', the $Cr_{23}C_6$ carbide predominates. We might therefore well consider our earlier $M_{23}C_6$ carbide (see *Figure 4.6*) as the modified 'chromium version' of the present A_3BC carbide and its predecessor upon electron abstraction. Both are face-centred cube-based (although $Cr_{23}C_6$ is so much more complex), and both clearly at, or near, the formal M_4C stoichiometry. The similar electronic structure (trivalency) of the Al and Cr atoms no doubt enables partial Cr $\rightarrow$ Al replacement*) in $Cr_{23}C_6$ (as in many Cr,Al compounds), which on passing from Group VI to VII and VIII develops into the Al/(Mn,Fe,Co,Ni) ordering of perovskite.

In further comment on *Table 4.13*, it should be added that all these perovskite isomorphs form complete or partial solid solutions. Thus, for example, Mn_3AlC dissolves both Fe (fully) and Cr (extensively), and Mn_3ZnC dissolves Al and Ga up to almost 10 atomic per cent. These Mn-base compounds (as some of the others) are interesting in showing a ferromagnetic transition and become strongly magnetic towards absolute zero.

It is of special interest here to note the existence of *interstitial-free* (i.e. intermetallic-compound-type) perovskite phases[358a]. These are Ce_3In, Ce_3Sn and Ce_3Pb. These have the ordered FCC Cu_3Au-type lattice (as naturally arises upon interstitial vacation), and the vacant sites can be *filled* by carbon up to Ce_3InC_x, Ce_3TiC_x etc., with *x* approaching 1. For the Group III, Rare Earth metals, as one partner the complete vacation is specially favoured, but in principle and to lesser degrees this should apply also to the Group IV and later isomorphs.

The formation of a double carbide in a system such as Fe–Al–C (important e.g. in cast irons) implies that graphitization becomes inhibited over a

* This could be ordered already in $(Cr,Al)_{23}C_6$, viz. in the various sites discussed earlier: a line of investigation which would be interesting. More recently a *perovskite-type carbide*, Ti_3AlC, has been synthesized[358b]: coexisting with and closely related to the 'H'-phase Ti_2AlC (see below). The spacing is $a = 4{\cdot}15_6$ Å, and it seems then likely that other perovskite isomorphs containing Group IV to VI metals occur (e.g. V_3AlC, and lending support to a possible Cr_3AlC). Clearly this Ti_3AlC is a simple deviant to a carbon-defective TiC, where the addition of aluminium promotes the NaCl $\rightarrow$ perovskite-type change, with decreased carbon. The structure of perovskite Ti_3InC (and some nitride isomorphs) has now also been determined[358a].

certain range of aluminium contents. This is well illustrated by *Figure 4.44*, and directly links up with the appearance of the ternary compound. Conversely, the fact of any such graphite inhibition, found empirically, can be taken as a good pointer to the existence of a double carbide. The same applies to Mn–Al–C, where the carbide $M_{23}C_6$ also prevents graphite formation, quite in line with its close relation to the present ternary carbide (as explained above). *Figure 4.42* also includes the Fe–Sn–C and Ni–Zn–C diagrams. The first brings out the austenite/carbide continuity especially well; the author also indicates areas of *spheroidal* and lamellar graphite forming, the former being promoted by the ternary carbide.

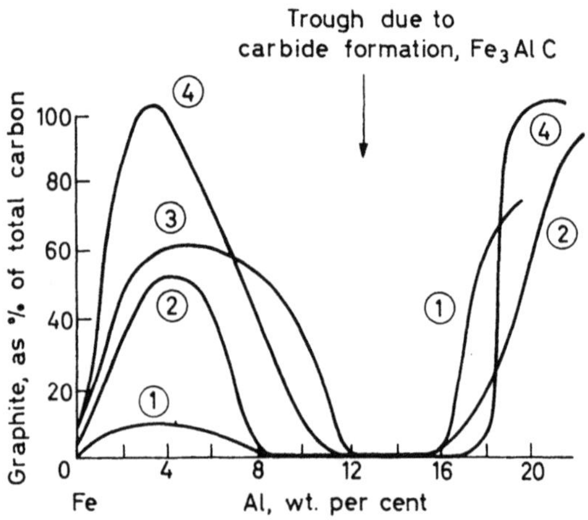

Figure 4.44. Graphitization diagram in Fe–Al–C *alloys, indicative of double carbide formation* (Löhberg and Schmidt[343])

(1)–(4) after various thermal histories

By courtesy of the Editor, *Archiv für das Eisenhüttenwesen*

One other point of principle ought to be made: the transition metals included in this double carbide are *only* those of Groups VII and VIII, i.e. those for which normally (without Al, etc.) carbide formation would have become weak or absent.

No double carbide with the Subgroup *B* metals is formed by the Group IV–V–VI metals, i.e. the self-sufficient strong carbide formers. The picture emerges, therefore, that the Al, Mg, etc. atoms are needed in the lattice as auxiliary in order to assist the iron-group metals also to become strong carbide formers! In fact, we may view in this sense the series of rock salt- and of perovskite-type carbide (in TiC, V_4C_3 and in Mn_3AlC, etc.) as direct structural descendants from each other with varying Group numbers.

Structurally, again we have in A_3BC a useful further instance to our earlier theme of FCC (matrix $\rightarrow$ carbide) transitions. This is associated, in the carbon-free *A–B* systems, with intermetallic compound formation and with superlattices. The ordering tendency is in fact 'inherited' by the carbide and becomes the more perfect one: from this viewpoint, the carbon atoms play the role of helping the pure intermetallic ordering, which may only be partial, to become complete. This is also reflected in the varying C deficiencies (index x, *Table 4.13*) which shows that the inter-metal system (e.g. Ni–Al) having the strongest tendency to superlattice formation are also those requiring the least interstitial carbon to stabilize the ternary carbide.

The *melting points* of the carbides[351] are close to those of the parent transition metals (i.e. 1,200–1,500°C, although there are little positive data); this certainly implies that they are less refractory than the Group IV–VI carbides.

The existence of these carbides is clearly of much practical application, for instance in cast iron metallography and in its effect on the mechanical properties of materials. Thus an improvement in properties of Fe–Al–C alloys at elevated temperatures has been reported[348], with increasing Fe_3AlC carbide content and an alleviation of the embrittlement which the superlattice Fe_3Al produces. An $Fe_3Al \rightarrow Fe_3AlC$ reaction through carbon addition can in this sense be beneficial.

Finally, some of these carbides have inherently interesting physical properties. The ferromagnetic properties referred to, in particular of Mn_3ZnC, Mn_3AlC and Ni_3ZnC, have been studied in detail by Butters and Myers[345], also in relation to structure and stoichiometry.

A more general survey of ternary carbides with one B-subgroup metal, including similar nitrides and borides, has been given by Stadelmaier[359].

A further, more recently discovered, group of carbides is *related* to the perovskites, but of general composition A_2BC where $A =$ a transition metal of Groups V or VI, $B =$ a non-transition metal, e.g. Cr_2AlC, V_2GeC, Cr_2GaC; *Table 4.14* lists these phases. The crystal structure is hexagonal, with $a = 2{\cdot}86$, $c = 12{\cdot}8_2$ Å$(c/a = 4{\cdot}8_3)$, for Cr_2AlC. One might well regard this group as a hexagonal analogue to the cubic perovskite structure. These are the so-called '*H phases*'[358]; for details of structure and electronic conditions of formation, the original papers should be consulted.

The structure of these phases is shown in *Figure 4.45* (upper part the unit cell, lower part planar sections and C environments).

There exists a further close structural link-up of this group with carbides of the β-Mn type, discussed earlier. Indeed, three new examples[358a] of such β-Mn-type phases have here been met, namely:

Mo_3Al_2C, $a = 6{\cdot}86_0$–$6{\cdot}86_6$ Å (Al-rich to Al-poor end of a homogeneity range)

Nb_3Al_2C, $a = 7{\cdot}06$ Å

Ta_3Al_2C, $a = 7{\cdot}03_9$ Å

No doubt these are only first prototypes of a larger isomorphous family A_aB_bX ($X =$ interstitial).

It may well be enquired whether at lower temperatures or with critical composition changes, these β-Mn type structures would not transform to the α-Mn type, as is the case in the π and χ phases and, of course, in Mn metal.

Another new development in this field[358c] is the occurrence of related η-*carbide*-type phases of transition metal/non-transition metal combinations, like Ti_2ZnC_x, Hf_2ZnC_x and Nb_2ZnC_x ($x < 1$) as well as of nitride analogues.

We should then consider: (*a*) the perovskite type of carbides; (*b*) the 'H' phases; (*c*) the β-Mn type of carbides (π phases); (*d*) the α-Mn type of carbides (χ phases); and (*e*) the η-carbides and their associates, as a family of closely related structural and electronic deviants able to accommodate a large variety of combinations of metals (transition or otherwise), of interstitials (to give e.g. also nitrides and oxides) and of vacancies (towards pure intermetallics). A further variant is a group of carbides of approximate

Table 4.14

Ternary 'H' *phases* A_2BC (Jeitschko *et al.*[355, 358])

Carbide*	Lattice spacings†, Å		
	a	*c*	*c/a*
Ti_2AlC	3·04	13·60	4·47
V_2AlC	2·913	13·14	4·51$_1$
Nb_2AlC	3·10$_3$	13·8$_3$	4·64
Ta_2AlC	3·07$_5$	13·8$_3$	4·49$_8$
Cr_2AlC	2·86$_0$	12·8$_2$	4·48$_3$
Ti_2SnC	3·16$_3$	13·67	4·22$_3$
Cr_2SnC	—	—	—
Zr_2SnC	3·34$_7$	14·5$_9$	4·35$_9$
Hf_2SnC	3·31$_2$	14·3$_9$	4·34$_4$
Nb_2SnC	3·24$_5$	13·77	4·24$_3$
Ti_2GaC	3·06$_4$	13·30$_5$	4·34$_2$
V_2GaC	2·93$_8$	12·8$_4$	4·37$_0$
Nb_2GaC	3·13	13·56$_3$	4·33$_2$
Ta_2GaC	3·10$_4$	13·57	4·37$_1$
Cr_2GaC	2·88$_6$	12·61$_0$	4·37$_0$
Mo_2GaC	3·0$_7$	13·1$_8$	4·36$_9$
Ti_2GeC	3·07$_9$	12·93	4·19$_9$
V_2GeC	3·00$_1$	12·25	4·08$_3$
Cr_2GeC	2·95$_4$	12·08	4·09$_1$
Ti_2InC	3·13$_2$	14·06	4·49$_0$
Zr_2InC	3·34$_7$	14·91	4·45$_5$
Hf_2InC	3·30$_7$	14·73	4·45$_5$
Nb_2InC	3·17$_2$	14·37	4·53$_0$
Ti_2PbC	3·20$_9$	13·81	4·30$_2$
Zr_2PbC	—	—	—
Hf_2PbC	—	—	—
Ti_2CdC	3·09$_9$	14·41	4·65$_3$
Ti_2TlC	3·15$_8$	13·98	4·42$_6$
Zr_2TlC	—	—	—
Hf_2TlC	—	—	—

* For some nitride analogues, see Chapter 5
† Some are liable to variations with homogeneity range

composition $A_{11}B_7C_5$ and of $Mn_5Si_3(D8_8)$-type structure, linking to the *silicides*, *carbosilicides* and *borides*, namely Nb_5GaC_x, $Ta_5Ga_3C_x$ and $Ta_5Ge_3C_x$ ($x < 1$). A further link extends towards the *oxide*, in that e.g. an analogue $Nb_5Ga_3O_x$ is now known[358a].

An exploration of further ternary systems such as Mo–Al–C and W–Al–C[358a,b] has indicated the probable existence of other instances of these carbides and of new modifications, and there is little doubt that many more such phases await discovery and possible applications.

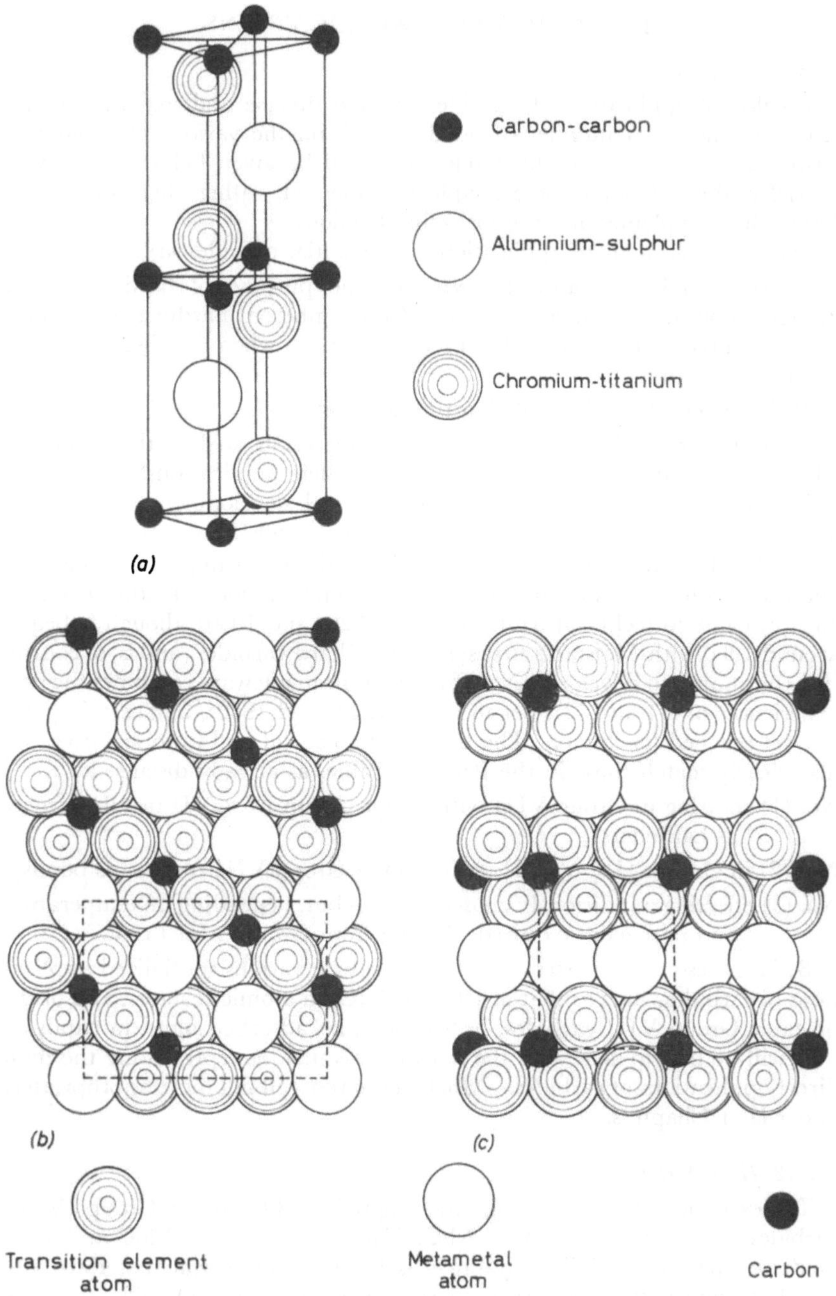

Figure 4.45. Structure of 'H' phase (Cr_2AlC *type*) (Jeitschko *et al.*[358c])

(*a*) Unit cell of H phase (Cr_2AlC or Ti_2SC type)
(*b*) Pseudohexagonal section of the perovskite carbide, showing arrangement of atoms (*thick lines*) in (110) plane. The layer of atoms behind (*thin lines*) can also partially be seen; from these two layers the whole crystal structure can be built up
(*c*) Section of the H phase represented to correspond with (*a*). In both figures the carbon atom is surrounded by six T atoms (octahedral environment)

By courtesy of the Editor, *Journal of less-common Metals*

4.13 CARBIDES: APPLICATIONS

4.13.1. General

The fields of application of carbides are too diverse and extensive to be dealt with here. A number of good reviews on the various technological aspects are available, to which references will be given below. However, a brief outline of points of principle is presented, rather than going into detail, thus supplementing the foregoing Sections.

The principal areas of application of interstitial carbides are:

(1) as secondary phases in *steels*, for the purpose of hardening and strengthening, not so much by virtue of their inherent hardness as by their mode of precipitation, and by their influence, when in dispersion, on dislocation locking.

(2) in cemented hard metals (cutting tools, etc.).

(3) in modern high-temperature-resistant materials, such as the refractory alloys ('superalloys') for jet engine and space-age applications*.

It is interesting to reflect that, in historical development, these three divisions correspond broadly to 'past', 'present' and 'future'.

No. (1) dates from antiquity, though with the main impetus in the 19th and early 20th centuries, and now fast developing on more scientific grounds. No. (2) came to technical fruition in the 1920s and 1930s, though it began around 1900 with Moissan's first synthesis of hard carbides (and is, of course, still continuing to develop). No. (3), which in many ways combines the best experiences of (1) and (2), began in the 1940s and 1950s, and certainly has not yet reached its climax. The three technological and historical divisions also clearly match those in the Periodic Table, as already discussed:

No. (1) centring in Group VIII with minor Group IV to VII metals (largely as hardeners);

No. (2) in Groups [IV, V, VI]—with minor Group VIII metals (as bonds);

No. (3) effecting a judiciously chosen overlap but, depending on temperature and properties required, shifting from Group VIII to IV–VI.

Besides these three classes, there are of course various specialized applications of carbides, such as that of UC and PuC, in atomic reactors, and of the unstable Group VIII carbides, or 'carbide-like' adsorbed films, in catalysis.

It will be best to deal with the hard metals first, as representing the most direct applications of carbides. Much of the rest will fall more appropriately under later Chapters.

4.13.2. Hard Metals

The extreme hardness and high melting points of the stable Group IV–VI carbides (TiC, ZrC, HfC, VC, NbC, TaC, Cr_3C_2, Mo_2C/MoC and WC) made *all* these potentially attractive as bases for industrial materials, e.g. to attain high cutting speeds or to resist wear to tools. Qualifications lie, however, in their brittleness and inability to self-sinter (except at highest

* These still include some advanced high-alloy steels but more as an incidental feature. Generically they are better referred to as Fe-based alloys, along with Co- and Ni-based alloys of similar standing and temperature ceiling in use: that is, up to ≈ 1,000°C, whilst for higher temperatures Group IV–VI-based alloys must take over.

temperatures). Most of their practical uses imply the need to attain a degree of toughness without detriment to hardness. This means effective bond metals which are more ductile but themselves require fair inherent strength and high melting point. A restrictive condition here is that they must, when liquid, have good wettability in contact with the carbides. Also they must themselves not be carbide formers—yet a partial (preferably temperature-sensitive) solid solubility between metal and carbide is beneficial and promotes bond strength. The condition for a suitable binder is one of well chosen balance. It should have a significant solubility in the carbide when liquid (this helps wetting) and on sintering, but should precipitate on cooling, so that the carbide portion is as pure as possible, though well bonded. It also must not form a ternary compound.

Since in service (e.g. during cutting)* high but transient and localized temperatures occur, a requirement is good oxidation resistance and thermal shock resistance at high temperatures, also implying good thermal conductivity to dissipate heat rapidly. Further obviously important properties are density, Young's modulus, bending strength and relative thermal expansion of carbide and binder phases. All these, often conflicting, conditions have greatly narrowed the practical choice of phase combinations, and intense research efforts have gone into determining the optimum balance for a given problem. Further restrictions are imposed by considerations of manufacture. The avoidance e.g. of interstitial impurities (notably oxygen) in the raw-carbide powders is important, and powder-metallurgical techniques (rather than melting) are the only ones feasible.

All these matters are determined by the properties of the single carbides listed e.g. in *Table 4.1*, and by the crystallographic and phase-diagram relationships described (*Figures 4.1, 3, 10, 28*, etc.). The degrees of intersolubility between the carbides, and between carbides and metals, as a function of temperature, and possible compound formation are controlling factors of particular relevance.

The most widely used and successful combination evolved to fulfil these requirements is that of WC–TiC base, with cobalt as the binder. The systems concerned are those shown in *Figures 4.10, 4.20* and *4.23* (Ti–W–C, Co–W–C and WC–TiC–Co). Rautala and Norton's Co–W–C phase diagram clearly implies the liability for the embrittling η-carbides (Co_3W_3C) to form for insufficient carbon or after specific heat treatments (by peritectic reaction below $\approx$ 1,350°C), instead of the (W,Ti)C and Co(WC) solid solutions aimed at. Furthermore, the simple W–C and Ti–C systems (*Figure 4.1*) imply the risk of possible W_2C formation upon carbon abstraction, with consequent increased embrittlement and oxidizability and, on the other hand, of defect lattice formation in TiC_{1-x}. The latter in itself is not too harmful, but it promotes the easy entry of lattice oxygen, and also embrittlement. Excess carbon, on the other hand, is also damaging, owing to the presence then of free graphite (a possibility arising particularly from impure or

* Practical requirements in cutting tools are hardness, wear resistance and strength. They must be harder than the work-piece material, e.g. a steel. They must not wear, i.e. must maintain their cutting edge—which does not just imply hardness but hardness allied to a degree of toughness and good intercrystalline bonding. Lastly, the impact and (transient) high-temperature-resistant properties referred to and fine grain size are also important factors.

incompletely reacted raw carbides used), which can prevent effective bonding of Co to WC.

It is worth while to elaborate the last point slightly further, as it provides an attractive illustration of the interplay of interstitial compounds. TiC and TiO are isomorphous, forming complete solid solutions. This is turned to advantage in hard metals and is one of the reasons for the inclusion of TiC. Thus it has been found[272] by x-ray analysis that, for instance during operational use as cutting tools, a thin surface film of TiO is formed on the TiC substrate. This is protective and physically isolates the tool from the work-piece metal (e.g. a steel); it prevents local welding of the hard metal/steel couple which used to be a main factor in tool wear of the earlier (TiC-free) WC–Co materials. WC can form no such protective oxide: in fact, WO_3 volatilizes. It requires the TiC compound with its tenacious and refractory layer, sealed by an outermost film of TiO_2 (rutile), to produce what may be called a local 'lubricating action' at high temperature. The tenaciousness is aided by the TiC–TiO solid solution which provides at the tool surface a steady carbide/oxide transition and an effective anchoring of the oxide layer. (This actually has its converse aspect in that one difficulty in the manufacture of titanium carbide, in the first place, from the Ti oxide, hydride or metal is to obtain carburization to completely pure TiC; however, this can be accelerated by the presence of chlorine.)

This fundamental point applies similarly to TiC-base alloys (see below) and promotes their oxidation resistance in gas-turbine uses at high temperatures. It also enters in TiC-base 'Cermets' (e.g.[360]).

Within the basic Co–WC–TiC system, there have in practice been many variations in composition. Such adjustment in Co/WC/TiC ratio in fact underlies their adaptability to different applications: depending on whether emphasis is laid e.g. on high hardness, impact strength, toughness, very-high-temperature performance for short times or long endurance under more moderate conditions. The 'grades' of cemented carbide tool materials commercially available can be adjusted to these different needs; e.g. the Co content normally varies between 6 and 13 per cent. A diagram illustrating the properties of WC–TiC–Co-base hard-metals[361] is shown in *Figure 4.46*.

The WC–TiC–Co combination[263,362–5]* is, of course, only a basis upon which many variants can be, and have been, built to adjust structure and properties by additions of both other carbides and other bond metals. Regarding the latter, Fe and Ni as well as stainless steel-type alloys (Fe–Ni–Cr) have all been tried as bonds but, except for specialized uses, cobalt has still generally proved superior, as strengthener and toughener.

Tantalum carbide (TaC) addition has brought important improvements. This enters into the cubic TiC–TaC solid solution (*Figure 4.3*) and is of two-fold benefit: by raising the melting point and by reducing the degree of oxygen solution and of oxidation in service. In this one may see a direct connection with the narrowing of the homogeneity ranges from TiC_{1-x} to TaC_{1-x} (*Figure 4.19*). TaC, as the Group V carbide, is the less prone to dissolve oxygen, apart from being the more refractory. Also, since Ta

* Gurland and Norton[366] consider the basic mechanical properties (strength and fracture) of such an aggregate of hard brittle particles in a soft, ductile matrix, with varying ratio, grain size and grain shape in wider principle.

already adjoins W in Group VI and is commensurate with it in atomic size, this offers increasing intersolubility with WC and ability to bond with it. The phase diagram TiC–TaC–WC in *Figure 4.28,* which contains a number of commercially important compositions, may again be referred to. These

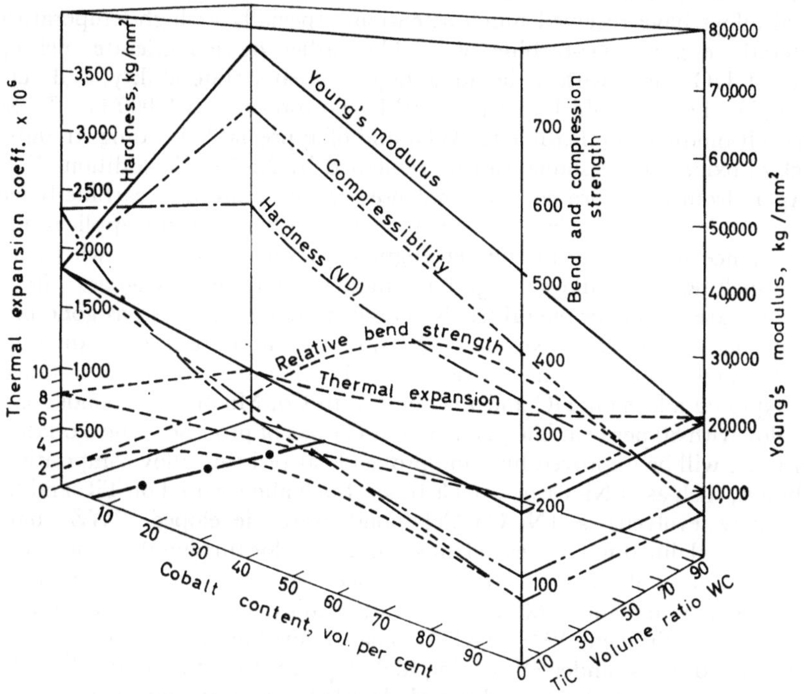

Figure 4.46. Some hard-metal properties as a function of composition (Ballhausen[361])

By courtesy of the Editor, *Stahl und Eisen*

generally lie in the two-phase field [Ti,Ta,W] C + WC, and we may note that this is the area of cubic → hexagonal transition and of maximum facility to stacking faults and dislocation locking. These lattice defects become the more interwoven through the rapid recession of the phase boundary with falling temperature[273]. Another attraction of Ta was that it greatly improved the bond strength and the toughness of the hard metal, owing partly to its grain-refining effect. Recent improvements on TiC–WC–TaC materials with Co binder have been achieved e.g. by Russian workers[367], including higher hardness, melting point and strength than without TaC.

Other additives of potential value are NbC (acting similarly to TaC), MoC, ZrC, HfC, VC and Cr_3C_2, each acting in its own special way of modifying structure and properties of the product, not necessarily good. Vanadium and chromium carbides already tend to introduce substantial reduction in refractoriness, but a measure of ductility. *Hafnium carbide**

* HfC is clearly one of the most interesting technologically, because of its extreme stability, and much applied research is now proceeding on this from the aspect of high-temperature physical properties (e.g.[96]). The non-stoichiometry of HfC_{1-x} is here important, and indeed the strict 1:1 composition does not appear to exist.

is of particular recent interest, being the most stable, high-melting and hardest of all carbides. It has shown appreciable experimental promise[368], notably in HfC–WC–TiC/Co combinations (see also intersolubilities, *Figure 4.22*).

In further practical developments pure (tungsten-free) TiC-based hard-metal alloys have received much attention, especially as high-temperature materials, e.g. for gas-turbine use. The rather more moderate melting point of TiC brings with it the advantage of greater sinterability, and very dense TiC bodies could be prepared[369] by hot-pressing at 3,000°C. These were self-binding (not feasible for WC) and of appreciable bending strength. Lack of oxidation resistance can be remedied by Ni–Co–Cr additions[370].

Apart from the more conventional uses as tool materials, carbide-based hard metals now gain importance in very-high-temperature applications, for instance in components for jet engines, rocket nozzles, etc. where the lower-melting 'superalloys' (e.g. Nimonics, special steels, Co-base alloys, Stellites) are no longer operable. Research in many quarters has gone into countering the inherent weakness of lower oxidation resistance, ductility and thermal shock resistance of carbides. The Plansee school, for instance, has explored alloys on TiC and ZrC base with Co, Mo or W as bonds; e.g. an alloy with 80 per cent TiC, 20 per cent Co was of promise. The oxidation resistance will be improved by using proven heat-resisting alloy compositions as bonds, such as of Ni–Cr or Co–Cr base. Such alloys based on TiC and (as ultimately evolved) a [Ni–Co–Cr] bond were developed ('WZ' hard metals[370]), claimed to be of excellent strength e.g. for turbine disc and blade uses, as well as of good oxidation resistance. By adjusting the proportion TiC/binder within the 75/25 to 35/65 range, alloys can be made adaptable to varying requirements[371–5]. TiC–TaC base alloys have also been developed with Co and Ni as binders[266], having high rupture strength and good oxidation resistance. TiC-base hard metals bonded with different heat-resisting materials, including martensitic steels, have opened further promising avenues[376]. Japanese research has led to e.g. a TiC Cermet with a Ni–Cr–Mo alloy binder[377].

Variants of TiC–TaC–NbC/Ni and TiC–Cr_3C_2/Co alloys have also been proposed[378–9]; such chromium carbide-bearing alloys are of particular interest in that a thin film of Cr_2O_3 would act protectively against oxidation. This has led to a new type of heat-resisting and tool material with nickel as binder, which has high wear and corrosion resistance. Their main fields of application are at more moderate or room temperature, the chemical use against acids and sodium hydroxide being a principal attraction[380]. A special advantage of Cr carbide is that its thermal expansion coefficient is comparable with that of iron and steel, useful e.g. in precision gauge applications and for hard metal inserts or linings to steel matrices, where wear resistance is also an important quality. The same attraction, incidentally, applies to the η-carbide Fe_4W_2C, when used as hard metal (investigated in detail by the author[2i]). More generally, as we move to carbide compositions involving the Group VII–VIII metals, their applicability by no means ceases but shifts in emphasis to some more specialist fields.

A wide technical and patent literature exists on this subject and that of carbide-base alloys for use up to 1,500°C.

As to techniques of hard-metal manufacture, references must suffice (e.g.[381]). However, an advance of particular interest is that of *infiltration* (as distinct from normal sintering of carbide/bond metal powder mixtures), where a pre-prepared carbide skeleton is subjected to capillary filling by the liquid metal*.

Obviously, carbide applications should not be viewed in isolation, and as will follow in the later Sections, the other interstitial compounds—nitrides, borides, silicides—find similar uses, sometimes in combination with carbides. The variant of carbide–oxide combinations would lead us into the field of metal ceramics. Here, for instance, TiC–Al_2O_3 and TiC–MgO have been tried (cf. Chapter 10).

Further industrial applications of carbide-based hard metals, apart from cutting tools (tipped lathe tools, milling cutters, drills, boring bits in the mining industry†), lie, for example, in extrusion dies, wire dies, hot-press dies, wear-resisting parts and hard surface facings. Regarding the latter, we may refer to the 'flame-plating' process, where a hard-face coating can be deposited onto surfaces of ordinary metal substrates, even though much lower melting (e.g. Fe, Al, Cu). WC + Co is again used, but the facing product consists not of these two phases, as in sintered products, but of an η-carbide, Co_3W_3C (*Figure 4.10*), as the stable compound.

The literature on the methods and mechanisms of bonding and sintering of carbides is itself extensive. Hot-pressing for high strength is an important feature, and much investigation has gone into attaining good bonding of the carbides, yet minimizing the binder metal. Thus, for example, 'fugitive binders' have been used whose main function is to promote solid solution at the carbide interfaces and local wettability (e.g.[382]). The binder metal itself (Co or Ni), or its oxide, is later removed by higher-temperature vacuum sintering.

High-speed steels which, among steels, are the nearest approach to cemented hard metals, are restricted in machining use to a local maximum temperature of about 300–400°C, i.e. that dictated by softening through martensite decomposition or, for higher temperatures, through austenite formation. The innate η-carbide, whilst very hard, is even at 30 per cent in the high-speed steel too small in amount to be the operative constituent compared with the steel matrix, unless artificially enriched, when it does act as hard carbide.

4.13.3. Carbides in Steels and in High-Temperature Alloys

In contrast to hard metals, where the emphasis is on immutability, hardness and refractoriness of the carbide, that in steel carbides lies largely in their mutability and even instability. In plain carbon steels, metastable cementite, Fe_3C, is clearly the most important: but upon this, alloying with the other

---

* Is it too venturesome to recall for a moment our more philosophical viewpoint (see Appendix) and compare such infiltrated hard-metal skeletons, on a macro scale, with the interstitial alloys themselves on the atomic scale? 'Infiltration' into the receptive transition metal lattice by C or N atoms strengthens its bond, and the analogy of atomic diffusion into vacant lattice sites, even though they may sometimes have to be created by a lattice distortion, to capillary action is one which holds in several respects.

† Where we might reflect that the hard carbide is put to work against its own interstitial element when in mass!

transition elements plays its variations, as already described. This is to stabilize, by Group VII (Mn), VI, V, IV metal additions, the $M_{23}C_6$, the cubic, hexagonal and double carbides, or to *de*stabilize, by Co or Ni additions, any carbides.

The subject of carbides in steels and in 'Superalloys' is so extensive and varied that they will require separate consideration. More important, however, is that the 'applications', if this is the right word, of carbides as microconstituents and precipitants in steel and alloy matrices are not unique to carbides: functionally many intermetallic compounds, nitrides, borides and other phases are of equal status as hardeners, strengtheners or embrittlers (although carbides, no doubt, are the most important). Often all these phases act in concert and in intricately interdependent sequences when in dispersion for a given steel or alloy, dissolving, decomposing, re-forming, transforming or precipitating, dependent on temperature, thermal history and concentration. In fact, in creep-resisting alloys the very complexity and 'staggered' nature of these phase equilibria at different temperature levels and for different thermal treatments is an essential factor.

Furthermore, some of the most influential functions of carbides as microconstituents occur not in their full state of precipitation but in statu nascendi, where their identity and distinction from the parent phase is still poorly defined. This is where the effect on dislocation movement is most pronounced. Lattice coherence, fit or misfit of spacings with those of the matrix, interfacial strains and effect on stacking faults and carbide-like lattice configurations previous to formation are here some of the more important factors.

These aspects have largely been developed already in Chapters 1 and 2, and in the initial parts of the present Chapter, and will be again referred to later.

4.13.4. *Various other Carbide Applications*

Whilst the uses of interstitial carbides in hard metals and as constituents in steels, etc. are the outstanding ones, there are various side applications.

Thus, in refractory alloys of Cr, Mo, W, Nb and Ta bases, small controlled quantities of carbides can act as dispersion strengtheners, similarly in Ti and Zr base alloys (see e.g.[383]). Here the non-solubility or very limited solubility of C in these metals can be directly utilized, and fine-particle dispersions, e.g. of Ta_2C in Ta, of TiC in Ti and in Mo would be of considerable interest. They could persist as such in service up to very high temperatures and act in reinforcement. If, as found in tungsten[151], the solubility becomes significant at very high temperatures ($>2{,}000$°C) but negligible at lower ones, precipitation and dispersion-hardening by W_2C particles becomes possible.

In the lamp industry the carbonization of W filaments in carbon-bearing gas has even in early days[88] enabled rigid tungsten carbide layers to be formed for mechanical strengthening and to prevent 'sagging'. Similarly, Ta_2C (or TaC) were formed on tantalum, forming a more effective and refractory light source than Ta itself (compare the earlier discussion on melting point differences).

Altogether, techniques have been devised to manufacture thin and moderately ductile carbide surface coatings on metals. In this connection, the surface hardening of steels by carburization is, of course, well established by many techniques. It forms one obvious aspect of steel applications of carbides, but here their effect is clearly integrated with that of martensite. Much of what will be said on nitrides (Chapter 5) also applies to carbides.

An interesting example of a further application is the use of carbides (as well as borides) in reduction cells in aluminium production by electrolysis[384]. In these, graphite had been used for many years in cell construction, but with the drawback of inefficiency by too high electrical resistance. Transition-metal carbides and borides have here greatly helped, because not only is their resistance lower but also they are stable and unreactive towards molten Al or the electrolyte at the high temperatures. The field of related applications in industrial production processes and plant is wide indeed.

The application of uranium carbides in nuclear energy reactor elements has already been referred to, and ThC, NpC and PuC come under the same heading. The solid solution of UC with other Group IV, V, VI carbides is here also relevant, because such additions are capable of stabilizing UC. In this special field of nuclear energy applications of carbides, two more recent Symposia[208,210] provide a source of useful information (see also Bibliography). There has also been much work on effects in (U,Zr)C nuclear fuel element particles which had been coated with pyrolytic graphite[305] and on UC fuel evaluation[386]. The use of UC–ZrC solid solutions and their stability towards graphite exemplifies a direct and elegant application of constitutional results to a practical problem, i.e. that of powder metallurgical fuel element design (see also the U–Zr–C system, *Figure 4.30*). A balance is here arrived at in compromising the active properties of UC with the more refractory ones of ZrC, where upon diffusion with the graphite coating, the formation of inter-layer UC_2 can be inhibited from a sufficient ZrC content.

The unstable carbides of Group VIII (Fe_3C, Co_3C, Ni_3C) are of interest in catalytic reactions[198]. And even the normally 'non-existent' carbides of the Pt-type metals can play a part during catalysis in form of thin carbide-like lattice configurations on surfaces and interfaces. Again, we should remember that, under very high *pressures*, some new carbides tend to become stabilized, a fact which very probably operates e.g. during diamond synthesis, where Group VIII metals act as catalysts. It might well be envisaged that such (what might be called) 'carbides quenched from high pressures', if only transiently stable, can be indirectly utilized in several technological directions; indeed, cementite can itself be regarded as a high-pressure product formed within the constraint of the steel matrix lattice.

SELECTED LITERATURE

I. *Binary Metal-Carbon Systems* (cf. Section 4.6)

The *thermal expansion* of the Group IV carbides has been determined[73,74] and, more recently, that of Group IV and V carbides[387]; in studying the *melting-point characteristics*[388] in the systems (Ti, Zr, Hf, Nb, Ta,)–C, eutectics were confirmed, also in the ternaries combining some of these metals. The *Hall coefficient* and low-temperature resistivity of TiC were investigated[70] as well as its *frictional properties* and deformation at high temperatures[78]; this paper gives much detail on the more technological properties of TiC for use in ceramics.

The *superconductive properties* of the Group V *hemi*carbides (V_2C, Nb_2C, Ta_2C) have also been discussed[389].

For some *field-emission microscopy* work on TaC, see [110].

Kelley (e.g.[145]) has also contributed much *thermodynamic data* of the refractory carbides more generally.

A recent critical review on the Group IV, V, VI and actinide binary carbide systems has been given by Storms[261], whilst Richardson's paper on the *thermodynamics* of carbides and carbon[390] now has become a classic.

The proceedings of a symposium held at A.E.R.E., Harwell, 1963 (*Carbides in Nuclear Energy*[208]) include further information on the U, Np, Pu systems, on physical and chemical properties of U, Pu, and Th carbides and their technological use and behaviour, specifically on U–C *vaporization*[208c], high-temperature property data[208d] and thermal conductivity[208e].

For a discussion of high-temperature thermodynamic properties of the carbides of Th, U and Pu, see[391].

In connection with the Rare Earth carbides, *neutron diffraction* work[252] on the La-carbides, as well as on YC_2, (R.E.)C_2 and (R.E.)$_2C_3$ has elucidated these structures. A review of the *magneto-chemistry* and x-ray absorption of the R.E. carbides was given by Vickery[241] who also published an earlier chemical survey[251] (see also [250]).

II. *Ternary Carbide Systems* (cf. Section 4.10)

The following papers represent further work on systems between pseudo-binary pairs of carbides or on ternaries M_1–M_2–carbon; the sequence is from Group IV to VIIIA. TiC–HfC, ZrC–HfC and HfC-base systems with Group V and VI carbides[269]; Ti–Mo–C and Mo–W–C[287]; Ti–Mo–C[287]; Zr–Cr–C[392a,b]; Zr–Mo–C[138, 392c]; Ti–W–C[393,394]; Zr–W–C[395]; VC–WC[396]; NbC–TaC and ZrC–NbC[397]; Hf–Ta–C[398]; Ta–W–C[399]; (TiC, TaC, WC)–Cr_7C_3[277].

For the carbide constitutions *underlying hard metals*: WC–TiC–TaC–Co, see also [400].

The converse case of the solubility *in cobalt* of TiC, NbC, TaC, Mo_2C, WC and their combinations has more recently been studied[401]; for dilatometric and structure studies on WC–Co alloys, see also [402].

Uranium carbide-based mixed systems have been extensively investigated in relation to atomic reactor materials, e.g.

- UC–PuC[403] (only preliminary at that stage); [279,280].
- U–Pu–C[404] (full survey); [208h] (especially U_2C_3–Pu_2C_3 section); property data[208].
- UC–HfC[405] (complete solubility confirmed at high temperatures, but with possible miscibility gap at lower ones).
- U–Zr–C and U–Nb–C[208b] (melting points, crystallographic and property data on ternary phases UMC_2).

Kinetics and diffusion rates of reactions (at 400–600°C) between U carbides and added metals: Ti, Zr, Th, V, Nb, Cr, Mo, Fe, Al and stainless steel[406]; U–Nb–C, U–W–C, U–Re–C and U–Y–C[208c]: (structure, hardness, metallography); U–Cr–C, U–Ni–C, U–Fe–Cr–C[208j] (phase equilibria); U–Fe–C, Pu–Fe–C and (limited) U–Pu–Fe–C[208k]: (equilibria extensively studied); UC–ZrC, UC–NbC and UC–TaC[278] (continuous series of solutions, with good conformance to Vegard's law); UC,UC_2- (and UN-) based systems[407].

For inter-carbide *lattice parameter* measurements (see e.g. *Figure 4.22*), cf. [268,408–410].

Further Plansee (Kieffer, Benesovsky, etc.) and Vienna (Nowotny) studies have been devoted to Cr–W–C[386]; U–Zr–C, U–Hf–C, U–Th–C, U–Nb–C, U–Ta–C, V–Mo–C, V–W–C, Nb–Mo–C, Ta–W–C[411,412]; HfC + other MC[270]; V–Mo–C[413]; Zr–Th–C[414a]; Hf–Ta–C[398]; Th–Mo–C and U–Mo–C[414a,c].

Some of the pseudo-binary monocarbide systems had been presented in earlier work[408].

The equilibria of UC_2 with UC on the one hand and, with carbon, of ZrC, HfC, NbC and TaC on the other, were given in[414c].

The study on HfC[270] also presents more general *thermodynamic* work on ternary carbide equilibria, including calculations of their enthalpies of formation and of the carbide–carbide and carbide–metal solubilities[411,415]. Partial reviews are contained in other papers by Nowotny[416].

III. *Hard metals applications and development* (cf. Section 4.13.2)

Review books or papers: on *metal ceramics* ('*Cermets*'), including *carbide hard metals*[91,263, 417–24]; the earlier classic[263a] was succeeded by[263b]; with more emphasis on the technological side of carbide hard metals[421]. A handbook on hard-metal tools [423] presents a recent survey of the *high-temperature chemistry* of the refractory carbides, whilst [424] reviews the technology and properties of refractory materials, including carbides, from the aspect of rocket propulsion uses.

As to some special *properties* of hard carbides: *thermal expansion*[425]; *hardness* at high temperatures[426] (also borides); *thermophysical* properties[427, 428a]; *thermodynamic stabilities*[428b]; *electrical properties*[428c]; various high-temperature properties[418]; structure and properties of WC–Co-base alloys[429]; *vapour pressures and vaporization rates*[430]; *non-stoichiometry* and its implications in hard carbides[431]; a survey of *carbide solid solution, properties* and *production* [432].

Becker's early work[88], industrially orientated, with evident success, covered the refractory carbides more generally and was basic to the subject, as were Moissan's numerous publications (around 1900).

IV. *High-Temperature Steels and 'Superalloys', with carbides as frequent constituents: developments* (cf. Section 4.13.3)

Some further literature on the *function of alloying elements in steels*[433] (a classic treatise); *stainless steels*[307]; *tool steels*[434]; *high-strength steels*[45,435]; *precipitation in alloy steels* (generally, but largely carbides[315,436], the latter containing e.g. reviews of work of Quarrell, Honeycombe, Nutting, McLean, etc., on particular aspects in this field; *processes during creep*[437]; *high-temperature gas turbine alloys*[438] and other *high-temperature materials*[439–441]

Note.—The '*Nimonic*' type of Ni–Cr-based alloys depends substantially on intermetallic (Ni_3Al, etc.), not so much on carbide precipitation; compare, however, [442].

On *austenitic* (e.g. 'G18B' type) and *ferritic* heat-resisting steels, for gas-turbines, etc.[2j,438,443,444].

Precipitation mechanisms in *iron, cobalt* and *nickel-based* refractory alloys[445–450].

Cobalt-based alloys (e.g. 'Vitallium' type)[334,451–453]; the journal 'Cobalt', published by the Cobalt Information Centre at Brussels), contains many other papers in this field.

Nickel-based alloys[454–456] (typical of many); see also [442].

General structural metallurgy of steels[25,213,457,458] and other papers, e.g. from Nutting's, Honeycombe's, Grant's and Cohen's (MIT) schools.

V. *Atomic Energy Applications of Carbides* (cf. Section 4.13.4).

The Proceedings of the 1963 Harwell Conference[208] provide 74 papers on uranium and allied carbide applications; Vol. 2 gives the more technological aspects: preparation and fabrication, irradiation behaviour, while Vol. 1 deals with basic properties. Similar information is contained in a (1963) East Hartford (U.S.A.) Conference[210]; the 4th (1961) Plansee Symposium[232] and the (1958) United Nations Symposium[459]; besides carbides of uranium, etc. these also include nitrides, oxides, borides, etc. A review of properties of uranium carbides, including temperature variations[460], is useful also for retrieving earlier references, while a later one[461] surveys the thermophysical properties of uranium carbide at high temperatures. An AIME Symposium[428] also includes papers concerning carbides.

REFERENCES

1 Westgren, A. and Phragmén, G. (*a*) *Trans. Am. Soc. Steel Treat.* 13 (1928) 539; (*b*) *Jernkont. Annlr* 117 (1933) 1; (*c*) *J. Iron Steel Inst.* (1922) I 241; (*d*) *Z. Phys.* 33 (1925) 777; (*e*) *Z. anorg. allg. Chem.* 187 (1930) 401; (*f*) *ibid.* 156 (1926) 27

2 Goldschmidt, H. J. *J. Iron Steel Inst.* (*a*) 170 (1952) 189; (*b*) 160 (1948) 345; (*c*) 194 (1960) 169; (*d*) (unpublished); (*e*) *Metallurgia* 56 (1957) 17; (*f*) Symposium on High Temperature Steels and Alloys for Gas Turbines, *Spec. Rep. Iron Steel Inst.* 43 (1951); (*g*) *Research, Lond.* 10 (1957) 289; (*h*) *J. less-common Metals* 2 (1960) 138; (*i*) *Brit. Patent 668847*, 1952; (*j*) *Metallurgia* 47 (1953) 215

3 Kuo, K. (*a*) *Jernkont. Annlr* 137 (1953) 141; (*b*) *J. Iron Steel Inst.* 173 (1953) 368; (*c*) *ibid.* 184 (1956) 288; (*d*) *J. Metals* 8 (1956) 97; (*e*) *Acta metall.* 1 (1953) 301, 611

4 Lipson, H. and Petch, N. J. *J. Iron Steel Inst.* 142 (1940) II 95

5 Ridal, K. A. and Quarrell, A. G. *J. Iron Steel Inst.* 195 (1960) 307; 200 (1962) 359; Honeycombe, R. W. K. In 'Metallography 1963', *Spec. Rep. Iron Steel Inst.* 80 (1964) 245

6 Beech, J. and Warrington, D. H. *J. Iron Steel Inst.* 204 (1966) 460

7 Kislyakova, E. N. *Zh. fiz. Khim.* 17 (1943) 108

8 Kiessling, R. *Proc. int. Symp. Reactivity of Solids*, Gothenburg 1952, 1065

9 Nevitt, M. V. In *Electronic Structure and Alloy Chemistry of the Transition Elements* (Ed. Beck, P. A.), New York (Interscience) 1963

10 Orell, F. L. and Fontana, M. A. *Trans. Am. Soc. Metals* 47 (1955) 554

11 Rostoker, W. *J. Inst. Metals* 4 (1952) 209

12 Taylor, A. and Sachs, K. *Nature, Lond.* 169 (1952) 411

13 Reiffenstein, E. *et al.*, *Mh. Chem.* 96 (1965) 1543

14 Rieger, W. *et al.*, *Mh. Chem.* 96 (1965) 232

15 Karlsson, N. *Mh. Chem.* 168 (1951) 558

16 Westgren, A. (*a*) *Nature, Lond.* 132 (1933) 480; *Jernkont. Annlr* 117 (1933) 501; (*b*) *ibid.* 118 (1935) 231

17 Hellborn, K. and Westgren, A. *Svensk kem. Tidskr.* 45 (1933) 141

18 Crafts, W. and Lamont, J. L. *Trans. Am. Inst. Min. metall. Engrs* 188 (1950) 561

19 Adelskjöld, V., Sundelin, A. and Westgren, A. *Z. anorg. allg. Chem.* 212 (1933) 401

20 Dyson, D. J. and Andrews, K. W. *J. Iron Steel Inst.* 202 (1964) 325

21 Cadek, J. *et al.*, *Kovove Mater.* 1 (1963) 83

[22] Rautala, P. and Norton, J. T. *J. Metals, N.Y.* 4 (1952) 1045; *Plansee Proc.* (1952) 303
[23] Schönberg, N. (*a*) *Acta metall.* 2 (1954) 837; (*b*) *Acta chem. scand.* 8 (1954) 624
[24] Hendricks, H. B. *Z. Kristallogr. Kristallgeom.* 74 (1930) 534
[25] Andrews, K. W. (*a*) *Acta crystallogr.* 16 (1963) 68; (*b*) *Acta metall.* 11 (1963) 939; (*c*) 12 (1964) 921; (*d*) *Nature, Lond.* 164 (1949) 1015
[26] Vertman, A. A. *et al.*, *Izv. Akad. Nauk SSSR, Otdel. tekh. nauk, Metallurgiya Topl.* 6 (1960) 123
[27] Chipman, J. *et al.*, *Trans. Am. Soc. Metals* 44 (1952) 1215
[28] Hägg, G. *Z. Kristallogr. Kristallgeom.* 89 (1934) 92
[29] Jack, K. H. (*a*) *Nature, Lond.* 158 (1946) 60; (*b*) *Proc. R. Soc.* A194 (1948) 41, 56; *Acta crystallogr.* 3 (1950) 392
[30] Roberts, E. C. *Trans. Am. Soc. Metals* 50 (1958) 123
[31] Lyashchenko, B. G. and Sorokin, L. M. *Kristallografya* 8 (1963) 382
[32] Gardin, A. I. *Dokl. Akad. Nauk SSSR* 146 (1962) 1068
[33] Kudielka, H. and Möller, H. *Z. Kristallogr. Kristallgeom.* 118 (1963) 213
[34] Keh, A. S. *Acta metall.* 11 (1963) 1101
[35] Radcliffe, S. V. *et al.*, *J. Iron Steel Inst.* 201 (1963) 143
[36] Hilliard, J. E. *Trans. Am. Inst. Min. metall. Engrs* 227 (1963) 429
[37] Pitsch, W. *Acta metall.* 10 (1962) 897
[38] Hale, K. F. and McLean, D. *J. Iron Steel Inst.* 201 (1963) 337
[39] Lipschutz, M. E. and Anders, E. *Geochim. Cosmochim. Acta* 24 (1961) 83; *Science, N.Y.* 134 (1962) 2095
[40] Bundy, F. P. *J. chem. Phys.* 38 (1963) 631
[41] De Carli, P. S. and Jamieson, J. C. *Science, N.Y.* 133 (1961) 1821
[42] Alder, B. J. and Christian, R. H. *Phys. Rev. Lett.* 7 (1961) 367
[43] *American Society of Metals Handbook*, 1948
[44] Andrews, K. W. and Hughes, H. *J. Iron Steel Inst.* 193 (1959) 304
[45] 'High Strength Steels' (Symposium), *Spec. Rep. Iron Steel Inst.* 76, 1962
[46] Hofer, L. J. E. and Peebles, W. C. *J. Am. chem. Soc.* 69 (1947) 893
[47] Clarke, J. and Jack, K. H. *Chemy Ind.* 46 (1951) 1004
[48] Trillat, J. J. and Oketani, S. *Acta crystallogr.* 5 (1952) 469
[49] Löhberg, K. *Arch. EisenhüttWes.* 32 (1961) 409
[50] Oketani, S. and Nagakura, S. *J. phys. Soc. Japan* 17 (1962) 235
[51] Nagakura, S. *J. phys. Soc. Japan* 14 (1959) 186; 16 (1961) 1213
[52] Dilewijns, J. *Cobalt* 11 (1961) 28
[53] Johnston, W. D. *et al.*, *J. phys. Chem., Ithaka* 64 (1960) 1720
[54] Hofer, L. J. E. and Cohn, E. M. *J. Am. chem. Soc.* 71 (1949) 189; *J. chem. Phys.* 218 (1950) 766
[55] Belorus, M. V. and Cherepin, V. T. *Fizika Metall.* 12 (1961) 685
[56] Apayev, B. A. *Soviet Phys. Metals Metallogr.* 12 (1962) 95
[57] Senetaeur, J. P. and Fruchart, R. *C.R. hebd. Séanc. Acad. Sci., Paris* 256 (1963) 3114
[58] Palatnik, L. S. *et al.*, *Soviet Phys. Metals Metallogr.* 11 (1961) 81
[59] McMullin, J. G. *et al.*, *Trans. Am. Soc. Metals* 46 (1954) 799
[60] Kasper, J. S. *Acta metall.* 2 (1954) 456
[61] Öhman, E. (*a*) *Metallwirt. Metallwiss., Metalltech.* 9 (1930) 825; (*b*) *Jernkont. Annlr* 128 (1944) 13; (*c*) *J. Iron Steel Inst.* 123 (1931) 445
[62] Preston, G. D. *Phil. Mag.* 52 (1928) 1198
[63] Bradley, A. J. *Phil. Mag.* 50 (1925) 1018
[64] Koh, P. K. *Trans. Am. Inst. metall. Engrs* 5 (1953) 339
[65] Ver Snyder, F. L. and Beattie, H. J. *Trans. Am. Soc. Metals* 47 (1954) 211
[66] Cadoff, I. and Nielsen, J. P. *J. Metals, N.Y.* 189 (1953) 248
[67] Ehrlich, P. *Z. anorg. allg. Chem.* 249 (1949) 1
[68] Jaffee, R. I., Ogden, H. R. and Maykuth, D. J. *Trans. Am. Inst. Min. metall. Engrs* 188 (1950) 1261
[69] Robins, D. A., In *The Physical Chemistry of Metallic Solutions and Intermetallic Compounds*, N.P.L. Symposium 9, London (H.M. Stat. Office) 1959
[70] Piper, J. *J. appl. Phys.* 33 (1962) 2394
[71] Bikerdike, R. L. and Hughes, G. *J. less-common Metals* 1 (1959) 42
[72] Williams, W. S. and Schaal, R. D. *J. appl. Phys.* 33 (1962) 955
[73] Krikorian, N. H. *et al.*, *J. electrochem. Soc.* 110 (1963) 587
[74] Houska, C. R. *Physics Chem. Solids* 25 (1964) 359
[75] Meerson, G. A. *et al.*, *Zh. priklad. Khim., Leningr.* 13 (1940) 66; 14 (1941) 291
[76] Taylor, R. E. *J. Am. Ceram. Soc.* 44 (1961) 525
[77] Hoch, M. and Vardi, J. *J. Am. Ceram. Soc.* 46 (1963) 245
[78] Brookes, C. A., In *Special Ceramics*, p. 221, New York (Academic Press) 1962
[79] Norton, J. T. and Lewis, R. K. *NASA Rep. CR 321*, 1963
[80] Hollox, G. E. and Smallman, R. E. *Proc. Br. ceram. Soc.* 1 (1964) 211

[81] Raman, S. and Ramachandran, G. N. *Curr. Sci.* 31 (1962) 321
[82] Samsonov, G. V. and Rozinova, N. S. *Izv. Sekt. fiz.-khim. Analiza Inst. Obshchei neorg. Khim.* 27 (1956) 126
[83] Benesovsky, F. and Rudy, E. (*a*) *Planseeber. PulvMetall.* 8 (1960) 66; (*b*) 9 (1961) 65; (*c*) *Metall* 14 (1960) 875
[84] Umanskii, Ya. S. *et al.*, *Izv. Akad. Nauk SSSR, Ser. Fiz.* 15 (1951) 24
[85] Kovalski, A. E. and Makarendo, T. G., quoted [91]
[86] Shaffer, P. T. B. *J. Am. Ceram. Soc.* 46 (1963) 177
[87] Lepie, M. P. *Trans. Br. ceram. Soc.* 63 (1964) 431
[88] Becker, R., e.g. *Phys. Z.* 34 (1935) 185; *Z. Phys.* 31 (1925) 268; *Hochschmelzende Hartstoffe und ihre technische Verwendung*, Berlin (Verlag Chemie) 1933
[89] Norton, J. T. and Mowry, A. L. *J. Metals, N.Y.* 1 (1949) 133
[90] Westrum, E. F. and Feick, G. *J. chem. Engng Data* 8 (1963) 176
[91] Samsonov, G. V. and Umanski, Ya. S. *NASA tech. Transl. F102*, 1962
[92] Nowotny, H. *et al.*, *Planseeber. PulvMetall.* 7 (1959) 79
[93] Nowotny, H. *Adv. X-Ray Analysis* 5 (1962) 13
[94] Cotter, D. G. and Kohn, J. A. *J. Am. Ceram. Soc.* 37 (1954) 415
[95] Glaser, F. W. *et al.*, *J. Metals* 5 (1953) 1119
[96] Sanders, W. A. *et al.*, In *High-temperature Materials*, Ed. Ault *et al.*, 2, 469, New York (Interscience) 1963
[97] Rostoker, W. and Yamamoto, A. *Trans. Am. Soc. Metals* 46 (1954) 1136
[98] Gurevich, M. A. and Ormont, B. F. *Dokl. Akad. Nauk SSSR* 96 (1954) 1165
[99] Storms, E. K. and McNeal, R. J. *J. phys. Chem., Ithaka* 66 (1962) 1401
[100] Fujishiro, S. and Gokcen, N. A. (*a*) *J. electrochem. Soc.* 109 (1962) 835; (*b*) *Trans. metall. Soc. A.I.M.E.* 221 (1961) 275
[101] Volkova, N. N. and Geld, P. V. *Tsvet. Metalty, Mosk.* 5 (1963) 89
[102] Mah, A. D. *U.S. Bur. Mines, Rep. Invest. 6177*, 1963
[103] Bowman, A. L. *et al.*, *Acta crystallogr.* 19 (1965) 6
[104] Storms, E. K. and Krikorian, N. H. *J. phys. Chem., Ithaka* (*a*) 63 (1959) 1747; (*b*) 64 (1960) 1471
[105] Kempter, C. P., Storms, E. K. and Fries, R. J. *J. chem. Phys.* 33 (1960) 1873
[106] Elliott, R. P. *Trans. Am. Soc. Metals* 53 (1961) 13; *Rep. U.S. atom. Energy Comm. TID 7625*
[107] Kimura, H. and Sasaki, V. *Trans. Japan Inst. Metals* 2 (1963) 98
[108] Brauer, G. and Lesser, R. *Z. Metallk.* 50 (1959) 8
[109] Brauer, G. *et al.*, *Z. anorg. allg. Chem.* 277 (1954) 249
[110] Huber, E. J. *et al.*, *J. phys. Chem., Ithaka* 65 (1961) 1846; 67 (1962) 793
[111] Kovattschenko, M. S. and Samsonov, G. V. *Dokl. Akad. Nauk Ukr. SSR* (1961) 1478
[112] Geld, P. V. and Kusenko, F. G. *Izv. Akad. Nauk SSSR, Otdel. tekh. nauk, Metallurgiya Topl.* (1960) 79
[113] Bowman, A. L. *J. phys. Chem., Ithaka* 65 (1961) 1596
[114] Pochon, M. L. *et al.*, In *Reactive Metals*, Ed. Clough, W. R., p. 327, New York (Interscience) 1959
[115] Smirnova, V. I. and Ormont, B. F. *Dokl. Akad. Nauk SSSR* 96 (1954) 557
[116] Cooper, J. R. and Hansler, R. L. *J. chem. Phys.* 39 (1963) 248
[117] Klein, R. and Leder, L. B. *J. chem. Phys.* 38 (1963) 1863
[118] Kosolapova, T. Ya. and Samsonov, G. V. *Khim. Zh.* 28 (1962) 931
[119] Ellinger, I. H. *Trans. Am. Soc. Metals* 31 (1943) 89
[120] Burgers, W. G. and Basart, J. C. M. *Z. anorg. allg. Chem.* 216 (1934) 209
[121] Kovalskii, A. E. and Makarenko, T. G., quoted [91]
[122] Frieman, E. and Sauerwald, F. *Z. anorg. allg. Chem.* 203 (1931) 64
[123] Hatsuta, K. *Technology Rep. Tôhoku Univ.* 10 (1932) 680
[124] Bloom, D. S. and Grant, N. J. *J. Metals* 2 (1950) 41
[125] Kelley, K. K. *Tech. Pap. Bur. Mines, Wash.* 662 (1944); *Ind. Engng Chem. analyt. Edn* 36 (1944) 865
[126] Ruff, O. and Föhr, T. *Z. anorg. allg. Chem.* 104 (1918) 27
[127] Lvov, S. N. *et al.*, *Soviet Phys. Metals Metallogr.* 2 (1961) 138
[128] Meinhardt, D. and Krisement, O. *Z. Naturf.* 15 (1960) 880
[129] Vintaikin, F. Z. *Fizika Metall.* 15 (1963) 144
[130] Nowotny, H. and Kieffer, R. *Z. anorg. allg. Chem.* 267 (1952) 261
[131] Kuo, K. and Hägg, G. *Nature, Lond.* 170 (1952) 245
[132] Nowotny, H. *et al.*, *Mh. Chem.* 85 (1954) 255
[133] Lander, J. J. and Germer, L. H. *Trans. Am. Inst. Min. metall. Engrs* 175 (1948) 660
[134] Sykes, W. P., van Horn, K. R. and Tucker, C. M. *Trans. Am. Inst. Min. metall. Engrs* 117 (1935) 173
[135] Rudy, E. and E. and Benesovsky, F. *Planseeber. PulvMetall.* 10 (1962) 42
[136] Tuttija, H. *Sci. Pap. Inst. phys. chem. Res., Tokyo* 19 (1932) 384
[137] Hoyt, S. L. *Trans. Am. Inst. Min. metall. Engrs* 89 (1930) 9

[138] Wallace, T. C. *et al.*, *J. phys. Chem., Ithaka* 67 (1963) 796
[139] Clougherty, E. V. *et al.*, *Nature, Lond.* 191 (1961) 1194
[140] Eremenko, V. N. and Velikanova, T. Ya. *Poroshk. metall.* (1965) 41; *Sov. Powder Metall. Metal Ceram.* 1 (1965) 25
[141] Kaye, G. *Nature, Lond.* 195 (1962) 1195
[142] Parthé, E. and Sadagopan, V. *Acta crystallogr.* 16 (1963) 202
[143] Skaupy, F. *Z. Elektrochem.* 33 (1927) 512
[144] Sykes, W. P. *Trans. Am. Soc. Steel Treat.* 18 (1930) 968
[145] Kelley, K. K. *U.S. Bur. Mines Bull.* 407, 1937; *J. Am. chem. Soc.* 62 (1940) 818
[146] Ivensen, A. E. *et al.*, *Soviet Phys. Metals Metallogr.* 12 (1962) 126
[147] Doloff, L. I. and Sara, R. V. *OTS Rep. AD* 278009 (1961); Sara, R. V. *J. Am. Ceram. Soc.* 48 (1965) 251
[148] Gleiser, W. C. M. and Chipman, J. *Trans. metall. Soc. A.I.M.E.* 227 (1962) 1278
[149] Leciejewicz, J. (*a*) *J. less-common Metals* 7 (1964) 318; (*b*) *Acta crystallogr.* 14 (1961) 200
[150] Butorina, L. N. and Pinsker, Z. G. *Kristallografiya* 5 (1960) 585; *Soviet Phys. Crystallogr.* 5 (1961) 560
[151] Goldschmidt, H. J. and Brand, J. A. *J. less-common Metals* 5 (1963) 181
[152] Isobe, M. *Sci. Rep. Res. Insts Tôhoku Univ.* 3 (1951) 468
[153] Jacobson, B. and Westgren, A. *Z. phys. Chem.* B20 (1933) 362
[154] Vogel, R. and Döring, W. *Arch. EisenhüttWes.* 9 (1935) 247
[155] Kuo, K. and Persson, L. E. *J. Iron Steel Inst.* 178 (1954) 39
[156] Gokcen, N. A. and Fujishido, S. *Trans. Met. Soc. A.I.M.E.* 227 (1963) 542
[157] Hughes, J. E. *J. less-common Metals* 1 (1959) 377
[158] Trzebiatowski, W. *Z. anorg. allg. Chem.* 233 (1937) 376
[159] Epstein, S. *The Alloys of Iron and Carbon*: *Constitution*, New York (McGraw-Hill) 1936
[160] (e.g.) Wever, F., Rose, A. and Peter, W. *Arch. EisenhüttWes.* 21 (1950) 367
[161] Esser, H. and Müller, G. *Arch. EisenhüttWes.* 7 (1933) 265
[162] Petch, N. J. *J. Iron Steel Inst.* 149 (1944) 143P
[163] Krainer, H. (*a*) *Arch. EisenhüttWes.* 21 (1950) 119; (*b*) *ibid.* 39; (*c*) 22 (1951) 53
[164] Lipson, H. and Parker, A. M. B. *J. Iron Steel Inst.* 149 (1944) 123P
[165] Genot, M. *Mém. scient. Revue Métall.* 60 (1963) 416
[166] Hale, K. F. and McLean, D. *J. Iron Steel Inst.* 201 (1963) 337
[167] Hume-Rothery, W. *et al.*, *J. Iron Steel Inst.* 145 (1942) 143P
[168] Horstmann, D. *The* Fe–C *constitutional diagram and the principles of heat treatment of* Fe–C *Alloys*, Düsseldorf (Verlag Stahleisen) 1961
[169] Johnston, W. D. *et al.*, *J. phys. Chem., Ithaka* 64 (1960) 1720
[170] McGrawth, J. and Rawlings, R. *Acta metall.* 12 (1964) 958
[171] Lyschenko, B. G. and Sorokin, L. M. *Kristallografiya* 8 (1963) 382
[172] (*a*) Rist, A. and Chipman, J.; (*b*) Cordier, J. A. and Kozokevitch, P. P., *In The Physical Chemistry of Steel Making*, Ed. Elliot, J. F., New York (Wiley) and London (Chapman & Hall) 1956
[173] Keh, A. S. *Acta metall.* 11 (1963) 1101
[174] Leslie, W. C., Fisher, R. M. and Sen, N. *Acta metall.* 7 (1959) 632
[175] Elliott, R. P. *Constitution of Binary Alloys. First Supplement*, New York (McGraw-Hill) 1965 (cf. [293])
[176] Meger, W. E. *Metallwirt., Metallwiss., Metalltech.* 17 (1938) 413, 465
[177] Takeda, S. *Honda Anniversary Vol.*, p. 864, Tôhoku Univ., 1936
[178] Bahr, M. H. and Jessen, V. *Ber. dtsch. chem. Ges.* 63 (1930) 2226
[179] Drain, J. *Annls Chim.* 8 (1953) 900
[180] Nagakura, S. *J. phys. Soc. Japan* 14 (1959) 189; 16 (1961) 1213
[181] Dilewijns, J. *Cobalt* 11 (1961) 28
[182] Kohlhaas, R. and Meyer, W. F. *Metallwirt., Metallwiss., Metalltech.* 17 (1938) 786
[183] Bernier, R. and Michel, A. *Bull. Soc. chim. Fr.* 11 (1949) 199
[184] Bernier, R. *Annls Chim.* 6 (1951) 104
[185] Oketani, S. *et al.*, *Nippon Kink. Gakk.* 18 (1954) 325
[186] Morrogh, H. and Williams, W. J. *J. Iron Steel Inst.* 155 (1947) 341
[187] Browing, L. C. and Emmett, P. H. *J. Am. chem. Soc.* 74 (1952) 1680
[188] Veertman, A. H. *et al.*, *Dokl. Akad. Nauk SSSR* 148 (1963) 342
[189] Meyer, G. and Sheffer, C. E. *J. Am. chem. Soc.* 75 (1953) 486
[190] Tebboth, J. A. *J. Soc. chem. Ind., Lond.* 67 (1948) 65
[191] Hofer, L. J. E. *et al.*, *J. phys. Colloid Chem.* 54 (1950) 1161
[192] Lander, J. J. *et al.*, *J. appl. Phys.* 23 (1952) 1305
[193] Siderenko, R. A. *Soviet Phys. Metals Metallogr.* 8 (1959) 114
[194] Collier, L. J. *et al.*, *J. Faraday Soc.* 30 (1934) 581
[195] König, H. *Naturwissenschaften* 38 (1951) 154
[196] Kempter, C. P. *J. chem. Phys.* 41 (1964) 1515

[197] Jeantet, B. and Knapton, A. G. *Planseeber. Pulv Metall.* 12 (1964) 12; Knapton, A. G., Personal communication
[198] Nadler, M. R. and Kempter, C. P. *J. phys. Chem., Ithaka* 64 (1960) 1468
[199] Raub, E. and Falkenburg, G. *Z. Metallk.* 55 (1964) 186, 190
[200] Kempter, C. P. and Nadler, M. R. *J. chem. Phys.* 33 (1960) 1580
[201] Wilhelm, H. A. and Chiotti, P. *Trans. Am. Soc. Metals* 42 (1950) 1295
[202] Wilhelm, H. A. *et al., J. chem. Soc.* (1949) Suppl. 2, 318
[203] Chiotti, P. *J. electrochem. Soc.* 101 (1954) 567 (and earlier papers)
[204] Peterson, D. T. *Trans. Am. Soc. Metals* 53 (1961) 765
[205] Kempter, C. P. and Krikorian, N. H. *J. less-common Metals* 4 (1962) 244
[206] Brett, N. H., Law, D. and Livey, D. T. *J. inorg. nucl. Chem.* 13 (1960) 44
[207] Bates, L. F. and Unstead, P. B. *Br. J. appl. Phys.* 15 (1964) 543
[208] (*a*) *Carbides in Nuclear Energy*, Ed. Russell, L. E. *et al.*, London (Macmillan) 1964; (*b*) Farr, J. D. and Bowman, A. L., p. 184; (*c*) Chubb, W. and Keller, D. L., p. 208; (*d*) Heeney, J. *et al.*, p. 69; (*e*) Alcock, D. C. *et al.*, p. 330; (*f*) De Crescente, H. A. and Miller, A. D., p. 342; (*g*) Wheeler, H. J., p. 358; (*h*) Dalton, J. T., p. 77; (*i*) Stahl, D. and Strasser, A., p. 373; (*j*) Briggs, Dutta, K. S. and White, J., p. 231; (*k*) Nichols, J. L. and Marples, J. A. C., p. 246
[209] Hill, N. A. and Cavin, O. B. *Rep. Oak Ridge natn. Lab. 3588*, 1964
[210] 4th Uranium Carbide Conference, East Hartford, Conn. *Rep. U.S. At. Energy Comm. TID 7676*, 1963
[211] Sharma, B. D. *Acta crystallogr.* 16 (1963) 322
[212] Atoda, T. *et al., Sci. Pap. Inst. phys. chem. Res., Tokyo* 55 (1961) 217
[213] Chang, R. *et al., Conf. on Materials Science and Technology for Advanced Appl., Am. Soc. Metals* (1962) 533
[214] Graham, L. J. *et al., J. appl. Phys.* 34 (1963) 1572
[215] Grossmann, L. N. *Bull. Am. Ceram. Soc.* 42 (1963) 265
[216] Mallett, M. W. *J. electrochem. Soc.* 98 (1951) 1718
[217] Rundle, R. E. *et al., J. Am. chem. Soc.* 70 (1948) 99
[218] Burdick, M. D. *et al., Bur. Stand. J. Res.* 54 (1955) 217
[219] Litz, L. M. *et al., J. Am. chem. Soc.* 70 (1949) 1718
[220] Esch, U. and Schneider, A. *Z. anorg. allg. Chem.* 257 (1948) 254
[221] Gillam, E. *Acta crystallogr.* 15 (1962) 1183
[222] Willis, B. T. M. (*a*) *Nature, Lond.* 167 (1963) 755; (*b*) *Proc. R. Soc.* A279 (1963) 122; (*c*) *Proc. Br. ceram. Soc.* 1 (1964) 9
[223] Kieffer, R., Benesovsky, F. and Nowotny, H. *Planseeber. PulvMetall.* 5 (1957) 33
[224] Levinson, L. S. *J. chem. Phys.* 38 (1963) 2105
[225] Brown, D. J. and Stobo, J. J. *Plansee Proc.* (1961) 279
[226] Chiotti, P. *J. Am. Ceram. Soc.* 35 (1952) 123
[227] Hardy, G. F. and Hulm, J. K. *Phys. Rev.* 93 (1954) 1010
[228] Chang, R. *Acta crystallogr.* 14 (1961) 1097
[229] Wilson, W. B. *J. Am. Ceram. Soc.* 43 (1960) 77
[230] Rough, F. A. and Bauer, A. A. *The Constitution of Uranium and Thorium Alloys*, Reading, Mass. (Addison-Wesley) 1958;
Rough, F. A. and Chubb, W. *Battelle Mem. Inst. Rep. BMI 370* (1959), *1441* (1960); Chubb, W. and Phillips, W. M. *Trans. Am. Soc. Metals* 53 (1961) 465; Bredig, M. A. and Chubb, W., *ibid.* 938
[231] Blumenthal, B. *Rep. Argonne natn. Lab.* 5958, 1959
[232] *Nucl. Engng* 5 (Aug. 1960), data sheet facing p. 354
[233] *Powder Metallurgy in Atomic Engineering* (4th Plansee Seminar), Ed. Benesovsky, F., 1962; (*a*) Murray, P., p. 325; (*b*) Benesovsky, F., p. 357
[234] Sheft, I. and Fried, S. F. *J. Am. chem. Soc.* 75 (1953) 1236
[235] Zachariasen, W. H. *Acta crystallogr.* 2 (1949) 388; 5 (1952) 17
[236] Coffinberry, A. S. and Ellinger, F. H. *Rep. U.S. atom. Energy Comm. A/Conf.* 8 P826, 1955
[237] Kruger, O. L. *J. Am. Ceram. Soc.* 46 (1963) 80
[238] Mulford, R. N. R. *et al., Proc. 2nd Int. Conf. Pluton. Metall.*, Grenoble 1960, p. 301, London (Cleaver-Hume) 1961
[239] Rosen, S. *et al., J. nucl. Mater.* 10 (1963) 90
[240] Pascard, R. *Plansee Proc.* (1961) 387
[241] Vickery, R. C. *et al., J. Am. chem. Soc.* 60 (1959) 498, 503, 505
[242] Vickery, R. C. *Chemistry of the Lanthanides*, London (Butterworth) 1953
[243] Friedrich, E. and Sittig, L. *Z. anorg. allg. Chem.* 144 (1925) 186
[244] Nowotny, H. and Auer-Welsbach, H. *Mh. Chem.* 92 (1961) 789
[245] Spedding, F. H. *et al., Trans. Am. Inst. Min. metall. Engs* 215 (1959) 192
[246] Carlsson, O. N. *et al.*, In 'Symposium on Newer Metals', *Spec. tech. Publs Am. Soc. Test. Mater.* 272, p. 144, 1960
[247] Brewer, L. and Krikorian, O. *J. electrochem. Soc.* 103 (1956) 38

[248] Stackelberg, M. v. *Z. phys. Chem.* B9 (1930) 437; *Z. Elektrochemie* 37 (1931) 542
[249] Warf, J. C. and Palenik, G. *OTS Rep. AD 236014*, 1960; PB 160016
[250] Gschneidner, K. A. *Rare Earth Alloys*, Princeton (Van Norstrand) 1961
[251] Spedding, F. H. and Daane, A. H. *The Rare Earths*, New York (Wiley) 1961
[252] (*a*) Atoji, M. *et al.*, *J. Am. chem. Soc.* 80 (1958) 1804; (*b*) Atoji, M. *J. chem. Phys.* 35 (1961) 1950; (*c*) Atoji, M. and Williams, D. C. *ibid.* 1960
[253] Spedding, F. H., Gschneidner, K. A. and Daane, A. H. *J. Am. chem. Soc.* 80 (1958) 4499
[254] Williams, J. *et al.*, *J. less-common Metals* 2 (1960) 352
[255] Griffiths, L. B. *Phil. Mag.* 7 (1962) 827
[256] Hunt, E. B. and Rundle, R. E. *J. Am. chem. Soc.* 73 (1951) 4172
[257] Gaines, G. B., Sims, C. T. and Jaffee, R. I. *J. electrochem. Soc.* 106 (1959) 881
[258] Kaye, G. *Nature, Lond.* 195 (1962) 1195
[259] Rudy, E. and E. and Benesovsky, F. *Mh. Chem.* 93 (1962) 522
[260] Mallet, M. W. *et al.*, *J. electrochem. Soc.* 98 (1951) 505; 99 (1952) 197
[261] Storms, E. K. *Los Alamos Sci. Lab.*, *Univ. Calif. Rep. LA 2942*, 1964
[262] Nowotny, H. and Glenk, G. *Metallforschung* 2 (1947) 265
[263] Schwarzkopf, P. and Kieffer, R. (*a*) *Refractory Hard Metals*; (*b*) *Cemented Carbides*, London (Macmillan) 1953 and 1960
[264] Andrieux, J. L. and Weiss, G. *Bull. Soc. chim. Fr.* 15 (1948) 598
[265] McKenna, P. M. *Metal Prog.* 36 (1939) 152
[266] Redmond, J. C. and Smith, E. N. *J. Metals, N.Y.* 185 (1949) 987
[267] Moissan, H. and Hoffmann, M. K. *C.R. hebd. Séanc. Acad. Sci., Paris* 138 (1904) 1358
[268] Norton, J. T. and Mowry, A. L. *J. Metals* 3 (1951) 923
[269] Nowotny, H., Kieffer, R. *et al.*, *Mh. Chem.* 90 (1959) 86, 669
[270] Nowotny, H. *et al.*, *Planseeber. PulvMetall.* 7 (1959) 79
[271] Molkov, L. P. and Vicker, I. V. *Vest. Metalloprom.* 8 (1936) 75
[272] Metcalfe, A. G. *J. Inst. Metals* 73 (1946) 591
[273] Nowotny, H., Kieffer, R. and Knotek, O. *Berg- u. hüttenm. Mh.* 96 (1951) 6
[274] Hinnüber, J. and Kinna, W. *Arch. EisenhüttWes.* 29 (1958) 391
[275] Rudy, E., Benesovsky, F. and Sedlatschek, K. *Mh. Chem.* 92 (1961) 841
[276] Kurnakov, N. N. and Troneva, M. Ya. *Zh. neorg. Khim.* 7 (1962) 157
[277] Rüdiger, O. (*a*) *Tech. Mitt. Krupp* 44 (1956) 136; (*b*) *Metall* 7 (1953) 967
[278] Brownlee, L. *et al.*, *J. Inst. Metals* 87 (1958) 58
[279] Rosen, S., Nevitt, M. V. and Mitchell, H. W. *J. nucl. Mater.* 9 (1963) 137
[280] Pascard, R. *Plansee Proc.* (1961) 387
[281] Nowotny, H., Kieffer, R., Benesovsky, F. and Laube, E. *Mh. Chem.* 88 (1957) 336; 89 (1958) 692
[282] Cromer, D. T. *et al.*, *Acta crystallogr.* 17 (1964) 272
[283] Rudy, E. *et al.*, *Mh. Chem.* 91 (1960) 176; *Planseeber. PulvMetall.* 8 (1960) 72
[284] Brownlee, L. D. and Whitehead, K. *Metrop. Vickers Res. Rep.* 4933, 1952
[285] Tombriel, F. *Plansee Proc.* (1955) 205
[286] Stecher, P., Benesovsky, F. and Nowotny, H. *Planseeber. PulvMetall.* 12 (1964) 89
[287] Albert, H. J. and Norton, J. T. *Planseeber. PulvMetall.* 4 (1956) 2
[288] Barta, J., Briggs, G. and White, J. *J. nucl. Mater.* 3 (1961) 322
[289] Baldock, P., McLaren, T. R. *et al.*, *J. nucl. Mater.* 5 (1962) 257
[290] Anselin, F., Calais, D. and Dean, G. *C.R. hebd. Séanc. Acad. Sci., Paris* 257 (1963) 3916
[291] Havell, R. F. and Baskin, J. *J. electrochem. Soc.* 108 (1961) 1068
[292] Kuzma, Ya. B., Lakh, V. I. *et al.*, *Poroshk. metall. Akad. Nauk Ukr. SSR* 3 (1963) 40
[293] Hansen, M. *Constitution of Binary Alloys*, New York (McGraw-Hill) 1958
[294] Pearson, W. B. *Handbook of Lattice Spacings and Structures of Alloys*, London (Pergamon) 1958
[295] Schenck, H. *et al.*, *Arch. EisenhüttWes.* 34 (1963) 37
[296] Craighead, C. M. *et al.*, *J. Metals* 188 (1950) 485, 514, 532
[297] Tofaute, W. and Büttinghaus, A. *Arch. EisenhüttWes.* 12 (1938) 33; *Metallurgist, Lond.* 11 (1938) 190
[298] Northcott, L. *Spec. Rep. Iron Steel Inst.* 24 (1939) 107
[299] Vogel, R. and Löhberg, K. *Arch. EisenhüttWes.* 7 (1934) 473
[300] Svechnikov, U. N. and Shurin, A. K. *Sb. nauch. Rab. Inst. Metallofiz., Kiev* 16 (1962) 124
[301] Vogel, R. and Martin, E. *Arch. EisenhüttWes.* 4 (1931) 487
[302] Wever, F. *et al.*, *Mitt. K.-Wilhelm-Inst. Eisenforsch. Düsseld.* 18 (1936) 229
[303] Maurer, E. *et al.*, *Arch. EisenhüttWes.* 13 (1940) 337
[304] Desch, C. H. *Rep. Br. elect. all. Ind. Res. Ass. J/E/T 125*, 1940
[305] Eggers, H. and Peter, W. *Mitt. K.-Wilhelm-Inst. Eisenforsch. Düsseld.* 20 (1938) 205
[306] Tofaute, W., Küttner, C. and Büttinghaus, A. *Arch. EisenhüttWes.* 9 (1936) 607
[307] Monypenny, J. H. G. *Stainless Iron and Steel* (2 vols.), London (Chapman & Hall) 1951, 1954
[308] Westgren, A., Phragmén, G. and Negresco, T. *J. Iron Steel Inst.* (1928) 386
[309] Belotskii, A. V. and Grindnev, V. N. *Izv. vyssh. ucheb. Zaved. Khim. Met.* 12 (1961) 107

310 Shukov, J. *J. phys. Chem., Ithaka* 36 (1962) 2787
311 Pepperhoff, H. *et al., Arch. EisenhüttWes.* 33 (1962) 611
312 Koch, W. and Kolbe-Rhode, H. *Z. anorg. allg. Chem.* 319 (1963) 312
313 Takeda, S. *Technology Rep. Tôhoku Univ.* 9 (1930) 627; 10 (1931) 42
314 Sato, T. *et al., Nippon Kink. Gakk.* 6 (1960) 395
315 'Precipitation Processes in Steels', Ed. Quarrell, A. G. *Spec. Rep. Iron Steel Inst.* 64, 1959
316 Umanski, Ya S. and Cebotarev, N. T. *Izv. Akad. Nauk SSSR Ser. Fiz.* 15 (1951) 24
317 Bennek, H. and Holzscheiter, C. G. *Arch. EisenhüttWes.* 9 (1935) 193
318 Bain, E. C. *et al., Trans. Am. Inst. Min. metall. Engrs* 100 (1932) 228
319 Walters, F. M. and Wells, C. *Trans. Am. Soc. Steel Treat.* 21 (1933) 830; *Trans. Am. Soc. Metals* 23 (1935) 751
320 Gensamer, M. *Trans. Am. Soc. Steel Treat.* 21 (1933) 1028
321 Kuznetsov, V. G. and Evseeva, N. N. *Zh. obshch. Khim.* 12 (1939) 405
322 Wever, F. and Matthieu, K. *Mitt. K.-Wilhelm-Inst. Eisenforsch. Düsseld.* 22 (1940) 9
323 Butler, J. F., McCable, C. L. and Paxton, H. W. *J. Metals* 221 (1961) 479
324 Vogel, R. and Sundermann, W. *Arch. EisenhüttWes.* 6 (1932) 35
325 Coutsouradis, D. and Habraken, L. *Cobalt* 4 (1959) 3
326 Bain, E. C. *Trans. Am. Inst. Min. metall. Engrs* 100 (1932) 13
327 Lange, H. and Mathieu, K. *Mitt. K.-Wilhelm-Inst. Eisenforsch. Düsseld.* 20 (1938) 175
328 Söhnchen, E. and Piwowarsky, E. *Arch. EisenhüttWes.* 5 (1931) 111
329 Andrew, J. H. *et al., Spec. Rep. Iron Steel Inst.* 23 (1938) 5
330 Samuel, P. L. G., Finch, L. G. and Rait, J. R. *Nature, Lond.* 175 (1955) 37
331 Steinitz, R., quoting thesis by Stover, E. R. *J. Metals* 5 (1953) 891
332 Zelikman, A. N. and Bernstein, O. S. *Sb. nauch. Trud. mosk. Inst. tsvet. Metall. Zolota* 23 (1952) 48
333 Stover, E. R. and Wulff, J. *Trans. Am. Inst. Min. metall. Engrs* 215 (1959) 127
334 Lane, J. R. and Grant, N. T. *Trans. Am. Soc. Metals* 44 (1952) 113
335 Wever, F. and Hashimoto, U. *Mitt. K.-Wilhelm-Inst. Eisenforsch. Düsseld.* 11 (1929) 319
336 Gurland, J. and Norton, J. T. *J. Metals* 4 (1952) 1051
337 Sandford, E. T. and Trent, E. M. *Spec. Rep. Iron Steel Inst.* 38 (1947) 84
338 Chaporova, I. N. and Shchetillina, E. A. *Izv. Akad. Nauk SSSR, Metallurgiya Topl.* (1959) 91 and (1961) 126
339 Dawihl, W. and Schröter, K. *Z. Metallk.* 41 (1950) 231
340 Krikorian, N. H. *et al., J. phys. Chem.* 64 (1960) 1517
341 Edwards, A. R. and Raine, T. *Plansee Proc.* (1952) 232
342 Morral, F. R. *J. Iron Steel Inst.* 130 (1934) 419
343 Löhberg, K. and Schmidt, J. *Arch. EisenhüttWes.* 11 (1938) 607
344 Huetter, L. T. and Stadelmaier, H. H. *Z. Metallk.* 50 (1959) 199
345 Butters, R. G. and Myers, H. P. *Phil. Mag.* 46 (1955) 132, 895
346 Morgan, E. R. *J. Metals* 6 (1954) 983
347 Swanson, M. L. *Can. J. Phys.* 40 (1962) 719
348 Waller, J. M. *OTS Rep. AD 239436*, 1960; *PB 153673*
349 Huetter, L. J., Stadelmaier, H. H. and Fraker, A. C. *Metall* 14 (1960) 113
350 Scheil, E. and Huetter, L. J. *Z. Metallk.* 44 (1953) 387
351 Stadelmaier, H. H. and Huetter, L. J. *Acta metall.* 7 (1959) 415
352 Stadelmaier, H. H. and Duplessis, J. J. *Metall* 15 (1961) 763
353 Stadelmaier, H. H. and Hardy, W. K. *Z. Metallk.* 52 (1961) 391
354 Stadelmaier, H. H. and Hammad, F. H. *Metall* 15 (1961) 124
355 (*a*) Jeitschko, W., Nowotny, H. and Benesovsky, F. *J. less-common Metals* 7 (1964) 133; (*b*) Nowotny, H., Jeitschko, W. and Benesovsky, F. *Planseeber. PulvMetall.* 12 (1964) 31
356 Stadelmaier, H. H. and Waller, T. M. *Metall* 15 (1961) 125
357 Huetter, L. J., Stadelmaier, H. H. and Hardy, W. K. *Trans. metall. Soc. A.I.M.E.* 218 (1960) 859
358 Jeitschko, W. *et al., Mh. Chem.* (*a*) 95 (1964) 30; (*b*) 94 (1963) 247, 332, 672, 1201; (*c*) 95 (1964) 133, 156, 178, 319, 432, 1040
359 Stadelmaier, H. H. *OTS Rep. AD 265251*, 1961; *Z. Metallk.* 52 (1961) 758
360 McBride, L., Greenhouse, H. M. and Shevlin, T. S. *J. Am. Ceram. Soc.* 35 (1952) 82
361 Ballhausen, C. *Stahl Eisen* 72 (1952) 489
362 Brownlee, L. D. *et al., Spec. Rep. Iron Steel Inst.* 38 (1947) 73
363 Kreimer, G. S. *et al., Izv. Akad. Nauk SSSR, Otdel. tekh. nauk, Metallurgiya Topl.* 3 (1962) 947
364 Doeg, H. H. *Jl S. Afr. Inst. Min. Metall.* (July, 1960) 663
365 Funke, V. H. *et al., Poroshk. metall. Akad. Nauk Ukr. SSR* 2 (1962) 66
366 Gurland, J. and Norton, J. T. *Plansee Proc.* (1956) 99
367 Rybalchenko, R. V. *et al., Zh. neorg. Khim.* 6 (1961) 2517
368 Kieffer, R., Benesovsky, F. and Messmer, K. *Metall* 13 (1959) 919; 6 (1952) 243
369 Glaser, F. W. and Ivanick, W. *J. Metals* 4 (1952) 387

[370] Kieffer, R. and Kölbl, F. *Planseeber. PulvMetall.* 1 (1953) 17
[371] Benesovsky, F. and Kieffer, R. *Neue Hütte* 2 (1957) 545
[372] Pfaffinger, K. *Planseeber. PulvMetall.* 3 (1955) 17
[373] Havekotte, W. L. *Metal Prog.* 64 (1953) 67
[374] Agte, C. and Blum, G. *Neue Hütte* 6 (1961) 139
[375] Engel, W. J. *Metal Prog.* 59 (1951) 664
[376] Epner, M. *Metal Treat. Drop Forg.* 12 (1961) 12, 40
[377] Sugiyama, M. and Suzuki, H. *J. Japan Soc. Powd. Metall.* 8 (1961) 152
[378] Trent, E. M. *et al., Metallurgia* 42 (1950) 111
[379] Roach, J. O. *J. electrochem. Soc.* 98 (1951) 160
[380] Kennedy, J. D. *Steel* 131 (1952) 92
[381] Kieffer, R. and Benesovsky, F. *Hartstoffe*, Vienna (Springer) 1963
[382] Forster, L. S. *et al., J. Am. Ceram. Soc* 33 (1949) 27
[383] Semchyshen, M. J. and Barr, H. N. *J. Metals, N.Y.* 13 (1961) 669
[384] Ransley, C. E. *J. Metals, N.Y.* 14 (1962) 129
[385] Benesovsky, F. *et al., Planseeber. PulvMetall.* 11 (1963) 146
[386] Rough, F. A. and Chubb, W. *Battelle Mem. Inst. Rep. BMI 1488*, 1961
[387] Houska, C. R. *J. Am. Ceram. Soc.* 47 (1964) 310
[388] Portnoi, K. J. *et al., Izv. Akad. Nauk SSSR, Otdel. tekh. nauk, Metallurgiya Topl.* (1961) 147
[389] George, A. L. *et al., Phys. Rev.* 129 (1963) 1524
[390] Richardson, F. D. *J. Iron Steel Inst.* 175 (1953) 33
[391] Eick, H. A. *et al., OTS Rep. TID 17413*
[392] (*a*) Fedorov, T. F. and Kuz'ma, Yu. B. *Poroshk. metall. Akad. Nauk Ukr. SSR* 3 (1965) 75; (*b*) Yeremenko, V. N. and Velikenova, T. Y. 11 (1961) 20; (*c*) Fedorov, T. F. *et al.* 3 (1965); *Soviet Powder Metall. Metal Ceram.* 3 (1965) 27
[393] Nowotny, H. *et al., Z. Metallk.* 44 (1954) 97
[394] Hinnüber, J. and Kinna, W. *Stahl Eisen* 82 (1962) 31
[395] Kuz'ma, Yu B. *et al., Poroshk. metall. Akad. Nauk Ukr. SSR* 2 (1965) 22. *Soviet Powder Metall. Metal Ceram.* (1965) 2126
[396] Gurevitch, M. A. and Ormont, B. F. *Metallov. term. Obrab. Metall.* (1958) 7
[397] Samsonov, G. V., Neshpor, V. S. and Strelnikova, N. S. *Dopov. Akad. Nauk URSR* (1958) 838
[398] Rudy, E. and Nowotny, H. *Mh. Chem.* 94 (1963) 507
[399] Rudy, E. *et al., Mh. Chem.* 93 (1962) 1176
[400] Brownlee, L. D., Edwards, R. and Raine, T. *Spec. Rep. Iron Steel Inst.* 58 (1956) 302
[401] Shchetilina, E. A. *et al., Izv. Akad. Nauk SSSR, Met. Gorn. Delo* (1964) 142
[402] Dawihl, W. and Altmayer, G. *Z. Metallk.* 54 (1963) 645
[403] Lied, R. C. and White, G. D. *J. Am. Ceram. Soc.* 45 (1962) 149
[404] Rosen, S., Nevitt, M. V. and Barker, J. J. *J. nucl. Mater.* 9 (1963) 128
[405] Krikorian, N. H. *et al., J. electrochem. Soc.* 110 (1963) 560
[406] Katz, S. *J. nucl. Mater.* 6 (1962) 172
[407] Briggs, G., Guha, J., Barta, J. and White, J. *Trans. Br. Ceram. Soc.* 62 (1963) 221
[408] Nowotny, H. and Kieffer, R. *Metallforschung* 2 (1947) 257
[409] Duwez, P. and Odell, F. *J. electrochem. Soc.* 97 (1950) 299
[410] Kovalskii, A. E. and Umanskii, Ya. S. *Zh. fiz. Khim.* 20 (1946) 769, 929; Brutcher transl. 2173, 2332
[411] Rudy, E. *Planseeber. PulvMetall.* 10 (1962) 32; *Z. Metallk.* 54 (1963) 112, 213; Brutcher transl. 5632
[412] Benesovsky, F., Rudy, E. and E. *Planseeber. Pulv Metall.* 10 (1962) 42
[413] Rudy, E., Benesovsky, F. and Rudy, E. *Mh. Chem.* 93 (1962) 693
[414] Rudy, E. and Benesovsky, F. *Mh. Chem.* (*a*) 93 (1962) 1279; (*b*) 94 (1963) 85; (*c*) *ibid.* 204
[415] Rudy, E. and Chang, Y. A. *5th Plansee Seminar*, Paper 40, 1964
[416] Nowotny, H. (*a*) *Bull. Soc. chim. Fr.* (1960) 1881; (*b*) in *Electronic Structure and Alloy Chemistry of the Transition Elements*, Ed. Beck, P. A., p. 179, New York (Interscience) 1963
[417] Norton, J. T. *J. Metals, N.Y.* 3 (1951) 966
[418] Glenny, E. and Taylor, T. A. *Powder Metall.* 1 (1958) 2, 189
[419] Murray, P. *Powder Metall.* 5 (1960) 64
[420] Kieffer, R. and Benesovsky, F. *Metall* 6 (1952) 171
[421] Dawihl, W. and Dinglinger, E. *Handbuch der Hartmetall-Werkzeuge*, Berlin (Springer) 1953, 1956
[422] Merz, A. In Eisenkolb, F. *Fortschritte der Pulvermetallurgie*, Vol. 2, Ch. 10, Berlin (Akademie-Verlag) 1963
[423] Bowman, M. G. *5th Plansee Seminar*, Paper No. 42, 1964
[424] Robinson, J. W. *OTS Rep. AD 290247*, 1962
[425] Harrington, L. C. and Rowe, G. H. *Proc. Am. Soc. Test. Mater.* 63 (1953) 620, 633
[426] Fitzgerald, L. M. *J. less-common Metals* 5 (1963) 356

REFERENCES

[427] Eick, H. A. *et al.*, *Proc. Symp. Thermodynamics of Nuclear Materials*, Vienna, 1962
[428] (*a*) Alcock, C. B. *et al.*; (*b*) Kaufman, L. and Sarney, A.; *A.I.M.E. Symp. on Compounds of Interest in Nuclear Reactor Technology*, Boulder (Colorado) 1964; (*c*) Piper, J. *Rep. AD 435624, NASA 464–1954*, 1964
[429] Bystroya, K. A. *et al.*, *Izv. Akad. Nauk SSSR, Otdel. tekh. nauk, Metallurgiya Topl.* 4 (1960) 124
[430] Bolgar, A. S. *et al.*, *Izv. Akad. Nauk SSSR, Otdel. tekh. nauk, Metallurgiya Topl.* 5 (1961) 142
[431] Anderson, J. S. *Nat. Phys. Lab. Symp. 9*, paper 7A, London (H.M. Stat. Office) 1959
[432] Samsonov, G. V. and Paderno, V. N. *Planseeber. PulvMetall.* 12 (1964) 19
[433] Bain, E. C. *The Function of Alloying Elements in Steels*, Ohio (Am. Soc. Metals) 1939
[434] Roberts, G. A., Hamaker, J. C. and Johnson, A. R. *Tool Steels*, Ohio (Am. Soc. Metals) 1962
[435] Cohen, M. 'On the development of High Strength Steels' (14th Hatfield Memorial Lecture), *J. Iron Steel Inst.* 201 (1963) 833
[436] 'Metallography 1963', Ed. Quarrell, A. G. *Spec. Rep. Iron Steel Inst. 80*, 1964
[437] 'Structural Processes in Creep', *Spec. Rep. Iron Steel Inst.* 70, 1961
[438] 'High Temperature Steels and Alloys for Gas Turbines', *Spec. Rep. Iron Steel Inst. 43*, 1951: (*a*) Harris, G. T. and Bailey, W. H., 160; (*b*) Harris, G. T. and Child, H. C., 167
[439] *High Temperature Materials I*, Ed. Campbell, I., New York (Interscience) 1959
[440] *High Temperature Materials II*, Ed. Ault, G. M., New York (Interscience) 1963
[441] AGARD Symposium *Refractory Metals and Alloys*, Ed. Semchyshen, M. and Perlmutter, I., London (Pergamon) 1963
[442] Betteridge, W. *The Nimonic Alloys*, London (Arnold) 1959
[443] Oliver, D. A. and Harris, G. T. *Metallurgia* 34 (1946) 293; *Trans. Inst. mar. Engrs* 59 (1947) 79; *Jl W. Scott. Iron Steel Inst.* 54 (1946) 97
[444] Harris, G. T. and Johnson, W. *Metallurgia* 51 (1955) 17
[445] Hagel, W. C. and Beattie, H. J. *Trans. Am. Soc. Metals* 49 (1956) 978
[446] Beattie, H. J. and Ver Snyder, F. L. *Trans. Am. Soc. Metals* 45 (1952) 397; 49 (1956) 883; *Nature, Lond.* 178 (1956) 209
[447] Beattie, H. J. *J. Metals, N.Y.* 14 (1962) 77
[448] Zwell, L., Fasiska, E. J. and v. Gemmingen, F. *Trans. metall. Soc. A.I.M.E.* 224 (1962) 198
[449] Mahla, E. M. and Nielsen, N. A. *Trans. Am. Soc. Metals* 43 (1950) 290
[450] Westbrook, J. H. *J. Metals* 9 (1957) 898
[451] Urbain, M. *Cobalt* 12 (1961) 11
[452] Lux, B. and Bollman, W. *Cobalt* 12 (1961) 32
[453] Habraken, L. and Coutsouradis, O. *Cobalt* 12 (1961) 3
[454] Barker, J. F., Couts, W. H. and Morris, R. J. *Metal Prog.* 78 (1960) 94
[455] Radavick, J. F. and Pennington, W. J. *Metal Prog.* 79 (1961) 94
[456] Henry, G., Plateau, J. *et al.*, *C.r. hebd. Séanc. Acad. Sci., Paris* 252 (1961) 2550
[457] Poynter, V. M. *Metal Prog.* 82 (1962) 111
[458] Smith, E. and Nutting, J. *J. Iron Steel Inst.* 187 (1957) 314
[459] United Nations *Peaceful Uses of Atomic Energy*, Vol. VI, Geneva, United Nations, 1958
[460] Strasser, A. *Nucl. Engng, Lond.* 5 (1960) 353
[461] Grossman, L. N. *J. Am. Ceram. Soc.* 46 (1963) 264

5

NITRIDES

5.1 SOME GENERAL ASPECTS

The transition metal nitrides are in many respects closely allied to the carbides: in structure, properties (metallic nature, high melting points and hardness in the early Groups) and in their fields of practical application. Also their stability and occurrence within the frame of the Periodic System is similar to those of the carbides. There are, however, important differences, largely bound up with the tri-valency of the nitrogen atom against quadri-valency of carbon which entails rather weaker bonds from metal to non-metal. The difference in atomic size is relatively slight, and the curious ambiguity in relative size of the interstitial N and C atoms has been referred to already (Chapter 2): this presents the N atom as the smaller for Group IV, V, VI nitrides (N = 0·69, C = 0·77 Å radius), but with an ascendancy in size over C on passing to Group VII–VIII nitrides. The latter is best shown in Jack's work on the iron nitrides and carbides, to be discussed below.

The close structural link between the nitrides and their metallic parent solid solutions is also a reflection of the relatively small size of the interstitial N atom. Hägg's rule is well fulfilled for the nitrides; the critical size ratio is, however, approached in Group VIII, with important practical consequences.

Thermodynamically, a further difference is rooted in the vapour pressure of nitrogen itself which, for the higher nitrides, tends to evolve in free form (as the stable graphite forms in carbides); this produces nascent gaseous nitrogen enclaves, first in atomic, then in molecular form, held under high pressure at dislocations, grain boundaries or voids. At surfaces it expresses itself as adsorbed layers. Thus, for instance, the truly *stable* Fe–N system is between α- and γ-iron solid solutions and gaseous *nitrogen*, without nitrides (just as Fe-graphite, not cementite). However, this does not preclude the occurrence of metastable nitrides which, though strictly speaking transient, are indeed persistent; the latter proves in practice to be of great usefulness. The hardness of nitrided layers in steels can be directly related to lattice strains derived from the basic instability of the nitrogen atom in the metal lattice.

We might observe that this factor is closely interlinked with the seemingly variable atomic size of the nitrogen atom in various nitrides, referred to above. The strongly bonded N atom, e.g. in TiN, might be envisaged as *in fact* the smaller, and the loosening of the bond on passing towards iron nitrides as the gradual freeing of the nitrogen atom to approach its natural free volume. In other words, the nitrogen atom can be regarded as *actually* being compressed by extreme pressures, whether exerted by the metal bond in the nitride crystal or externally. (It might well be interesting to investigate whether under very high pressures the N atom in Group VII–VIII nitrides does not attain the size it has for lower Group numbers, i.e. become again smaller than the carbon atom.)

5.2 BINARY NITRIDES: SURVEY

Table 5.1 collates the data on the transition metal nitrides, as far as can be assembled from many sources. They include compositions, crystal structures and various properties of the nitrides of the Groups IV–VIII transition metals. In addition, we have, towards the earlier Groups, the nitrides of scandium, yttrium, the Rare Earths (e.g. CeN) and Alkaline Earths, which are very stable and of high melting point. On the other hand, at the later Groups of nitride-forming block, the nitrides cease to exist or, better, 'fade away' with cobalt and nickel, as do the carbides, although highly unstable Cu_3N and Ag_3N have been reported. Pressure dependence is here a vital factor, however, and indeed at this front new nitride structures can be rendered stable by extreme pressures.

We also have the very stable light-atom (non-transition-metal) nitrides of Si, Al, Mg, B and Be, as well as possibly of the alkali metals—certainly lithium nitride—which can be noted here only in passing*.

As to properties, the advantage of hardness of the nitrides is often vitiated by their extreme brittleness; that of high melting point by the fact that the oxidation resistance at high temperatures is much poorer than that of silicides, borides and carbides. The dissociation pressure at high temperatures, particularly from Group VI onwards, also limits their uses, even though it leads to alternative applications (see Section 5.6). Nitrides do come into their own, however, as additions to other interstitial compounds, such as in carbonitrides, and as subsidiary phases in high-temperature materials. Indeed, taking a wider view, we might see in the silicon and boron nitrides a fulfilment of the function of nitrogen in maintaining, when joined with silicon or boron as 'partner interstitials', strong and stable bonds at high temperatures, which it quite fails to achieve in the metallic nitrides.

Figure 5.1 shows a survey of the Group IV–VIII nitride systems, arranged according to the Periodic Table, which may be compared with those shown for the carbides, borides, silicides, oxides and hydrides. For the following comments it may help to consult these diagrams and *Table 5.1.*

The *Group IV* metals (Ti, Zr, and probably Hf) stand out by the high primary solid solubility of nitrogen in the hexagonal (α-)form, which it stabilizes to very high temperatures†. The peritectic temperatures attained (2,350°C

* The hardness and valuable refractory properties of silicon nitride and boron nitride should be recalled, i.e. the 'inter-interstitial phases' (Chapter 2). Although fundamentally so different, their practical applications are closely akin and complementary to the metallic nitrides such as TiN.

† The stabilization of α-Ti by nitrogen is so extensive, approaching 25 at. per cent N at 1,400°C, that the phase forms what might be regarded as a HCP 'pseudo-nitride' developing from the primary solution. There is a lattice expansion[1] to $a = 3{\cdot}03_3$, $c = 5{\cdot}09_2$ Å (from $a = 2{\cdot}9504$; $c = 4{\cdot}6833$ Å for Ti), and with atomic *ordering* around 20 per cent Ni. In apparent conflict, the solubility of nitrogen in α-Ti has also been reported[2] as 17 at. per cent N at 900°C, with *random* N-atom distribution. This then switches[3] to the tetragonal anti-rutile type structure of Ti_2N. Both order or disorder effects may, however, prove to be compatible, depending on heat treatment and degree of equilibration; it would seem a rewarding subject for neutron diffraction study.

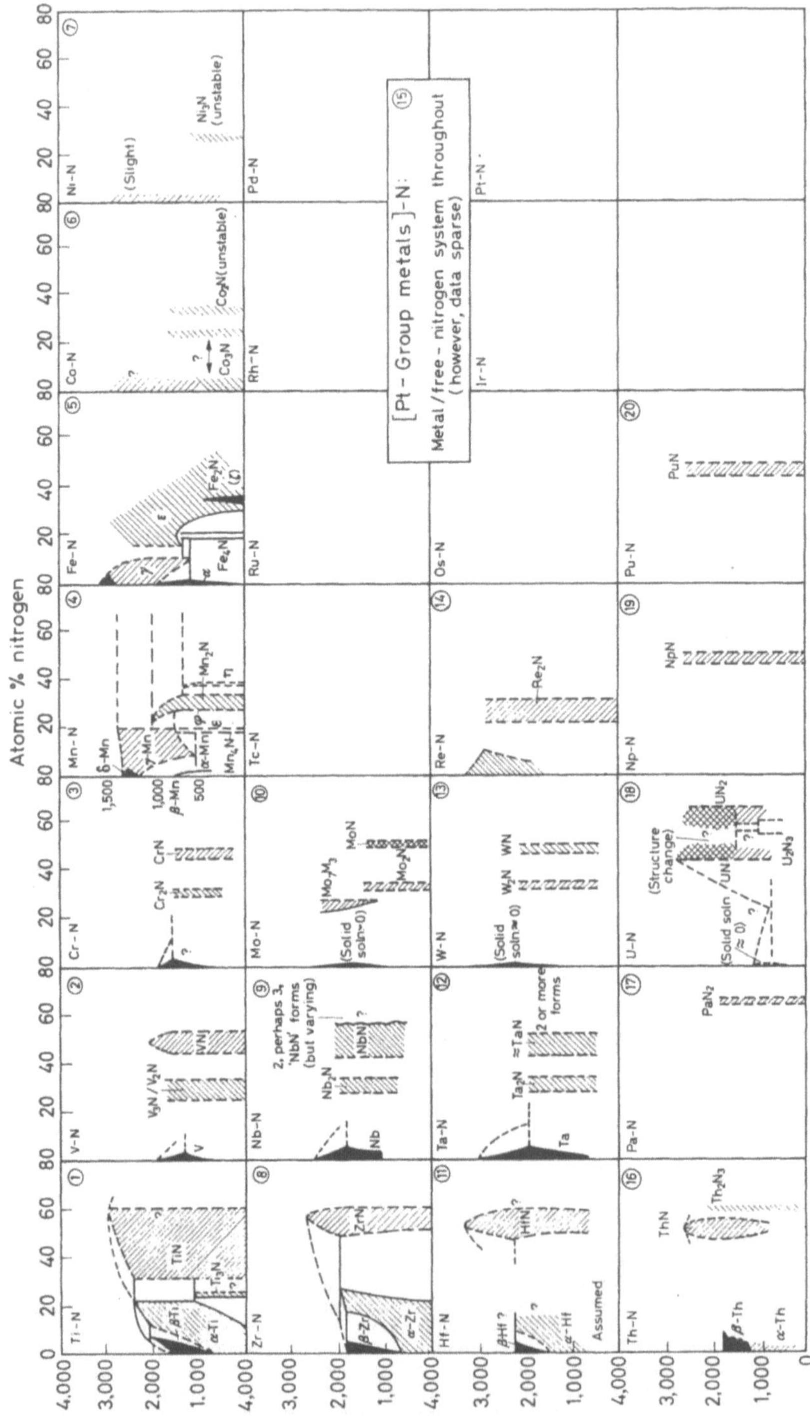

Figure 5.1. The pattern of binary metal–nitrogen systems

for Ti–N) considerably exceed the melting point of the metal. These nitrogen-saturated α-Ti and α-Zr solutions may indeed be viewed already as pseudo-nitrides, particularly as they also structurally anticipate, in Group V, the true hexagonal nitrides, i.e. $V_{3-2}N$, Nb_2N, Ta_2N. In fact, the N solubility limits (cf. Chapter 3) in α-Ti and α-Zr approach this stoichiometry. With increasing N content in any one system, we have the transformation from hexagonal to cubic close-packing in α-(Ti,N) → TiN, $V_2N \rightarrow VN$; α-(Zr,N) → ZrN and α-(Hf,N) → HfN; and, later, in $Cr_2N \rightarrow CrN$.

However, a complication begins in the case of the larger *Group V* atoms, niobium and tantalum. The change-over in hexagonal → cubic stacking here fails to materialize, and the symmetry remains of hexagonal type: in NbN* and (N-deficient) TaN the *simple* hexagonal WC-type of structure. The occurrence of the latter is thus brought forward from Group VI for the carbides, to Group V for the nitrides. Besides the WC type, Nb and Ta mononitrides adopt further hexagonal forms such as the 'B–8' type in NbN and 'B–35' in TaN.

The essentially *metallic* nature of the bonding in TiN is still similar to that in TiC but contrasts to the *ionic* bonding in isostructural TiO. This has been reaffirmed recently both electrochemically[4], and from the aspect of atomic and ionic sizes[5], but with some lively later controversy[6]. Nitrogen is able to dissolve in TiO both as an ion and as a covalent atom but, conversely, oxygen can dissolve in TiN only as an ion.

A study of resistivity and microhardness of TiN_{1-x} as a function of its non-stoichiometry[7] showed already a distinct tendency towards *ionic* bonding, as against metallic in the carbides: increasingly so the more nitrogen-defective the titanium nitride becomes. The conclusion can be generalized to other Group IV and V nitrides and there is, in fact, a part-way stage of mononitrides between the monocarbides and monoxides.

The wide homogeneity ranges of all the Group IV and V nitrides should be noted; the various polymorphs observed are not necessarily bound to the same stoichiometry, but this can vary slightly. Thus for instance $NbN_{0.95}$ ($B\text{–}8_1$ type) undergoes a modification to the simple stacking repeat, so as to give twice the unit cell of 'NbN'. A third form of NbN, of γ-MoC type, has also been reported, and much of what has been said on the MoC-problem (cf. Chapter 4) can be applied to NbN. Tantalum 'mononitride' can variate to a whole series of closely related structures[8], probably of slightly varying N content†. The critical effect of trace impurities, e.g. O or C, to steer these structures is also important; much of the earlier work on nitride equilibria has probably suffered from inadvertent oxygen or carbon contamination.

This impurity effect has also cast some doubt on the vanadium–nitrogen system. Here a body-centred tetragonal phase reported around 16 at. per cent N proved invalid, being really oxygen-bearing. In the Ta–N system,

* For niobium, there have been suggestions of the tetragonally distorted B1 structure, in the form of a compound reported around $Nb_{14}N_{11}$ (see *Table 5.1*); however, this was later suspected to be stabilized by oxygen impurity, i.e. ≈ $Nb_{14}(N,O)_{11}$.

† The Ta–N system affords a very instructive example of a change in stacking sequence. We have here, in series, compounds of *close-packed*, *simple* and complex *'mixed'* packing. Thus a whole variety of new nitrides has now been discovered in the refractory metals, where the stacking changes can be 'rung' almost at will by isomorphous replacements (alloying) or heat treatment. The *iron* carbonitrides provide a modified extension of the same feature to the later Groups.

Table 5.1. Summary of

Base metal	*Nitride* ($\longleftrightarrow$ = *homogeneity range*)	*Crystal structure type; spacings, Å* *M = molecules per unit cell*	*Density*	*Melting point, °C*
Group IV				
Ti	Ti_3N (possibly)	tetragonal; $a=4{\cdot}92$, $c=5{\cdot}16$		peritectic below 1,050?
	Ti_2N	tetragonal; $a=4{\cdot}9452$, $c=3{\cdot}0342$ ('anti-rutile' type)		
	TiN $\longleftrightarrow$	cubic Bl 4·221 (at $TiN_{0\cdot45}$) –4·244 (at TiN) $M=4$	5·43	2,950 (or 2,350?)
Zr	ZrN $\longleftrightarrow$	Bl 4·537 (at Zr end) –4·577 (at $ZrN_{1\cdot0}$); $M=4$	7·3	2,980
Hf	HfN $\longleftrightarrow$	Bl 4·524 (at $HfN_{0\cdot74}$) –4·518 (at $HfN_{1\cdot0}$) $M=4$	14·0	3,310
Th	ThN $\longleftrightarrow$	Bl 5·1593 $M=4$	11·56	2,630
	Th_2N_3	hexagonal (D5$_4$ type) $a=3{\cdot}883$ $c=6{\cdot}187$ $M=1$	10·40	?
Group V				
V	(V_3N?)V_2N $\longleftrightarrow$	hex. C.P. N superlattice $a=2{\cdot}823$, $c=4{\cdot}545$ (at $VN_{0\cdot48}$) $M=1$ –$a=2{\cdot}839$, $c=4{\cdot}560$ (at $VN_{0\cdot50}$)		
	VN $\longleftrightarrow$	cubic Bl $a=4{\cdot}066$ (at $V_{0\cdot72}$) –$a=4{\cdot}139$ at $VN_{1\cdot0}$ $M=4$	6·1	2,050 (or 2,360?)
Nb	Nb_2N $\longleftrightarrow$	hex. C.P. $a=3{\cdot}056$, $c=4{\cdot}957$ (at Nb end) –$a=3{\cdot}056$, $c=4{\cdot}995$ (at N end) $M=1$	8·31	2,400
	$Nb_{14}N_{11}$ (?) $\longleftrightarrow$	tetr. distorted Bl $a=4{\cdot}385$, $c=4{\cdot}310$ (at Nb end) –$a=4{\cdot}386$, $c=4{\cdot}331$ (at N end)		
	$NbN_{0\cdot9}$ to $NbN_{1\cdot0}$	cubic Bl $a=4{\cdot}381$–$4{\cdot}392$ $M=4$		
	NbN $\longleftrightarrow$	simple hexagonal $a=2{\cdot}950$, $c=2{\cdot}772$ (at Nb end) $M=1$ WC type $a=2{\cdot}958$, $c=2{\cdot}779$ (at N end)	7·3	2,050
	NbN (2nd form) $\longleftrightarrow$ ($NbN_{0\cdot95}$)	hex. (B–8$_1$ type) $a=2{\cdot}968$, $c=5{\cdot}535$ $M=2$		

Note—Considerable limits of error in melting point, hardness and other property figures must be *transition metal nitrides*

Heat of formation, kcal/g atom at R.T., 1 atm	*Hardness* (microhardness, kg/mm^2)	*Other properties: electrical resistivity,* ρ, mΩ.cm *Young's modulus,* E (kg/mm^2); *superconductivity,* SPCD: *upper temp. limit*	*Remarks*
			TiN has wide homogeneity range, reported to $TiN_{0\cdot06}$ at 900°C; maximum spacing at 50 per cent
−80·4	8–9 Moh (1,200 at $TiN_{0\cdot6}$ to 1,900 at $TiN_{0\cdot97}$ (50 g load)	$\rho=113$ at $TiN_{0\cdot56}$–27 at TiN $E=8{,}000$ SPCD < 5·6°K	
−87·3	+8 (1,510, 100 g Knoop)	$\rho=160-21$, dep. on stoichiometry SPCD < 10°K or higher	ZrN: homogeneity range wide but less than TiN
−88·2	1,640 (for $HfN_{0\cdot86}$)	$\rho=33$ for $HfN_{0\cdot86}$	HfN: homogeneity range less than for ZrN (see *Table 5.4*)
−310			
	1,900 (for $VN_{0\cdot34}$; 50 g load)	$\rho=123$ for $VN_{0\cdot34}$	V-richer BC tetr. phase reported ≈ 16 at. per cent N, but probably oxide
−52	1,520 (for $VN_{0\cdot74}$; 50 g load)	$\rho=85$ for $VN_{1\cdot0}$ SPCD < 3°K	note spacing maximum at 50 per cent N
−61·1	1,720 (for $NbN_{0\cdot46}$; 50 g load)		
−59	1,396 (for $NbN_{0\cdot96}$; 50 g load)	$\rho=200$, but down to 40 dep. on stoich. SPCD < 16°K (note exceptionally high superconductive temperature)	note that Bl structure, in contrast to TaN, can still be stabilized in some work by O_2 impurity 'NbN' allotropes are closely related: 2nd form=2 × unit cell; a 3rd form reported of γ'–MoN type

emphasized, particularly due to possible impurities (e.g. O_2) or uncertain stoichiometry.

Base metal	*Nitride* (↔ = *homogeneity range*)	*Crystal structure type; spacings,* Å			*Density*	*Melting point,* °C
Group V—cont.						
Ta	Ta_2N ↔	hex. C.P.	a=3·048 c=4·915 (at $TaN_{0\cdot41}$) –a=3·042 c=4·923 (at $TaN_{0\cdot50}$)	M=1	15·86	
	$TaN_{0\cdot8}$ to $_{0\cdot9}$ ↔	simple hex. (WC-type)	a=2·938 c=2·883	M=1		
	TaN ↔	hex. (B–35 type)	a=5·1913 c=2·9100 (at $TaN_{1\cdot0}$)	M=3	14·34	3,090
Pa	PaN_2	cubic, Cl, CaF_2 type (as UN_2)		M=4		
Group VI						
Cr	Cr_2N ↔ (only small range)	hex. C.P. N superlattice	a=4·759 (Cr end) c=4·438 –a=4·805 (N end) c=4·479			
	CrN ↔	cubic, Bl	a=4·149	M=4	6·1	1,450 (dec.)
Mo	≈Mo_7N_3 ↔ (H.T. phase 'β' about 28 atomic % N)	F.C. tetrag.	a=4·188 c=4·024	M=4		stable only above 600°C
	Mo_2N ↔	cubic, Bl (N superlattice?)	a=4·165 (Mo end) –a=4·169 (N end)	M=4		dec. to Mo + N above 895°C (for 1 atm)
	MoN ↔	hex. with N superlattice	a=5·725 c=5·608	M=8		
W	W_2N ↔	cubic, Bl	a=4·126	M=4		
	WN ↔	simple hex. (WC type)	a=2·893 c=2·826	M=2		
	WN_2	rhombohedral	a=2·89 c=16·40	M=3		
U	UN	cubic, Bl	a=4·8897	M=4	14·32	2,650 amended to 2,850°C above 2·5 atm
two forms:	U_2N_3 ↔	BCC (D–53) Mn_2O_3 type	a=10·68—10·58 at $UN_{1\cdot75}$	M=16	11·24	transformation point ≈ 1,120°C
	U_2N_3 ↔	trigonal (D–52) La_2O_3 type	a=3·700 c=5·826			
	UN_2 ↔	cubic, Cl CaF_2 type	a=5·301	M=4	11·73	

Heat of formation, kcal/g atom at R.T., 1 atm	*Hardness* (microhardness, kg/mm²)	*Other properties: electrical resistivity,* ρ, mΩ.cm *Young's modulus,* E (kg/mm²); *superconductivity,* SPCD*: upper temp. limit*	*Remarks*
−64	1,220 (for $TaN_{0\cdot58}$; 50 g load)	$\rho=263$ (for $TaN_{0\cdot45}$) no SPCD	a series of high-metal Ta 'nitrides' also known, including $TaN_{0\cdot05}$ which is N superlattice in Ta solution
−59·0	1,060 (for $TaN_{1\cdot0}$; 50 g load)	$\rho=128$ (for $TaN_{1\cdot0}$) no SPCD	a FCC (Bl) TaN, not existing at atm. pressure, produced by high pressures at high temperatures
−31	1,570 (for $CrN_{0\cdot47}$; 50 g load)	$\rho=79$ (for $CrN_{0\cdot50}$) no SPCD	
−30	1,083 (for $CrN_{0\cdot99}$; 50 g load)	$\rho=640$ (for $CrN_{0\cdot93}$) no SPCD	
		ρ unknown	above ≈ 600°C
−19·5		SPCD < 5°K	
	unknown (brace spanning from "ρ unknown" to "ρ=9")	SPCD < 12°K	a second hexagonal form of MoN, found by electron diffraction, probably only metastable, at surfaces
−17		ρ unknown ↕	another (simple cubic) W_2N form reported but not confirmed (probably due to oxygen)
		$\rho=9$	note N-deficiency in WN; dec. to W_2N *in vacuo.* A 'WN_2' reported earlier was recently confirmed, akin already to UN_2
	580 (100 g Knoop)	$\rho=208$	no 'MoN_2' analogue UN chemically more stable than UC U–UN eutectic at 1,130°C U_2N_3–UN_2 partial solution, pressure-dependent $U_2N_3 \rightarrow UN + UN_2$ at high pressures

Base metal	*Nitride (⟷ = homogeneity range)*	*Crystal structure type; spacings, Å*	*Density*	*Melting point, °C*
Group VII				
Mn	Mn_4N ⟷ (small)	FCC, Fe_4N type with superlattice (linking with γ–Mn) $a=3{\cdot}852$ (Mn end) –3·868 (N end) $M=1$		
	Mn_5N_2/Mn_2N ⟷ (wide)	hex. C.P. $a=2{\cdot}779$, $c=4{\cdot}529$ (Mn end) (N atoms random) –$a=2{\cdot}834$, $c=4{\cdot}537$ (N end) $M=1$		decomposes above ≈ 900°C
	Mn_3N_2 ⟷ (small, defective) ≈ 38 at. per cent N	FC tetrag. $a=4{\cdot}202$, $c=4{\cdot}039$ (Mn end) (N atoms ordered) –$a=4{\cdot}215$, $c=4{\cdot}137$ (N end)		
	Mn_6N_5 ⟷	FC tetrag. $a=4{\cdot}2234$, $c=4{\cdot}1086$ (Mn end) –$a=4{\cdot}2145$, $c=4{\cdot}1486$ (N end)		
Re	Re_3N/Re_2N (metastable)	FCC metal lattice $a=3{\cdot}93$		
Np	NpN	cubic, Bl $a=4{\cdot}897$ $M=4$	14·2	
Group VIII				
Fe	Fe_4N (γ'phase)	FCC, with N superlattice $a=3{\cdot}791$ (Fe end) –3·801 (N end) $M=1$	6·6	forms from below ≈ 700°C
	ε-phase Fe_3N/Fe_2N ⟷ (wide range)	HCP, with N superlattice $a=2{\cdot}660$, $c=4{\cdot}344$ (Fe end) –$a=2{\cdot}764$, $c=4{\cdot}420$ (N end) (but many other data available)	varying	
	Fe_2N (ζ-phase)	orthorhombic, N superlattice $a=5{\cdot}511$, $b=8{\cdot}29$, $c=4{\cdot}425$	6·4	forms from ε below ≈ 500°C
Ru, Os	no data but stable nitrides unlikely			
Pu	PuN ⟷	cubic, Bl $a=4{\cdot}906$ $M=4$	14·2	
Co	Co_3N ⟷ (metastable)	HCP, with N superlattice $a=2{\cdot}663$, $c=4{\cdot}360$ (Co end) –$a=2{\cdot}66$, $c=4{\cdot}358$ (N end)	7·92	
	Co_2N	orthorhombic (as Co_2C) $a=2{\cdot}848$, $b=4{\cdot}636$, $c=4{\cdot}338$ $M=2$	6·3	
Ni	Ni_3N ⟷ (metastable)	HCP with N superlattice (ε-Fe–N type) $a=2{\cdot}668$, $c=4{\cdot}312$ $M=2$		decomposes at ≈ 400–500°C (dep. on pressure)
Pt	group metals (Rh, Ir, Pd, Pt): no nitrides but nitride-like surface-adsorbed structures possible			

Heat of formation, kcal/g atom at R.T., 1 atm	*Hardness* (microhardness, kg/mm²)	*Other properties: electrical resistivity, ρ,* mΩ.cm *Young's modulus, E* (kg/mm²); *superconductivity,* SPCD: *upper temp. limit*	*Remarks*
−30·3		ferromagnetic (Curie point 470°C)	extensive solid solubility of N in Mn; γ-Mn stabilized FC tetr. on quenching
−48·2			Mn_4N may merge with γ-Mn at higher temperatures range 28·4–34·6 per cent N for Mn_2N note *ordered* N in Mn_4N, *random* N in Mn_2N and Mn_3N_2
			Mn_6N_5 is a new phase, reported to form peritectoidally below ≈ 580°C; no Mn nitrides stable below 400°C?
≈ 0			
−2·6	very hard but associated with N/martensite, etc.; heat-treatment-dependent	ferromagnetic (Curie point 480°C)	extensive γ-Fe solubility of N Fe–N equilibria highly pressure-dependent; intermediate stages: N-austenite/N-martensite/Fe_4N and ε-transformations
−0·9			
(−95?)			
+2			extensive solubility of N in α and β-Co but no stable nitrides
0·2			no 'Ni_2N' (as Co_2N) reported but possible

a similar tetragonal distortion of the BC cube has recently been reported[9], produced by quenching the BC primary solid solutions from high temperature. Whether this ought to be regarded as a genuine nitride is arguable, and certainly it is a metastable condition which can be annealed out. Similar metal-rich 'semi-nitrides' can almost certainly be formed from other Group IV–VI primary solutions though, as mentioned, the issue is complicated by possible impurities.

The kinetics of the Ta → N reaction at 800–1,300°C, from initial N solution to the actual nitrides[10], revealed an intricate series of phases, depending on pressure/temperature/time conditions. These are '$Ta_{20}N$' (essentially a slightly distorted primary solution phase occurring at grain interfaces with a BCC superlattice), Ta_2N, $Ta_{1\cdot25-1\cdot11}N$ and TaN. The appearance of such, probably metastable, very high-metal phases, distinct from the matrix at grain boundaries, is an important effect, typical of the nitriding kinetics of the refractory metals more generally. We will later see it similarly developed in form of new high-metal *oxides*, such as the 'NbO_x', 'TaO_y', etc. (Chapter 8). Such dilute nitrogen alloys are also liable to *precipitation-hardening* effects: e.g. on Cr–N alloys this was studied in detail[11], and work on similar lines would be worth pursuing particularly in V–N, Ta–N and Nb–N (and O) alloys.

The differences between these various nitride modifications are perhaps much better terms 'modulations', as they merely involve changes in stacking sequence. Thus stacking faults within nitride single crystals are certain to be common, depending on local lattice defects, vacancies and impurity atoms. It is doubtful, in fact, whether these structure modulations are not so closely entwined within a given nitride crystal (particularly near dislocations and sub-grain boundaries) that identification of a 'pure' and structurally perfect phase becomes impossible.

Whilst for the M_2N nitrides the homogeneity ranges mean essentially a nitrogen-defect lattice, there can be both M and N excess for the mononitrides. Characteristic spacing changes associated with this have been instanced already for TiN_{1-x} (Chapter 2). In *Figure 5.2* they are set against those in VN_{1-x}.

Pressure is a particularly critical factor in deciding the nitride structures in Group V. For instance, in *tantalum* mononitride, TaN, the (normally non-existant) cubic form *can* be produced[12c], at about 45 kilobars, 1,400°C, creating isomorphy with the Group IV nitrides (TiN, etc.). Thus, viewed in compositio n-temperature-pressure space, it is highly probable that e.g. a high-pressure TaN_{1-x} is capable of forming a complete series of solid solutions with the normal-pressure TiN_{1-x}, ZrN_{1-x} or HfN_{1-x} nitrides; the same, no doubt holds for NbN_{1-x} and VN_{1-x} at high pressures.

Passing to the *Group VI* nitrides, these have already lesser stability and lower heats of formation than those of Groups V and IV. We may note here, incidentally, that the large conflicts in the melting points of 'pure' chromium reported in earlier work were later attributed to nitrogen, both in solution and as nitride. Similarly, a presumed 'hexagonal Cr' was in fact proved to be chromium nitride.

Extensive researches have been made on the effect of nitrogen on Cr (e.g.[12,13]) and, in particular, the precipitation mechanism of Cr_2N in the Cr matrix.

The changeover from hexagonal to cubic nitride, as in $V_2N \rightarrow VN$, continues in chromium; however, for Mo and W it becomes seemingly reversed, Mo_2N and W_2N being cubic close-packed, MoN and WN hexagonal—but simple not close-packed. The occurrence, for molybdenum,

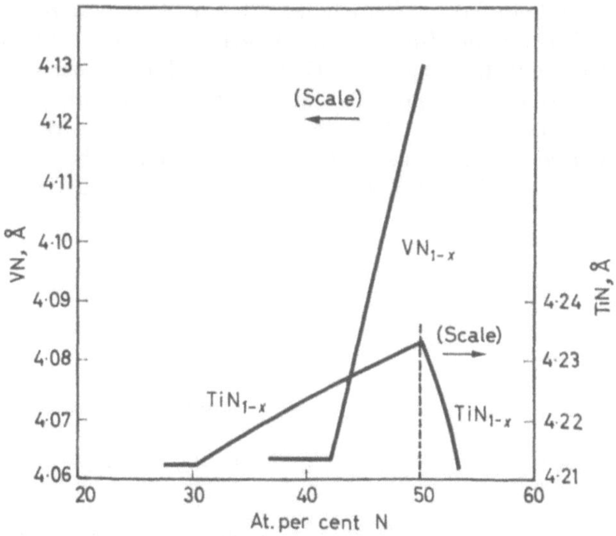

Figure 5.2. Change of lattice spacing in titanium and vanadium mononitrides with N *content in the non-stoichiometric ranges* (TiN[12a], VN[12b])

of the metal-richer high-temperature 'β' phase (see *Figure 5.1*) near 28 at. per cent N is of interest. Being face-centred tetragonal, it may be regarded as a stage preceding the FC cubic one in Mo_2N, the lattice having become distorted through excessive nitrogen absences—unless indeed this β form is, at temperature, also cubic, becoming tetragonal only on quenching (like γ-Mn). No tungsten analogue to this has been reported, but it may well exist. According to more recent electron diffraction work[14], MoN possesses a second hexagonal modification, differing slightly from the first in N atom positions and lattice dimensions. This may well be classed as an extension to the NbN and TaN modulations and could also occur in WN. The $W_2N \rightarrow WN$ transition again represents a straightforward change in stacking sequence through added nitrogen. It is interesting to recall that tungsten carbide (WC, its simple hexagonal isomorph) decomposed at very high temperatures to $W_2C + C$; WN does so too (and much more readily) in its own characteristic manner, i.e. *in vacuo*, to $W_2N + N_2$. Thus we have

	Simple hexagonal	
	WN ↙ ↘ WC	
	Cubic close-packed	*Hexagonal close-packed*
	W_2N	W_2C
	$+$*gas*	$+$*graphite*
variable:	*pressure*	*temperature*

The occurrence of the *di*nitride for tungsten is new in the evolution along Groups IV, V, VI, and this is the highest nitride met with so far, other than for actinides. There is no MoN_2 (except conceivably under high pressures?), and WN_2 may be regarded as a lead-over to uranium, with its UN_2. (The parallel of W$\longleftrightarrow$U carbides (Chapter 4) may here be recalled.) The structure is rhombohedral, with hexagonal layers in three-fold stacking, but a considerable distance apart (5·47 Å). The N atoms occupy the tetrahedral holes between W atoms, with an appreciable distortion.

The bond types and crystal chemistry of tungsten nitrides have been discussed in a group of Russian papers[15,16]. The schematic diagram evolved is shown in *Figure 5.3*, and the W—N phases and their structures

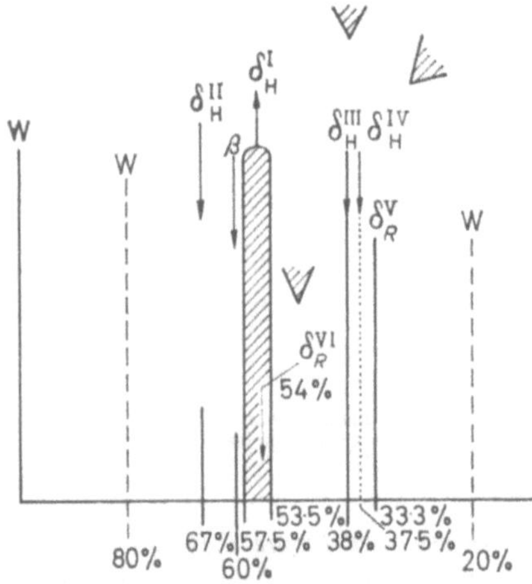

Figure 5.3. Phase dispositions in the W–N system[15]

Cf. *Table 5.2*; temperatures left open; shaded triangles = high-temperature phases of variable composition

By courtesy of the American Institute of Physics

are more closely described in *Table 5.2.* There is a whole series (6) of seemingly separate tungsten nitrides which are, however, structurally very closely related stages of varying bond types, where metallic bonds begin to be superposed by covalent ones. Whilst in the First Long Period nitrides (Ti, V, Cr) the co-ordination around the metal is octahedral, it tends to degenerate in the Second and Third Long Periods (Nb, Mo, Ta, W) to prismatic: a complex sequence arises, with three types of distinct metal positions, entailing variations in stacking sequence.

In the wider context of interstitial alloys, some of these structures are already comparable to transition-metal boride and to the molybdenum disulphide (MoS_2) structures. The borderline position, as far as interstitial bonding is concerned, of tungsten is here again well brought out; it is also reflected by the increasing complexity of phase sequences in W–C$\rightarrow$W–N$\rightarrow$ W–O (compared diagrams); the partial covalency in the nitrides develops further in the oxides.

Before leaving the more refractory nitrides, we turn briefly to the *actinon nitrides* (for composition and details, see *Table 5.1* and *Figure 5.1*). The nitrides of Th, (Pa?), U, Np and Pu are all high-melting, although their decomposition is strongly pressure-dependent.

Table 5.2. Tungsten nitrides[15]

By courtesy of the American Institute of Physics

Phase Symbol	Lattice constants Å	Space group	Position of W normal	Position of W defect	Position of N	Approx. formula and theor. density, g/cm³	Minimal distances, Å W—N	Minimal distances, Å N—N
1	2	3	4	5	6	7	8	9
δ_{II}^{I}	a=2·885 c=15·30—15·46 c/a=5·3—5·35	$P6_3/mmc$—D_{6h}^{4}	W_I *at* 2 (c): $\frac{1}{3}\frac{2}{3}\frac{1}{4}$, $\frac{2}{3}\frac{1}{3}\frac{3}{4}$	W_{II} *at* 4 (f): $\frac{1}{3}\frac{2}{3}z$, $\frac{2}{3}\frac{1}{3}z$, $\frac{2}{3}\frac{1}{3}\frac{1}{2}+z$, $\frac{1}{3}\frac{2}{3}\frac{1}{2}-z$, z=0·06	N *at* 4 (f): z=−0·165	$W_{1.15}N$—$W_{1.8}N$ ρ=13·6—15·7	W_I—N 2·16 W_{II}—N 2·26	2·60— 2·62
δ_{H}^{II}	a= 2·89 c=22·85 c/a= 7·9	$P\bar{3}$—C_{3i}^{1}	W_I *at* 2 (c): 00z, 00$\bar{z}$, z_1=0·0607 W_{II} *at* 2 (d): $\frac{1}{3}\frac{2}{3}$, z, $\frac{2}{3}\frac{1}{3}\bar{z}$, $z_2=\frac{1}{3}-z_1$; W_{III} *at* 2 (d): $z_3=\frac{1}{3}+z_1$		N_I *at* I (b): 00½ N_{II} *at* 2 (d) z=0·154	W_2N; ρ=12·0	W_I—N 2·88 W_{II}—N 3·09 W_{III}—N 2·91	2·89
δ_{H}^{III}	a= 2·87 c=11·00 c/a= 3·81	$P6_3/mmc$—D_{6h}^{4}	W_I *at* 2 (b): 00¼, 00¾	W_{II} *at* 2 (a): 000 00½	N *at* 4 (f): z=0·126	$W_{0.64}N$; ρ=11·0	W_I—N 2·15 W_{II}—N 2·16	2·75
δ_{H}^{IV}	a= 2·89 c=10·8 c/a= 3·73	$P6_3$—C_6^6	W *at* 2 (b): $\frac{1}{3}\frac{2}{3}z$, $\frac{2}{3}\frac{1}{3}\frac{1}{2}+z$ z=0·625	W_{II} *at* 2 (b): z=0·375	N_I *at* 2 (a): 00z, 00½−z, z=0·25 N_{II} *at* 2 (b) z=0·25	$W_{0.6}N$; ρ=10·63	W_I—N 2·16 W_{II}—N	1·67
δ_{R}^{V}	a= 2·89 c=16·4 c/a= 5·67	$R\bar{3}m$—D_{3d}^{5}	W *at* 3 (a): 000, $\frac{1}{3}\frac{2}{3}\frac{1}{3}\frac{2}{3}\frac{1}{3}\frac{2}{3}$		N *at* 6 (c): (000, $\frac{1}{3}\frac{2}{3}\frac{1}{3}$, $\frac{2}{3}\frac{1}{3}\frac{2}{3}$)+00$z$, 00$z$ z=0·1785	$W_{0.5}N$; ρ=9·0	W—N: 2·915 3·03 3·04	1·71
δ_{R}^{VI}	a= 2·89 c=23·35 c/a= 8·07	$R\bar{3}m$—D_{3d}^{5}	W_I *at* 3 (a): 000, $\frac{1}{3}\frac{2}{3}\frac{1}{3}\frac{2}{3}\frac{1}{3}\frac{2}{3}$	W_{II} *at* 6 (c): z=0·120	N *at* 6 (c): z=0·277	$W_{1.17}N$; ρ=13·6	W_I—N 2·13 W_{II}—N 2·12	3·12

Like the corresponding carbides, these nitrides show close similarities to the Rare Earth nitrides, in composition (all at or near 50 at. per cent N), chemically (e.g. hydrating tendency) and in structure. The mononitrides ThN, UN, NpN and PuN are all of cubic close-packed structure and (where known) high melting point. N addition converts the hexagonal (La_2O_3-type) structure in Th_2N_3 to the BCC (Mn_2O_3-type) in U_2N_3*. Perhaps the most interesting effect, however, is the maintenance in UN_2 of the close-packed cubic metal lattice, quite as in UN but this time with the CaF_2- instead of NaCl-type structure: the only difference is that the N atoms have changed places from the octahedral to the more numerous but less spacious tetrahedral positions. UN_2 can in fact only be obtained under high nitrogen pressure. It is probable (though uncertain, and no doubt pressure-dependent) that at high temperatures a mutual solid solution $UN \leftrightarrow UN_2$ could be formed, with both these sites partially occupied. U_2N_3 decomposes peritecticoidally at higher temperatures to enable this to occur. The analogous position in the U–C system should again be recalled (see e.g. *Figure 4.2* as against *Figure 5.1*). The CaF_2 structure still also occurs in protactinium dinitride, PaN_2, but there seems no Th analogue.

The 'bareness' of the Np–N and Pu–N diagrams in *Figure 5.1* does not imply that no nitrides other than *M*N occur but simply that no data are available.

An interesting point arises from melting-point observations on UN. A previously accepted value of 2,630°C was more recently amended to 2,850°C, if measured under high *pressures*. It is suggested that the earlier low value applies to the partly decomposed mononitride, a finding which may more generally throw doubt on earlier melting-point (and other thermodynamic) data for such nitrides of critical stability. This raises the point of principle that any property data on nitrides for the later transition metals must increasingly be conditional on high N pressures for their correctness. Similar reasoning will apply later to the oxides as it did, with less emphasis, to carbides.

Passing to *Groups VII and VIII*, the growing instability, yet persistence, of nitrides is the dominant feature for the 1st Long Period. In the 2nd and 3rd Long Period (Pt Group metals) this develops into the abrupt breakdown of nitrides altogether, except possibly at very high pressures. The behaviour trend is quite similar to that of the carbide. A cubic rhenium nitride, Re_2N, has been reported[17], isomorphous to W_2N, whilst there is no rhenium carbide.

The evolution of nitride equilibria along the series Ti $\rightarrow$ V $\rightarrow$ Cr $\rightarrow$ Mn $\rightarrow$ Fe $\rightarrow$ Co $\rightarrow$ Ni is largely self-explanatory from *Figure 5.1*. The greater complexity of the Mn– and Fe–N systems is directly associated with their technological role. Basically it is brought about by the increasingly transitory and unstable location of the N atom in the metal host lattice of the nitrides concerned, and the blurred distinction between primary solution and actual nitride. (Nitrogen martensite, and associated effects, are some manifestations of this; see below.) The stabilization of austenite and of its FCC Mn, Ni analogues is linked with wide solubilities for the N atom

* The isomorphy of some of these M_2O_3 nitrides with known *oxides* should here be noted which, in the carbides, cannot yet be realized.

which, upon reaching nitrogen saturation, 'continues', though in the modified form of the (still FCC but ordered) Mn_4N and Fe_4N; they derive for manganese directly from FCC γ, for iron from the hexagonal ε phase. This ε-nitride in itself has a very wide homogeneity range for nitrogen (see diagram) and may be rather better classed as a 'semi-nitride' in the sense of a continuation of a primary solution with changed stacking sequence. This theme had already been a familiar one in the earlier stable nitrides of Groups IV–VI but now develops into that of integration with the matrix solid solutions. A further implication of these, fundamentally unstable, states is the considerable pressure dependence of the Fe–N alloy kinetics and the evolution of free nitrogen, e.g. *in vacuo*.

For Co and Ni, the nitrides Co_3N, Ni_3N and Co_2N decompose at quite moderate temperatures or under reduced pressures. Structurally the transitions Co $\rightarrow$ Co_3N and Ni $\rightarrow$ Ni_3N still follow the change of cubic $\rightarrow$ hexagonal stacking sequence with ordered N atoms. The orthorhombic Ni_2N is isomorphous with the carbide, Co_2C (cf. Chapter 4). A phase which had at one time been thought to be a 'hexagonal form of nickel' has been shown[18] to be really the nickel nitride, Ni_3N, isomorphous with ε-Fe_3N. It is worth adding that an unstable *copper* nitride, 'Cu_3N', also exists, probably of the ReO_3-type structure, but decomposes above about 300°C (at at. pressure).

The structure of Mn_4N, including the magnetic arrangement, has been shown from neutron diffraction work[19] to have the FCC Mn lattice centred by N atoms, and similarly that of Fe_4N[20].

The hexagonal close-packed structure in ε(Fe,N) can really be regarded as *rectifying* an apparent anomaly amongst the pure transition metals[21] (noted in the theory of their crystal structures[22]) of metallic iron not yet having adopted this structure, as in fact Ru and Os, of the same Group, have. With regard to electron orbitals, the nitrogen atom is an effective agent able to absorb the 3*d* electrons from the iron atoms, thus allowing 'iron' to adopt the HCP structure, as in Ru and Os. We might add that nitrogen does the same to *manganese*, equalizing this element structurally with Tc and Re. The stabilization also of the FCC structures by nitrogen in Fe and Mn is consistent with its later appearance in elemental Co, Rh, Ir, Ni, Pd and Pt.

Altogether the nitrides are distinguished by extensive homogeneity ranges (cf. Section 5.4) and by the fact that, as e.g. in NbN and TaN, closely allied allotropes, mainly involving a multiplication of the unit cell through a slight distortion, are not rigidly bound to the same stoichiometry: divergencies in N content which, indeed, cause those in structure.

The vanishing stability of nitrides in order Cr $\rightarrow$ Mn $\rightarrow$ Fe $\rightarrow$ Co $\rightarrow$ Ni is reflected in the increasing difficulty of their *preparation*. Thus the Group IV–VI nitrides can all be made simply by nitriding the metal in N_2 gas proper, but those of Mn, Fe, Co, Ni only in ammonia (or in N_2 under high pressure, with difficulty). In other words, it requires nitrogen atoms in atomic, not molecular form to allow them to enter the metal lattice interstitially. (From our 'philosophical' viewpoint, we may express this by saying that in NH_3, one interstitial atom, hydrogen, is made to act as an aid temporarily to hold the other, nitrogen, in a largely bonded state, suitable to permit its infiltration into the metal lattice. This only arises for the tighter metal lattices of Fe, etc., not for the more open lattices of Ti, etc.)

In general, the nitride lattices, as far as the metal atoms are concerned, are either the same as, or closely related to, the cubic or hexagonal close-packed pure metal lattices (as discussed already in Chapters 1 and 2). The nitrogen atoms in the Fe, Co and Ni nitrides tend to be *ordered* rather than random as in the more refractory nitrides. Whilst mutual N—N atom bonds do not yet arise in the nitride nor C—C bonds in the carbide lattices, as B—B and Si—Si bonds do in borides and silicides, such regular alignment ('ordering') in nitrides may perhaps be considered as energetically the initial step in the approach to non-metal bond formation.

The decomposition of nitrides, on heating or under reduced pressure, may in this sense be regarded as a breakdown of the (already weak) metal–nitrogen bond. The approach to this state and segregation of N atoms in lattice clusters, and eventually the N—N bonding to give molecular nitrogen, develops into nascent gas evolution at dislocations, grain boundaries, voids and surfaces—to leave behind the changed stacking sequence of the lower nitride or of the primary solid solution.

The kinetics of this nitrogen evolution has been studied, e.g. on Fe–N alloys of ε-(HCP) structure, in some of Jack's[18] now classical work. Here the wide homogeneity field (see *Figure 5.1*) allows nitrogen loss without major structure changes, apart from those in lattice dimensions. The rate of free nitrogen evolution derived involves three consecutive processes: (1) diffusion of N atoms through the lattice to voids or surfaces; (2) combinations of N atoms to give adsorbed N molecules; and (3) desorption of the latter; of these, stage (2) is the total rate-determining one, being far slower than the diffusion rate.

The significance of Jack's work on the structures and transformation kinetics in iron nitrides, even though referred to before (Chapter 2), is such as to require some elaboration. The Fe–N phase diagram gives only an incomplete picture, and the intermediate stages not appearing in an equilibrium diagram (although implied by it) are equally important. Thus, apart from the structures of the Fe_4N (γ'-phase)[23], Fe_4N/Fe_2N ($=\varepsilon$ solution range), and the Fe_2N (ζ, low-temperature form), we have the so-called 'nitrogen martensite' formed by quenching (α''-stage) and tempering (α''-stage) and the 'nitrogen austenite'.

α'-Martensite is body-centred tetragonal and fully analogous to carbon martensite, with N atoms placed anisotropically along the cube edge, to strain the Fe lattice. However, upon tempering, distinct differences occur from the case of carbon, to give a structure intermediate to N-martensite and Fe_4N (which for carbon had only the very different analogue of cementite). The α'' phase has an enlarged unit cell covering 8 distorted BCC unit cells of the parent martensite, of dimensions $a=5{\cdot}72$, $c=6{\cdot}29$ Å. The ideal formula is $Fe_{16}N_2$, with the following atom positions ($z=0{\cdot}06$):

16 Fe at $0, 0, \bar{z};\ \frac{1}{2}, \frac{1}{2}, \frac{1}{2} - z;\ \frac{1}{2}, \frac{1}{2}, z;\ 0, 0, \frac{1}{2}+z;\ \frac{1}{2}, 0, 0;\ 0, \frac{1}{2}, 0;$
$\frac{1}{2}, 0, \frac{1}{2};\ 0, \frac{1}{2}, \frac{1}{2};\ \pm(\frac{1}{4}, \frac{1}{4}, \frac{1}{4});\ \pm(\frac{1}{4}, \frac{1}{4}, \frac{3}{4});\ \pm(\frac{1}{4}, \frac{3}{4}, \frac{1}{4});\ \pm(\frac{3}{4}, \frac{1}{4}, \frac{1}{4})$

2N at $0, 0, \frac{1}{4}$ and $\frac{1}{2}, \frac{1}{2}, \frac{3}{4}$

The nitrogen atoms are thus in rare but ordered positions, occupying $\frac{1}{24}$ of the number of octahedral interstices; however, up to half the N atom sites may be vacant, to give an (α'') homogeneity range from Fe_8N to $Fe_{16}N$.

Figure 5.4 illustrates the structure of α''-martensite, also viewed as a distorted FCC pseudo-austenite lattice, in conjunction with the γ, Fe_4N and α' structures. The Fe_4N (γ') phase, as mentioned, is truly FCC, the only difference from N austenite being N-atom ordering at ($\frac{1}{2}$, $\frac{1}{2}$, $\frac{1}{2}$) positions.

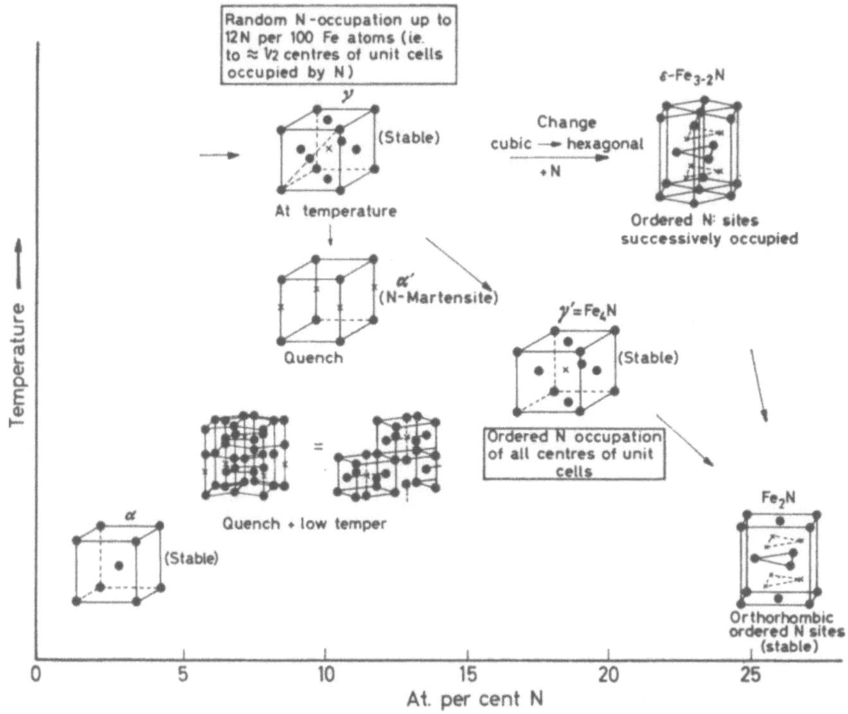

Figure 5.4. Sequence of iron–nitrogen structures

In the hexagonal ε phase, the wide composition range implies a lattice-dimensional change as shown[24] in *Figure 5.5.* At ε compositions around Fe_3N and Fe_2N, there exist distinct superlattices[23,25] with a unit cell: $a' = \sqrt{3}a$; $c' = c$ where a and c are the HCP Fe atom spacings.

However, the superlattice persists in modified fashion throughout the ε range. Jack finds that the basic structural unit in ε, allowing for all stages of ordered occupation, is $a' = 2\sqrt{3}a$; $c' = c$. The HCP Fe–N structure has later been linked with the idea of hybridized bonding orbitals[21].

The low-temperature ζ-Fe_2N phase of orthorhombic unit cell can be considered as a structural distortion of the ε lattice. Some anisotropic Fe atom displacement is caused by an N atom rearrangement; this is more closely packed in ζ, having filled previously available ordered lattice vacancies. Interpreting ζ as of inherited pseudo-hexagonal unit cell, this acquires the new dimensions $a' = 4\sqrt{3}a$; $c' = c$.

To summarize, we have consecutive changes in unit cell size, caused by ordering within the hexagonal Fe atom lattice, which range from $a^1 = \sqrt{3}a$ to $4\sqrt{3}a$; the repeat along the c axis remains unaltered. The presence of ordered vacancies in these nitride lattices is in itself an important factor, as far as diffusion infiltration by other interstitials and physical properties are concerned. These effects have had a profound technological importance in the understanding, for instance, of the surface-hardening process by nitriding; we will refer to this again below (Section 5.6).

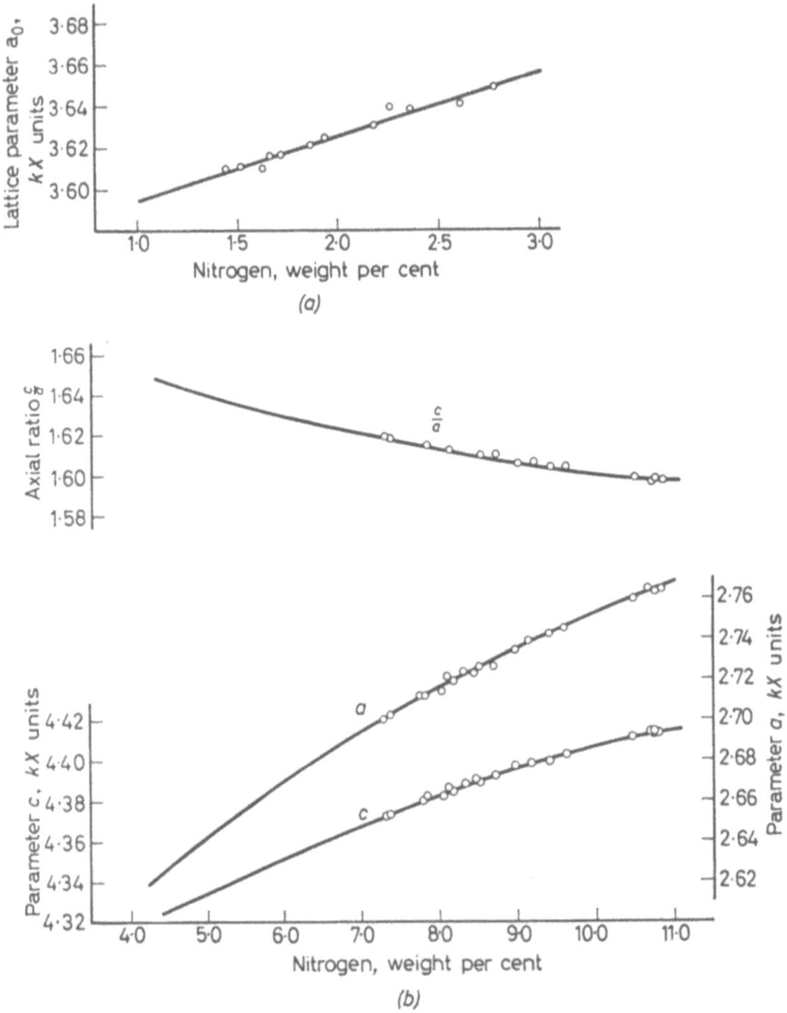

Figure 5.5. Lattice spacing variations in: (a) nitrogen austenite (FCC); (b) HCP Fe–N *ε-phase, with nitrogen content*[24]

By courtesy of the Metallurgical Society of AIME

In Mn_4N and Fe_4N the magnetic structure, studied by neutron diffraction[19,26], shows that generally the increase in nitrogen vacancies decreases the saturation moment. Also the moments of the FCC *metal* atom lattice critically vary with interstitial N occupancy. The N superlattice in Fe_4N has also been observed from x-ray intensities.

The precipitation mechanism of iron nitride, Fe_4N, in ferrite has been the subject of a great deal of work. For example, electron microscopy and electron diffraction have shown a strong preferred orientation relationship during precipitation of a (metastable) very-low-nitride 'Fe_8N' in platelets along (100) ferrite matrix planes[27,28]. Fe_4N also precipitates in lamellar form on either austenite or ferrite lattice planes. Such very-low-'nitrides' are, however, almost certainly transient stages, in this case perhaps an iron-supersaturated form of γ^1-(Fe_4N) which, upon annealing, would gain interstitial nitrogen.

The electronic configurations of Fe and of N atoms in Fe_4N and in $(Fe,Ni)_3N$ have also been studied, using Mössbauer spectra[29]. This work indicates that the N atoms at the body centres donate electrons to the Fe atoms at the face centres. It also shows Fe/Ni atom ordering upon adding nickel to Fe_4N, in that the Ni replace the Fe atoms at the *corners* preferentially to those at the face centres (see also Fe–Ni–N entry in *Table 5.5*, p. 239). We might see in this the accentuation, through N addition, of an ordering tendency already incipient in Fe_3Ni.

This raises the much wider question whether, in principle, the addition of electron-donor interstitials would promote ordering in transition-metal binary alloys which otherwise would remain disordered. It arises in both the FCC and BCC metal lattices, and much evidence seems to accumulate to substantiate this possibility.

Di*cobalt* nitride, Co_2N is isomorphous[30] with Co_2C, having the orthorhombic cell with the interstitial atoms located in octahedral holes within the Co lattice. The striking ascendency of size of nitrogen atom over that of the carbon atom is well demonstrated here: the volume per cobalt atom is 14·2 Å^3 in Co_2N, 14·03 Å^3 in Co_2C. The Fe and Co nitrides, Fe_2N and Co_2N, though both orthorhombic, are not isostructural; this is due to a different anisotropic location of the N atoms within the (basically) hexagonal close-packed metal lattice, causing different degrees of distortion. On the other hand, Co_3N and (ε-)Fe_3N are isostructural (HCP metal lattice with the superlattice qualifications described above). Systematically, we might thus note that increasing addition of interstitial nitrogen to both Fe and Co (1) stabilizes the HCP* structure, after the initial cubic CP deviation in γ- and γ'-(Fe,N), producing superlattices; (2) distorts this lattice by 'overloading' (supersaturating) with nitrogen atoms in a manner which differs between Fe and Co. This 'overloading' is literally true because these nitrides are in fact quite unstable and evolve free nitrogen readily.

Regarding the manganese–nitrogen system, an amendment of the phase equilibria has been proposed[31] (*Figure 5.6*). (Mn amalgams were here used as preparatory material for the alloys.) Whilst confirming the Mn_4N, Mn_5N_2 and Mn_3N_2 phases (with details as in *Table 5.1*), a higher face-centred tetragonal nitride around Mn_6N_5 was added which forms peritectoidally at low temperatures. Except for some features of the structural sequence, resemblance of this Mn–N diagram to that of Fe–N is not yet apparent. The *low*-temperature nature of the nitrides higher than Mn_4N is interesting, however,

* It may well be possible that hexagonal α–Co and ε–Co_3N (both isostructural) form continuous solid solutions, although it is not proved.

as is the direct Mn_4N/gas equilibrium at higher temperatures, evolving in the later Group to the γ-Fe/gas and Co/gas equilibria as the true ones. Even so, the Mn–N system appears to require re-examination to the same degree of detail as Jack's Fe–N, including experimental comparison also using neutron diffraction.

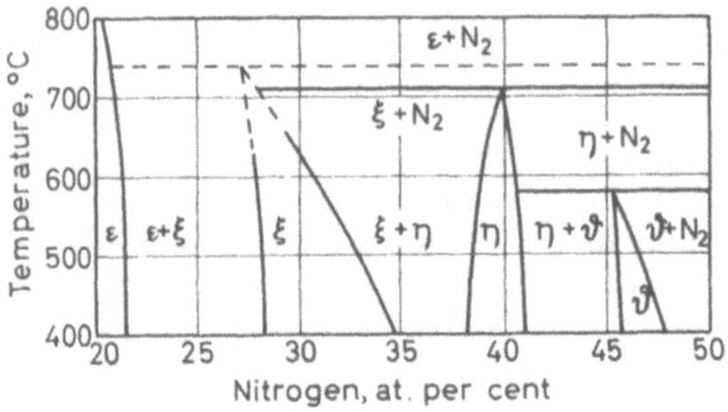

Figure 5.6. Portion of the Mn–N *phase diagram*[31]

By courtesy of the Editor, *Zeitschrift für Metallkunde*

5.3 FRINGE CASE NITRIDES

In addition to the transition-metal nitrides, some 'fringe cases' should also briefly be considered, namely the nitrides of (*a*) the Rare Earth, Group III and earlier transition metals, (*b*) the non-transition metals, notably aluminium and the light elements and (*c*) Cu, Ag and the Pt-group metals (i.e. the left, top and right borders of the transition-metal block in the Periodic Table).

(*a*) Scandium, like its neighbour titanium, still forms a very stable mononitride, ScN, of cubic (NaCl-type) structure[32] and melting point 2,650°C (TiN = 2,950°C). No data on yttrium nitride appear to be known, although very probably it is isomorphous and also refractory, as is true for the Rare Earth nitrides (*Table 5.3*). All are of NaCl-type structure.

The *Alkaline Earth* nitrides, as the carbides, strongly hydrolyse. Be_3N_2, Mg_3N_2 and the high-temperature form of Ca_3N_2 all have the cubic D_5 type of structure, with a large unit cell containing 16 molecules (spacings in *Table 5.3*). Sr_3N_2 and Ba_3N_2 are also mutually isomorphous, but of so far uncertain structure. Of the *alkali* metals, the only positive knowledge is on the hexagonal Li_3N.

(*b*) The very stable *aluminium* nitride, AlN, as well as GaN and InN, are hexagonal*, of the 'wurtzite' (ZnS)-type structure, while silicon nitride, Si_3N_4, which has gained much importance through its special refractory properties and resilience at high temperature, has two crystalline (α and β) forms (*Table 5.3*). Aluminium nitride and boron nitride, BN, stand out not only by their stability but also by low sublimation and decomposition pressures up to very high temperatures[33] (>1,800°C), as does magnesium nitride, Mg_3N_2. This contrasts sharply with the transition-metal nitrides, such as TiN and ZrN, as to their much higher sublimation and decomposition pressures and rates. A basic point of interest here is that in TiN to ZrN the defect lattices

* These nitrides fall outside our scope of interstitial compounds, but some references to AlN, BN and Si_3N_4 are included in the Bibliography.

Table 5.3

Some 'fringe' nitrides, with lattice spacings (Å)

NaCl *type*		Cubic D_5 *type*		
Rare Earths and Group III		*Alkaline Earth Group*		*Alkali Group*
ScN*	4·45	Be_3N_2	8·15	Li_3N hexagonal
LaN	$5{\cdot}30_1$	Mg_3N_2	9·97	$a = 3{\cdot}665$
CeN	5·02	$Ca_3N_2\alpha$	11·42	$c = 3{\cdot}890$
PrN	$5{\cdot}16_7$	β hexagonal		1 mol. unit cell (unstable)
NdN	$5{\cdot}15_1$	Sr_3N_2, Ba_3N_2 } as β-Ca_3N_2; structure undetermined		(Na_3N, K_3N, etc.)
PmN	—			
SmN	$5{\cdot}04_6$			
EuN	$5{\cdot}01_4$			
GdN	$4{\cdot}99_9$	*Miscellaneous*		
TbN	$4{\cdot}93_3$	AlN hexagonal β–4 (wurtzite) type	$a = 3{\cdot}110$	
DyN	$4{\cdot}90_5$		$c = 4{\cdot}975$, 2 mol./unit cell	
HoN	$4{\cdot}87_4$	GaN isomorphous	$a = 3{\cdot}19$ $c = 5{\cdot}18$	
ErN	$4{\cdot}83_9$	InN isomorphous	$a = 3{\cdot}54$ $c = 5{\cdot}70$	
TmN	$4{\cdot}80_9$	ZnN cubic D_5		
YbN	$4{\cdot}78_6$	(as Mg_3N_2, etc.)	$a = 9{\cdot}763$	
LuN	$4{\cdot}76_6$			
		Si_3N_4 two hexagonal forms:		
		α: $a = 7{\cdot}753$ $c = 5{\cdot}618$		
		β: $a = 7{\cdot}606$ $c = 2{\cdot}909$		
		Ge_3N_4 isomorphous		
		BN hexagonal, graphite-like		
		Cu_3N cubic, anti-ReO_3-type (?)		
		Ag_3N FC cubic $a = 4{\cdot}379$		

* YN probably same structure

allow nitrogen more readily to escape and aggregate to molecules than the more rigidly stoichiometric lattices of the non-transition-metal nitrides.

The *heats of formation* (at 25°C) of AlN, BN and Si_3N_4, derived from their dissociation pressures[34,35], are −76·1, −59·8 and −176·3 kcal/mole, respectively; they may be compared with those of *Table 5.1.*

(*c*) The unstable later nitrides Cu_3N and Ag_3N (*Table 5.3*) link well with the interstitial series described. Structurally they are close relations to the FCC metal lattices of γ-Fe, Fe_4N and Cu and Ag proper, with ordered N atoms in interstices, except for a slight controversy on Cu_3N. No nitrides occur for the *Pt-group metals*, but for rhenium a W_2N analogue, Re_2N, is known. A reservation ought to be made even for the later Groups, as far as high pressures and thin surface films are concerned. The data here are sparse.

5.4 NON-STOICHIOMETRY IN NITRIDES

The transition-metal nitrides of cubic and hexagonal close-packed structures are, as mentioned, particularly prone to defect lattice formation. It is worth while trying to compare the relevant data available in this respect, as for the carbides (Chapter 4). However, a difficulty is caused by their insufficiency and by the possible presence of interstitials other than nitrogen as impurities in some of the work, always to be remembered.

Figure 5.7 and *Table 5.4* collate the data available (cf. those of the carbides, *Figures 4.3, 4.19*). The values refer to the maximum ranges within the given lattice, whatever the temperature; pressure dependence must clearly affect the data, but they generally apply to atmospheric pressure.

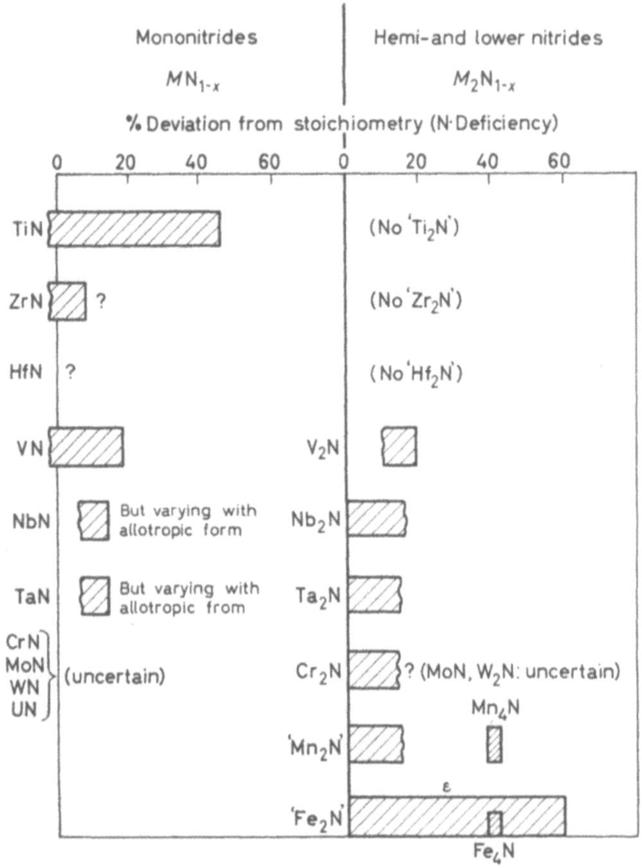

Figure 5.7. Non-stoichiometry in transition metal nitrides (Collation of maximum ranges available)

No quantitative rule can yet emerge from this, and, if anything, *Figure 5.7* may serve as a 'programme' for further research to define these boundaries systematically. Nevertheless, a number of interesting features arise. The high FCC lattice defection in Group IV* decreases towards Group V where it is, in fact, partially but less extensively transferred from the mono- to the hemi-nitrides. The former begin to depart from the equiatomic composition towards nitrogen deficiency, in association with the more complex structure changes. It is noted that, in contrast, to the carbides, the Group IV and V mononitrides also are capable of dissolving some

* At any rate for TiN; the ZrN and HfN limits are insufficiently known but probably also wide.

excess nitrogen, but the extent to which this occurs is ill-defined. This lack of definition of the N-rich boundary of the N-rich nitrides reflects, in fact, a problem of principle, in that the very loose interstitial lattice binding for N-atom saturation may well be considered as a transition stage from true atomic lattice nitrogen to atom pairing to give N_2 molecules, perhaps segregating at grain boundaries, and ultimately to mere adsorption of free gas.

Table 5.4

Non-stoichiometric ranges of nitrides (at. per cent N)

System	*Maximum ranges within given close-packed structure*		
Group IV			
Ti–N	27–52 (?) for $TiN_{0.38}$–$TiN_{1.2}$		
Zr–N	46 (?) – > 50 for $ZrN_{0.55}$–$ZrN_{1.0}$ at 1,500°C		
Hf–N	$HfN_{0.74}$–> $HfN_{1.1}$		
Th–N	not known but finite		
Group V			
V–N	27–30 (?) for V_2N $VN_{0.48}$–$VN_{0.50}$	41·5– >51 for VN $VN_{0.72}$–$VN_{1.0}$	
Nb–N	28·5–33·3 for Nb_2N $NbN_{0.40}$–$NbN_{0.50}$	44·5–47·5 (?) for ε–NbN (but varying with structure amongst the polymorphic forms)	
Ta–N	29·1–33·3 for Ta_2N $TaN_{0.30}$–$TaN_{0.50}$ (at 2,000°C)	44·5–47·5 (?) for TaN $TaN_{0.98}$–$TaN_{1.0}$ (but varying with structure; 50·0 reported for ε–TaN)	
Group VI			
Cr–N	32 (or 27·5 ?)–33·3 for Cr_2N	CrN not known	
Mo–N	~ 28 for H.T. β form; 30–33·3 for Mo_2N	MoN small, uncertain	
W–N	not exactly known, but complex series of finite non-stoichiometries; cf. *Figure 5.3*		
U–N	uncertain but considerable at high temps., small at low temps. (see *Figure 5.1*)		
Group VII			
Mn–N	20·0–21·4 for Mn_4N	28·4–34·6 for Mn_2N	~ 38–39 for Mn_3N_2
Group VIII			
Fe–N	19·0–20·6 for Fe_4N	~ 14–33 for ε–Fe_3N/Fe_2N	
Co–N	not known		
Ni–N	not known		

The pressure sensitivity of any N-rich boundary must in any case be extreme, and it may even be queried whether the observed N excess in TiN, ZrN and VN is not a transitional rather than a true equilibrium condition. This would still be consistent with the break in lattice spacings shown in *Figure 5.2*.

On passing to Groups VII and VIII, the mononitrides (or their equivalent) cease completely in favour of gaseous equilibrium—in continuation of the former trend. The formation of metal-richer nitrides in iron and manganese is accompanied by the strong recovery of non-stoichiometric ranges, particularly for the hexagonal ε phase. In this sense we might consider the wide

γ-Fe solid solution (N austenite) as a direct extension of this non-stoichiometric range to the face-centred cube, with a deviation to the ordered structure of γ'-(Fe_4N,Mn_4N) which can only exist over a relatively narrow homogeneity range.

The wide homogeneity range of 'TiN' (cf. *Figure 5.2*) has been confirmed[36] as extending from 30–50 at. per cent N, and the bond type begins to change within this range from metallic towards ionic bonding, with increasing N content[37]. This is revealed by non-linear electrical resistance changes with Ti/N ratio in TiN_{1-x}, in contrast to linearity for Ti *carbides* (TiC_{1-x}). For the stoichiometric composition, a linear dependence on temperature is attained, but as the vacancies increase, anomalies appear, taking the form of pronounced resistence maxima which become displaced with varying nitrogen contents. This variation is attributed to one in degree of metallic/ionic bonding with varying lattice defection.

The general point to note is again the indication that the character of bonding is already more ionic in TiN than in TiC, where it is still substantially metallic—part-way towards TiO.

5.5 TWO-METAL NITRIDE SYSTEMS

From the compositions, structures and stabilities of the single-metal nitrides described above, the trends of mixed nitride equilibria can often be reasonably foreseen and checked with observation. However, this does not excuse the sparsity of the systematic work published hitherto. In *Table 5.5* the information available has been collated.

The extensive homogeneity ranges of the cubic or hexagonal close-packed binary nitrides give rise to similar ternary regions, provided that size ratios and electron valencies of the partner metals are favourable. Thus e.g. TiN, ZrN, NbN, TaN (with their N-defect ranges) are all intersoluble; where the structure change cubic $\rightarrow$ hexagonal makes complete solutions impossible, as in TiN–TaN and ZrN–TaN, it is presumed (but would provide an interesting field for further electron-microscopic investigation) that the saturated two-phase assembly can consist of networks of closely interwoven stacking-fault sequences in dispersed (Ti,Ta)N and (Ta,Ti)N lattice domains rather than of sharply defined pure cubic and hexagonal crystals*. Lattice spacing changes have been measured[38] for many such mononitride pairs; the resulting curves are shown in *Figure 5.8*. It should here be noted, however, that owing probably to oxygen impurity, NbN and TaN were also determined as cubic, in contrast to later findings[39a]. Nevertheless, these curves are valuable for practical use, and the mere fact that trace impurities can render the nitride pairs isomorphous and entirely intersoluble again, lends point to the general theme.

The VN–ZrN system forms an exception, as here the metal atom size ratio is unfavourable for any appreciable intersolubility. This was quite similar in the corresponding carbides. It implies that in a ternary system such as TiN–ZrN–VN or ZrN–NbN–VN a miscibility bay should occur, in contrast to complete solution systems such as TiN–VN–NbN.

* Unless the special conditions of manufacture, e.g. by sintering from the pure nitrides, favour this separate condition, and even then, the effect would arise at the interface.

Table 5.5. Two-metal nitride relations

CSS = continuous, *PSS* = partial solid solutions; *TC* = ternary compound

System	*Solid solutions or compounds*	*Structure; remarks*	*References*
Ti-Zr-N	TiN-ZrN *CSS*	Bl (i.e. Cubic, NaCl type)	38
Ti-Hf-N	presumed similar (TiN-HfN *CSS*)		38
Ti-V-N	TiN-VN *CSS*	Bl	38
Ti-Nb-N	TiN-NbN *CSS**	Bl	38
Ti-Ta-N	TiN-TaN *PSS* no *TC* in the above systems	Bl to approx. 10 at. per cent Ta then 2-phase Ti(Ta)N + TaN	39b
Ti-*M*-N	no data available *M* = Cr, Mo, W, U, Mn, Fe and Pt group metals		
Ti-Co-N	*TC* at $Ti_7Co_3N_{10}$ (i.e. '*MN*' stoichiometry)	WC type (simple hex.) $a = 2{\cdot}94$, $c = 2{\cdot}90$ Å	39b
Ti-Ni-N	*TC* at $Ti_2Ni_3N_{10}$ (i.e. '*MN*' stoichiometry)	WC type	39b
Zr-Hf-N	not determined, but *CSS* ZrN-HfN in little doubt	Bl	
Zr-V-N	ZrN-VN *PSS* (only very slight)	Bl	38
Zr-Nb-N	ZrN-NbN *CSS**)	Bl	38
Zr-Ta-N	not determined but probably similar to Zr-Nb-N		
Zr-*M*-N	*M* = Cr, Mo, W, U, Mn, Fe, Co, Ni and Pt group metals no data available		
Hf-*M*-N	*M* = any other metal: no data available		
Th-*M*-N	*M* = any other metal: no data available		
V-Nb-N	VN-NbN *CSS**	Bl	38
V-Ta-N	VN-TaN *PSS*: no *TC*	Bl to approx. 25 at. per cent Ta	39b
V-*M*-N	*M* = Cr, Mo, W, U, Mn, Fe, Co, Ni and Pt group metals: no data available		
Nb-*M*-N	*M* = Ta (though (Nb,Ta)N solutions certain), Cr, Mo, W, U, Mn, Fe, Co, Ni and Pt group metals: no data available		
Ta-Cr-N	TaN-CrN *PSS*	Bl(CrN-end) to approx. 25 at. per cent (no data at TaN end)	39b
	TC '*X*'	unknown structure; coexisting with TaN; phase unconfirmed	39b
Ta-Mn-N	Two *TC's*: (1) same as phase '*X*' (Ta, Cr, N) above; (2) near Ta_3MnN_4	isomorphous to '*X*' hexagonal; no solubility of Ta in Mn_5N_3 nor of Mn in ε-TaN, but some, to 10 at. per cent Mn in ε-TaN (causing anomalous expansion)	39b
Ta-Fe-N	*TC* near $Ta_4Fe_2N_5$	hexagonal; related to Ta_3MnN_4 Ta insoluble in Fe nitrides	39b
Ta-Co-N	*TC* $Ta_4Co_2N_5$	isomorphous to $Ta_4Fe_2N_5$	39b
Ta-Ni-N	*TC* $Ta_4Ni_2N_5$	isomorphous to $Ta_4Fe_2N_5$ (i.e. Fe/Co/Ni completely replaceable in this phase)	39b
Ta-*M*-N	*M* = Mo, W, U or Pt group metals: no data available		

* Conflict: complete solution unlikely, as pure NbN is HCP (possibly due to oxygen content); extensive solubility certain, however.

Table 5.5.—continued

CSS = continuous, *PSS* = partial solid solutions; *TC* = ternary compound

System	*Solid solutions or compounds*	*Structure; remarks*	*References*
Cr–Mo–N	*TC* '$CrMoN_x$'	hexagonal; possibly binary-based ternary	40
Cr–Mn–N	$(Cr,Mn)_{2\,(or\,3)}N$	hexagonal, with Cr_2N-type superlattice; probably Cr_2N solid solution, not *TC*	41
Cr–Fe–N	$(Cr,Fe)_2N$	hexagonal solid solution in ε-Fe_2N; no *TC* (much work exists on primary solutions of N in ferrite and austenite)	
Cr–*M*–N	*M* = W, U, Co, Ni, Pt group metals: no data available		
Mo–*M*–N only Mo–(Mn, Fe, Co, Ni)–N and	*TC* ≈ $Mo_3Fe_7N_4$ (typically)	β-Mn type; likely to be representative of other [Group VI + (VII, VIII)A] metal nitrides	42
Mo–Co–N	*TC* $Mo_8Co_2N_{10}$ (i.e. '*M*N' stoichiometry)	hexagonal WC type; possibly intersoluble with MoN	39b
Mo–Ni–N	*TC* $Mo_8Ni_2N_{10}$	WC type, as above	39b
W–*M*–N	*M* = U, Mn, Fe, Co, Ni, Pt group metals: no data available		
U–*M*–N	*M* = any other metal: no data available		
Mn–Fe–N	$Mn_4N \longleftrightarrow Fe_4N$ *CSS*	FCC, ordered	43
Mn–*M*–N	*M* = Co, Ni Pt group metals: no data available		
Fe–Ni–N	Fe_4N and ε-$Fe_{3-2}N$ extensive *PSS* $(Fe,Ni)_4N$, $(Fe,Ni)_{3-2}N$; probably *CSS* if Ni-end member stabilized	FCC and HCP; in direct γ-matrix, co-existence M_4N and ε series form two-phase region [probably similar for (Fe, Co, N) and (Co, Ni, N)]	44
	'*TC*' Fe_3NiN	cubic Fe_4N type but distinctive in having Fe and Ni in *ordered* positions (apart from ordered N atoms)	45
	TC FeNiN (?)	of tetragonal cell, ordered Fe and Ni, $a = 2{\cdot}830$, $c = 3{\cdot}713$ Å	46
		in the section $Fe_{4-x}Ni_xN$, with $0 < x < 4$, the γ′–Fe_4N structure was found up to $x = 3{\cdot}33$, followed by a duplex equilibrium, then by FeNiN	46
Fe–Pt–N Fe–Pd–N	*TC* Fe_3PtN	FCC Fe_4N type phase but *ordered* Fe and Pt positions; also an analogue Fe_3PdN, of cubic perovskite type note $Fe_4N \rightarrow Fe_3NiN \rightarrow Fe_3PdN \rightarrow Fe_3PtN$ continuous series	54a
Fe–*M*–N	*M* = Co and most Pt group metals; no data available		
Co–Ni–N Ni–[Pt group]–N	no data available		

The *lower* nitrides, e.g. V_2N, Nb_2N, Ta_2N, no doubt also form complete mutual solid solutions, as do the carbides, although there seems to be no experimental verification. Again, the dependence on stoichiometry should be critical, and nitrogen excess in local regions would, in a system such as V–Nb–N, cause a sensitive switch-over in stacking sequence (see *Table 5.1*), to produce e.g. the simple hexagonal from the close-packed arrangement.

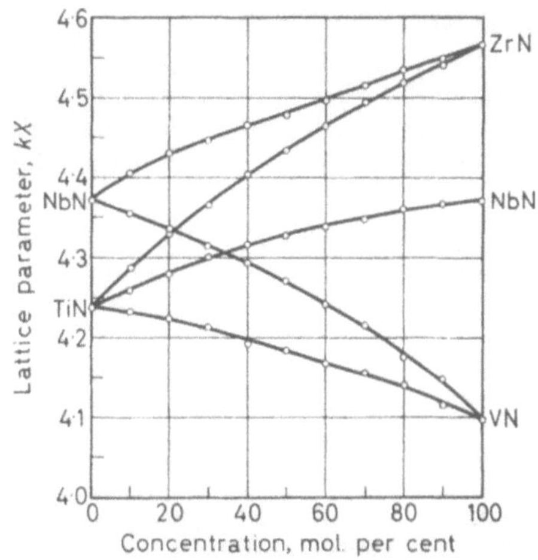

Figure 5.8. Lattice spacings versus composition, for pairs of transition metal nitrides[38]

By courtesy of the Electrochemical Society, New York

It is here of particular interest that a *ternary* compound of the WC-type structure has been observed in the series TiN–'CoN' and TiN–'NiN', namely at about $Ti_{0{\cdot}7}Co_{0{\cdot}3}N$ and $Ti_{0{\cdot}7}Ni_{0{\cdot}3}N$. The virtually identical lattice dimensions and densities are

$$a = 2{\cdot}94,\ c = 2{\cdot}90\ \text{Å}\ (\text{density},\ 2{\cdot}74)$$
$$a = 2{\cdot}94,\ c = 2{\cdot}89\ \text{Å}\ (\text{density},\ 2{\cdot}74),$$

for the Co and Ni compounds, respectively. Mononitrides of Co and Ni do not of course exist, and these metals should here be regarded as mainly playing the role of small-atom electron donors to TiN; effectively they produce an equivalent to NbN or TaN, which themselves have the WC structure. It seems well possible that Group VIII metals more generally, e.g. Ru, Pd or other members of the Pt family, would also be capable of producing this structure when added to Group IV (Ti,Zr,Hf) nitrides; however, this is not proven. The composition ratio, found at about 7:3, in this compound of combined Group IV + VIII metals would almost suggestively produce the electronic balance similar to that in the Group V and VI nitrides which have this structure.

Very little is known on mixed Group (IV–VI) and (V–VI) nitride systems. It can only be surmised that CrN forms at least partial solutions with its cubic isomorph TiN, with VN probably complete ones, with ZrN most likely none. The position for the hexagonal pairs, Cr_2N–V_2N, etc. would be similar. The displacement, in Mo_2N and W_2N, of the cubic NaCl-type structure to the new 2:1 stoichiometry would not prevent their continuous solutions with TiN, ZrN, HfN and, when cubic-stabilized, with NbN and TaN. No doubt the nitrogen deficiency would be progressive, and major changes in superlattices would be involved. In fact, the metal-rich boundaries of the 'TiN' and 'ZrN' regions (see *Figure 5.1*) are themselves far from the M_2N stoichiometry.

It is also highly likely that NbN, TaN, MoN and WN, all of simple hexagonal structure and favourable size ratio, form continuous solutions, not only with one another and with WC, but also with the above (Ti,Co)N and (Ti,Ni)N compounds.

Ternary nitrides begin to appear[39b] in Ta–Cr–N, Ta–Mn–N and Ta–(Fe,Co,Ni)–N. *Table 5.5* gives a broad survey as far as is possible, and *Table 5.6* summarizes the crystallographic and other lattice data. An important point to note is again their hexagonal structure, this time of a stacking type intermediate to hexagonal and cubic close-packing. Thus even the *ternary* compounds present nothing fundamentally new but only a new adaptation to the same basic feature.

Table 5.6. Ternary nitride structures[39b]

System	*Compound*	*Structure and spacings*, Å; *density*
Ta–Cr–N	undetermined '*X*' phase $(Ta,Cr)_aN_b$	unknown
Ta–Mn–N	*X* phase $(Ta,Mn)_aN_b$	unknown
	Ta_3MnN_4	hexagonal: $a = 3{\cdot}023$, $c = 10{\cdot}49$*; density (calc.) = 13·04
Ta–Fe–N	$Ta_2MN_{2\cdot5}$ (M = Fe,Co,Ni; completely intersoluble)	hexagonal, similar to Ta_3MnN_4*
Ta–Co–N	$Ta_2FeN_{2\cdot50}$	$a = 5{\cdot}156$, $c = 10{\cdot}31$; 12·70
Ta–Ni–N	$Ta_2CoN_{2\cdot50}$	$a = 5{\cdot}161$, $c = 10{\cdot}30$; 12·74
	$Ta_2NiN_{2\cdot45}$	$a = 5{\cdot}168$, $c = 10{\cdot}36$; 12·60
	Small $M \longleftrightarrow$ N homogeneity ranges only	

*Stacking sequence ABACABAC . . . , i.e. mixtures of hexagonal and cubic close packings; ordered Ta and Mn atoms

Although only these *tantalum*-base compounds have been reported[39b], the existence of niobium analogues seems at least highly probable. The occurrence of true ternary nitrides in other (unexplored) systems is also likely but can only be surmised. The so-called (WC-type) ternary compounds $Mo_{0\cdot8}Co_{0\cdot2}N$ and $Mo_{0\cdot8}Ni_{0\cdot2}N$ reported are in all likelihood solid solutions in MoN which already have this structure.

'η-Nitrides', M_6N, in analogy to η-carbides, have also been reported[47], namely Mn_3Ti_3N and Fe_3Ti_3N, with no doubt other members existing. Indeed, this family has recently been extended[48] to include phases such as Nb_3Zn_3N, where lattice defection of type M_6N_{1-x} is likely, as well as variable metal atom ratios (as for η-carbides).

These *are* certainly true ternary compounds of distinctive structure, and other such nitrides, in line with their carbide analogues, are probable. The diagonal relationship for favourable ternary phase formation in direction $\swarrow\!\!\nearrow$ within the transition-metal block of the Periodic Table, established for carbides and silicides, is likely also to emerge for the nitrides; however, the data are yet too sparse to be conclusive.

Among mixed Group VII/VIII nitride systems (Mn–Fe–N, Fe–Ni–N, etc.), mutual solid solutions between isomorphs are general. Particularly important in alloy steels are clearly the Mn_4N/Fe_4N (γ') solutions and their close relations to nitrogen austenite, as well as those based on the ε-Fe_3N phase. This dissolves large amounts, if not the whole, of Co and Ni, with the proviso, however, that they are liable to have become unstable, except for higher pressures, before this limit is reached. Superlattice formation is important, as indeed recurs already in the metallic Fe–Co and Fe–Ni systems themselves, but now projected to the nitrides. Thus the γ'-phase Fe_4N adopts an ordered metal atom structure in Fe_3NiN and Fe_3PtN[45], recalling the superlattices in Fe_3Ni and Fe_3Pt. An ordered-phase 'FeCoN' might be enquired into. A phase 'FeNiN' of primitive tetragonal unit cell and ordered Fe and Ni atoms (as in FeNi) has also been reported[46] which probably is in direct derivation from FeNi itself, but distorted by anisotropic nitrogen insertion.

The absence of data on nitrides in important alloy systems such as Fe–Ti–N, Fe–Cr–N or Fe–Co–N is surprising. (There exists much work on the effect of nitrogen on the primary α- and γ-solid solutions in alloy steel matrix phases and in nitrogen martensites, e.g. Fe–Cr–N, Fe–Cr–Mn–N, but this is on the subject of primary solutions (Chapter 3) rather than nitrides.)

The high stability of *aluminium* nitride, AlN, gives the Fe–Al–N system some special interest, also in its practical application. Even a trace of nitrogen in Fe–Al alloys produces AlN, and it has been observed that nitrogen, if in adequate quantity, diverts all Al from Fe solid solution to form aluminium nitride[49] which then equilibrates Fe_4N and ε-(Fe,N). Although not a strictly interstitial compound, AlN can, in its metallographic effect of nitride stabilization, be placed in a similar category as TiN (to which it is chemically also closely related).

The substitutional series $Fe_4N \rightarrow Fe_3NiN \rightarrow Fe_3PdN \rightarrow Fe_3PtN$ is particularly interesting from the systematic aspect, as here we have a progressive ordering of metal and interstitial nitrogen atoms, eventually to give the *perovskite* ($CaTiO_3$) type of structure (to be discussed more fully in Chapter 8). In contrast to the oxide perovskites, the nitride and carbide ones are 'anti-structures' containing the N or C atoms in the body-centred position of the FCC metal lattice[50].

On the subject of M–M'–N systems, where M is a transition metal but M' is not, *lithium*-bearing alloys, Co–Li–N, Ni–Li–N and Ti–Li–N are of interest. Hexagonal compounds have been found around $CoLi_5N_2$ and $NiLi_5N_2$, which represent limits of solid solutions of Co and Ni in Li_3N[51], and a cubic phase, $TiLi_5N_3$, is also known[52]. (There seems to be no similar work on other alkali-bearing nitride systems, e.g. with Na or K.)

The so-called ternary 'H' phase (composition $M_2M'X$) found among the carbides and discussed already in Chapter 4, has its *nitride* analogues. A compound Ti_2AlN is known[53], of the same hexagonal structure and of lattice dimensions $a = 2{\cdot}99_4$, $c = 13{\cdot}61$ Å, $c/a = 4{\cdot}54_4$, i.e. smaller than those for the carbide, Ti_2AlC, with which no doubt it is intersoluble. A degree of non-stoichiometry, Ti_2AlN_{1-x}, is likely. Analogues 'Ti_2GeN' and 'Ti_2SnN' could not be prepared, but it has been possible[53a] to obtain the phases Ti_2GaN and Ti_2InN; Zr_2InN, Zr_2TlN and Hf_2InN, as well as V_2GaN and V_2GeN.

There is little doubt that further nitride analogues of [Group IV, V, (VI?) A/B Group metal] combinations will be forthcoming.

Double nitrides of *β-manganese*-type structures, as incarbides (cf. Chapter 4), have been reported[42] in the systems Mo–Mn–N, Mo–Fe–N, Mo–Co–N and Mo–Ni–N. A typical formula is '$Mo_{13}Fe_7N_4$', but with probable deviations and homogeneity ranges; the metal atoms are in the definite ordered positions characteristic of the β-Mn structure (cf. *Figure 4.16*). To this we can now add the compounds[53b] V_3Zn_2N, V_3Ga_2N and Nb_3Al_2N (with other [Group IV, V, VI A/B Group metal] combinations probably to be expected).

The Stadelmaier-type double carbides of *perovskite* structure (with FCC ordered metal structure) have, as mentioned, their counterparts amongst the nitrides[54b,55], such as, typically[53b], Fe_3(Mg and Al)N, Ni_3AlN (also with Co?), $(Mn,Fe,Co,Ni)_3ZnN$, $(Fe,Co)_3(Ga,Ge,Sn)N$, $(Fe,Co,Ni)_3InN$, Co_2SnN and $Ti_3(In,Tl)N$. Considerable variations in stoichiometry occur not only in respect to nitrogen deficiency but also to metal proportions in the FCC lattice of perovskite.

Thus the *chief double nitrides* so far observed are of the following types: η-carbide (M_6N); β-Mn type (M_5N); H phase (M_4N) and perovskite (M_4N). Other groups of double nitride may eventually well have to be added (some of $Cr_{23}C_6$ or α-Mn types?).

Apart from the survey of more systematic ternary systems above, there exists much work relevant to more specific industrial problems of special alloy developments, such as the effect of alloying metals on the nitrogen solubility in γ-, α- or liquid iron; the precipitation of the more refractory nitrides from γ-Fe, e.g. of NbN[56] and VN[57] or the structure of nitrided surface layers.

Other than iron base, the possibility of precipitation-hardening by nitride nucleation within two-*refractory metals* systems is also clearly implied. As an example, precipitation-hardening in Nb–Hf–N alloys has recently been studied[58], and we may say more generally that the addition of an *earlier* transition metal (e.g. of Groups III, IV, V) to a later one (e.g. of Groups V, VI, VII or VIII) in the presence of nitrogen should lead to such potential hardening effects. Precipitation of the early-Group nitrides, dispersion strengthening and scavenging can thus be attained for nitrides as for carbides and oxides.

This leads directly to Applications.

5.6 APPLICATIONS OF NITRIDES

The practical applications of interstitial nitrides lie in areas similar to those of carbides but are rather more limited. This limitation is, basically, closely connected with their structural peculiarities and lesser stability; on the other hand, these give the nitrides certain subtle advantages in other directions, such as the hardening effects in the iron-group metals, bound up with the very fact of instability. In high-temperature ceramics or cemented hard metals, nitrides have found only small and specialized use*, despite their

* Excepting materials of the Si_4N_3 and BN class.

high melting points, because oxidation resistance is generally poor and brittleness severe. However, nitrides are often unavoidably present in solid solution in carbide-based hard metals, and not necessarily to disadvantage. The relatively high dissociation tendency of nitrides at low pressures and higher temperatures is disadvantageous; they are also difficult to sinter. We might attribute their poor oxidation resistance in part to the defect lattices and non-stoichiometric ranges which invite ready lattice solubility for oxygen. This accommodation of oxygen can occur by both vacancy filling and substitution of nitrogen atoms, the latter being helped by the looser bonding compared with carbides; this is aggravated by the higher nitrogen vapour pressure in the nitrides.

The stable *Group IV nitrides* (TiN, ZrN and HfN) have been put to use as refractory bodies[59], e.g. for the manufacture of crucibles from finely divided nitride powders. These proved of interest for melting metals such as cerium, beryllium and iron, which would be readily oxidizable from oxide ceramics (according to Campbell[60], this formed at the time part of the Manhattan project). The crucibles were very hard and had good thermal shock resistance —again a property that can be associated with the more loosely packed lattice. In air the material ignites, so that use in inert gas, nitrogen or (at lower temperatures where vapour pressure remains inappreciable) *in vacuo* is essential. Both TiN and ZrN crucibles thus proved moderately successful, and porosity to which sintered TiN bodies were prone could be reduced by small rutile (TiO_2) additions. It seems likely (although no information is available) that ScN, YN and Rare-Earth nitrides could be put to similar uses on a small scale.

The *Group V* and *VI* nitrides, being much less stable, are ruled out for such ceramic application. In fact, no direct uses seem to have appeared for VN, NbN, TaN, CrN, MoN, WN, etc., except as secondary metallurgical phases in the development of refractory alloys or of special steels, and as dispersion strengtheners similar to carbides and oxides. In this context titanium nitride, TiN, generally as carbonitride Ti(C,N) solid solution, is well known as a microconstituent in steels to which titanium had been added. This, however, can hardly be called a 'use' of nitrides but rather one of the stronger nitride- and carbide-forming element to scavenge nitrogen otherwise undesirably distributed. This applies even more potently to yttrium and the rare earths which, in trace additions, bond any solution nitrogen in alloys to such an extent as to act as effective purifiers, indirectly promoting ductility, e.g. in certain refractory alloys.

In Group VI metals, this aspect manifests itself strongly in the brittleness of 'pure' Cr, Mo and W which according to now well accepted theory is caused by interstitial content. This is most notably expressed in chromium, where nitrogen and Cr_2N nucleation cause the inherent brittleness*; much intense research has been proceeding, particularly in Australia[61–65], to combat this and to lower the ductile/brittle transition temperature by removing or neutralizing trace nitrogen. In Cr, Mo and W base alloys, too, controlled and intentional nitride hardening is a further possibility of

* Whether or not nitrogen is necessarily responsible for Cr brittleness is still controversial, however, and recent work[65] points to the existence of other causes as well.

development for high-temperature application: although nitrides, owing to their instability, are less effective as dispersion hardeners than carbides and oxides.

The dispersion strengthening of chromium by nitrogen and Cr_2N formation is one of the more positive aspects, including the possibility of internal nitriding within the chromium mass[66]. Small additions to chromium of earlier transition metals like Ce, Y, Ti, Zr or Ta have been found beneficial in lowering the ductile/brittle transition temperature, due clearly to the formation of the more stable nitrides[67]. In *molybdenum*-based alloys it was also possible to achieve an improvement of creep rates by nitriding[68].

Again, small amounts of nitrogen in titanium- and zirconium-base alloys tend to stabilize the hexagonal α-form rather than BCC β, with undesirable effects. We may regard this, in fact, as the precursor to nitride formation (cf. Chapter 2), which in Group V develops into V_2N, Nb_2N and Ta_2N, i.e. the actual hexagonal nitrides. Also, nitrogen in titanium-based alloys can affect properties greatly, first by solid solution, then by nitride nucleation; thus hardness maxima can be caused by transitory stabilization of the so-called 'ω' phase[69], a metastable constituent intermediate to β- and α-Ti. Although this is not a true nitride, it represents a structural distortion derived from the BCC, stabilized in the first instance by addition of Group V to Group IV metals, e.g. Nb or V to Ti or Zr, but further so by insterstitial nitrogen.

Probably the most important application, however, of interstitial nitrides arises not with the stable ones of Groups IV–VI but with the metastable and unstable nitrides of Group VIII, i.e. in the nitriding of steels and other iron alloys. We need here only recall Jack's work on the iron nitrides[18,23] described above, with its wide implications in regard to surface-hardening of alloy steels as well as internal hardening effects. Nitriding is done generally at moderate temperatures (about 500°C) in ammonia. The hardness can arise firstly through the nucleation of hard nitrides of the alloying elements, e.g. Cr_2N, Mn_4N, Mo_2N or AlN, and secondly through the formation of nitrogen austenite and martensite in the various stages of their structural development (γ, α' and α'' phases, see above). For higher nitrogen contents, this then leads to the Fe_4N (γ') and the ε-Fe_3N–Fe_2N phases, with their implied cubic/hexagonal transformations. The carburizing and carbonitriding processes are closely allied, and their structural mechanism described in Jack's investigations now forms an integral part of the whole technology surrounding them. (Further references to the nitriding process are given in the Bibliography.) Very hard wear-resisting layers could also be attained on nitrided surfaces of steels which had previously been coated with molybdenum or tungsten, i.e. containing substantially the Mo or W nitrides*. The structure of nitrided steels had already been studied in early x-ray work[70] showing, from the highly diffuse reflections, that extreme lattice strains and distortions were set up in the matrix, as well as a thin outer nitride layer. Both of these accounted for the high hardness, the lattice distortions probably being the more important than the actual nitrides. In

* The point of principle here arising is to utilize the more stable *earlier*-Group nitrides in hardened layers on steels, as a variant to, or in support of, the metastable Fe-based nitrides.

retrospect, however, it may be said that the two effects are really inseparable, and this and other earlier work is mainly superseded.

In heat-resisting steels small amounts of nitrogen are also able to improve high-temperature creep resistance. It was, therefore, considered possible to use low-nitrogen additions economically as a substitute for more expensive alloy elements in producing effective creep resistance. This should not be attributed to free nitrides but rather to their incipience in matrix solid solutions, i.e. it again amounts to the creation of Cottrell atmospheres and to effective dislocation pinning (cf. Chapter 3). Nevertheless, nitride lamellae are already liable to nucleate and would still contribute to this until embrittlement sets in. Basically, their role is similar to that of carbides, but with certain differences implied in the rather higher mobility of the N atom and a partial tendency to ionic bonding.

In a high-*chromium* stainless steel, the morphology of nitrides has shown the hexagonal close-packed $(Cr,Fe)_2N$ nitride to be essentially responsible for hardenability and to appear in coherent needle-like precipitates.

Nitrogen addition can be made to produce high strength and toughness in steels; but nitrogen without further alloy additions can also cause brittleness through iron nitride formation on aging. An important development by which such brittleness could be prevented is simultaneous aluminium and vanadium addition. This introduces particles of aluminium and vanadium nitrides to give dispersion hardening and to inhibit grain growth[71].

The practical effects of such special alloy nitrides on steel properties have been studied extensively. Thus, for instance, in low-Ti- or Zr-bearing steels, an extraordinary increase in creep strength could be achieved by the slightest amounts of nitrides[72]. These were finely dispersed TiN solid solutions with TiC or ZrN. A similar behaviour occurs, as mentioned, through fine-particle formation, AlN, in aluminium-bearing steels. The crystal-growth-inhibiting effect is pronounced e.g. on austenite crystallites, but such grain-growth prevention through dispersed nitrides is quite a general attribute in alloys, capable of much future development. The effect occurs already in mild steel but does so on an enhanced scale in the higher-alloy steels containing the stronger nitride formers: for instance, it was possible to improve properties of high-speed steels, such as resistance to temper softening and tool performance during machining.

The valuable strengthening effects of trace nitrogen on steels and ferrous alloys of various kinds have been greatly developed in recent years. Grain-refining effects by VN inclusions in *silicon* iron have also been attained[73]. Where silicon is prominent in irons, the formation of stable *silicon nitride*, Si_3N_4, has been observed, somewhat in analogy to AlN in Fe–Al alloys. It is known from extracts of silicon-bearing steels that Si_3N_4 can occur there (in both its allotropic α and β forms) quite competitively to metallic nitrides or carbonitrides.

The introduction of small quantities of titanium with nitrogen and aluminium into alloy steels not normally bearing these elements, with consequent grain refinement and increase in strength through nucleation of fine particles of TiN or AlN, can be illustrated by further examples[74–76].

Finally, on the subject of applications of interstitial nitrides, we should refer to the very *unstable* nitrides ending the series in Group VIII. Those of Fe, Co and Ni have been used in *catalysts* of ammonia syntheses, such as in the Fischer–Tropsch process. Here again, from the more systematic viewpoint it may be seen that even the very light and transient bonding of nitrogen in the Fe, Co or Ni lattices can be utilized to 'hold' N in atomic form for subsequent donation in further reaction*, e.g. to produce ammonia.

Thus, looking at the overall picture, the crystal chemistry that characterizes the nitrides is well reflected in their applications:

(1) The stable Group III and IV nitrides being used (*a*) physically in ceramic applications, but already with the caution of possible nitrogen loss at high temperatures (or replacement by oxygen and other interstitials where present, or by vacancies: non-stoichiometric ranges), and (*b*) dispersed as hardening constituents or as nitrogen-scavengers within Group V–VIII bulk alloys;

(2) the semi-stable Group V and VI nitrides having only very limited applications and, principally, deleterious effects (e.g. Cr embrittlement), although being of some positive application for strengthening when dispersed in steels;

(3) the metastable and unstable Group VII and VIII nitrides, chiefly those of iron, being of extreme importance in steel hardening, where a major part is played by the intermediate lattice distortions and transformations (e.g. N martensite, Fe_4N and ε-Fe_3N phases) and by the intricately interwoven non-stoichiometric ranges of cubic and hexagonal lattices (with their implied stacking-fault and dislocation locking);

(4) the quite unstable and almost fleetingly existing Co and Ni (possibly Cu, Ag and Pt-group) nitrides used in catalysts where the metal bonding of N atoms at surfaces and interfaces of other kinds (crystal boundaries, even at dislocations) is just sufficient for a 'holding' action in the atomic state.

The very stable silicon and boron nitrides (Si_3N_4, BN) fall outside this discussion, being non-metallic and non-interstitial. Nevertheless, they should be mentioned (see Bibliography) because their recent important applications (in the high-temperature ceramics and hard-materials fields, also as minor dispersed phases) are complementary to those of the metallic nitrides. It has also been found that aluminium nitride, AlN, can be grown in form of 'whiskers' of considerable strength (e.g. in an arc-image furnace), exemplifying just one aspect of future utilization of such nitrides as fibre-type strengtheners in materials. To a significant extent, aluminium nitride on its own is comparable to SiC (carborundum), both electronically and in properties and fields of application, e.g. as high-temperature ceramics. Particularly important ceramic applications have been found for silicon nitride as material of considerable strength and oxidation resistance up to high temperatures, with a remarkable degree of resilience.

Furthermore, it is interesting to note that the *metal*-nitride stability becomes so reduced in the later transition-metal Groups that for instance in low-carbon steels containing small amounts of N with *boron* or silicon, it is the *boron nitride*, BN, or *silicon nitride*, Si_3N_4, which forms rather than a free-metal nitride.

* This is the converse to the use of NH_3 for preparing the more stable nitrides.

Thus these nitrides find two-fold applications: as independent special ceramic phases and as subsidiary precipitates in alloy matrices. For instance, a silicon nitride phase has been reported in commercial silicon-killed steels[76], attributed the composition 'SiN' (different in structure to the normal Si_3N_4) and isomorphous to AlN. This raises the important point that the nitride morphology and structure can thus be highly sensitive to the environment in which the phases form, e.g. to coherence of lattice matching with the parent solution, as distinct from free growth. The same is true also for the formation of refractory metal nitrides, and further metastable structures may offer scope for unusual new property variations.

SELECTED LITERATURE

I *Binary Metal–Nitrogen Systems*

Kiessling, R. and Liu, Y. H. *Trans. Am. Inst. Min. metall. Engrs* 191 (1951) 639 [Cr–N, W–N]
Rudy, E. and Benesovsky, F. *Mh. Chem.* 92 (1961) 415 [Ti–N; Zr–N; Hf–N]
Samsonov, G. V. and Verkhoglyadova, T. S. *Dokl. Akad. Nauk SSSR* 142 (1962) 608 [Hf–N, V–N, Ta–N, Cr–N]
Hardness and electrical resistance
Schönberg, N. *Acta chem. scand.* 8 (1954) 627 [Zr–N]; *ibid.* 213 [V–N, Cr–N]; *ibid.* 208 [Nb–N]; *ibid.* 620 [Ta–N]; *ibid.* 199, 204 [Mo–N, W–N]

Ti–N[1–7,36–7]

Vainshtein, E. E. *et al., Fizika metall.* 12 (1961) 360
Palty, A. E. *et al., Trans. Am. Soc. Metals* 46 (1954) 312
Brager, A. *Acta phys.-chim. URSS* 10 (1939) 593; 11 (1939) 617

Zr–N

Domagala, R. F. *et al., Trans. metall. Soc. A.I.M.E.* 206 (1956) 98
Baker, T. W. *Acta crystallogr.* 11 (1958) 300
Nowotny, H. *et al.*[1]

Hf–N

Nowotny, H. *et al., Radex Rdsch.* 6 (1960) 367

V–N

Brauer, G. and Schnell, W. D. *J. less-common Metals* 7 (1964) 23
Hahn, H. *Z. anorg. allg. Chem.* 258 (1949) 58
Rostoker, W. and Yamamoto, A. *Trans. Am. Soc. Metals* 46 (1954) 1136
Mah, A. D. *U.S. Bur. Mines Rep.* RI 6177 (1963)
Heats of formation
Brager, A. *Acta phys.-chim. URSS* 14 (1941) 297

Nb–N

Cost, J. A. and Wert, C. A. *USAF Rep. AD 273232*, 1962
Thermodynamic study including primary solubilities and Nb_2N equilibria
Brauer, G. *Z. Elektrochem.* 46 (1940) 397
Brauer, G. and Jander, J. *Z. anorg. allg. Chem.* 270 (1952) 160; Brauer, G. and Esselborn, R. *ibid.* 308 (1961) 53; 309 (1961) 151
Elliott, R. P. and Komjathy, S. *Columbium Metallurgy*, p. 367, New York (Interscience) 1961
Mah, A. D. *J. Am. chem. Soc.* 80 (1958) 3872
Heats of formation

Ta–N[9,10,12c]

Brauer, G. and Zapp, K. H. *Z. anorg. Chem.* 277 (1954) 129
Pemsler, J. P. *J. electrochem. Soc.* 108 (1961) 744
Benesovsky, F. *3rd Plansee Seminar*, 1958, p. 291

Cr–N[11,13]

Verkhoglyadova, T. S. *et al., Poroshkovaya Met. Izv. Akad. Nauk Ukr. SSR* 1 (1961) 9
Corliss, L. M. *et al., Phys. Rev.* 117 (1960) 939
Eriksson, S. *Jernkont. Annlr* 118 (1934) 530

Blix, R. *Z. phys. Chem.* 83 (1929) 229
Miyagawa, O. and Okamoto, M. *Nippon Kink. Gakk.* 23 (1959) 568
Servi, I. S. and Forgeng, W. D. *J. Inst. Metals* 88 (1959) 164

Mo–N

Troitskoya, N. V. and Pinsker, Z. G.[14]; *Kristallografiya* 4 (1959) 30; *Soviet Phys. Crystallogr.* 6 (1961) 34

W–N[15,16]

Kiessling, R. and Peterson, L. *Acta metall.* 2 (1954) 675
Neugebauer, J. *et al., Z. anorg. allg. Chem.* 302 (1959) 50

Mn–N[19,31]

Baratshvidi, I. B. and Fedotoff, V. P. *Dokl. Akad. Nauk. SSSR* 139 (1961) 1354
Zwicker, U. *Z. Metallk.* 42 (1951) 274
Juza, R. *et al., Z. Elektrochem.* 61 (1957) 804

Fe–N[18,20,21,23–5,27–9]

Bose, B. N. and Hawkes, M. F. *J. Metals, N.Y.* 2 (1950) 307
Pitsch, W.[28]
Precipitation effects of nitrided steels
Burdese, A. *Metallurgia ital.* 49 (1957) 195

Co–N[30]

Terao, N. *mém. Scient. Revue métall.* 57 (1960) 96

Ni–N[18]

Juza, R. and Sachsze, W. *Z. anorg. allg. Chem.* 251 (1943) 201
Trillat, J. J. *et al., Bull. Soc. chim. Fr.* (1957) 804
Terao, N. *Naturwissenschaften* 45 (1958) 620; 46 (1959) 204

Pt group metals–N

Nitrides instable

Actinides–N

Th–N[59]

Rundle, R. E. *Acta crystallogr.* 1 (1948) 180; Zachariasen, W. H. *ibid.* 388
Street, R. S. and Waters, T. N. *Rep. AERE* (Harwell) M1115 (1962)

U–N

Olsen, W. H. and Mulford, R. N. R. *J. phys. Chem., Ithaka* 67 (1963) 952
Rundle, R. E. *et al., J. Am. chem. Soc.* 70 (1948) 99
Kempter, C. P. *et al., Analyt. Chem.* 31 (1959) 156
Mueller, M. H. and Knott, H. W. *Acta crystallogr.* 10 (1958) 751
Williams, J. and Sambell, R. A. *J. less-common Metals* 1 (1959) 217
Katz, J. J. and Rabinowitch, E. *The Chemistry of Uranium*, New York (McGraw-Hill) 1951
Trzebiatowski, W. *et al., Bull. Acad. Polon. Sci., Ser. Sci. Chim.* 10 (1962) 385
Speidel, G. O. *et al., Bull. Am. Ceram. Soc.* 49 (1963) 265
Hardness and resistance
Evans, P. E. and Davies, T. J. *J. nucl. Mater.* 10 (1963) 43
Benz, R. and Bowman, M. G. Paper at *Am. Chem. Soc. Meeting*, Chicago 1964 (cited by Storms, cf. Section IV)
Vaughan, D. A. *J. Metals, N.Y.* 8 (1956) 78

Pu–N

Olsen, W. H. and Mulford, R. N. R. *J. phys. Chem., Ithaka* 68 (1964) 1048
Rand, M. H. and Street, R. S. *Rep. AERE* (Harwell) *M*-973 (1962)
Anselin, F. *et al.* in *Carbides in Nuclear Energy*, Ed. Russell, L. E. *et al.*, Harwell Symposium, London (Macmillan) 1963
Zachariasen, W. H. *The Transuranium Elements*, New York (McGraw-Hill) 1949
Ellinger, F. H. In *The Metal Plutonium*, p. 281 (Eds. Coffinberry and Miner), Chicago Univ. Press, 1961

(Sc, Y and Rare Earths)–N

e.g. Gschneidner, K. A. *Rare Earth Alloys*, Princeton (Van Norstrand) 1961
Spedding, F. H. and Daane, A. H. *The Rare Earths*, New York (Wiley) 1961

REFERENCES

Si_3N_4

Popper, P. and Ruddleston, S. N. *Trans. Br. Ceram. Soc.* 60 (1961) 603
Parr, N. L. *et al.*, In *Special Ceramics*, London (Heywood) 1960; *Powder Metall.* 8 (1961) 145, 152, 164, 196
Oliver, D. A., In *High Temperatures in Aeronautics*, p. 327, Oxford (Pergamon) 1962
Reviews, technology, properties and structure

AlN

Taylor, K. M. and Camille, L. *J. electrochem. Soc.* 107 (1960) 308
Long, G. and Foster, L. M. *J. Am. Ceram. Soc.* 42 (1959) 53

BN

Finlay, G. R. and Fetterley, G. H. *Bull. Am. Ceram. Soc.* 31 (1952) 141
Wentorff, R. H. *J. chem. Phys.* 26 (1957) 956
Milledge, H. J. *et al.*, *Nature, Lond.* 184 (1959) Suppl. 10, 715; Vickery, R. C. *ibid.* 268
Pease, R. S. *Acta crystallogr.* 5 (1952) 356

II. *Ternary Metal–Nitrogen Systems* (cf. Section 5.5)

Duwez, P. and Odell, F.[38]
Solution series between mononitrides
Schönberg, N.[39b]
Kotelnikov, R. B. *Zh. neorg. Khim.* 3 (1958) 841
Nitride solution series
Jeitschko, W. *et al.*[53a] [Ti–Al–N]; [53b] [Ti–(Ge, Sn)–N, V–(Ga, Zn)–N, Nb–Al–N]
Begley, R. T. *et al.*[58] [Nb–Hf–N]
Smith, R. P.[56]; Tayeb, E. and Parlee, N. A. D.[57] [NbN, VN precipitation from γ-Fe]
Kuo, K.[47]; Stadelmaier, H. H. and Meussner, R. A.[48] (η-carbide-type ternary nitrides]
Evans, D. A. and Jack, K. H.[42] [Mo–*M*–N, *M* = Mn, Fe, Co, Ni]
Wiener, G. W. and Berger, J. A.[45] [Fe–Ni–N, Fe–Pt–N]
Arnott, R. J. and Wold, A.[46]; Schueter, W. *Stahl Eisen* 72 (1952) 197 [Fe–Ni–N]
Stadelmaier, H. H. and Fraker, A. C.[54a] [Fe–Ni–N, Fe–Pd–N, Fe–Pt–N]
Norton, J. T.[49] [Fe–Al–N]
Stadelmaier, H. H. and colleagues[54,55] [Fe–Mg–N, etc.]
Sachsze, W. and Juza, R.[51]; Juza, R. *et al.*[52] [(Co,Ni)–Li–N]

III. *Applications of Nitrides* (cf. Section 5.6)

Fry, A. *Stahl Eisen* 43 (1923) 12; Lehrer, E. *Z. Elektrochem.* 36 (1930) 383; Eisenhut, O. and Kaupp, E. *ibid.* 392; Djikstra, L. J. *Trans. Am. Inst. Min. metall. Engrs* 185 (1949) 252; Jack, K. H.[18,23]
Earlier papers on nitriding, still of interest
On 'Progressing in nitriding': Leeming, W. *Metal Prog.* 85 (1964) 86; Dashfield, D. A. *ibid.* 88; Jones, C. K. and Martin, S. W. *ibid.* 94

IV. *General reviews including nitrides*

Storms, E. K. 'A critical review of refractories', *U.S. atom. Energy Comm. Los Alamos Rep. LA-2942*, 1964
Samsonov, G. V. and Umanskii, Ya S. 'Hard Compounds of refractory metals', *NASA Techn. Trl. F. 102*, 1962
Campbell, I. E. *High Temperature Technology*, New York (Wiley), 1962
Kieffer, R. and Schwarzkopf, P. *Refractory Hard Metals*, p. 223, New York (MacMillan) 1953
Kieffer, R. and Benesovsky, F. *Hard Materials*, p. 286, Vienna (Springer) 1963
Pearson, J. and Ende, U. J. C., 'Thermodynamics of nitrides and nitrogen', *J. Iron Steel Inst.* 175 (1953) 52
Milliken, M. T. and Senkin, J. E. 'Some kinetic and thermodynamic properties of the refractory metal borides and nitrides. An annotated bibliography', *Univ. Calif. Radiat. Lab. Rep. 7559*, 1963

REFERENCES

1 Nowotny, H. *et al.*, *Mh. Chem.* 92 (1961) 403
2 Holmberg, B. *Actà chem. scand.* 16 (1962) 1255
3 Yurko, G. A. *et al.*, *Acta crystallogr.* 15 (1962) 1309
4 Phillip, W. *Acta metall.* 10 (1962) 583
5 Pearson, W. B. *Acta metall.* 10 (1962) 123

[6] James, W. J. and Straumanis, M. E. *Acta metall.* 12 (1964) 739; Philipp, W., *ibid.* 740
[7] Samsonov, G. V. and Verkhoglyadova, T. S. *Dokl. Akad. Nauk SSSR* 138 (1961) 342
[8] Schönberg, N. *Acta chem. scand.* 8 (1954) 199, 204
[9] Gebhardt, E. *et al.*, *Z. Metallk* 52 (1961) 464
[10] Östhagen, K. and Kofstad, P. *J. less-common Metals* 5 (1963) 7
[11] Weaver, C. W. *Acta metall.* 10 (1962) 1151
[12] (*a*) Ehrlich, P. *Z. anorg. allg. Chem.* 259 (1949) 1; (*b*) Hahn, H. 258 (1949) 58; (*c*) Giardini, A. A. and Kohn, J. A., Paper at Amer. Crystallogr. Ass. Conf., Boulder 1961
[13] Sully, A. H. *Chromium*, London (Butterworths) 1954
[14] Troitskaya, N. V. and Pinsker, Z. G. *Kristallografiya* 6 (1961) 43
[15] Khitrova, V. I. and Pinsker, Z. G. *Soviet Phys. Crystallogr.* 6 (1962) 712
[16] Khitrova, V. I. *Kristallografiya* 6 (1961) 549
[17] Hahn, H. and Konrad, A. *Z. anorg. allg. Chem.* 264 (1950) 174
[18] Jack, K. H. *Acta crystallogr.* 13 (1950) 392
[19] Takei, W. J. and Shirane, G. *Phys. Rev.* 119 (1960) 122
[20] Frazer, B. C. In *Pittsburgh Diffraction Conf.* 1956
[21] Hume-Rothery, W. *Phil. Mag.* 7 (1962) 1955
[22] Altmann, S. L., Coulsen, C. A. and Hume-Rothery, W. *Proc. R. Soc.* A240 (1957) 145
[23] Jack, K. H. *Proc. R. Soc. A.* 195 (1948) 34; 208 (1951) 216; *Acta crystallogr.* 15 (1952) 404
[24] Paranjpe, V. E., Cohen, M. *et al.*, *J. Metals. N.Y.* 188 (950) 261
[25] Hendricks, S. B. and Kosting, P. R. *Z. Kristallogr. Kristallgeom.* 74 (1930) 511
[26] Takei, W. J. *et al.*, *Phys. Rev.* 125 (1962) 1893
[27] Booker, G. R. *Acta metall.* 9 (1961) 590
[28] Pitsch, W. *Arch. EisenhüttWes.* 32 (1961) 573
[29] Shirane, G. *et al.*, *Phys. Rev.* 126 (1962) 49
[30] Clarke, J. and Jack, K. H. *Chemy Ind.* 46 (1951) 1004
[31] Lihl, F. *et al.*, *Z. Metallk.* 53 (1962) 715
[32] Becker, K. and Ebert, F. *Z. Phys.* 31 (1925) 268
[33] Dreger, L. H. *et al.*, *J. phys. Chem., Ithaka* 66 (1962) 1556
[34] Hildenbrand, D. L. and Hall, W. F. *J. phys. Chem., Ithaka* 67 (1963) 888
[35] Pehlke, R. D. and Elliott, J. F. *Trans. metall. Soc. A.I.M.E.* 215 (1959) 781
[36] Samsonov, G. V. and Umanskii, Ya. S. *NASA Rep. TT-F-102*, 1962
[37] Samsonov, G. V. and Verkhoglyadova, T. S. *Dokl. Akad. Nauk SSSR* 138 (1961) 342
[38] Duwez, P. and Odell, F. *J. electrochem. Soc.* 97 (1950) 299
[39] Schönberg, N. (*a*) *Acta chem. scand.* 8 (1954) 199, 208; (*b*) *ibid.* 213
[40] Beattie, H. J. and Ver Snyder, F. L. *Trans. Am. Soc. Metals*, 45 (1950) 397
[41] Sully, A. H., Hume-Rothery, W. and Pearson, W. B. *J. Inst. Metals* 76 (1950) 719; cf. 75 (1949) 195
[42] Evans, D. A. and Jack, K. H. *Acta crystallogr.* 10 (1957) 762
[43] Hahn, H. *F.I.A.T. Rev. inorg. Chem.* 1939–46, 11 (1948) 70
[44] Hahn, H. and Mühlberg, H. *Z. anorg. allg. Chem.* 259 (1949) 721
[45] Wiener, G. W. and Berger, J. A. *J. Metals* 7 (1955) 360
[46] Arnott, R. J. and Wold, A. *Physics Chem. Solids* 15 (1960) 152
[47] Kuo, K. *Acta metall.* 1 (1953) 301, 611
[48] Stadelmaier, H. H. and Meussner, R. A. *Mh. Chem.* 96 (1965) 228
[49] Norton, J. T. *Trans. Am. Inst. Min. metall. Engrs* 113 (1934) 262
[50] Stadelmaier, H. H. *Z. Metall.* 52 (1962) 758
[51] Sachsze, W. and Juza, R. *Z. anorg. allg. Chem.* 259 (1949) 278
[52] Juza, R. *et al.*, *Z. anorg. allg. Chem.* 273 (1953) 48
[53] Jeitschko, W., Nowotny, H. and Benesovsky, F. *Mh. Chem.* (*a*) 94 (1963) 1198; (*b*) 95 (1964) 436
[54] Stadelmaier, H. H. and Fraker, A. C. (*a*) *Trans. metall. Soc. A.I.M.E.* 218 (1960) 571; (*b*) *Z. Metallk.* 53 (1962) 48
[55] Stadelmaier, H. H. and Yun, T. S. *Z. Metallk.* 52 (1961) 477
[56] Smith, R. P. *Trans. metall. Soc. A.I.M.E.* 224 (1962) 190
[57] Tayeb, E. and Parlee, N. A. D. *Trans. metall. Soc. A.I.M.E.* 227 (1963) 929
[58] Begley, R. T. *et al.*, *Plansee Seminar* 5 (1964) Paper No. 20
[59] Chiotti, P. *J. Am. Ceram. Soc.* 35 (1952) 123
[60] Campbell, I. E. *High-Temperature Technology*, New York (Wiley) 1962
[61] Sully, A. H. *et al.*, *J. Inst. Metals* 81 (1953) 569, 585
[62] *Ductile Chromium*, e.g. pp. 169 and 196, Symposium, Cleveland (Am. Soc. Metals) 1957
[63] Wood, W. A. *et al.*, *Trans. metall. Soc. A.I.M.E.* 230 (1964) 837
[64] Wilms, G. R. *J. less-common Metals* 6 (1964) 169 and other papers
[65] Bullen, F. P. *et al.*, *Phil. Mag.* 101 (1964) 803
[66] Ryan, N. E. and Landau, C. S. *J. Aust. Inst. Metals* 8 (1963) 273
[67] Henderson, F. *et al.*, *J. Inst. Metals* 92 (1963) 111

REFERENCES

[68] Mukherjee, A. K. and Martin, J. W. *J. less-common Metals* 5 (1963) 403
[69] Sperner, F. and Meixner, J. *Z. Metallk.* 50 (1959) 691
[70] Wood, W. A. *Phil. Mag.* 7 (1933) 719
[71] *Metalwkg Prod.* 107 (1963) 76
[72] Rapatz, F. and Frehser, J. *Stahl Eisen* 69 (1949) 605
[73] Fiedler, H. C. *Trans. Am. Inst. Min. metall. Engrs* 221 (1961) 1201
[74] Kubik, A. *et al., Radex Rdsch.* 6 (1961) 739
[75] Peter, W. and Wiesenecker, I. *Stahl Eisen* 82 (1962) 23
[76] Kremnev, L. S. and Geller, Ya. A. *Izv. vÿssh. ucheb. Zaved. Khim. Met.* 9 (1961) 129
[77] Arrowsmith, J. M. *J. Iron Steel Inst.* 202 (1963) 699

6

BORIDES

6.1 GENERAL SURVEY

The place of borides amongst interstitial compounds is a singular one because, unlike the smaller interstitial atoms (C, N, H, O), boron is of a size such that direct B–B bonds can already exert themselves in the lattice, the metal–interstitial and metal–metal bonds being no longer exclusive. Furthermore, this implies that we are on the borderline of Hägg's rule and that even for the lower borides the basic metal lattice can undergo major modifications. This tendency will become more accentuated in the silicides (Chapter 7) which in many respects are related to the borides, both in properties and structure.

The identity and structures of the transition metal borides have been indicated in *Figure 2.1* together with other interstitial compounds. For the lower borides such as M_2B (Al_2Cu structure in Fe_2B), the B atoms are still in isolation, and the M–B bonds substantially prevail. The bonds evolve, however, with increasing boron content, from (1) isolated B atoms in M_2B to (2) B–B chain linkage, (3) double chains, (4) network and (5) framework structures of B atoms interpenetrating the metal lattice: that is, from 0–1- to 2–3-dimensional linkage of boron atoms. Nevertheless, the metal–metal bond still persists throughout, and the essentially metallic character of the borides, hardness, high melting points and conductivity are some of their important features.

The radius of the boron atom, assessed from the interatomic distances in phases in which direct B–B contact predominates, such as CaB_6 or UB_{12}, is approximately 0·87 Å. However, its variability with co-ordination number and environment must be stressed, giving a possible radius range of at least $\pm$ 0·1 Å. (For its relations to the H, C, N, O, Si and other atoms see also *Figure 2.1.*)

In view of the great increase of interest in borides in recent times, both basic and practical, which is due to their special characteristics, some background publications[1–8] are given and a note on some of the main centres of work is appended (see Bibliography).

Table 6.1 summarizes the structure of the borides and some of their properties as far as known—which is a considerable qualification.

The stoichiometric series $M_2B \rightarrow MB \rightarrow M_3B_4 \rightarrow MB_2 \rightarrow MB_6 \rightarrow MB_{12}$ clearly represents the main stations in the above bonding progression from isolated B atoms to those building their 3-dimensional framework. *Figure 6.1* shows this sequence of B atom configurations. Starting with the *high-metal* borides, the M_2B type are all of $CuAl_2$ (C–16)-type structure or closely related to it and contain sheets of metal tetrahedra with B atoms located interstitially between these. The boron atoms thus are arranged in rows within the metal lattice, with a B–B distance just exceeding twice the radius of the boron atom. However, although in this direction B–B bonding is not yet quite achieved, these aligned rows may be viewed already as a near approach to such a bonding. The members all belong to Groups VI–VIII, i.e. the metals of the smaller atomic sizes (Ni_2B, Co_2B, Fe_2B, Mn_2B, Cr_2B, Mo_2B and W_2B). The Mn_4B (Cr_4B) structure (Kiessling's 'δ'-Mn–B) is very similar to that of Mn_2B; in fact, the metal lattice may be considered as almost the same, except for a slight readjustment due to B deficiency. Mn_4B possesses a significant homogeneity range. On the boron-richer side to Cr_2B, we have

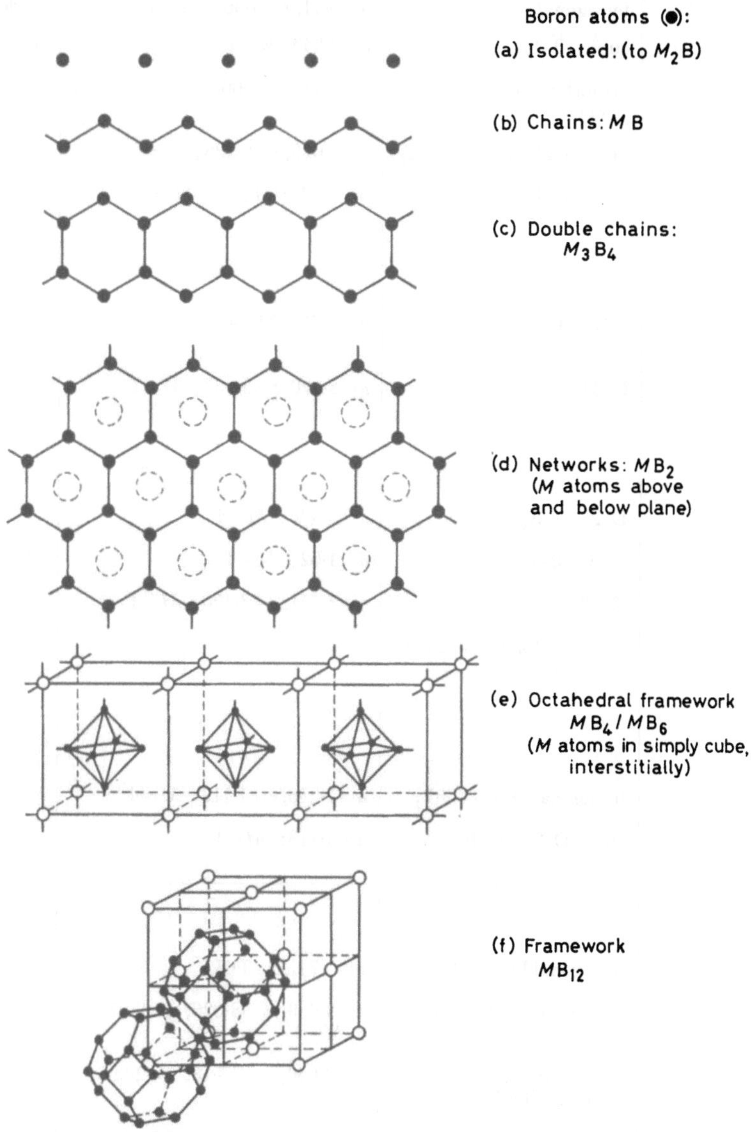

Figure 6.1. Boron atom configurations in borides

Table 6.1. Structure and properties of transition-metal borides

Base metal	*Borides formed*	*Structure type (Strukturbericht nomenclature)*	*Lattice dimensions, Å; no. of molecules per unit cell, M*	*Melting point, °C*
Group IV				
Ti	Ti_2B	tetragonal	$a=6{\cdot}11$, $c=4{\cdot}56$; $M=4$	dec. 2,250
	TiB	cubic (B–1)	$a=4{\cdot}24$; $M=4$	dec. ~ 1,900?
	$TiB_{1{\cdot}1}$ (?)	orthorhombic (B–27= FeB)	$a=6{\cdot}14$, $b=3{\cdot}06$, $c=4{\cdot}56$; $M=4$	
	TiB_2	hexagonal (C–32=AlB_2)	$a=3{\cdot}028$, $c=3{\cdot}228$; $M=1$	2,980
	Ti_2B_5	hexagonal (D–8_h=W_2B_5)	$a=2{\cdot}98$, $c=13{\cdot}98$; $M=1$	dec. 2,100?
Zr	ZrB	cubic (B–1)	$a=4{\cdot}65$; $M=4$	dec ~ 1,250?
	ZrB_2	C–32	$a=3{\cdot}170$, $c=3{\cdot}533$; $M=1$	3,040
	ZrB_{12}	D–2_f=UB_{12}	$a=7{\cdot}408$; $M=4$	2,680
Hf	HfB	cubic (B–1)	$a=1{\cdot}62$; $M=4$	dec. peritectic ~ 2,900
	HfB_2	C–32	$a=3{\cdot}141$, $c=3{\cdot}470$; $M=1$	3,250
	(HfB_{12}?)	(checked for but not confirmed)		
Th	ThB_4	tetragonal D–1_e=ThB_4	$a=7{\cdot}256$, $c=4{\cdot}113$; $M=4$	>2,500
	ThB_6	cubic D–2_1=CaB_6	$a=4{\cdot}113$; $M=1$	2,195
Group V				
V	V_2B (?)			~2,050
	V_3B_2 (±)	tetragonal D–5a=U_3Si_2	$a=5{\cdot}74_6$, $c=3{\cdot}03_2$	~1,900
	VB	orthorhombic B_f=CrB	$a=3{\cdot}058$, $b=8{\cdot}026$, $c=2{\cdot}97$; $M=1$	dec. ~2,300
	V_3B_4	orthorhombic D–7_b=Ta_3B_4	$a=3{\cdot}030$, $b=13{\cdot}18$, $c=2{\cdot}986$; $M=2$	~2,400 (2,100)
	VB_2	C–32	$a=3{\cdot}006$, $c=3{\cdot}056$; $M=1$	

(In order of Periodic System and of increasing B content)

Heat of formation, kcal/g atom [] *value for monocarbides*	*Hardness,* kg/cm^2 at 30–50 g load * micro; [ ] *value for monocarbides*	*Density (theoretical)*	*Other properties: electrical resistivity, ρ, μΩ. cm; thermal conductivity, κ, cal/cm/sec/°C thermal expansion coefficient, β,/°C*	*Remarks*
			$\rho=40$	Ti_2B high-temperature phase only
~−24 [−21·9]	3,370 [3,200]* (+9 Moh)	4·52	$\rho=10$ at 50°C, rising to 60 at 1,000°C (linear); $\kappa=0{\cdot}14$ at 50°C, 0·11 at 1,000°C; $\beta=6{\cdot}39\times10^{-6}$ (≈20°–600°C [fairly constant]; mean?)	
	(+9 Moh)	5·7		stable 800°–1,200°C only?
<−26 [−24·1]	2,300 [2,600]* (8 Moh)	6·1	$\rho=9$–11; $\kappa=0{\cdot}055$ (at 200°C); $\beta=6{\cdot}3\times10^{-6}$ (0°–200°C mean); $7{\cdot}3\times10^{-6}$ (0°–1,400°C mean)	
	92 Rockwell A	3·7		
		12·8		HfB of orthorhombic FeB-type structure also reported; sensitive conditions of change-over cubic B–1↔cubic B–27 for the Group IV monoborides observed; B–27 can probably be stabilized by small-atom solutions
		11·2	$\rho=10$; $\beta=5{\cdot}3\times10^{-6}$ (0°–1,000°C?)	
		8·45		as UB_4
		6·08		structure of ThB_6 related to ZrB_{12} and UB_{12}
	(+9 Moh)	5·44	$\rho=35$–40	
	2,070 (2,800)* (8–9 Moh)	5·10	$\rho=16$	

Base metal	*Borides formed*	*Structure type (Strukturbericht nomenclature)*	*Lattice dimensions, Å; no. of molecules per unit cell, M*	*Melting point, °C*
Group V cont.				
Nb	Nb_3B (?)		no details known	
	Nb_2B (?)			
	Nb_3B_2	D–5a = U_3Si_2	$a=6{\cdot}18_5$, $c=3{\cdot}28_1$	dec. 1,800
	NbB	orthorhombic B_f = CrB	$a=3{\cdot}298$, $b=8{\cdot}724$, $c=3{\cdot}166$; $M=2$	~2,300
	Nb_3B_4	orthorhombic D–7_b = Ta_3B_4	$a=3{\cdot}305$, $b=14{\cdot}08$, $c=3{\cdot}137$; $M=2$	dec. ~2,700
	NbB_2	C–32	from $a=3{\cdot}110$, $c=3{\cdot}085$ (Nb end) to $a=3{\cdot}085$, $c=3{\cdot}311$ (B end); $M=1$	dec. ~ >2,900
Ta	Ta_3B (?)	as Nb_3B	no details known	
	Ta_2B	tetragonal C–16 = $CuAl_2$	$a=5{\cdot}778$, $c=4{\cdot}864$	dec. ~1,800
	Ta_3B_2	D–5a = U_3Si_2	$a=6{\cdot}18_4$, $c=3{\cdot}28_7$	dec. ~2,100
	TaB	B_f = CrB	$a=3{\cdot}276$, $b=8{\cdot}669$, $c=3{\cdot}157$; $M=4$	~2,400
	Ta_3B_4	orthorhombic $D7_1$ = Ta_3B_4	$a=3{\cdot}29$, $b=14{\cdot}0$, $c=3{\cdot}13$; $M=2$	dec. 2,650
	TaB_2	C–32	from $a=3{\cdot}099$, $c=3{\cdot}224$ (Ta end) to $a=3{\cdot}057$, $c=3{\cdot}291$ (B end); $M=1$	~3,200
Pa	?			
Group VI				
Cr	Cr_4B ('δ') (or Cr_2B? (second form conflicting formulae reported)	orthorhombic D–1_f = Mn_4B	$a=14{\cdot}71$, $b=7{\cdot}38$, $c=4{\cdot}26$; $M=8$	dec. ~1,700?
	Cr_2B	C–16 = $CuAl_2$	$a=5{\cdot}18_5$, $c=4{\cdot}31_6$; $M=4$	dec. ~1,850?
	Cr_5B_3 (or Cr_3B_2)	tetragonal D–8 = CrB	$a=5{\cdot}44$, $c=10{\cdot}07$; $M=4$	1,960
	CrB	B_c = CrB	$a=2{\cdot}969$, $b=7{\cdot}858$, $c=2{\cdot}932$; $M=1$	~2,050?
	Cr_3B_4	D–7_b = Ta_3B_4	$a=2{\cdot}986$, $b=13{\cdot}020$, $c=2{\cdot}952$; $M=1$	
	CrB_2	C–32	$a=2{\cdot}969$, $c=3{\cdot}066$; $M=1$	~1,850

Heat of formation, kcal/g atom [ ] *value for monocarbides*	*Hardness,* kg/cm^2 at 30–50 g load * micro; [ ] *value for monocarbides*	*Density* (*theoretical*)	*Other properties: electrical resistivity,* ρ, $\mu\Omega$. cm; *thermal conductivity,* κ, cal/cm/sec/°C *thermal expansion coefficient,* β,/°C	*Remarks*
	(+9 Moh)	7·5	ρ=64·5	
<−12 [15]	2,200 [2,400]* (+8 Moh)	7·21	ρ=32; κ=0·040; $\beta=5{\cdot}3\times10^{-6}$ (0°–1,000°C?)	
				Ta_3B stable only at very high temperatures alternatively reported as $Ta_{2\cdot4}B$
		14·9	ρ=100	alternatively reported as $Ta_{1\cdot6}B$
		13·69		
<−17·5 [−19·25]	2,500 [1,800]* (+9 Moh)	12·60	ρ=68; κ=0·026	
		6·24		
		6·53		
	(+9 Moh)	6·12	ρ=64	
	(8·5 Moh)	6·11		
<−10	1,800 [1,300, Cr_3C_2?]*	5·6	ρ=20; $\beta=11{\cdot}0\times10^{-6}$/C (0°–600°C?)	

Base metal	*Borides formed*	*Structure type (Strukturbericht nomenclature)*	*Lattice dimensions, Å; no. of molecules per unit cell, M*	*Melting point, °C*
Group VI cont.				
Mo	Mo_2B	C–16	$a=5{\cdot}543$, $c=4{\cdot}735$; $M=4$	dec. ~2,000
	Mo_3B_2	tetragonal Cr_3B_2 type		2,070
	α-MoB } (HR)	tetragonal B_g=MoB	$a=3{\cdot}110$, $c=16{\cdot}9_5$; $M=8$	α→β, ~2,000
	β-MoB } (HR)	orthorhombic B_f=CrB	$a=3{\cdot}16$, $b=8{\cdot}61$, $c=3{\cdot}08$; $M=8$	~2,080
	MoB_2 (HR)	C–32 (high-temp. form only)	$a=3{\cdot}06$, $c=3{\cdot}10$; $M=1$	~2,100
	Mo_2B_5(HR)	hexagonal D–8_1=Mo_2B_5	$a=3{\cdot}01$, $c=20{\cdot}93$; $M=3$	decomp. to Mo_2B_5 below ~1,600°C
	MoB_4	tetragonal D–1_e=ThB_4	$a=6{\cdot}34$, $c=4{\cdot}50$; $M=4$	dec. ~1,600 to MoB_2
W	W_2B	C–16	$a=5{\cdot}564$, $c=4{\cdot}740$; $M=4$	~2,770
	α-WB	B_g=MoB	$a=3{\cdot}115$, $c=16{\cdot}93$; $M=8$	α→β, ~1,950
	β-WB	B_f=CrB	$a=3{\cdot}19$, $b=8{\cdot}40$, $c=3{\cdot}07$; $M=4$	2,920
	WB_2 (?) }	C–32	$a=6{\cdot}35$–$8{\cdot}24$, $c=16{\cdot}4$–$15{\cdot}6$; $M=1$	~2,220
	W_2B_5 }	D–8_h=W_2B_5	$a=2{\cdot}982$, $c=13{\cdot}87$; $M=2$	~2,300
	WB_4	tetragonal D–1_e=ThB_4	$a=6{\cdot}34$, $c=4{\cdot}50$; $M=4$	dec. ~1,600°C to lower borides
U	UB_2	C–32	$a=3{\cdot}1314$, $c=3{\cdot}9857$ (U end) to $a=3{\cdot}1293$, $c=3{\cdot}9893$ (B end); $M=1$	~2,400
	UB_4	D–1_e=ThB_4	$a=7{\cdot}075$, $c=3{\cdot}979$; $M=4$	~2,500?
	UB_{12}	D–2_f=UB_{12}	$a=7{\cdot}468$; $M=4$	~2,200?
Group VII				
Mn	Mn_4B ('δ')	orthorhombic D–1_f=Mn_4B	$a=14{\cdot}53$, $b=7{\cdot}292$, $c=4{\cdot}209$; $M=8$	
	Mn_2B	tetragonal C–16	$a=5{\cdot}148$, $c=4{\cdot}208$; $M=4$	
	MnB	orthorhombic B–27=FeB	$a=5{\cdot}560$, $b=2{\cdot}977$, $c=4{\cdot}145$; $M=4$	
	Mn_3B_4	orthorhombic D–7_b=Ta_3B_4	$a=3{\cdot}009$, $c=3{\cdot}039$; $M=1$	
	MnB_2	C–32	$a=3{\cdot}009$, $c=3{\cdot}039$; $M=1$	

Heat of formation, kcal/g atom [ ] *value for monocarbides*	*Hardness,* kg/cm² at 30–50 g load * micro; [ ] *value for monocarbides*	*Density (theoretical)*	*Other properties: electrical resistivity, ρ, μΩ. cm; thermal conductivity, κ, cal/cm/sec/°C thermal expansion coefficient, β,/°C*	*Remarks*
	1,660 (100 g load) (8–9 Moh)	9·31	$\rho=40$	
				Mo_3B_2 high-temperature phase only, above 1,850°C
	1,570 (100 g load) (8 Moh)	8·77	$\rho=45$	
			$\rho=25$	
<-8 [$-1\cdot0$]	2,330* (1,570) (+9 Moh)	7·78	$\rho=45$	
	2,350	7·48	$\rho=25$	
		4·8 (meas.)		
		16·72		
	9 Moh	16·0		
-6 to -3 [$-4\cdot2$]		13·6		
	2,660 [2,400]* (+9 Moh)	13·1	$\rho=21$	'WB_2', of C–32 structure, probably superseded
		8·3 (meas.)		
		12·71	$\beta=8\times10^{-6}$ (//c axis) $=9\times10^{-6}$ (//a axis)	
		9·38		
		5·82		
		no details known		

Base metal	*Borides formed*	*Structure type (Strukturbericht nomenclature)*	*Lattice dimensions, Å; no. of molecules per unit cell, M*	*Melting point, °C*
Group VII cont.				
Tc	?			
Re	Re_3B	orthorhombic 'Re_3B type'	$a=2{\cdot}890$, $b=9{\cdot}313$, $c=7{\cdot}25$	
	Re_7B_3	$D\text{–}10_2=Th_7Fe_3$	$a=7{\cdot}504$, $c=4{\cdot}882$	
	ReB_2 or ReB_3	hexagonal 'ReB_3 type'	$a=2{\cdot}900$, $c=7{\cdot}478$; $M=2$	
Group VIII				
Fe	Fe_2B	C–16	$a=5{\cdot}109$, $c=4{\cdot}249$; $M=4$	dec. 1,389
	FeB	B–27=FeB	$a=5{\cdot}506$, $b=2{\cdot}952$, $c=4{\cdot}061$; $M=4$	~1,550
Ru	Ru_7B_3	hexagonal $D\text{–}10_2=Th_7Fe_2$	$a=7{\cdot}46_2$, $c=4{\cdot}71_3$	peritectic dec. 1,600
	$Ru_{11}B_8$	orthorhombic	$a=11{\cdot}61$, $b=11{\cdot}34$, $c=2{\cdot}84$	
	~RuB	BC cubic	$a=7{\cdot}02$	1,500
	$\sim Ru_2B_3$	('complex structure')		1,550
	$\sim RuB_2$	orthorhombic or hexagonal cp	$a=4{\cdot}6443$, $b=2{\cdot}8668$, $c=4{\cdot}0449$; $M=1$ $a\simeq 3$ Å; $c/a\sim 1{\cdot}10$	
Os	~OsB	cubic (?)	$a=7{\cdot}04$	
	$\sim OsB_2$			
	$\sim OsB_{2+}$	orthorhombic (as RuB_2)	$a=4{\cdot}6832$, $b=2{\cdot}8717$, $c=0{\cdot}4761$; $M=1$	
Pu	PuB	B–1	$a=4{\cdot}92$	
	PuB_2	C–32	$a=3{\cdot}18$, $c=3{\cdot}90$	
	PuB_4	$D\text{–}1_e=ThB_4$	$a=7{\cdot}10$, $c=4{\cdot}014$	
	PuB_6	$D\text{–}2_1=CaB_6$	$a=4{\cdot}115\text{–}4{\cdot}140$	
Co	Co_3B	$DO_{11}=Fe_3C$	$a=5{\cdot}225$, $b=6{\cdot}629$, $c=4{\cdot}408$	
	Co_2B	C–16	$a=3{\cdot}016$, $c=4{\cdot}220$; $M=4$	
	CoB	B–27=FeB	$a=5{\cdot}253$, $b=3{\cdot}043$, $c=3{\cdot}956$; $M=4$	1,265
Rh	Rh_7B_3	$D\text{–}10_2=Th_7Fe_3$	$a=7{\cdot}47_1$, $c=4{\cdot}77_7$	
	Rh_2B	orthorhombic, $C\text{–}23=Co_2Si$		
	$\sim RhB_{1\cdot 1}$	B–8='anti-NiAs'	$a=3{\cdot}30_9$, $c=4{\cdot}22_4$	
	RhB_2			

Heat of formation, kcal/g atom [ ] *value for monocarbides*	*Hardness,* kg/cm² at 30–50 g load * micro; [ ] *value for monocarbides*	*Density (theoretical)*	*Other properties: electrical resistivity,* ρ, $\mu\Omega$. cm; *thermal conductivity,* κ, cal/cm/sec/°C *thermal expansion coefficient,* β,/°C	*Remarks*
		no details known		
		no details known		
		no details known		

Base metal	*Borides formed*	*Structure type (Strukturbericht nomenclature)*	*Lattice dimensions, Å; no. of molecules per unit cell, M*	*Melting point, °C*
Group VIII cont.				
Ir	Ir_3B_2 (?)			
	IrB_{1-2}	$C_c = ThSi_2$	$a = 2{\cdot}81_6$, $c = 10{\cdot}26$	
	IrB_2 (?)			
Ni	Ni_3B	$DO_{11} = Fe_3C$	$a = 5{\cdot}211$, $b = 6{\cdot}619$, $c = 4{\cdot}389$	
	Ni_2B	C–16	$a = 4{\cdot}990$, $c = 4{\cdot}245$; $M = 4$	~1,230
	Ni_4B_3 ('*o*' form)	orthorhombic '*o*-Ni_4B_3'	$a = 11{\cdot}953$, $b = 2{\cdot}981$, $c = 6{\cdot}569$ (Ni end) to $a = 11{\cdot}973$, $b = 2{\cdot}985$, $c = 6{\cdot}584$ (B end)	~1,160?
	Ni_4B_3 ('*m*' form)	monoclinic '*m*-Ni_4B_3 type'	$a = 6{\cdot}430$, $b = 4{\cdot}882$, $c = 7{\cdot}818$ $\beta = 103°18'$	dec. 1,020
	NiB	$B_f = CrB$	$a = 2{\cdot}925$, $b = 7{\cdot}396$, $c = 2{\cdot}966$; $M = 4$	
	Ni_2B_3?			
Pd	Pd_3B	$DO_{11} = Fe_3C$	$a = 5{\cdot}46$, $b = 7{\cdot}57$, $c = 4{\cdot}85$	
	Pd_5B_2	monoclinic	$a = 12{\cdot}79$, $b = 4{\cdot}95$, $c = 5{\cdot}47$; $\beta = 97°$	
	Pd_3B_2 (?)			
Pt	PtB	hexagonal B–8 = 'anti-NiAs'	$a = 3{\cdot}35_8$, $c = 4{\cdot}05_8$	

the Cr_5B_3 (U_3Si_2)-type structure as a further variant, in which partial B–B bonding is already realized. Thus the trio $Cr_4B \rightarrow Cr_2B \rightarrow Cr_{1{\cdot}67}B$ presents a breakthrough to boron–boron contact amongst closely related structures.

Size factor rather than electron valency is the principal factor governing the boride structures, which often makes simple geometrical considerations, such as based on availability of lattice space and on interstitial size, well applicable. This also allows to predict reasonably the extent of metal–metal, of boron–silicon and, with reservations, of boron–carbon substitution.

It is useful to remember that $CuAl_2$ is the precipitation-hardening phase in aluminium–copper alloys and to compare relations between the pairs Al/Al_2Cu and M/M_2B (where M = Ni, Co, γ-Fe or γ-Mn). Both represent isostructural precipitations from a FCC parent phase; in the former case, this occurs via a much investigated series of metastable stages, $\theta \rightarrow \theta' \rightarrow \theta''$ modifications giving rise to Guinier–Preston zone effects. It may well be asked whether the borides M_2B have not an analogous structural relation to the parent metal, and whether a corresponding lattice mechanism would apply. The primary solubility of B in Ni, Co, γ-Fe and γ-Mn is appreciable (cf. *Figure 3.1*) and also recedes with decreasing temperature, so that similar intermediate stages are liable to occur. In Mn_4B we may perhaps even see a stabilized version of the Guinier–Preston zone-type structure—a point worth examining.

Heat of formation kcal/g atom [ ] *value for monocarbides*	*Hardness* kg/cm² at 30–50 g load * micro; [ ] *value for monocarbides*	*Density* (*theoretical*)	*Other properties: electrical resistivity, ρ,* μΩ. cm; *thermal conductivity, κ,* cal/cm/sec/°C *thermal expansion coefficient, β,*/°C	*Remarks*
		no details known		
		no details known		

In the next higher class, the monoborides *M*B, zigzag chains of B atoms directly bonded pervade the metal lattice in channels; each boron is centrally surrounded by trigonal metal prisms. We have here to distinguish 3 closely related sub-types in structure, represented by FeB, CrB and MoB (see *Table 6.1*), in which the metal prisms are connected in varying manners. The FeB and CrB types are tied to the *M*B composition, whilst the MoB type shows a small homogeneity range on either side, associated with a change in unit cell axes; this degree of tolerance for the boron chains to contain B vacancies or B excess[1] is facilitated by the larger Mo and W metal atoms.

The B 'double chains' in the M_3B_4 borides (*Figure 6.1c*) may be regarded as an intermediate stage towards network formation, finally accomplished in MB_2.

These evolving hexagonal boron nets (*Figure 6.1d*) then give rise to the AlB_2 (C–32)-type structure in MB_2 (or to those related to it); we might regard this as alternating near close-packed metal and boron layers where the latter may become slightly puckered. In the Mo and W borides, a certain variant to MB_2 occurs at compositions Mo_2B_5 ($=M_{0\cdot8}B_2$), with a rather more complex though still closely related structure. The main difference is again one in stacking sequence, namely metal-sheet layering of type AAAA in MB_2 (simple hexagonal as in WC), AABBAABBAA in W_2B_5 and AABBCCAABBCC in Mo_2B_5. The boron-atom network fits into the interstices in varying modes.

Lastly, the high-B borides (MB_4, MB_6, MB_{12}) may be considered as three-dimensional boron frameworks containing the metal atoms interstitially. They become the more prominent for the rare earth and alkaline earth (CaB_6-type) borides, as with the actinides, e.g. UB_{12}.

In CaB_6 the boron atom octahedra—a three-dimensional evolution from the hexagonal networks—are indicated, within their simple cubic metal lattice, in *Figure 6.1e*.

Stable borides are not unique to the transition metals; they are also formed e.g. by the alkaline earths (CaB_6, SrB_6, BaB_6, BeB_2, MgB_2), aluminium (AlB_2, AlB_{10}, AlB_{12}), scandium, yttrium and the rare earths, some isostructural to the present ones. It is arguable, however, whether these phases qualify as 'interstitial compounds'; indeed, some of the transition-metal borides transgress the borderline of this concept, the further with increasing boron content. The three-dimensional boron framework such as in AlB_{12} may, in fact, justly be termed 'reversed interstitials', in that the metal atoms are interposed in the large holes within this framework (cf. *Figure 6.1e*). This also explains the observed increase in *homogeneity width* for the higher boron phases in the M–B systems, due to latitude in M–atom and B–atom occupancy. It is of interest that in one of its forms AlB_{12} is already the same as the rhombohedral modification of pure boron itself, showing that, crystallographically at least, the presence of metal, though accommodated, is not essential. This is corroborated by the indication of a small but finite solubility e.g. of Nb in elemental boron itself, and by the behaviour of other MB_{12} (and higher) borides. Indeed, this is one of the difficulties in preparing high-purity boron.

Hägg's rule, valid for the carbides, nitrides and hydrides and originally formulated[2] to include the borides, begins to fail for the latter and even more for silicon, owing to the development of direct boron (silicon) linkages. However, it still applies for the metal-richer borides. *Table 6.2* gives the

Table 6.2

Atomic radius ratios r_B/r_{metal}

Ti 0·59	V 0·64	Cr 0·69	Mn 0·66	Fe 0·69	Co 0·69	Ni 0·70
Zr 0·54	Nb 0·59	Mo 0·62	Tc 0·65	Ru 0·65	Rh 0·65	Pd 0·64
Hf 0·55	Ta 0·59	W 0·61	Re 0·63	Os 0·65	Ir 0·64	Pt 0·63
Th 0·48	Pa —	U 0·55	Np —	Pu —		

radius ratios r_B/r_M, calculated under the assumption of $r_B=0{\cdot}87$ Å. Hägg's critical ratio (0·59) thus still divides the field broadly as to where the simpler boride structures are possible and where not, with strong qualifications, however.

The lesser importance of the precise metal atom species is also manifested by the wide isomorphous replacements possible for metals greatly differing in atomic size, e.g. in W_2B, Fe_2B, Ni_2B, all of the same C–16 structure.

Surveying the position, the principal new feature of borides, compared with the previous interstitial compounds, is the existence of B–B bonds. As long as the boron atoms remain isolated, as in M_2B, the borides are still of Hägg type and quite analogous to the smaller atom phases (hydrides, carbides, nitrides). However, as soon as the B–B link is established and strengthened, from MB upwards, the radius-ratio criterion ($r_B/r_M < 0{\cdot}59$) becomes less significant, the more important factor being the evolving* independent boron bond (B chains→double chains→hexagonal networks→3-dimensional frameworks). Initially the boron lattice fits into the hexagonal metal atom lattice in varying stacking sequence, but later causes radical displacements. This independent boron bond also helps to explain how stable borides still can exist for the iron and platinum Group metals (Fe, Co, Ni), whilst carbides and nitrides become metastable, unstable or do not exist.

Nevertheless, the stability of borides decreases with increasing atomic number along each Long Period, and a 'copper boride' no longer occurs: the Cu–B system (probably also Ag–B and Au–B) is a simple eutectic one, occupying a position that is for the carbides displaced to the earlier Groups VII and VIII.

Considering the MB_2 phases it may be noted from *Table 6.1* that their lattice spacing decreases in the sequence $ZrB_2 \rightarrow NbB_2 \rightarrow TaB_2 \rightarrow TiB_2 \rightarrow VB_2$, which follows naturally that of metal atom sizes, and such that the c/a ratio decreases only slightly.

The B–B bond formation by covalent chains and nets implies that electrons are freed for transfer from boron to the metal atom lattice which clearly still favours the metallic properties of the borides (high hardness, melting points, conduction, etc.), and in this sense still makes them of 'interstitial compound type', comparable to the carbides.

The bonding energies of the borides, calculated[8] in comparison with those of carbides, nitrides and silicides from heats of formation and other thermodynamic data, show a maximum throughout at Ti in the first Long Period and at (Nb, Ta) in the second and third Long Periods. This decrease in bonding energy in the order of Groups IV→V→VI, etc. persists for all interstitial compounds but becomes less and less pronounced in the sequence nitrides→carbides→borides→silicides.

The decrease in bonding energy within a Period, therefore, progresses inversely to the atomic diameter of the interstitial atom and to the increase in degree of direct non-metal bonding, which links up well the thermodynamic with the crystallographic and chemical observations. We might

* Amongst carbides, the nearest analogy is the Cr_3C_2 structure where C–C chains are incipient but not quite established.

place 'hydrides' at the head of this series, recalling that here the decrease of bond energy within the Period is the most rapid one.

The relation between the metallic base elements and the lower-B interstitial borides may be considered in slightly greater detail. The metal–metal co-ordination in the borides is still quite similar to that in the parent metal, namely 11 or 12. This may be exemplified by Ni_3B, of cementite structure, which has two kinds of environment to each nickel atom, at distances[1] (Å)

Environment I: 2·43; 2 at 2·46; 2 at 2·50; 2·55, 2·57, 2·58, 2·71; 2 at 2·74, i.e. 11 neighbours, *mean distance* 2·57

Environment II: 2 each at 2·50, 2·55, 2·57, 2·58, 2·71 and 2·79, i.e. 12 neighbours, *mean distance* 2·62.

In the elemental Ni lattice (FCC) the *M–M* distance (12 neighbours) is 2·49 Å. Although, on passing from *M* to M_3C, lattice expansion and distortion through B interstition are thus appreciable (as applies also to the transition of austenite $\rightarrow Fe_3C$, cf. Chapter 4), yet the basic relationship is still preserved. Unstable Ni_3C, which is isomorphous, has much shorter *M–M* distances than Ni_3B; continuous carbon $\rightarrow$ boron replacement is here still possible, accompanied by rapid stabilization of nickel–cementite through boron. The iron atom, on the other hand, is already too large for a 'Fe_3B' cementite to occur; instead, the first Fe–B compound becomes Fe_2B, of $CuAl_2$-type structure, which is also related to the FCC of γ-Fe, though differing (see Chapter 4 and above).

In a system such as Ni–Fe–B, therefore, the balanced additions to Fe of Ni plus B can stabilize the cementite structure and become similar to the effect of carbon alone. On the other hand, upon Ni deficiency the transformation to the $CuAl_2$ structure would occur:

$$\begin{bmatrix} \text{(Ni, Fe)}_3\text{B} \\ \text{(cementite type)} \end{bmatrix} \longleftrightarrow \begin{bmatrix} \text{(Fe, Ni)}_2\text{B} \\ \text{(Al}_2\text{Cu type)} \end{bmatrix} \longleftrightarrow \begin{bmatrix} (\gamma\text{-Fe, Ni}) \\ \text{(FCC)} \end{bmatrix}$$

This reaction will then be modified by carbon addition to stabilize the cementite structure and free the nickel to also stabilize austenite. A wide initial C–B replacement in $Fe_3(C,B)$ is possible. This has a bearing on the important and often improving effects of boron additions to steel; the cobalt analogue $Co_3(B,C)$ occupies here a position intermediate to those of Fe and Ni. Here also arises the more general point of a still relatively close relation of the metal-richer borides to the carbides which is no longer continued for the silicides, further exemplified by the occurrence of the wide boron solutions possible in the $M_{23}C_6$- and M_7C_3-type carbides.

Thus in Fe_3C an extensive C $\longleftrightarrow$ B exchange can take place (e.g.[9]). Also the cubic $M_{23}C_6$ carbides have boron-bearing isomorphs[10,11]. A ternary compound occurs in the Fe–C–B system, of this structure and approximate composition $Fe_{23}C_3B_3$ with, however, a considerable homogeneity range; i.e. B $\longleftrightarrow$ C replacement can occur (at 800°C) from about $Fe_{80\cdot5}C_{13}B_{6\cdot5}$ to $Fe_{80\cdot5}C_8B_{11\cdot5}$, with simultaneous lattice-spacing increase. The phase is stable and melts congruently; it equilibrates austenite (at 800°C) and the 'borocementite' $Fe_3(C,B)$ on the other side. The latter compound, however, is metastable[10], decomposing to ternary carbide + graphite upon annealing.

Although this is a 'mixed' interstitial compound and will be referred to in Chapter 10, its special link with the carbide types established has to be emphasized. As we know, chromium and manganese stabilize $M_{23}C_6$ from cementite amongst the carbides, so that Cr or Mn addition no doubt would tend to produce a complete solution $Fe_{23}C_3B_3 \rightarrow (Cr,Fe)_{23}C_6$, i.e. lower the boron required for this structure, freeing the iron boride itself. This is the counterpart (on the other side of Fe) to the case of Ni with B stabilizing cementite*.

The close relation to the carbides also persists through the platinum group borides, such as Ru_7B_3 and Rh_7B_3 (of Cr_7C_3-type), Pd_5B_2 (of Mn_5C_2-type) and Pd_3B (of Fe_3C-type, as Ni_3B); the diagonal shift within the Periodic Table may be noted.

Amongst the MB borides (see *Table 6.1*), the FeB-, CrB- and MoB-type boron-chain structures are, as mentioned, all closely related. The *two* Ni_4B_3 types can be regarded as structural variants[12] to NiB with, however, a partial tendency for the boron bonding to revert to that in M_2B. In MB, each metal atom has an environment of 10 other M and of 7 B atoms. The structure evolves for the platinum group metals to that of PtB and '$RhB_{\sim 1.1}$', where the metal atoms become nearly hexagonal close-packed, with B atoms occupying the interstices. Alternatively, these monoborides (and monosilicides) have been interpreted as of the 'nickel arsenide' class (Chapter 2).

Passing to the MB_2 group (prototype AlB_2), where the double chains and nets of boron atoms evolve, the Ta_3B_4 and Mn_3B_4 types of atomic environment are sufficiently similar to those in MB_2 to be regarded as its modulations[3]. Diborides cease to occur for Mo, W and Re but are replaced by W_2B_5, Mo_2B_5 and ReB_3—still similar, but with a somewhat unusual boron environment, which may be regarded as the beginning of a break-through to the three-dimensional boron bonding of the high-boron compounds established in the ThB_4, CaB_6 and $AlB_{12}(UB_{12})$ types, each with its now appreciable range of homogeneity. The octahedral framework of boron atoms is sufficiently rigid and strongly bonded to contain the metal atoms in their voids without large variations in MB distances. AlB_{12}, as mentioned, already adopts the structure of an elemental boron modification.

From the crystallographic and chemical viewpoint, the borides thus occupy a unique position in which many of their features for lower-boron contents still relate to the carbides, whilst their relation to the silicides becomes more firmly established in the medium boron range. In this wider context we may also mention the aluminides, recalling the Periodic Table square

```
  |    |
—B—C—
  |    |
—Al—Si—
  |    |
```

The aluminides, although they no longer qualify as interstitial compounds, the borides and silicides being on the borderline already, form an interesting

* The idea of $Cr_{23}C_6$-type borides has recently been extended further to new *two-metal* borides (see below).

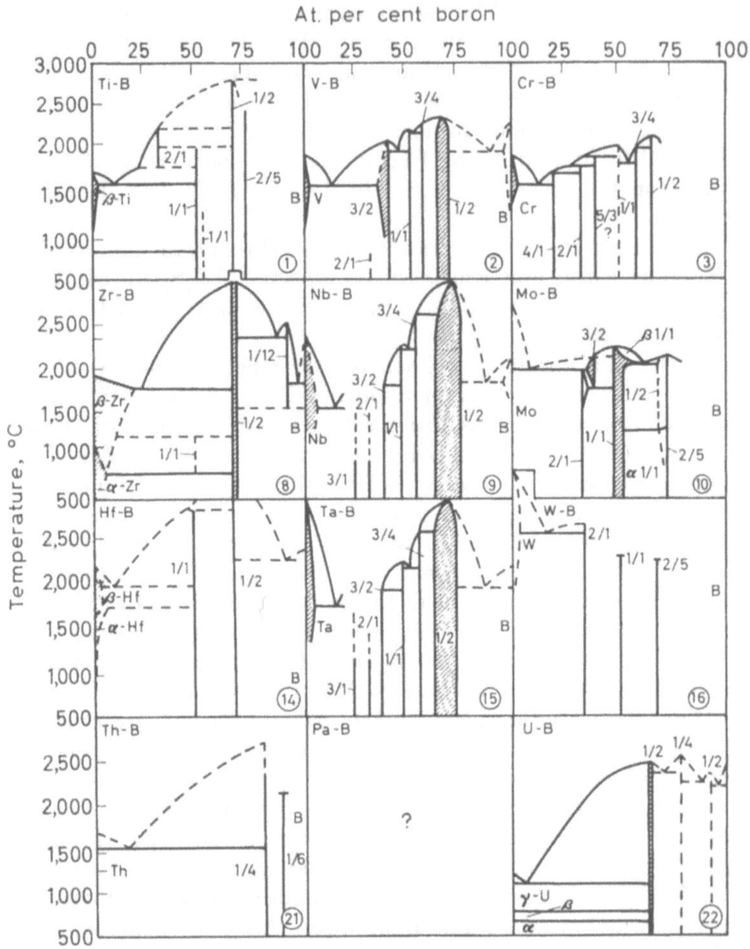

Figure 6.2. The pattern of binary metal–boron systems

series of isomorphous link structures from interstitial to true intermetallic compounds.

The borides of the *platinum group metals* are particularly interesting, because here we have for the first time stable interstitial compounds of the noble metals, although their energies of formation are still lower than those for Group IV, V, VI borides. A characteristic feature is the multiplicity of these Pt-metals borides and their variability in structure, as shown in *Table 6.1.* Many additional and unidentified ones exist[13,14], with the consequent complexity of the phase diagrams concerned by no means all resolved. The occurrence here of relatively low eutectic temperatures[15] is also a relevant feature in this context. All this points to a still relative instability of bondings in the (Ru, Rh, Pd, Os, Ir, Pt)–B structures; it leads to a tendency of these phases if not to decompose, at least to contain boron atoms which are 'uncomfortable' in their metal surrounds and seeking alternative environment. Ready transformations upon relatively small changes in temperature,

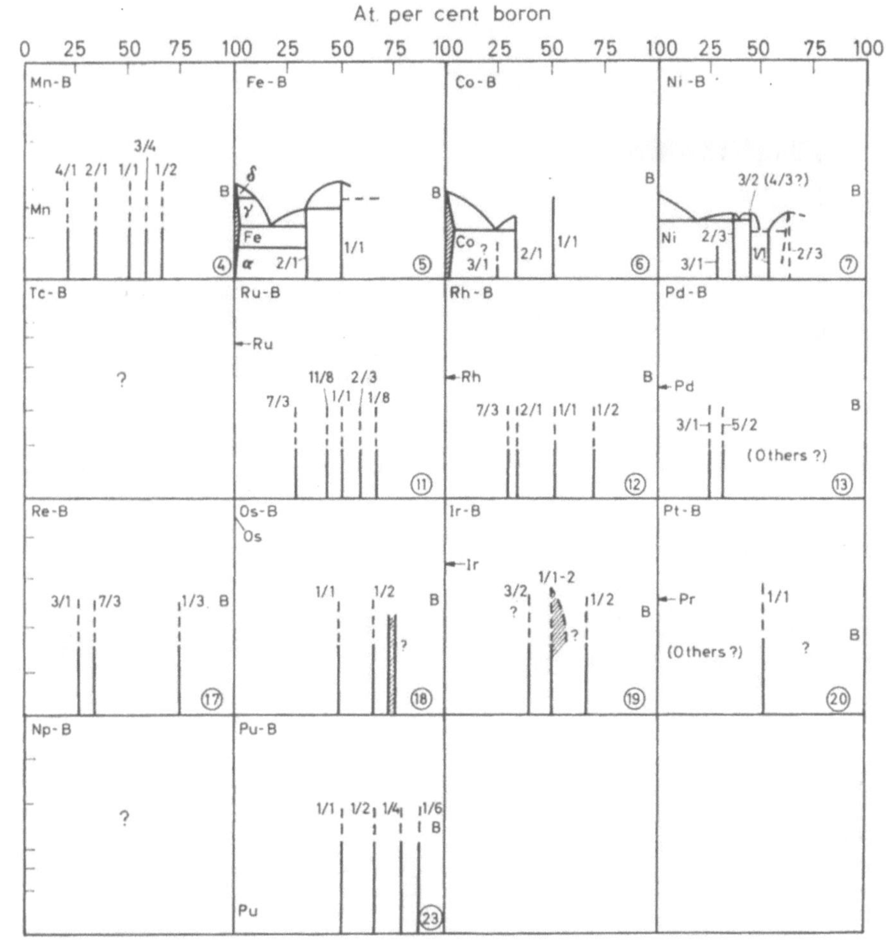

Figure 6.2b

composition and probably also pressure are thus a feature. The heats of formation of these borides have not yet been determined, but they have been estimated[14] as being of the order of 10 kg/g atom.

Figure 6.2 shows as complete a survey as could be prepared of the transition metal boron phase diagrams, arranged according to the Periodic Table to illustrate the principal features (cf. *Table 6.2*) and the trends with Period and Column. For example may be noted (1) the extremely high melting point for the diborides, concentrating in ZrB_2, HfB_2, NbB_2 and TaB_2 all exceeding 3,000°C; (2) the phases limited to high-temperature ranges only; (3) the wide variations in eutectic temperatures and homogeneity ranges, and (4) not least important, the gaps in our present knowledge. *Figure 6.2* will also serve as a data chart for the approximate eutectic temperatures and compositions involved.

Figure 6.3 shows the areas of structurally isomorphous (or near-isomorphous) phases, superimposed upon the 'elemental grid' in the Periodic

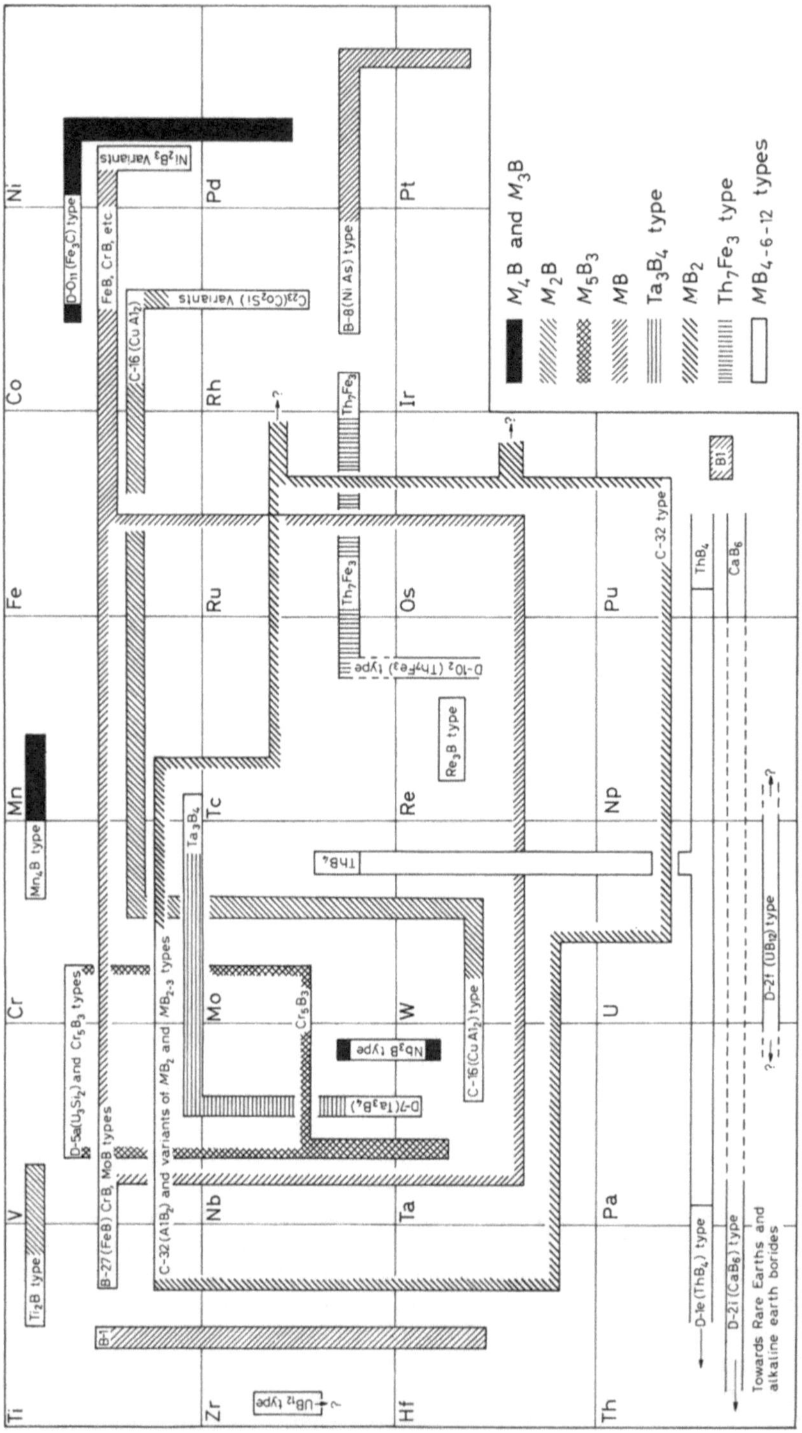

Figure 6.3. Fields of boride structures within the Periodic Table

Table. This may be used as a guide to possible or likely solid-solution formation amongst the borides, and together with the corresponding diagram for the silicides, between borides and silicides.

6.2 INDIVIDUAL (SINGLE-METAL) BORIDE SYSTEMS

After this, more general, introduction, additional points of individual borides will be considered (to be read in conjunction with *Figure 6.2* and *Table 6.1*).

Ti–B—The identity of a separate phase '$TiB_{1.1}$' is still controversial. There has also been an argument regarding TiB. This was first thought[16] to have the ZnS-type structure; however, the NaCl type has since been established[17] beyond doubt, TiB being isomorphous with titanium carbide. The possibility of a complete interstitial solubility [TiC–TiN–TiO–TiB] thus arises, making titanium unique in this respect. Ti_2B is a high-temperature phase only, and the twin reactions $Ti_2B \rightarrow TiB + \text{liq.}$ and $TiB \rightarrow Ti_2B + TiB_2$ should be noted. This implies that, over a certain limited high-temperature range, we have a rapid change-over in boron bonding from isolated to chain to network type.

Zr–B and Hf–B— ZrB is a high-temperature phase only (although this requires verification), the eutectic Zr/ZrB_2 being dominant[18]. In the hafnium–boron system[19], included in *Figure 6.2*, a higher boride, HfB_{12}, analogous to ZrB_{12}, seems at least probable, but not yet confirmed.

V–B—The existence of a compound 'V_2B' is controversial; it was claimed[20] to be isomorphous with Nb_2B and Ta_2B, but later its existence was doubted. Small amounts of impurity in preparation (such as O, N, C, Si) may certainly make a critical difference to the stability or otherwise of given structure types amongst the borides. This impurity effect can be particularly sensitive for the metal-richer borides, as also occurs in mixed systems (Chapter 10).

Nb–B and Ta–B—Whilst Ta_2B ($CuAl_2$-type) is firmly established, the existence of Nb_2B is still doubtful, the Nowotny school denying, others affirming it. The fact of such a difference between the behaviour of Nb and Ta, usually so similar, vis-à-vis boron is of interest in itself. In more recent studies on the Ta–B system[21], establishing the $Ta/Ta_2BTa_2B_3/TaB$ equilibria, it was thought that the composition of 'Ta_2B' should be altered to $Ta_{2.4}B$, that of '$Ta_{1.5}B$' (Ta_3B_2) to $Ta_{1.6}B$.

The possibility of some non-stoichiometry must be borne in mind for most borides, and more precise studies will be necessary to refine the compositions and often even to verify the existence of some of the phases in question.

The Nb_3B and Ta_3B phases[22] require substantiation.

Cr–B, Mo–B and W–B—The group VI borides are relatively well established, except for the lack of equilibrium data for tungsten (see *Figure 6.2*). The atomic radii of Mo and W are such as to take us, for boron, slightly above the borderline of Hägg's rule, and the lower and medium-range borides are still of relatively simple structure.

The existence[23] of two new tetraborides, MoB_4 and WB_4, may be recorded. They are of tetragonal UB_4-type structure, though of larger axial ratio. These phases occurred, however, always in conjunction with M_2B_5 and dissociated above 1,600°C. Thus these boron network structures, hitherto regarded as characteristic only of the actinides (UB_4, ThB_4), cancontinue 'up the groups', i.e. to the lighter metals. (The occurrence of, so far hypothetical, tetraborides of rhenium and the platinum family seems very possible.)

Th–B, U–B and Pu–B—The actinon borides are identified in *Table 6.1* and in *Figure 6.2.* The absence of lower borides (probable direct eutectics metal$\leftarrow\rightarrow MB_2$), a noteworthy feature, means that the least amount of boron added to U or Th immediately builds a structure with highly developed B–B linkage, U and Th being unable to accommodate atoms in isolation or in one- or two-dimensional linkage.

Some interesting date are available on the *thermal expansion* of UB_2. (The thermal expansion coefficients of borides in general are where known included in *Table 6.1*). The UB_2 expansion has been determined[24] along with that of β-USi_2 which is isostructural, both being of AlB_2 type. The coefficients are given in *Table 6.3* together with the bond strengths.

Table 6.3

Expansion coefficients, β (per °C) *and bond lengths* (Å) *in* UB_2 *and* β-USi_2 (From Beckman and Kiessling[24]; for lattice spacings, cf. *Tables 6.1* and *7.1*)

Compound	*Direction in lattice, along*	*Mean coefficient** × 10^6	*Bond lengths* U–U *distance within* — *between sheets*		B–B(Si–Si) *distance within sheets*†	U–B(Si) *distance between sheets*
UB_2	a-axis	9	3·14	3·99	1·81	2·69
	c-axis	8				
	U–B bond direction	6				
USi_2	a-axis	57				
	c-axis	−26	3·85	4·06	2·22	8·01
	U–Si bond direction	16				

* Over range 20°–205°C
† No bond between B sheets

As discussed earlier, the UB_2 (AlB_2-type) lattice may be considered as stacking of hexagonal sheets of U atoms between the hexagonal B (or Si) atom network. In UB_2, the U–U distances between the sheets (3·14 Å) are still within the range of metallic bonding; in U metal itself (α, β, γ-U), the distances range between 2·75 and 3·30 Å. In UB_2, the B–B distances are 5 per cent larger than twice the boron radius, the U–B distance 11 per cent larger than the sum of metal+boron radius, and the thermal expansion of UB_2 is well related to this bonding. The fact that the coefficient is so similar in the different bond directions (*Table 6.3*) would indicate that the lattice coheres by bonds of the same nature. *Within* the U sheets we have direct metal bonding, but *between* these sheets the U–U distance ($\sim$4 Å) is too large for this. The interposed B atoms in fact *help* cementing the U atoms together[24] by forming U–B–U bonds (in addition to the B–B bonds within the boron nets) and transferring electrons from B atoms to the unfilled

6*d* levels of the U atoms. In other words, the near-isotropy in expansion reflects a special type of metal–metal bond in which boron acts as an intermediary. In the silicide (USi_2), on the other hand, the situation becomes significantly different, since for the larger Si atom this type of bonding is no longer possible. This then results in the extraordinarily large and anisotropic expansions shown (see also Chapter 7).

Mn–B, Fe–B, Co–B, Ni–B—Several points have to be added to the earlier comments and to the data of *Table 6.1* and *Figure 6.2*. Thermodynamic data on the manganese borides are still quite insufficient. The two iron borides, Fe_2B and FeB, and the isomorphies and appearance of the various boride structures on passing to the adjoining elements (i.e. $\rightarrow$ Cr, Mn and $\rightarrow$ Co, Ni) are of obvious importance in considering effects of boron in steel, such as the above of balancing the formation of the cementite and the $CuAl_2$-type lattices.

The Fe–B system has been extensively studied, and stable Fe_2B is always the phase equilibrating with the primary α, γ and δ solid solutions. However, there are considerable disagreements otherwise, and *Figure 6.4* shows, in more detail than *Figure 6.1*, three versions of the system[9,25,26]. A renewed investigation, particularly of solubilities, may be justified, and varying purity of the iron and boron used may well underlie the differences shown.

Although Fe_2B is the equilibrium boride, the actual borides in low-boron steels have been demonstrated, after certain conditions of heat treatment, as *both* Fe_2B and FeB, segregated at the grain boundaries of the ferritic and (in alloy steels) austenitic matrix. This liability to metastable coexistence of Fe_2B and FeB (e.g.[27]) can be important in affecting properties. It is a direct reflection of the transitional state of bonding in iron, where either of these structures can be stabilized or destabilized by third-element additions, trace carbon, for instance, playing a critical rôle.

The phase Co_3B and a complete solid solution $Co_3B \longleftrightarrow Ni_3B$ of cementite structure has been affirmed[28]. The two closely related Ni_4B_3 allotropes, referred to[12] as '*o*' and 'm' forms (viz. orthorhombic and monoclinic), represent transition stages between that of boron-atom isolations in Ni_2B and of chain bonds of NiB. In *o*-Ni_4B_3, $\frac{1}{3}$ of the B atoms remain isolated and $\frac{2}{3}$ form chain bonds, whilst in the m-form *all* B atoms have already moved into contacting positions; but the Ni positions have not yet appreciably changed, as they will do in NiB. The overall series $Ni_3B \rightarrow Ni_2B \rightarrow$ *o*-$Ni_4B_3 \rightarrow$ m-$Ni_3B_4 \rightarrow$ NiB demonstrates perhaps most strikingly the progression in boron bonds. Furthermore, since the NiB structure is, amongst the carbides, closely related to Cr_3C_2, that of Ni_3B to cementite, it would appear probable that the total series Ni $\rightarrow Ni_3B \rightarrow \ldots \rightarrow$ NiB amongst the borides represents a transition, to some extent analogous to the series austenite $\rightarrow M_{23}C_6 \rightarrow M_3C \rightarrow M_7C_3 \rightarrow M_3C_2$ (where $M =$ Fe,Mn,Cr) amongst the carbides; i.e. the effect becomes displaced to the lower atomic numbers by three elements along the 1st Long Period.

This is also reflected in the fact that, throughout the Group VII and VIII borides, extensive lattice substitutions of B by C atoms are possible. Examples are the occurrence of compound[10] $Fe_{23}(C,B)_6$, referred to already, isomorphous with the $Cr_{23}C_6$ carbides, of Pd_3B and Ni_3B, with Fe_3C structure, and of Pd_5B_2, with[29] Mn_5C_2, which is particularly striking; hence it is worth

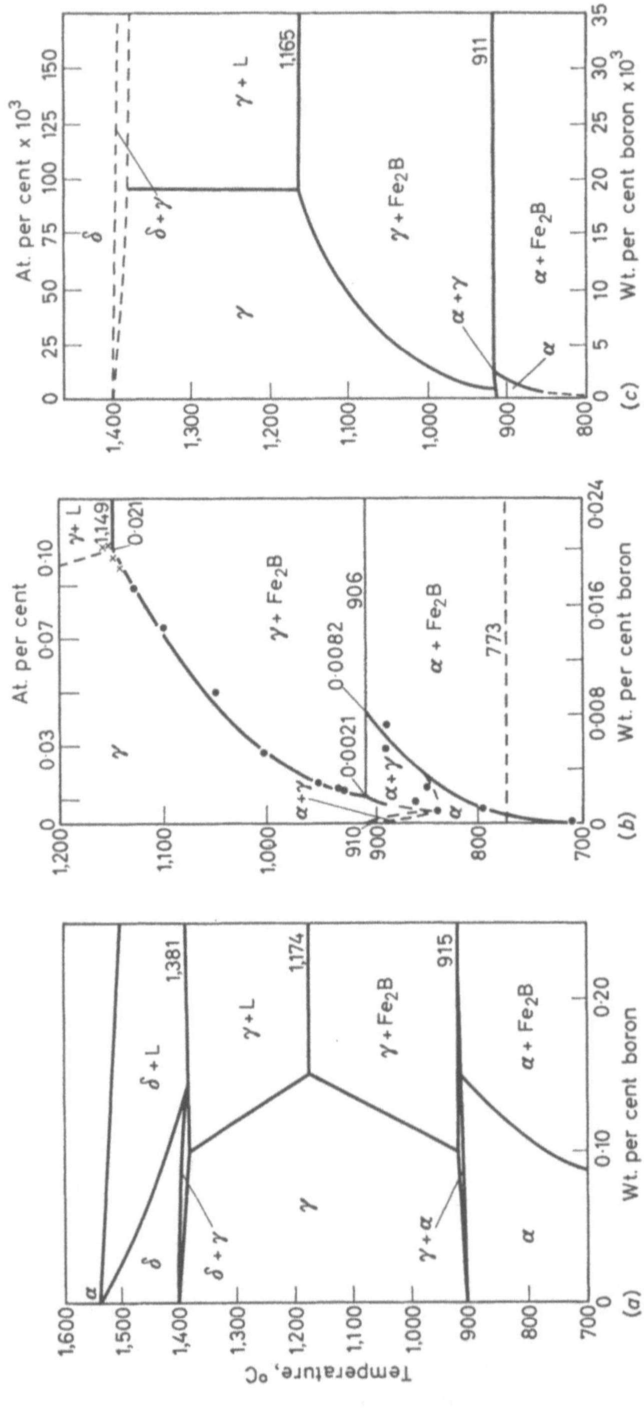

Figure 6.4. *The* Fe–B *equilibrium diagram* (Fe *end*); (From Nicholson[9b] [*a, c*] and McBride, Spretmak and Speiser[26] [*b*])

By courtesy of the Editors, *Journal of Metals, New York* and *Transactions of the American Society of Metals* while to examine the question of cementite isomorphy a little further. *Table 6.4* summarizes compounds now known to have this structure[13], including carbides, borides, a Pd silicide and phosphide as well as $NiAl_3$ (which, however, is a special case that needs reconsideration).

Table 6.4

Phases with cementite structure (After Aronssen and Rundqvist[14])

Phase	Lattice spacing, Å *a*	*b*	*c*
Mn_3C	5·080	6·772	4·530
$Fe_3C(\rightarrow(Fe,Cr_3C)$	5·0890	6·7428	4·5230
Co_3B	5·225	6·629	4·408
Ni_3B	5·211	6·619	4·389
Pd_3P	5·645	7·558	5·071 Pd end (hom. range)
	5·980	7·440	5·164 P end (hom. range)
Pd_3Si	5·735	7·555	5·260
$Fe_3B_{0\cdot8}C_{0\cdot2}$	5·4052	6·6685	4·4500
$Fe_3(B,Si)$	5·299	6·674	4·454 Fe end (hom. range)
	5·363	6·660	4·458 B end (hom. range)
$NiAl_3$	6·6115	7·3664	4·8118
Re_3B,Ru_7B_3 structure related (spacings see *Table 6.1*)			

The fact of interest is that $Fe_3(C,B)$ can remain stable up to 80 per cent boron substitution which then becomes 100 per cent in Co_3B and Ni_3B. Also, some silicon can act as a promotor to stabilize a 'Fe_3B' boride, which seems consistent if we consider silicon as a strong 'lattice widener' (or 'prop') assisting boron to enter into a cementite lattice.

Also structurally related to the cementite group are the Re_3B and Ru_7B_3 types. The latter links on to the Cr_7C_3 group amongst the carbides.

This leads up appropriately to the noble-metal borides, Re–B and [Pt group]–B (cf. earlier comments and *Figures 6.1, 6.2* and *Table 6.1*). The hexagonal layer structure of ReB_2 (now preferred to 'ReB_3') has been determined in detail[30].

The Ru_7B_3 structure applies also to Re_7B_3 and Rh_7B_3. The next higher rhenium boride, ReB_3, is more closely related to the W_2B_5 and Mo_2B_5 types in the preceding Group VI, the difference being essentially one of stacking sequence of the hexagonal metal and boron networks[31]. The Ru–B alloys[32] obey the same principle but tend already towards iron group features.

It may be surmised (though not being known) that *technetium* analogues TcB_3 and Tc_7B_3 exist, with, however, a tendency to the Mn–B structures. The 'anti-NiAs' structure of PtB becomes modified in Rh–B, with a small change in stoichiometry such that the B environment resembles already that in the AlB_2 type; the existence of a similar palladium boride, PdB_{1-2}, would seem probable, though it has not been reported yet.

Again, the change from NiB to PtB is a simple one of stacking change from the hexagonal NiAs to the CrB type. (For discussions of these phases, also in relation to silicides, see Bibliography.)

Figure 6.2 tends to show that much further work is necessary on the identity and thermal equilibrium data of the Pt-metal/boron systems.

Rare Earth borides—For completeness, it is noted that the Rare Earths and Group III metals also form stable MB_4- and MB_6-type borides. These link on to the respective Alkaline Earth and Actinide boride isomorphs. The hexaborides (structure: *Figure 6.2*) are already very refractory. Thus typical melting points[33] are for LaB_6, 2,210°C, CeB_6, 2,190°C, compared with CaB_6 and SrB_6, 2,235°C, BaB_6, 2,270°C and ThB_6, 2,195°C. The constancy of these values, which persist in all the other hexaborides, is remarkable; it sharply contrasts with the considerably larger variations in melting point for other, metal-richer, isomorphous boride families, e.g. the diborides (cf. *Table 6.1*, *Figure 6.2*). This is clearly associated with the dominating B atom framework structure, accommodating the *metal* atoms interstitially and making the metal type almost unimportant. It also is reflected by the relative lattice dimensions—LaB_6, 4·156 Å, CeB_6, 4·141, CaB_6, 4·145, SrB_6, 4·19, BaB_6, 4·268 Å—the differences being much smaller than e.g. for the diboride and carbide series of the same metals.

The Rare Earth hexaborides (cubic, CsCl-type structure) are remarkable for their quite metallic behaviour, as good electrical and thermal conductors (e.g. LaB_6 27 $\mu\Omega$ cm, better than metallic La, 59), hard, of fair strength and, although brittle, machinable with care[34,35]. The tetragonal tetraborides, on the other hand, are of remarkably less metallic nature than the hexaborides. This is due to essentially *ionic* bonding, whilst in MB_6 *metallic* bonding is recovered, as also reflected by the lattice dimensions. No metal-richer borides than MB_4 appear to exist for the Rare Earths, though another type, 'MB_x', has been reported[35] which, however, is strongly suspected to be due to carbon impurity (possible formula MB_2C). The difficulty of producing *pure* borides uncontaminated by C, O or other interstitials, increases with decreasing B content; the Rare Earth case only exemplifies the more general position in borides. Some earlier reported 'diborides' later proved, in fact, to be oxyborides. Apart from the practical difficulties of experimental exclusion, this can also well be explained on the structural ground of boron linkages as discussed. The latter become more sensitive to third-atom interference at the earlier stages preceding the establishment of three-dimensional boron frameworks which are almost self-contained.

6.3 TWO-METAL BORIDE SYSTEMS

From the above survey it follows naturally that, where two binary borides are isomorphous and the size and electronic factors not too divergent, complete series of solid solutions are formed. Hume-Rothery's 15 per cent rule can be applied as a first guide[36]. In this respect, *Figures 6.2* and *6.3* are largely self-explanatory, and only a brief indication of these ternary systems remains to be added. Practical interest attaches to the metal–boride equilibria largely because of the problem of finding a suitable metallic bond for the established borides (e.g. in cemented borides and to allow impregnation procedures).

In *Table 6.5* available data have been assembled, again in the order of the Periodic Table.

Table 6.5

Inter-boride systems: solid solutions and double borides of independent structure (cf. p. 282)

(*CSS* = continuous solid solution series; *Cpd* = compound; *H.R.* = homogeneity range)

System	*Solid solution series or compound formed*	*Structure*	*References; remarks*
Ti–Zr–B	TiB_2–ZrB_2	*CSS:* C–32 (AlB_2 type) throughout	[35,37] (for M_{VA}–M_{VIA}–B ternary systems, cf. [39])
Ti–Hf–B	TiB_2–HfB_2		
Zr–Hf–B	ZrB_2–HfB_2		
Ti–V–B	TiB_2–VB_2		
Ti–Nb–B	TiB_2–NbB_2		
Ti–Ta–B	TiB_2–TaB_2		
Ti–Cr–B	TiB_2–CrB_2		
Ti–Mo–B	TiB_2–MoB_2		
Zr–Nb–B	ZrB_2–NbB_2		
Zr–Ta–B	ZrB_2–TaB_2		
Zr–Mo–B	ZrB_2–MoB_2		
V–Cr–B	VB_2–CrB_2		
Nb–Cr–B	NbB_2–CrB_2		
Ta–Cr–B	TaB_2–CrB_2		
Ta–Fe–B	*Cpd:* $FeTaB_2$ Fe_5TaB_2	structure not determined	[39] cf. *Table 6.6*
Ta–Co–B	$CoTaB_2$		[39,40]
	Co_5TaB_2	cubic, $Cr_{23}C_6$ type	
Ta–Ni–B	$NiTaB_2$	isomorph to $CoTaB_2$	[39]
	Ni_5TaB_2	isomorph to Co_5TaB_2	
Cr–Mo–B	CrB_2–MoB_2		[35]
Cr–Mn–B	Cr_2B–Mn_2B	*CSS:* C–16 type	[41]; no doubt also Cr_4B–Mn_4B, etc. solutions
	CrB–MnB	*CSS:* B–27 type	
Cr–Fe–B	Cr_2B (in '$CuAl_2$' form)–Fe_2B, CrB–FeB partial solid soln. probable		
Cr–Co–B	Cr_2CoB_4?	Cr_3B_4 type?	[28] (phase diagram up to 50 per cent B)
Cr–Ni–B	Cr_2NiB_4		
Mo–Fe–B	$FeMo_2B_4$	orthorhombic, Cr_3B_4 type ←	[42]
Mo–Co–B	$CoMo_2B_2$	—	none of the *binaries* has the Cr_3B_4-type structure, but (Mo+Fe), (Mo+Co), (Mo+Ni) can substitute Cr
	$CoMo_2B_4$	orthorhombic, Cr_3B_4 type ←	
Mo–Ni–B	$NiMo_2B_2$	—	[43]
	$NiMo_2B_4$	orthorhombic, Cr_3B_4 type ←	
Mn–Fe–B	Mn_2B–Fe_2B	*CSS:* C–16 type	[43] cf. *Figure 6.5*
	MnB–FeB	*CSS:* B–27 type	
Mn–Co–B	Mn_2B–Co_2B	*CSS:* C–16 type	
	MnB–CoB	*CSS:* B–27 type	
Mn–Ni–B	Mn_2B–Ni_2B	*CSS:* C–16 type	
	MnB–NiB	*CSS:* B–27 type	
Fe–Co–B	Co_2B–Fe_2B	*CSS:* C–16 type	
	CoB–FeB	*CSS:* B–27 type	
Th–Na–B	$ThB_6 \rightarrow (NaB_6)$	cubic D–2 type	[44]; extensive solubility to at least 7·7 at. per cent Na, density[45] 5·85
Mo–Al–B	*Cpd:* $Mo_7Al_6B_7$	orthorhombic; $a = 6{\cdot}34$, $b = 7{\cdot}03$, $c = 5{\cdot}76$ Å; $M = 1$	
Co–Al–B	*Cpd:* $Co_{20}Al_3B_{6-12}$	cubic, $Cr_{23}C_6$ type; $a = 10{\cdot}50$ Å	[46b]

System	*Solid solution series or compound formed*	*Structure*	*References; remarks*
Table 6.5–cont.			
Ni–Al–B	*Cpd:* $Ni_{20}Al_3B_{6-12}$	cubic, $Cr_{23}C_6$ type; $a = 10{\cdot}55$ Å; wide *H.R.*	46c
Co–Mg–B	(no ternary phase found but probable, similar to Ni–Mg–B)		
Ni–Mg–B	Ni_7MgB_2; modified[46d] to $Ni_{19\cdot5}Mg_{3\cdot5}B_6$ (the complex Ni corner of the Ni–Mg–B system also determined)	$Cr_{23}C_6$ type; $a = 10{\cdot}49/10{\cdot}55$ Å; wide *H.R.*	11, 46e (Co–Mg–B to Co–Zn–B)
Co–Zn–B	no ternary compound found, but high solubility of Zn in Co_2B		
Ni–Zn–B	$Ni_{20}Zn_3B$ to $Ni_{19}Zn_4B_8$ and 3 other compounds, of unknown structure	$Cr_{23}C_6$ type; $a = 10{\cdot}50/10{\cdot}55$ Å	46f
Co–Ga–B	Co_7GaB_2	$a = 10{\cdot}52$ Å	11
Ni–Ga–B	Ni_7GaB_2	$Cr_{23}C_6$ type, $a = 10{\cdot}53$ Å	
Co–Ge–B	no ternary compound found, but possible		
Ni–Ge–B	Ni_7GeB_2	$a = 10{\cdot}53$ Å	11
Co–Sn–B	$Co_{21}Sn_2B_6$	$Cr_{23}C_6$ type $a = 10{\cdot}60$ Å	
Ni–Sn–B	$Ni_{21}Sn_2B_{5-9}$	$Cr_{23}C_6$ type $a = 10{\cdot}58_4/10{\cdot}59_8$	46a
Co–In–B	no ternary compound found but possible		
Ni–In–B	Ni_7InB_2	$Cr_{23}C_6$ type; $a = 10{\cdot}60$ Å	46g

*No information appears available on the following systems**

Ti–W–B	Zr–Mn–B	Ta–Mo–B
Ti–Mn–B	Zr–Fe–B, Zr–Co–B, Zr–Ni–B (except on Fe, Co, Ni additions to ZrB_2 (structure and properties))	Ta–W–B
Ti–Fe–B		Ta–Mn–B
Ti–Co–B		U-bearing ternaries
Ti–Ni–B	Hf-bearing ternaries[47]	Th-bearing ternaries
Zr–V–B	V–Nb–B	Pt group metal-bearing ternaries, etc.
Zr–Cr–B	V–Ta–B	
Zr–W–B (except kinetics of Zr–W–B reactions: *possible* ternary compound[49])	V–Group [VI VIII Metals]–B	
	Nb–[Group VI VIII Metals]–B though Nb analogues to the above Ta double borides are probable	

* For a summary on Stadelmaier's investigations into two transition-metal–boron systems, cf. [48] and Bibliography.

The time factor for attaining equilibrium is important. The system TiB_2–ZrB_2, for instance, required for the attainment of a complete solid solution[36] more than 21 h at 1,500°C but only ½ h at 2,650°C, so that in some investigations there could easily have been imperfect equilibrium.

Within complete diboride solid solutions MB_2–$M'B_2$, Vegard's law applies to a fair approximation. However, deviations do occur, and it was thought[50] that spacing changes along the *c* dimension are governed by the *larger* metal atom, those along *a* by the smaller one.

In work[41] on the distribution of equilibria in the systems Mn–Cr–B, Mn–Fe–B, Mn–Co–B, Mn–Ni–B and Co–Fe–B for the range $(Mn–M)_2B$—

(Mn–*M*)B (*M*=Cr, Fe, Co, Ni) [i.e. that of the $CuAl_2$$\leftarrow\rightarrow$FeB types], it was found that each homologous series forms complete solutions, and the tie lines and partitions of metal atoms between the two-phase diboride+

Table 6.6

Additional data on some double borides[39]

Phase	*Reaction temperature,* °C	*Eutectic with* Fe, Co, Ni (°C)	*Micro-hardness*	*Density*
$FeTaB_2$	(decompose to TaB_2+metal	—	1,568–1,714	—
$CoTaB_2$	vapour above 1,400°C)	—	1,486–1,632	—
$NiTaB_2$	150	—	1,598–1,725 (hardness range due to anisotropy)	10·1
Fe_5TaB_2	1,175	1,130	—	10·3
Co_5TaB_2	1,170	1,120	—	10·1
Ni_5TaB_2	1,180	1,090	—	10·5

monoboride series could be determined. This is shown in *Figure 6.5* for Mn–Co–B, the tie lines here always showing Mn enrichment in *M*B compared with MB_2 and, more generally, the *lower* group metal (with the higher boron affinity) tending to segregate into the monoborides. Similar conditions are

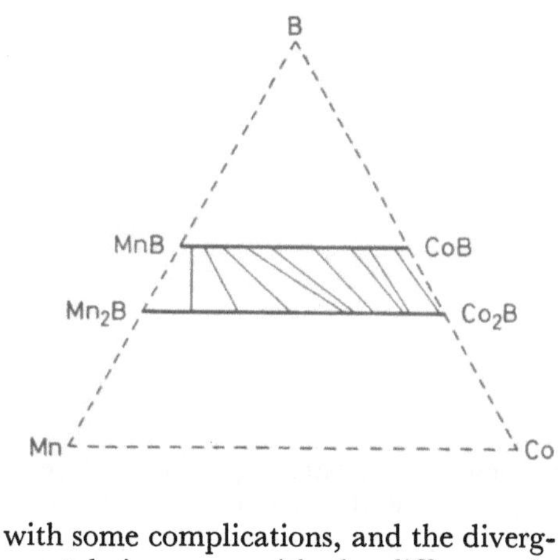

Figure 6.5. Tie lines in the Co–Mn–B *System* (After Hägg and Kiessling[41])

By courtesy of the Editor, *Journal of the Institute of Metals*

found in the other systems, though with some complications, and the divergence in partition between the two metals increases with the difference in atomic number.

Powder-metallurgical reaction[46a] of ZrB_2 with Fe,Co and Ni produced compacts of very high strength and hardness, some containing FeB, Co_3B and Ni_3B besides the initial ZrB_2. The sections ZrB_2–Fe, ZrB_2–Co and ZrB_2–Ni are not pseudo-binary, and clearly some boron from ZrB_2, despite its high stability, is partitioned off to form the lower borides of the Group VIII metals. This again sharply contrasts with the carbides and reflects the marked difference in bond type and stability of the later carbides and borides.

The formation of *double borides* of independent structure, analogous to the double carbides (e.g. η-carbide), is possible for combinations of Group VI + Iron Group metals (see *Table 6.5*). Thus there are compounds Cr_2NiB_4, Mo_2NiB_4, Mo_2CoB_4, etc., of great hardness and refractoriness. It is probable, though not yet known, that tungsten analogues exist, as well as double borides of [Group IV, V]–[Group VII, VIII] combinations. The compounds '(Fe,Co,Ni)TaB_2' and '$(Fe,Co,Ni)_5TaB_2$' in fact form a eutectic[39] with metallic Fe, Co and Ni, resulting in a quasi-binary system of the type shown for Ni in *Figure 6.6.*

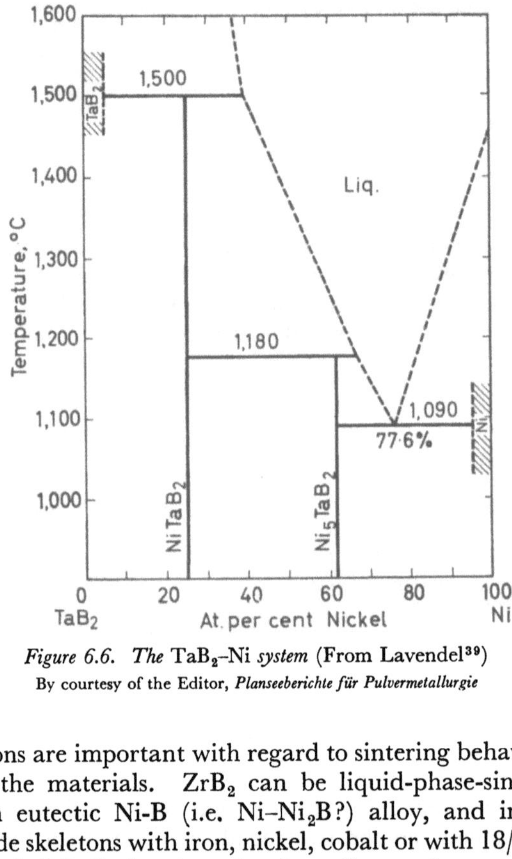

Figure 6.6. The TaB_2–Ni *system* (From Lavendel[39])
By courtesy of the Editor, *Planseeberichte für Pulvermetallurgie*

These relations are important with regard to sintering behaviour and good wettability of the materials. ZrB_2 can be liquid-phase-sintered[51] by the addition of an eutectic Ni-B (i.e. Ni–Ni_2B?) alloy, and impregnation is feasible of boride skeletons with iron, nickel, cobalt or with 18/8-type stainless steel, 'Nimonics' (Ni–Cr base) and other alloys. Low-melting eutectics between TiB_2 and Fe, Co, Ni have also been reported[52], and further work[51] included, as impregnants for TiB_2 and ZrB_2, many types of lower-melting alloys especially centred around Group VIII and the eutectics between these metals and borides and silicides. Nevertheless, the eventual equilibrium tends even here often towards stable borides (see above), and the fundamental problem is to inhibit this final equilibrium by any means, such as by only keeping the effective diffusion temperature low. The theme of alloying TiB_2, ZrB_2 and TaB_2 with Fe, Co, Ni, etc. has been pursued in many other studies.

The property variations occurring for series of double-boride solid solutions are of similar interest. A diboride series TiB_2–CrB_2 (both of C–32 structure) evolves compositions of oxidation resistance at higher temperatures superior to each of the end members[53].

Chromium borides as a base have received much attention in earlier work[54]; an alloy of CrB with *nickel* showed good strength and high-temperature oxidation resistance.

The ternary system Th–U–B[55] was confirmed to show a complete series of solid solutions $ThB_4 \rightarrow UB_4$ with *a* and *c* spacings decreasing in this direction and a marked positive deviation from Vegard's law (*Figure 6.7*).

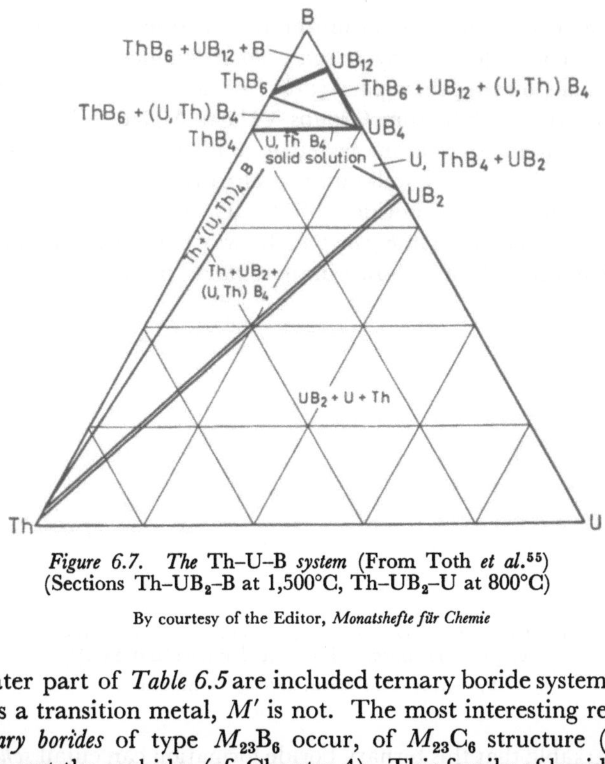

Figure 6.7. The Th–U–B *system* (From Toth *et al.*[55])
(Sections Th–UB_2–B at 1,500°C, Th–UB_2–U at 800°C)

By courtesy of the Editor, *Monatshefte für Chemie*

In the later part of *Table 6.5* are included ternary boride systems *M*–*M'*–B where *M* is a transition metal, *M'* is not. The most interesting result is that stable *ternary borides* of type $M_{23}B_6$ occur, of $M_{23}C_6$ structure ($D8_4$), well known amongst the carbides (cf. Chapter 4). This family of borides includes *M* = Co and Ni, *M'* = the B sub-group metals (Al, Mg, Zn, Ga, Ge, Sn, In). However, the compound stabilities vary considerably, as between Co and Ni (see pairs in *Table 6.5*), and wide homogeneity ranges for Ni give way to small ones (or non-existence) for Co; iron analogues of this structure no longer exist, but they do in the mixed Fe-borocarbides, $Fe_{23}B_3C_3$, as described. The double borides all melt congruently, and the higher the valency of the *M'* metal, the lower the *M* content.

These double borides characteristically equilibrate with the FCC Ni- and Co-based solid solutions (of Al, Zn, etc.), as was the case for the double carbides. This is quite in the same sense as the $M_{23}C_6$ carbides, e.g. (Fe, Cr)$_{23}C_6$, equilibrate austenite and precipitate from it, as in stainless steels.

However, for the borides the extra electronic balance furnished by the non-transition metals is required. As we have seen in the carbides, the same balance there creates the ternary perovskite-type structure, as another variant also developing from the FCC matrix. It may also be recalled that amongst the ternary carbides the $M_{23}C_6$ structure 'takes over' from the perovskite ones on progressing from (Ni, Co, Fe) to Mn and Cr as bases. Thus, in the triangular relation

$$\begin{array}{ccc} \text{FCC} & \longleftrightarrow & \text{perovskite} \\ & \nwarrow\searrow \quad \swarrow\nearrow & \\ & M_{23}C_6 & \end{array}$$

each side amounts to a mode of adjusting the electron balance replacements. These can be controlled (*a*) by boron/carbon balance (with nitrogen and other interstitials playing their part); (*b*) by transition/non-transition metal balance; and (*c*) by shift from Groups VIII to VII to VI as base metals. Hence, the suitable path of atom exchange to maintain isomorphy can, from present knowledge, be fairly well outlined already.

The areas of ternary borides of $Cr_{23}C_6$ type in systems (Co,Ni)–(Al,Sn,Zn)–B are indicated in *Figure 6.8*. By way of illustration, the evolution of some of these equilibria is shown in more detail in *Figure 6.9*.

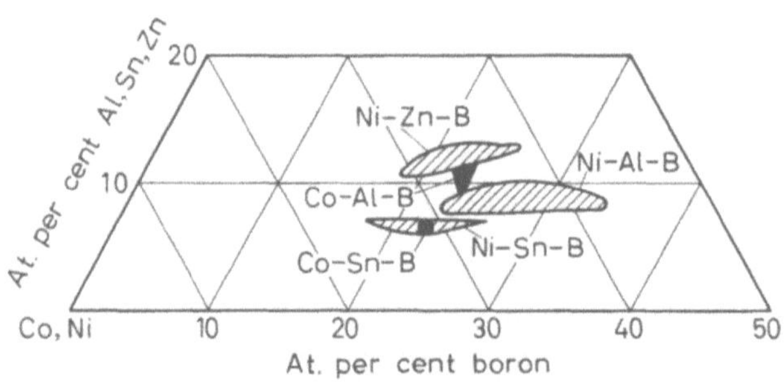

Figure 6.8. Homogeneity fields (at 800°C) *of ternary borides of* $Cr_{23}C_6$*-type structure* (From Stadelmaier and Yun[11])

By courtesy of the Editor, *Zeitschrift für Metallkunde*

It is remarkable that this ternary boride formation can entail *spheroidization* of eutectics, notably the Ni–Al eutectic in Ni–Al–B[56].

It is clear from *Table 6.5* that large gaps in our knowledge on the ternary boride systems still exist, and these are listed at its end to invite investigation. In many cases, the established isomorphous relations and surrounding binary systems make at least an outline prediction of the equilibria possible, but with notable pitfalls.

An excellent review[57] of the ternary borides of $Cr_{23}C_6$-type structure describes the precise lattice positions preferred by the transition- and non-transition-metal atoms in the unit cell of $(M,M')_{23}B_6$ (cf. *Figure 4.6*). From evaluating the atomic distances found, it arose that, upon substituting C by B and the metal by the non-transition metal (M'), the lattice has gained a certain degree of ionization, i.e. the metals will become positively charged. The *M′* atoms will avoid boron atoms as neighbours, and there is a certain indication of boron *pairing* where B becomes excessive in $Ni_{20}Al_3B_{12}$ as against $Ni_{20}Al_3B_6$ and its analogues.

For mixed interstitial compounds involving boron (*M*–B–C, *M*–B–N, *M*–B–Si), which are of great importance, cf. Chapter 10.

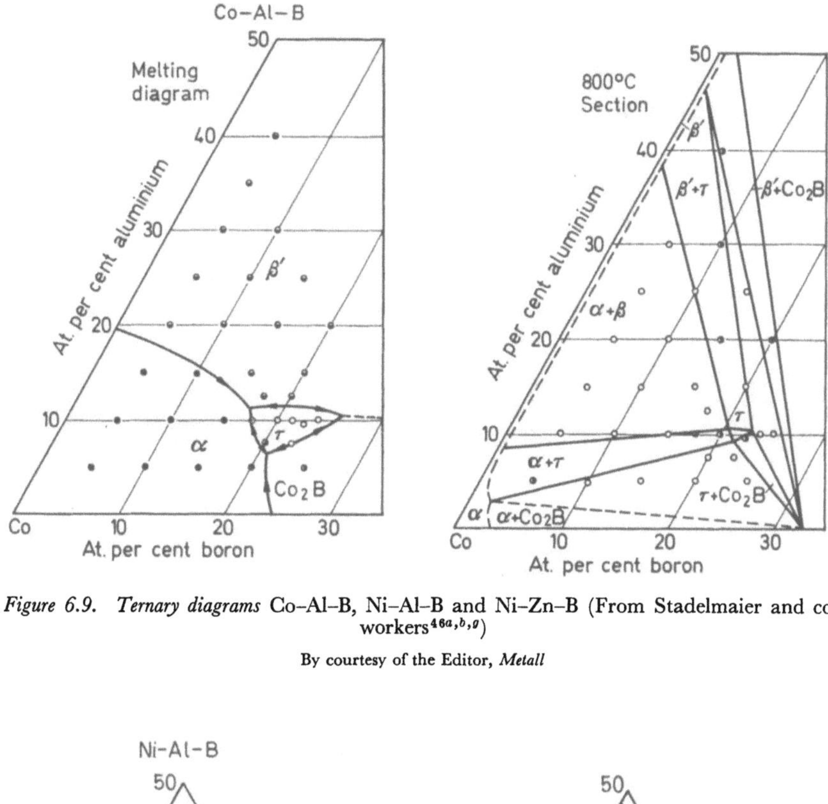

Figure 6.9. Ternary diagrams Co–Al–B, Ni–Al–B and Ni–Zn–B (From Stadelmaier and co-workers[46a,b,g])

By courtesy of the Editor, *Metall*

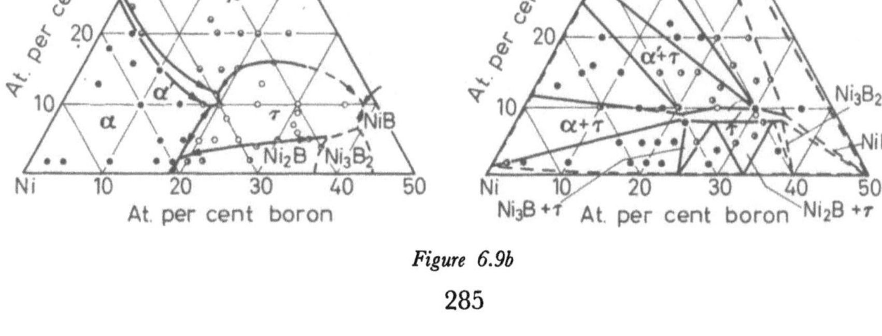

Figure 6.9b

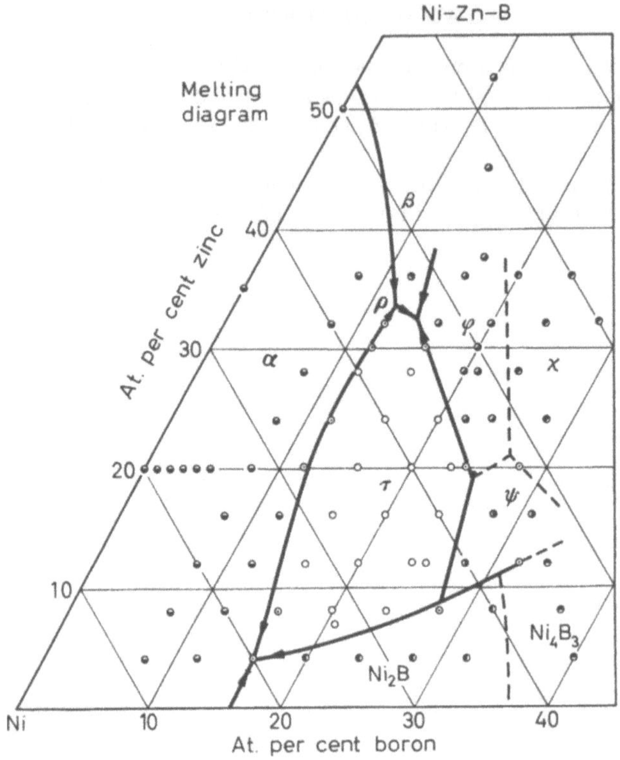
Ni–Zn–B
Melting diagram
At. per cent zinc
50
40
30
20
10
β
ρ
φ
α
χ
τ
ψ
Ni₄B₃
Ni₂B
Ni
10
20
30
40
At. per cent boron

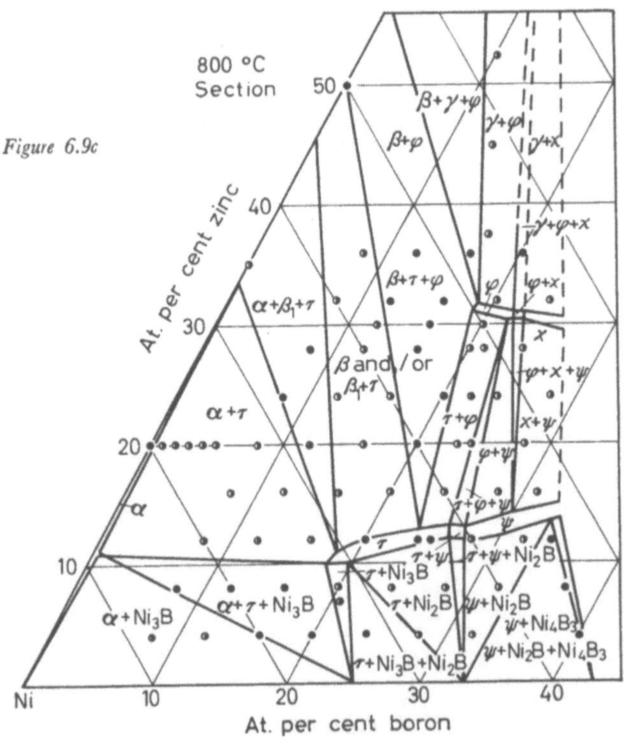
800 °C Section
At. per cent zinc
50
40
30
20
10
β+γ+φ
γ+φ
β+φ
γ+χ
γ+φ+χ
β+τ+φ
φ
φ+χ
α+β₁+τ
χ
β and/or β₁+τ
φ+χ+ψ
α+τ
τ+φ
χ+ψ
φ+ψ
τ+φ+ψ
α
ψ
τ
τ+ψ
τ+ψ+Ni₂B
τ+Ni₃B
α+Ni₃B
α+τ+Ni₃B
τ+Ni₂B
ψ+Ni₂B
ψ+Ni₄B₃
τ+Ni₃B+Ni₂B
ψ+Ni₂B+Ni₄B₃
Ni
10
20
30
40
At. per cent boron

Figure 6.9c

New isomorphs

The number of $Cr_{23}C_6$-type double-boride isomorphs is continuously increasing. The following new representatives have been reported[58] (referred to as 'τ' phases):

$Co_{20}Ti_3B_6$;	$Ni_{20}Ti_3B_6$	$Co_{21}Mo_2B_6$	?
$Co_{21}Zr_2B_6$;	$Ni_{20}Zr_3B_6$	$Co_{21}W_2B_6$	?
$Co_{21}Hf_2B_6$;	$Ni_{20}Hf_3B_6$	$Co_{21}U_2B_6$;	$Ni_{21}U_2B_6$
$Co_{21}V_2B_6$;	$Ni_{21}V_2B_6$	?	$Ni_{21}Mn_2B_6$
$Co_{21}Nb_2B_6$;	$Ni_{21}Nb_2B_6$	$Co_{21}In_2B_6$	?
$Co_{21}Ta_2B_6$	$Ni_{21}Ta_2B_6$	$Co_{21}Sb_2B_6$;	$Ni_{21}Sb_2B_6$

but no corresponding phases occur in the systems [Ti, Zr, Hf, V, Cr, Mo, W, Mn]–Fe–B, [Cr,Mo,W]–Ni–B and [Cr,Mn]–Co–B.

All these compounds possess a fair homogeneity area, and the above formulae are therefore only approximative.

Thus we have substantially $M_{21}M'_{2-3}B_6$ where M = Co,Ni; M' = Group IV, V, VI and some B sub-group metals, but with a definite selectiveness in electron balance. Further, as yet undiscovered members of this family can readily be predicted, with the platinum-group metals and rhenium very likely to enter.

The $Cr_{23}C_6$ family appears much more extensive amongst the borides than the carbides, and of probably the greater general stability; but the exclusion, upon present evidence, of *iron* analogues (as in $Fe_{21}Mo_2C_6$) amongst the borides is notable.

Seen more widely, we ought to regard these as just typical points of one continuous single-phase region in the transition metal–boron–carbon space in which the pure borides and pure carbides just form end members. Very probably this will also extend to nitrides and oxides, i.e. N and O atoms acting as substituents in the $Cr_{23}C_6$ (as for the η-carbide) structure, though apparently not yet reported. A limited solubility for silicon and aluminium also seems likely. In practical terms, it means e.g. that in stainless and other steels, where $M_{23}C_6$ carbide occurs, this could be decidedly stabilized by boron additions, if accompanied by suitable metal combinations.

Further, complex, two-metal borides are continuously reported[59] so that the scope of *Table 6.5* is now one of rapid growth. Such phases link on to a reported compound Mo_2FeB_2 which, in consonance with the M_3B_2 stoichiometry, had the U_3Si_2 type of crystal structure, the smaller boron atom substituting silicon, simultaneously with the smaller metal atoms. Again, the beginning of an extended region in *borosilicide* space seems to evolve here, by *smaller-atom* substitution, just as $Cr_{23}C_6$ served as prototype for *large-atom* substitution. In the systems studied, M–M'–B (where M = Mo,W; M' = Fe,Co,Ni), the further compounds

Mo_2CoB_2, Mo_2NiB_2
W_2FeB_2, W_2CoB_2 and W_2NiB_2

were found, related to, but representing structural distortions as from Mo_2FeB_2. Other new phases here were the M_2B analogues MoCoB and WCoB. New data have also been reported on mixed *aluminium*-boride systems in which ternary phases such as $(Nb,Al)B_2$, $(Ta,Al)B_2$ and MoBAl occur; it is remarkable that a small amount of Al added to MoB is able to stabilize the CrB-type structure.

6.4 APPLICATIONS

Some properties of the borides other than crystallographic are included in *Table 6.1.* In their practical applications, borides were latecomers on the scene, but the scope is extensive, rapidly growing and still far from being adequately explored.

Some principal attractions are their great *hardness* (unfortunately with extreme brittleness) and high *melting points.* The latter are comparable to, though somewhat lower than, those of the corresponding carbides and nitrides, but they exceed some of those of the silicides. Borides, however, as silicides, stand out by their *chemical stability and inertness,* implying often good oxidation resistance at high temperatures. This renders them suitable as refractory alloy constituents and in protective coatings. A further attraction in this context is their relatively good creep resistance. The high thermal emission properties of the borides also have found their use. The thermal expansions, where known, are mostly comparable with those of the silicides. Some borides, notably CrB and alloys based on it, likewise have shown good thermal shock resistance.

Owing to their resistance to oxidation and creep, borides have also found use in atomic reactor components, where resistance to both high temperatures and chemical attack becomes important. An example here is again the chromium monoboride[7] which, with a 15 per cent nickel binder, resists oxidation up to 950°C, and this property is attractively coupled with only a relatively minor loss of hardness on increasing temperature. This compares favourably e.g. with tungsten carbide-base alloys. Bonded uranium borides have also been applied in this field.

Much detailed basic research (including microstructure and hardness, etc.[60]) has more recently been performed in connection with the newer high-temperature *ceramic applications* of the early-Group diborides such as TiB_2. The properties of TiB_2, ZrB_2, HfB_2, NbB_2 and TaB_2 indicated[61] encouraging oxidation resistance, especially for ZrB_2 and HfB_2.

Further development in boride applications consisted in admixing silicides and other compounds. Thus, for instance, a refractory boride, 'Boride Z', has been reported, containing ZrB_2 plus $MoSi_2$, which is oxidation-resistant in air up to 1,100°C and has high electrical conductivity and thermal shock resistance.

An attractive application for borides lies in fields where their interconnecting rôle to other compounds serves well, such as in heterogeneous assemblies like special ceramics, metal ceramics, transition joints or protective coatings for high-temperature alloys. This applies e.g. to borides in combination with already established silicide systems where they add bonding and adhesion characteristics and a degree of fluxing (vitrifying action). An important feature here is inherent in the extensive isomorphous structure relationships between the metal borides and silicides (expressed at a glance in *Figures 6.3* and *7.2* with their distinctive variation) which, following the path of favourable atomic sizes and Group relations, permits selective bridging solid solutions between such phases in adjoining parts of the matrix; simultaneous variations of metal replacement in such borosilicide series are then possible (Chapter 10). In one example[62], the addition of aluminium boride, AlB_2, to Nb_5Si_3, a compound of poor oxidation resistance, was able to improve it greatly. Aluminium boride, though based on a non-transition metal, nevertheless enters in this context, both because of its isomorphy with known transition metal (C–32 type) borides and because of the large extent to which with silicides Si → Al replacement is possible and sometimes advantageous (Chapter 7); as a third factor, we may add that of control of fusion point and vitrification.

Viewed in broad principle, we should regard the elemental group B, Si, Al and O as an integral whole, the leading factor being the crystal *lattices* they form with the transition metals, whilst the particular atoms to which these are hosts are the more secondary variables.

The *melting points* of the Group IV → V → VI diborides have been plotted, in conjunction with those of the disilicides, in *Figure 6.10* which reveals several marked tendencies, clear from the diagram.

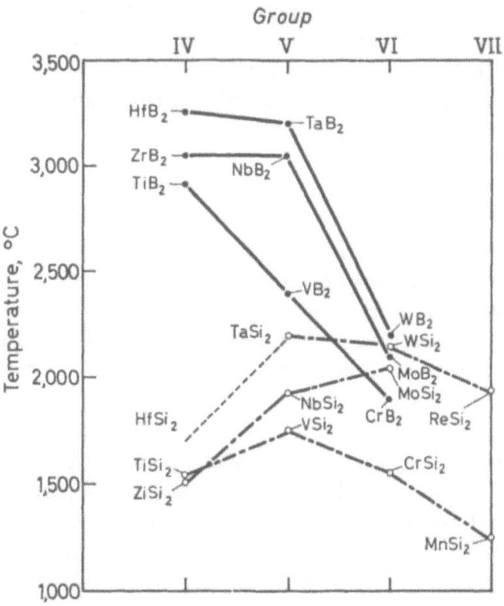

Figure 6.10. Melting-point trends amongst borides and silicides

In line with their metallic character, the borides are good electrical and thermal conductors. Some, such as NbB, MoB and W_2B, become super-conducting at relatively high temperatures[63].

The *hardness* of certain borides is so extreme that they have been considered as bases for hard-metal developments, with some difficulties, however, due to their brittleness. *Molybdenum boride* cemented with nickel[43] (which involves the ternary boride, Mo_2NiB_2, referred to above) has cutting qualities quite comparable with those of tungsten carbide base but with the added attraction of less cost and weight. Such boride alloys are in ascendance not only as cutting tools, becoming to some extent competitive with high-speed steels, but also as corrosion- and abrasion-resistant materials in chemical, grinding, mixing, etc. plant.

A condition throughout, however, is the ability of effectively cementing borides, made the more important by their brittleness. The latter, in contrast to, say, the carbides, is directly associated with their more complex structures and the weakening of the metal—metal bond in favour of the direct bond between the non-metal atoms. The choice of an effective (ductile) bonding material, to correspond e.g. to cobalt or nickel for the carbides, is here much more problematic, however, because these metals form stable borides themselves, so that embrittlement can occur on sintering—a difficulty increased even more in the silicides (Chapter 7). The advantage of the borides (and silicides) of inertness in high-temperature applications is also partly vitiated by this embrittlement. Further difficulties are created by the need to resort, as bond metals for the highest-temperature uses, to the more refractory metals (e.g. of the Groups IV–VI, rhenium or possibly the platinum family) which, in turn, tend to react to form their *own* boride. Nevertheless, some limited success has accompanied the use of ZrB_2 as hard metal, sintered with Fe, Co or Ni as binders[46a,64].

As a secondary additive to conventional cemented carbide-type materials, some refractory borides have also found interest, e.g. TiB_2 added to tungsten carbide[65].

Before the bonding problem can really be considered as solved, much more knowledge of the complex metal–boron systems is required (constitution, thermal data, especially eutectics and heats of formation related to properties); the situation is similar in the silicides (Chapter 7) and the mixed silicoborides (Chapter 10).

Refractory *molybdenum borides* and other Group VI borides have been used as brazes to join refractory metal (Mo, W, Nb, Ta) components. The relatively moderate melting temperature ($\sim$ 2,000°C) of the borides and eutectics can there be turned to considerable advantage[64]. Other fields of application are in corrosion- and abrasion-resisting parts and in cutting tools, where e.g. Mo_2B and Mo_2NiB_2 become prominent constituents. It seems probable that, as cutting materials, such *boride-base composites* may gain a place intermediate to high-speed steels and carbide-based hard metals but with possible developments well rivalling the latter. In metal ceramics and coatings on molybdenum, etc., Mo borides have also found application, and ZrB_2 has been applied in the hard-metal industry for cutting tools, etc., for which TiB_2 has also been tried, with some degree of promise[66].

The chemical corrosion resistance of borides, whilst generally fair, is not as good as that of comparable carbides and silicides.

As hard linings for tools, e.g. zirconium boride onto steels has been used. Spraying and sintering of hard linings of borides onto surfaces is well feasible, e.g. with Cr boride[67].

Boriding as a surface treatment of steels in the same sense as carburizing and nitriding has given successful results in improving wear resistance[68–9]. Such boride diffusion coatings have also been extended from alloy steels to Ni- and Co-base alloys, and this is certainly not the limit.

Much technical development work has been done on the *impregnation* on boride 'skeletons', using lower melting metals and alloys as impregnants[51].

Zirconium boride, ZrB_2, has found special interest in practice, owing to its exceptional stability (to a more limited extent also TiB_2 and, potentially, HfB_2). For example, a high-strength, high-temperature-resisting material has been developed on ZrB_2 as basis, by powdermetallurgical means which allows to produce very hard components, also resistant to corrosion. As a bond, the process uses not a metal (owing to the difficulty mentioned above) but *boron carbide*. ZrB_2 alloys bonded with Ni additions—in some variants, by adding other low-melting metals, e.g. Al, Zn, Cu, Sn, Pb—have found attractive uses due to high oxidation and thermal shock resistance, e.g. in turbine blades and fuel nozzles. Zirconium diboride has a valuable resistance to reaction with molten salts; materials of this base can be used as refractory linings in the conventional sense, as tool-tip materials[70] and ceramic furnace components[71]. The good electrical conductivity of ZrB_2, together with its wear resistance, has opened new uses, e.g. in sliding contacts; and there are many more general applications in the electronics industry (cf. [7] and Bibliography). The electrical properties of borides of Groups IV–VI (surveyed in conjunction with carbides and nitrides[72]) show other useful potentialities, especially at high temperatures and in unusual environments.

The high refractoriness of TaB and TaB_2 and their non-reactivity with e.g. molten uranium and calcium have made the borides, as well as ZrB_2, suitable as *crucible materials*[73] for melting these metals; this typifies another potential line of boride applications, comparable with some ceramic oxides.

The chemical resistance of Cr, Mn, Fe, Co and Ni borides, in particular to acids, has also been found attractive[74]. Their corrosion resistance, combined with electric and thermal conductivity and relatively good mechanical strength, makes them suitable materials for cathodes and for container construction in aluminium reduction cells[75]. TiB_2 has proved particularly attractive. However, in these, as in other applications of borides, small amounts of impurities can have a radical effect on properties and on sintering behaviour, as is well understandable from the structural characteristics explained above, in particular the sensitive alteration of bonding type.

Massive TiB_2 and other diboride shapes (cylinders, etc.) have proved of value[76] because of their relatively high mechanical strength allied with resilience and heat resistance maintained up to about 2,000°C.

Quite a different field of application of some later-Group borides lies in their *catalytic* action, such as that of nickel boride as a catalyst for hydrogeneration[77].

In fine-particle dispersion strengthening, transition-metal borides now play their parts along with oxides and some carbides. Here again refractoriness and non-solubility in the matrix material is a condition. In heat-resistant alloys of nickel, cobalt, iron, etc. base, the Group IV and V borides (e.g. ZrB_2, ThB_2) have proved of interest. Titanium alloys of high modulus which contain dispersed borides have also been developed[78], and the feasibility of this method clearly extends generally to other refractory base metals (Zr, Nb, W, etc.).

A large new field of interest is now developing particularly with regard to the effect of boron on precipitation in Nb, Cr, Ni, Mo alloy steels[79], both ferritic and austenitic; that on mechanical properties such as creep strength can likewise be highly beneficial. Much special research has been devoted to the effect of small boron additions on the 'superalloys'; for instance, two new double borides have been extracted and determined in detail[80] from such alloys ('Incoloy'), of M_3B_2 type and tetragonal structure. Chemically, the metal is here chiefly Mo or Ti, with less Fe, Ni and Al, and structurally these compounds are interesting because they contain an isolated boron-atom *pair*—an intermediate stage to the isolated single B atoms and the chain zigzag structure described earlier. Such pairs recur amongst the silicides (Chapter 7). The particular borides are accompanied by TiC; it is itself refractory and here goes into matrix solution at high temperatures, precipitating at lower ones. The point of principle, however, which is typified here is the potentiality that many new double or multiple borides are able to form in such alloys, of varied structure and composition, which function as strengtheners (along with carbides), precipitating 'relay-fashion' over wide temperature ranges.

The non-metallic boron carbides, boron nitride and boron silicide fall outside the present scope, but their own valuable ceramic properties have found application in similar fields (see Bibliography).

The properties of the *Rare Earth* hexaborides referred to (high conductivity, etc.) are certain to open for them new and special fields of application. To their relative strength and refractoriness we should add favourable nuclear characteristics, owing to the high neutron capture cross-section of the metals. The same applies to the other R.E. hard-metal compounds (carbides, silicides) but with special emphasis on the borides in view of the combination of further beneficial properties. In a sense we can almost envisage the boron-atom framework in the B-richer borides (*Figure 6.1*) as a 'conveniently constructed atomic stand' to hold the metal atoms in space during service!

The Group III and Rare Earth borides have recently gained interest also as additives to refractory materials, e.g. with regard to the high-temperature thermo-emissivities of ScB_2 and TmB_6 (determined[81] along with those of some carbides such as HfC and NbC).

The borides are perhaps among the most versatile of all the interstitial compounds as to applications (basically owing to the critical size and electron condition of the boron atom), though more so potentially at present than in actual fact—not so much on their own but when joined in composites and alloys, and often as mere additives and 'modulants', e.g. in precipitation mechanisms.

SELECTED ADDITIONAL LITERATURE

Basic (structural; classification; reviews)

Much of our structural knowledge is due to the work of Kiessling[1] and the Swedish school derived from Hägg's classical investigations[2], continued more recently by Aronsson[3]. Further contributions, constitutional as well as on applications, originated from Plansee/Vienna and the allied Yonkers (U.S.A.) schools (Nowotny, Schwarzkopf, Kieffer and associates[4,5], etc.) Brewer's centre[18,76] in California produced e.g. bonding and thermodynamic researches on borides, whilst in Russia the work centres chiefly around the work of Samsonov, Umanskii and their teams (e.g.[6,7]).

Thompson, R. and Wood, A. A. R. *Chem. Engng, Albany* 166 (1963) 51
Review

Belikov, A. M. and Umanskii, Ya. S. *Soviet Phys. Crystallogr.* 4 (1960) 643
Anisotropy

(*a*) Brewer, L. and Krikorian, O.[8]; (*b*) Johnson, R. W. and Daane, A. *J. chem. Phys.* 38 (1963) 425; (*c*) Silver, A. H. and Kushida, U. T. *ibid.* 865
Electron structure and bonds

Kaufman, L.[61a]; Kaufman, L. and Clougherty, E. V.[61b]
Thermodynamic properties: detailed review of Group IV and V diborides for high-temperature applications

Mezaki, R. *et al.*, *Symp. Thermodynamics of Nuclear Materials, Int. atom. Energy Ag.*, p. 775, Vienna, 1963
Thermodynamic properties of Group IV–VI borides; high-temperature heat contents and entropies

Boron, ed. Kohn, J. A. *et al.*, New York (Plenum Press) 1960
Mainly the element B: containing, however, limited papers on borides

Norton, J. T. *et al.*, *J. Metals, N.Y.* 1 (1947) 749
Earlier work mainly on diboride structures

Samsonov, G. V. *Dokl. Akad. Nauk SSSR* 86 (1952) 329
Microhardness

Specific Systems and Uses

Rudy, E. and Benesovsky, F. *Mh. Chem.* 92 (1961) 415
Zr–B and Hf–B binaries; Zr–B–N and Hf–B–N ternaries

Westrum, E. F. and Feick, G. *J. chem. Engng Data* 8 (1963) 193
ZrB_2: heat capacity and other thermodynamic properties

Gorelik, S. S. *et al.*, *Izv. vyssh. ucheb. Zaved. Khim. Met.* 5 (1962) 143
Ti and Zr borides; recrystallization temperatures compared with those of Ti and W carbides, with and without Co bonding

Kovalchenko, M. S. *et al.*, (*a*) *Dopov. Akad. Nauk URSR* (1958) 740; (*b*) *Izv. Akad. Nauk SSSR, Otdel. tekh. nauk, Metallurgiya Topl.* 2 (1960) 115
(*a*) Mo–ZrB_2 system; (*b*) metal–boride combinations

Fyodorov, T. F. *et al.*, *Poroshk. Met., Izd. Akad. Nauk Ukr. SSR* 2 (1962) 49
Cr-TiB_2 system

Kisly, P. S. *et al. Poroshk. Met. Izd. Akad. Nauk Ukr. SSR* 2 (1962) 53
Cr borides; mainly physical properties

(*a*) Kieffer, R. and Benesovsky, F. *Powder Metall.* 1–2 (1958) 145; (*b*) Funke, V. F., Judkovski, S. I. and Samsonov, G. V. *Izv. Akad. Nauk USSR, Otdel. tekh. nauk, Metallurgiya Topl.* 4 (1962) 126; (*c*) Tyrrell, M. E. and Houck, C. W. *U.S. Bur. Mines, Rep. Invest.* 6262 (1963); cf. [60]
Iron-group metals as binders for Group IV–VIA borides: technological but related to constitution, e.g. ternary borides

Cadeville, M. C. and Meyer, A. J. P. *C.R. hebd. Séanc. Acad. Sci., Paris* 255 (1962) 3391
Iron-group borides: Curie point and magnetic moments

Aronsson, B., Stenberg, E. and Aselius, J.[31]; *Nature, Lond.* 183 (1959) 1318; 195 (1962) 377
Ni and Pt-group borides; structural changes

Aronsson, B. *Acta chem. scand.* 17 (1963) 2036
Newer data on RuB_2, OsB_2, $IrB_{1.35}$ structures; general discussion of boron-richer borides, MB–MB_3

Buddery, J. H. and Welch, A. J. E.[14]
Pt-metal–boride structures; some of this work now superseded

(*a*) Samsonov, G. V. *High Temperature Compounds of Rare Earth Metals with Non-metals*, New York, (Plenum Press) 1965; (*b*) Binder, I.[34]; (*c*) Post, B. *et al.*, *J. Am. chem. Soc.* 78 (1956) 1800; (*d*) Spedding, F. H. and Daane, A. H. *The Rare Earths*, New York (Wiley) 1961
Surveys of Rare Earth borides

Johnson, R. W. and Daane, A. H. *J. phys. Chem., Ithaka* 65 (1961) 909
Investigation of the La–B system

L'vov *et al.*[72]
Thermal conductivities of Rare Earth and Alkaline Earth borides

Ti–B Ogden, H. R. and Jaffee, R. I. *J. Metals, N.Y.* 3 (1951) 335
Ti–B; Zr–B Glaser, F. W. *J. Metals, N.Y.* 4 (1952) 391; 5 (1953) 1117
Zr–B; Ta–B; Mn–B Kiessling, R. *Acta chem. scand.* 3 (1949) 90, 603; 4 (1950) 146

V–B; Nb–B; Ta–B	Nowotny, H., Benesovsky, F. and Kieffer, R. *Z. Metallk.* (*a*) 50 (1959) 258; (*b*) ibid. 417
Nb–B, Cr–B, etc.	Andersson, L. H. and Kiessling, R.[18]
Cr–B, W–B	Post, B. and Glaser, F. W. *J. chem. Phys.* 20 (1952) 1050
CrB_2, etc.	Hodkin, D. C. *Rep. Prog. Chem.* 46 (1949) 68
Mo–B	Weiss, G. *Annls Chim.* 1 (1946) 446; Steinitz, R. *et al., J. Metals, N.Y.* 4 (1952) 983
Fe–B	Leak, G. M. and Thomas, W. R. *Nature, Lond.* 176 (1955) 29
Co–B	Fruchart, R. *C.r. hebd. Séanc. Acad. Sci., Paris* 256 (1963) 3304
Ru–B	Aronsson, B. *Acta chem. scand.* 13 (1959) 109
U–B	Matterson, K. J. *et al., Plansee Proc.* 9 (1961) 329

Applications

Campbell, J. *Mater. Meth.* 31 (1950) 59
Ni-bonded ZrB_2 and other metal/boride couples

Hamjian, H. and Lidman, W. *J. Am. Ceram. Soc.* 35 (1952) 144
B_4C as bond for borides

Adv. Mater. Technol., Winter 1963
'Boride Z' as hard metal, ZrB_2-based

Richardson, L. *J. electrochem. Soc.* 101 (1954) 2230
Borides in high-temperature technology

Campbell, I. E. in *High Temperature Technology*, London (Chapman & Hall) 1956
Review

Schwarzkopf, P. and Kieffer, R., in *Refractory Hard-Metals*, New York (Macmillan) 1953

Kieffer, R. and Benesovsky, F., in *Hartstoffe*; *Hartmetalle*, Vienna and New York (Springer) 1963, 1965

Milliken, M. T. and Senkin, J. E. 'Some kinetic and thermodynamic properties of the refractory metal borides [and nitrides]'. *Univ. Calif. Radiat. Lab. Rep.* 7559 (1963)
Bibliography

REFERENCES

1 Kiessling, R. *Fortschr. chem. Forsch.* 3 (1954) 41; *J. electrochem. Soc.* 98 (1951) 166; *Acta chem. scand.* 4 (1950) 209
2 Hägg, G. *Z. phys. Chem.* B 6 (1929) 221; B 12 (1931) 33
3 Aronsson, B. *Ark. Kemi* 16 (1960) 379; with Lundstrom, T. and Rundqvist, S. *Borides, Silicides and Phosphides*, London (Methuen) 1965
4 Nowotny, H., in *Adv. X-Ray Analysis*, ed. Müller, Vol. 5, New York (Plenum Press) 1962; *Electronic Structure and Alloy Chemistry of the Transition Elements*, ed. Beck, New York (Interscience) 1963
5 Post, B. and Glaser, F. W. *Refractory Binary Borides*, New York (Interscience) 1963
6 Samsonov, G. V. and Markovskii, L. Ya. *Usp. Khim.* 25 (1956) 190, 234
7 Samsonov, G. V. and Umanskii, Ya. S. *NASA tech. Transl.* TL-102, 1962
8 Brewer, L. and Krikorian, O. *Univ. Calif. Radiat. Lab. Rep.* 2544 (1954)
9 Nicholson, M. E. *J. Metals, N.Y.* (*a*) 4 (1952) 148; (*b*) 6 (1954) 185; (*c*) 9 (1957) 1
10 Carroll, K. G. *et al., Nature, Lond.* 174 (1954) 978
11 Stadelmaier, H. H. and Yun, T. S. *Z. Metallk.* 53 (1962) 754
12 Rundqvist, S. *Acta chem. scand.* 13 (1959) 1193
13 Aronsson, B. and Rundqvist, S. *Platin. Metals Rev.* 5 (1961) 93
14 Buddery, J. H. and Welch, A. J. E. *Nature, Lond.* 167 (1951) 362
15 Reinacher, G. *Rev. Metall., Paris* 54 (1957) 321
16 Ehrlich, P. *Z. anorg. allg. Chem.* 259 (1949) 1
17 Andersson, L. H. and Kiessling, R. *Acta chem. scand.* 4 (1950) 160
18 Glaser, F. W. and Post, B. *J. Metals, N.Y.* 5 (1957) 1117
19 Nowotny, H. *et al., Mh. Chem.* 91 (1960) 348
20 Kudielka, H., Nowotny, H. and Findeisen, G. *Mh. Chem.* 88 (1958) 1048
21 Leitnaker, J. M. *et al., J. electrochem. Soc.* 108 (1961) 568
22 Nowotny, H., Benesovsky, F. and Kieffer, R. *Z. Metallk.* 50 (1959) 417
23 Chrétien, A. and Helgorsky, J. *C.r. hebd. Séanc. Acad. Sci., Paris* 252 (1961) 742
24 Beckman, G. and Kiessling, R. *Nature, Lond.* 178 (1956) 1341
25 Wever, F. and Müller, A. *Mitt. K. Wilhelm-Inst. Eisenforsch. Düsseld.* 11 (1930) 193
26 McBride, C. C., Spretnak, J. W. and Speiser, R. *Trans. Am. Soc. Metals* 46 (1954) 499
27 Östberg, G. *Nature, Lond.* 200 (1963) 1311
28 (*a*) Kolomytsev, P. T., with Zhukovskii, N. *Proc. Akad. Sci. USSR* 124 (1959) 151; (*b*) 130 (1960) 121

REFERENCES

[29] Stenberg, G. E. *Acta chem. scand.* 15 (1961) 861
[30] La Placa, S. and Post, B. *Acta crystallogr.* 15 (1962) 97
[31] Aronsson, B., Stenberg, G. E. and Aselius, J. *Acta chem. scand.* 14 (1960) 733
[32] Obrowski, W. *Metall* 17 (1963) 108
[33] Samsonov, G. V. *Usp. Khim.* 28 (1959) 189
[34] Binder, I. *J. Am. Ceram. Soc.* 43 (1960) 287
[35] Post, B. *et al.*, *Acta metall.* 2 (1954) 20
[36] Schwarzkopf, P. and Glaser, F. W. *Z. Metallk.* 44 (1953) 353
[37] Glaser, F. W. and Ivanich, W. *Powder Metall. Bull.* 6 (1953) 126
[38] Rudy, E., Benesovsky, F. and Toth, L. *Z. Metallk.* 54 (1963) 345
[39] Lavendel, H. W. *Planseeber. PulvMetall.* 9 (1961) 80
[40] Stadelmaier, H. H. and Hofer, G. *Metall* 18 (1964) 460
[41] Hägg, G. and Kiessling, R. *J. Inst. Metals* 8 (1952) 57
[42] Post, B., Pipitz, E. and Herz, H. *Powder Metall. Bull.* 7 (1954) 149
[43] Steinitz, R. and Binder, I. *Powder Metall. Bull.* 6 (1953) 123
[44] Blum, P. and Bertaut, F. *Acta crystallogr.* 7 (1954) 81
[45] Halla, F. and Thury, W. *Z. anorg. allg. Chem.* 249 (1942) 229
[46] Stadelmaier, H. H. *Metall* (*a*) 16 (1962) 793; (*b*) and Gregg, R. A. *ibid.* 407; (*c*) and Fraker, A. C. *ibid.* 212; (*d*) with Schöbel, J. D. and Sagmüller, J. R. 18 (1964) 23; (*e*) and Burgess, R. E. 17 (1964) 781; (*f*) and Jordan, L. T. 16 (1962) 752; (*g*) *Z. Metallk.* 531 (1962) 719
[47] Funke, V. F. and Yudkovsky, S. I. *Izv. Akad. Nauk SSSR, Otdel. tekh. nauk, Metallurgiya Topl.* (1962) 126
[48] Stadelmaier, H. H. *et al.*, *Metall* 18 (1964) 460, 1065; 19 (1965) 121
[49] Chrétien, A. and Helgorski, J. *C.r. hebd. Séanc. Acad. Sci., Paris* 253 (1961) 2693
[50] Glaser, F. W. *et al.*, *J. Metals, N.Y.* 5 (1953) 1119
[51] Kieffer, R., Benesovsky, F. and Honak, E. R. *Z. anorg. allg. Chem.* 268 (1952) 191
[52] Samsonov, G. V. *Metallov. term. Obrab. Metall.* 1 (1958) 35
[53] Portnoi, K. I., Samsonov, G. V. and Frolova, K. I. *Planseeber. PulvMetall.* 8 (1961) 195
[54] Sindeband, S. J. *Trans. Am. Inst. Min. metall. Engrs* 185 (1949) 198
[55] Toth, L. E., Nowotny, H., Benesovsky, F. and Rudy, E. *Mh. Chem.* 92 (1961) 945
[56] Schöbel, J. D. and Stadelmaier, H. H. *Z. Metallk.* 33 (1962) 326
[57] (*a*) Stadelmaier, H. H., with Hofer, G. *et al.*, *Metall* 18 (1964) 460, 1065, 1285; (*b*) with Draughn, R. A. and Hofer, G. *Z. Metallk.* 54 (1963) 640
[58] Ganglberger, E., Nowotny, H. and Benesovsky, F. *Mh. Chem.* 96 (1965) 1144
[59] Rieger, W., Nowotny, H. and Benesovsky, F. *Mh. Chem.* (*a*) 96 (1965) 844; (*b*) 95 (1964) 1502; (*c*) Schöbel, J. D. and Stadelmaier, H. H. *Metall* 19 (1965) 715; 20 (1966) 31; (*d*) Stadelmaier, H. H. and Avery, J. G. *Z. Metallk.* 56 (1965) 195
[60] Lynch, C. T. *et al.*, Paper at *A.I.M.E. Symp.* Cleveland, 1963
[61] (*a*) Kaufman, L., with Clougherty, E. V. *5th Plansee Seminar*, 1964; (*b*) Paper at *A.I.M.E. Symp.*, Cleveland, 1963
[62] Goldschmidt, H. J. *J. Inst. Metals* 87 (1959) 235
[63] Matthias, B. T. *et al.*, *Rev. mod. Phys.* 35 (1963) 1; Raub, C. J. *et al.*, *Jnl Phys. Chem. Solids* 24 (1963) 1093
[64] Everhardt, J. *Mater. Meth.* 40 (1954) 90
[65] Tyrrell, M. E. and Farrier, G. M. *U.S. Bur. Mines, Rep. Invest.* 6095 (1962)
[66] Yudkovski, S. I. *et al.*, *Vest. Machinost.* 42 (1962) 44
[67] Renninger, H. *Werkstoffe Korros., Mannheim* 13 (1962) 269
[68] Minkevic, A. N. *Hart tech. Mitt.* 17 (1962) 141
[69] Ornig, H. and Schaaber, O. *Hart tech. Mitt.* 17 (1962) 131
[70] Brewer, R. C. *Engrs' Dig., Lond.* 20 (1959) 205
[71] Brewer, L. *et al.*, *J. Am. Ceram. Soc.* 34 (1951) 173
[72] Lvov, S. N., Nemchenko, V. F. and Samsonov, G. V. *Dokl. Akad. Nauk SSSR* 135 (1960) 577
[73] Jenkins, I. L. and Keen, N. J. *J. less-common Metals* 4 (1962) 387
[74] Markovskii, L. Ya. and Bezurh, E. T. *Zh. prikl. Khim. Leningr.* 35 (1962) 491
[75] Ransley, C. E. *J. Metals, N.Y.* 14 (1962) 129
[76] Mandorf, V. *et al.*, in *High Temperature Materials II*, ed. Ault *et al.*, New York (Interscience) 1963
[77] Joseph, N. *Ind. Engng Chem. analyt. Edn* 44 (1952) 1006
[78] Brown, A. R. G. *et al.*, *J. Inst. Metals* 91 (1963) 161
[79] Bungardt, K. and Lennartz, G. *Arch. EisenhüttWes.* 34 (1963) 531
[80] Beattie, H. J. *Acta crystallogr.* 11 (1958) 607
[81] Samsonov, G. V. *et al.*, *Ukr. fiz. Zh.* 8 (1963) 700

7

SILICIDES

7.1 BASIC AND GENERAL

With a radius of the silicon atom as large as 1·17 Å*, most silicides no longer strictly qualify as 'interstitial compounds' but rather occupy a hybrid position, intermediate to interstitial and intermetallic compounds. Nevertheless, their inclusion here is justified on several grounds. Firstly, there is their close relationship to the borides (Chapter 6), structurally, in properties and in areas of application. Also silicon still *does* form genuine interstitial compounds when in association with the larger transition-metal atoms of Groups IV and V; and even in the higher silicides, the new Si–Si bond does not entirely destroy the insterstitial character. (Amongst primary solid solutions (Chapter 3) we had seen already that the Si atom can enter the metal lattice either substitutionally or interstitially—an ambiguity also transposed to the compounds and depending on similar factors.)

Silicon might well be viewed as a 'potential' interstitial atom, in the sense that the electronically homologous carbon is an 'actual' one, and inhibited only from realization by being just too large. The close relationship between the homologous Si and C atoms (both in Group IVB; four valency electrons; their part in silico-organic chemistry, etc.) gives special point to a comparison between silicides and *carbides*, and close relations will be found between them. This aspect of silicides forming an evolution from carbides is also supported by the intermediate silicocarbides known (Chapter 10).

In many of their features the silicides in fact already behave like true intermetallic compounds, in others as a hybrid between these and interstitials like carbides, the borides occupying an intermediary position.

Stable silicides are formed by the same group of metals as for borides, but additionally this time by *copper* (boride formation had ceased with nickel in the 1st Long Period) and, at the opposite end, by the Group IIIA metals (Sc,Y), Rare Earth, Alkaline Earth and alkalis. On the other hand, aluminium forms no silicides (contrast AlB_2), although the AlB_2-type structure does occur amongst the silicides. Incidentally, it may be observed how, viewing the Periodic Table as a whole, the elemental area over which hydrides → nitrides → carbides → borides and → silicides can be stable, expands progressively in this order, i.e. clearly paralleling the atomic size of the non-metal.

Table 7.1 (p. 304) shows the transition-metal silicides occurring between Groups IV and VIII, their structures and salient properties. The distribution of structure (to be viewed in conjunction with those of the borides, *Figure 6.3*) may be represented as in *Figure 7.1.* The transition metal–Si

* Si–Si bond radius; but it can range up to 1·34 Å (e.g. in pure silicon), depending on the co-ordination number and structure concerned.

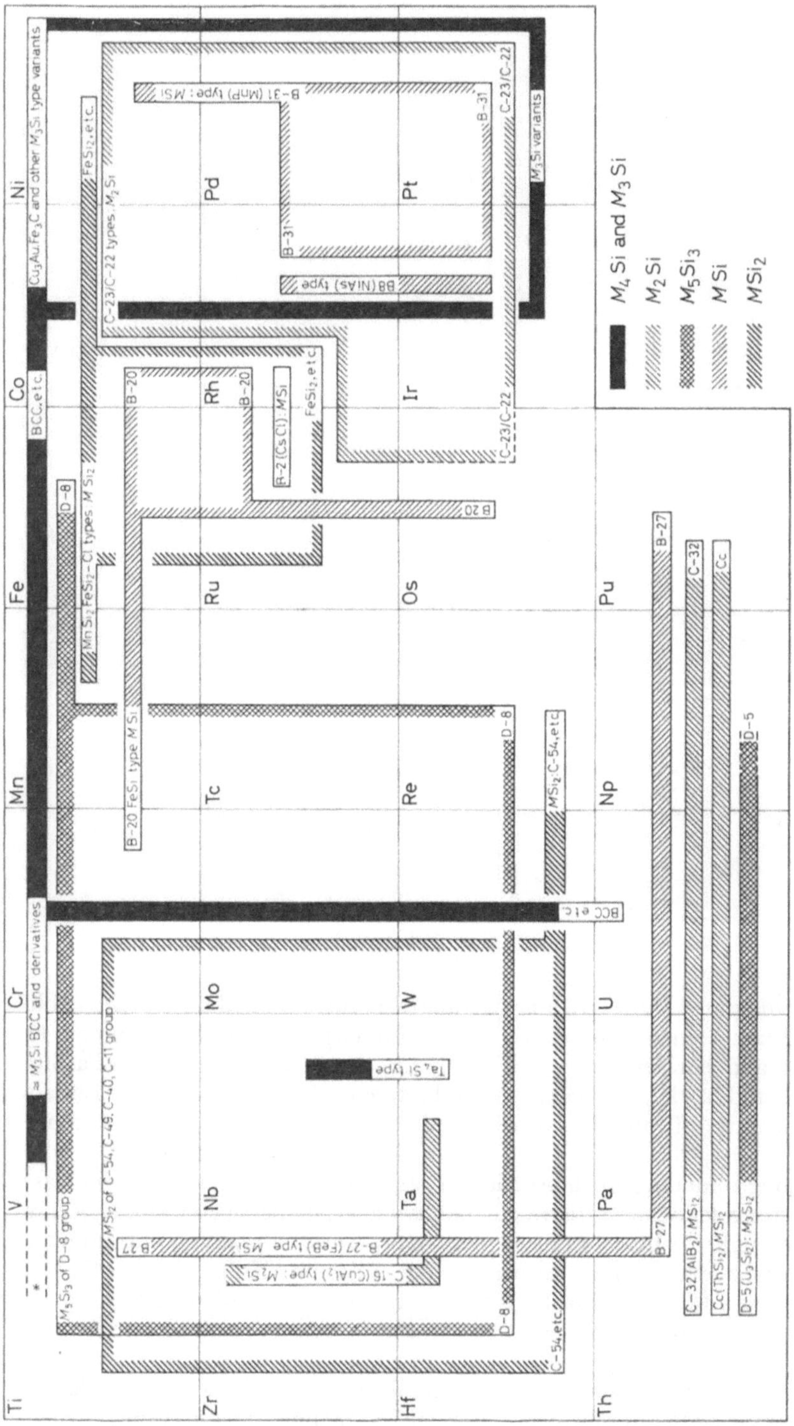

* Extensive Si solid solutions in BCC β-Ti and β-Zr : parent to M_5 Si$_3$

Figure 7.1. Fields of silicide structures within the Periodic Table

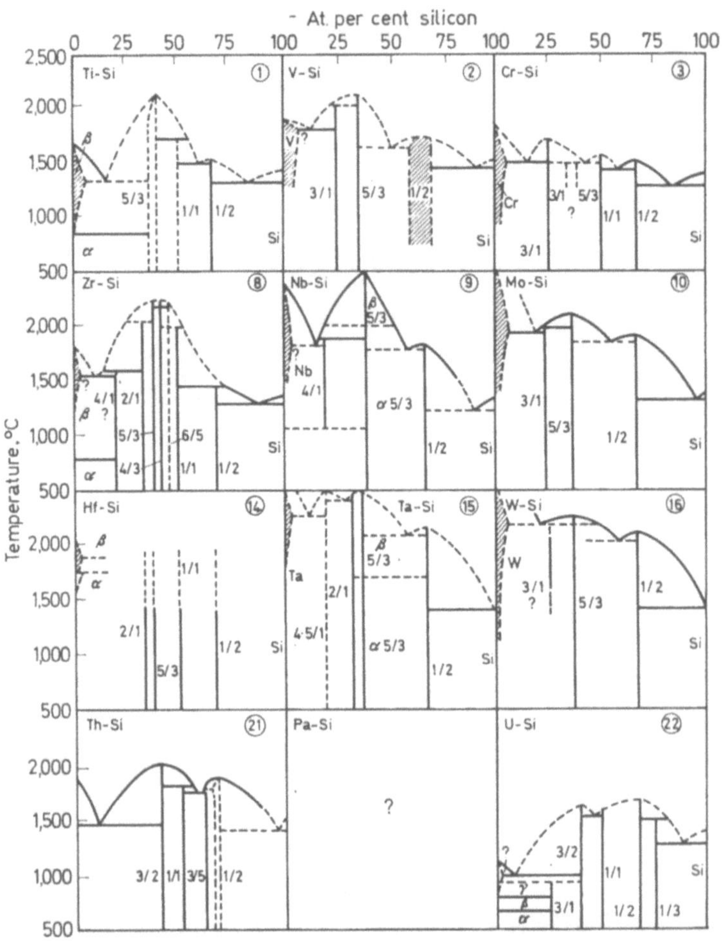

Figure 7.2. The pattern of metal–silicon systems

binary diagrams (*Figure 7.2*) are arranged according to the Periodic Table to facilitate comparison. The several isomorphous families common to silicides and borides should be noted, e.g. the (FeB–USi), (Cr_5B_3–Ta_5Si_3) groups, and a detailed comparison between *Tables 7.1* and *6.1* reveals many features of interest, not least suggesting where solid solutions could form.

Although bonding in the silicides is still essentially metallic, an increasingly strong tendency to covalent bonds between the silicon atoms occurs for the higher silicides; disilicides, MSi_2, show already semiconductor properties, as does, at the next higher stage, silicon itself. The lower silicides, such as the M_3Si phases of β-W type, are structurally still closely akin to, and derivable from, the parent metal ones, but more complex structures appear with increasing silicon.

The sequence of progressive non-metal to non-metal bonding, established amongst the *borides*, is repeated, in a modified manner, amongst the silicides.

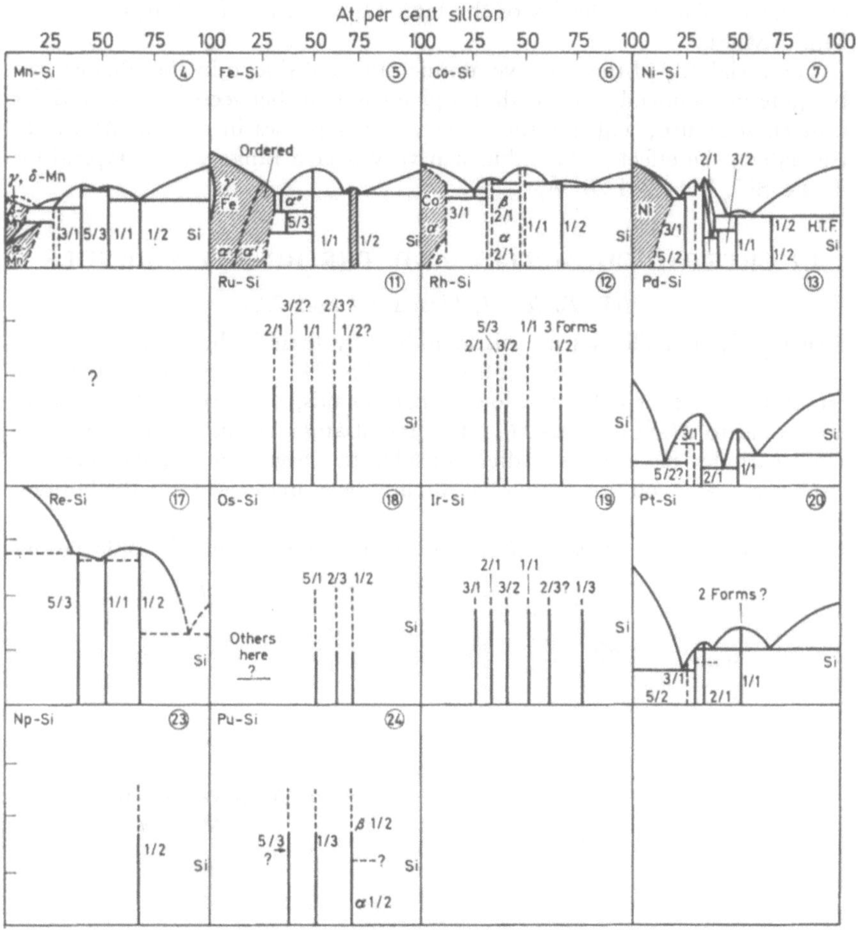

Figure 7.2b

Thus, with increasing Si addition to the transition metal[1,2], we have

(1) *isolated* Si atoms bonded to metals only (e.g. Cr_3Si, U_3Si types; also still in CrSi, FeSi, CoSi);

(2) *pairs of Si atoms*, still isolated (e.g. U_3Si_2, $FeSi_2$);

(3) one-dimensional (zigzag) *chains* of Si atoms (e.g. Co_2Si, Mn_3Si_2, and Mo_5Si_3 generally);

(4) two-dimensional, densely packed, *layers* of both Si and metal atoms (e.g. $MoSi_2$, $NbSi_2$, $TiSi_2$, etc.; cf. *Figure 7.4* below);

(5) three-dimensional *frameworks* of Si atoms (e.g. $ThSi_2$ and Rare Earth disilicides).

(For a more detailed discussion of this classification see Bibliography.)

The truly metallic nature of the Group IV–VI silicides has also been proven e.g. through detailed magnetic susceptibility and electric resistivity measurements, the latter always with a positive temperature coefficient[3]. Nevertheless, there are indications for at least partial covalent bonding exerting itself in the silicides of the later Groups[4], e.g. for Mn_3Si, Mn_5Si_3, MnSi, $MnSi_2$.

The variability of the effective atomic radius of silicon in the silicides can be quite pronounced, even within a given system, between systems and for a given structure; e.g. for the ternary Laves phases in systems *M*–*M'*–Si (see later), the effective Si radius can vary over a range which, typical for Ti–Fe–Si [5], is 1·22–1·281 Å.

7.2 CRYSTAL CHEMISTRY AND THE BINARY SILICIDES

(cf. *Table 7.1*, *Figures 7.1* and *7.2*)

Structurally, the silicides are distinguished by their higher degree of complexity than that in carbides, nitrides and borides. This is in line with the establishment, except for the low Si compounds, of direct Si–Si bonding, additional to *M*–Si and *M*–*M'*. The metal-atom lattices, unlike those in most carbides or nitrides, differ considerably from the typical metallic structures—an evolution which has already been under way for the borides. The silicon-poorer structures, such as those of β–W (W_3O) type in V_3Si, Cr_3Si and Mo_3Si (see *Figure 7.3*) can still be regarded essentially as deviations

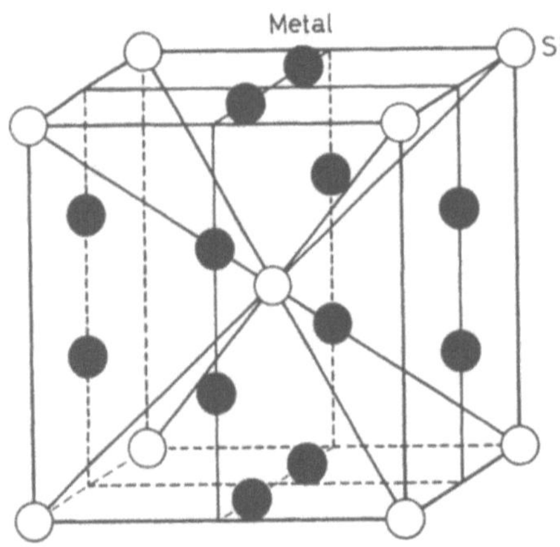

Figure 7.3. β–W-type structure: Cr_3Si, Mo_3Si, *etc.*

from the BCC parent metal; it is here particularly interesting to note how, for Mn_3Si (i.e. next Group), this indeed develops into the BCC itself (disordered)* and in Fe_3Si into the (Fe_3Al-type) superlattice within the primary Fe–Si solution. Another evolution of this kind occurs in the CsCl-type BCC superlattice structure (as in FeAl) found in the monosilicides RuSi and RhSi.

* We may also note that a limited silicon addition has the effect of structurally 'normalizing' manganese, i.e. of eliminating the complex structures of *elemental* (α,β)-Mn to stabilize the BCC of its two neighbours, Cr and α-Fe. Also the magnetic structure of Fe_3Si and Fe_3Si_2, when studied by means of the Mossbauer effect[6], has revealed two types of Fe atom which, as in the neighbouring manganese, are distinguished by their lattice sites and magnetic behaviour.

A 'Co_3Si' analogue does not appear to have been reported, although its existence would seem logical.

A further important feature is the tendency to polymorphic *changes*, such as in the M_5Si_3 family (e.g. α,β-Nb_5Si_3 involving the so-called 'T–1' and 'T–2' structures[7]) and in the Pt group silicides. To the extensive disilicide family of C–54/C–40/C–49/C–11-type structures and their composition-sensitive transformations we shall return below.

The trimetal silicide, M_3Si, structures are of interest also in some of their later-Group variants, such as Ni_3Si which adopts from the face-centred cube of Ni the Cu_3Au-type superlattice. Thus, overall, we have the following M_3Si evolutions with varying atomic number:

β-W *type* ⟶ BCC *disordered* ⟶ BCC *ordered* ⟶ Cu_3Au-*type order**

β-W type	BCC disordered	BCC ordered	Cu_3Au-type order
(in Groups V, VI: V,Cr,Mo,W?) → Nb_4Si,Ta_4Si variants; ↓ U_3Si type (BC tetragonal variant: U,Ir)	(in Group VII: Mn)	(in Group VIII: Fe,Co?)	(in Group VIII: Ni) evolves to Fe_3C type in Pd_3Si,Pt_3Si type (near-cubic)

The V_3Si (β-W-type) structure, incidentally, is also adopted by the germanide V_3Ge with which complete solid solutions are formed[8].

For copper, a phase Cu_5Si of β-Mn-type structure begins to occur which is so far unique amongst silicides but could be the nucleus of discovering other isotypic low silicides. (The link-up is relevant to the β-Mn type and 'χ' phases (Chapter 4) which can also be C- or N-bearing.)

Quite a number of questions concerning the lower silicides would arise from this, such as: Does a *cobalt* phase, 'Co_3Si', exist, of an ordered type intermediate to the known types of Fe_3Si (BCC) and Ni_3Si (Cu_3Au)?—Seeing that carbon can stabilize certain intermetallic compound structures like that of α- and β-Mn type, can Si (and B) do the same, and under what conditions?—What precisely is the mechanism of Si stabilizing the Laves phases to become ternary, for which there are now many examples?

The non-occurrence[9] in the W–Si system of a 'W_3Si' analogue to Mo_3Si would seem anomalous and, although a 'W_3Si' phase has been noted[10], produced by heating at very high temperatures, it may have been due to extraneous factors (oxygen content?); the absence of a 'W_3Si' has been confirmed[11,12], also in unpublished work by the writer. The stability conditions for the low-Si equilibria in the Groups IV–VI metals are highly critical, as reflected e.g. by the problem surrounding Nb_4Si and Ta_4Si (or $Ta_{4.5}Si$). The type of conflict met with amongst the metal-rich silicides may be illustrated on the Nb–Si system[13].

Whilst the existence of the higher silicides, $NbSi_2$ and Nb_5Si_3, is generally agreed, *in addition* the phase Nb_4Si was found[13,14], isomorphous with the known Ta_4Si. On the other hand, a phase 'Nb_2Si' had also been reported[15], and its possible existence could not be excluded. Then, however, a direct $Nb_5Si_3 \leftrightarrow$ Nb equilibrium was supported[16], without Nb_4Si, but later[17] again the occurrence of Nb_4Si. We were able to confirm the existence of Nb_4Si and its isomorphy to Ta_4Si; but in contrast to the latter, Nb_4Si appears as a high-temperature phase only, decomposing at lower temperatures into Nb + Nb_5Si_3. This reaction, probably eutectoidal, can be extremely sluggish but the rate is temperature-dependent. Thus alloys which were Nb_4Si as-melted (m.p. = 1,950°C), remained so after annealing at temperatures down to 1,200°C for 3 h but became Nb + α-Nb_5Si_3 after long-time anneals at 1,000° and 1,100°C. These findings in principle agreed with those of

* The inversion of this 'anti-Cu_3Au' structure in USi_3 should be noted.

Knapton[14]. An unusual effect found was that Nb_4Si tended to become stabilized in the presence of *excess* Nb but decomposed, after otherwise identical treatment, for excess Si to Nb_5Si_3 + Nb.

Thus the Nb_4Si problem is highly complex, and there are even indications of a possible second high-temperature form. In this connection, the existence of (α/β) allotropy in the associated Nb_5Si_3 becomes important. The latter transformation occurred around 2,000°C, although the β-form can be readily retained metastably at lower temperatures. Thus the presence, at the reaction temperature, of either α or β-Nb_5Si_3 could itself affect the rate of formation or decomposition of Nb_4Si.

It is noted that the melting point of Nb_4Si, 1,930°C, and of the $\beta \rightarrow \alpha$ transformation in Nb_5Si_3, 1,900–2,100°C, are of the same order, so that Nb_4Si could crystallize from Nb-rich liquid + either α- or β-Nb_5Si_3. Which of these alternatives operates may sensitively determine the formation of the Nb_4Si structure. In this connection, the observation[14] is significant that Nb_4Si, once decomposed, cannot be re-formed by solid reaction but only by re-melting. Also, although the existence[15] of a reported Nb_2Si phase between Nb_4Si and Nb_5Si_3 is now improbable, it gives additional point to the uncertainties in this part of the system.

Niobium should be considered within the frame of the other Group IV–VI metals more generally. *Table 7.1* and *Figure 7.1* show how some of the low-Si silicides form peritectically, so that their nucleation can be inhibited by heat treatment; others are high-temperature phases only; yet others may persist but be unstable. In the light of the electron structure of the transition-metal compounds generally, Nb in conjunction with Si is quite liable to show a varying kinship, leaning towards the behaviour of its neighbours V, Ta, Zr or Mo, depending on treatment, so that various metastable equilibria might readily occur.

Another phase reported[18a], Nb_3Si, is of considerable interest, since it becomes superconducting below 1·5°K, as other A_3B compounds, and is of the ordered Cu_3Au-type structure. It would imply an extension of the M_3Si branch of the structure-field diagram (*Figure 7.1*) to include Nb_3Si, and possibly also a 'Ta_3Si' not yet reported. Furthermore, we know that A_3B compounds of the β-W-type structure (e.g. Nb_3Sn) have a high superconductive transition temperature. The Nb/Si atomic size ratio just fails to achieve this structure in Nb_3Si but, as we know, in neighbouring Mo_3Si and V_3Si it is attained. An attractive proposition, therefore, might be to see where in a pseudo-binary system such as $Nb_3Si \leftrightarrow Mo_3Si$ or $Nb_3Si \leftrightarrow V_3Si$ the structural transition from Cu_3Au- to β-W-type would occur, and whether by moderate Mo (V) addition this could be accompanied by a raised superconductive transition temperature.

This could apply more generally to M_3Si-based systems, including ternary metal substitutions, or indeed boron for silicon, in consonance with a smaller metal atom. The replacement of Si by Sn in Nb_3Si would also be a natural one, in view of their equivalency and the larger size of the Sn atom, which must cause the $Nb_3Si \rightarrow Nb_3Sn$ (= $Cu_3Au \rightarrow \beta$-W-type) transformation at a certain ratio.

The equilibrium temperatures of Nb_3Si in the Nb–Si diagram and the relation to Nb_4Si are still insufficiently clear, and the phase cannot yet be allowed for in *Figures 7.1* and *7.2.*

There is now some doubt as to the validity[18a] of the Cu_3Au structure for Nb_3Si in the *pure* state, as it is likely to have been stabilized by interstitial contamination (oxygen?). Nevertheless, the ready formation of this structure is important and provides another example of the highly critical nature of the low silicides. Their sensitivity, structurally and property-wise, to trace impurities cannot be too much emphasized; the so-called 'Nowotny phases' ($D\text{–}8_8$ structure) provide a further example, being really ternary silicides. That compounds 'Nb_3Si' and Ta_3Si in fact do exist has been confirmed by Schubert[18b], who found them isomorphous with Ti_3P, and by the writer's work (unpublished). Similarly, it has been shown[19] that intermetallic compounds of the β-W (Cr_3Si)-type structure can transform to the Cu_3Au type by the introduction of O, C or N atoms.

The existence of Nb_3Si of Cu_3Au structure, even though interstitial-stabilized, raises a number of points of principle. There could well be a polymorphous change to the β-W type, either in Nb_3Si or in any of its solid solutions with adjoining 'A_3B' compounds, creating a special interest in studying systems such as $Nb_3Si \rightarrow V_3Si$, Ta_3Si, Mo_3Si, Nb_3Sn, etc.

Amongst the M_3Si structures, we might especially note the occurrence of that of cementite (Fe_3C), as discussed for the borides (see *Table 6.1*). The only silicide actually reported[20] so far is Pd_3Si, though others may exist. The monotonic sequence Ni_3Si (= ordered Cu_3Au type) → Pd_3Si (Fe_3C type) → Pt_3Si (monoclinic but almost cubic structure) and the transition between the three is worth following up, as each, in its own way, still retains some of the characteristics of the face-centred cubic metal parent. A precipitation of Pd_3Si from Pd, as Fe_3C from austenite and by a similar lattice mechanism, is entirely possible. (Conversely, the question of possible *transient* superlattice formation preparatory to $\gamma \rightarrow Fe_3C$ transformation might be investigated.)

The fact of Pd_3Si having the cementite structure at all raises the question of possible solid-solution formation with Fe_3C *itself* and its stabilization perhaps by Pd and Si addition (e.g. in steels or cast irons). Also could there be other Fe_3C-type silicides or, indeed, borides amongst the noble metals, e.g. a Rh_3Si capable of dissolving in Fe_3C and stabilizing it?

The tetragonal M_5Si_3 (or M_3Si_2) phases are all either isomorphous or closely related. Characteristic is a liability for *pairs* of Si atoms to occur within the metal lattice (e.g.[21]) in U_3Si_2. Another is the structural sensitivity of this group to impurities. The slightest addition to M_5Si_3 (such as Nb_5Si_3) of carbon or other interstitials (B, N, O) can cause transformation to the '$D\text{–}8_8$' structure. The latter, in fact, holds for the pure *binary* silicide in Group IV, but not in Groups V and VI where it becomes a *ternary* phase, requiring second interstitials[22].

In this intermediate silicon range (from M_2Si to MSi_2), the silicides are most closely related to the borides; therefore, much of what was said (Chapter 6) in this connection about the latter with regard to stacking sequences of metal-atom layers and interposition of non-metals applies to the silicides. Whilst in the M_2Si structures the M atom layers contain only metal, with Si filling the interspaces, in M_5Si_3 structures, such as α- and β-Nb_5Si_3 and Mn_5Si_3, these metal layers also begin to take up Si atoms, but substitutionally. Thus we have, as a point of principle, the situation of a mixed interstitial *and* substitutional solid solution of Si in these silicides. In fact, some ternary M–Si–B compounds also adopt the M_5Si_3-type structures; e.g. that of W_2Si_3 occurs in $Fe_{4.8}Si_2B$, revealing the division yet more clearly in that in this silicoboride the Si atom replaces metal, the B atom the formerly interstitial Si. Also, as mentioned, a small amount of extra C, B, or N sensitivity steers the M_5Si_3 type to another structure to give the Nowotny phases, $D\text{–}8_8$, where again the smaller C atom has entered interstitially and displaced the Si atom entirely into metal sites.

Amongst the M_5Si_3 silicides, Nb_5Si_3 and Ta_5Si_3 again occupy a crossroads position by existing in two allotropic (α,β) forms, each representing adjoining silicide and boride families, α-Nb_5Si_3 and Ta_5Si_3 being isomorphous with Cr_5B_3, β-Nb_5Si_3 and Ta_5Si_3 with V_5Si_3, Mo_5Si_3 and W_5Si_3. Controversy surrounded a presumed 'Nb_5Si_3' of the $D\text{–}8_8$ structure[23], later shown[14] to be due to carbon impurity.

In the *monosilicides*, MSi (*Table 7.1, Figure 7.2*), we have, as in the borides, the FeB, CrB and MoB sub-forms with, this time, Si–Si zigzag chains,

Table 7.1. Transition-metal

Metal base	*Silicides*	*Structure type*	*Lattice dimensions, Å (M=number of molecules per unit cell)*	*Melting point, °C*
Group IV				
Ti	Ti_5Si_3	hexagonal, Mn_5Si_3 type (D–8_8)	a=7·448, c=5·114; M=2	2,120
	TiSi	orthorhombic, FeB type (B–27)	a=6·49_2, b=3·67_8, c=4·97_0; M=4	Peritectic below 1,760
	$TiSi_2$	orthorhombic, $TiSi_2$ type (C–54) (ordered)	a=8·253, b=4·783, c=8·540; M=8	1,540
Zr	[Zr_4Si?] Zr_2Si	tetragonal, $CuAl_2$ type (C–16)	a=6·6120, c=5·2943; M=4	2,110? (dec.)
	Zr_5Si_3 (=Zr_3Si_2?)	hexagonal, Mn_5Si_3 (D–8_8)	a=7·85_4, c=5·53_5; M=2	2,210 (dec.)
	ZrSi	orthorhombic, FeB type (B–27)	a=6·982, b=3·786, c=5·302; M=4	2,095 (dec.)
	$ZrSi_2$	orthorhombic, $ZrSi_2$ type (C–49) 'Zr_4Si' reported as hexagonal (A–3)—probably Zr solution? 'Zr_4Si_3' and 'Zr_6Si_5' reported but uncertain	a=3·72, b=14·76, c=3·67; M=4	~1,600
Hf	Hf_2Si	tetragonal, C–16	a=6·48, c=5·21; M=4	
	Hf_5Si_3(?)	hexagonal (D–8_8)	a=6·999, c=3·671	
	HfSi	B–27=FeB	a=6·85_5, b=3·75_3, c=3·19_1; M=4	
	$HfSi_2$	C–49=$ZrSi_2$ Hf_5Si_3 probable but unconfirmed	a=3·677, b=14·550 c=3·649; M=4	~1,700?
Th	Th_3Si_2	tetragonal, U_3Si_2 type (D–5a)	a=7·835, c=4·154	~1,850
	ThSi	B–27=FeB	a=7·88, b=4·15, c=5·89; M=4	
	α-$ThSi_2$ (low temp.)	tetragonal, $ThSi_2$ type (C_c)	a=4·134, c=14·375; M=3	

silicides: structures and properties

Heat of formation (298°K), (kcal/g atom)	*Hardness* (kg/mm²) at 50 g, unless otherwise stated)	*Density* (x-ray)	*Other properties: Electrical resistivity, ρ,* μΩ cm, at R.T.; expansion coeff., β superconductivity (SPCD): upper temp. limit	*Remarks*
−46·2 (or −73) for ⅓ Ti_5Si_3	986 (100 g)	4·32	SPCD < 1·2°K	
−31·0 for 1 TiSi	1,039 (100 g)	4·21		
−16·1 for ½ $TiSi_2$	618 (100 g)	4·4	$\rho = 16{\cdot}7$	
−50 (or −74) for 1 Zr_2Si	2,220	5·99		
−46 (or −72) for ⅓ Zr_5Si_3	2,250	5·90		$D8_8$ structure in Zr_5Si_3 and Hf_5Si_3 requires small C, O or N for stabilization; probably also applies to Ti_5Si_3
−37 (or −58) for 1 ZrSi	2,100	5·65	$\rho = 49{\cdot}4$ SPCD < 12°K	
−19 (or −30·5) for ½ $ZrSi_2$	1,060	4·9	$\rho = 106{\cdot}2$	
−73 for ⅓ Hf_5Si_3				
	930	8·03		
		9·80		
		9·08		
	1,120	7·79		

Metal base	*Silicides*	*Structure type*	*Lattice dimensions, Å (M=number of molecules per unit cell)*	*Melting point, °C*
Group IV cont.				
Th	β-$ThSi_2$ (Th_2Si_3) (high temp.)	hexagonal, AlB_2 type (C–32) phases Th_6Si_{11} and Th_3Si_5 (two polymorphs?) also reported α- → β-$ThSi_2$ transition ≈ 1,700°C	$a=3{\cdot}985$, $c=4{\cdot}220$; $M=1$	
Group V				
V	V_3Si	cubic, β-W type (A–15)	$a=4{\cdot}722$; $M=2$	peritectic below ~2,060
	V_5Si_3	tetragonal, W_5Si_3 type ($D–8_m$) '*T1*'	$a=9{\cdot}42_9$, $c=4{\cdot}75_7$; $M=4$	2,150
	VSi_2	hexagonal, $CrSi_2$ type (C–40)	$a=4{\cdot}571$, $c=6{\cdot}372$; $M=3$	1,750
Nb	Nb_4Si [Nb_3Si?]	hexagonal, ε-Fe_3N (Ta_4Si) type	$a=3{\cdot}59$, $c=4{\cdot}46$; $M=2$	peritectic below ~1,950
	α-Nb_5Si_3 (low temp.)	tetragonal, Cr_5B_3 type (D_8) '*T2*'	$a=6{\cdot}57_0$, $c=11{\cdot}88_4$; $M=4$	
	β-Nb_5Si_3 (high temp.)	tetragonal, W_5Si_3 type ($D–8_m$) '*T1*'	$a=10{\cdot}01_8$, $c=5{\cdot}07_2$; $M=4$	2,480
	$NbSi_2$	C–40: $CrSi_2$ type	$a=4{\cdot}795$, $c=6{\cdot}589$; $M=3$	1,930
Ta	Ta_4Si or $Ta_{4.5}Si$ [Ta_3Si?]	hexagonal, (ordered, $D0_{19}$type)	$a=6{\cdot}09_3$, $c=4{\cdot}90_9$; $M=2$	2,510
	Ta_2Si	tetragonal, C–16= $CuAl_2$ type	$a=6{\cdot}157$, $c=5{\cdot}03_9$; $M=4$	2,460
	α-Ta_5Si_3 (low temp.)	$D8_m=W_5Si_3$ '*T1*'	$a=9{\cdot}8_8$, $c=5{\cdot}0_6$; $M=4$	
	β-Ta_5Si_3 (high temp.)	Cr_5B_3 type ($D–8_1$) '*T2*'	$a=6{\cdot}51_6$, $c=11{\cdot}87_3$; $M=4$	2,500
	$TaSi_2$	C–40=$CrSi_2$	$a=4{\cdot}783$, $c=6{\cdot}565$; $M=3$	2,200
Pa	?			
Group VI				
Cr	Cr_3Si (hom. range)	A–15=β-W type	$a=4{\cdot}564$; $M=2$	1,750

Heat of formation (298°K), (kcal/g atom)	*Hardness* (kg/mm² at 50 g, unless otherwise stated)	*Density* (x-ray)	*Other properties: Electrical resistivity, ρ,* μΩ cm, at R.T.; expansion coeff., *β* superconductivity (SPCD): upper temp. limit	*Remarks*
−20·8 for ½ $ThSi_2$				
−37 for 1 V_2Si	1,430–1,560	5·67 (meas.)	SPCD < 17°K	
	1,350–1,510 890–960	4·8		
< −7	1,090 (100 g)	4·7	$\rho = 13{\cdot}3$ SPCD < 1·2°K	
−21 (or −47) for ⅓ Nb_5Si_3	830–820 (100 g) 700 (100 g)	8·01 6·56		
		6·26		
< −8 or −15 (−11 to −28) for ½ $NbSi_2$	1,080 (100 g)	5·7	$\rho = 6{\cdot}3$ SPCD < 12°K	Nb_4Si, Nb_3Si, etc. problematic Nb_4Si high-temp. phase only but retainable an earlier 'Nb_2Si' redundant, but Nb_3Si exists D–8_8 structure in Nb_5Si_3 and Ta_5Si_3 when C-stabilized probably also applies to V_5Si_3 a *third* Nb_5Si_3 form existing—viz. Mn_5Si_3 isomorph?
−32·2 for 4½ $Ta_{4.5}Si$	(1,000–1,200, 40 kg)	12·86		
−30·7 (or −25 to −30) for 1 Ta_2Si	(1,200–1,500, 40 kg)	13·54		
−28·8 (or −50) for ⅓ Ta_5Si_3	(1,200–1,500, 40 kg)	13·06		
−13·9 for ½ $TaSi_2$	1,410 (1,000–1,200, 40 kg)	9·1	$\rho = 8{\cdot}5$ SPCD < 2·2°K	
('Cr_2Si' −35) for 1 Cr_2Si?	1,005	6·5	$\rho = 45{\cdot}5$	

Metal base	*Silicides*	*Structure type*	*Lattice dimensions, Å (M=number of molecules per unit cell)*	*Melting point, °C*
Group VI cont.				
Cr	Cr_5Si_3 (hom. range) (Cr_3Si_2)	D–8_m=W_5Si_4 type ('*TI*')	a=9·170, c=4·636; M=4	1,600
	CrSi	cubic (B–20=FeSi type)	a=4·629; M=4	1,630
	$CrSi_2$ (hom. range)	C–40=$CrSi_2$	a=4·431, c=6·364; M=3	1,550
Mo	Mo_3Si	A–15=β-W type	a=4·890; M=2	2,030
	Mo_5Si_3 Mo_3Si_2	B.C. tetragonal, D–8_m=W_5Si_3 type ('*TI*')	a=9·642, c=4·905; M=4	2,190
	$MoSi_2$	C–11b ($MoSi_2$ type)	a=3·202, c=7·85$_2$; M=2	2,050
W	[W_3Si(?)]	A–15	a=4·910; M=2	
	W_5Si/W_3Si_2	D–8_m=W_5Si_3 type ('*TI*')	a=9·605, c=4·964; M=4	2,350
	WSi_2	C–11b=$MoSi_2$	a=3·211, c=7·868; M=2	2,165
U	U_3Si	B.C. tetragonal, U_3Si type	a=6·029, c=8·696; M=4	
	U_5Si_3	tetragonal, U_3Si_2 (D–5a)	a=7·3299; c=3·9004; M=2	
	USi	B–27=FeB	a=7·65, b=3·90; c=5·65; M=4	
	β-USi_2(U_2Si_3?)	C–32=AlB_2	a=3·84$_3$, c=4·069; M=1	
	α-USi_2	C_c=$ThSi_2$	a=3·98, c=13·74; M=4	
	γ-USi_2(hom. range)	cubic, Cu_3Au type ($L1_2$)	a=4·053–4·035$_3$; M=1	≈1,700

Heat of formation (298°K) (kcal/g atom)	*Hardness* (kg/mm²) at 50 g, unless otherwise stated)	*Density* (x-ray)	*Other properties: Electrical resistivity*, ρ, $\mu\Omega$ cm, at R.T.; expansion coeff., β superconductivity (SPCD): upper temp. limit	*Remarks*
	1,280	5·9	$\rho=114$	D–8_8 structure in Cr_5Si_3 stabilized by carbon addition; note exceptionally high ρ for $CrSi_2$; homogeneity range reported as $CrSi_{1\cdot9}$ to $CrSi_{2\cdot6}$
	1,005		$\rho=143$	
>-8	1,130 (100)	4·4	$\rho=1{,}420$	
-21 (or -24) for 1 Mo_3Si	1,300 (100)	8·97		
-22 (or $-23\cdot4$) for $\frac{1}{3}$ Mo_5Si_3?	1,700 (100)	7·4		
$-13\cdot6$ (or $-15\cdot7$) for $\frac{1}{2}$ $MoSi_2$	1,300 (100)	6·2	$\beta=5\cdot1\times10^{-6}$ $\rho=21\cdot8$ therm. cond. 0·075 cal/sec/cm/°C	
		16·2	SPCD < 1·3°K	'W_3Si' now disproved; interstitial-stabilized
-15 for $\frac{1}{3}$ W_5Si_3	770	12·2	SPCD < 2·8°K	
$-11\cdot2$ for $\frac{1}{2}$ WSi_2	1,080	9·3	$\rho=38\cdot2$ $\beta=11\cdot3$ (*c* axis), 7·1 (*a* axis) SPCD < 12·2°C	
		15·60		
		12·21		
		10·4	$\beta=57$ (*a* axis), -26 (*c* axis) $\times10^{-6}$	USi_2 dec. above ≈450°C to USi_3+ '$USi_{1\cdot88}$'; γ-USi_2=$U_{1\cdot07}Si_{2\cdot14}$, in terms of atoms per (cubic) unit cell
		9·25		
		8·98		
		8·30 for USi_3		

Metal base	*Silicides*	*Structure type*	*Lattice dimensions, Å (M=number of molecules per unit cell)*	*Melting point, °C*
Group VII				
Mn	Mn_3Si	BCC (A–2), not ordered	$a=2{\cdot}857$; $M=\frac{1}{2}$	
	Mn_5Si_3	Mn_5Si_3 (D–8_8)	$a=6{\cdot}912$, $c=4{\cdot}812$; $M=2$	1,280
	MnSi	B–20=FeSi	$a=4{\cdot}557$; $M=4$	1,275
	$MnSi_2$	tetragonal	$a=5{\cdot}524$, $c=17{\cdot}457$	
Tc	unknown but presumed similar to silicides of Mn and Re, e.g. TcSi of B–20 type			
Re	Re_5Si_3	D–8_m=W_5Si_3 type ('*TI*')	$a=9{\cdot}53$, $c=4{\cdot}81$; $M=4$	1,920
	ReSi	B–20=FeSi BC tetragonal	$a=4{\cdot}775$; $M=4$	peritectic below 1,860?
	$ReSi_2$	C–11b=$MoSi_2$	$a=3{\cdot}129$, $c=7{\cdot}674$; $M=2$	1,930
Np	$NpSi_2$	C_c=$ThSi_2$	$a=3{\cdot}97$, $c=13{\cdot}70$; $M=4$	
Group VIII				
Fe	Fe_3Si	BCC (A-2) disordered–ordered	$a=2{\cdot}8277$; $M=2$	
	Fe_5Si_3 (Fe_3Si_2?)	D–8_8=Mn_5Si_3 type	$a=6{\cdot}7552$, $c=4{\cdot}7174$; $M=2$	H.T. phase 820–1,050
	FeSi	B–20=FeSi	$a=4{\cdot}488$–$4{\cdot}483$ (hom. range); $M=4$	1,415
	$FeSi_{2+x}$	tetragonal, $FeSi_2$ type	$a=2{\cdot}698$–$2{\cdot}683$, $c=5{\cdot}141$–$5{\cdot}127$ (hom. range); $M=1$	1,220
Ru	Ru_2Si	'anti-$PbCl_2$' (C–23)	$a=5{\cdot}27_9$, $b=4{\cdot}00_5$, $c=7{\cdot}41_8$; $M=4$	
	Ru_3Si_2(?)			
	RuSi	CsCl type (B–2) and B–20=FeSi type	$a=2{\cdot}909$; $M=1$ $a=4{\cdot}70_3$; $M=4$	low temp. form high temp. form
	Ru_2Si_3	tetragonal (slightly distorted)	$a=5{\cdot}53$, $c=4{\cdot}47$	
	$RuSi_2$(?)	$RuGe_2$ type		
Os	OsSi	CsCl type (B–2) and B–20=FeSi	$a=2{\cdot}960$; $M=1$ $a=4{\cdot}729$; $M=4$	
	Os_2Si_3	tetragonal	$a=5{\cdot}58$, $c=4{\cdot}48$	

Heat of formation (298°K), (kcal/g atom)	*Hardness* (kg/mm^2 at 50 g, unless otherwise stated)	*Density* (x-ray)	*Other properties: Electrical resistivity*, ρ, $\mu\Omega$ cm, at R.T.; expansion coeff., β superconductivity (SPCD): upper temp. limit	*Remarks*
				a new high-Mn silicide $Mn_{4\cdot5}Si$ recently detected
-30 for $\frac{1}{8}$ Mn_5Si_3				Mn_3Si: martensite-type transformation occurs upon quenching
				Mn_5Si_3: D–8_8 structure here genuine (without C, etc.)
		5·3		'$MnSi_2$' possibly a Si solution in Mn_3Si_5; semiconductor properties
				a phase '$MnSi_{1\cdot67-1\cdot73}$' reported (at 950°C), eutectic with Si and melting incongruently
				occurrence of 'Re_3Si' probable but not known
$> -9\cdot8$ for $\frac{1}{8}$ Re_5Si_3				
$-10\cdot2$ (or $-11\cdot6$) for 1 ReSi		13·1		
$-8\cdot3$ (or $-12\cdot5$) for $\frac{1}{2}$ $ReSi_2$	≈1,500		$\rho=100$	
		9·03		
('Fe_2Si' $-9\cdot2$)			$\rho=55$	Fe_3Si: superlattice over wide range of α-Fe–Si solid solution (devolving into separate phases in Mn_3Si and Cr_3Si)
$-4\cdot6$			$\rho=170$	
$-9\cdot6$		6·16	$\rho=240$	FeSi: several modifications (α',α'')
$-5\cdot6$	1,075	4·74 (meas.)	$\rho=350$	$FeSi_{2+x}$ becomes stoichiometric below ≈960°C (eutectoid/peritectoid reaction)
				'Os_3Si' probable but not known

Metal base	*Silicides*	*Structure type*	*Lattice dimensions, Å (M=number of molecules per unit cell)*	*Melting point, °C*
Group VIII cont.				
Os	$OsSi_2$	monoclinic	complex, similar to $OsGe_2$ structure	
	$OsSi_3$(?)			
Pu	Pu_5Si_3(?)			
	Pu_3Si_2(?)			
	PuSi	B–27=FeB type	a=7·933, b=3·847, c=5·727; M=4	
	(Pu_2Si_3?)			
	α-$PuSi_2$	C_c=$ThSi_2$	a=3·967, c=13·72; M=4	
	β-$PuSi_2$	C–32=AlB_2 (pseudo)	a=3·884, c=4·082; M=1	
Co	Co_3Si	?		1,210
	Co_2Si	'anti-$PbCl_2$' (C–23)	a=4·918, b=3·738, c=7·109; M=4	1,332
	CoSi	B–20=FeSi	a=4·447; M=4	1,460
	$CoSi_2$	C–1=CaF_2 type	a=5·367; M=4	1,326
Rh	Rh_2Si	'anti-$PbCl_2$' (C–23)	a=5·40$_8$, b=3·93$_0$, c=7·38$_3$; M=4	
	Rh_5Si_3	orthorhombic, Rh_5Ge_3 type	a=5·30$_9$, b=10·07$_4$, c=3·88$_8$	
	Rh_3Si_{2+x}	B–8=NiAs type	a=3·94, c=5·047; M=2	
	RhSi	orthorhombic; MnP type (B–31): H.T. form	a=5·53, b=3·063, c=6·36$_2$	
		B–20=FeSi: L.T. form	a=4·675; M=4	
	$RhSi_2$ (Rh_2Si_3?)	B–2=CsCl type (with trace Al)	a=2·963	
Ir	Ir_3Si	U_3Si type (BC tetr.)	a=5·22$_2$, c=7·954; M=4	
	Ir_2Si	'anti-$PbCl_2$' (C–23)	a=5·28$_4$, b=3·98$_9$, c=7·615; M=1	
	Ir_3Si_2	NiAs=B–8	a=3·96$_3$, c=5·12$_6$; M=2	
	IrSi Ir_2Si_3(?) $IrSi_2$(?)	MnP=$B_3$31	a=5·58, b=3·211, c=6·273	
	$IrSi_3$	hexagonal?	a=4·350, c=6·610	

Heat of formation (298°K), (kcal/g atom)	*Hardness* (kg/mm²) at 50 g, unless otherwise stated)	*Density* (x-ray)	*Other properties: Electrical resistivity, ρ,* μΩ cm, at R.T.; expansion coeff., β superconductivity (SPCD): upper temp. limit	*Remarks*
				Ru_2Si_3, Os_2Si_3: ranges $MSi_{1.8\pm0.3}$; $OsSi_3$: range $OsSi_{3\pm0.5}$
		10·15		
		9·12		Pu–$PuSi_3$ eutectic?
		9·18		
				Co_3Si probably exists at high temperatures only, 1,170–1,210°C Co_2Si: two forms CoSi: homogeneity range $CoSi_{0.9-1.09}$
−9·2				
−12·0				
−8·2	890	4·94	ρ=64·8	
				Rh_3Si_{2+x}: high-temp. phase only, >700°C
		8·5		

Metal base	*Silicides*	*Structure type*	*Lattice dimensions, Å (M = number of molecules per unit cell)*	*Melting point, °C*
Group VIII cont.				
Ni	Ni_3Si	ordered, $Ll_2=Cu_3Au$	$a=3{\cdot}5040$; $M=1$	peritectic below 1,165
	Ni_5Si_2	trigonal	$a=6{\cdot}670$, $c=12{\cdot}33$ (at Ni end) $=6{\cdot}670$, $c=12{\cdot}26$ (at Si end)	1,282
	Ni_2Si (low temp.)	'anti-$PbCl_2$' (C–23)	$a=4{\cdot}99$, $b=3{\cdot}72$, $c=7{\cdot}08$; $M=4$	
	Ni_2Si (high temp.)	hexagonal Ni_2In ($B–8_2$)	$a=3{\cdot}805$, $c=4{\cdot}908$; $M=2$	1,318
	Ni_3Si_2	orthorhombic	$a=12{\cdot}229$, $b=10{\cdot}05$, $c=6{\cdot}924$; $M=16$	peritectic below 845
	NiSi	B–31=MnP	$a=5{\cdot}18$, $b=3{\cdot}34$, $c=5{\cdot}62$; $M=4$	992
	$NiSi_2$	C–1=CaF_2	$a=5{\cdot}406$; $M=4$	peritectic below 993
Pd	Pd_3Si	Fe_3C type (DO_{11})	$a=5{\cdot}735$, $b=7{\cdot}555$, $c=5{\cdot}260$; $M=4$?	
	Pd_2Si	Fe_2P type (C–22)	$a=6{\cdot}497$, $c=3{\cdot}432$ (Pd end) $=6{\cdot}528$, $=3{\cdot}437$ (Si end); $M=3$	1,250
	PdSi	B–31=MnP	$a=5{\cdot}599$, $b=3{\cdot}381$, $c=6{\cdot}133$; $M=4$	1,100
Pt	Pt_3Si	monoclinic	$a=7{\cdot}71_6$, $b=7{\cdot}77_3$, $c=7{\cdot}77_8$; $\beta=88{\cdot}15°$	peritectic below 986
	Pt_2Si (high temp.?)	trigonal, C–22=Fe_2P type	$a=6{\cdot}43_6$, $c=3{\cdot}56_9$; $M=3$	1,100
	Pt_2Si (low temp.?)	tetragonal (or C–16 type?)	$a=5{\cdot}55_4$, $c=5{\cdot}92_3$; $M=3$	
	PtSi	B–31=MnP	$a=5{\cdot}59_5$, $b=3{\cdot}60_3$, $c=5{\cdot}93_2$; $M=4$	1,229

whilst for the Pt Group the new FeSi-, MnP- and NiAs-types structure appear.

There is a further important link-up to the *aluminides* to which some of these silicide structure groups also extend, e.g. the ordered CsCl type in RuSi to FeAl, the W_5Si_3 type to Zr_5Al_3. We might consider this essentially as a size-factor effect, but electron concentration must enter as well. (This relation develops, for the lower silicides, structurally into 'anti-types', as in the isomorphous Pd_3Si and $NiAl_3$.) This partial substitution of Si by Al is known in many silicides (e.g. [24]) with, however, in general a modulation of the silicide-base structures (see also below).

The frequent isomorphy with *germanides* should also be noted; this follows perhaps naturally from the common valency, but owing to the larger size of the Ge atom is possible only where the metalloid is no longer interstitial. Considerable Si → Ge substitutions in certain silicides

Heat of formation (298°K) (kcal/g atom)	*Hardness* (kg/mm²) at 50 g, unless otherwise stated)	*Density* (x-ray)	*Other properties; Electrical resistivity, ρ,* μΩ cm at R.T.; expansion coeff., β superconductivity (SPCD): upper temp. limit	*Remarks*
('Ni_2Si' −5·3)		7·905 (meas.)		
				Ni_5Si_2: note small homogeneity range with c-axis expansion only
−10·5		7·23		
				Ni_2Si: H.T. form has extensive solubility for Si
−10·7				
−10·3				
	1,020			$NiSi_2$: possible H.T. form exists
				'Pd_5Si_2' reported but uncertain Pd_2Si: small homogeneity range
				Pt_3Si *almost* cubic, from lattice dimensions 'Pt_5Si_2' reported but = Pt_3Si?

seem probable. A further structural link-up of the Group VIII silicides exists with the *phosphides*, in particular those of the NiAs and MnP class[25].

The *disilicides*, MSi_2, are probably the most important with regard to applications, though not necessarily the most stable. They are represented mainly by the five families of $TiSi_2$ (C–54)*, $ZrSi_2$ (C–49), $ThSi_2$ (C–32), $CrSi_2$ (C–40) and $MoSi_2$ (C–11) type and evolve towards Group VIII in a variety of fairly simple forms. The cubic CaF_2 (calcium fluoride, C–1) class in $CoSi_2$ and $NiSi_2$ is attained via the tetragonal $MnSi_2$ and $FeSi_2$ variants,

* Structural types in *Strukturbericht* nomenclature.

whilst in the platinum group metals the disilicides variate to MSi_3 or cease to be stable altogether (although in some cases this needs confirmation; cf. *Figure 7.1* and *Table 7.1*).

In *uranium* disilicide, the existence of its three polymorphic forms, of $ThSi_2 \rightarrow AlB_2 \rightarrow Cu_3Au$ type with increasing temperature, places uranium in a remarkable junctional position, each form being distinctive of other disilicide areas. Also have we here a decreasingly rigid stiochiometry, due either to lattice vacancies in Si sites (cf. β-$USi_2 \rightarrow USi_{1.5}$) or to further $M \rightarrow Si$ substitutions (cf. γ-$USi_2 \rightarrow USi_3$).

The coefficient of expansion of β-USi_2, as distinct from its UB_2 isomorph, has been discussed in Chapter 6. The very large and anisotropic expansion of USi_2 has been attributed[26] to the facts that no direct bonding force exists between the U atoms in the sheets and that the hexagonal network of Si atoms parallel to the layers is considerably compressed. The lattice is held together by Si–Si and U–Si–U bonds. Much crystallographic data are now known[27] on USi_2 and the associated compound range ($USi_{1.88}$ to USi_3). A decomposition of USi_2 into $USi_3 + USi_{1.88}$ occurs above 1,100°C, i.e. a division into structurally related Si-excess and Si-deficiency domains. A similar dimorphism has been reported[28] for $ThSi_2$.

The members of the $TiSi_2/CrSi_2/MoSi_2$-type series, all closely related, can again be described in terms of a simple variation in hexagonal stacking sequence of metal and silicon layers, from 4- to 3- to 2-fold repeat (in this order for the above three types); the $ZrSi_2$ type is intermediate to these and the $ThSi_2$ structure. Wallbaum's[29] original representation of one such layer is shown in *Figure 7.4*. Amongst the *iron* group disilicides, the change from

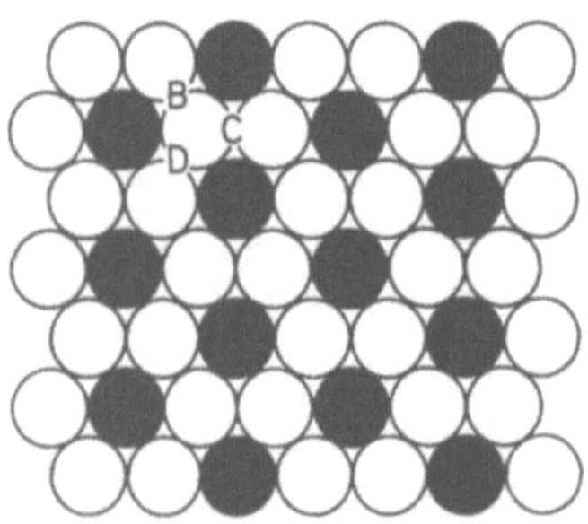

Figure 7.4. Atomic layering in disilicides (from Wallbaum[29])

By courtesy of the Editor, *Zeitschrift für Metallkunde*

the CaF_2 type for $NiSi_2$ and $CoSi_2$ to the $FeSi_2$ type implies an interesting transition from the viewpoint of interstitial alloys[30], as shown in *Figure 7.5*: the normal FCC metal lattice in $NiSi_2$ and $CoSi_2$, where a Si atom centres each octant, transforms in $FeSi_2$ to a structure of uniplanar metal–atom layers with Si forming adjacent pairs. This continues to evolve, from $FeSi_2$ through $MnSi_2 \rightarrow CrSi_2 \rightarrow VSi_2 \rightarrow TiSi_2$ to the Wallbaum-type hexagonal layer lattices. All these points directly bear on the *two*-metal silicide series (below).

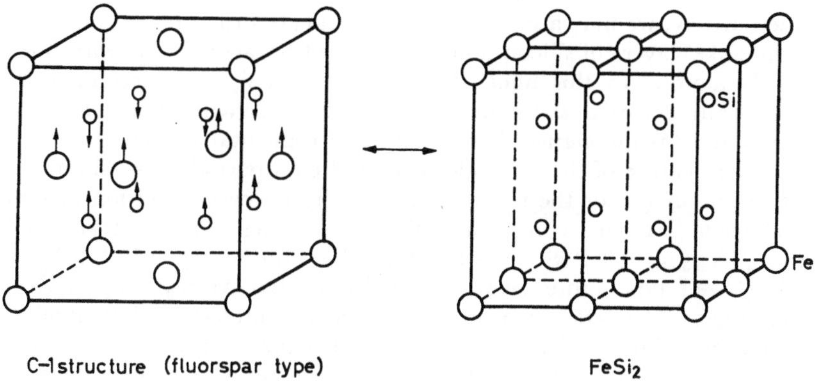

Figure 7.5. Structure relation of fluorspar-type to $FeSi_2$*-type silicide* (From Wittman, Burger and Nowotny[30]). (Small arrows indicate atom movements for transformation)

By courtesy of the Editor, *Monatshefte für Chemie*

The non-stoichiometric, Si-excess, $FeSi_2$ phase (termed ζ) undergoes a eutectoid transformation[31] at 917·5°C, resulting in an apparently true stoichiometric compound $FeSi_2$ below this temperature. The true $FeSi_2$ phase, in turn, decomposes peritectoidally to FeSi + ζ at a higher temperature (967·5° C, as shown in the FeSi–Si diagram[31] (*Figure 7.6*). Considerable

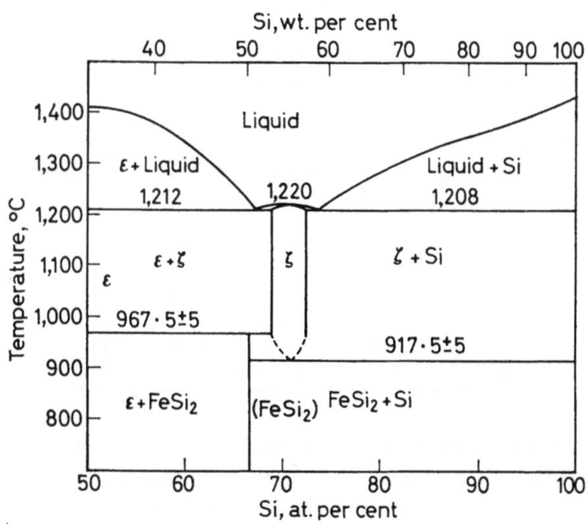

Figure 7.6. The FeSi–Si *diagram* (From Holdhus[31])

By courtesy of the Iron and Steel Institute, London

volume changes are linked with these transformations. Whether structurally this restoration of stoichiometry means the perfect occupation of all Fe and Si lattice sites in $FeSi_2$ (see *Figure 7.5*) or a balance of vacancies is an open question inviting investigation.

Alternatively, we may also interpret the disilicide structures as containing 'graphite-like' layers of silicon atoms in form of hexagonal nets with interposed M atoms. In this light, the above $\zeta \rightarrow FeSi_2 + Si$ decomposition clearly implies one with *excess* silicon to (*a*) a structure with such *graphite*-like Si layers and (*b*) the *diamond*-like structure of elemental silicon.

A further feature of the disilicides is that often a partial Si → Al replacement can occur (as for the monosilicides) with, however, a certain amount of distortion. Ternary compounds of the following ranges are known[24]: $TiSi_{1\cdot7-1\cdot4}Al_{0\cdot3-0\cdot6}$ of $ZrSi_2$ (C–49) type but distorted; $Mo(Si,Al)_2$ of $NbSi_2$ (C–40) type (note: *not* that of MSi_2) and $MoSi_{0\cdot7}Al_{1\cdot3}$ of $ZrSi_2$ (C–49) type. The latter two seem of particular interest in that small Al replacement in $MoSi_2$ (Group VI) alters its structure to give, electronically, a *Group V* equivalent, and on further Al replacement, a *Group VI* equivalent structure. This may well be correlated to a change in electron concentration involved by $Si^{4+} \rightarrow Al^{3+}$ substitution.

There are no silicides beyond the MSi_3 stoichiometry: in sharp contrast to the borides, Si analogues to MB_4, MB_6 and MB_{12} do not exist. Free silicon directly equilibrates with the di- (or, at most, tri-) silicides; its diamond-type structure might in fact be envisaged as inherent already in some of the disilicides.

An osmium trisilicide, $OsSi_3$, again with a spatial Si framework, has been reported[32]; altogether, the highest silicides are attained in the later Groups, VIII and the Pt metals. The $OsSi_3$ phase is highly resistant to chemical attack, and probably this applies to other Pt-Group silicides (see below, 'Applications').

An important fact is that in the silicides the metal–silicon distances tend to become smaller with increasing Group number[1]. The M–Si contacts, e.g. in the iron- and platinum-group silicides, are very close (less than the sum of the nominal atomic radii) but distant e.g. for the actinide silicides. This is strikingly illustrated by the pair U_3Si–Ir_3Si where, despite the isomorphy of the two, we have

	Shortest M–Si distance, Å	*Radius sum*, Å
Ir_3Si	2·4	2·53 (i.e. ≈ + 0·1)
U_3Si	2·93	2·7 (i.e. ≈ − 0·2)

Thus there is a considerable 'elastic factor' in M–Si distance. However, the tendency throughout to close Si–Si contacts should be noted; some, as in $FeSi_2$, are even closer than in elemental silicon. This sharply contrasts with e.g. the interstitial carbides and nitrides where, on the contrary, the tendency is mostly towards maximum C–C and N–N distance.

A bond classification similar to that of Kiessling for the borides (Chapter 6) may also be applied to the silicides (*Table 7.2*).

Amongst the *Rare Earth* silicides, such as those of cerium, many are isomorphous with, and similar in properties (notably refractoriness) to, the preceding families, in particular those of Groups IV and the actinides. Many of the Rare Earth disilicides are likewise isomorphous with $ThSi_2$

Table 7.2. Classification of silicides by Si–Si *bonds* (After Aronsson[1])

Isolated (or near-isolated) Si *atoms*	*Paired* Si *atoms*	*Chain* Si	*Network* Si	3-*Dimensional* Si *framework*
all M_3Si *all* M_2Si (no doubt also in Nb_4Si,$Ta_{4.5}Si$, though not yet proved) *some* M_5Si_3(M_3Si_2): Ti_5Si_3,Zr_5Si_3,Mn_5Si_3,Fe_5Si_3, Rh_3Si_2,Ir_3Si_2 *some* MSi: RuSi,CrSi,MnSi, ReSi,RuSi,OsSi,RhSi,IrSi, NiSi,PdSi,PtSi	U_3Si_2,Th_3Si_2	*some* M_5Si_3: V_5Si_3,α-Nb_5Si_3,α-Ta_5Si_3, Re_5Si_3, Rh_5Si_3 *some* MSi: ZrSi,Hf Si,USi,PuSi	*some* M_5Si_3: Cr_5Si_3,Mo_5Si_3,W_5Si_3, β-Nb_5Si_3,β-Ta_5Si_3 *some* $M$$Si_2$: β-$ThSi_2$,β-USi_2,β-$PuSi_2$	*all* $M$$Si_2$ (except those in preceding column) *all* $M$$Si_3$ silicon, elemental*

* Re silicon proper: though obvious, this classing of Si is not trivial, since the diamond lattice of elemental Si evolves from that of the disilicides, in which it is already incipient; some of the physical properties of Si are also closely related to those of the silicides. In the borides this becomes the more pronounced, in the MB_{12} structures and the quasi-solutions of metal in elemented boron (see Chapter 6); the metal atom entry strengthens the bond in both cases. It is interesting to note that in the high-C *carbides* this evolves to the layer lattice of *graphite* instead of the diamond lattice for silicon (which does also form for C under special high-pressure conditions).

(tetragonal), although in some cases distorted towards orthorhombic symmetry. *Table* 7.3 summarizes some structural and properties details for Group III and Rare Earth metals.

Table 7.3

*Rare Earth disilicides** (From Binder[33], etc.)

Silicide	*Symmetry* (T=tetragonal, O=orthorhombic)	*Lattice constants*, Å *a*	*b*	*c*	*Density*	*Other properties* *El. resistivity*, ρ, $\mu\Omega$ cm	*Melting point*, °C
$ScSi_2$†	?						
YSi_2	O	4·04	3·95	13·33	4·5	—	1,520
$LaSi_2$	T	4·31	—	13·80	5·0	273	1,520
$CeSi_2$	T	4·27	—	13·88	—		
$PrSi_2$	T	4·20	—	13·76	—		
$NdSi_2$	O	4·18	4·15	13·56	4·7	349	1,520
$SmSi_2$	O	4·105	4·035	13·46	—		
$EuSi_2$	T	4·29	—	13·66	—		
$GdSi_2$	O	4·09	4·01	13·44	6·4	263	1,540
$DySi_2$	O	4·04	3·95	13·33	6·8		

* Compare also the related $ThSi_2$ and USi_2 in *Table 7.1*.

† For scandium, a compound Sc_5Si_3 has also been reported[34]. Its crystal structure is again of the D–8_8 (Mn_5Si_3) type with $a = 7{\cdot}861$, $c = 5{\cdot}853$ Å; similarly for yttrium[35], Y_5Si_3. It is striking to note this recovery in Group III and IV of the D–8_8 structure, which had already occurred in Group VII and lapsed in V and VI where C or B were required to stabilize it.

The germanides, Sc_5Ge_3, La_5Ge_3 and Ce_5Ge_3, are also isomorphous, and the existence of stable 5/3 analogues amongst all Rare Earth silicides is probable.

The constancy of melting point (where known) in *Table 7.3* should be noted and the moderate refractoriness. Rare Earth silicides of stoichiometries other than MSi_2 have been reported[33] (MSi, M_5Si_3), as well as structure surveys[36,37]. Close relations exist again with the Rare Earth borides.

The transition-metal silicides are mostly good metallic conductors, as shown e.g. by the Hall coefficients[34]. Only $CrSi_2$ and $ReSi_2$ are impurity semi-conductors. Certain silicides of the Fe group metals (Mn, Fe, Co, Ni) are also ferromagnetic[38].

The *heats of formation* of a number of Group IV–VI silicides as reported by Robins and Jenkins[39] for Ti_5Si_3, TiSi, $TiSi_2$; Zr_5Si_3, ZrSi, $ZrSi_2$; $ThSi_2$; V_2Si; Ta_5Si_3, $TaSi_2$; $MoSi_2$ and WSi_2, showed a maximum in Group IV and generally lesser stability of the *di*silicides as against the lower silicides. By correlating this with the metal bond lengths, the replacement could be shown of metal–metal bonds in the metal by strong metal–silicon bonding in the silicides, also in line with the maintenance of metallic character. Again, we might read this as effectively the complete and genuine conversion of the Si atom into a metal atom *when bonded* in the silicides; for C in carbides, N in nitrides and O in oxides this is partially also true, subject to various provisos described earlier, and it again applies for the hydrides (Chapter 9).

The bonding energies of silicides of composition M_3Si_2 (see *Figure 8.1*, p. 350), together with those of the non-silicide refractory phases[40], prove to be remarkably similar for the different types compared with, however, a shift in maximum and height with Group and Period. The silicides show relatively the lowest bonding energies, with their maxima in Groups IV and V. An important point there is the similarity of bond-energy curves for the parent transition metals *themselves* to those of their silicides, borides, carbides and nitrides. This may be seen as strong supplementary evidence of the similar bond types in these phases, namely metallic, in tune with our original theme (cf. Chapters 1 and 2). *Figure 7.7* compares the *ratios* of bonding energies of silicides, carbides, nitrides[40] to those of the metal (see also *Figure 6.9*).

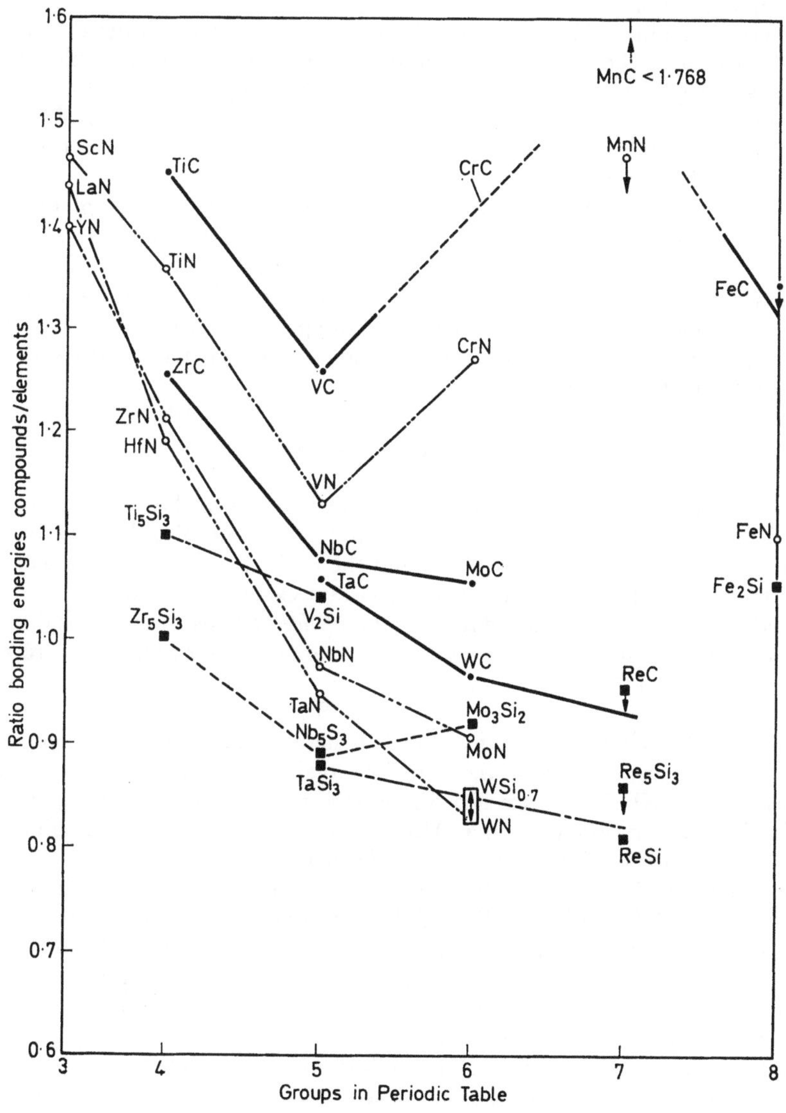

Figure 7.7. Ratio of bonding energies of silicides, carbides and nitrides to those of the metals (After Brewer and Krikorian[40])

By courtesy of the Electrochemical Society, New York

A comparison of the binary silicide and boride systems (cf. *Figures 6.2* and *7.2*, *6.3* and *7.1*) reveals a number of further basic differences and common factors, in conformance with their structural relations: for instance, *Figure 7.2* shows a remarkable absence of homogeneity ranges for each of the silicides except VSi_2. This contrasts to the borides and, it is felt, ought not yet to be finally accepted. Nevertheless, the capacity for lattice defection in the silicides is lower (also than in the carbides, nitrides and oxides), a circumstance which profoundly affects their properties, e.g. in coatings and their applications.

Referring again to *Figure 7.2*, much active current work on these binary silicide systems makes a constant revision of such a collation desirable; for instance, a slightly modified version of the V–Si diagram shown has been derived[41], and other changes and additions, especially for the low-Si ends or refractory metal–silicon systems, must be expected. (For individual *M*–Si systems see Bibliography.)

At the 'late end' of the transition-metal block, whilst for the platinum group metals silicides are still persistent (in contrast to carbides, etc.), they cease to occur[42] for Ag and Au with which Si forms plain eutectics. This contrasts, however, with Cu, which still does form a low-silicide $\approx Cu_5Si$ (with an appreciable Si/Cu homogeneity range, substitutional) and three forms[43], all modulations of the hexagonal close-packed structure, in stacking change from FCC copper.

7.3 TWO-METAL SILICIDES

From the single-metal silicides so far described, many of the two- (or multiple-) metal silicides systems can fairly confidently be predicted on the basis of crystal chemistry and isomorphies, and the limited experimental observations tend to confirm this, with some important exceptions, however.

Table 7.4 summarizes such data on *inter-silicide* systems as are available*. The question of solid solution or ternary compound formation is here of particular interest. Work has concentrated less on overall systems (M–M'–Si) than on definite homologous series such as MSi_2–$M'Si_2$, M_5Si_3–M'_5Si_3, etc.; the work of the Kieffer–Nowotny schools is here probably the most comprehensive. A few overall investigations are available, however, such as on Fe–Mo–Si, Fe–Nb–Si and Cr–Nb–Si. Furthermore, there are various more 'ad hoc' studies on systems such as $MSi_2 + M'$ in relation to finding suitable practical metal *bonds* for composite silicide materials.

As discussed, the mixed disilicide systems are governed by the four main types of C–54($TiSi_2$), C–49($ZrSi_2$), C–40($CrSi_2$) and C–11($MoSi_2$) structure and by the variants represented by the Fe, Pt group and actinide silicides. Thus *Table 7.4* includes the MSi_2, M_5Si_3, etc. isomorphs forming continuous solid solutions, and the near- or non-isomorphs forming limited ones. It also shows where ternary compounds are formed. An outstanding feature is that double silicides of a previously established *single*-metal type can appear in a series of differing terminal structures. This is the case e.g. in the

* As a side effect, the Table shows the many gaps still existing in our knowledge of two-silicide systems, even though they are now rapidly filled in view of the practical applications of these materials, e.g. for high-temperature protection.

Table 7.4

Ternary silicide systems, M_1–M_2–Si
(in order of Periodic System; transition metals only)

System	*Solid solutions or compounds*	*Structure: remarks*	*References*
Ti–Zr–Si	Ti_5Si_3–Zr_5Si_3 *CSS**	← $D\text{-}8_8$ →	23
	$TiSi_2$–$ZrSi_2$ *PSS**	C-54 → (5%) (45%) ← C-49†	45
Ti–Hf–Si	probably similar to Ti–Zr–Si; no data		
Ti–V–Si	$TiSi_2$–VSi_2 *PSS*	C-54 → (5%) (85%) ← C-40	45
Ti–Nb–Si	certainly similar to Ti–Ta–Si; no data		
Ti–Ta–Si	Ti_5Si_3–Ta_5Si_3 *PSS*	C-54 → (Nil) (50%) ← C-40	
	$TiSi_2$–$TaSi_2$ *PSS*		46
Ti–Cr–Si	$TiSi_2$–$CrSi_2$ *PSS*	C-54 → (7%) (88) ← C-40	45
Ti–Mo–Si	Ti_5Si_3–Mo_5Si_3 *PSS*	$D\text{-}8_8$ → (50%) (slight) ← $D\text{-}8_m$	23
	$TiSi_2$–*CPD*–$MoSi_2$	C-54 → C-40, 7–60% $MoSi_2$ ← C-11 Mo reduces $TiSi_2$ to Ti_5Si_3	
Ti–W–Si	Ti_5Si_3–W_5Si_3 *PSS*	$D\text{-}8_8$ → (25%) (50%) ← $D\text{-}8_m$	45
	$TiSi_2$–*CPD*–WSi_2	C-54 →← C-40 →← 58% ← C-11 C-40, 5–42% WSi_2	23
Ti–Co–Si	ternary '*G*' and '*E*' phases		
Ti–Ni–Si	6 ternary silicides including Laves, '*G*' and '*E*' phases		50
Ti–Mn–Si	ternary Laves, etc. phases		5, 48
Ti–Re–Si	Ti–Si_2–*CPD*–$ReSi_2$	C-54 → C-40 ← C-11	49
Ti–*M*–Si, where *M* = Th, U, Np, Pu, Fe and Pt Group metals; no data available			
Zr–Hf–Si‡	*CSS* between homologues not in doubt; no data		
Zr–V–Si	$ZrSi_2$–VSi_2 mutually insoluble		45
Hf–V–Si	probably similar to Zr–V–Si; no data		
Zr–Nb–Si	Zr_5Si_3–Nb_5Si_3	*CSS* doubted as not isomorphous but *PSS* certain	
Hf–Nb–Si	similar?		
Zr–Ta–Si	$ZrSi_2$–$TaSi_2$ *PSS*	C-49 → (Nil) (49%) ← C-40	46
Hf–Ta–Si	probably similar; no data		
Zr–Cr–Si	$ZrSi_2$–$CrSi_2$ almost insoluble		46
Hf–Cr–Si	probably similar; no data		
Zr–Mo–Si	$ZrSi_2$–$MoSi_2$ *PSS*	C-49 → (5%) (Nil) ← C-11	46
Hf–Mo–Si	probably similar; no data		
Zr–W–Si Hf–W–Si	probably similar to Zr–Mo–Si, Hf–Mo–Si '*E*'-phases		
Zr–Co–Si Zr–Ni–Si Hf–Ni–Si Zr–Fe–Si Hf–Fe–Si	'*G*' and '*E*' phases		

* *CSS* = complete series of solid solutions
PSS = partial series of solid solutions
CPD = a ternary compound occurs.
† Percentages mean mol. per cent of opposite silicide; at 1,300°C unless otherwise stated (but qualified equilibrium).
‡ Zr and Hf, Nb and Ta, Mo and W tabulated in pairs.

System	*Solid solutions or compounds*	*Structure: remarks*	*References*
Table 7.4 cont.			
Zr–*M*–Si and Hf–*M*–Si, where *M* = Th, U, Mn, Re, Np, Pu and Pt Group metals: no data			
Th–*M*–Si	no data available		
V–Nb–Si	V_3Si (A–15 β-W-type) solutions to about 32% Nb_3Si: then $(Nb_3V)_5Si_3$ + (Nb,V) equilibria (rôle of Nb_4Si not clear)		45
	VSi_2–$NbSi_2$ *CSS*	← C–40 →	52
V–Ta–Si	V_3Si (A–15 solutions to < 5% 'Ta_3Si': then probably as for Nb		45
	V_5Si_3–Ta_5Si_3 *CSS* VSi_2–$TaSi_2$: *CSS* probable but no data	← D–8 →	
V–Cr–Si	V_3Si–Cr_3Si *CSS*	← A–15 →	
	VSi_2–$CrSi_2$ *CSS* probable, but no data		
V–Mo–Si	V_3Si–Mo_3Si *CSS*	← A–15 →	
	VSi_2–$MoSi_2$ *PSS*	C–40 → (64%) (4%) ← C–11	52
	V–$MoSi_2$ section: ternary compound suggested		53
V–W–Si	probably similar to V–Mo–Si; no data		
V–Mn–Si	Si strong '*G*' stabilizer		
	V_3Si–Mn_3Si *PSS*	← A–15 →	
V–Fe–Si	ternary compounds $V_3Fe_5Si_2$ of α-Mn type		48, 54
	and extensive Si-stabilized σ phase from VFe		51
V–Co–Si	ternary compounds: Laves phase ($MgZn_2$ type)		42
	$V_{2.8}Co_{5.2}Si_2$ of α-Mn type		55
	$V_{1.5}Co_6Si_{2.5}$ CsCl type		54
	extensive Si σ phase from VCo		
V–Ni–Si	ternary Laves phase ($MgZn_2$ type) at $V_5Ni_6Si_4$		54
	extensive σ phase from VNi, also Laves, '*G*' and '*E*' phases		5
V–*M*–Si where *M* = U, Re, Np, Pu and Pt-Group metals: no data			
Nb–Ta–Si	*CSS* $(Nb,Ta)_4Si$, $(Nb,Ta)_5Si_3$ and $(Nb,Ta)Si_2$ certain; no data		
Nb–Cr–Si	5 ternary phases at about $Cr_5Nb_3Si_2$ (Laves phase); $Cr_2Nb_4Si_4$; $Cr_4Nb_3Si_4$; $Cr_9Nb_3Si_8$; $CrNb_3Si_{10}$; some have wide homogeneity ranges; Nb_5Si_3–Cr_5Si_3 slightly soluble, form compound which is β-Nb_5Si_3-stabilized by Cr; $NbSi_2$–$CrSi_2$ insoluble		56
Ta–Cr–Si	$TaSi_2$–$CrSi_2$ *PSS*	C–40 → (40%) (12%) ← C–40 (note miscibility gap between isomorphs)	45
Nb–Mo–Si	*CSS* between homologues probable; no data		
Ta–Mo–Si	Ta_5Si_3–Mo_5Si_3 *CSS* (?)	← $D\text{–}8_8$ → (but carbon-stabilized?)	45
	$TaSi_2$–$MoSi_2$ *PSS*	C–40 → (56%) (16%) ← C–11	46
	Nb_5Si_3–W_3Si_3 *PSS*		57
Nb–W–Si	$NbSi_2$–WSi_2 *PSS* (extensive in $NbSi_2$, nil in WSi_2)		57

System	Solid solutions or compounds	Structure: remarks	References
Table 7.4 cont.			
Ta–W–Si	'Ta_4Si and Ta_2Si stable in the presence of W'		
	$TaSi_2$–WSi_2 *PSS*	C-40 → (30%) (24%) ← C-11	46
Nb–Fe–Si	at least 6 ternary phases; extensive Laves-phase solutions for Si; '*E*' phase occurs		58
	Nb_5Si_3–Fe_5Si_3 insoluble; form compounds		
	$NbSi_2$–$FeSi_2$ insoluble		
Ta–Fe–Si	probably similar to Nb–Fe–Si; no data		
Nb–Co–Si, Ta–Co–Si, Nb–Ni–Si, Ta–Ni–Si	Laves, '*G*' and '*E*' phases		
Nb–*M*–Si, Ta–*M*–Si	where *M* = U, Mn, Re, Np, Pu and Pt Group metals: no data		
Cr–Mo–Si	Cr_3Si–Mo_3Si *CSS*	← A-15 →	45, 59
	$CrSi_2$–$MoSi_2$ *PSS*	C-40 → (32%) (46%) ← C-11	
Cr–W–Si	Cr_3Si dissolves up to 20% 'W_3Si' at 1,500°C; actual W_3Si not found		60
	$CrSi_2$–WSi_2 *PSS*	C-40 → (16%) (70%) ← C-11	45
	Cr_5Si_3–W_5Si_3 *CSS*		60
Cr–Mn–Si	Si stabilizes σ phase (from Mn_3Cr); large solubility of Mn in Cr_3Si		48, 55
Cr–Fe–Si	probable B-20 (CrSi–FeSi) *CSS*		62
	$CrSi_2FeSi_2$ *PSS*, simple eutectic system		62–3
Cr–Co–Si	Si σ phase extensive (from Cr_3Co_2)		55
Cr–Ni–Si	ternary silicide Cr_2Ni_3Si;		64
	ternary σ phase Cr_6Ni_3Si (hom. range)		55
Cr–*M*–Si, where *M* = U, Re, Np, Pu and Pt Group metals:	no data		
Mo–W–Si‡	Mo_5Si_3–W_5Si_3 *CSS*		59
	$MoSi_2$–WSi_2 *CSS*		65
	Mo_3Si decomposes to (W,Mo) + $(W,Mo)_5Si_3$ upon W addition		
Mo–Mn–Si, W–Mn–Si	ternary Laves phase		5
Mo–Re–Si	$MoSi_2$–$ReSi_2$ *CSS*		66
Mo–Fe–Si	ternary silicides, Fe_3Mo_3Si, Fe_3Mo_2Si and Fe_2Mo_2Si		67
W–Fe–Si	ternary silicides FeWSi and FeW_2Si reported; pseudo-binary section W–FeSi, containing FeWSi *CSS*		68
	Fe_3Si_2–FeWSi. αFe equilibrates Fe_3Si_2, FeWSi and Fe_3W_2; ternary Laves phase		48
Mo–Co–Si, W–Co–Si	ternary Laves phases ($MgZn_2$ type)		48
	'$MoSi_2$–Co form eutectic system'		53
Mo–Ni–Si	'$MoSi_2$–Ni form eutectic system'; ternary Laves phase		53
W–Ni–Si	ternary Laves phase		

‡ Zr and Hf, Nb and Ta, Mo and W tabulated in pairs.

System	*Solid solutions or compounds*	*Structure: remarks*	*References*
Table 7.4 cont.			
Mo–*M*–Si, W–*M*–Si	where *M* = U, Mn, Re, Np, Pu and Pt Group metals: no data except for Mo–U–Si where new ternary phase reported		69
U–*M*–Si, Mn–*M*–Si, Re–*M*–Si, Np–*M*–Si, Pu–*M*–Si and [Pt group metals]–*M*–Si	where *M* = Group VII and VIII metals: little data except as follows:		
Mn–Fe–Si**	Mn_3Si–Fe_3Si *CSS*, Mn_5Si_3–Fe_5Si_3 *CSS*, MnSi–FeSi *CSS* (highly complex equilibria)		70–1
	ternary silicide extending from $Fe_{6\cdot0}Mn_{74\cdot5}Si_{19\cdot5}$ to $Fe_{0\cdot8}Mn_{80\cdot2}Si_{19\cdot6}$		72
Mn–Co–Si	ternary compound $MnCo_2Si$ of Laves phase ($MgZn_2$ type structure)		54
Mn–Ni–Si	'*G*' phase; Si strongly stabilizes *β*-(Mn,Ni); several ternary silicides		
Fe–Co–Si**	FeSi–CoSi *CSS*	→ B–20 →	30
	$FeSi_2$–$CoSi_2$ nearly insoluble (seems abnormal)		
Fe–Ni–Si**	FeSi–NiSi *PSS*	B–20 → (50%) (10%) ← B–31	30
	$FeSi_2$–$NiSi_2$ *PSS*	$FeSi_2$ → (5%) (30%) ← C–1	73 (Fe-rich alloys only)
Co–Ni–Si**	CoSi–NiSi *PSS*	B–20 → (50%) (20%) ← B–31	30
	$CoSi_2$–$NiSi_2$ *CSS*	← C–1 →	

** The knowledge available in these systems chiefly concerns primary metal solution fields (α, γ, etc.), not silicides.

$TiSi_2 \longleftrightarrow MoSi_2$ system (C–54 ⟷ C–11) which contains a ternary phase* $(Ti,Mo)Si_2$ of C–40 structure ($CrSi_2$-like); other such ternary compounds, some having wide homogeneity regions, are noted in the Table. In consequence, these new ternary phases are able to form complete solid solutions with their single-metal analogues, e.g. $CrSi_2 \longleftrightarrow (Ti,Mo)Si_2$. As an example, *Figure 7.8* shows the two systems[46] $CrSi_2$–WSi_2–$TiSi_2$ and $TaSi_2$–$MoSi_2$–$TiSi_2$. The lattice dimensions (*a* and *c* axes, atomic volumes) change steadily within the extensive single-phase fields thus defined.

As to the quantitative solution limits mentioned in *Table 7·4*, the precise figures should not be stressed, as they depend on the specific heat-treating conditions employed—hot-pressing, usually annealing at 1,300°C—and equilibrium may not necessarily have been attained. However, the values may at least be taken as indicative.

The M_5Si_3 phases where isomorphous also form complete solutions. However, the D–8_8-type structure requires, in the case of Group V and VI metals, a small amount of *carbon* for this to be attained, otherwise the change is to the 'T_1' and 'T_2' types above. An important point to note is that a

* The ternary compound compositions range from about $Ti_{0\cdot8}Mo_{0\cdot2}Si_2$ to $Ti_{0\cdot4}Mo_{0\cdot6}Si_2$, and for tungsten from $Ti_{0\cdot9}W_{0\cdot1}Si_2$ to $Ti_{0\cdot6}W_{0\cdot4}Si_2$.

trace of extra interstitial such as carbon, but also nitrogen, oxygen and boron, can serve sensitively to control the degree of mutual solid solubility of these multi-silicide series, whether to advantage or detriment (see also Chapter 10).

The M_3Si silicides of β-W type also permit extensive mutual solid solutions, but again sensitively subject to interstitial presence or absence.

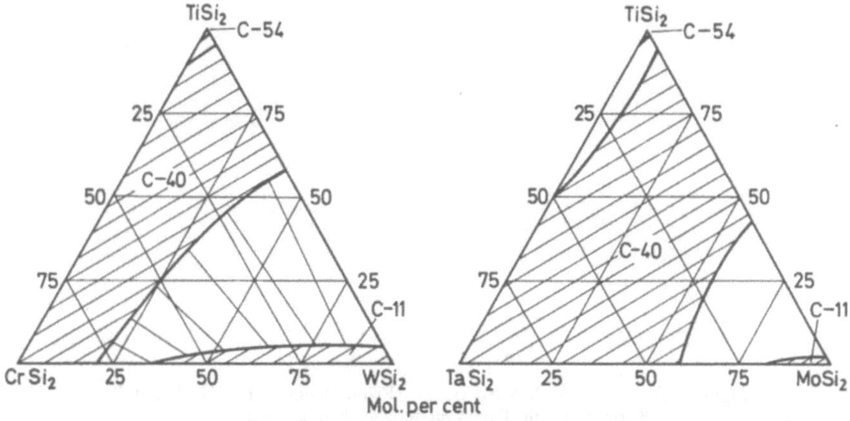

Figure 7.8. Sections through the systems $TiSi_2$–$CrSi_2$–WSi_2 *and* $TiSi_2$–$TaSi_2$–$MoSi_2$ (From Kudielka and Nowotny[46])

By courtesy of the Editor, *Monatshefte für Chemie*

The mono- and disilicide series of iron-group[30] base again show wide isomorphous ranges, FeSi–CoSi–NiSi, subject to changes from the B–20 to the B–31 type (cf. *Table 7.4*), and $FeSi_2$–$CoSi_2$–$NiSi_2$ with those from the Cl (fluorspar) to the $FeSi_2$ type, leading to the structural changes referred to already (see *Figure 7.5*). Thus also it is reasonable that CrSi, MnSi, FeSi and CoSi, all isomorphous and of B–20-type structure, form a continuous mutual series of solid solutions[74]. However, when we come to nickel, CrSi can only dissolve up to 40 mol. per cent NiSi (B–31 structure) and MnSi up to 70 per cent NiSi, whilst the converse solubilities, i.e. *in* NiSi, are very small.

Considering combined systems between the Group IV–VI and the *iron group* (VIII) silicides, the information is as yet insufficient to gain an overall picture; *Table 7.4* gives some details on this subject. The dominant feature is *ternary compound formation.* In the Fe–Mo–Si system[67] shown in *Figure 7.9*, three such phases have been established: Fe_2MoSi_2, Fe_3Mo_2Si and Fe_2Mo_2Si (the latter two, however, being solid solutions of Si in the Fe–Mo intermetallic compounds Fe_3Mo_2 and FeMo of μ and σ structures), and the Fe–Nb–Si and Cr–Nb–Si systems[56,58] (*Figures 7.10* and *7.11*) show a number of new compounds, and some highly complex equilibria. A notable feature is the stabilization by silicon of some well-known intermetallic compound structures. The *Laves phase* (of hexagonal $MgZn_2$ type) appears as a ternary compound ('ρ') in Cr–Nb–Si which in Fe–Nb–Si takes the form of an extensive Si solid solution based on Fe_2Nb. Ternary Si–Laves phases have also been reported[47] as Mo_2Co_3Si, W_2Co_3Si and $V_4Co_5Si_3$, as '$VCo_{1.1}Si_{0.9}$' (no doubt the same as the last-named) and[54] as '$VNi_{1.2}Si_{0.8}$'. Extensive

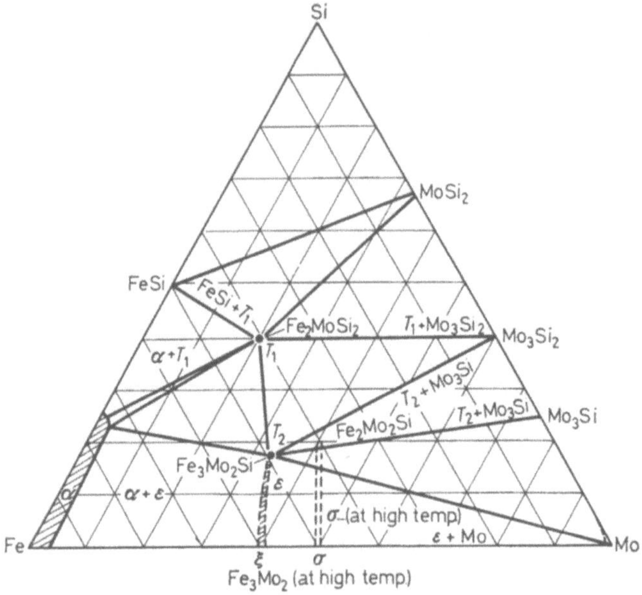

Figure 7.9. The Fe–Mo–Si *system* (From Vogel and Gerhardt[67])
By courtesy of the Editor, *Archiv für das Eisenhüttenwesen*

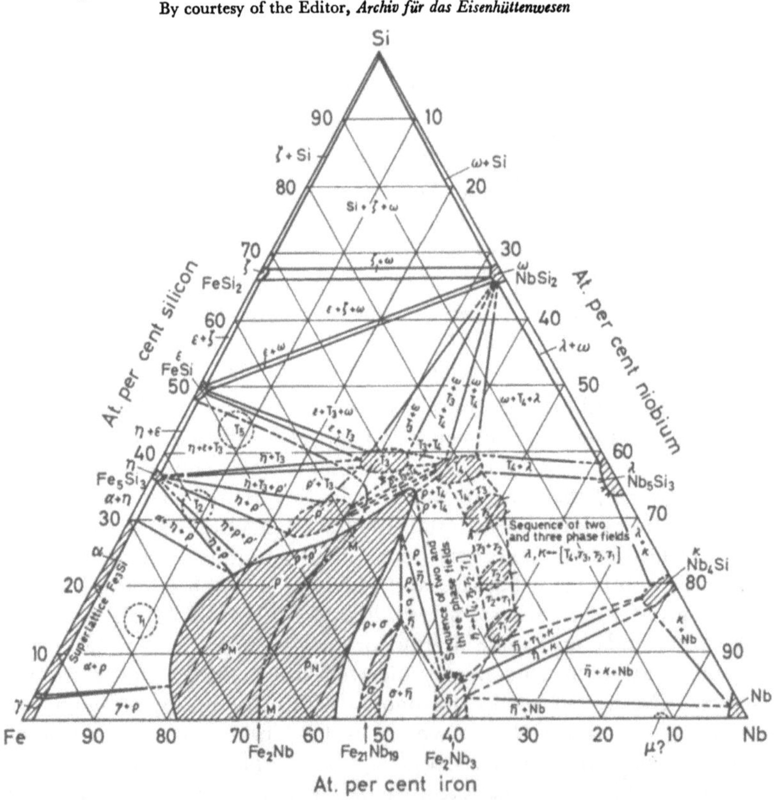

Figure 7.10. The Fe–Nb–Si *system,* 1,000°C section (From Goldschmidt[56])
By courtesy of the Iron and Steel Institute, London

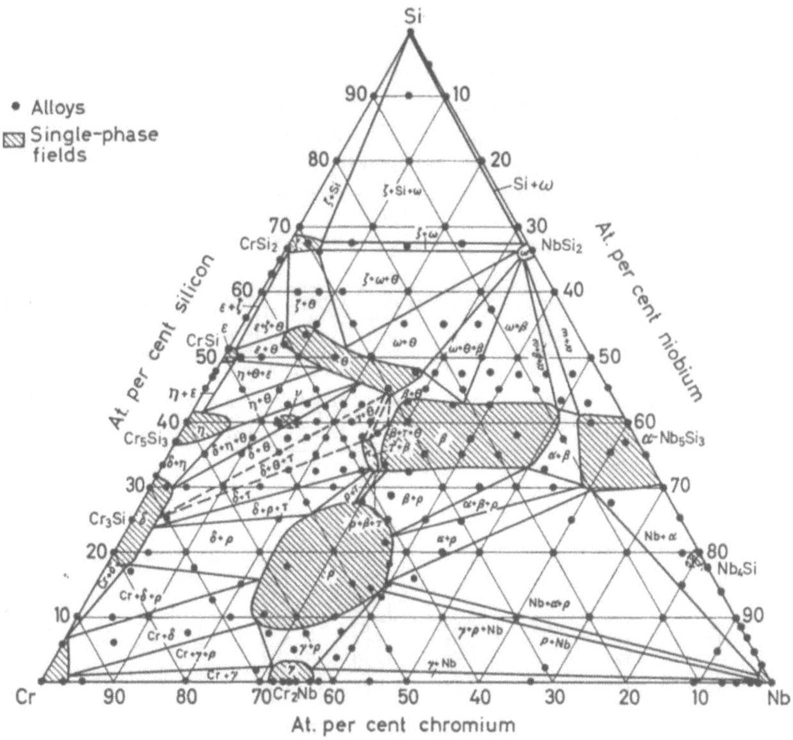

Figure 7.11. The Cr–Nb–Si *system,* 1,000°C section (From Goldschmidt and Brand[58])
By courtesy of the Editor, *Journal of less-common Metals*

Si-bearing σ phases occur[55] in the systems V–Fe–Si, V–Co–Si, V–Ni–Si and Cr–Co–Si, Cr–Ni–Si and Cr–Mn–Si as well as[58] in Fe–Nb–Si. Ternary silicides of *α-manganese* type* are known[54] in $V_3Fe_5Si_2$ and $V_{2\cdot8}Co_{5\cdot2}Si_2$. A silicide Ni_4MoSi_2 has also been reported[39], whilst in Ni–Ti–Si six silicides have been observed[50], of approximate formulae NiTiSi, Ni_5Ti_6Si, $Ni_7Ti_2Si_3$, Ni_3Ti_2Si, Ni_4TiSi_3 and $Ni_4Ti_4Si_7$.

More recently[51] considerable extensions took place in this field, a further group of ternary, so-called 'G' ($Ti_6Ni_{16}Si$)- and 'E' (TiNiSi)-type silicides and germanides being found. The 'G' phases are FC cubic with 116 atoms per unit cell and varying spacings: for $Ti_6Ni_{16}Si_7$, $a = 11\cdot187$ Å. The 'E' phases are orthorhombic with 30–36 atoms per unit cell (uncertain) and $a = 7\cdot02$, $b = 5\cdot18$, $c = 11\cdot11$ Å. A full list of 'G' and 'E' phases and their spacings is given in *Table 7.5*. The first to detect and define the 'G' silicide were Beattie and co-workers[76]; with Gladyshevskii[54,75], they showed the structure to be isomorphous with Mg_6Mn_{23}, the intermetallic compounds. The general feature throughout, considering the transition-metal silicides only, is the combination of Group III–V and Group VIII elements, with apparently a recession of these structures for Groups VI and VII, although this may not be final. Writing the 'G' compounds generally as $A_6B_{16}C_7$,

* We may here remember the *carbon*-stabilized α-Mn (χ) phases (Chapter 4)—the parallel with the silicon-stabilized ones seems important and worth further investigation.

Table 7.5

The ternary silicides (and germanides) of 'G'- and 'E'-phase types

($G = A_6B_{16}X_7$, $E = ABX$ where $X =$ Si or Ge); including lattice dimensions, Å: figures under E phase = a, b and c axes, under G phase = cube edge (From Spiegel, Bardos and Beck[51]; (*a*)[47]; (*b*)[54]; (*c*)[75])

G Phases

Silicides B \ A	Sc	Ti	Zr	Hf	V	Nb	Ta	Cr	Mo	W	Mn
Ni	11·429 (*a*)	11·187 11·22 (*b*)	11·423 11·47 (*b*)	11·39 (*b*)	11·153 11·13 (*b*)	11·249 11·24 (*b*)	11·215 11·22 (*b*)	no	no (*b*)	no (*b*)	11·15 (*b*)
Co	11·438 (*a*)	11·201	11·417	exp.	no	11·235	11·198	no?	no	no	no
Fe	no?	no	no	no?	no	no	no	no	no	no	no
Germanides											
Ni	11·663 (*a*)	11·421	11·689	exp.	no	11·504	exp.	no?	no	no?	11·41 (*a*)
Co		no	11·625	exp.	no	11·476	exp.	no?	no?	no?	no
Fe	no?	no?	no	no?	no?	no?	no?	no?	no?	no?	no?

E Phases

Silicides B \ A	Ti	Zr	Hf	V	Nb	Ta	Cr	Mo	W
Fe	no	7·33 5·46 11·47	exp.	no	7·11 5·29 11·25	7·07 5·25 11·18	no (*c*)	no (*c*)	no (*c*)
Co	7·04 5·20 11·04	7·27 5·44 11·47	exp.	6·85 5·04 10·78	7·04 5·24 11·18	6·98 5·26 11·11	no?	no (*c*)	no (*c*)
Ni	7·02 5·18 11·11	7·35 5·44 11·47	exp.	6·87 5·05 10·91	7·07 5·25 11·18	7·04 5·23 11·11	no?	no (*c*)	no (*c*)
Germanides									
Fe	no	no	no?	no	no	7·27 5·30 11·47	no?	no?	no?
Co	no	7·41 5·52 11·55	exp.	7·02 5·16 11·11	7·18 5·33 11·25	7·16 5·30 11·32	no?	no?	no?
Ni	7·16 5·27 11·32	7·39 5·51 11·70	exp.	7·04 5·14 11·19	7·20 5·38 11·25	exp.	no?	no	no?

no = G or E phase absent at temperatures investigated
no? = probably absent, by analogy
exp. = expected but not yet verified

the Ti group elements are the most effective A metals, Ni the most effective B metal. However, there seems little doubt that the compositions shown are merely first representatives of a much wider family amongst transition-metal double compounds, including possibly the Pt group as B metals; borides and carbides as well as silicides might further be included. The transition from the G phase to the *η-carbide*-type family (Chapter 4) is suggested as a further fruitful subject of study, as the two are structurally closely related. The G phase might almost be considered as the silicon 'version' of the η-carbide type. A further fascinating aspect is to seek the link-up to isomorphous (Si-free) pure intermetallics of G-related type, similarly as is known for the η-carbides.

The *'E' silicides* concentrate at a composition 'ABC' (= ABSi or ABGe), and extension of work in the much wider framework of alloy chemistry, particularly to the Laves, α-Mn and σ phases occurring in the binaries of the same ternary systems, would be of interest.

Generally, all these ternary silicides are of high hardness and good oxidation resistance, although variably so. A particularly interesting case of interplay of silicide structure in two-metal systems occurs in Cr–Nb–Si, where the very-high-temperature β form of Nb_5Si_3 is stabilized to low temperatures by Cr additions (β field in *Figure 7.10*) to give a seemingly pure ternary silicide; this links, however, to the binary one with raised temperature.

Some further work on M_1–M_2-silicon systems[40] showed that, if M_1 = Ta, M_2 = Mo or W, then $Ta_3Si_{0.6}$ and Ta_2Si equilibrated free Mo and W; if M_1 = Ta, M_2 = Ti, a join occurs between Ta_2Si and Ti_5Si_3, with a solubility and expansion of the Ti_5Si_3 lattice by Ta; if M_1 = Ti, M_2 =Mo, the latter reduces $TiSi_2$ to Ti_5Si_3, and a join Ti_5Si_3–$MoSi_2$ is possible; if M_1 = Mo, M_2 = W, then Mo_3Si is stable in the presence of tungsten (i.e. supporting the non-formation of a 'W_3Si' analogue); and if M_1 = Zr, M_2 = Nb, a join occurs between Zr_5Si_3 and Nb, and complete inter-solubility of Zr_5Si_3 with Nb_5Si_3 (Mn_5Si_3-type structure series).

The relative stability ranking of the silicides, according to Krikorian's thermodynamic studies, is in the decreasing order Ti_5Si_3, $TaSi_{0.6}$, Ta_2Si, $TaSi_{0.2}$ > Mo_3Si > $WSi_{0.7}$; Zr_5Si_3, Zr_2Si, Zr_4Si > $NbSi_{0.55}$.

Extensive study has in recent years been proceeding on ternary *alumino-silicide* systems of the transition metals (*M*–Si–Al), in connection with their likely high-temperature oxidation resistance and use for protective coatings. (Ti–Al–Si and Mo–Al–Si had been mentioned above.) Here again, ternary compound formation is common, but also considerable Si $\longleftrightarrow$ Al interchangeability, giving pseudo-binary silicide–aluminide sections with complete or large solution ranges. Much of this work is once more due to the Plansee School. The systems[77] (Ti,Zr,Hf)–Al–Si are shown in conjunction in *Figure 7.12* where the trend of equilibria should be self-explanatory. Particularly interesting is the occurrence, in the shape of pure *ternary* compounds, of structures which otherwise are known as binary aluminides or silicides. This continues through Groups IV, V and VIA. For example, the 'C–49' structure of $ZrSi_2$ and $HfSi_2$ recurs in $Ti(Si,Al)_2$; the 'C–54' structure of $TiSi_2$ in $Nb(Si,Al)_2$ and $Mo(Al,Si)_2$, and the 'C–40' structure of VSi_2, $NbSi_2$, $TaSi_2$ and $CrSi_2$ in *another* compound, $Mo(Al,Si)_2$, of different Al/Si ratio, and in $W(Si,Al)_2$. (So far, the 'C–11b' structure of $MoSi_2$ and WSi_2 has not yet found its ternary counterpart, but this is not excluded and could possibly occur for the Pt group.)

Note also the almost complete (D–8_8-type) solid-solution range from Zr_5Si_3 to Zr_5Al_3 (lattice expansion with increasing Al/Si ratio): 'almost' because *pure* Zr_5Si_3 of D–8_8 type does not exist but, as discussed, becomes stabilized by small amounts of extra interstitials B, C, O or N; these become unnecessary in the presence of aluminium. This develops strikingly for hafnium (see diagram) where we have the D–8_8 isomorph in form of a pure *ternary* compound only, of wide homogeneity range.

In Groups V and VI, the (V,Nb,Cr,Mo)–Al–Si systems have been investigated[78]. The systems for Nb, Cr, Mo are included in *Figure 7.12* to

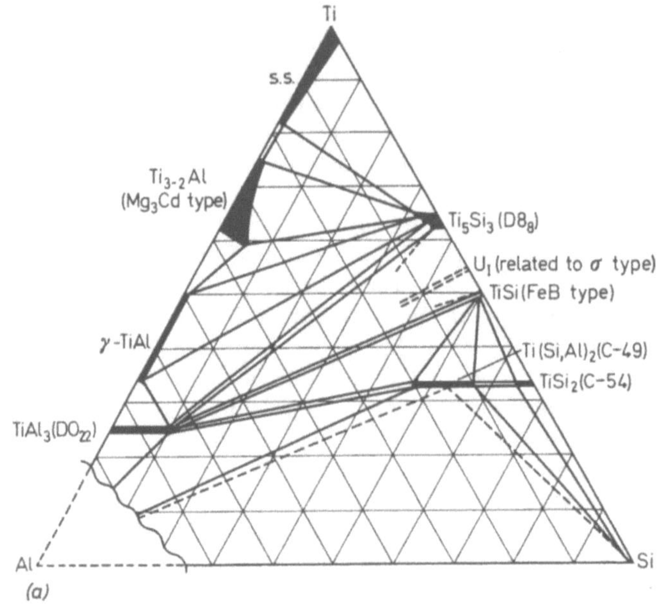

Figure 7.12. Phase equilibria M–Si–Al where *M* = Ti, Zr, Hf; Nb, Cr, Mo (*a–c* from Schob, Nowotny and Benesovsky[77]; *d–f* from Brukl, Nowotny and Benesovsky[78])

By courtesy of the Editors, ***Planseeberichte für Pulvermetallurgie*** and ***Monatshefte für Chemie***

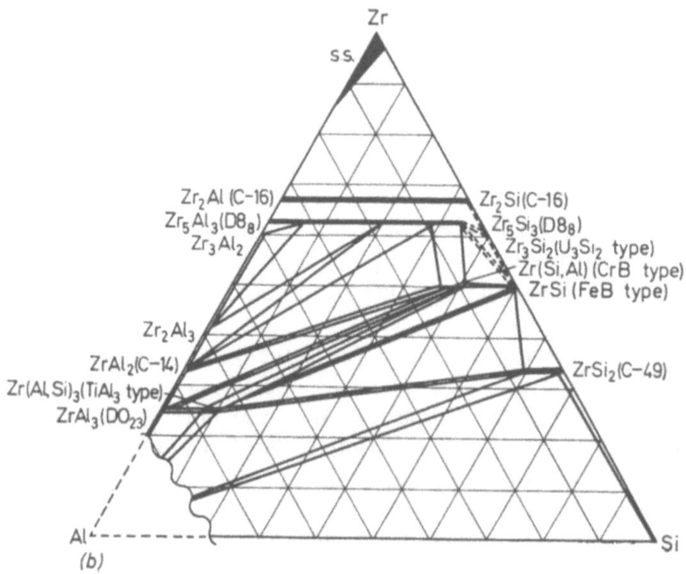

Figure 7.12b

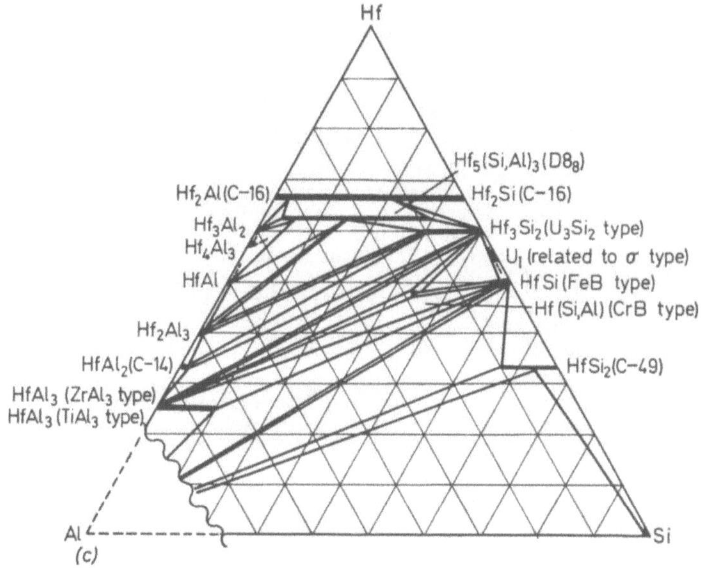

Figure 7.12c

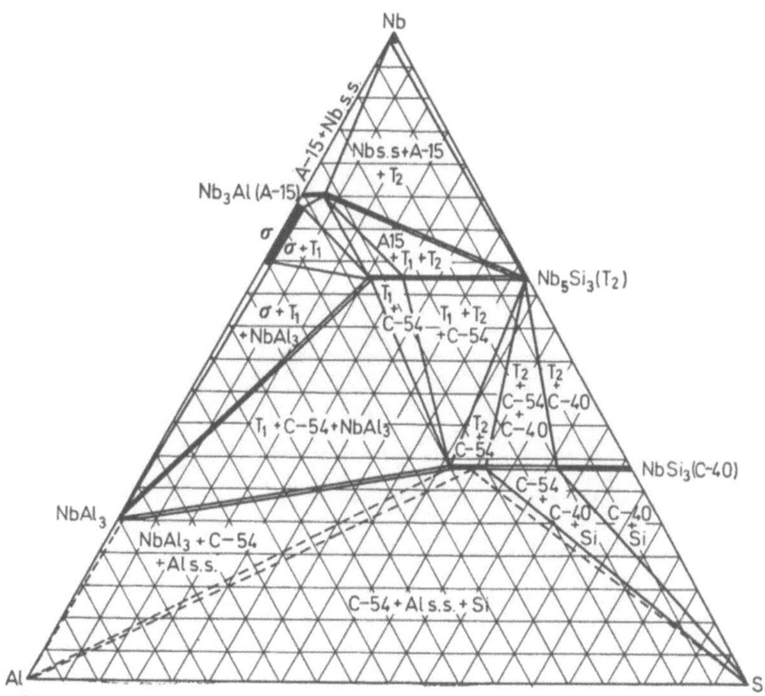

Figure 7.12d

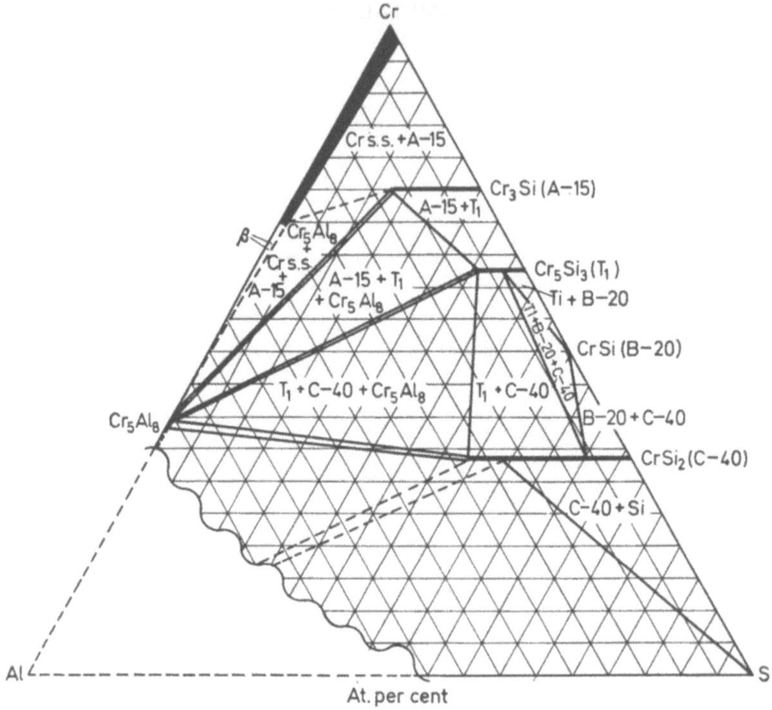

Figure 7.12e

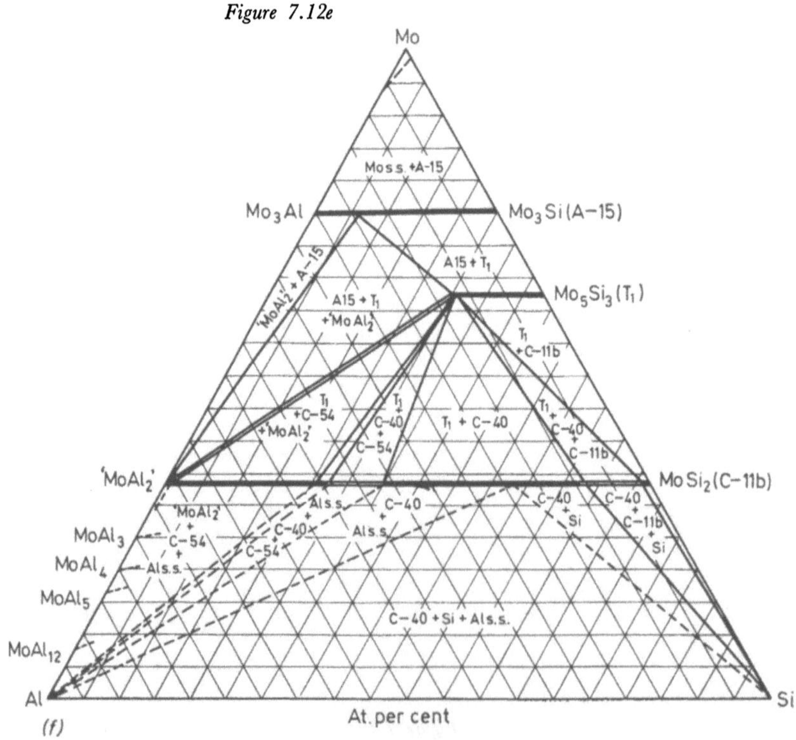

Figure 7.12f

compare with those of Group IV above. The ternary compounds should again be noted. The phase $Cr_4Al_{13}Si_4$ (with homogeneity range) has face-centred cubic structure, whilst a second Cr–Al–Si compound reported really evolved from $CrSi_2$ by Al solution.

Al-rich compounds in the Cr–Al–Si, Mn–Al–Si and Fe–Al–Si systems had been reported[79] earlier; the structure of Mn_3SiAl_9 was found[80] to be similar to $CoAl_5$, with vacancies. Clearly we here transgress from interstitial to intermetallic compounds, but this highlights the 'bridge' position of silicon in this respect. For further references to *M*–Al–Si systems see Bibliography.

Many of the phases referred to have a direct bearing on the oxidation resistance and protectiveness of the compounds concerned (see below, Applications). It may be noted here, however, that good oxidation resistance depends on thin protective *oxide* films forming (SiO_2 on $MoSi_2 \rightarrow$ mullite, etc.; $MoSi_2(Al)$?), which links the present *M*–Al–Si systems directly to the whole field of silico-aluminates (mineral structures). Here Si–Al ion substitution and metal interstition play, of course, a major part which must divolve from the structure relations of the *metallic* substrate compounds or solutions.

Against Si $\leftrightarrow$ Al exchange we might set Si $\leftrightarrow$ B exchange: both aluminium and boron, being of Group IIIB, would divert valency electrons from silicon. However, Al is larger, B smaller than Si in atomic size, so that silicoborides truly remain within our terms of reference (see Chapter 10), and both are of interest as potential modulants of oxidation resistance.

Germanium, being an immediate homologue to silicon, has also received some recent attention, from the viewpoints of both chemical structure and of applications (see Bibliography). In a system such as W–Ge, in contrast to W–Si (and to Mo–Ge, which does form germanides analogous to the Mo-silicides: Mo_3Ge of β-W structure, Mo_5Ge_3, Mo_2Ge_3 and $MoGe_2$), compounds no longer form, but there is a simple eutectic. Germanium substitutes, however, extensively in W-silicides and, in the Mo–W–Ge system[81–2], tungsten in the Mo-germanides. Again, the V–Ge system[7] proves akin to V–Si, though with some structural deviations.

Much valuable work on ternary silicide systems has also emerged from the practical interest in bonding silicides (e.g. $MoSi_2$) with other metals. This includes wettability and impregnation studies[83], bonding by Fe, Ni, Co, Cr and stainless steel[84]; bond by Ni, Co, and Pt, where the critical position of ternary silicide or solid-solution formation immediately affects performance in high-temperature uses. Thus double silicides can be severely embrittling[85]. A basic condition is that the bond metal should (*a*) be stable towards the silicide but (*b*) nevertheless have good wettability (to some extent a conflicting condition, limiting the choice of partner metals), quite apart from the requirements of thermal expansion, strength, shock resistance and refractoriness.

Many multi-silicide systems are still unexplored (see *Table 7.4*) and a full correlatory survey is not yet possible at this stage. However, as for borides and carbides, indications of a pattern emerge in which the $\nearrow$ and $\searrow$ diagonal combinations in the Periodic Table favour *double silicides* and *solid solutions*, respectively.

The interesting idea of 'pseudo-silicides*' has also been formulated[86]. These are intermetallic compounds not containing silicon themselves but based on structures established for the silicides and maintained by electronically balanced replacement of elements, under the restriction of atomic sizes. Examples of such pseudo-silicides are $Hf_5(In_{0\cdot5}Sb_{0\cdot5})_3$ of $D\text{–}8_8$ structure, and $Mo(Al,P)_2$ of C–40 structure (for a fuller discussion, the original paper[86] is worth referring to).

7.4 PROPERTIES AND APPLICATIONS OF SILICIDES

Some physical and chemical properties of the silicides are included in *Table 7.1*. All transition-metal silicides are metallic in character, and probably the most important properties in practice are their high melting points, in common with borides and carbides, and their stability and chemical inertness against oxidation and other reactions, maintained up to high temperatures. The latter is utilized, for instance, in *protective layers* upon alloys for high-temperature service. All silicides are hard and brittle, and this brittleness and liability to form microcracks often militates against such applications succeeding and would require effective ductile bonding means. Their use in hard metals is also thus restricted.

The electrical and thermal conductivities of silicides are high, and the coefficient of thermal expansion is in some cases, notably for $MoSi_2$, comparable with that of the substrate metals, such as molybdenum- or niobium-base alloys, to which silicide layers could be applied as protectants. Several silicides, e.g. V_3Si, Mo_3Si, W_5Si_3 and $ThSi_2$, have gained interest as *superconductors*, V_3Si having one of the highest superconductive transition temperatures[87] known, 17° abs.

The high thermal conductivities of silicides assist their thermal shock resistance and thus, their attraction as *high-temperature materials*. Hardness and strength at high temperatures, though only moderate, are adaptable by alloy additions or composite layerings, graded from alloy substrate to the exposed surfaces; a measure of ductility, and especially the oxidation resistance, all weigh favourably for silicides. To this must be added their adaptability in expansion coefficients.

The *melting points* (*Table 7.1*) are best surveyed in form of the binary diagrams (*Figure 7.2*). In *Figure 6.9* the melting points of the Group IV–VI disilicides and diborides had been compared graphically. The very different trends with increasing atomic number along Period and Group are self-explanatory. Of note are the relatively low values for the Group IV silicides, in contrast to the Group IV borides, but the strong silicide ascendancy towards Group VI. Many silicides melt congruently, and the high *eutectic* temperatures should be noted, in contrast to those in the boride systems. A remarkable feature, in fact, is the often relatively small difference between eutectic and melting point and between melting points of different silicides, reflected by the more 'undulating' nature of the melting diagrams in *Figure 7.2*. A detailed comparison with *Figures 4.1*, *5.1* and *6.2* will be instructive.

* This can, it is suggested, be fruitfully extended to 'pseudo-borides', 'pseudo-carbides' and 'pseudo-nitrides', a theme in harmony with one of the basic ideas of this book.

The emergence of the M_5Si_3 silicides as of higher melting point than the disilicides contrasts with the boride analogues. Nevertheless, it is the disilicides, notably $TiSi_2$, $CrSi_2$, $MoSi_2$ and WSi_2, which exhibit the most attractive refractory properties. These depend on quite a complex of factors other than melting point, such as the ability (1) to form thin, continuous and self-healing surface films of vitreous silica; (2) to form adhesive bonds to the alloys substrate; (3) to have thermal shock resistance and thermal expansions adaptable to both base-metal and admixed phases (which implies also good thermal conductivity), and (4) to form effective diffusion barriers to other embrittling interstitials entering the substrate alloys. Also the *di*silicides can more readily be sintered and are somewhat softer than the lower silicides and, with the aid of suitable matrices, can attain a small degree of ductility.

The mechanism of protection, for instance of $MoSi_2$, may be envisaged as follows. The highly volatile metal oxide, MoO_3 (of melting point 795°C and of high pressure already at much lower temperatures), tends to disappear, leaving a film of amorphous silica which becomes increasingly more viscous with time and adhering to the $MoSi_2$ matrix to protect it. For other silicides, such as WSi_2, $NbSi_2$, $TiSi_2$, $CrSi_2$, etc., the stable metallic oxides are by nature less volatile. Whether or not an effective protection can arise, depends then on (1) their interreaction with silica and the physical properties of any resulting metal silicate supplied by the substrate; (2) the atomic volume ratios of both the oxides and the silicide concerned; (3) the eutectics in the metaloxide–SiO_2 system; and (4) the ability to vitrify and for the SiO_2 to remain vitrified. This complex interplay of conditions makes it difficult to prescribe a universally valid mechanism, and none has yet been arrived at; a transient rôle of silicon suboxide, SiO, formation is also likely to participate.

Returning for a moment to the 'basic philosophy' of interstitial compounds and their interplay, we may note that a property like unreactiveness and resistance to oxidation, which seems 'inherent' in the silicides, is really not so. It depends on their ability to coat themselves with oxides (mainly silica, even though in very thin layers), and thus the property is ultimately the function of *another* interstitial compound, namely really that saturated in the element (oxygen) from which it is to protect!

In view of the critical nature of the metal–silicon binary systems (cf. *Figure 7.1*), it is easy, by a slight change in composition or during the advance with time in service exposure of the material at temperature, to depart from the MSi_2 stoichiometry, i.e. to introduce either the low-melting siliceous eutectic or, on the other side, the M_5Si_3 phases with their very different properties.

The superior scale resistance of the disilicides over other compositions (e.g. M_5Si_3, M_3Si) may be seen instructively in *Figure 7.13* which shows the degree of oxidation of Mo–Si and Ta–Si alloys as a function of composition[83,88]. The microhardness of the disilicides, ranging from 800 to 1,300 kg/mm^2, is rather less than that of other refractory interstitial compounds.

Regarding oxidation protection by silicide coatings (in air), the following comparative 'ranking'[89] may be of interest (testing time: order of hours); it should be emphasized that this can be greatly time-dependent, according to the eventual breakdown of a protective layer.

Oxidation-resistant to

$ZrSi_2$, $NbSi_2$: 800°–1,100°C

$TaSi_2$: ~1,400°C (ductility slightly higher than for $NbSi_2$)

$CrSi_2$: nearly melting point (1,570°C)

$MoSi_2$, WSi_2: above 1,700°C

To this may be added rhenium disilicide, $ReSi_2$, which has also found more recent application in high-temperature protection*. $ReSi_2$ is of more recent interest with regard to oxidation resistance also in the alloyed state. The continuous solid solution it forms with $MoSi_2$ contains, at 50 per cent

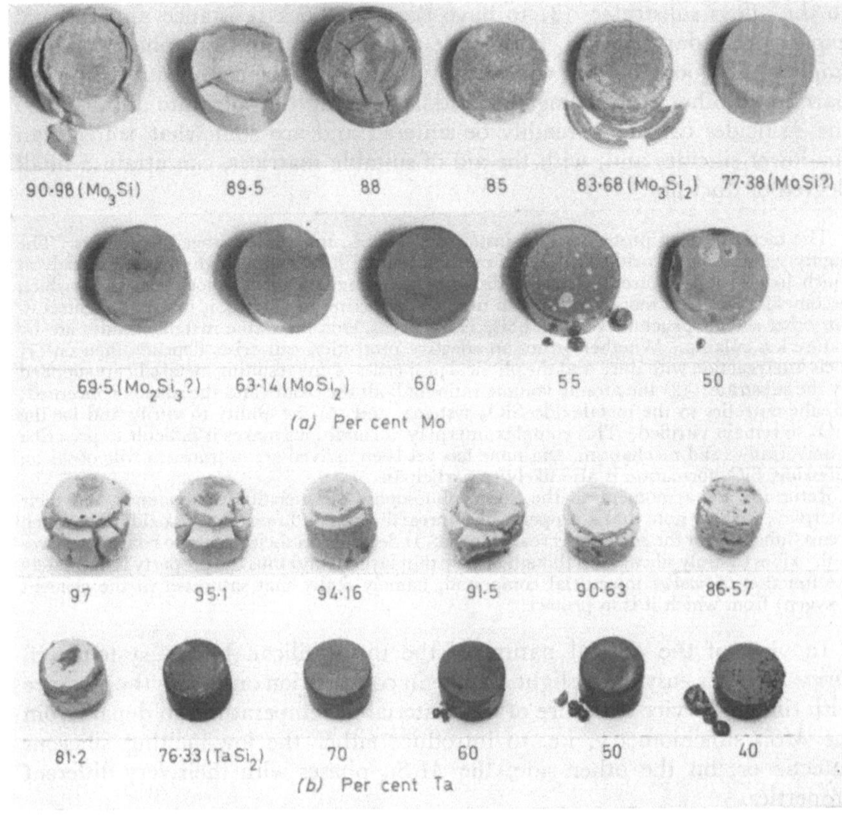

Figure 7.13. (*a*) *Actual* Mo–Si *alloys,* (*b*) Ta–Si *alloys after oxidation at* 1,500°C *for* 1 *h* (From Kieffer and Benesovsky[88])

By courtesy of the Iron and Steel Institute, London

$ReSi_2$–$MoSi_2$, a composition of extreme oxidation resistance at high temperatures[90b]. The system $TiSi_2$–$ReSi_2$ also develops an intermediate compound at 10 per cent $ReSi_2$ of the $CrSi_2$ (C–40)-type structure[49], which, as $CrSi_2$ itself, might attract unusual attention in view of its properties. The same applies e.g. to the mixed (Ti,Mo)Si_2 compound. The broader principle then arises to variate the mutual alloying of the transition-metal silicides (and

* For $ReSi_2$, any weight change when held in air at 1,400°C ceased after 30 min, due to a thin impervious oxide film[90a] (a behaviour shown, in principle, by all disilicides). $ReSi_2$, incidentally, is valuable as a high-temperature semiconductor, and it has also found electrical applications up to 1,700°C. However, there seems a conflict, in that *poor* oxidation resistance for all rhenium silicides has been reported[91] alternatively.

borides) such that ternary phases would be stabilized of a new and independent structure (controlled by electron relations), unlike that of the end members but isomorphic to that of other known single silicides, themselves of proven merit.

Thus among the disilicides themselves there are marked gradations in protective performance, in part connected with a slight ductility possessed by $NbSi_2$, $TaSi_2$, $MoSi_2$ and WSi_2 and not by the Group IV silicides, with matching of lattice volumes, eutectic and melting temperatures (cf. melting points, *Table 7.1*). VSi_2 and $NbSi_2$ are not nearly as scale-resistant as the Group VI silicides. It is notable that an improvement in protectiveness from lower to higher temperature by silicides can in some cases be attained, aided by a degree of *sintering* of the oxide overlayer.

Adjustments of oxidation-resistance properties can be achieved by blending the various transition-metal (Group IV–VII) disilicides with one another and (in the sense of our earlier remarks on ternary silicide systems) with aluminium, boron and other B Group elements. Aluminium addition to $MoSi_2$, for instance, has been claimed to be beneficial in that the brittleness and thermal shock cracking can be lessened.

One established means of producing the silicide on a material surface, rather than by direct deposition and diffusion processes, is by *siliconizing* the metal. A process of halogenide–hydrogen deposition can be used (gaseous diffusion) of the type (for molybdenum): $Mo + 2SiCl_4$ (vapour) $+ 4H_2 \rightarrow MoSi_2 + 8HCl$. The hydrogen here acts as the carrier gas, saturated with the chloride (or, in more recent developments, fluoride) vapour, and a well adhesive and uniform $MoSi_2$ layer is formed. This remains effective up to about 1,800°C, WSi_2 similarly up to about 2,000°C. It has been shown on silicided Mo and W wires[89] bearing adhesive and non-porous layers. Seen on a microscale, a fine layer of Mo_5Si_3 is formed *between* Mo base and $MoSi_2$, varying in thickness with temperature and time. Mo_3Si, on the other hand, if formed is deleterious, being as oxidizable as the metal; it needs protection by a further top layer of $MoSi_2$. Such siliciding processes have been usefully applied in practice not only to extended surfaces and to wires but also to powder particles for subsequent sintering. This brought about an important new technological development in preparing powder-metallurgical alloys, say of Mo–Si base, which can be used e.g. in producing *heater elements* for furnaces to operate in air to 1,700°C or more[89c]; in this connection, it is relevant that the electrical resistivity can be greatly temperature-dependent: for hot-pressed $MoSi_2$, 21 $\mu\Omega$.cm at room temperature rising[92] to about 80 at 1,500°C.

Developments of $MoSi_2$-base heating elements were largely on a powder metallurgical basis: deliberate admixtures of SiO_2 and Al_2O_3 as well as metallic constituents succeeded in giving good electrical resistivity, density, strength and 'life' at temperature[93]. The oxide admixture helps in particular to increase the, otherwise rather too low, resistivity of the silicides.

In these applications, a difficulty implicit in the brittleness of silicides is that of crack formation and damage of the protective oxide film by only slight mechanical work, and much thought has gone into various means of self-healing, etc. The existence of such a thin film of vitreous silica[94] may be demonstrated by *Figure 7.14*. The easy damage to this layer and to open exposure of $MoSi_2$ (which itself is quite readily oxidized, if the SiO_2 barrier is absent, to give MoO_3 vapour) can here be envisaged.

Valuable heat-resisting properties have also been claimed[95] for manganese silicides, '$MnSi_2$' (Mn_3Si_5).

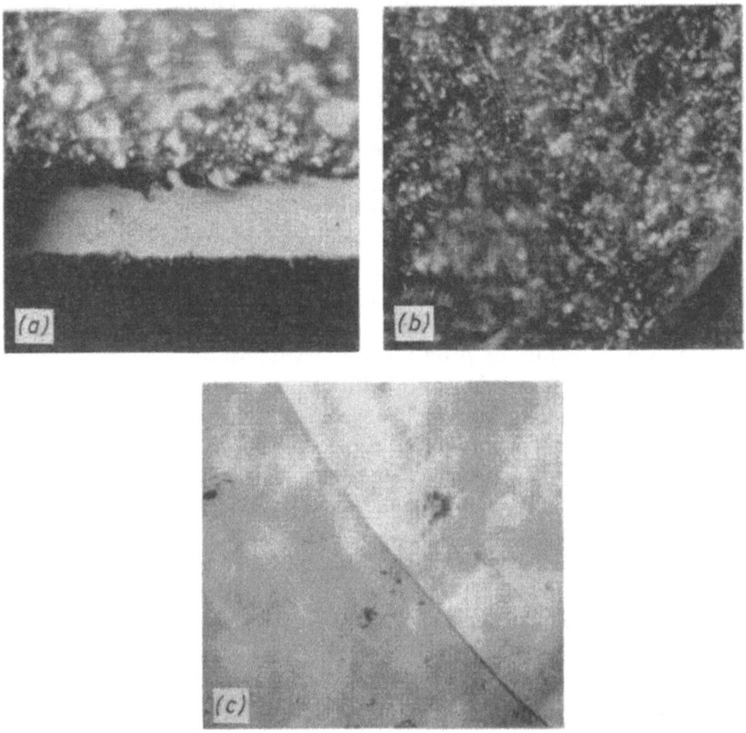

Figure 7.14. Silica 'glass' film developed on outer surface of oxidized $MoSi_2$ (After Fitzer and Rubisch[94])

(*a*) Section normal to surface; (*b*) focused on substrate; (*c*) focused on surface

By courtesy of the Editor, *Elektrowärme*

The practical fields of application of silicides extend not only to protective coatings on gas-turbine components, rocket nozzles, re-entry cones in 'space-age' applications (with their very different temperature/time requirements, often short times only at extreme temperatures), but further to covering molybdenum wires or tapes in furnace-heater windings. Silicides also play their part as integral micro-constituents in certain steels and 'super-alloys' for refractory uses, either sintered or melted. Here their peculiar midway position between intermetallic and interstitial compounds, such as carbides, can be utilized, with the feasibility to attain e.g. some degree of ductility. Much specialized research is proceeding to determine the optimum combinations of materials to fit the many problems requiring solution here. (See also σ- and Laves-phase stabilization by Si, described earlier, which can have adverse effects in metals, such as embrittling. This leads up to the more indirect applications of silicides mentioned, i.e. to their occurrence as significant secondary constituents and precipitates in heat-resisting steels and other alloys, such as Beattie's[76] complex silicide and phases in the Fe–Nb–Si, Cr–Nb–Si and Ni–Ti–Si systems. These 'working silicides' in practical alloys are not generally binary ones nor pure silicides, in that they can also harbour other interstitials like C, N or B.

The importance of silicides in technological developments of the future can hardly be overemphasized. For instance, in aerospace research, silicide coatings have been found effective to protect refractory Group IV–VII metals from oxidation even at 1,650°C, for a time sufficiently long to make a large difference during re-entry; and at temperatures in the 1,000–1,400°C range there are good prospects similarly to protect constructional components for long-time service use.

Attempts have also been made to produce high-temperature *thermocouples*[96] from silicide pairs, e.g. of $CrSi_2$–$CoSi_2$.

The ability of *platinum* and Pt-group metals to form silicides is marked. This can be of severe disadvantage in causing silicon embrittlement, for instance in Pt/Pt–Rh thermocouples[97] and similar components operating in siliceous contacts, particularly with silica ceramics used in sulphur-bearing atmospheres.

Chemically, silicides also possess good *acid resistance* to the mineral acids, and even, in the case of the Ti- and Mo-silicides, to hydrofluoric acid and to alkalis[2,98].

The electrical properties of the silicides have been described e.g. for a manganese silicide[95] (cited as $MnSi_{1.67-1.73}$—defect $MnSi_2$?), for Ti_5Si_3, TiSi and $TiSi_2$[99] and many other silicides (Hall constants[38]). The high-temperature semi-conductor properties of $ReSi_2$ have been mentioned already. Other silicides possessing favourable semiconductor properties are MnSi, FeSi, CoSi and NiSi, with, however, critical differences which well may relate to the crystallographic ones discussed earlier. Semiconductor properties of $CrSi_2$ and $MnSi_2$ have also been evaluated[100].

The possibility exists of *pairs* of silicides for thermocouples and in thermo-element generators[96] where the thermoelectric properties e.g. of Cr and Co silicides as a function of temperature and of degree of stoichiometry lead to silicide couples of considerable value.

The important question of 'cementing' silicides suffers from the difficulty pointed out for the borides, in that any likely bond metal, which must be fairly high melting, may itself ultimately form a brittle silicide: again bringing back the obvious importance of knowing the basic two-metal systems (*Tables 7.1* and *7.3*), the eutectics formed and other basic properties.

(The mixed silicides, e.g. silicoborides, silicocarbides, etc. of the transition metals will be discussed in Chapter 10.)

SELECTED LITERATURE

I. *General and Basic* (classification; structural aspects; reviews)
Aronsson, B.[1]; Samsonov, G. V. and Umanskii, Ya. S.[2]; Samsonov, G. V.[98]
Kiessling, R. *Fortschr. chem. Forsch.* 3 (1954) 41
Laves, F. and Wallbaum, H. J. *Z. Kristallogr. Kristallgeom.* 101 (1939) 78

Wallbaum, H. J.[29]
Structure of disilicides: early work, now classical
Nowotny, H., in *Electron structure and alloy chemistry of the transition elements*, ed. Beck, p. 179, New York (Interscience) 1963
Nowotny, H. and Parthé, E. *Planseeber. PulvMetall.* 2 (1954) 34
Nowotny, H. *2nd Plansee Seminar* (1956) 166
Parthé, E. *et al.*[52]
Structures of Group IV–VI silicides
Schwarzkopf, P. and Kieffer, R. *Refractory Materials*, London (MacMillan) 1953
Grinthal, R. D.[85]
Kieffer, R. and Benesovsky, F. (*a*) *Hartstoffe*, Vienna (Springer) 1963; (*b*) *Spec. Rep. Iron Steel Inst.* 58 (1954) 292; (*c*) *Powder Metall.* 1/2 (1958) 145
Pietrokowsky, P. and Duwez, P. *J. Metals, N.Y.* 3 (1951) 772
Robins, D. A.[3]
Bonding; physical aspects; magnetic properties
Robins, D. A. and Jenkins, I.[39]
Heats of formation
Schubert, K. *Kristallstrukturen zweikomponentiger Phasen*, Berlin (Springer) 1964
Crystal chemistry
Aronsson, B., Lundström, I. and Rundqvist, S.[25]
Schubert, K. and Esslinger, P. *Z. Metallk.* 48 (1957) 193
Templeton, C. and Dauben, C. *Acta crystallogr.* 3 (1950) 261
Brewer, L. and Krikorian, O.
Brewer, L. *et al., J. Am. Ceram. Soc.* 33 (1950) 291[16,40]; *Univ. Calif. Radiat. Lab. Rep.* 2888 (1955)
Structural, thermodynamic: heats of formation; bond energies
Searcy, A. W.[84]
Heats of formation
Hansen, M. [and Elliott, R. P.] *Constitution of Binary Alloys; First Supplement*, New York (McGraw-Hill) 1958, 1965
Pearson, W. B. *A Handbook of Lattice Spacings and Structures of Metals and Alloys*, London (Pergamon) 1958
Binary and ternary silicides
Neshpor, V. S., Samsonov, G. V. and coworkers[38,90]; *Kristallografiya* 6 (1961) 466; *Ogneupory* 28 (1963) 134
Electron structure; physical properties; (electrical; thermal expansion)

II. *Binary Metal*–Si *Systems*

(Phase diagrams; structure, constitutional aspects, Group IVA–VIIIA)

Ti-Si Hansen, M. *et al., Trans. Am. Soc. Metals* 44 (1952) 518
Phase diagram
Brewer, L. and Krikorian, O.[16]; Schachner, H. *et al.*[23]
Brukl, C., Nowotny, H., Schob, O. and Benesovsky, F. *Mh. Chem.* 92 (1961) 781
Structure

Zr-Si Lundin, C. E., McPherson, D. J. and Hansen, M. *Trans. Am. Soc. Metals* 45 (1953) 901
Brewer, L. and Krikorian, O.[40]
Schob, O., Nowotny, H. and Benesovsky, F. *Mh. Chem.* 92 (1961) 1218; Schachner, H. *et al.* 84 (1953) 677; 85 (1954) 1140; Nowotny, H., Lux, B. and Kudielka, H. 87 (1957) 447

Hf-Si Post, B., Glaser, F. W. and Moskowitz, D. *J. chem. Phys.* 22 (1954) 1264
Smith, J. F. and Bailey, D. M. *Acta crystallogr.* 10 (1957) 341
Nowotny, H. *et al., Mh. Chem.* 89 (1958) 701
Schob, O. *et al.* (as Zr–Si)
Nowotny, H., Braun, H. and Benesovsky, F. *Radex Rdsch.* (1960) 367

Th-Si Brown, A. *Acta crystallogr.* 14 (1961) 860
Brown, A. and Norreys, J. J. *J. Inst. Metals* 89 (1960) 238; *Nature, Lond.* 183 (1959) 673
$ThSi_2$ and its dimorphism
Stecher, P. *et al., Mh. Chem.* 94 (1963) 549
Jacobson, E. L. *et al., J. Am. chem. Soc.* 78 (1956) 4850

V-Si Schachner, H. *et al.*[23]; Parthé, E. *et al.*[52]
Kieffer, R., Benesovsky, F. and Schmid, H. *Z. Metallk.* 47 (1956) 247
Pearson, W. B. and Hume-Rothery, W. *J. Inst. Metals* 80 (1952) 652
Rostoker, W. and Yamamoto, A. *Trans. Am. Soc. Metals* 46 (1954) 1136
Golutvin, Yu. M. and Kozlovskaya, T. M. *Zh. fiz. Khim.* 34 (1960) 2350; 36 (1962) 362
Changes in enthalpy and heat capacity

Nb-Si Knapton, A. G.[14]
Overall diagram
Goldschmidt, H. J.[13]
Alyamovsky, S. J. *et al.*, *Russ. J. inorg. Chem.* 7 (1962) 432
Samsonov, G. V. *et al.*, *Zh. neorg. Khim.* 3 (1958) 868
Schubert, K. *et al.*[18b]
Nb_3Si of Ti_3P-type structure
Galasso, F. and Pyle, J.[18a]
Nb_3Si of Cu_3Au-type structure
cf.[23,35]

Ta-Si Kieffer, R. *et al.*, *Z. Metallk.* 44 (1953) 243
Nowotny, H. *Mh. Chem.* 84 (1953) 1; with Brukl, C. and Benesovsky, F. 92 (1961) 116
Brewer, L. *et al.*, *J. Amer. Ceram. Soc.* 33 (1950) 291
cf. [14,18b,35]

Pa-Si —

Cr-Si Borén, B. *Ark. Kemi* 11A (1933) 2
Parthé, E. *et al.*[22,52]; Dauben, C. H. *et al.*[44]; Sakata, T. and Tokushima, T.[96]
Kieffer, R., Benesovsky, F. and Schroth, H. *Z. Metallk.* 44 (1953) 437
Kurnakov, N. N.[63]; *Dokl. Akad. Nauk SSSR* 26 (1940) 362; 34 (1942) 110
Guzeva, L. N. and Ovechkin, B. I. *Izv. Akad. Nauk SSSR, Otd. tekh. nauk, Metall. Topl.* (1957) 27
Goldschmidt, H. J. and Brand, J. A.[56]
Phase diagram

Mo-Si Templeten, D. H. and Dauben, C. H. *Acta crystallogr.* 3 (1950) 261
Searcy, A. W. and Tharp, A. G. *J. phys. Chem., Ithaka* 64 (1960) 1539
Thermodynamic data
Kieffer, R. and Cerwenka, E. *Z. Metallk.* 43 (1952) 101
Nowotny, H. *et al.*[65]; *Mh. Chem.* 85 (1954) 255; Amberg, S. 91 (1960) 412
cf.[22,44]

W-Si Kieffer, R. *et al.*[8]; Efimenko, L. N. *et al.*[12]
Blanchard, R. and Cueilleron, J. *C.r. hebd. Séanc. Acad. Sci., Paris* 244 (1957) 1782
Obrowsky, W. *J. Inst. Metals* 89 (1960) 79
Aronsson, B. *Acta chem. scand.* 9 (1955) 1107
cf.[44,52]

U-Si Kaufmann, A. R. *et al.*, in *The Chemistry of Uranium*, ed. Katz and Rabinowitch, New York (McGraw-Hill) 1951; *Trans. Am. Inst. Min. metall. Engrs* 209 (1957) 23
Zachariasen, W. H.[21]; *Acta crystallogr.* 1 (1948) 265
Brown, A. and Norreys, J. J.[27]; [and as Th–Si]
USi_2, USi_3 reactions, structure and non-stoichiometry
Brauer, G. and Haag, H. *Z. anorg. allg. Chem.* 259 (1949) 197; 267 (1951) 198, 211

Mn-Si Vogel, R. and Bedarf, H. *Arch. EisenhüttWes.* 7 (1933) 423
Laves, F. *Z. Kristallogr. Kristallgeom.* 89 (1934) 189
Senataeur, J. P. and Fruchart, R. *C.r. hebd. Séanc. Acad. Sci., Paris* 258 (1964) 1524
Kuzma, Yu. B. and Gladyshevskii, E. I. *Zh. neorg. Khim.* 9 (1964) 674
Low-Mn silicides
Davydov, K. N. *et al.*, *Fizika Metall.* 2 (1956) 192; 12 (1961) 424; *Phys. Metals Metallogr., N.Y.* 12 (1961) 168
Mn_3Si martensite-type transformation
Korshunov, V. A. *et al.*, *Fizika Metall.* 11 (1961) 945; 12 (1961) 277; *Phys. Metals Metallogr. N.Y.* 11 (1961) 118; 12 (1961) 101
$MnSi_2$ non-stoichiometry; new phase $MnSi_{1\cdot 67-1\cdot 73}$ (?)
Aronseon, B.[1]
Mn_3Si ordering as Fe_3Si
Dudkin, L. D. and Kuznetsova, E. S. *Dokl. Akad. Nauk SSSR* 141 (1961) 94
Golutvin, Ya. M. *et al.*[4]
Heats of formation and heat capacities
Burger, K. O. *et al.*[74]
Mn_3Si_5
Schwamma, O., Nowotny, H. and Wittmann, A. *Mh. Chem.* 94 (1963) 681, 924
Structure of $MnSi_{1\cdot 7}$, etc.
Gupta, K. P. *Trans. Am. Inst. Min. metall. Engrs* 230 (1964) 253
New phase around Mn_9Si_2

Re-Si Searcy, A. W. and McNees, R. A. *J. Am. chem. Soc.* 75 (1953) 1578; 77 (1955) 5290
Knapton, A. G.[91]
Neshpor, V. S. and Samsonov, G. V.[90]

Np-Si Zachariasen, W. H.[21]
Structure of $NpSi_2$, etc.

Fe-Si (extensive earlier work; cf. Hansen, Pearson above, etc.)
Siderenko, F. A. *et al., Fizika Metall.* 8 (1959) 465, 735; 9 (1960) 861; *Phys. Metals Metallogr., N.Y.* 8 (1959) 34, 88; 9 (1960) 56
Siderenko, F. A. and Geld, P. V. *Izv. vyȳssh. ucheb. Zaved. Khim. Met.* 7 (1963) 140
ε-FeSi
Aronsson, B. *Acta chem. scand.* 14 (1960) 1414
Holdhus, H.[31]
$FeSi_2$, non-stoichiometry
Seyboldt, A. U. *Trans. Am. Inst. Min. metall. Engrs* 212 (1958) 161
Abrikosov, N. K. *Izv. Sekt. fiz.-khim. Analiza Inst., Obshch. neorg. Khim.* 27 (1956) 157

Pu-Si Elliott, R. O. and Larson, A. C.; Schonfield, F. W., in *The Metal Plutonium*, ed. Coffinberry and Miner, Chicago Univ. Press, 1961
Coffinberry, A. S. and Waldron, M. B., in *Metallurgy and Fuels*, Progress in Nuclear Energy, Ser. 5, Vol. 1, Oxford (Pergamon) 1956
Zachariasen, W. H.[21]
Structure
Runnalls, O. J. C. and Boucher, R. R. *Acta crystallogr.* 8 (1955) 592

Co-Si Vogel, R. and Rosenthal, K. *Arch. EisenhüttWes.* 7 (1934) 689
Pfisterer, H. and Schubert, K. *Z. Metallk.* 41 (1950) 433; Köster, W. and Schmid, E. 29 (1937) 232
Bertaut, F. and Blum, P. *C.r. hebd. Séanc. Acad. Sci., Paris* 231 (1950) 626
Geller, S. *Acta crystallogr.* 8 (1955) 83
Sakata, T. and Tokushima, T.[96] [cf. Cr–Si]
Also homogeneity range

Ni-Si Wittman, A. *et al.*[30]
Pilström, G. *Acta chem. scand.* 15 (1961) 893
Structures of Ni_3Si_2 and Ni_5Si_2
Iwase, K. and Okamoto, M. *Honda Anniv. Vol., Sci. Rep. Res. Insts Tohôku Univ.* (1936) 777
Osawa, A. and Okamoto, M. *Nippon Kink. Gakk.* 2 (1938) 378
Schubert, K. and Pfisterer, H. *Naturwissenschaften* 37 (1950) 112; *Z. Metallk.* 41 (1950) 438
Lashko, N. I. *Dokl. Akad. Nauk SSSR* 81 (1951) 605

Pt group metals-Si Buddery, J. H. and Welsh, A. J. E. *Nature, Lond.* 167 (1951) 362; Aronsson, B., Åselius, J. and Stenberg, E. 183 (1959) 318
Structure comparison, incl. borides
Aronsson, B. and Åselius, J. *Acta chem. scand.* 15 (1961) 1571
Ru_2Si
Aronsson, B. and Nylund, A. *Acta chem. scand.* 14 (1960) 1011
Schwamma, O. *et al.* [cf. Mn–Si]
Finnie, L. N.[32]; with Searcy, A. W. *Acta crystallogr.* 12 (1959) 260
Ru_2Si_3 and Os_2Si_3 homogeneity ranges; phase $OsSi_3$, etc.
Bhan, S. and Schubert, K. *Z. Metallk.* 51 (1960) 327
Rh_5Si_3, etc.
Gohle, R. and Schubert, K. *Z. Metallk.* 55 (1964) 503
Pt_7Si_3, etc.
Schubert, K. *et al., Naturwissenschaften* 47 (1960) 303
Pfisterer, H. and Schubert, K. *Z. Metallk.* 41 (1950) 358
Anderko, K. and Schubert, K. *Naturwissenschaften* 39 (1952) 351; *Z. Metallk.* 44 (1953) 307
Reinacher, G. *Rev. Métall., Paris* 54 (1957) 321
Rao, N. K. and Winterhager, H. *Trans. Indian Inst. Metals* 10 (1956) 139
Voronov, N. M. *Izv. Sekt. Platiny̆, Inst. Obshch. neorg. Khim.* 13 (1936) 145
Engström, I. *Acta chem. scand.* 17 (1963) 775

(Sc, Y, Rare Earths)-Si Brauer, G. and Haag, H. *Z. anorg. allg. Chem.* 267 (1951) 198
Rare Earth disilicides, structures
Perri, J. A. *et al., J. phys. Chem., Ithaka* 63 (1959) 616, 2073; Tharp, A. G. 66 (1962) 758

(Sc, Y, Rare Earths)-Si *cont.*
Matthias, B. I. *et al.*, *Phys. Rev.* 112 (1958) 89
Sc silicides, incl. ScSi and $ScSi_2$
Parthé, E. *Acta crystallogr.* 12 (1952) 559; 13 (1960) 868; cf.[34,35,52]
Structures, etc.
Brown, A. (quoted[25])
Ce–Si system; structures
Mayer, I. P. *et al.*, *J. phys. Chem., Ithaka* 66 (1962) 693
Lundin, C. E.[37]
Gschneidner, K. A. *Rare Earth Alloys*, Princeton (Van Norstrand) 1961
Reviews
Smith, G. S. *et al.*, *Nature, Lond.* 210 (1966) 1148
Binder, I. *J. Am. Ceram. Soc.* 43 (1960) 287
Y and R.E. silicides and borides; structure, properties

For *individual binary silicides* see also I (Hansen; Elliott; Pearson; Aronsson *et al.*; Kieffer and Benesovsky; Samsonov).

III. *Ternary Systems, M_1–M_2*–Si (cf. *Table 7.4*)

Al-bearing:

Ti-Al-Si, Zr-Al-Si, Hf-Al-Si, Mo-Al-Si
Brukl, C. *et al.*, *Mh. Chem.* 92 (1961) 781; Schob, O., Nowotny, H. and Benesovsky, F. *ibid.* 1218
Structure of TiSi, $Ti(Al_2Si)_2$, $Mo(Al_2Si)_2$

V-Al-Si
Gerhardt, E. and Joseph, G. *Z. Metallk.* 52 (1961) 310
Equilibria (Al corner only)

Nb-Al-Si
Nowotny, H., Benesovsky, F. and Brukl, C. *Mh. Chem.* 92 (1961) 193

Ta-Al-Si, W-Al-Si
Nowotny, H., Brukl, C. and Benesovsky, F. *Mh. Chem.* 92 (1961) 116
Equilibria in the range 20–80 Si, 10–75 Al (at. per cent)

Cr-Al-Si, Mo-Al-Si, V-Al-Si, Nb-Al-Si
Brukl, C. *et al.*[75]
Summary

Mo-Al-Si
Holleck, H. *et al.*, *Mh. Chem.* 93 (1962) 1075
Phases $Mo_3(Al,Si)$, MoAlSi, $Mo(Al,Si)_2$, including thermal expansion

U-Al-Si
Petzow, G. and Kvernes, I. *Z. Metallk.* 52 (1961) 693
U_3Si_2–UAl_2 section (= eutectic pseudo-binary)

Co-Al-Si
Burger, K. O. *et al.*[74]
Ternary compounds, $Co_3Al_3Si_4$ (Ir_3Ge_7 type) and Co_2AlSi_2 (Ni_2Al_3 type)

Nowotny, H. *Aluminium* 37 (1961) 580
Brief appraisal of *M*–Al–Si systems with reference to crystal chemistry of aluminides
Pötzschke, M. and Schubert, K. *Z. Metallk.* 53 (1962) 548
Transition-metal aluminide structures as analogous to silicides
Edshammer, L. E. *Acta chem. scand.* 16 (1962) 30 [Zr, Hf–Al]; Boller, H. *et al.*, *Mh. Chem.* 92 (1961) 330 [Hf–Al], etc.

Ge-bearing:

Th-Si-Ge
Stecher, P. *et al.*, *Mh. Chem.* 94 (1963) 549
Th_3Si_2–Th_3Ge_2 and $ThSi_2$–$ThGe_2$ pseudo-binaries (complete solid solutions)

U-Si-Ge
Brixner, L. H. *J. inorg. nucl. Chem.* 25 (1963) 783

V-Si-Ge
Holleck, H., Benesovsky, F. and Nowotny, H. *Mh. Chem.* 96 (1965) 570

Nb-Si-Ge, Ta-Si-Ge
Brixner, L. H. *J. inorg. nucl. Chem.* 25 (1963) 257
Compounds, structures, solubilities; thermodynamic properties

W-Si-Ge
Stecher, P. *et al.*[81]
Overall equilibria
Nowotny, H.[82]

Silicon-stabilized intermetallic compounds, 'electron phases' (of transition metals); work from P. A. Beck's (Urbana) school:

Spiegel, F. X., Bardos, D. and Beck, P. A.[51]
'*G*' and '*E*' phases
Bardos, D. I. *et al.*[5]
Ternary Laves phases in *M*–*M'*–Si where *M* = [Ti, V, Nb, Ta, Mo, W], *M'* = [Mn, Fe, Co, Ni]; structure and electronic relations
Bardos, D. I. and Beck, P. A. *Trans. metall. Soc. A.I.M.E.* 236 (1966) 64
α-Mn, σ and other type phases in V–(Mn, Fe, Co, Ni)–Si, Cr–Mn–Si and V–Mn–Fe–Si systems; structures and equilibria

Gupta, K. P. *et al.*[55]
σ-phases, spacings, etc.
Gupta, K. P. *Trans. metall. Soc. A.I.M.E.* 230 (1964) 253
Ternary phases in Fe–Mn–Si
Bardos, D. I. *et al.*[48]
β-Mn, σ, etc. phases in Fe–Mn–Si, Co–Mn–Si, Mn–Ni–Si, Mn–Ti–Si, Mn–V–Si, Mn–Cr–Si; corresponding systems with Ge and Sn instead of Si; equilibria and electronic relations
Bardos, A. M. and D. I. and Beck, P. A. *Trans. metall. Soc. A.I.M.E.* 227 (1963) 991
Ternary Si-bearing Laves phases $M\,(M'_{1-x}Si_x)_2$ in M–M'–Si; M and M' as before

Applications

Schwarzkopf, P. and Kieffer, R.; Kieffer, R. and Benesovsky, F. (cf. I)
Wehrmann, R., in *High Temperature Technology*, ed. Campbell, I. E., p. 151, New York (Wiley) 1956
Samsonov, G. V.[98]; Umanskii, S. Ya.[2]
Fitzer, E.[89]
Siliciding of W and Mo
Verkhoborin, L. F. *et al.*[11]
Siliciding of W, Mo, Ta
Campbell, I. *et al.*, *J. electrochem. Soc.* 96 (1949) 218; Beidler, E. *et al.* 98 (1951) 21
Siliciding
Todd, G. and Parry, E. *Nature, Lond.* 203 (1964) 967
$MoSi_2$ protection
Robins, D. A. and Jenkins, I.[39]
Stabilities of silicides in refractories
Krier, C. A. and Blocher, J. M. *Electrochem. Technol.* 1 (1963) 137
Silicides in coatings, etc.
Wlodek, S. T. *J. electrochem. Soc.* 108 (1961) 177
Cox, A. R. and Brown, R. *J. less-common Metals* 6 (1964) 51
Kieffer, R., Benesovsky, F. and Konopicky, C. *Ber. dt. keram. Ges.* 31 (1954) 223
Brukl, C., Benesovsky, F. and Nowotny, H. *Planseeber. PulvMetall.* 11 (1963) 20
Mo–Al–Si oxidation

REFERENCES

1 Aronsson, B. *Ark. Kemi* 16 (1960) 379
2 Samsonov, G. V. and Umanskii, Ya. S. *NASA tech. Trans. F-102*, 1962
3 Robins, D. A. *Powder Metall.* 1–2 (1958) 172
4 Golutvin, Ya. M. *et al.*, *Zh. fiz. Khim.* 37 (1963) 1362
5 Bardos, D. I., Gupta, K. P. and Beck, P. A. *Trans. metall. Soc. A.I.M.E.* 221 (1961) 1087
6 Shinjo, T. *et al.*, *J. phys. Soc. Japan* 18 (1963) 797
7 Nowotny, H. *et al.*, *2nd Plansee Seminar* (1955) 166
8 Holleck, H., Nowotny, H. and Benesovsky, F. *Mh. Chem.* 94 (1963) 497
9 Kieffer, R., Benesovsky, F. and Gallistl, E. *Z. Metallk.* 43 (1952) 284
10 Matyushenko, N. N. *et al.*, *Fizika Metall.* 8 (1959) 879
11 Verkhoborin, L. F. *et al.*, *Phys. Metals Metallogr., N.Y.* 13 (1962) 67
12 Efimenko, L. N. *et al.*, *Izv. Akad. Nauk. SSSR, Otdel. tekh. nauk, Metallurgiya Topl.* (1965) 4, 163
13 Goldschmidt, H. J. *J. Iron Steel Inst.* 194 (1960) 169
14 Knapton, A. G. *Nature, Lond.* 175 (1955) 730
15 Brauer, G. and Scheele, W. *Z. anorg. allg. Chem.* 24 (1948) 103
16 Brewer, L. and Krikorian, O. *Univ. Calif. Radiat. Lab. Rep.* 2544 (1954)
17 McMurty, C. H. and Crandall, W. B. *Amer. Ceram. Soc. Bull.* 13 (1956) 13; private communication
18 (*a*) Galasso, F. and Pyle, J. *Acta crystallogr.* 16 (1963) 228; (*b*) Schubert, K. *et al.* *Naturwissenschaften* 51 (1964) 506
19 Philipsborn, H. and Laves, F. *Acta crystallogr.* 17 (1964) 219
20 Aronsson, B. and Rundqvist, S. *Acta crystallogr.* 15 (1962) 878
21 Zachariasen, W. H. *Acta crystallogr.* 2 (1949) 94
22 Parthé, E., Schachner, E. and Nowotny, H. *Mh. Chem.* 86 (1955) 183
23 Schachner, E., Cerwenka, E. and Nowotny, H. *Mh. Chem.* 85 (1954) 245
24 Brukl, C., Nowotny, H. and Benesovsky, F. *Mh. Chem.* 92 (1961) 967
25 Aronsson, B., Lundström, T. and Rundqvist, S. *Borides, Silicides and Phosphides*, London (Methuen) 1961
26 Beckman, G. and Kiessling, R. *Nature, Lond.* 178 (1956) 1341
27 Brown, A. and Norreys, T. J. *Nature, Lond.* 191 (1961) 61
28 Brown, A. *Acta crystallogr.* 14 (1961) 860

[29] Wallbaum, H. J. *Z. Metallk.* 13 (1941) 378
[30] Wittmann, A., Burger, K. O. and Nowotny, H. *Mh. Chem.* 92 (1961) 961
[31] Holdhus, H. *J. Iron Steel Inst.* 200 (1962) 1024
[32] Finnie, L. N. *J. less-common Metals* 4 (1962) 24
[33] Perri, J. A., Binder, I. and Post, B. *J. phys. Chem., Ithaka* 63 (1959) 616
[34] Arbuckle, J. and Parthé, E. *Acta crystallogr.* 15 (1962) 1205
[35] Parthé, E., Lux, B. and Nowotny, H. *Mh. Chem.* 86 (1955) 859
[36] Mayer, I. P., Banks, E. and Post, B. *J. phys. Chem., Ithaka* 66 (1962) 693
[37] Lundin, C. E. in *The Rare Earths*, ed. Spedding and Daane, New York (Wiley) 1961
[38] Neshpor, V. S. and Samsonov, G. V. *Dokl. Akad. Nauk SSSR* 134 (1960) 1337
[39] Robins, D. A. and Jenkins, I. *2nd Plansee Seminar* (1955) 187
[40] Brewer, L. and Krikorian, O. *J. electrochem. Soc.* 103 (1956) 38
[41] Rostoker, W. and Yamamoto, A. *Trans. Am. Soc. Metals* 46 (1954) 1136
[42] Hager, P. *Trans. metall. Soc. A.I.M.E.* 227 (1963) 1000
[43] Foley, J. H. and Raynor, G. V. *Trans. Faraday Soc.* 57 (1961) 51
[44] Dauben, C. H. *et al., J. phys. Chem., Ithaka* 60 (1956) 443
[45] Nowotny, H. *et al., Mh. Chem.* 85 (1954) 241
[46] Kudielka, H. and Nowotny, H. *Mh. Chem.* 87 (1956) 471
[47] Dwight, E., Connor, R. A. and Downey, J. W., cited[51]
[48] Bardos, D. I. *et al., Trans. metall. Soc. A.I.M.E.* 236 (1966) 40
[49] Duffin, W. J., Parthé, E. and Norton, J. T. *Acta crystallogr.* 17 (1964) 450
[50] Westbrook, J. H. *et al., G.E.C. (U.S.A.) Rep.* 58RL/2117 (1958)
[51] Spiegel, F. X., Bardos, D. I. and Beck, P. A. *Trans. metall. Soc. A.I.M.E.* 227 (1963) 575
[52] Parthé, E., Nowotny, H. and Schmid, H. *Mh. Chem.* 86 (1955) 385
[53] (*a*) Savitskii, E. M. *et al., Russ. J. inorg. Chem.* 7 (1962) 573; (*b*) Savitskii, E. M. and Baron, V. V. *Trudy Inst. Metall. A.A. Baikova* 5 (1960) 156
[54] Gladyshevskii, E. I. *et al., Soviet Phys. Crystallogr.* (*a*) 6 (1961) 207; (*b*) *ibid.* 615
[55] Gupta, K. P., Rajan, N. S. and Beck, P. A. *Trans. metall. Soc. A.I.M.E* 218 (1960) 617
[56] Goldschmidt, H. J. and Brand, J. A. *J. less-common Metals* 3 (1961) 34
[57] Dokukina, N. V. and Shamrai, F. J. *Poroshk. Met., Izd. Akad. Nauk Ukr. SSR* 2 (1962) 32
[58] Goldschmidt, H. J. *J. Iron Steel Inst.* 194 (1960) 169
[59] Nowotny, H. *et al., Mh. Chem.* 84 (1953) 579
[60] Kieffer, R., Schob, O. *et al., Mh. Chem.* 93 (1962) 517
[61] Anderson, A. G. and Jette, E. R. *Trans. Am. Soc. Metals* 24 (1936) 375
[62] Dubrovskaya, L. B. and Geld, P. V. *Zh. neorg. Khim.* 7 (1962) 145; *Russ. J. inorg. Chem.* 7 (1962) 73
[63] Kurnakov, N. N. (*a*) *Dokl. Akad. Nauk SSSR* 34 (1942) 158; (*b*) *Izv. Sekt. fiz.-khim., Analiza Inst. Obshch. neorg. Khim.* 17 (1949) 209
[64] Guard, R. W. and Smith, E. A. *J. Inst. Metals* 88 (1959) 285, 373
[65] Nowotny, H., Kieffer, R. and Schachner, H. *Mh. Chem.* 83 (1952) 1243
[66] Neshpor, V. S. and Samsonov, G. V. *Fizika Metall.* 18 (1964) 187
[67] Vogel, R. and Gerhardt, R. D. *Arch. EisenhüttWes.* 32 (1961) 47
[68] Vogel, R. and Töpker, H. *Arch. EisenhüttWes.* 13 (1939) 183
[69] Sikrica, M. and Ban, Z. *Proc. Int. Conf. nucl. Mater., Prague* 1963
[70] Vogel, R. and Bedarf, H. *Arch. EisenhüttWes.* 10 (1936) 581
[71] Jaenicke, E. *Arch. EisenhüttWes.* 11 (1937) 409
[72] Gupta, K. P. *Trans. metall. Soc. A.I.M.E.* 230 (1964) 253
[73] Greiner, E. S. and Jette, E. R. *Trans. Am. Inst. Min. metall. Engrs* 152 (1943) 48
[74] Burger, K., Wittmann, A. and Nowotny, H. *Mh. Chem.* 93 (1962) 9
[75] Gladyshevskii, E. I. and Kuzma, Ya. B. *Zh. strukt. Khim.* 66 (1960) 57
[76] Beattie, H. J., with Ver Snyder, F. L. *Nature, Lond.* 178 (1956) 208; with Hagel, W. C. *J. Metals, N.Y.* 209 (1957) 911
[77] Schob, O., Nowotny, H. and Benesovsky, F. *Planseeber. PulvMetall.* 10 (1962) 65
[78] Brukl, C., Nowotny, H. and Benesovsky, F. *Mh. Chem.* 92 (1961) 967
[79] Pratt, J. N. and Raynor, G. V. *Proc. R. Soc.* A205 (1951) 103; *J. Inst. Metals* 79 (1951) 211
[80] Robinson, K. *Acta crystallogr.* 5 (1952) 397
[81] Stecher, P., Benesovsky, F. and Nowotny, H. *Mh. Chem.* 94 (1963) 1154
[82] Nowotny, H. *Mh. Chem.* 92 (1961) 365
[83] Kieffer, R. and Benesovsky, F. *Powder Metall.* 1/2 (1958) 145
[84] Searcy, A. W. *J. Am. Ceram. Soc.* 40 (1957) 431
[85] Grinthal, R. D. *J. electrochem. Soc.* 107 (1960) 59
[86] Boller, H. and Parthé, E. *J. Metals, N.Y.* 12 (1962) 90; *Acta crystallogr.* 16 (1963) 830
[87] Hardy, G. and Hulm, J. *Phys. Rev.* 89 (1953) 884
[88] Kieffer, R. and Benesovsky, F. (*a*) *Spec. Rep. Iron Steel Inst.* 58 (1954) 292

[89] Fitzer, E. (*a*) *Berg-u. hüttenm. Mh.* 97 (1952) 81; (*b*) *1st Plansee Seminar* (1953) 244; (*c*) *2nd Plansee Seminar* (1956) 56
[90] Neshpor, V. S. and Samsonov, G. V. *Fizika Metall.* 11 (1961) 638
[91] Knapton, A. G. *3rd Plansee Seminar*, 1958 (1959) 412
[92] Glaser, F. W. *J. appl. Phys.* 22 (1951) 103
[93] Tamura, K. and Kawaguchi, T. *J. Japan Soc. Powd. Metall.* 8 (1961) 145
[94] Fitzer, E. and Rubisch, O. *Elektrowärme* 5 (1958) 163; Fitzer, E. *et al.* 6 (1958) 1
[95] Korshunov, V. A. and Gel'd, P. V. (*a*) *Fizika Metall.* 11 (1961) 945; *Soviet Phys. Metals Metallogr.* 11 (1961) 118; (*b*) Korshunov, V. A. *et al.* 12 (1962) 101
[96] Sakata, T. and Tokushiva, T. *Trans. Japan Inst. Metals* 5 (1963) 34
[97] Goldschmidt, H. J. and Land, T. *J. Iron Steel Inst.* 155 (1947) 221
[98] Samsonov, G. V. *Silicides and their Technical Applications*, Kiev (Ukr. Akad. Sci.) 1959
[99] Neshpor, V. S. *et al., Ukr. fiz. Zh.* 5 (1960) 839
[100] Dudkin, L. D. and Koznolsova, E. S. *Poroshk. Met., Izd. Akad. Nauk Ukr. SSR* 2 (1962) 31

8

OXIDES

8.1 INTRODUCTION

The topic of oxides is so large and varied that this account has to concentrate on aspects relevant to the systematic pattern of interstitial compounds. Also, publications on oxides in more specific fields (ceramics, minerals, semi-conductors, ferroelectrics, etc.) go beyond our scope, but references will be given. The criterion of metallic character must be guiding, i.e. those oxides are most important in which metal–metal bonds are still dominant and the oxygen atom is interstitial. However, an outlook to the ionic oxides should form part of the picture.

Seen in a wider context, oxides should not be ranked as of higher basic importance than carbides, nitrides, borides, silicides or hydrides. Their special role is mainly incidental to our living in the particular atmosphere of this planet in which molecular oxygen happens to be ubiquitous and inescapable, as well as in gaseous form, and, in contrast to nitrogen, to be of greater reactivity with most metals. Indeed, modern vacuum technology largely owes its existence to the need for inhibiting some of the phases of this Chapter. In practice, a further consequence of our being surrounded by oxygen is that oxides have their two physical types of occurrence: (*a*) in bulk (ceramics, minerals, etc.), (*b*) two-dimensionally, chiefly in scale layers and films upon metals. The latter can be either propagative (in corrosion) or self-inhibiting (in protective coatings). Again this leads to several fundamental issues regarding the position of oxides in the nature of things. The truly interstitial oxides are in general only the lower-oxygen ones of the transition metals. The oxygen atom is a strong electron absorber, so that the composition range of metallic bonding remains small or limited, and the tendency is for compounds of partly (and ultimately wholly) ionic or covalent bonding to be formed. Thus oxides like Al_2O_3, MgO, CaO, SrO, etc. are all non-metallic; furthermore, their base metals do not as a rule form any metallic lower oxides (except in some cases unstably, e.g. in thin films or under exceptional reducing conditions). In contrast, for the transition-metal systems such as Ti–, Nb–, W– or Fe–oxygen, the oxides lower than TiO_2, Nb_2O_5, WO_3 or Fe_2O_3 become important. They can readily stabilize under reducing conditions or in metal contact, e.g. during oxidation processes.

In any one metal–oxygen system we shall try to delineate the compositional boundaries between true interstitial and non-metallic compounds and note any transition range of metallic character. This is obviously an area of semi-conductors. *Non-stoichiometry* enters prominently in oxide structures, as it did in the carbides and nitrides. Viewed in the light of the atom sequence $B \rightarrow C \rightarrow N \rightarrow O \rightarrow F$ along the first Short Period, the oxygen atom already has a considerable tendency towards ionic bonding, as have the halides, and a higher degree of electron exchange with the metal orbitals than in the nitrides. The position of oxides, intermediate between nitrides and halides, is therefore clearly reflected electronically as well as in the resulting properties. The 'atomic size' of oxygen is in consequence also a variable concept: depending on the structure in which it is incorporated, it becomes ill defined and blurred with its 'ionic size' under varying valencies. However, this does not alter the fact of its being a 'small atom' in Hägg's sense and capable of forming true interstitial compounds with the transition metals. We might merely instance the isomorphous series (TiC/TiN/TiO) where C, N and O atoms are exchangeable. The eventual *ionic* radius of O^{--} is much larger, ≈1·4 Å (e.g. in CaO), compared with the atomic radius of oxygen, e.g. in TiO, of about 0·7 Å.

The bonding energies in interstitial oxides, together with those of carbides, nitrides, borides, silicides, some halides and the metals, are shown in

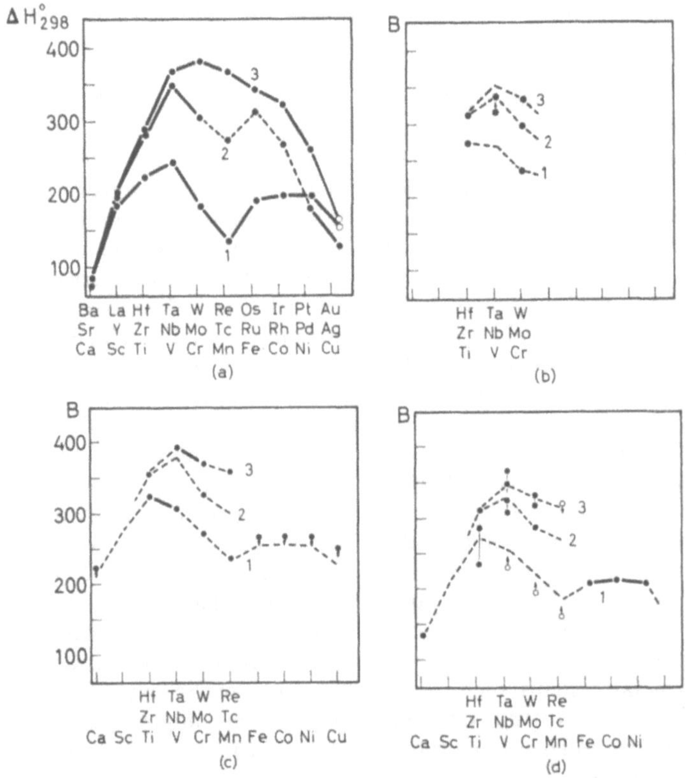

Figure 8.1. Comparison of heats of formation and bond energies
(After Brewer and Krikorian[1] and Kiessling[2])

(*a*) Heats of formation of transition metals of the 1st–3rd Long Periods (= 1–3) at room temperature; (*b*)–(*h*) Variation of bonding energies, *B* (kcal) of some transition-metal compounds with Group and Period: (*b*) = *monoborides;* (*c*) = *monocarbides;* (*d*) = *silicides,* M_3Si_2 (calculated on basis of 1¼ moles of M_3Si_2 unit); (*e*) *mononitrides;* (*f*) *monoxides* (compared with monosulphides for the 1st Long Period; (*g*) *dioxides* (calculated on basis of ⅔ mole of MO_2); (*h*) *difluorides* and *dichlorides* for the 1st Long Period

By courtesy of the Institute of Metals, London, The Electrochemical Society, New York, and the University of California Lawrence Radiation Laboratory, Berkeley

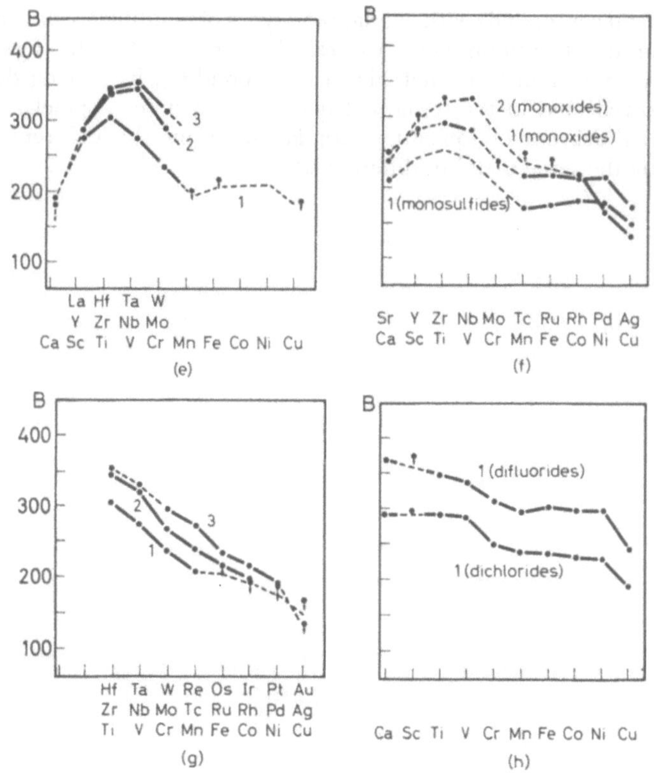

Figure 8.1e, f, g, h

Figure 8.1. The similarity of the monoxide curves with those of the carbides, etc. stands out and also the fact that the type of bonding throughout resembles that in the metals[1]. This contrasts with the basically different bonding energy curves for the transition-metal *di*oxides, which already resemble the fluorides with their ionic-type bonding.

Generally, the lowest stable interstitial oxides are the monoxides, from TiO to NiO in Groups IV to VIII and equivalents in the higher Periods. Whilst the lower oxides such as W_3O, Ta_4O, Ti_2O, etc. are also established, they are often transient or metastable, requiring metal contact, reducing atmospheres or low oxygen pressures for their existence. Our knowledge of important aspects of the low oxides is still very incomplete, although recent researches into the oxidation kinetics of the refractory metals have brought significant advances in constitutional knowledge.

The transition-metal oxides are presented in *Table 8.1* (p. 356) and, by way of phase-diagram sequence, in *Figure 8.2*. These survey the *single-metal/* oxygen compounds and phase diagrams, as far as it was possible to assemble them.

In addition, there exist numerous *double* oxides, e.g. those of η-carbide-type structure, A_3B_3O, which are again metallic and interstitial in character and can indeed form solid solutions with their carbide and nitride analogues. Double oxides will be dealt with later, but one point of principle is that two- or multiple-metal compounds with overlapping types of bonding are particularly favoured in the transition-metal oxides. The reason for this again lies in the ready transition from metallic to ionic bonding, in varying degrees. This is also reflected in the ternary structures formed (e.g. spinels), where such varying degrees of ionization are favoured by the different metal species or by the same metal in different sites.

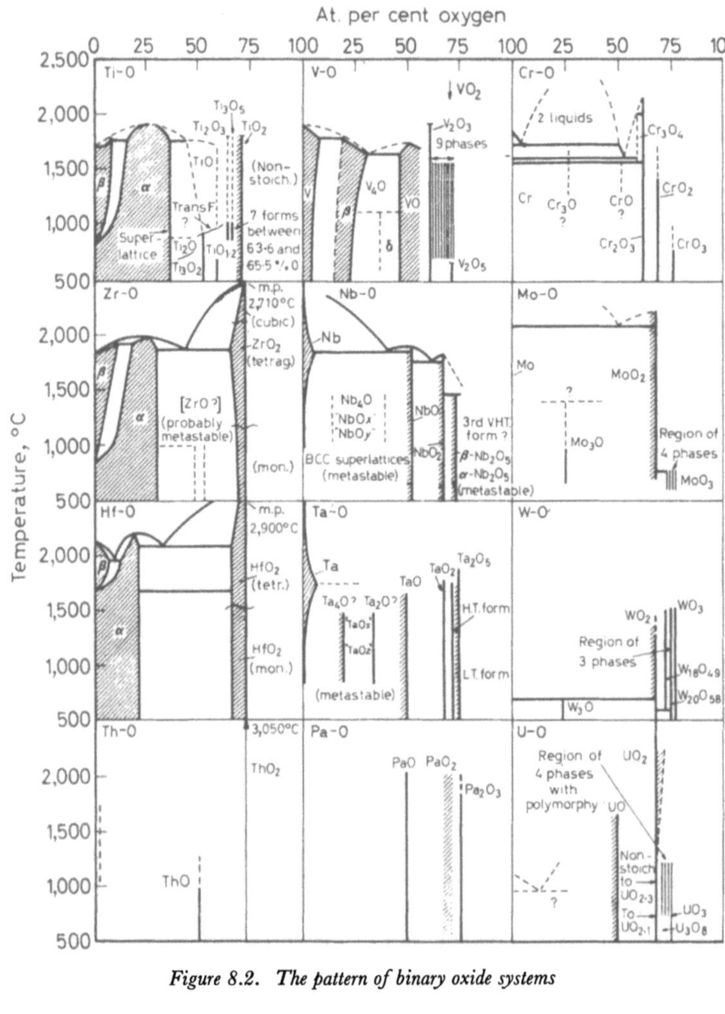

Figure 8.2. The pattern of binary oxide systems

The tendency to depart from the metallic bond and for oxygen to ionize is incipient already in the metal-rich compounds. Ionization can be initiated by small variations in external factors, such as temperature, pressure, oxygen availability, etc. The possibility exists even of *local* fluctuations in type of bonding (i.e. of degree of electron sharing between metal and oxygen), within different lattice regions, or from one crystallite to another, for instance near surfaces or dislocations. More macroscopically it can occur, e.g. within a layered scale aggregate. Anisotropic variations in bond type can also be induced.

The interstitial atom acts as electron donor to the *d* orbital of the metal atom (cf. Chapter 2). This still applies to the oxygen atom but is qualified by its inclination towards electron *absorption* being more readily realized.

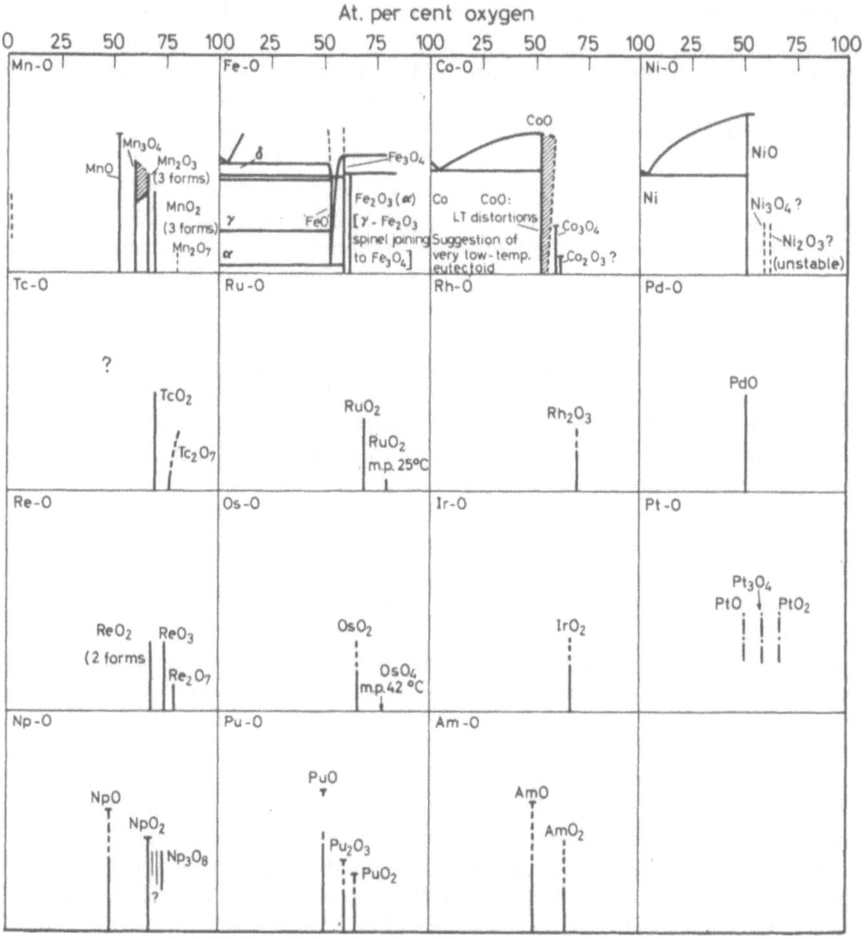

Figure 8.2b

Whilst, therefore, in silicides and borides it is largely atomic size which can prejudice the interstitial character of the compound, in the oxides, where size of the O atom is quite commensurate with C or N, it is the electronic structure.

A further aspect is that, along any one Period in the Periodic Table, the metal/non-metal bond strength decreases as the *d* orbital becomes filled. Therefore, hybrid-type bonding can increasingly arise, coupled with the possible development of unsymmetrical bonding forces in the structure;

this has been viewed[2] as a deviation of the interstitial atoms from spherical shape, and much work on low-temperature distortions in FeO, CoO, NiO and the spinels, from cubic to tetragonal or orthorhombic symmetry, is a case in point[3]. The spinel-type oxides (Fe_3O_4, $FeO{\cdot}Cr_2O_3$, etc.) occupy a particularly critical position in this respect, because metallic and ionic bonding (to di- and trivalency) are here closely intermixed. (These and other such cases, such as the perovskites, will be discussed below.)

Non-stoichiometry is as important amongst interstitial oxides as for the carbides and nitrides, but in the light of the bond differences, electronically different modes of defect can occur. Thus, excess of (1) O, due to cation vacancies; (2) metals, due to anion vacancies; (3) interstitial cations; (4) interstitial anions, all become possible contingencies in FeO, CoO, NiO and, continuing along the 1st Long Period, Cu_2O. An important consequence is that electrons can be trapped near a vacancy; in Cu_2O, for example, a cupric Cu^{++} ion can be trapped in a positive hole. This subject impinges on the immense field of semiconductivity and the occurrence of N- and P-type conductivity related to electronic or positive hole type of defects where oxides form only a part (cf. Bibliography and also that to Chapter 2).

Further implications lie in the oxidation and tarnishing reactions on metals. Thus, in a layered assembly $M/MO/O_2$ (M = metal), diffusion through MO can proceed either by M or by O migration. The rate-determining factor is in general (e.g. in the Cu–O system) the diffusion of M ions and its dependence on vacancies and electron bond type. The latter varies critically within sequences of oxide phases in any one scale-layer assembly (such as $Cu/Cu_2O/CuO$ or $Fe/FeO/Fe_3O_4/Fe_2O_3$). This is more sensitive in oxides than in carbides and other interstitial compounds where, despite drastic structural change between phases, at least the type of bonding remains substantially similar.

A further example may illustrate the delicate balance which controls the physical nature of some of the oxides. High-purity nickel oxide, NiO, is an insulator, yet the introduction of traces of impurity ions can render it conducting. Thus odd Ni vacancies which occur may be filled e.g. by lithium ions, with a rapid (100- or 1,000-fold) increase of conductivity. This effectively changes the local lattice stoichiometry very slightly, in the sense of local Li_2O configurations (Li^+ for Ni^{++}), i.e. a tendency towards increased O ion vacancies in the neighbourhood. Conversely, a Cr^{+++} ion impurity added to NiO would attract oxygen, causing O vacancies further afield; the effect is to create a P- instead of N-type semiconductor. The position becomes rather less critical on passing to the earlier transition metals, e.g. from Cu_2O and NiO towards the equivalent oxides FeO,VO, TiO, and to phases MO_{1-x} where metallic bonding becomes governing. The critical bond position, however, occurs in these systems too, but it is transposed towards the higher oxides (e.g. Fe_3O_4, TiO_2).

These defect lattice aspects obviously also impinge on the kinetics of any other types of reaction involving oxide layers, not merely in plain oxidation and reduction processes but also in those affecting surface mobility, catalytic reactions, adsorption rates, chemisorption, etc. They have an effect on sintering, tarnishing, semiconductivity, magneto-chemical behaviour, on the precipitation of second phases in oxide systems and on dispersion strengthening by oxides within an alloy matrix. One of the important points is the potential 'trigger action' of traces of impurity (in their role of filling or creating vacant lattice sites) and thus releasing a chain of reactions inhibited without them.

The following discussion of the transition-metal oxides will proceed from the single to the two (and higher) metal/oxygen systems, being accompanied by a brief survey of structures and properties*.

8.2 SURVEY OF BINARY OXIDES

In *Figure 8.2* and *Table 8.1* a complete survey of the binary Group IV to VIII oxide systems is given, as far as information is available. However, the gaps in our knowledge are still conspicuous, and it is indeed one object to reveal these and to suggest lines of further research required.

The following is the overall picture in proceeding regularly from Group IV to VIII (it may help to keep *Figure 8.2* and *Table 8.1* at hand).

8.2.1. Suboxides, from primary (*M*,O) *solutions to monoxides*

The stabilization of the hexagonal modifications of α-Ti, Zr and (probably) Hf by oxygen is of particular interest, because not only is the primary oxygen solubility very extensive (to about 30 atomic per cent O) but the melting point is increased with oxygen as from the pure metal to attain a maximum (≈1900°C, 25 per cent O for Ti; cf. *Figure 8.2*). Although a solid solution, this might be effectively considered an 'oxide', i.e. a true, high-melting interstitial compound of HCP structure, of compositions such as Ti_3O or Zr_4O. The fact that it joins by solid solution to the pure metal is incidental, though important enough in oxidation kinetics. The phase diagrams of Ti–O and Zr–O almost 'spell out' the gettering action, and the neighbouring V–O and Nb–O systems in Group V the lesser effectiveness of these metals.

Furthermore, these pseudo-oxides are quite comparable with the earlier hexagonal hemicarbides and nitrides, e.g. V_2C, V_2N, Nb_2C, Mo_2C, etc., and it seems probable that continuous series of solid solutions can occur with these carbides and nitrides, such as Ti_3O–Nb_2C, provided that size and electronic factors are favourable. Of the lower hafnium oxides, little is known, but similarity to Zr may be presumed (e.g. an 'Hf_4O' solid solution with α-Hf). Ordering effects are here quite likely, such as a superlattice in Ti_3O, but still little investigated.

These lower oxides show the following further characteristics. In Group IV, the α-Ti/TiO, α-Zr/ZrO_2 and α-Hf/HfO_2 equilibria hold good. However, the compounds Ti_2O and Ti_3O_2 have been reported, the latter forming

* In consequence of the circumstance referred to above, geology is mainly concerned with oxides. However, whilst in nature two of the potential interstitial atoms, O and Si, are substantially uniformly dispersed in the earth's crust, the carbon atom tends to occur more highly segregated, such as in coal deposits and in limestone. This in turn is brought about by the incidence of life. It leads back to earlier considerations (Chapter 2) and the fact that C, O, N and H (i.e. our principal species of interstitial atom) *amongst themselves* combine in form of organic and biological structures. Whereas, upon decomposition, O, N and H all largely disperse, carbon remains either segregated in coal or diamond or, still combined, as oil deposits or carbonates.

Table 8.1 Interstitial oxides: systematic survey of occurrence, structure and some properties of single-metal oxides

System	Oxides ($\leftrightarrow$ = homogeneity range)	Structure: Type	Structure: Lattice dimensions, Å; *M* = molecules per unit cell	Density	Melting point[4a], °C	Heat of formation kcal/mole at R.T.	Other properties; remarks (ρ = el. resistivity (Ω·cm); H = hardness; β = expansion coefficient; *SC* = semi-conductor; *NC* = non-conducting at R.T.; *MC* = metallic conductor; *HR* = homogeneity range ($\leftrightarrow$))
Group IV Ti–O	α-(Ti,O) solution (pseudo-oxide $\leftrightarrow$ 'Ti_3O')	*HCP*	a = 2·950–2·959, c = 4·683–4·81 at O saturation	4·5 (decreasing with O)	1,900 (max. at 25 a/o O)	—	H and ρ increase with O in α-(Ti,O) solution; O is strong α stabilizer; interstitial solubility to ≈ 35 a/o O
	Ti_2O $\leftrightarrow$	hex. superlattice (NiAs type)					
	Ti_3O_2	tetragonal (?) or hexagonal	a = 5·333, c = 6·645; M = 2 a = 5·790, c = 4·80; M = 2				forms peritectoidally below ≈925°C; tetragonal cell amended to hexagonal
	TiO $\leftrightarrow$	cubic (B1)	4·180; M = 1 (but varying within *HR*)		2,020	−135	stoichiometric TiO disproportionates into *two* defect 'TiO' types; Ti and O vacancies *HR*: ≈ 44–45 at. per cent O; tetragonal low-temp. form also reported: distortion from B1; metallic *and* ionic bonding in TiO; ρ ≈ 300 at R.T.
	Ti_2O_3	rhombohedral D–5 type (Al_2O_3)	a = 5·45, α = 59·1°; M = 2	3·6	1,800	−375	homologous series of 7 crystalline forms between 63·5 and 65·6 a/o O
	Ti_3O_5	monoclinic	a = 9·44, b = 3·80, c = 9·75, β = 92·1°; M = 4		1,640	−587	two monoclinic forms of Ti_3O_5 reported; one similar in structure to $Fe_{0\cdot2}TiO_2$
	TiO_2 (3 forms) $\leftrightarrow$	tetragonal (C–4) (for rutile; also anatase and brookite structures)	a = 4·5937, c = 2·958; M = 4	4·2	1,850	−219	H = 6 Moh; β = 7·5 × 10^{-6} (0–600°C) *HR* to lower O: to about 1 per cent vacation of O positions. $TiO_{1\cdot983}$–$TiO_{2\cdot000}$ *SC* ρ = 1·2 × 10^4 at 800°C, to 8·5 at 1,200°C, to ρ > 10^{14} at R.T. conductivity in TiO_2 can also be induced by vacuum treatment and by hydrogen impregnation
Zr–O	α-(Zr,O) solution (pseudo-oxide $\leftrightarrow$ 'Zr_3O')	*HCP*	a = 3·229–3·240, c = 5·143–5·187 at O saturation	6·5 (decreasing with O)	1,975 (max. at 25 a/o O)	—	H and ρ increase with O in α-(Zr,O) solution; O is strong α stabilizer; interstitial solubility to 29 a/o O

Group IV cont.	ZrO ↔	cubic (B1)	4·584–4·620				ZrO probably metastable; other suboxides improbable, certainly not stable. α-(Zr,O)–ZrO_2 eutectic at 1,900°C and 43·5 per cent O
	ZrO_2 ↔	monoclinic (low-temp.)	a = 5·17, b = 5·26, c = 5·30, β = 80·17°	5·6	2,690	−259	H = 6·5 Moh; β = 11·0 × 10^{-6} (0–1,000°C) *HR* to lower O, viz. to 63 per cent O from ZrO_2 at eutectic; strong thermal hysteresis in 900–1,100°C range
		tetragonal (high-temp.); both are fluorspar-type derivatives; *cubic* form needs stabilizer addition, e.g. MgO, CaO, but also reported as genuine very-high-temperature form	a = 5·08, c = 5·17	6·1			*NC* ρ = 10^6 at 385°C
Hf–O	α-(Hf,O) solution (pseudo-oxide to ≈20·5 at. per cent O at 1,350°C)	*HCP*	a = 3·194–3·214, c = 5·052–5·101 at O saturation				H and ρ increase with O in α-(Hf,O) solution
	HfO_2 ↔	monoclinic (low-temp.)	a = 5·12, b = 5·18, c = 5·25; β = 98°	9·7	2,780	−272	β = 6·5 × 10^{-6} (0–400°C)
		tetragonal above 1,870°C (as ZrO_2); cubic fluorspar type uncertain		10·0			*NC* ρ = 5 × 10^9 at 400°C *Note:* no data on lower Hf-oxides (HfO?) or on O solution in α-Hf, but latter certainly high; α-stabilization by O; probably HfO_2 only oxide
Th–O	ThO_2	cubic (C1) fluorspar type	5·5840	9·7	3,300	−293	H = 6·5 Moh; β = 9·3 × 10^{-6} *NC* (0–1,000°C) ρ = 2·6 × 10^7 at 550°C to 1·5 × 10^4 at 1,220°C no oxides lower than ThO_2
Group V V–O	($V_{10}O$? to) V_4O ↔	BC tetragonal (Ta_4O analogue); O atoms in octahedral positions	a = 2·989–2·942, c = 3·263–3·512		uncertain		Note expansion by O along c, contraction along a axis tetragonal distortion to primary V; *HR* 15–22 a/o O

(*continued on page 358*)

System	Oxides (↔ = homogeneity range)	Structure: Type	Structure: Lattice dimensions, Å	Density	Melting point[4a], °C	Heat of formation, kcal/mole at R.T.	Other properties; remarks (ρ = el. resistivity (Ω·cm); H = hardness; β = expansion coefficient; SC = semi-conductor; NC = non-conducting at R.T.; MC = metallic conductor; HR = homogeneity range (↔))
Group V cont.	$V_{13}O_7$	hexagonal?					forms peritectoidally below 1,140°C
	VO ↔	cubic (B1)	{4·042 at $VO_{0·75}$; 4·093 at $VO_{1·00}$; 4·128 at $VO_{1·20}$	6·1–5·5	≈1,800	−106	V and O vacancies throughout; *HR* $VO_{0·75}$–$VO_{1·20}$
	V_2O_3	rhombohedral D-5 type (Al_2O_3)	a = 5·467; α = 53·74°; M = 2	4·9	1,980	—300	*MC* ρ = 5·5 × 10^{-3} at R.T.; *HR* $VO_{1·4}$–$VO_{1·53}$
	(then series:) V_3O_5	monoclinic	a = 9·98, b = 6·53, c = 9·835; β = 138·80°; M = 4				
	$VO_{1·75}$–VO_2	phase series of related structures, leading up to VO_2					homologous series $VO_{1·5,\,1·67,\,1·75,\,1·80,\,1·84,\,2·0;\,2·17,\,2·5}$
	VO_2 ↔	monoclinic but near-rutile	a = 5·743, b = 4·517, c = 5·375; β = 122·61°; M = 4	4·4	1,550	−176	transforms to true rutile at 70°C; near-isomorph to MoO_2; note critical TiO_2–VO_2–NbO_2–MoO_2 relations
	V_6O_{13}	monoclinic	a = 11·90, b = 3·671, c = 10·122; β = 100·87°; M = 2	3·9			V_6O_{13} dec. → VO_2 + V_2O_5
	V_2O_5 ↔	orthorhombic	a = 11·519, b = 4·573, c = 3·564; M = 2	3·4	670	−383	*HR* V_2O_5–$V_2O_{4·8}$ (similar to Nb_2O_5)
Nb–O	Nb_4O; Nb_2O (unstable) ('NbO_x', 'NbO_z')	BC tetragonal	a = 3·36, c = 3·25		dec. 350°C (?)		Nb_2O formed during Nb oxidation: tetragonal distortion of primary Nb; probably an 'Nb_4O' sub-oxide, similar to V_4O and Ta_4O, also occurs
	NbO ↔	cubic (B1) (vacancy ordering)	4·211	7·3	1,945	−116	*HR* probable for NbO but not determined; sub-oxides 'NbO_x' and 'NbO_z' form during Nb oxidation kinetics; x = 1/2, z uncertain
	NbO_2 ↔	tetragonal (rutile-related)	a_r = 4·846, c_r = 2·993 (for rutile sub-cell); M = 4	5·9	1,915	−194	*MC*; NbO_2 has 'supercell' of rutile, with $a = 2\sqrt{a_r}$, $c = 2c_r$ and 32 NbO_2 units; ρ = 0·09 at R.T.; an 'Nb_2O_3' sometimes referred to is probably NbO_2
	Nb_2O_5 (2 forms, possibly 3)	α 'low temp.' β 'high-temp.' Nb_2O_5	structures uncertain; α orthorhombic but also reported monoclinic	4·6	1,460 (1,495?)	−463	*SC* (?); *HR* Nb_2O_5–$Nb_2O_{4·8}$ near 1,400°C; α-Nb_2O_5 is metastable, β stable at all temps.; a third very-high-temp. form reported; also structural variants to β

Group V cont. Ta–O	Ta_4O (Kofstad's 'TaO_y')	orthorhombic (distorted *BCC* of Ta)	$a = 7{\cdot}194–7{\cdot}238$, $b = 3{\cdot}266–3{\cdot}273$, $c = 3{\cdot}204–3{\cdot}238$				other lower-Ta oxides may occur, but still problematic suboxides 'TaO_x' and 'TaO_z' form transiently during oxidation kinetics; $y = \frac{1}{4}$ z uncertain but $\geq \frac{1}{2}$
	Ta_2O (probably Kofstad's 'TaO_z')	orthorhombic					unconfirmed; note increasing uncertainty on lower oxides down Groups IV and V
	TaO ↔	cubic (B1)	$a = 4{\cdot}422–4{\cdot}439$				
	TaO_2 ↔	tetragonal (rutile-related)	$a = 4{\cdot}709$, $c = 3{\cdot}065$; $M = 2$	10·4			
	$TaO_{2{\cdot}25}$	structure unknown					
	Ta_2O_5 (2 forms, possibly 3) ↔	low Ta_2O_5 (α) orthorhombic; high Ta_2O_5 (θ) tetragonal	$a = 6{\cdot}180$, $b = 43{\cdot}93$, $c = 3{\cdot}890$; $M = 12$ $a = 3{\cdot}80$, $c = 35{\cdot}60$; $M = 6$	8·3 8·5	1,890	−499	*SC* (?); transforms ≈ 1,320°C (?); defect structure of Ta_2O_5 reported; small *HR* exists; Ta–O diagram very incompletely known
Pa–O	PaO	cubic (B1)	4·961	13·4			
	PaO_2 ↔	cubic (Cl, fluorspar type)	5·505				
	Pa_2O_5	orthorhombic, as low-temp. Ta_2O_5	$a = 6{\cdot}92$, $b = 4{\cdot}02$, $c = 4{\cdot}18$				
Group VI Cr–O	Cr_3O	cubic (A–15, β-W type)	4·544; $M = 3$				unstable
	CrO (possibly)	hexagonal, wurtzite type?					unconfirmed; unstable if existent; the stable system probably Cr–Cr_2O_3 equilibrium, except for high-temp. Cr_3O_4
	Cr_3O_4	distorted spinel tetragonal	$a = 8{\cdot}72$, $c = 7{\cdot}50$				true spinel can be stabilized, e.g. to FeO·Cr_2O_3
	Cr_2O_3	rhombohedral D–5_1 (Al_2O_3) type	$a = 5{\cdot}35$; $\alpha = 55{\cdot}0°$; $M = 2$	5·25	2,275 (dec. before melting)	−270	*SC*; $\rho = 1{,}300$ at 350°C, 23 at 1,200°C; $\beta = 9{\cdot}6 \times 10^{-6}$ (0–1,400°C); antiferromagnetic; structure distortion at low temps.
	CrO_2	tetragonal (rutile type)	$a = 4{\cdot}41$, $c = 2{\cdot}91$; $M = 2$	4·90		−139	liquid immiscibility
	CrO_3	orthorhombic	$a = 5{\cdot}743$, $b = 8{\cdot}557$, $c = 4{\cdot}789$; $M = 4$	2·82	(dec.)	−137	

(*continued on page 360*)

System	Oxides (↔ = homogeneity range)	Structure: Type	Structure: Lattice dimensions, Å	Density	Melting point[4a], °C	Heat of formation, kcal/mole at R.T.	Other properties; remarks (ρ = el. resistivity (Ω·cm); H = hardness; β = expansion coefficient; *SC* = semi-conductor; *NC* = non-conducting at R.T.; *MC* = metallic conductor; *HR* = homogeneity range (↔))
Group VI cont. Mo–O	Mo_3O	cubic (A–15, β-W type)	5·549; $M = 3$	8·66			mestastable
	MoO_2	monoclinic (distorted rutile)	$a = 5{\cdot}610$, $b = 4{\cdot}843$, $c = 5{\cdot}526$; $\beta = 119{\cdot}62°$; $M = 4$	6·5		−141	Mo and MoO_2 form eutectic at ≈2,100°C
	then homologous series $Mo_{2\cdot75}$, $MoO_{2\cdot88}$, $MoO_{2\cdot89}$ of related structures, leading to MoO_3; general formula: Mo_xO_{3x-1}						
	MoO_3	orthorhombic	$a = 13{\cdot}825$, $b = 3{\cdot}694$, $c = 3{\cdot}954$; $M = 4$	4·71	795	−180	note very low m.p., high vapour pressure; MoO_2 and MoO_3 begin to volatilize well below melting; form eutectic at 778°C; MoO_3 transformation at ≈650°C suspected; Mo–O diagram very incompletely known
W–O	W_3O	cubic (A–15, β-W type)	5·036; $M = 2$	14·8			decomposes → W + WO_2 above 725°C
	WO_2	monoclinic (distorted rutile as MoO_2)	$a = 5{\cdot}560$, $b = 4{\cdot}884$, $c = 5{\cdot}546$	?	1,270 (or > 1,700)	−135	
	then homologous series $WO_{2\cdot65/2\cdot75}$ ↔	$WO_{2\cdot88/2\cdot92}$ ↔ of related structures, leading to					homologous series with appreciable *HR* for each member; related to ReO_3
	WO_3 (3 forms)	monoclinic (low-temp.)	$a = 3{\cdot}85$, $b = 7{\cdot}54$, $c = 7{\cdot}29$; $\beta = 90{\cdot}3°$	7·2	1,470 (dec. before melting)	−200	ferroelectric properties
		orthorhombic (med. temp.)					change mon. ↔ orth. about 300°C
		tetragonal (high-temp.) (distorted ReO_3 type)	$a = 5{\cdot}250$, $c = 3{\cdot}915$ (at 770°C)				WO_2 and WO_3 begin to volatilize well below melting
U–O	UO	cubic (B1)	4·92	14·1			

Group VI cont.	UO_2 ↔	cubic (Cl, fluorspar type)	5·458; $M = 4$	10·91	2,230 or 2,880 (preferred)	−255	$SC = 3{\cdot}8 \times 10^4$ at 20°C *HR* from UO_2 to $UO_{2\cdot2}$ with decreasing spacing; new interstitial O sites giving $UO_{2\cdot1}$ at 600°, $UO_{2\cdot2}$ at 1,050°C
	$UO_{2\cdot2}/UO_{2\cdot3}$ ↔	tetragonal (distorted fluorspar type)	link to UO_2, but *c* axis increasing, *a* axis decreasing with O content; ordering of O atoms			−1080	
	homologous series $UO_{2\cdot33}$	$UO_{2\cdot43}$, $UO_{2\cdot5}$, $UO_{2\cdot66}$	(U_3O_8) to UO_3	11·05			all closely related structures, liable to non-stoichiometry and temp. variations
	U_3O_8	orthorhombic and hexagonal forms		10·06		−845	
	UO_3 (6 known forms)	trigonal, orthorhombic, etc.; γ-UO_3 tetragonal	$a = 6{\cdot}89$, $c = 19{\cdot}94$; $\alpha = \beta = 90°$, $\gamma = 90{\cdot}34°$; $M = 4$	8·35		−291	U_3O_8 dec. above 750°C to UO_2; *NC*?
Group VII Mn–O	MnO	cubic (B1)	$4{\cdot}435_5$	5·4	1,785	−93	*SC*; $\rho = 10^8$ at 20°C = 5·5 MoH; antiferromagnetic at −151°C; ordering and distortion to rhombohedral at low temps.
	Mn_3O_4	tetragonal (spinel-related); also cubic spinel at H.T.?	$a = 8{\cdot}140$, $c = 9{\cdot}435$; $M = 8$ $a = 8{\cdot}7$?	4·8	1,560	−332	'hausmannite' (distorted spinel)
	Mn_2O_3 (3 forms)	cubic (D–5_3 type) tetragonal BC tetragonal	9·408; $M = 16$ $a = 8{\cdot}1$, $c = 9{\cdot}4$ $a = 8{\cdot}85$, $c = 9{\cdot}95$ (?)			−233	= α-Fe_2O_3 analogue, but cubic variant = Mn_3O_4 structure with Mn vacancies (γ-Fe_2O_3 analogue)
	MnO_2 (3 forms)	tetragonal (rutile type) orthorhombic orthorhombic	$a = 4{\cdot}395$, $c = 2{\cdot}86$; $M = 2$ $a = 9{\cdot}27$, $b = 2{\cdot}866$, $c = 4{\cdot}533$ $a = 9{\cdot}36$, $b = 2{\cdot}85$, $c = 4{\cdot}43$	4·83		−125	'pyrolusite' 'ramsdellite' related to ramsdellite; synthetic
	Mn_2O_7					−174	
Tc–O	TcO_2	monoclinic (rutile variant as MoO_2)	$a = 5{\cdot}53$, $b = 4{\cdot}79$, $c = 5{\cdot}53$; $\beta = 120°$; $M = 4$			−104	other technetium oxides, e.g. TcO_3 (ReO_3 analogue?) probably exist but not reported
	Tc_2O_7				119	−266	
Re–O	ReO_2 (2 forms)	monoclinic (rutile variant as MoO_2) orthorhombic	$a = 5{\cdot}562$, $b = 4{\cdot}838$, $c = 5{\cdot}561$; $\beta = 12°$; $M = 4$ $a = 4{\cdot}810$, $b = 5{\cdot}64$, $c = 4{\cdot}601$; $M = 4$			−101	

(*continued on page 362*)

System	Oxides ($\leftrightarrow$ = homogeneity range)	Structure: Type	Structure: Lattice dimensions, Å	Density	Melting point[4a], °C	Heat of formation, kcal/mole at R.T.	Other properties; remarks (ρ = el. resistivity (Ω·cm); H = hardness; β = expansion coefficient; *SC* = semi-conductor; *NC* = non-conducting at R.T.; *MC* = metallic conductor; *HR* = homogeneity range ($\leftrightarrow$))
Group VII cont.							
	Re_2O_7	orthorhombic	$a = 15{\cdot}25$, $b = 5{\cdot}48$, $c = 12{\cdot}5$; $M = 8$		296	−300	
	ReO_3 $\leftrightarrow$	cubic (D–O_9 type)	3·73; $M = 1$	7·43	dec.	−147	structural relation to WO_3 and to perovskites; *HR*
Np–O	NpO	cubic (B1)	5·00	11·11			NpO–NpO_2–Np_3O_8 similar to those in uranium
	NpO_2	cubic (Cl)	5·425			−246	
	Np_3O_8	orthorhombic (U_3O_8 type)					probably further variants in this range, as for U
Group VIII Fe–O	'FeO' $\leftrightarrow$	cubic (B1)	4·3088 (Fe end), 4·2800 (O end) (at R.T., as retained)	5·6–5·7	1,370–1,390	−65	'wüstite'; *HR* (metal-deficient lattice): 51·4° to peritectoid 570°C; 53·3 at. per cent O at 1,000°C; antiferromagnetic; rhombohedral low-temp. distortion becomes tetragonal at higher O; two varieties of wüstite, of different O content indicated; $\beta = 12{\cdot}2$ (100–1,000°C)
	Fe_3O_4 $\leftrightarrow$	cubic (H–1_1) spinel type	8·3940; $M = 8$		1,590 (dec.)	−265	'magnetite', ferromagnetic; transforms to rhombohedral at −158°C; small *HR*
	Fe_2O_3 (2 forms) $\leftrightarrow$	rhombohedral (α) (D–5, Al_2O_3 type)	5·427; $\alpha = 55{\cdot}26°$; $M = 2$	5·25	peritectic, dec. 1,457°C	−195	'haematite'; antiferromagnetic transition at 675°C, with changes in lattice expansion; magnetic transformations reported at −40°, −7° and (Néel point) + 726°C
		cubic (γ) spinel $\leftrightarrow$ type (a third form suggested)	8·322 *kX*				Fe vacancies in spinel; in γ–Fe_2O_3 *ordered* in octahedral sites; *HR*; metastable; ferromagnetic $\beta = 14{\cdot}9$ (20–900°C) Curie point of $Fe_2O_3 \approx 640°C$
Ru–O	RuO_2	tetragonal (rutile type)	$a = 4{\cdot}51$, $c = 3{\cdot}11$; $M = 2$			−53	

Group VIII cont.							
Os–O	OsO_2	tetragonal (rutile type)	$a = 4{\cdot}51$, $c = 3{\cdot}19$; $M = 2$				
	OsO_4	monoclinic	$a = 8{\cdot}66$, $b = 4{\cdot}52$, $c = 4{\cdot}75$; $\beta = 117{\cdot}9°$; $M = 2$		42°C	−80	
Pu–O	PuO	cubic (B1)	4·948	13·9			$PuO–PuO_2$ relations similar to those for U and Np, but intermediate distortions
	Pu_2O_3 (2 forms)	BC cubic ($D–5_3$ type)	11·04; $M = 16$	10·2			
		hexagonal (La_2O_3 type)	$a = 3{\cdot}841$, $c = 5{\cdot}958$	11·47			
	PuO_2	cubic (Cl fluorspar type)	5·396	11·46		−246	
Co–O	CoO	cubic (B1)	4·2581	6·5	1,810	−58	*SC*; $\rho = 100$ at 300°C; antiferromagnetic below 10°C, with tetragonal distortion, increasing towards L.T.; Co–CoO eutectic at 1,451°C, similar to Ni–NiO; $-\beta = 15{\cdot}0$ (20–900°C)
	Co_3O_4	cubic (H11) spinel type	8·09; $M = 8$	6·17	dec. 900	−197	close thermal relations to CoO to which it transforms as $Co_{1-x}O$ $\beta = 13{\cdot}6$ (20–900°C)
	Co_2O_3 (possibly)	hexagonal (haematite type)	$a = 4{\cdot}64$, $c = 5{\cdot}74$	5·34			unconfirmed: metastable or occurring only as hydrate?
Rh–O	Rh_2O_3	rhombohedral ($D–5_1$, Al_2O_3 type)	5·45; $\alpha = 55{\cdot}72°$; $M = 2$	8·2		−69	Rh_2O and RhO reported but not certain
Ir–O	IrO_2	tetragonal (rutile type)	$a = 4{\cdot}49$, $c = 3{\cdot}14$; $M = 2$		dec. 900	−46	
Ni–O	NiO	cubic (B1); below 150°C distorted to rhombohedral	4·1946 $a = 2{\cdot}9518$ $\alpha = 60{\cdot}07°$ (R.T.)	7·4	1,960	−58	*NC* → *SC*; $\rho = 6{,}700$ at 600°C, 240 at 1,250°C; eutectic Ni–NiO at 1,438°C; $H = 5{\cdot}5$ Moh; $\beta = 17{\cdot}1$ (20°–1,000°C); antiferromagnetic; synthetic ferromagnetic NiO with Ni vacancies also reported; monoclinic, i.e. *artificial HR* (not normally occurring, but similar to CoO and FeO); 'Ni_3O_4', 'Ni_2O_3' and 'NiO_2', occasionally reported, very doubtful
Pd–O	PdO	tetragonal	$a = 3{\cdot}03$, $c = 5{\cdot}33$; $M = 2$	8·3		−21	PdO conceivably becomes cubic at high temps. (as NiO), but not known
Pt–O	PtO	tetragonal, as PdO	$a = 3{\cdot}04$, $c = 5{\cdot}34$; $M = 2$				
	Pt_3O_4	BC cubic	6·266; $M = 2$	8·8			all metastable
	PtO_2	hexagonal	$a = 3{\cdot}08$, $c = 4{\cdot}19$			+41	

peritectoidally at 925°C where the hemioxide, Ti_2O, is in the nature of a superlattice phase evolving from the α-(Ti, O) solution[4b,5]. The ordering process leading up to this is already noticeable at the composition 'Ti_3O', in the form of an increase in axial ratio, i.e. the 'pseudo-compound' composition of maximum melting point (see above). This would lead from HCP α-Ti, through the successive oxygen occupation of octahedral holes, ideally to an NiAs-type structure in 'TiO'. However, the occupation ceases when the anti-CdI_2-type structure of Ti_2O is attained as it transforms to the cubic symmetry of (O-deficient) TiO. This is a stacking change of Ti atoms clearly dictated by the need to find new O sites; at high temperatures this is direct, whilst at low temperatures the process goes via the intermediate stage of tetragonal Ti_3O_2. Nothing seems to be known of any equivalent process in Zr or Hf, but the point would be worth studying, particularly the problem of a metastable cubic 'ZrO' which may evolve before ZrO_2.

In Groups V and VI the lower oxides gain considerably in prominence. The body-centred tetragonal 'V_4O' (with homogeneity range) is clearly a distortion of the parent vanadium structure which in the similar Ta_4O evolves to still lower (orthorhombic) symmetry. In niobium, an equivalent process of lower-Nb oxides has been indicated by Norman's and Kofstad's studies on Nb oxidation (see later).

As to the lower oxide equilibria for Nb and Ta, the position is still far from clear, particularly whether they should be regarded as unstable 'kinetic' phases which only form in the course of reaction and strictly have no room in the equilibrium diagram. The phases $V_{13}O_2$ (peritectic, hexagonal), the controversial 'Nb_2O' and 'Ta_2O' and their relation to the reported 'NbO_x', 'NbO_z', 'TaO_y', 'TaO_z' should also be clarified. Pressure sensitivity of these equilibria will be strong, and though unstable at atmospheric or lower pressures, such structures may well become stabilized at high pressures and low temperatures. The issue also arises whether in Group V at very low temperatures an intermediate hexagonal analogue to α-Ti and α-Zr could conceivably arise, stabilized by oxygen, without excluding a possible true new polymorph to V, Nb or Ta on approaching the absolute zero temperature.

Passing to Group VI (Cr, Mo, W), the M_3O phases of 'β-W'-type structure (cubic A–15 type) evolve. β-W was believed originally to be a second form of tungsten, but proved[6] to be an oxide, W_3O*. This structure is again derived from the BCC parent by O-atom addition, as discussed already in connection with the silicide isomorphs Cr_3Si and Mo_3Si (*Figure 7.3*). Mo_3O and W_3O are firmly established (except for the precise stoichiometry), but 'Cr_3O' is speculative, as is the issue whether these phases are really metastable or stable. Thermodynamically, the question is complicated by competing higher-oxide formation superimposed on extreme temperature and pressure sensitivity.

In summary, the oxides lower than MO evolve in Groups IV → V → VI from (1) the wide hexagonal close-packed α-(Ti, O)-type solid solution, of 'pseudo-compound' status, to (2) the BC tetragonal and orthorhombic

* This formula is, in later light, debatable, but that 'β-W' is a low oxide is beyond doubt.

V_4O, Nb_4O (?), Ta_4O (?) types, with superlattice tendency, to (3) the cubic W_3O type which is a genuine new phase.

No oxides lower than *MO* exist in any of the later transition-metal systems, viz. of the Groups VII, VIII, nor of the actinide metals. Their possible transient formation, e.g. at surfaces or adjoining dispersed oxide particles, ought not to be excluded, however. This applies up to nickel. In the adjacent copper, we have Cu_2O, and for silver, Ag_2O. These, and not CuO, may be regarded as the direct descendants of the cubic close-packed *MO* structures, the lower stoichiometry corresponding to a defect lattice.

8.2.2. Monoxides

TiO forms a prototype for the large family of cubic NaCl-type oxides, with its variants. Extensive non-stoichiometry can occur, on both sides of the equi-atomic composition, due both to oxygen and metal vacancies; a similar situation applies in Group V to 'VO'. An analogue 'ZrO' has been reported, but its existence is uncertain; zirconia is certainly the only stable oxide.

The α-Ti/TiO equilibrium again represents a transition from hexagonal to cubic close packing of metal atoms. Seen more broadly, the β-Ti → (Ti, O) → TiO series is another instance of the familiar BCC/HCP/FCC transition, in the new mode of oxygen content acting as controlling parameter.

Taking an overall view of the occurrence of close-packed metallic monoxides, whether stable or metastable, the following pattern of isomorphs arises (for equilibria see *Figure 8.2*):

TiO	VO	CrO?	MnO	FeO	CoO	NiO	(Cu_2O)*
ZrO	NbO	(Mo) none	(Tc) none?	(Ru) none	(Rh) none	PdO	(Ag_2O)*
HfO?	TaO	(W) none	(Re) none	(Os) none	(Ir) none	PtO?	
(Th) none	PaO	UO	NpO	PuO	AmO	CmO	

(also: MgO, CaO, SrO, BaO, ScO, YO, [Rare-Earth] O, outside or bordering the transition metals)

* For Cu and Ag, the logical descendants to the *MO* structure lie not at Cu_2O and AgO but at the hemioxides.

Thus many variants develop from 'TiO'. Along the First Long Period (TiO → NiO), the type of bonding gradually changes from metallic to ionic. This is accompanied by the emergence of semi-conductivity, anti-ferromagnetism and, structurally, anisotropy (distortions from cubic symmetry). Simultaneously, the homogeneity ranges narrow, although a limited number of lattice vacancies remain a feature throughout. A noteworthy minimum in stability occurs at 'CrO', the existence of which is not even certain. The latter minimum is, however, accompanied by the ascendance to the highly stable and refractory Cr_2O_3 and the initiation of spinels, both associated with a marked change in bonding type. This is another instance of the borderline position of chromium (observed in so many other respects already, e.g. in the carbides).

One remarkable feature is *melting-point* behaviour for the monoxides as the bonding changes along this series, with a trough at FeO, viz. TiO 2020 (135), VO 1800 (106), MnO 1785 (93), FeO 1390 (65), CoO 1810 (58),

NiO 1960°C (58); the heats of formation (in brackets), on the other hand, steadily decrease.

Down Groups IV and V (excluding the actinides), the monoxide stability decreases, followed in Groups VI–VIII by the large area of omissions, but the termination is a slight recovery for the NiO Group—unstable PdO and PtO.

On the heavy-element side, the area where *M*O is absent is, however, firmly contained by the stable actinon monoxides. No 'ThO' has been reported, but the structure becomes stable for UO and the transuranic oxides (cf. Chapter 2). In this series, metallic bonding is regained. The lattice dimensions of these oxides are shown, in comparison with the NaCl-type carbides, nitrides, etc., in *Figure 4.3.*

The wide *non-stoichiometric* ranges in TiO and VO do not continue in NbO, TaO and the other cubic monoxides. However, though more restricted, they assume their own fields of importance, e.g. in $UO \leftrightarrow UO_2$ (nuclear energy applications), in FeO and in the semi-conducting types (NiO, etc.).

CrO can be produced transiently by aluminothermic reaction, namely $3Cr_2O_3 + 2Al \rightarrow 6CrO + Al_2O_3$. This is a simple and direct example of the principle of utilizing the higher heat of formation of one oxide to reduce

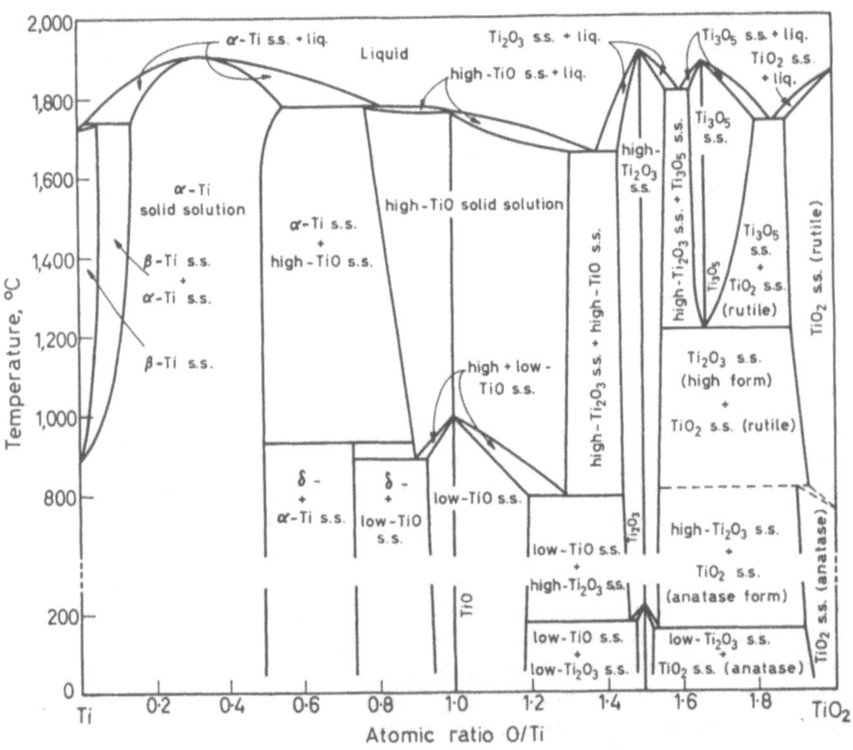

Figure 8.3. The Ti–O *system: 'A not-impossible diagram'* (De Vriess and Roy[8])

(s.s. = solid solution)

By courtesy of the American Ceramic Society

another, even though the initial and final oxides are isomorphs (hematite type) and represent in part just an exchange of cations in the same lattice. Divalent chromium could more stably be built into a spinel lattice ($CrO \cdot Al_2O_3$), but we may regard the aluminothermic 'CrO' as a metastable retention product of its (probably stable) high-temperature form. (The Bibliography gives some basic literature on the TiO → NiO monoxide series.)

The monoxides clearly are in a critical position, as regards the change-over from the non-metallic to the metallic state.

8.2.3. Range between MO and MO_2 (e.g. TiO ←→ TiO_2)

These are marked by the highest degree of complexity. In Ti–O we have, besides the better defined Ti_2O_3 and Ti_3O_5, no less (possibly more) than seven intermediate forms[7a]. *Figure 8.3* shows a suggested 'possible' Ti–O diagram[8] (alternative to that included in *Figure 8.2*).

The changes extend to the subsequent systems, as far as $MO_{2\cdot5}$ in Group V and MO_3 in Group VI; the manner in which this happens is summarized in *Table 8.2*.

Table 8.2. Series of phases from MO to MO_3 in oxides of Group IV–VI elements

MO	*Intermediate series*	*MO_2*	*$MO_{2\cdot5}$*	*MO_3*
TiO	$TiO_{1\cdot5}$ $TiO_{1\cdot67}$ $TiO_{1\cdot75, 1\cdot8, 1\cdot83,}$ $TiO_{1\cdot86, 1\cdot875, 1\cdot88, 1\cdot90}$*	TiO_2†	– – – – – – – –	– – – –
(ZrO)	– – – – – – – –	ZrO_2†	– – – – – – – –	– – – –
(HfO)	– – – – – – – –	HfO_2†	– – – – – – – –	– – – –
		ThO_2		
VO	$VO_{1\cdot5}$ $VO_{1\cdot67}$ $VO_{1\cdot75, 1\cdot8, 1\cdot84}$	VO_2	$VO_{2\cdot17}$‡, $VO_{2\cdot5}$ – – – – – –	– – – –
NbO }	(straight *MO*–*MO_2* transition possible but uncertain; some intermediate stages thought probable)	{ NbO_2	possible series $NbO_{2-2\cdot5}$ $NbO_{2\cdot5}$† – – – – – –	– – – –
TaO }		{ TaO_2	$TaO_{2\cdot25}$‡ $TaO_{2\cdot5}$† – – – – – Possible series $TaO_{2-2\cdot5}$?	– – – –
CrO?	$CrO_{1\cdot3}$ $CrO_{1\cdot5}$ – – – – –	– CrO_2	– – – – – – – –	– CrO_3
	– – – – – – – –	– MoO_2	$MoO_{2\cdot75, 2\cdot88, 2\cdot89}$	MoO_3
	– – – – – – – –	– WO_2	$WO_{2\cdot65 \leftrightarrow 2\cdot75}$, $WO_{2\cdot88 \leftrightarrow 2\cdot92}$	WO_3†
UO		UO_2– $UO_{2\cdot2-2\cdot3}$, $UO_{2\cdot33, 2\cdot43}$	$UO_{2\cdot5}$ $UO_{2\cdot66}$†	

* Representing a series Ti_nO_{2n-1} where $n = 4$–10; further extension to phases with $n > 10$, $\sim n = \infty$, may be possible; for *lower* oxygen phases, TiO, $TiO_{1\cdot5}$ and $TiO_{1\cdot67}$ arithmetically precede this series, viz. $n = 1$, 2 and 3—For vanadium, with some omissions, the case is similar.

† = Several polymorphs.

‡ A homologous series $Nb_2O_5 \rightarrow NbO_2$ and $Ta_2O_5 \rightarrow TaO_2$, though not reported, could well occur, initially closely related to structures of β-Nb_2O_5 and β-Ta_2O_5 (Roth, private communication). For V, a similar series $V_2O_5 \rightarrow VO_2$ is possible. Further such series may occur when metals are interchanged in ternary systems, e.g. $VO_2 - Nb_2O_5$, etc.

This multiple homologous series is transposed to increasing oxygen contents from Groups IV to VI. It is accompanied by increasing structural complexity and instability, and by a decline in melting points. The latter is perhaps most conspicuous in V_2O_5 (670°C) and MoO_3 (795°C), which is of direct practical consequence in the oxidation of these metals. It is significant that the multiple oxide series is followed by the highest-oxygen material of low melting point, as it indicates the increasing difficulty of oxygen atoms to find suitable lattice sites before escaping as a free gas. It is doubtlessly associated with the change in bonding from metallic to covalent and the correlated loss of metallic properties. Viewed from a different angle, these homologous series are perhaps the nearest substitute for non-stoichiometric ranges, as actually realized in Nb_2O_5 ($\rightarrow Nb_2O_{4\cdot 8}$) and in TiO_2 ($\rightarrow TiO_{2-x}$). The difference stems from the capability of a given stoichiometric compound to tolerate a slight oxygen deficiency without some distortion of the remaining lattice.

8.2.4. Dioxides

The MO_2 phases show a simple and systematic interrelation within the transition block. Broadly they fall into one of two principal classes: those of *rutile* structure, with TiO_2 as prototype, and those of *fluorspar* structure, with ThO_2 as prototype (and their respective variants). Each occupies a definite and continuous area within the transition-metal block, as shown in *Table 8.3.*

*Table 8.3**

TiO_2	– VO_2	– CrO_2	– MnO_2	– FeO_2(?)		
ZrO_2	– NbO_2	– MoO_2	– TcO_2	– RuO_2	←	*Rutile type*
HfO_2	– TaO_2	– WO_2	– ReO_2	– OsO_2	– IrO_2	
ThO_2	– PaO_2	– UO_2	– NpO_2	– PuO_2	– AmO_2	

↗ *Fluorspar type* (ZrO_2 and HfO_2 when stabilized)

* For details in structure variations, etc., see *Table 8.1.*

The tetragonal TiO_2 and the cubic ThO_2 structures are compared in *Figure 8.4.* The deviations amongst the rutiles are appreciable; enlarged unit cells can evolve, with distortions, e.g. from tetragonal to orthorhombic symmetry. Non-stoichiometric ranges are common, with O vacancies

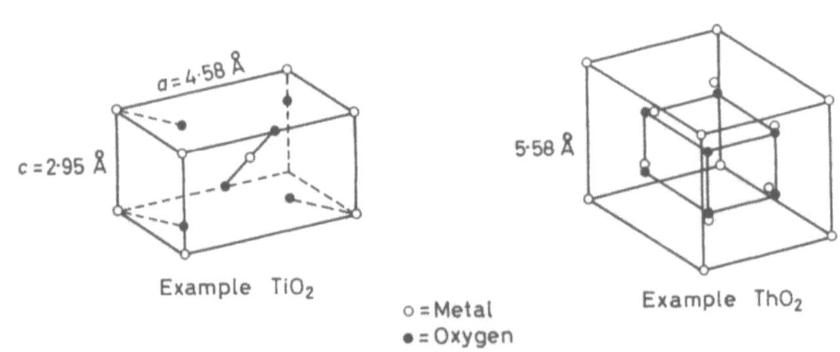

Figure 8.4. Rutile- and fluorspar-type oxide structures

occurring in both types. TiO_2, ZrO_2 and HfO_2 each have several polymorphic forms (rutile, anatase and brookite* in TiO_2; monoclinic, tetragonal and, with certain stabilizers, cubic in ZrO_2), but these can be described as fairly simple derivatives from the rutile and fluorspar structures, respectively.

ZrO_2, with its borderline position, has been proved conclusively by x-ray high-temperature work[9] to possess a truly cubic CaF_2-type form above 2,285°C; it is quenchable only at extreme rates, but the transformation temperature is lowered by MgO addition[10]. This links with the well known 'stabilization' of zirconia (cf. Section 8.4). It seems likely that a similar situation occurs in HfO_2 (pure), but so far only the monoclinic and tetragonal forms are known. The defect structure (non-stoichiometry) e.g. of ZrO_2 and HfO_2 is known to influence greatly the electrical conductivity of these compounds. This is revealed by a critical dependence on oxygen pressure, which has been attributed to the transport of interstitials *coupled* with oxygen vacancies—faster than that of each defect independently[11a].

Most dioxides are preceded, for lower O contents, by the cubic rocksalt type of oxides in which the metal atom lattice is already the same as in the fluorspar-type dioxides; only the oxygen positions change over from octahedral to tetrahedral sites. The pair $UO \longleftrightarrow UO_2$ is one example where, significantly, negligible intersolubility occurs, indicating that both oxygen sites in the same lattice† are mutually exclusive. In contrast, UO_2 forms a pronounced oxygen-excess lattice, up to $UO_{2.3}$. It is noteworthy that the multiple structure changes (see *Table 8.2*) intervening between MO and MO_2 occur only where MO_2 has the *rutile* structure, whilst for the fluorspar types, MO and MO_2 are always in direct equilibrium. The latter is consistent with the similarity of the two metal lattices, but the former means that, for the NaCl type → rutile transformation to occur, the lattice mechanism is a much more complex one which can only proceed in discreet stages as each new small dose of oxygen atoms is added.

The fact of isomorphy or near-isomorphy of the rutile family ($TiO_2 \rightarrow VO_2 \rightarrow NbO_2 \rightarrow MoO_2$, etc.) implies ready mutual solid solutions amongst these, subject to atomic size and electron factors being favourable. Such solid solutions are indeed observed, but with the complication that slight deviations in oxygen content are sufficient to stabilize other phases such as Nb_2O_5, MoO_3, etc. Basically, the attainment of such structural isomorphs by alloy addition has an immediate bearing on scaling problems and is one of the objects in designing oxidation-resisting alloys (see Section 8.5).

The delineation between metallic and ionic bonding becomes critical amongst the rutile- and fluorite-type oxides. In general they are insulators; the ceramic nature of ThO_2 and of stabilized zirconia need only be mentioned, although even here ZrO_2 is noted amongst refractories as of relatively high conductivity. TiO_2 is a semiconductor, as are other rutiles, and a strong temperature dependence of conductivity is a characteristic feature (see

* It may be noted that anatase and brookite, although occurring as natural minerals, are not equilibrium forms of TiO_2, and that by oxidation of Ti metal only rutile can be produced.

† This is in remarkable contrast to the carbides, where we have seen that $UC \leftrightarrow UC_2$ are largely intersoluble (cf. *Figure 4.1*). It demonstrates again the essential difference in bonding of C and O atoms to uranium.

Table 8.1). Thus the electrical resistivity of TiO_2 is 12,000 Ω cm at 800°C and 8·5 Ω cm at 1,200°C, as against a value $>10^{14}$ at room temperature. Anisotropy in physical properties and in diffusion rates is associated with the tetragonal unit cell, and oxides of this type can in some cases be considered as having varying types of bonding, depending on their crystallographic direction, as distinct from the directed bond (cf. Chapter 2).

Rutile, TiO_2, has been the subject of much specialized research from the aspects of point defects, electrical conductivity and mechanical properties. The potential oxygen deficiency is considerable, and it is possible, for instance by vacuum reduction at elevated temperature, to render the phase conducting. Conduction can also be produced by atomic *hydrogen* in the lattice. A particularly fruitful field of study has been the effect of neutron irradiation on TiO_2, as indeed on all the present oxides (UO_2, spinels, etc.).

The effect of radiation consists in the production of new vacancies and interstitials, both anionic and cationic, as well as, where vacancies already exist, their transient filling or increase in number. Energy is stored in the lattice as atoms are knocked out to form such defects, and can be released with heat effects upon healing, as O and metal ions return to their normal positions. It is particularly interesting that, if ceramic oxides, which inherently already possess polymorphic form, are irradiated, phase transformations can be induced; thus monoclinic ZrO_2 can be converted to the cubic form, though not necessarily stably. (On a wider view, this would seem to offer attractive possibilities in many areas of ceramic science, such as conversion between the forms of silica or of iron oxides, and thus a limited control of properties as desired.) Metallic oxidation and corrosion can also be strongly influenced by irradiation which generally has an accelerating effect on diffusion rates of vacancies across an oxide layer, protective or otherwise; moreover, this effect is highly temperature-dependent (cf. Bibliography).

The work of Smallman and co-workers[12] on the non-stoichiometry and vacancy distribution in TiO_2 should here be referred to; they studied thin rutile films by electron microscopy and showed that, when these were heated, the original stacking faults could be annealed out except where retarded by defects; oxygen atoms are fed into dislocation loops, and this produces metal interstitials and vacancy clustering. The process yields an increase in mechanical strength and in ductility, as well as developing colour centres in the O-deficient rutile. The plastic deformation of TiO_2 in relation to dislocation movement and slip systems has also been studied[12], as has the dislocation system in rutile (using an etchpit method)[13].

It is interesting to compare the non-stoichiometries of TiO_2 and UO_2. Whilst in TiO_2 (rutile) we have oxygen deficiency which, upon saturation with metal, leads to the homologous series of *lower* oxides, in UO_2 (fluorspar type, Willis[14]) we have oxygen excess which then also culminates in a homologous series. The two cases may be regarded as an 'inversion' of each other, and the interstitial and point-defect phenomena, involving as they do anion and cation excess, respectively, are to a large extent complementary.

There is one further important point of principle regarding the dioxides. Whilst their occurrence terminates in the First Long Period at Group VII (Mn), they continue, though unstably, to the platinum group metals in the

Second and Third Long Periods (cf. *Figure 8.2*), at the same time tending towards the structural deviations detailed in *Table 8.1.*

An insight into the much more universal formation of given oxide structures with varying *pressure* is exemplified by silica, SiO_2, which can adopt the *rutile* form at high pressures[15]. Effects like this are clearly of great consequence in practice, if the question is one of isomorphy and of creating intersolubilities amongst oxides which otherwise are incompatible. (For individual dioxides see below, Section 8.3.)

8.2.5. Progression to hematites, (M_2O_3)

The general decrease in refractoriness of oxides formed from Group IV towards higher Groups is evident from *Figure 8.2* and *Table 8.1*, as well as from vapour pressure, heat of formation and other thermodynamic data. Since the oxygen-*richest* oxides in each system progress from MO_2 to M_2O_5 to MO_3 for Groups IV–VI, the properties of these oxides are of special interest: effects such as the catastrophic oxidation rates of Mo and V at high temperatures are directly related to the low melting points and high vapour pressures of MoO_3 and V_2O_5. The ascendence of the M_2O_3 oxides of hematite-type structure must be considered in this light.

This structure occurs already in Ti_2O_3 (indeed earlier, in Sc_2O_3 of Group III) and becomes increasingly stable in V_2O_3 and Cr_2O_3, where it culminates in the provision of good refractory properties for this oxide; Cr_2O_3 is the most important in that (when in thin-film form*) it gives to chromium metal and stainless steel their characteristic oxidation resistance. The structure continues to Fe_2O_3 and, possibly, to an unstable Co_2O_3. No 'Ni_2O_3' exists; Mn_2O_3 is anomalous (see below). The structure is generally restricted to the First Long Period, although a Rh_2O_3 of hematite structure has been reported.

8.2.6. Spinels

The evolution of hematites, M_2O_3, is closely linked, for the later transition metals, with that of the spinels, of which Fe_3O_4 (= $Fe^{++}O{\cdot}Fe_2{}^{+++}O_3$) is probably the most prominent representative. Its cubic unit cell is shown in *Figure 8.5* in conjunction with an equivalent pseudo-unit cell of FeO. The two are closely related[16b], insofar as the spinel structure contains two kinds of iron atom: (*A*) those forming the gross FCC framework repeated in 'FeO'; (*B*) the remaining Fe ions, with oxygen in the cell interiors distributed as in *Figure 8.5*, which have altered their positions as from 'FeO'. All *A* sites and certain *B* sites are occupied by Fe^{++} ions, whilst all Fe^{+++} ions occupy *B* sites. This is reflected in their coordination with oxygen. It should be noted that the oxygen atoms in the spinel structure are again as close-packed as possible. (See also Section 8.4 and Bibliography.)

In a systematic survey, the spinels begin to appear at a (reputed) Cr_3O_4 which, however, is unconfirmed and would presuppose the transient existence of divalent Cr^{++} ions (the same doubt applies to a 'CrO'). However,

* The shielding by stable Cr_2O_3 and possibly by spinel in thin film form on Cr helps to render low-melting CrO_3 innocuous. In contrast, in molybdenum the MoO_3 does become effective, because an equivalent shielding 'Mo_2O_3' can no longer exist. (This, incidentally, has been demonstrated[16a] in an oxidation series on Cr–Mo alloys which were oxidation-resistant up to Mo contents at which the Cr_2O_3 film broke down—when catastrophic oxidation set in.)

Cr-bearing spinels do become strongly stabilized by Group VII and VIII ions (Fe^{++}, etc.) to give $FeO{\cdot}Cr_2O_3$ (chromite), $CoO{\cdot}Cr_2O_3$, etc. (as well as by other divalent partners, cf. $MgO{\cdot}Cr_2O_3$). Each of these, along with Cr_2O_3, contributes to protective films on stainless steels and heat-resisting alloys, whilst e.g. $MgO{\cdot}Cr_2O_3$ and solid solutions enter basic refractories and chrome ores effectively.

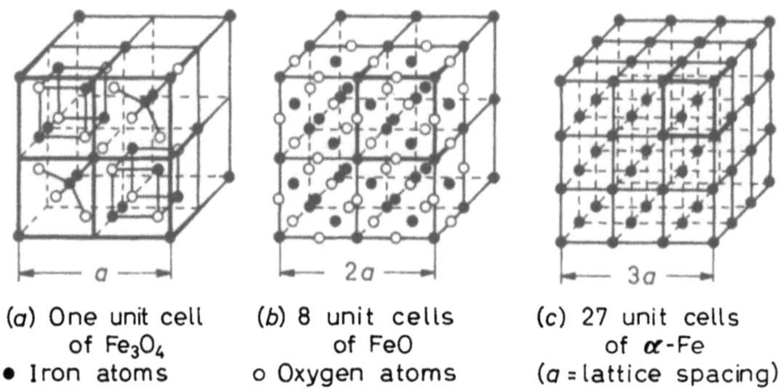

Figure 8·5. The Fe_3O_4, FeO *and* α-Fe *structures: comparable pseudo-unit cells* (After Goldschmidt[16b])

(Back layers not filled in, for clarity)

By courtesy of the Iron and Steel Institute, London

The knowledge of spinels, their structure and magnetic and electrical (semiconductor) properties, is extensive; these aspects will be referred to later. In a systematic context, the spinels are critically poised on the borderline between metallic and ionic bonding, and both bond types may be said to hybridize in them. The electron balance can vary with composition, degree of non-stoichiometry and, for a given composition, with temperature. In the extreme, spinels include the purely ceramic (non-metallic) types ($MgO{\cdot}Al_2O_3$).

Quite in general, spinels should be viewed in conjunction with the MO and M_2O_3 structures (cf. Sections 8.2.2 and 8.2.5). Three classes of oxide formation should here be differentiated amongst the single metals, namely of

(a) spinels (M_3O_4) plus MO plus M_2O_3, as for Fe

(b) MO only, as for Ni

(c) M_2O_3 only, as for Cr.

For cobalt, and for more than one metal present, this can evolve into a more complex intermediate set of phase assemblies and ionic conditions. The nature of spinels as basically ternary compounds should be re-emphasized, with the binary ones as singularities. A list of spinels is thus included under ternary spinels (*Table 8.6*, p. 397).

The liability to form lattice vacancies of metal and oxygen sites is pronounced in spinels. It obviously becomes important in diffusion processes, remembering that spinels are frequently integral in oxide scale assemblies. The occurrence of γ-Fe_2O_3 (and γ-Al_2O_3), still of spinel type, exemplifies how far the defect lattice can extend. γ-Fe_2O_3 forms solid solutions with

Fe_3O_4, with progressive ferrous ion loss, to become a polymorph to hematite. A similar effect can occur amongst Mn oxides ($Mn_3O_4 \rightarrow Mn_2O_3$).

Basically, the spinels ought to be classified as double oxides, even though in special cases they become single oxides, as in Fe_3O_4 which, however, is effectively still a 'double oxide' of ferric and ferrous ions.

In the Co–O system, the CoO/Co_3O_4 relationship becomes particularly closely integrated, in line with Co_3O_4 being the last stable spinel along the First Long Period, and already with a ready mutation to CoO. Thus Co_3O_4 spinel[16c] becomes unstable at high temperatures and transformed to the NaCl-type 'CoO' structure, with oxygen excess, i.e. metal vacancies as in 'FeO'. This can be reconverted sluggishly to spinel at lower temperatures, the composition, however, remaining unchanged during these lattice changes. This feature of sensitive thermal convertibility from the FCC to the spinel lattice, i.e. from $Co_{1-x}O$ to $Co_{0.75}O$, has been tentatively taken over in the Co–O phase diagram shown in *Figure 8.2*. The instability of 'Co_2O_3' (hematite structure) falls into the same picture, which amounts to the disappearance of Co^{+++} ions also promoting the spinel $\rightarrow$ 'CoO' conversion. In the neighbouring Ni–O system the position has clearly resolved itself, NiO remaining the only oxide.

In the γ-M_2O_3 oxides of spinel type, the formation of *stacking faults* has been observed e.g. in γ-Al_2O_3[17], as a first step to transforming into the hematite type. Quite generally, this effect is as important amongst transforming ionic oxides as amongst the more metallic interstitial compounds.

Much specialized investigation has been made of the magnetization of Fe_3O_4, from its role as the old 'loadstone' to recent geophysical work on its orientation relative to the earth's magnetic field, which helped to trace the historical changes in rock morphology, and its magnetic structure[18].

Cation deficiency in spinel structures has also been studied extensively after Verwey's[19] and Barth's[20] pioneering work. In the transition $Fe_3O_4 \rightarrow$ γ-Fe_2O_3 it has been doubted[21] whether γ-Fe_2O_3 really is of true spinel structure. A kind of superlattice is found in which 3 unit cells are stacked linearly ($c/a = 3$), with positions of the 8 vacancies ($=2\frac{2}{3}$ per unit cell) determined by a fourfold screw axis with translation $c/4$.

8.2.7. *The structural trinity of phases $MO/MO{\cdot}M_2O_3/M_2O_3$*

The structural trinity of these phases dominates, with varying emphasis, the whole series from Cr to Ni and the alloys based on these metals. It governs their oxidation processes, and for iron and steel the assembly $FeO/Fe_3O_4/Fe_2O_3$ is a familiar one in scales and corrosion products. The development of stainless steels and of heat-resisting alloys of Fe, Co or Ni base depends in the last resort on the correct structure type of the protective oxide film formed, and the correct adjustment of the compositional balance within this oxide phase aggregate, e.g. the Cr/Ni ratio in M_2O_3 or spinels. Corrosion resistance in relation to the substrate also depends on favourable fit, volume ratio and relative expansions. *Figure 8.6* summarizes the occurrence of MO and spinel-type phases and relates it to ionic radii. The diagram helps to show their compatibility for given compositions of the two oxide types, e.g. in relation to oxide layers and ceramic composites.

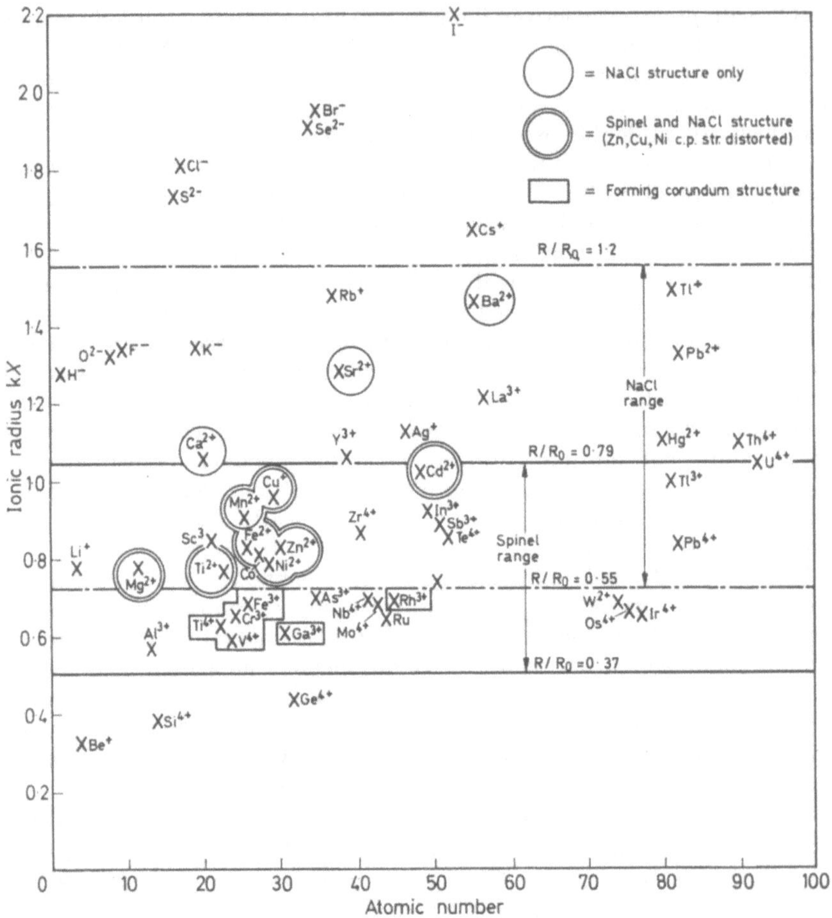

Figure 8.6. Occurrence of MO (NaCl *type*) *and* M_3O_4 (*spinel type*) *oxides as a function of ionic radii of the metals and of atomic number*

8.2.8. Highest oxides

It may be observed that the oxides of greatest oxygen concentration, in any one system, are generally liable to sensitive polymorphous changes. In themselves, they change systematically with Group and Period. Thus, for example, in TiO_2 we have the stable rutile with two subsidiary forms, anatase and brookite. In ZrO_2 the three—cubic, tetragonal and monoclinic—forms are subject to strong phase hysteresis on heating and cooling; this may also occur in HfO_2, but need not because HfO_2 already shows characteristics akin to the Rare Earth oxides. In the neighbouring Nb_2O_5, where the α and β forms were first thought also to be low- and high-temperature states, this is no longer correct: α is in fact metastable, β stable throughout, and the transformation is irreversible (*Figure 8.7*). For V_2O_5 and MoO_3, the tendency to weak bonding expresses itself in a low melting point, in other cases in metastability, easy decomposition and critical oxygen pressure of the higher oxides. In Co this is illustrated by the ambiguous identity of the highest oxide (Co, Co_3O_4 or Co_2O_3). (For rhenium and tungsten trioxides, ReO_3 and WO_3, and their relation to the perovskites and 'tetragonal bronzes', see Section 8.4.3.)

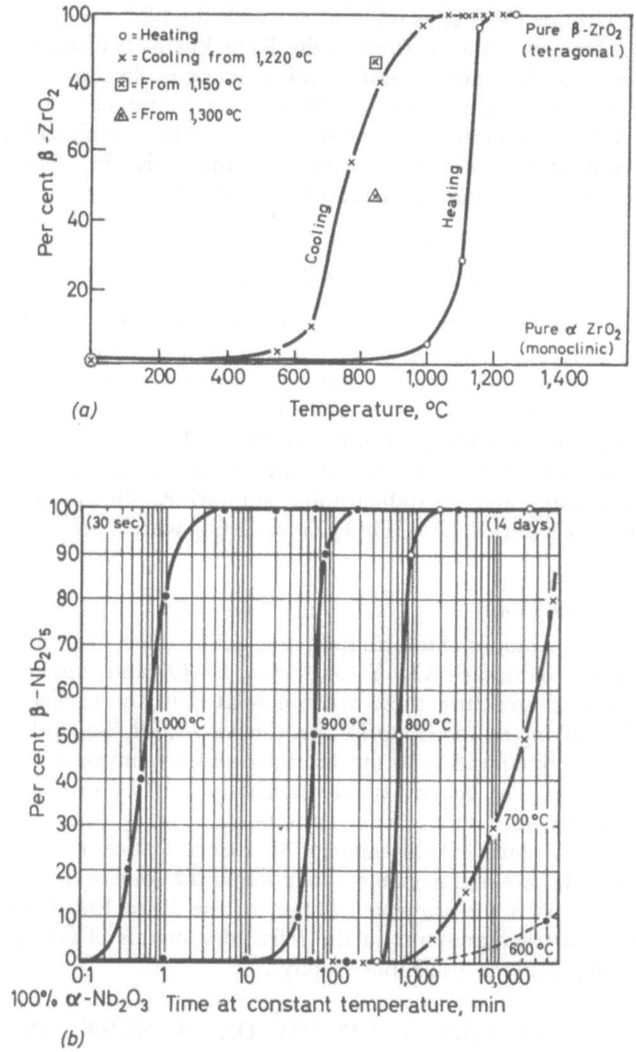

Figure 8.7. (a) Phase hysteresis in ZrO_2 (α to β form); *(b) 'non-phase hysteresis' and kinetics of permanent transformation in* Nb_2O_5 (α to β form) ***with time at constant temperature*** (After Goldschmidt[16d,g])
By courtesy of The Institute of Metals, London and Plenum Publishing Corporation, New York

8.2.9. Actinon oxides

Here the dioxides are again highly refractory. Starting with ThO_2 (m.p. 3,050°C—highest of any oxide), the melting point decreases with increasing atomic number. The series is dominated by the $MO \leftrightarrow MO_2$ structure relations described. It continues, from uranium upwards, to higher oxides, viz. M_3O_8 and MO_3, partly in the form of non-stoichiometric ranges, partly in that of the multiple homologous series.

The highest oxides, ThO_2, Pa_2O_5 and UO_3, formally follow the valency positions in Groups IV–VI, as is consistent with the filling of the incomplete 6*d* orbitals. However, the relation to the Rare Earths is already close, and the electronic configurations and binding energies become modified as compared with the lighter-element Periods: a filling of the 5*f* orbitals, leading to the conception of the actinides as '5*f*' or combined '5*f*–6*d*' elements. A multiplicity of valency conditions thus arises[22a]. The build-up of such alternative electron valencies occurs as follows:

Ac 1; Th 4; Pa 4, 5; U 6, 4, 3, 5;

Np 4, 3, 6, 5; Pu 4, 3, 6, 5; Am 3, 4, 6, 5

We may consider the actinides in the context of '6*d*' and '5*f*' elements, as of dual orbital filling (this duality[23] leads to a corresponding 'triality' in the Rare Earth oxides). The fugitive nature of valency states in the actinides is reflected directly in the oxides formed and in the ease of acquiring different structures and stoichiometries. On the other hand, the close relation of the actinide oxides to the other Group IV–VI oxides still persists, as reflected in the isomorphies and crystallographic similarities shown in *Table 8.1*, e.g. the fluorspar series $UO_2/ThO_2/ZrO_2/CeO_2$ which are largely intersoluble.

8.2.10. Rare Earth oxides

A considerable amount of information now exists concerning the Group III and Rare Earth oxides, Sc_2O_3, Y_2O_3, CeO_2, La_2O_3 and numerous others. In properties and structure these are still related to the earlier transition-metal oxides, linking on to Group IV, i.e. the Ti, Zr, Hf, Th oxides, but they also occupy a bridge position to the actinides and, to some extent, the Alkaline Earth oxides. The general stoichiometries are MO (NaCl type), M_2O_3 (cubic, hexagonal and monoclinic variants) and MO_2 (CaF_2 type), with deviations and a considerable capacity to defect lattice formation. The Rare Earth oxides are highly stable—so much so that one of their most significant applications is the scavenger action of the Rare Earths for oxygen, when present in trace amounts in the refractory metals of Groups IV–VI, e.g. V, Nb, W, etc. (see also Bibliography).

8.3 INDIVIDUAL BINARY OXIDE SYSTEMS

The above survey may be supplemented by data on individual systems.

Ti–O[24–31]—In the primary HCP α-Ti–O solid solution (our 'pseudo-oxide' above), in which interstitial oxygen expands the lattice with marked anisotropy (*a* and *c* axes, cf. *Table 8.1*), the radius[32a] of the O atom is about 0·66 Å, i.e. the same as that derived by Pauling for oxygen in the covalent phases. Nevertheless, the bonding is still metallic and remains so beyond compound TiO. On the Ti_2O phase, discussed already, and the low-temperature phase Ti_3O_2 little further data are available[24] except that the tetragonal Ti_3O_2 unit cell has been amended to a hexagonal one, related to α-Ti from which it evolves probably by a change in stacking sequence. The intermediate Ti_2O phase is interesting in this context, as it too is hexagonal (NiAs type), and the whole low-O range in Ti–O would thus amount to a series of modifications upon the hexagonal α-Ti lattice, first via a superlattice and then by simple hexagonal ordering. After Ti_3O_2, this switches to the cubic symmetry of TiO, consistent with the tendency to regain close packing. In α(Ti → O) solutions and in TiO, low-temperature ordering has been confirmed[33–4], both of vacancies and of O atoms.

In 'TiO', the wide homogenity range shown[35] corresponds to additional Ti and O vacancies on each side of the strict $TiO_{1.0}$; this is really a 'ternary compound' (cf. Chapter 2) of Ti + O + vacancies and, in lattice terms, $Ti_{0.85}O_{0.85}$ is probably a more appropriate expression of the composition. There is a decrease of spacing with increasing oxygen resulting. Metallic bonding still applies in TiO, as has been shown by resistivity measurements[32a] (≈300 μΩ cm) and by structural arguments. Oxygen is in this case still a donor, not an absorber of electrons, as it will be at higher concentrations. A phase transformation occurs[28] in TiO between 900° and 1,000°C, with some likelihood of a tetragonal distortion at lower temperatures.

Also, according to Magneli[7], the non-stoichiometric range becomes drastically restricted, and an invariant form of precise composition $TiO_{1.00}$ *does* exist at lower temperatures, rather as indicated in the 'not-impossible' Ti–O diagram (*Figure 8.3*). This has a deformed NaCl-type structure, with vacancy ordering in both the anion and cation sub-lattices. A somewhat divergent result is that of Bright[31] who, in considerable work on the TiO range and its decomposition, found it actually impossible to prepare pure stoichiometric TiO, obtaining a duplex structure of two defect isomorphs with opposite vacancies.

These phase changes in TiO are still quite problematic and demand further study. A basic difficulty here is that of maintenance and identification of composition (Ti/O ratio), and of isolation in view of the multiple further phases liable to intervene.

The ionization of oxygen towards O^- and higher electronegativity does not really gain hold until the multiple series $Ti_2O_3 \ldots TiO_2$ commences (and indeed, we may say reversely, this is caused by it, as discussed). Still, even in TiO_2 full ionization is not quite accomplished yet. The anatase (tetragonal) and brookite (orthorhombic) forms of TiO_2, structurally different[36]—as against rutile—should probably not be considered as stable modifications of the pure oxide, although all three are found in nature.

The dislocations and stacking-fault formation in rutile formed upon oxidation from TiC, have been studied by the electron microscope work of Smallman's school[12d], also in relation to non-stoichiometry (TiO_{2-x}). Oxygen-denuded TiO_2 has increased strength, and pronounced clustering of vacancies can occur in the rutile structure. This work was extended[12a–c]), showing a definite plastic deformation in rutile single crystals, related to the crystallographic slip system, upon compression at different temperatures—also as a function of non-stoichiometry.

Other studies of oxides, exemplified on rutile but of general importance regarding their mechanical properties[37], showed that externally applied strains can both change the crystal symmetry and produce internal friction peaks, with a remarkable similarity to those in primary interstitial solid solutions. The peaks are in this case manifested by the movements of Ti ions from one type of site to another in the rutile structure. These are examples from an ever-growing literature in the rutile field, directed to semi-conductor or ceramics applications (cf. Bibliography).

Zr–O[38–40], Hf–O—The Zr–O diagram (*Figure 8.1*) and data (*Table 8.1*) summarize the position. After the extensive HCP α-Zr solution and its termination in what effectively amounts to a congruently melting compound ($\approx Zr_3O$), the only stable oxide is ZrO_2 with its polymorphs. There was some early controversy regarding the existence of suboxides, and a metastable 'ZrO' of NaCl-type structure has been claimed[41a]; this could, however, have been stabilized by hydrogen.

Recently the Zr–O system has been re-investigated[42] and reaffirmed, but with some quantitative amendments in solubilities; also a non-quenchable high-temperature form of ZrO_2, stable above 1,600°C, is suggested. As only metallographic and resistivity measurements were carried out, incidentally indicating that even here a small degree of semblance to metallic bonding is still retained, this high-temperature structure could not be identified; it may well be asked whether this is not the truly cubic (CaF_2-type) form of ZrO_2, otherwise stabilized by added oxides (MgO, CaO, etc.). This has been doubted as a low-temperature form[43] but indeed confirmed[44] above 2,200°C.

The lower oxides, 'Zr_2O_3' and 'ZrO', have been observed as transitional stages when zirconium carbide was prepared by carburization of ZrO_2, i.e. it needs the reducing action of *carbon* with its high terminal affinity for Zr to produce such lower oxides. The same occurs on TiC, VC, NbC and TaC formation, where the lower oxides can form transiently, but this does not of course mean that these are equilibrium phases.

For *hafnium*, HfO_2 in its three forms, analogous to ZrO_2, appears to be the only oxide existing. The high solubility and stabilization of α-Hf by O (as for Zr) may be seen in *Figure 8.2**. The phase hysteresis (monoclinic ⟷ tetragonal) in ZrO_2 and HfO_2 has been reinvestigated[46] and shown to occur in the 1,000–1,200°C range for ZrO_2, the 1,500–1,600°C range for HfO_2, with cooling reversals in the 970°–750°C and 1,550°–1,450°C ranges, respectively. The ZrO_2 and HfO_2 hysteresis effects and the diffusionless transformations implied have been aptly compared[44] with martensitic effects.

V–O[47–52]—The highly complex V–O system (*Figure 8.1*, *Table 8.1*) is still far from established, especially at high temperatures, despite the numerous and often conflicting data known.

The tetragonal low-oxide V_4O, which follows the BCC solid solution and can extend to much higher V contents, possibly to $V_{10}O$, has its isomorphs in the Nb and Ta oxides described below. It is a direct tetragonal distortion of the BCC lattice of V. Neutron diffraction[51] has shown that the O atoms enter the octahedral sites, initially randomly but with increasing degree of order (verified lower down Group V, for Nb). *Figure 8.8* indicates this

* Rudy and Stecher[45] established the β-Hf → O solubility as 0·90 per cent O (at 1,350°C but almost independent of temperature), also showing that the non-stoichiometric range HfO_{2-x} extended to $x \approx 0{\cdot}25$.

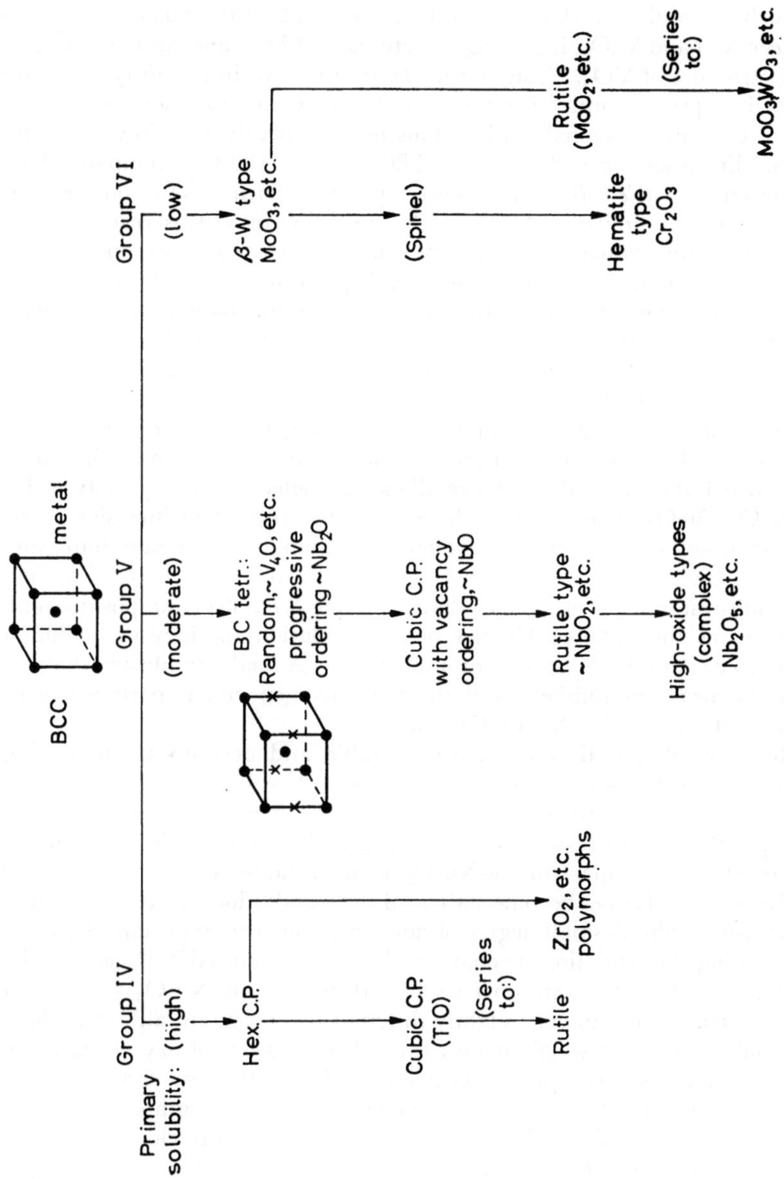

Figure 8.8. Evolution of oxide structures in Group IV, V and VI with increasing oxygen content (*schematic*)

sequence schematically, with extensions, to compare with Group IV and VI. A low-temperature phase, $V_{13}O_7$, has been claimed[32b] to form by peritectoidal reaction below 1,140°C, but this may need confirmation. Regarding the cubic 'VO' phase, with its wide homogeneity range, *Table 8.1* should be self-explanatory, along with our remarks on TiO and NbO below.

Much interesting work has been done on details of the homologous series between VO and V_2O_5, including structures[41b,49,53–5], and on the deformed rutile structure of VO_2. A significant feature is that, in the VO_2 phase, the monoclinic pseudo-rutile structure transforms at elevated temperatures to that of the *true* tetragonal rutile. This occurs already[56] at 70°C, with the lattice dimensions $a = 4{\cdot}530$, $c = 2{\cdot}869$ Å (at 180°C; compare TiO_2 dimensions in *Table 8.1*). The room-temperature form of VO_2, on the other hand, is already similar to MoO_2. The transition in VO_2 is reversible and associated with changes in resistivity, magnetic susceptibility and thermal expansion. Again this fits into the overall pattern of rutile transitions and represents one stage in progress from metallic to ionic bonding, which reaches a high degree of sensitivity in the rutiles of Groups V and VI. VO_2 lies critically on the borderline between the normal rutile of TiO_2 and the distorted group of MoO_2 type.

The structure of V_2O_5, in which the homologous series terminates, and the manner of its bonding are now fairly well established[57], and as in Nb_2O_5 (from which it differs), there is a small oxygen-defective homogeneity width.

Nb–O, Ta–O—These systems have been the subject of intensive recent investigations, in view of the attraction of niobium- and tantalum-based alloys as high-temperature materials, with the objective to render them oxidation-resistant. The factual data in *Table 8.1* reflect the position only partially, as the detailed kinetics of the oxidation mechanism, including metastable states, is of paramount importance. A final equilibrium diagram is by no means established, and the lower-oxygen end is particularly ill-defined (cf. *Figures 8.1*, Nb–O diagram).

The suboxides, if they exist, are unstable and decompose on heating, except at low pressures, into $Nb + Nb_2O_5$. The latter is the dominant, and macroscopically usually the only, oxide upon metal oxidation. Of its two (possibly three) forms, only β-Nb_2O_5 is the truly stable one, and the presumed 'low'-temperature α-Nb_2O_5 is metastable, though persistent[16d] (cf. *Figure 8.7*). It does become stabilized by Ta addition, to give its eventual Ta_2O_5 isomorph. A small degree of non-stoichiometry, important though in determining the oxidation mechanism, has been reported[58a,59], as reaching down to $Nb_2O_{4{\cdot}8}$; it is even conceivable that the various 'Nb_2O_5' polymorphs reported correspond to very slightly different compositions, although chemical analysis seems to weigh against this. The presence of oxygen vacancies in Nb_2O_5 has also been proved gravimetrically[11b] (for semi-conductor, etc. effects associated with this and related defects, see Bibliography).

The next lower oxide, NbO_2, has a tetragonal structure closely related to rutile[60]. Upon further reduction, this passes to the cubic (NaCl-type) of NbO which has a considerable capacity for defect lattice formation. An important feature here is that of vacancy ordering, the degree of which can be varied and controlled by added atoms. A number of unstable suboxides, NbO_x ($x = 0{\cdot}5, 0{\cdot}3, 0{\cdot}2$?) has been variously reported on approaching the primary Nb $\rightarrow$ O series of solid solutions. It is indeed debatable whether they can be sharply distinguished from this solution as new phases or whether they evolve from it by a continuous ordering process and simultaneous lattice distortion. Thus two tetragonal suboxides, 'NbO_x' and 'NbO_z', have been reported[61], where $x \approx 0{\cdot}5$ and z is an uncertain, probably variable, fraction. A TaO_z analogue to NbO_z also exists, whilst for tantalum a further suboxide TaO_y ($y = 0{\cdot}25$) has been found[61]; again, this may have an NbO_y analogue. All these oxides have metallic character and are in this sense truly interstitial compounds, as against a more ceramic Nb_2O_5. These phases are highly pressure- and temperature-dependent, as is their formation from the metal. Line broadening effects and spacing changes occur in BCC-parent Nb owing to O solution and partial precipitation in the 200°–400°C range, to a degree indicating considerable supersaturation, and culminating in a superlattice phase. An 'Nb_2O' phase, formed during oxidation of Nb powder[58b], was also shown to amount to a BC tetragonal distortion (spacings $a = 3{\cdot}36$, $c = 3{\cdot}25$ Å, against $a = 3{\cdot}30$ Å for Nb). Although experimental evidence is lacking, this could be structurally a direct analogue to V_4O, except for occupation of further interstitial sites and ordering.

This work on the low (metallic) oxides of Nb and Ta, their structures, morphology and transformations, has been considerably elaborated[61–2].

The various nominal formulae quoted (Nb:O = 4:1, 2:1, etc.) may not really be of much significance; more probably, they correspond to a gradual process of occupation in which vacancies can be filled or remain free. This can occur in an ordered or statistical manner, as Norman[61] has shown, with consequent varying degrees of tetragonal distortion. Furthermore, this process may take place homogeneously or heterogeneously over small lattice regions (manifested by x-ray line broadening, sharpness or line split), and during oxidation or reduction all are liable to occur at various stages of advance.

A decomposition $5Nb_2O \rightarrow 8Nb + Nb_2O_5$ occurs[58b] at about 350°C, for a fine oxidizing powder; however, this should probably not be regarded as an equilibrium (peritectoid-type) decomposition but rather as attached to particular experimental conditions. The resulting 'Nb_2O_5' is initially in an amorphous form. It is not at this stage clear how NbO and NbO_2 enter into the kinetic sequence; it seems at least likely that the transformation BCC $\rightarrow$ FCC (metal lattice) would occur with excess oxygen, as implied in Nb_2O and NbO.

Structurally there is a remarkable resemblance to the austenite $\leftrightarrow$ martensite $\leftrightarrow$ ferrite transition in Fe–C alloys, with oxygen taking the place of carbon atoms, concomitant with the larger-size metal atom. The $\gamma \leftrightarrow \alpha$-Fe-type transformation of course no longer occurs in Nb (and any Group V or VI metals) but, in line with the theme presented throughout this book, has been transposed to higher interstitial contents. The view of the Nb_2O- (or V_4O- or Ta_4O-)type stage as martensite-related could well prove fruitful to pursue, both from the property and metallographic aspects. In particular, the progressive *reduction* of niobium oxides, from the higher oxides (such as under the influence of sodium or other active reducing media) to the finely dispersed metal, could be considered as a near analogue to the $\gamma \rightarrow$ martensite $\rightarrow \alpha$ transition in Fe–C alloys.

Whilst reduction in hydrogen of Nb_2O_5 to the metal is impossible, it has been shown in early work[63] that it does occur in the presence of *nickel* (in form of powdered Ni carbonyl), to a degree where it is possible to produce metallic Ni–Nb alloys; this actually permitted a study of the Ni–Nb phase diagram using reduction-produced alloys. In this context the point of principle arises that, by altering the electronic balance through the introduction of a Group VIII metal atom*, Nb_2O_5 (and, certainly, V_2O_5, Ta_2O_5, etc.) can be rendered reducible by hydrogen where otherwise it would not be. There is evidence that sodium is more powerful and can achieve reduction to the metal, but perhaps Ni, etc. would accelerate this.

The NbO_2 structure is a rutile of multiple unit cell[7b] (cf. *Table 8.1*) and the phase is indeed stabilized by Ti addition. Single-crystal data have been able to define the structure more closely[60]. The reactions and metastable states observed during the kinetic processes of oxidation of niobium and its alloys clearly are intricately linked with the oxide structures found—and this, indeed, makes the determination of a true Nb–O or Ta–O equilibrium diagram so difficult (cf. Bibliography).

As regards Ta–O, all phases in *Table 8.1*, i.e. $Ta_4O - Ta_2O_5$, are now moderately well established, but again many kinetically doubtful points remain, as for Nb–O. There are indications[58b] of a BC tetragonal Ta_2O analogue to Nb_2O, whilst the BC orthorhombic 'Ta_4O' evolves from this on higher oxygen addition, thus suggesting a Ta_nO with $n < 2$. Nevertheless, the orthorhombic TaO_y and a tetragonal TaO_z (with $y = 0{\cdot}25$ and z uncertain) were lately reaffirmed[61]. The position has often been confused by the presence also of nitrogen, and a previously reported 'hexagonal Ta_2O' really corresponded[64] to an *oxynitride* of composition $Ta_2(O,N)$.

In much recent work on the oxidation kinetics of tantalum, both by rate curves and constitutionally, it was possible to confirm, after the initial BCC $Ta \rightarrow O$ solution with expansion, the two suboxides[61] 'TaO_y' and 'TaO_z' formed at lower and higher temperatures.

There is one important distinction between the Ta–O and Nb–O systems[65]. Whilst, upon mixing Nb and Nb_2O_5, the suboxides can readily be formed at $\approx$ 1,000°C *in vacuo*, Ta and Ta_2O_5 do not so react, and any suboxides are much more difficult to prepare synthetically. This, along with the very easy formation of lower oxides in vanadium in Group V, falls quite into line with the parallel case down Group IV, i.e. the multiple lower oxides for Ti and their absence in Zr and Hf.

In amplification of the niobium oxide structures—and this may typify other refractory metal–oxygen systems—it may be noted that, metastably, numerous forms can evolve, e.g. in thin films which are greatly dependent on environment. Thus there exist a tetragonal 'NbO_x' and a monoclinic Nb_2O_3, apart from the (Nb,O) solution, NbO, NbO_2 and Nb_2O_5. The important observations on sub-oxides formed during progressive oxidation (mainly of the Oslo School of Kofstad and associates) give a more kinetic

* There seems little doubt that Co, Fe or Pt-type metals could act similarly.

approach to this question, not necessarily an equilibrium one. This includes extensive work on the oxidation mechanism at varying temperatures and pressures. The indication is that oxidation of Ta and Nb involves (*a*) initial chemisorption of oxygen at the surface, followed by (*b*) diffusion and solid solution in the BCC matrix; then ordering of dissolved oxygen to give the Ta_4O, 'TaO_z' or 'NbO_x' platelets before the higher oxide nucleates.

Defect lattices and vacancies in Nb_2O_5 and Ta_2O_5 are often decisive in governing oxidation rates. For instance, Ta_2O_5 can change basically in electronic character with oxygen pressure: from a *p*-type semiconductor (with O vacancies) at 1 atm to *n* type (with O interstitials) at low pressures, the transition $p \leftrightarrow n$ type being a function of temperature. An even greater influence on oxidation rate lies of course in the polymorphic changes referred to. The improvements or otherwise through alloying elements added to Nb- or Ta-base alloys, e.g. W or Ti, consist essentially in their modifying action on both the phase structures and the vacancy/interstitial character of the phases concerned.

Cr–O—The direct equilibrium Cr–Cr_2O_3 is generally considered to be the true one. However, the metallic Cr_3O phase of β-W-type structure has been produced and analysed[41], although probably metastable. Similarly, a phase 'CrO' was claimed[66] but is unconfirmed. The question as to the existence of a pure Cr_3O_4 spinel (without FeO, etc.) is an open one, but it has been strongly supported as a high-temperature phase[67], distorting from cubic to tetragonal on quenching. The existence of both CrO_2 and Cr_3O_4 show this system as the point of confluence of the rutile and spinel areas. (For details of the Cr_3O_4, Cr_2O_3, CrO_2 and CrO_3 phases, see *Table 8.1.*)

Mo–O, W–O—Most of the knowledge of the crystal chemistry of the molybdenum and tungsten oxides we owe to the Uppsala school. The metallic Mo_3O phase could be prepared by reaction between Mo + MoO_2. W_3O, earlier considered as a polymorph of tungsten ('β-W'), was proved to be the oxide[68] and its structure analysed.

The whole series Cr_3O–Mo_3O–W_3O is distinguished by partial ionization already of the oxygen atoms, revealed by anomalous interatomic distances. However, essentially metallic bonding holds. Furthermore, the M_3O sequence (A–15 structure) is subject to an order/disorder process, Cr_3O showing superlattice, W_3O random distribution, with Mo_3O occupying an intermediate position. The evolution here from the above lower Group V oxides (Nb_2O, Ta_4O, etc.) should be noted, the W_3O-type oxides differing structurally much more from the BCC parent metals.

It has been held that structurally[41c] this Group VI series forms intermediate stages between the BCC lattice of (Cr, Mo, W) and the FCC. From a systematic viewpoint, this is the more interesting facet of the work as it foreshadows trends which are increasingly developing in Groups VII and VIII. Also, the M_3O oxides may perhaps be considered as the nearest approach to the cubic close-packed structures which in Group VI fail to materialize but do so on its left (TiO, VO, etc.) and right (MnO, FeO, etc.).

Upon oxygen addition, the monoclinically distorted rutiles of MoO_2 and WO_2 form the next stage, followed by the highly complex homologous series of structures described earlier. Kihlborg[69] has studied these equilibria in some more detail, and *Figure 8.9* shows his temperature/composition diagram[69b] for MoO_x phases up to $x = 3$. In this connection the *mixed* phases $(Mo,W)_{10}O_{29}$ and $(Mo,W)_{11}O_{32}$ are relevant, slightly deviating in structure from the single-metal Mo and W series[7c] and adding further to the multiplicity of structure variants and stoichiometries. The magnetic susceptibility of tungsten oxides has been determined[70] in the series WO_x, with x decreasing from 3 to 2. Viewing this series 'from the far end', i.e. as one of oxygen-deficient structures derived from the high oxide, electron readjustment in the (already ionic) lattice of WO_3 entails, upon oxygen

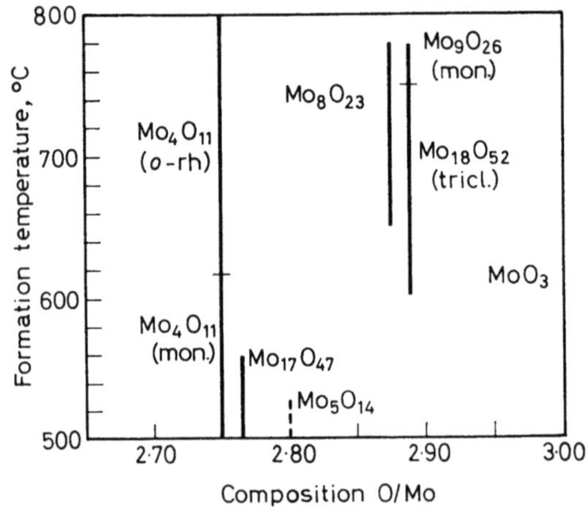

Figure 8.9. Homologous series of Mo *oxides: formation temperature against composition* (From Kihlborg[69b])

By courtesy of the Editor, *Arkiv för kemi, mineralogi och geologi*

abstraction, a successive distortion in structure (manifested in this multiple series). If we still had metallic bonding, the same series would no doubt have taken the form of a normal vacancy solid-solution range instead of the step-wise series, and it is especially interesting as an illustration of the development associated with that of ionic bonding.

It could be shown[7d] that WO_3, MoO_3 and the homologous series M_nO_{3n-1} are already structurally closely related to the much simpler ReO_3 lattice (cf. *Figure 8.18*, p. 414), with regular slight displacements of atoms. Further modulations take the form of the perovskite or tungsten bronze type series (see Section 8.4). These entirely non-metallic MO_3 oxides transcend our terms of reference (but see Bibliography).

U–O—The relationships between the cubic UO_2 and UO (there is no lower oxide) have been described already. The non-stoichiometry can extend up to $UO_{2\cdot35}$, but the limit is temperature-dependent: it has been reported as $UO_{2\cdot10}$ from about 600°C, as $UO_{2\cdot20}$ from about 1,050°C upwards. By neutron diffraction, Willis and associates[14] were able actually to locate the oxygen atoms in $UO_{2\cdot13}$ single crystals. Three interstitial oxygen positions have to be distinguished besides the normal ones: (*a*) as in the fluorspar structure of UO_2 (approx. $\frac{1}{4}\,\frac{1}{4}\,\frac{1}{4}$ if U at 000, but slightly displaced), additional sites; (*b*) at 0·5, 0·4, 0·4 (along cube edge) which displace the nearest oxygen atoms at (*a*) slightly from their sites to new positions (*c*). The largest holes, $\frac{1}{2}\,\frac{1}{2}\,\frac{1}{2}$ in the UO_2 structure, it may be noted, are not occupied; Willis *et al.* explain this as being due to the need for maintenance of the electronic balance between the two nearest U^{4+}, as in UO_2, by conversion to U^{5+}, the polarizing influence of the latter preventing occupation of the $\frac{1}{2}\,\frac{1}{2}\,\frac{1}{2}$ holes. Only a small percentage of sites (*b*) and (*c*) are actually filled, more with rising temperature, but all normal sites (*a*) are so occupied. The essential new point of principle is that we have not a simple metal substraction lattice, as in FeO, but one where an 'overload' of interstitial oxygen is accommodated by slight displacements of the original oxygen sites*.

This process is clearly preparatory to the actual structure distortion to tetragonal[71–2] which increases with O content in $UO_{2\cdot20-2\cdot30}$. This is then followed by the complex homologous series to UO_3 referred to. Electrical resistivity measurements[73] of non-stoichiometric UO_2 have shown a temperature dependence for UO_{2+x} between 500° and 450°C, consistent with the ionic nature of non-stoichiometric UO_2. *Figure 8.10* shows such electrical conductivity curves[74–5]; it also contains a more detailed delineation of the UO_2 + oxygen region of the phase diagram[75].

Towards the metal side, a *sub*-stoichiometry of UO_2 has been shown to exist[76] at high temperatures (above 1,800° C, produced *in vacuo*); on cooling, metallic uranium then precipitates. This metallic precipitate has been identified[77] as the β-form of U, i.e. the σ structure. Such degradation of anion-deficient UO_2 into the σ phase seems of particular interest, but is more likely to depend on aging temperature. A point of principle here is the potential precipitation-hardenability of an oxide by the metal nucleating from it and its variability if it can do so in form of its different allotropes.

Much special work on the UO_2 fine structure and its non-stoichiometry has been done in connection with atomic reactor researches, e.g. dislocation studies by electron microscopy and other techniques, combined with investigation of properties such as strength, creep behaviour, etc.[78]. Temperature dependence is clearly here structurally critical.

Mn–O—Manganese oxides occupy a key position in our pattern, as in so many other respects for Mn. Structurally they are slightly anomalous. The dioxide, MnO_2, is retained, as the last member of the rutile series, already with a tendency to polymorphs of lower symmetry (in orthorhombic 'ramsdellite'). On the other hand, the (MnO–Mn_3O_4–Mn_2O_3) series foreshadows the iron oxides, but still with a small structural difference (see *Table 8.1*). MnO, however, is already isomorphous to FeO. Again, these oxides are notable in their semiconductor aspects.

It is noteworthy that MnO, as CoO, NiO and most other cubic monoxides, is stable down to room temperature and below, whereas FeO forms its eutectoid at 570°C. We may see here clearly an effect similar to the $\gamma \rightarrow \alpha$ change in pure iron (viewing the *M* lattice in *M*O as the descendant of the FCC of austenite) and, when recalling that the corresponding metal-binary

* This neutron diffraction work by Willis, which deals with both UO_2 and ThO_2, represents, apart from the specific information provided, a good demonstration of the power of neutron diffraction to locate interstitial positions and movements.

systems, Fe–Mn and Fe–Ni, both show strong austenite stabilization to low temperatures, parallel to oxygen stabilizing MnO and NiO*.

Finally, the $Mn–O_2$ system is the first along the First Long Period with no oxides lower than *M*O. This again coincides with the commencement in Mn of the FC cubic phase in the pure metals, evolving to γ-Mn → γ-Fe → Co → Ni. It implies that, in the presence of oxygen, any intermediate structural stages are now 'unnecessary', because the FCC metal atom lattice in *M*O can be directly taken over from the pure metal. Where the latter was BCC (as in Ti, V and Cr), such intermediate oxide stages *were* necessary and may be regarded as preparatory to the FCC configuration.

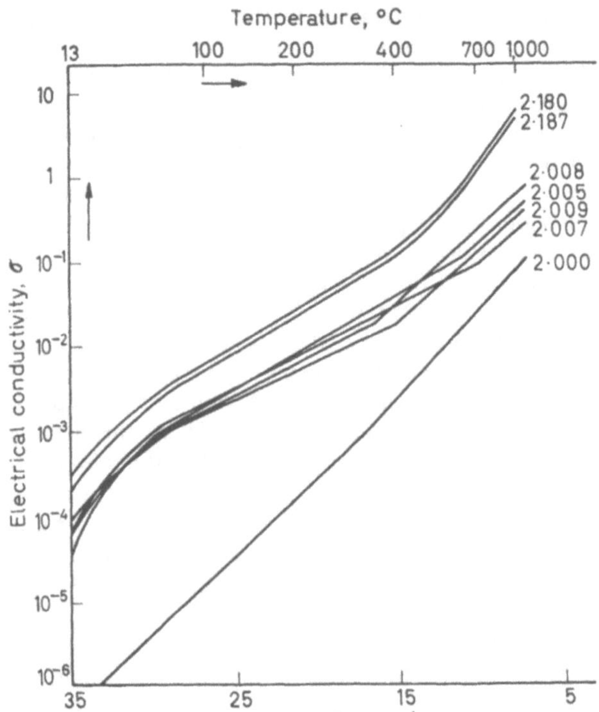

(a)

Figure 8.10. (a) Electrical resistivity as function of temperature, for UO_2 *of varying non-stoichiometry, from* UO_2 *to* ≈ $UO_{2\cdot2}$; *(b) The* 'UO_2+' *phase diagram* (From Brabers[74] and Belle and Austern[75])

By courtesy of United Nations Office at Geneva and Messrs. Wiley, New York

The termination of oxides lower than cubic *M*O from Group VII upwards is not only structurally natural but also bound up with the transition from metallic to ionic bonding, characteristic of the oxides. It contrasts with the carbides and nitrides, where Fe_3C, Fe_4N, etc. freely occur, with metallic bonding (but are also structurally related to the FCC, as are the monoxides).

* The eutectoid decomposition of γ-(Fe,C) in the iron–carbon system at 710°C may also be recalled, which in FeO would correspond to the 570°C eutectoid.

The cases of the lower borides and silicides show similarities but are complicated by the establishment of non-metal/non-metal bonds and by size-factor considerations (cf. Bibliography).

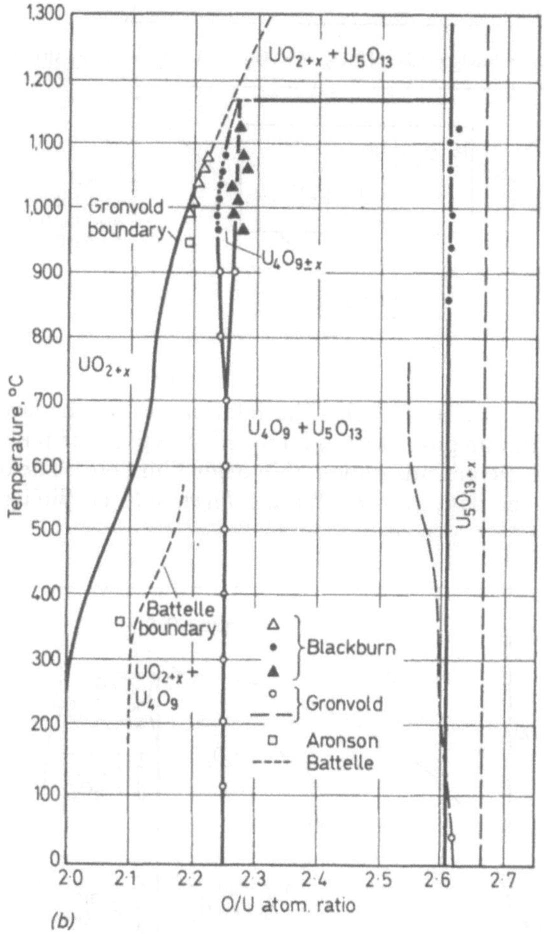

Figure 8.10b

Fe–O, Co–O, Ni–O—The literature on these oxide systems, and that on oxidation, scale structures, etc. is of course voluminous. In addition to the points made and the data in *Table 8.1* and *Figure 8.2*, see the Bibliography (with no claim to completeness).

Re–O—The pseudo-rutile structure of ReO_2 adopts a further (orthorhombic) modification[7e]. Oxygen excess transforms this into orthorhombic Re_2O_7, then into ReO_3 (cf. *Figure 8.18*, p. 414). The latter may be regarded as a simple structural parent to many other oxide forms, e.g. the tungsten oxides, the 'tetragonal bronzes' and perovskites (Section 8.4). Pure metallic rhenium oxides do not appear to be known (see also Bibliography).

Table 8.4

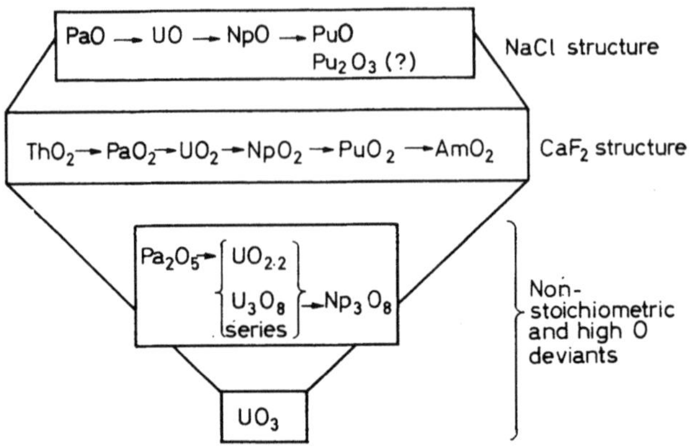

[Ru, Rh, Pd; Os, Ir, Pt]–O—The information on the largely unstable oxides of the platinum group metals is limited, except in the physico-chemical field of catalysts, oxygen adsorption, thin oxide films, etc. The oxides known are covered by the data in *Table 8.1* and *Figure 8.2* (cf. Bibliography).

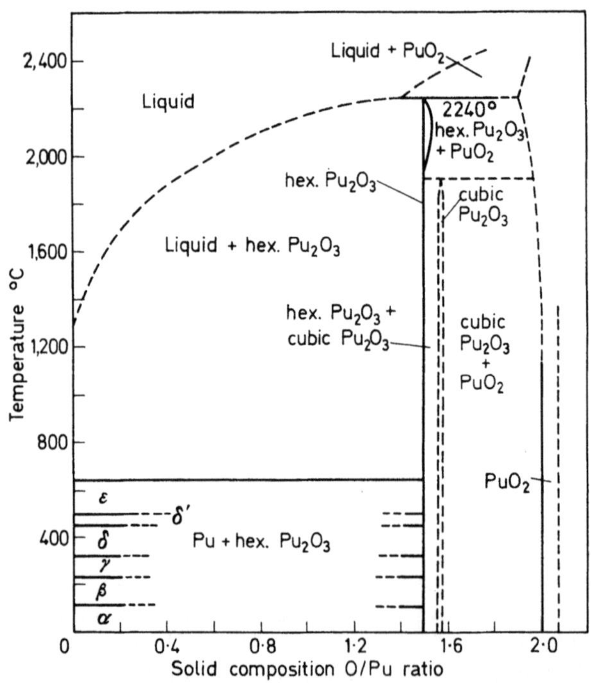

Figure 8.11. The Pu–O *phase diagram* (From Holley *et al.*[74b])
By courtesy of United Nations Office at Geneva

Actinon–O systems other than U–O—For thorium, ThO_2 is strictly the only oxide, and the fluorspar structure persists along the Period past UO_2 to the transuranic oxides. These have been largely investigated by Zachariasen[22] who gives the steadily decreasing lattice dimensions

$$\begin{array}{cccccc} ThO_2 \rightarrow & PaO_2 \rightarrow & UO_2 \rightarrow & NpO_2 \rightarrow & PuO_2 \rightarrow & AmO_2 \\ 5{\cdot}597 & 5{\cdot}507 & 5{\cdot}468 & 5{\cdot}436 & 5{\cdot}397 & 5{\cdot}388\ \text{Å} \end{array}$$

in line with the '5*f*' contraction, i.e. successive filling of 5*f* orbitals, by electrons, as the 4*f* orbitals are filled in the Rare Earth series. The lower and higher oxides (cf. Bibliography) discussed can be systematized as shown in *Table 8.4*.

Figure 8.11 shows the Pu–O diagram, supplementing that in *Figure 8.2*.

8.4 TWO (OR MULTIPLE) METAL OXIDES

8.4.1. Introduction

Upon combination of two or more metal atom species with oxygen, a great variety of new structure types can arise, either not known amongst the single metal oxides or showing major distortions from them. This is partly caused by the new factor of liability to ionic bonding which now becomes important, in contrast to the carbides and nitrides. The crystal structures and properties thus become sensitive to the valency and electronic relationship between the two metal partners and these and oxygen. In other words, the proportion of the participating atoms becoming cations and anions can critically vary in these systems with oxygen content, temperature and pressure. Relative atomic size, although obviously still paramount, thus becomes a more ill-defined concept, diffusing to that of ionic size (and often with variable valencies at that), depending on the system.

The single-metal oxides described above can form extensive mutual series of ternary solid solutions, if isomorphous and if size and valency factors are favourable. However, such solution series can be perturbed and broken into close homologous series of related structures. They are also subject to variable lattice defects, to superlattice and directional (anisotropy) effects which, in themselves, are temperature-, pressure- and concentration-sensitive. On the other hand, the flexibilities in substitution and in non-stoichiometry can be extensive, either without structure change or with one of a finer second-order. This fact is widely utilized in 'designing' new synthetic oxides with special properties, e.g. magnetic or electrical ceramics. In nature, mineralogy provides abundant examples of these isomorphous or near-isomorphous families, generally excluding lower oxides with metallic bonding which are 'man-made'.

In *Table 8.5* we have attempted to collate the information available on ternary metal oxide systems, M_1–M_2–O, composing the transition-metal block from Group IV to VIII in systematic order. There are still many gaps in this knowledge, and one purpose of this book is indeed to point these out and to suggest fruitful fields of research. This survey, being limited to the transition-metal frame, represents only a partial picture of the scope of the two-metal oxide systems concerned. It is characteristic of these oxides

Table 8.5. Two-metal oxide systems (inter-transition metal only)

Abbreviations:
PB = Pseudobinary system
CPD = Ternary compound
CSS = Complete series of solid solutions
PSS = Partial series of solid solutions
⊖ = Details not known; investigation required

System	*Principal compounds or solid solutions; special notes*
Ti–Zr–O	TiO_2–ZrO_2 pseudo-binary; *PSS* rutile (Ti → Zr)O_2 max. 20 per cent (Zr → Ti)O_2 (tetr.) max. 38 per cent *CPD* $TiZrO_4$; forms[79] *eutectic* with TiO_2 at 1,760°, *peritectic* with ZrO_2 at 1,820°C [structure of $TiZrO_4$ orthorhombic[80]; two polymorphous forms] Overall ternary not known ⊖; however, α-(Ti → O) and α(Zr → O) primary solutions no doubt continuous; then *PSS* of Zr in the lower Ti oxides, probably equilibrating ZrO_2 soln. and $TiZrO_2$ ⊖ Solubility of Zr in TiO probably extensive but must break down at some value ⊖
Ti–Hf–O	not known but presumed similar in many respects to Ti–Zr–O ⊖
Ti–Th–O	not known ⊖
Ti–V–O	intersolubilities of isomorphs, e.g. TiO–VO, Ti_2O_3–V_2O_3, etc. very likely but not studied ⊖, except *CSS* TiO_2–VO_2 (rutile series), but including[69] *CPD* $TiVO_4$; tetragonal, $a = 4{\cdot}58$, $c = 2{\cdot}95$ Å; $M = 1$
Ti–Nb–O	(1) *PB* TiO_2–Nb_2O_5; *PSS* extensive[81a]; *CPD* $TiO_2{\cdot}Nb_2O_5$ and $TiO_2{\cdot}3\ Nb_2O_5$; 3 eutectics Structure[87] of $TiNb_2O_6$: β-Nb_2O_5 derivative (2) *PB* TiO_2–NbO_2; *CSS* rutile series but with superlattice developing[16c,61] (3) *CSS* (TiO–NbO) and primary β-(Ti,O)–(Nb,O) in no doubt (though not studied as such?) ⊖ overall ternary not known ⊖
Ti–Ta–O	not known as such but probably similar in many respects to Ti–Nb–O ⊖ *CSS* between isomorphs (e.g. TiO_2–TaO_2, etc.) likely but requires study ⊖
Ti–Cr–O	Ti corner known[83]; α-(Ti,O) solution becomes destabilized through Cr for β two ternary *CPD*: Ti_3Cr_3O (cubic η-carbide type) and Ti_4Cr_4O (approx. tetragonal?); both decompose before melting ⊖
Ti–Mo–O Ti–W–O	(1) *PB* TiO_2–MoO_2 (WO_2) (*PSS* of rutiles) and TiO–MoO_3 (WO_2) probable (2) solubility of Mo and W in TiO small[61] (3) overall diagrams not known; no ternary compounds found[41d] but still possible TiO_2 and other multiple Ti oxides probably in direct equilibrium with Mo and W oxides ⊖
Ti–U–O	not known; *PB* TiO–UO and TiO_2–UO_2 probable, with multiple non-stoichiometric transition stages ⊖
Ti–Mn–O*	ternary little known, except (1) *CPD* Ti_3Mn_3O/Ti_4Mn_2O (η-oxide)[32c,41d] (2) on *PB* MnO–TiO_2: *CPD* $TiMnO_3$ (ilmenite type, Fe_2O_3-related) and $TiMn_2O_4$ (spinel)[84] ⊖
Ti–Re–O	ternary little known, except[41d] *CPD* TiReO, *HCP* (M_2C)-type structure but non-stoichiometry; $TiReO_{1-x}$ where possibly $0 < x < 1$, i.e. complete O vacancies ⊖
Ti–Fe–O*	*PB* TiO_2–FeO contains: (1) *CPD* $TiFeO_3$ (ilmenite = Fe_2O_3 deviant), $TiFe_2O_5$ and $TiFe_2O_4$ (pseudo-brookite = spinel deviant); magnetic structure of this series known

System	*Principal compounds or solid solutions; special notes*
Ti-Fe-O* *cont.*	(2) Ti_3Fe_3O/Ti_4Fe_2O (η-oxide[32c]) ⊖ order/disorder transformations in the system $FeTiO_3$–Fe_2O_3 linked with magnetic changes[85] Fe–Ti–O system at 1,200°C showed[86] equilibria and structure changes in the Fe_2O_3–Ti_3O_3, spinel ($Fe_nTi_{3-n}O_4$) and $Fe_pTi_{3-p}O_5$ series; also non-stoichiometry and reducibility as function of pressure
Ti-Co-O*	*PB* TiO_2–CoO contains (1) *CPD* $TiCoO_3$ (ilmenite type), $TiCo_2O_5$ and $TiCo_2O_4$ (spinel deviant) (2) Ti_3Co_3O/Ti_4Co_2 (η-oxide[32c]) ⊖
Ti-Ni-O*	(1) *PB* TiO_2-NiO with *CPD* analogous to TiO_2–CoO series probable (e.g. $TiNiO_3$ of ilmenite type); needs verification ⊖ (2) Ti_3Ni_3O/Ti_4Ni_2O (η-oxide[32c]) ⊖
Zr-Hf-O	ZrO_2–HfO_2 intersolubilities certain but detailed system not determined ⊖
Zr-Th-O	*PB* ZrO_2–ThO_2: 2 phases at lower temps.; but[87] above 2,000°C extensive *PSS* of ZrO_2 in ThO_2 to give coexisting tetragonal and cubic forms of 'ZrO_2' ⊖ a compound ZrO_2–ThO_2 reported; ZrO_2–ThO_2 system[88]
Zr-V-O	not known, except for ternary *CPD* $(Zr,V)O_2$ (CaF_2 type[41d]), i.e. cubic ZrO_2 form stabilized by V, then towards rutile of VO_2; small *PSS* $ZrO_2 \leftrightarrow V_2O_5$ likely ⊖
Zr-Nb-O	(1) *PB* ZrO_2–Nb_2O_5: limited solubilities; *CPD*[81a] $6ZrO_2{\cdot}Nb_2O_5$; peritectic at 1,675°C; cf.[16c] (2) lower-oxygen system unknown but *PSS* ZrO_2–NbO_2 certain, with rutile-type stabilization ⊖
Zr-Ta-O	not known as such but Zr–Nb–O resemblances probable ⊖
Zr-Cr-O	not known as such, except ternary *CPD*; $(Zr,Cr)O_2$ of CaF_2 structure[41d]; *PB* ZrO_2–Cr_2O_3 probable, with possible *CPD* ⊖
Zr-Mo-O Zr-W-O	not known as such, except *CPD* $(Zr,Mo)O_2$ and $(Zr,W)O_2$ of CaF_2 structure[41d] *PB* and *PSS* $ZrO_2 \rightarrow MoO_2$ and $ZrO_2 \rightarrow WO_2$ likely, also *CPD* $ZrO_2{\cdot}xMo(W)O_3$ ⦶ Y–W–O system: a study[89] showed that Y_2O_3 (*PB*) forms 5 compounds $WO_3{\cdot}x(YO_{1\cdot5})$ where $x = \frac{2}{3}, \frac{2}{1}, \frac{15}{4}, \frac{9}{2}$ and $\frac{6}{1}$
Zr-U-O	(1) *PB* ZrO_2–UO_2 (both isomorphs) contains large *PSS* (possibly *CSS*?), receding at lower temps., with miscibility bay[90]; non-stoichiometry of ZrO_2–UO_2 solutions[91]; also ZrO_2–UO_2 system[88] (2) oxygen addition to Zr–U binary stabilizes α-Zr and α-U, decomposes the stable ϵ-Zr_3U intermetallic compound to *HCP* (Zr,O) solution + U; later UO_2–ZrO_2 solution formed[92] ⊖
Zr-Mn-O	not known as such except ternary *CPD* $(Zr,Mn)O_2$ of CaF_2 structure[41d]; further ternary *CPD* probable ⊖
Zr-Re-O Zr-Np-O Zr-Pu-O	undetermined; *CPD* again likely, including large $(Zr,Pu)O_2$, etc. intersolutions of CaF_2 type ⊖
Zr-Fe-O	mainly unknown, except ZrO_2–FeO plain duplex system[93]; *CPD* $ZrO_2{\cdot}xFe_2O_3$ (ilmenite analogue?) seems possible ⊖
Zr-Co-O Zr-Ni-O	not known but ZrO_2–CoO (NiO) possibly plain duplex systems; ternary *CPD* probable ⊖
Hf-*M*-O (*M* = most other transition metals)	generally unknown except for likely similarities to ZrO_2-base systems and possible predictions from binary structures, etc. ⊖
Th-*M*-O (*M* = most other transition metals)	generally unknown except *CSS* ThO_2-UO_2; non-stoichiometry and thermodynamic properties[91] ⊖

System	*Principal compounds or solid solutions; special notes*
V–Nb–O	*PB* V_2O_5–Nb_2O_5 shows extensive *PSS* *CPD* $V_2O_5{\cdot}Nb_2O_5$ (m.p. 1,125°C) and $V_2O_4{\cdot}Nb_2O_5$ (m.p. 1,810°C) *PSS* of end isomorphs in V–O and Nb–O systems (see *Table 8.1*) likely; *CSS* VO_2–NbO_2 rutile series[60] ⊖
V–Ta–O	not known but *PB* V_2O_5–Ta_2O_5, etc. probably similar to above ⊖
V–Cr–O	not known except rutile–solution series VO_2–CrO_2 including $VCrO_4$ (rutile structure[94]); *CSS* V_2O_3–Cr_2O_3 and further ternary *CPD* probable ⊖
V–Mo–O V–W–O	not known except[60] *CSS* VO_2–MoO_2; probable *PB*: V_2O_5–MoO_3 (with *PSS* and very low-melting eutectic) ⊖ Ternary oxides still possible ⊖
V–U–O	not known ⊖
V–Mn–O	not known but *PB* VO_2–MnO_2, V_2O_3–Mn_2O_3 probable, with extensive solid solutions ⊖ spinel $(V,Mn)_3O_4$? ⊖
V–Re–O	not known except *CPD* $VReO_{1-x}$ $(1 > x > 0)$ of *HCP* structure[41c] ⊖
V–Np–O V–Pu–O	not known ⊖
V–Fe–O	not known overall, but *CPD* and extensive solid solutions of VO–FeO, spinel and hematite type occur ⊖ (1) in V_2O_3–Fe_2O_3 *PSS*, but transformation[95] towards V_2O_3 (magnetic properties and structures by neutron diffraction); ionic character of $VFeO_3$ (ilmenite type?)[96] ⊖ (2) spinels V_2FeO_4 and VFe_2O_4[96–7] (3) *CPD* $VFeO_4$ no longer a rutile, although $VRhO_4$ is one FeO, FeV_2O_4, V_2O_3 and VO are successively in equilibrium; first two soluble in Fe, last two not. Series of pesudo-ternaries Fe/V/VO, Fe/VO/V_2O_3, Fe/V_2O_3/FeV_2O_4, Fe/FeV_2O_4/FeO VO reduces FeO; cf.[98]
V–Co–O V–Ni–O	not known; probably similarities to V–Fe–O ⊖
Nb–Ta–O	(1) *PB* Nb_2O_5–Ta_2O_5; *CSS:* β-Nb_2O_5 and α/θ–Ta_2O_5; *CPD* $2Nb_2O_5{\cdot}Ta_2O_5$ reported[99,100] structure BC tetragonal, $a = 15{\cdot}77$, $c = 3{\cdot}84$ Å; $M = 4$, $\rho = 6{\cdot}76$ (2) *CSS* $(Nb,Ta)O_2$ and (Nb,Ta)O not in doubt nor NbO_x–TaO_x series[11b,61–2]
Nb–Cr–O	*PSS* β-$Nb_2O_5 \rightarrow Cr_2O_3$; extensive ternary rutile and hematite series rutile series $NbO_2 \rightarrow NbCrO_4 \rightarrow CrO_2$ studied[60] ⊖
Nb–Mo—O	*PSS* β-$Nb_2O_5 \rightarrow MoO_3$, up to limit, then ready MoO_3 volatalization; rutile series[60] MoO_2 ⊖
Nb–W–O	*PSS* β-$Nb_2O_5 \rightarrow WO_3$ very extensive *PB* Nb_2O_5–WO_3 contains *CPD* 'Y' structure[101] of $Nb_2O_5{\cdot}3WO_3$; again rutile series NbO_2–WO_2, probably *CSS*; perovskite- and tetragonal bronze-type series ⊖
Nb–U–O	not known but *PSS* $Nb_2O_5 \rightarrow UO_2$, and extensive $(Nb,U)O_2$ (rutile ↔ fluorspar) solution series probable ⊖
Nb–Mn–O	little known but extensive rutile- and hematite-type series likely; *CPD* $Nb_2Mn_4O_9$ of hematite structure[102] and Nb_2MnO_6 (columbite type); monoxide series NbO–MnO probably *PSS* ⊖
Nb–Re–O	not known but series Nb_2O_5 ↔ perovskite ↔ ReO_3 probable ⊖
Nb–Np–O Nb–Pu–O	not known but similarities to Nb–U–O probable ⊖

System	*Principal compounds or solid solutions; special notes*
Nb–Fe–O Nb–Co–O Nb–Ni–O	*PSS* β-$Nb_2O_5 \rightarrow Fe_2O_3$ β-$Nb_2O_5 \rightarrow Co_3O_4$ (CoO) β-$Nb_2O_5 \rightarrow NiO$ } occur then *CPD*: rutile ($NbFeO_4$, etc.), columbite (Nb_2FeO_6, etc.) and hematite[102] series ($Nb_2Fe_4O_9$, etc.) extensive; H.T./L.T. relations of rutile/columbite; strong non-stoichiometry. *PSS* NbO–FeO, NbO–CoO and NbO–NiO likely ⊖ Nb-bearing η-oxides almost certainly also exist (e.g. Nb_3Fe_3O, with non-stoichiometry?) but not reported ⊖
Nb–Pt metals–O	$NbRhO_4$ rutile known; others possibly occur with Pt-group metals ⊖
Ta–Cr–O Ta–Mo–O Ta–W–O Ta–U–O Ta–Mn–O Ta–Re–O Ta–Np(Pu)–O Ta–Fe–O Ta–Co–O Ta–Ni–O	(1) known only fragmentarily, except for probable similarities to Nb-base oxides (2) dominance of rutiles, columbites ($TaMn_2O_6$, $TaFe_2O_6$), perovskites, etc. (3) rutiles include $TaCrO_4$, $TaFeO_4$, $TaRhO_4$; 'trirutiles' (tapiolite type) Ta_2FeO_6, Ta_2CoO_6 and Ta_2NiO_6 (4) hematites occur, e.g.[102] in $Ta_2(Mn,Fe,Co)O_9$ (5) in Ta–Mn–O two ternary phases, Ta_3Mn_3O (η-carbide type) and $TaMn_2O_3$ (CoSn type) reported[41c] (6) Ta-bearing η-oxides almost certainly also exist (e.g. Ta_3Fe_3O, with non-stoichiometry?) but not reported ⊖
Cr–Mo–O Cr–W–O	largely unknown except (1) *CSS* Cr_3O–Mo_3O (W_3O) probable (2) *CSS* CrO_2–MoO_2 rutile series known[60] *PSS* CrO_2–WO_2 probable; trirutile Cr_2WO_6 reported [103] (3) *PSS* CrO_3–MoO_3 (WO_3) almost certain (4) no ternary metallic *CPD* found[41c] but a '$(Cr,Mo,W)_2O_3$ 'related to Cr_2O_3 (hematite type with structure distortion); its Cr/Mo(W) ratio > 1 ⊖
Cr–U–O	in Cr_2O_3–UO_2–O_2 partial system (1,200°–1,600°C) *CPD* Cr_2O_3, U_3O_8 and a new ternary phase, $CrUO_4$, reported[104] (orthorhombic, $a = 4{\cdot}86_8$, $b = 5{\cdot}04_8$, $c = 11{\cdot}78_5$ Å; $M = 4$); otherwise equilibria uncertain ⊖
Cr–Mn–O	not known as such but extensive $(Cr,Mn)_2O_3$ and spinel ($MnO{\cdot}Cr_2O_3$) series exist; MnO probably dissolves some Cr ⊖
Cr–Re–O	not known except *CPD* $CrReO_{1-x}$ of HCP structure[41c], where $1 > x > 0$ ⊖
Cr–Np(Pu)–O	not known ⊖
Cr–Fe–O	(1) tentative overall diagram[105]; cf.[67] (2) *CSS* Cr_2O_3–Fe_2O_3 (hematites); *CSS* of *spinels* $Fe_3O_4 \rightarrow FeO{\cdot}Cr_2O_3$ (chromite), continuing to a probable Cr_3O_4 at higher temperatures ⊖ (3) magnetic structure in Cr_2O_3–Fe_3O_2 system determined[106]
Cr–Co–O Cr–Ni–O	not determined overall—surprising in view of importance in oxide layers on alloys, etc.; however, spinels $CoO{\cdot}Cr_2O_3$ and $NiO{\cdot}Cr_2O_3$ exist, no doubt joining to 'Cr_3O_4' at higher temperatures; strong non-stoichiometries probable, to give spinels of M_2O_3 type (related to γ-Fe_2O_3); $NiO{\cdot}Cr_2O_3$ reported[107] as *only CPD* on *PB* Cr_2O_3–NiO ⊖
Mo–W–O	(1) *PB* MoO_3–WO_3 plain eutectic, *CPD* $Mo_{0{\cdot}5}W_{0{\cdot}5}O_3$ of different structure, which would probably mean double eutectic ⊖ (2) multiple homologous series of *CPD* (lower oxides) M_nO_{3n-1} where $n = 8 - 12, 14$, $M =$ Mo,W in different proportions[7c]; (cf. binary Mo–O and W–O phase series, *Table 8.1*) ⊖
Mo–U–O	not known ⊖

System	*Principal compounds or solid solutions; special notes*
Mo–Mn–O	not known overall but *CPD* Mo_3Mn_3O (η-oxide)[41c]; possible $MoMnO_4$ rutile? $\ominus$
Mo–Re–O Mo–Np(Pu)–O	not known $\ominus$
Mo–Fe–O Mo–Co–O Mo–Ni–O	overall equilibria not known but (1) *CPD* Mo_3Fe_3O, Mo_3Co_3O and Mo_3Ni_3O of η-carbide structure, with homogeneity ranges (2) *CPD* $\approx$ MoFeO and MoCoO of *HCP* (M_2C-type) structure (no 'MoNiO'); these are ideally oxide analogues of the hemicarbides, though compositions range over $MoFeO_{1-x}$, $MoCoO_{1-x}$ where $0 < x < 0{\cdot}6$ (deficiency as in carbides) $\ominus$ (3) *CPD* $CoMoO_4$ and $NiMoO_4$, of monoclinic symmetry, reported[108] (4) Mo–Ni double oxide of β-Mn structure[109], with no doubt further analogues
W–U–O	not known except double oxides reported[110]; but *PSS* WO_2–UO_2 probable, with non-stoichiometric effects $\ominus$
W–Mn–O	not known overall but *CPD* W_3Mn_3O (η-oxide)[41c]; *CPD* $WMnO_4$ (mineral 'huebnerite') nearest approach to rutile $\ominus$
W–Re–O	not known but *PB* series WO_3–ReO_3 (related to perovskite) should be of special structural interest $\ominus$
W–Np(Pu)–O	not known $\ominus$
W–Fe–O W–Co–O W–Ni–O	overall equilibria not known but (1) *CPD* W_3Fe_3O, W_3Co_3O and W_3Ni_3O, of η-carbide structure, with homogeneity ranges (2) *CPD* $\approx$ WFeO and WCoO of *HCP* (M_2C-type) structure (no 'WNiO'); cf. comment on Mo (3) *CPD* $WFeO_4$ (mineral 'ferberite', isomorph to $MnWO_4$; wolframite series) $\ominus$ Fe_2WO_6 columbite of two forms reported[103]
U–Mn–O U–Re–O	not known $\ominus$
U–Np(Pu)–O	extensive solution series between end isomorphs probable but systems not generally known $\ominus$ UO_2–PuO_2 system, *CSS* melting behaviour[111]
U–Fe–O U–Co–O U–Ni–O	not known $\ominus$
Mn–Re–O	not known overall except *CPD* $MnReO_{1-x}$, of *HCP* structure, where $1 > x > 0$
Mn–Np(Pu)–O	not known $\ominus$
Mn–Fe–O Mn–Co–O Mn–Ni–O	incompletely known but (1) *PB* MnO–FeO; *PSS* extensive but not *CSS*[112] (2) *PB* Mn_3O_4–$MnO{\cdot}Fe_2O_3$–Fe_3O_4, spinel series *CSS* at high temperatures (above 1,160°C), i.e. high/low hausmannite transformation. At lower temps. *PSS* still extensive[113] (to 55 per cent Mn_3O_4 in Fe_3O_4 at 400°C) (3) *PB* Mn_2O_3–Fe_2O_3 *PSS* but not *CSS*[114] (4) interlinkages of (1), (2), (3) not known; Fe–Mn–O system reviewed[115] from the viewpoint of slag constitutions; O solubilities in liquid Fe–Mn[116] (5) *PB* MnO–CoO and MnO–NiO; *PSS* probably extensive; *PB* Mn_3O_4–Co_3O_4; *PSS* probable $\ominus$ spinel system $Mn_3O_4 \leftrightarrow Co_3O_4$, cf.[117]

System	*Principal compounds or solid solutions; special notes*
Re–Np(Pu)–O	not known ⊖
Re–Fe–O Re–Co–O Re–Ni–O	not known except *CPD* $ReFeO_{1-x}$, and $ReCoO_{1-x}$ of *HCP* structure, where $1 > x > 0$ '$ReNiO_{1-x}$' analogue possible but not reported ⊖
(Np,Pu)–(Fe,Co,Ni)–O	not known ⊖
Fe–Co–O	*CSS* FeO–CoO; *CSS* Fe_3O_4–$CoO{\cdot}Fe_2O_3$–Co_3O_4[118] *PSS* $Fe_2O_3 \rightarrow$ 'Co_2O_3' ⊖
Fe–Ni–O	*CSS* FeO–NiO; *CSS* $Fe_2O_3 \rightarrow NiOFe_2O_3$ (solubility of oxygen in Fe–Ni but not compounds determined[119]) ⊖ reduction of NiO in contact with Fe[120]; (O-atom diffusion to Ni and γ-Fe)
Co–Ni–O	*CSS* CoO– NiO ⊖ overall systems Fe–Co–O, Fe–Ni–O, Co–Ni–O require study ⊖
	Some Rh and other mixed Pt oxides also exist; systems containing Pt-metal Group oxides, where stable, do not, however, appear to have been systematically examined ⊖

* To Ti–(Mn, Fe, Co, Ni–O):

(*a*) Ternary *CPD* ≈ $(Ti,Fe)O_2$, $(Ti,Co)O_2$ and $(Ti,Ni)O_2$ of *fluorite*-type structure (distinct from rutile)[11d]

(*b*) Series TiO → (MnO, FeO, CoO, NiO) of NaCl structure might be thought *PB*, with partial solutions; however, complicated by Ti–O multiple series and intermediate phases

that, in principle, their family membership should exceed the transition metal confines. Altogether, the boundary between, and the very concept of, transition metal and non-transition metal becomes much more fluid for oxides, due to the advent of ionization.

The spinel and perovskite families are ready examples. Thus one or both metal partners can, in place of Ni^{++}, Fe^{++}, Fe^{+++}, Cr^{+++}, V^{x+}, Ti^{x+}, etc. become Mg^{++}, Al^{+++}, Ca^{++}, Zn^{++}, the Rare Earth ions, etc., not to mention silicon and the almost limitless variation arising in silicates and aluminosilicates. Also we return again to our original theme which includes 'holes' and electrons as valid atom-equivalent species. Point defects of this sort acquire a special importance in ternary oxide structures. In fact, the knowledge of the transition-metal oxides containing as partner the alkali or alkali earth metals, aluminium, silicon, etc. (e.g. $NaWO_4$, $CaTiO_3$, etc.) was until recent times more extensive than that of the inter-transition metal oxides. Both groups should be considered to form an integral whole.

In parallel with *Table 8.5*, *Table 8.6* and *Figure 8.12* give a classification of the double oxides into their basic structural families. The subsidiary diagrams in *Figure 8.12* superimpose these families on the transition-metal grid of the Periodic Table. Any two atoms participating in the given structure are joined. The pattern of these joins in itself reveals points of interest—not the least being the possibility of predicting further, unreported, oxides.

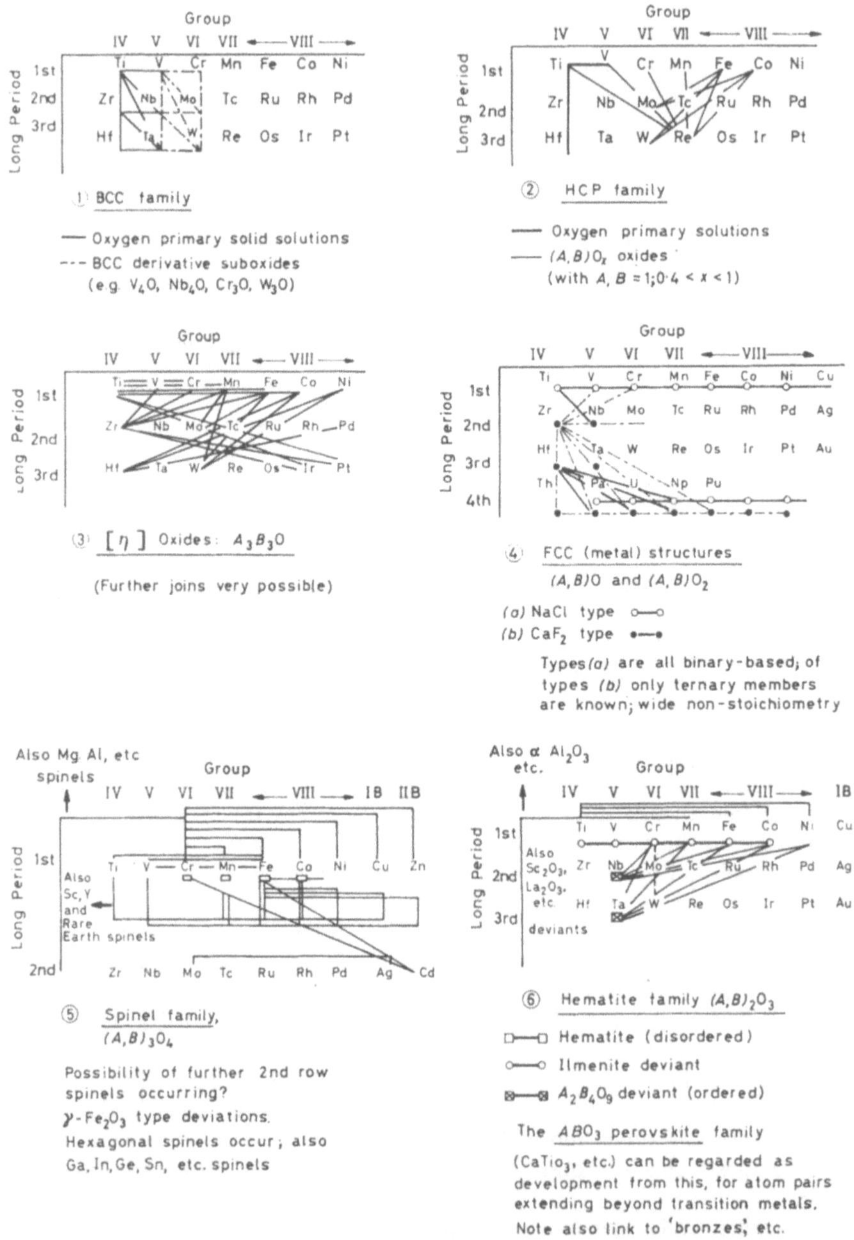

Figure 8.12. Some double oxides of the transition metals, grouped by isomorphous families and occurrences

Partner metal atoms joined; stoichiometries as stated (but with qualifications)

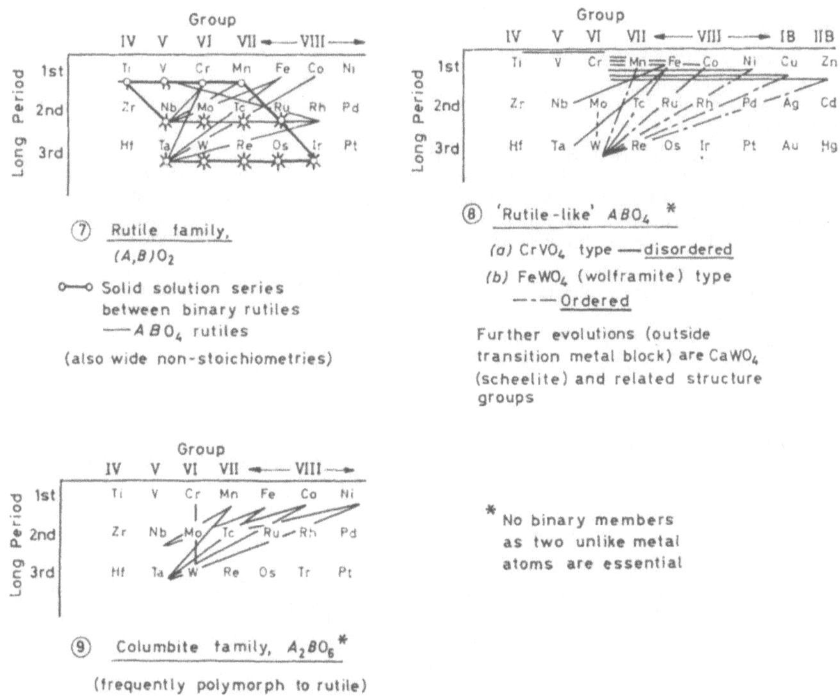

Figure 8.12b

8.4.2. *Survey of structure families in the double oxides*

The discussion of double oxides will broadly follow the order of increasing oxygen content. The starting point will be the BCC metal family, viz. the primary oxygen solutions in Ti, Zr, etc. which evolve in Groups V and VI to new suboxide structures derived from the BCC. The oxygen solid solubility in the later-Group BCC transition metals is negligible (see Chapter 3), and these metals are no longer prone to form such suboxides.

Throughout the discussion the occurrence of possible polymorphy should be remembered. The structures noted for given metal pairs are valid for some but not necessarily all temperatures in the solid. Conversely, more than one of the given structures could apply to similar compositions. This will occur frequently for the rutile, columbite and associated families.

Table 8.6. Classification of double oxides by structure families

⊖ = need of further research indicated

(1) *BCC family* (with O atoms interstitially)

1.1 Primary inter-solid solutions (strict BCC) [β-(Ti,O), β-(Zr,O), β-(Hf,O), (V,O), (Nb,O), (Ta,O)]

1.2 Lesser distortions of *BCC*: [$V_{10}O/V_4O$, Nb_4O, Ta_4O, 'NbO_x', 'NbO_y', 'TaO_y', 'TaO_z'] and intersolutions (including unstable suboxides) ⊖

1.3 Derivatives: Cr_3O, Mo_3O, W_3O (β-W type) and intersolutions

These suboxides cease at Group VII; little is known (although this could be very important) about intersolutions *between* these Group IV–VI suboxides and possible new mixed compounds formed ⊖

Table 8.6 cont.

(2) *HCP family* (with O atoms interstitially)

2.1. Primary inter-solid solutions [α-(Ti,O), α-(Zr,O), α-(Hf,O)]
Lesser distortions of *HCP*: $V_{13}O_7$ (possibly also an O-stabilized *HCP* form of Nb and Ta, though not reported) ⊖

2.2 Oxides $(A,B)O_x$ where $A:B \approx 1:1$, x ranges 0·4–1
A = Mo,W,Re; B = Fe,Co when A = Mo,W; B = Ti,V,Cr,Mn,Fe,Co when A = Re; this is all reported so far; however, probably many more representatives exist ⊖
non-stoichiometric range: $\approx A_2B_2O$ to ABO; 'ABO' is here the ideal lattice formula, comparable to M_2C in carbides of same (L'_3) structure but with non-stoichiometry

Notes: (*a*) The metal combinations hitherto established are those which, without oxygen, would favour σ and Laves phases; this may be universal but would need amplifying by further searches ⊖

(*b*) Link-up between 2.1 and 2.2 would be of interest: it seems likely that in a system like Ti–*A*–*B*–O (e.g. Ti–Mo–Fe–O), a continuous solution α-(Ti,O) ↔ (Mo,Fe)O would occur ⊖

(3) *η-Carbide-type oxide*, A_3B_3O

Cr_3Ti_3O, Mn_3Ti_3O, Fe_3Ti_3O, Co_3Ti_3O, Ni_3Ti_3O, Cu_3Ti_3O; Rh_3Ti_3O, Pd_3Ti_3O, Ir_3Ti_3O, Pt_3Ti_3O, V_3Zr_3O, Cr_3Zr_3O, Mn_3Zr_3O; Fe_3Zr_3O(?), $Rh_{1.8}Zr_{4.2}O$, Pd_2Zr_4O, $Ir_{1.8}Zr_{4.2}O$, $Pt_{1.6}Zr_{4.4}O$, Ni_2Hf_4O, Pd_2Hf_4O; Mn_3Ta_3O; Mn_3Mo_3O, Mn_3W_3O; Fe_3Mo_3O, Fe_3W_3O; Co_3Mo_3O, Co_3W_3O; Ni_3Mo_3O, Ni_3W_3O

Notes: (*a*) No doubt others exist, e.g. Fe_3Ta_3O, Fe_3Nb_3O, etc. ⊖

(*b*) The M_6O stoichiometry is not necessarily adhered to. Both A/B and metal/oxygen ratios can be variable; the latter to complete O deficiency. Typical ranges are:

$$Fe_2Ti_4O \leftrightarrow Fe_3Ti_3O$$
$$Fe_2Ti_4O \leftrightarrow FeTi_2$$

(*c*) The η-carbide structure amongst suboxides may not be the only one among oxides typified by known carbides and nitrides (e.g. do oxide isomorphs occur to $M_{23}C_6$, M_7C_3, Fe_3C or WC types?) ⊖

(4) *FCC family* (with O atoms interstitially)

4.1 NaCl type: [TiO, VO, (CrO?), MnO, FeO, CoO, NiO; Cu_2O; NbO, TaO; PaO, UO, NpO, PuO] primary inter-solutions (subject to size and electronic factors)
Note: No purely ternary NaCl-type oxides seem to have been found comparable with the ternary *HCP* oxides (ABO_x above) but could possibly exist or be predicted (exception see p. 406) ⊖

4.2 CaF_2 type: [ZrO_2 when stabilized, e.g. by Mg or Ca, HfO_2, ThO_2, PaO_2, UO_2, NpO_2 PuO_2, AmO_2, etc.] and their intersolutions where favourable; including non-stoichiometric ranges
also *ternary* oxides: $(Ti,Fe)O_2$, $(Ti,Co)O_2$, $(Ti,Ni)O_2$; $(Zr,V)O_2$, $(Zr,Cr)O_2$, $(Zr,Mo)O_2$, $(Zr,W)O_2$, $(Zr,Mn)O_2$ reported so far; others likely ⊖

(5) *Spinel family*, A_2BO_4

Iron spinels	*Cobalt spinels*	*Manganese spinels*
$MnFe_2O_4$	(Co_3O_4)	
Fe_3O_4	$NiCo_2O_4$	(Fe,Co,Ni,Cu,Zn,Cd)Mn_2O_4 compounds very likely to exist, though confirmation required ⊖
$CoFe_2O_4$‡	$CuCo_2O_4$†	
$NiFe_2O_4$	$ZnCo_2O_4$†	
$CuFe_2O_4$*	$CdCo_2O_4$	
$ZnFe_2O_4$	$(MgCo_2O_4?)$	
$CdFe_2O_4$		
$(MgFe_2O_4)$		

* Cubic at high temps. only, tetragonal at lower; possibly also for $ZnFe_2O_4$ (?)
† Possibly existing ⊖
‡ Possible inversion to $FeCo_2O_4$ (?)

Table 8.6 cont.

Chromium spinels	*Vanadium spinels*	*Titanium spinels*
$MnCr_2O_4$	MnV_2O_4	$MnTi_2O_4$
$FeCr_2O_4$ (chromite)	FeV_2O_4/VFe_2O_4	$FeTi_2O_4/TiFe_2O_4$
$CoCr_2O_4$	CoV_2O_4/VCr_2O_4	$CoTi_2O_4/TiCo_2O_4$ (?)
$NiCr_2O_4$	(NiV_2O_4/VNi_2O_4)†	$NiTi_2O_4$ (?)†
$CuCr_2O_4$	CuV_2O_4 (?)	$CuTi_2O_4$ (?)†
$ZnCr_2O_4$	ZnV_2O_4	$ZnTi_2O_4$
$CdCr_2O_4$	CdV_2O_4 (?)	($TiMg_2O_4$)
($MgCr_2O_4$)	(MgV_2O_4)	(others probable)

Aluminium spinels	*Miscellaneous*	
$MnAl_2O_4$	$ZnRh_2O_4$	$ZnAl_2O_4$
$FeAl_2O_4$	$MgRh_2O_4$	$MgIn_2O_4$
$CoAl_2O_4$	$GeNi_2O_4$	$MoAg_2O_4$
$NiAl_2O_4$	$SnCo_2O_4$	$MoNa_2O_4$
$CuAl_2O_4$	$SnZn_2O_4$	WNa_2O_4
$ZnAl_2O_4$	$SnMg_2O_4$	*silicon-* and *germanium*-bearing spinels also known (quaternary, to balance valencies), e.g. $(GeCr)LiO_4$, $(GeRh)LiO_4$ ⊖
$CdAl_2O_4$	$SnAl_2O_4$	
($MgAl_2O_4$) ('spinel' proper)	$ZnGa_2O_4$	
	$MgGa_2O_4$	
	$CdGa_2O_4$	

† Possibly existing ⊖

(6) *Hematite family,* $(A,B)_2O_3$

6.1 [Ti_2O_3, V_2O_3, Cr_2O_3, Mn_2O_3, Fe_2O_3, (Co_2O_3?), Al_2O_3] intersolutions *disordered*

6.2 Ilmenites: $MnTiO_3$, $FeTiO_3$, $CoTiO_3$, $NiTiO_3$, $MgTiO_3$, $CdTiO_3$, $ZnTiO_3$; probably others ⊖

6.3. e.g. niobates and tantalates of type A_2B_4O, viz.:
$Nb_2Mn_4O_9$, $Nb_2Fe_4O_9$, $Nb_2Co_4O_9$; probably $Nb_2Bi_4O_9$; $Nb_2Mg_4O_9$ and Ta analogues; near-hematite structure but formed by ordering of A and B ions; also $(Cr,Mo)_2O_3$, $(Cr,W)_2O_3$, slightly distorted
deviant M_2O_3 structures in Sc_2O_3 (cubic), La_2O_3, Ti_2O_3, As_2O_3 types could possibly occur as ternaries ⊖

(7) *Rutile family* (including 'trirutiles')

7.1 [TiO_2, VO_2, CrO_2, MnO_2, NbO_2, MoO_2, TcO_2 RuO_2, TaO_2, WO_2, ReO_2, OsO_2 IrO_2] and their intersolutions (where favourable)

Note: could TaO_2 possibly also have the fluorspar structure as a borderline dimorph? ⊖

7.2 $NbCrO_4$, $NbMnO_4$, $NbFeO_4$, $TaCrO_4$, $TaMnO_4$, $TaFeO_4$, $VRhO_4$, $NbRhO_4$, $TaRhO_4$; $TaGaO_4$; $AlSbO_4$, $CrSbO_4$, $FeSbO_4$, $RhSbO_4$, $(Nb,Cr)_aO_b$, $(Nb,Mn)_aO_b$, $(Nb,Fe)_aO_b$, etc. (similar for Ta), where $a:b = 1:4$ but can diverge widely; $MoMnO_4$?
the *trirutiles* include some of these but tend to dimorphy with the A_2BO_6 columbites [see (9)]

(8) *Rutile (1st stage) deviants*: $CrVO_4$ *and wolframite* ($FeWO_4$) *types*

8.1 $CrVO_4$ type: $CrVO_4$, [$CrMnO_4$*, $CrFeO_4$*], $CrCoO_4$, $CrNiO_4$, $CrCuO_4$, $CrZnO_4$; also $TiVO_4$ and $TiZrO_4$(?): *disordered*

* Not reported but considered very probable

8.2 $FeWO_4$ type: $WMnO_4$, $WFeO_4$ (wolframite), $WCoO_4$, $WNiO_4$, ($WCuO_4$(?), $WZnO_4$, $WMgO_4$, $NbFeO_4$, $TaFeO_4$, $NbInO_4$, $TaInO_4$ $NbScO_4$, $TaScO_4$: *ordered*; probably also Mo- (instead of W-) bearing members (not reported) and other extensions

Notes: (*a*) Types 8.1 and 8.2 can also occur as high and low polymorphs to similar compositions

(*b*) *monoclinic, distorted* $CoMoO_4$ and $NiMoO_4$ reported

(*c*) more distant relations to this group are the scheelite ($CaWO_4$), barite ($BaSO_4$), zircon ($ZrSiO_4$) families, etc. ⊖

'fergusonites' are $YTaO_4$ and numerous rare earth niobates and tantalates of ABO_4 type

Table 8.6 cont.

(9) *Columbite family* (*rutile 2nd stage deviants*)

Nb_2MnO_6, $NbFeO_6$, Nb_2CoO_6, Nb_2NiO_6, Ta_2MnO_6, Ta_2FeO_6 and $(Nb,Ta)_a(Mn,Fe,Co,Ni)_bO_c$ where a:b:c can deviate widely from 2:1:6; generally, complete intersolutions of these occur

Nb_2PbO_6, Ta_2PbO_6, Nb_2SnO_6, $Nb_2(Ca;Sr;Ba)O_6$, $Ta_2(Ca,Sr,Ba)O_6$, Fe_2WO_6 and Cr_2WO_6 columbite-like trirutiles also reported; further members certain ⊖

8.4.3. Further comments on classes of double oxides

Further comment will now be made on each of the classes shown in *Table 8.6*, in conjunction with *Figure 8.12.*

(1) BCC-*based solutions*—The only significant BCC solutions of oxygen occur in Groups IV and V; however, formation of the BCC-based suboxides in Groups V and VI should be viewed as one of continuing evolution. Thus, upon alloying oxygen-bearing Ti by the addition of Nb or W, the NbO_x, TaO_y, etc. (where x and y are small, probably variable, fractions) and the W_3O types of atomic redistribution and distortion of the unit cell should gradually become realized, and the intermediate stages of precipitation of suboxides become possible. However, very little is known about this, although so obviously important in deciding the resistance to oxidation in an initially thin-film structure.

The tetragonal phase VO_w ($w = 0{\cdot}10$–$0{\cdot}25$), still very closely related to the BCC metal solution, will evolve down Group V to the Kofstad-type suboxides NbO_x, NbO_y, TaO_z, etc. where x, y and z are likely to be of comparable order to w. Upon alloying, extensive, though not necessarily complete, substitution may be expected. (This represents a fruitful field of research on thin films on a metal substrate, using electron microscopy and diffraction methods.) Passing to Group VI, a main difference within the A–15 class of Cr_3O-, Mo_3O- and W_3O-type phases is one of order/disorder, another that of lattice vacancies; moreover, upon alloying, transitions in degree will become obvious.

On a more general plane, the W_3O-(A–15)-type phase is only one representative of a large family of intermetallic phases. Others are Mo_3Si, Cr_3Si, V_3Si (see Chapter 7) and numerous A_3B compounds where A belongs to Groups IV–VI, B to the Mn, Fe, Co, Ni or Cu Groups; B can also be Ge, Sn (e.g. Nb_3Sn), concomitant to Si, Al (e.g. Nb_3Al), Ga; As and Sb. Thus, in certain cases, solution formation and stabilization may be expected upon alloying. (The W_3O-type phases have been surveyed from the atomic size and electronic viewpoint[121–2].)

(2) HCP *oxides*—The principal *hexagonal close-packed* oxides are the oxygen-stabilized α-Ti, Zr and Hf solid solutions and their mutual solutions (the 'pseudo-oxides' Ti_3O, etc. discussed earlier). Upon Group V metal addition, after limited solutions, these could evolve into new oxides, closely akin to the carbides and nitrides, V_2C, Nb_2C, Nb_2N, etc. The reported '$V_{13}O_7$' is a near approach to this, and an analogue, though not reported, is possible for Nb or Ta. However, the most interesting new event is the occurrence of the true HCP double oxides

$$ABO_x \text{ with } A = (\text{Mo, W}) - B = (\text{Fe, Co})$$
$$= (\text{Re}) \quad - \quad = (\text{Ti,V,Cr,Mn,Fe,Co})$$

that is, combinations of the heavier Group VI–VII with First Long Period metals. Other members are very likely, including Tc-bearing ones. The stoichiometry is variable between $x = 0{\cdot}4$ and $1{\cdot}0$ in ABO_x, but 'ABO' will correspond to complete lattice occupation. This family of two-metal HCP oxides seems at present rather isolated from the single-metal HCP family, i.e. the primary α-Ti (oxygen) solid solutions and their isomorphs discussed earlier; however, as noted in *Table 8.6*, solution linkages between the two groups of compounds are likely, if the electron/atom and size ratios are kept balanced. Even a cursory inspection of diagram (2) in *Figure 8.12* suggests likely double oxides of this type, involving Nb, Ta and possibly some of the later Pt-group metals. For rhenium in particular, a link-up with the primary (HCP) Re metal is distinctly possible; this would imply an oxygen solid solution in Re that could be stabilized by the First Long Period metals*. On the other hand, the similarity should be noted of the metal pairs that form the HCP to those in the *η-carbides*, M_6C, e.g. $Fe_3W_3C \leftrightarrow (Fe,W)O_x$ $(0{\cdot}4 < x < 1)$.

Seen synoptically, it is evident (cf. *Table 8.6*, note 2.2*a*) that those metal combinations which, without interstitial, adopt the σ and Laves phases, will favour the η-carbide type structure, at low-carbon or nitrogen contents, and with low oxygen content the HCP structures. The relative size of the two unlike metal atoms here plays probably a more decisive part than the electronic factor, and the rules now fairly well established for Laves and σ-phase occurrence also seem likely to govern those of the present compounds.

It is instructive to represent these relations schematically as in *Figure 8.13*, showing the typical Mo–Fe–O and the Re–Fe–O ternary systems. Thus in

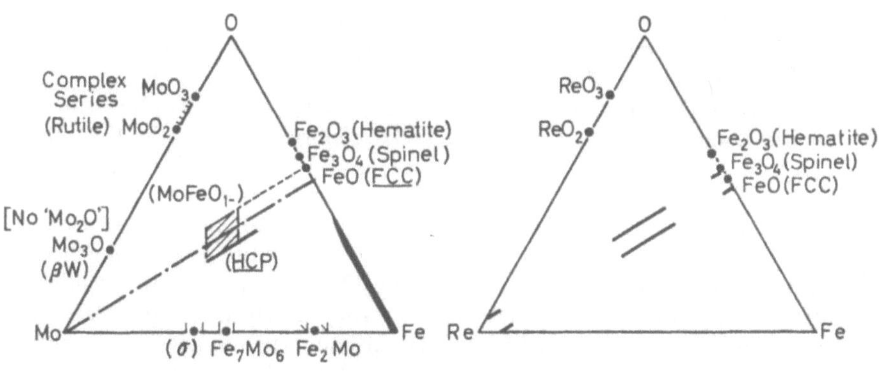

Figure 8.13. Ternary diagrams Mo–Fe–O *and* Re–Fe–O (*schematic*)

Mo–Fe–O the HCP compound M_2O occurs at about equi-atomic metal composition, with a fair homogeneity range. There is, of course, no 'Mo_2O' or 'Fe_2O' and no suggestion of a pseudo-binary section at M_2O level.

* Looking back at the introductory chapters where the *pure* metal structures were surveyed (cf. *Figures 1.1* and *2.4*), the present two quite separate HCP colonies of the early and the late transition metals, i.e. Ti, Zr, Hf in Group IV and Re, Os, Ir, Ru in Groups VII/VIII, are *linked* by way of low interstitial additions, in this case the HCP double oxides.

However, a most interesting section of this kind occurs at Mo ↔ FeO. The HCP double oxide clearly lies on this, and we have the structural series

$$BCC \rightarrow HCP \rightarrow FCC$$

(metal atoms, with O interstitial). This sequence is very familiar in many other systems discussed before, e.g. $V \rightarrow V_2C \rightarrow VC$, β-Ti → α-(Ti, C) → TiC, etc. We might well regard this as a new manifestation of the same theme —the HCP form as intermediate state between the BCC and FCC. Moreover, an equivalent to the 'FeO' homogeneity range (defect metal lattice) seems to recur in the HCP double oxide, but this point needs more study. *Figure 8.13* also brings out the stabilization by oxygen from σ and the Laves phase of the HCP oxide, whilst carbon additions give the η-carbide.

Similar pseudobinaries will occur in Mo–CoO, W–FeO, W–CoO systems and may evolve for NiO by formation of a HCP phase, 'WNiO' (?). The ternary boundaries of the HCP '*ABO*' homogeneity ranges, particularly the character and degree of non-stoichiometry, are at present unknown but have been tentatively indicated in *Figure 8.13* as inclined towards 'FeO'.

On the other hand, in the Re–Fe–O system (*Figure 8.13*), a Re–FeO pseudo-binary section contains the natural sequence (HCP Re) → (HCP ReFeO) → (FCC FeO), where the possibility of continuous solid solution between the first two exists, at least at high temperatures. A similar series would probably prevail between Re and CoO, MnO, (CrO), VO and TiO, with variable solution ranges, controllable by adjusting the proportions of these metals.

(3) *η-oxides*—As noted in *Table 8.6*, a general picture of the occurrence of η-oxides, M_6O, is still incomplete; but Nevitt[121] has made some useful additions to the work of Rostoker[32] and Schönberg[41]. Some obviously possible further η isomorphs are suggested in *Figure 8.12*, e.g. Nb-bearing and others involving Re and the Pt Group metals.

In contrast to the η-carbides and nitrides, however, η-oxides can be formed *between* 1st Long Period metals. Those so far reported all contain Ti; but this may be specific to the researches done, and V-base η-oxides are very possible (V_3Fe_3O, etc.?). The general combinations emerging are of the type A = 1st Long Period, B = 1st–3rd Long Period metal, as is clear from *Figure 8.12*, Graph 3. However, new work[121] has also now established 2nd–3rd Period combinations, such as Zr_4Pd_2O, $Zr_{4.4}Pt_{1.6}O$, with broader variations in metal ratios. The radius ratio R_A/R_B clearly is a critical factor and varies for given cases from 1·08 to 1·26, although this is not a sufficient condition, and the electron: atom ratio is important. The diagonal relation is quite similar to that for the η-carbides (see *Figure 4.2*). In particular, the high-speed steel carbide Fe_3W_3C has its Fe_4W_2O analogue, implying a possible ready solid solution $Fe_3W_3(C,O)$, with interesting practical consequences in a carbide/oxide transition. Furthermore, comparison of *Figure 8.12*, Graphs 2 and 3, shows certain coincidences of HCP and η-oxide families for the same metals, e.g. Fe–W, Co–W, etc. Noting that there is an approach in composition (HPC = $FeWO_{1-0.4}$; η = $FeWO_{0.17}$), the question of transformation of one to the other with varying oxygen and temperature seems to provide a fruitful field of enquiry; even a polymorphy for any one composition is conceivable. Similarly, starting from the pure HCP (α-Ti or α-Zr)-low-oxygen solid solutions, we may note a phase transition HCP → η-oxide, as the later transition metals (Mn, Fe, Co, etc.) are added in increasing amounts. Again, the capability should be noted of some of the η-oxides to lose their O atoms entirely without structure change, but in other cases with a change. Such cases are[32,121,123]:

η intermetallic (O-free)	*η-oxide*
Ti_2Cu	Ti_4Cu_2O
Ti_2Ni	Ti_4Ni_2O
[Ti_2Co η *and* Laves phase]	Ti_4Co_2O
[Ti_2Fe = BCC]	Ti_4Fe_2O/Ti_3Fe_3O
[Ti_2Mn = BCC (β-Ti solid solution)]	Ti_3Mn_3O

To this we may add a phase[16e] of η-Nb_3Fe_2, closely associated with a σ and Nb–Fe Laves phase, which almost certainly has an allied (Nb_4Fe_2O/Nb_3Fe_3O). The occurrence or otherwise of the oxygen-vacated η-oxide structure depends critically on the later transition metal present, i.e. gives way to the Laves, σ or BCC structures as the Group numbers approach each other. Again it may be noted that all the pure binary metal systems giving rise to η-oxides are *themselves* capable of transmuting to the Laves phase, in some cases also to the σ phase, but only if the size differences are sufficient (e.g. not for Cr–Fe, etc.). The latter pairs tend towards *spinels* when oxidizing.

Additional pure intermetallic η-types reported[121b] are ternary: $Hf_2Ni_{0.67}M_{0.33}$ where M = Ru, Re or Os. It seems likely that interstitial oxygen will in this case again dissolve. Attention should also be drawn to the possibility of three-metal η-oxides where two-metal oxides would not be stable—a situation quite generally liable to occur amongst the present oxide classes (e.g. HCP, FCC, etc.) and worth further study.

Figure 8.14 shows[121] some typical ranges of oxygen content for η-type phases in the four systems Ti–(Ni,Co,Fe,Mn)–O. We might consider the

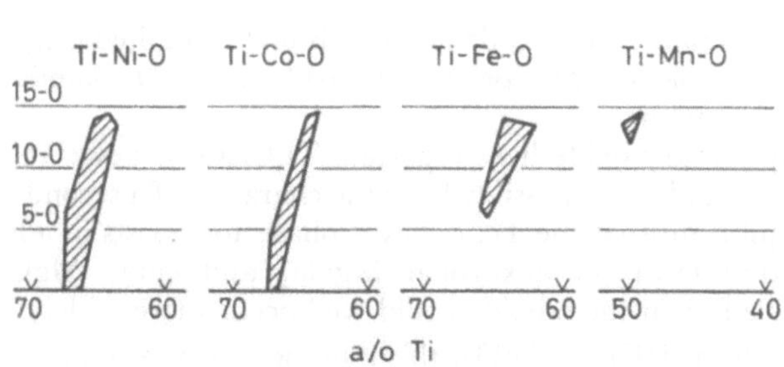

Figure 8.14. Ranges of concentration of oxygen in Ti_2Ni-*type* (η) *phases in the ternary systems* Ti–Ni–O, Ti–Co–O, Ti–Fe–O *and* Ti–Mn–O (From Nevitt[121a])

By courtesy of The Metallurgical Society of AIME

oxygen atom as a stabilizer of what is essentially an intermetallic compound between the transition metals, the more probable the closer the Group numbers. For Zr as one partner, the O atom is always necessary: it acts as an electron acceptor, as does C. A wider survey of Ti_2Ni- (i.e. η-)type phases with potential oxygen filling, based on Ti,Zr and Hf with the later 1st, 2nd and 3rd Long Period transition metals, has revealed many new isomorphs, included in *Table 8.5* and *Figure 8.12.* Some are intermetallics only, others require lattice oxygen, depending on relative atomic size and position in the Periodic Table.

We may note that all η-oxide members contain transition metals *only*, in contrast to most of the other oxides discussed.

Structurally, *Figure 4.6*, of the η-carbide structure, may be recalled; the O atoms would occupy the sixteen interstitial *d* positions then occupied by carbon, amongst the 96 metals. The vacation of the oxygen sites from M_6O occurs progressively towards the A_2B intermetallic, as *Figure 8.14* shows. However, as we know, A_2B stoichiometry is not rigid, and as in the η-carbides, can develop to AB: i.e. $A_4B_2O \leftrightarrow A_3B_3O$, with other metal-ratio variations. Significantly, the lattice spacings in η-Mo_3Mn_3O (11·07–11·20 Å) and in η-Mo_3Mn_3C (11·13 Å) are quite similar, showing clearly that the oxygen atom here is effectively still of much the same size as carbon, and atomic, not ionic. Metallic bonding still applies in η-oxides, without the breakthrough to ionization which characterizes the higher oxides.

In recapitulating the lower-oxide phases met so far, it may be useful to expand slightly on the Ti–Cr–O system, which will show several of the basic points. The Ti corner of this system[83] is shown in *Figure 8.15* for the temperature sections 1,400, 1,200, 1,000 and 800°C. The points to notice are:

(*a*) The BCC (β-Ti,O) solid-solution field continues along Ti–Cr, narrowing with decreasing temperature. More oxygen causes the BCC → HCP → FCC transition for high Ti, the Laves → η transition for higher Cr.

(*b*) The HCP (α-Ti,O) phase assumes at high temperatures the character of a *true oxide*, joining the α-Ti solution below 880°C. It becomes destabilized by Cr.

(*c*) The Laves phase ($TiCr_2$, denoted γ) is only a limited oxygen solvent, giving way to the η-carbide-type ternary oxide Ti_3Cr_3O, familiar from the ternary carbide systems.

The second ternary oxide here reported, Ti_4Cr_4O, is new; it closely adjoins the η phase and could possibly have the character of a second polymorph to it. It is significant that the $TiCr_2$ Laves phase, too, exists in a high- and a low-temperature form (of hexagonal $MgCu_2$ and cubic $MgCu_2$ types), and a similar effect in the η-oxide could well occur more widely.

(*d*) Viewed from HCP α-(Ti,O), this may be re-expressed by saying that Cr addition causes transformation

$$\text{HCP} \longrightarrow \text{Laves for } \textit{low} \text{ oxygen}$$
$$\searrow \eta \text{ for } \textit{higher} \text{ oxygen}$$

This evolves, for still higher oxygen, to the FCC → η change (TiO = NaCl structure), as we have observed earlier on a broader base (cf. *Table 8.5*).

(*e*) Recalling that η is itself isomorphous to the $TiNi_2$ type of intermetallic (cf. *Figure 8.12*, Graph 3), it emerges that oxygen addition to $TiCr_2$ has the same effect structurally as passing along the 1st Long Period amongst the intermetallic compounds:

$$\underset{\text{Laves}}{TiCr_2,\ TiMn_2,\ TiFe_2} \rightarrow \underset{\eta}{TiCo_2,\ TiNi_2}$$

$$\searrow \underset{\eta}{Ti_2Cr_3O,\ Ti_3Mn_3O,\ TiFe_2O}$$

This is effectively an equivalent between electron addition and oxygen addition, which is again a reflection of the fact that oxygen acts, in these

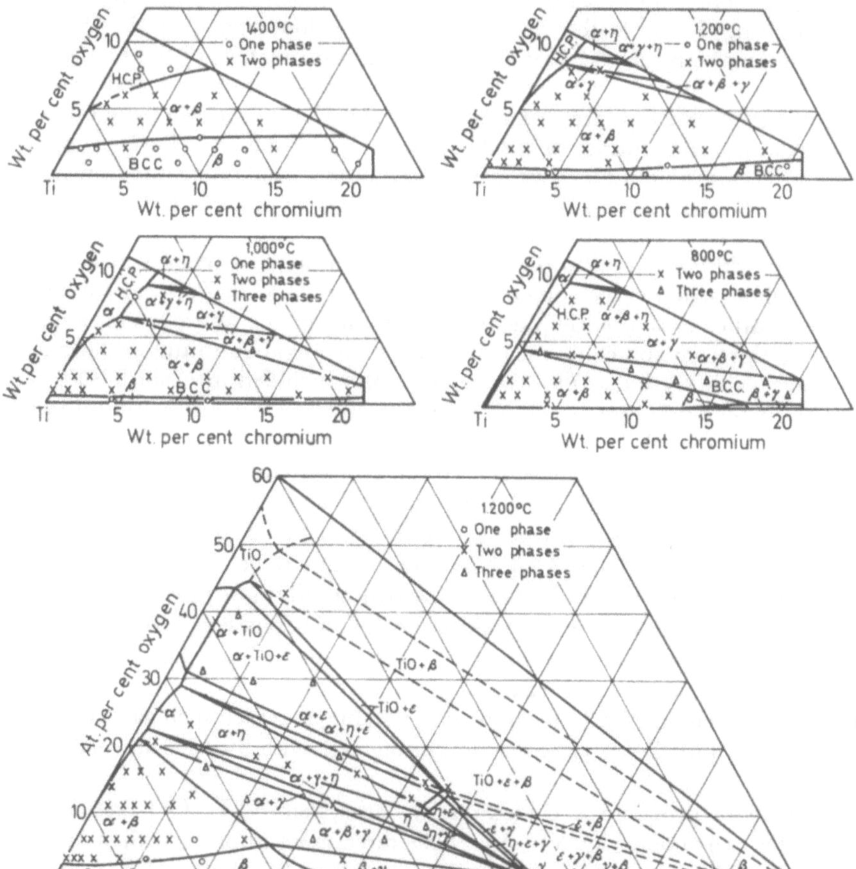

Figure 8.15. Ti–Cr–O *diagram at different temperature sections* (From Wang and Grant[83])
Ti-corners for 1,400°, 1,200°, 1,000° and 800°C; full diagram for 1,200°C showing evolution of structures
By courtesy of the Metallurgical Society of AIME

low oxides, as an electron donor, not as an electron absorber (as it does when ionization sets in at a later stage). It may well be suggested that there must be a situation of borderline alloys (e.g. Ti–Fe–Co–O) where the Laves and η phases become polymorphs, controlled by oxygen (and other interstitial) content.

(4) FCC-*based* oxides: NaCl and CaF_2 types—The FCC double oxides form the two major sub-groups of NaCl and CaF_2 types (monoxides or dioxides), depending on the oxygen sites in the metal lattice. The single-metal members shown in *Table 8.1* form extensive solid solutions, conditioned though by size ratio and electron factor. Thus in the *monoxides* the continuity of FeO/NiO/CoO/MgO solution is established; but MnO and the earlier transition-metal monoxides form only partial or very limited solutions with these. In borderline cases the temperature dependence can be critical. A complicating factor is the deviation from stoichiometry (e.g. in 'FeO') and the ready appearance, for the early transition metals, of the higher or lower oxides, in place e.g. of 'TiO'. No *ternary* transition metal oxides of NaCl-type structure are so far known which would not also have binary end members, in contrast to the HCP and CaF_2 families. However, such purely ternary oxides are conceivable for couples containing the 2nd and 3rd Long Period metals and would be worth searching for*. Defect lattices can introduce certain further anomalies, readily revealed by lattice spacings. Thus the addition, for example, to 'FeO_{1+x}' of a divalent metal which itself forms a more strictly stoichiometric monoxide, e.g. MgO or NiO, can initially cause a lattice *expansion*, despite the smaller size of the added metal atom, because this contributes to vacancy filling, not metal substitution. (Some spacings of binary monoxide systems were also included in *Figure 4.3.*)

The low-temperature rhombohedral distortions occurring in the binary cubic monoxides (NiO, CoO, etc.) will clearly apply to the mixed oxides in solution, although these have not been specifically studied. It is, however, of interest to consider the NiO–CuO system where around a composition $NiCuO_2$ a tetragonal superlattice can occur[125], with c/a varying with Ni/Cu ratio at 1,000°C; similar work has been published on Co–Cu–O. CuO is itself monoclinic, and it is Cu_2O which is structurally more nearly equivalent to NiO, so that we may well consider the $NiCuO_2$ phase as a distortion at high temperatures of a type which for NiO occurs at low ones, but here is engendered by copper and excess oxygen. Furthermore, extreme modifications of the cubic NaCl structure are notable in cases such as $CuFeO_2$ (delafossite) and $CuCrO_2$, which are hexagonal with 3 molecules per unit cell.

A point of principle is evidenced by the system V–Fe–O (*Table 8.5*). Here the two monoxides VO and FeO directly equilibrate with metallic Fe, both of NaCl type. However, in between it is not (V,Fe)O solid solutions that occur but, with increasing vanadium, a succession FeO–FeV_2O_4–V_2O_3–VO; evidently, the spinel and hematite structures intervene in the equilibrium. Clearly VO reduces FeO to form the higher mixed oxides in which the V ion preponderates. It is interesting to recall that in the binary

* For example, in *Figure 8.12*, Graph 4 shows the 1st and 4th Long Period *M*O quite isolated from each other, except for a tenuous link along Group V (NbO, TaO), and it is inviting to search for certain intermediate oxides of NaCl type which would be purely ternary and involve the 2nd and 3rd Long Period metals.

Ternary NaCl-type monoxides do occur, if one alkali-metal atom is involved, of atomic size similar to the transition-metal one, e.g. $Fe^{++}Li^{+}O_2$.

Cr–O system of Group VI (between V–O and Fe–O, see *Figure 8.2*), a binary 'CrO' is also destabilized in favour of the spinel and the very stable Cr_2O_3; one may well suggest that, electronically, the two systems become analogous, i.e. that V–Fe–O becomes 'Cr–O'-like. In Group VII, with Mn–O, the rocksalt-type oxide MnO has already recovered its stability, which could be reflected in a limited (Fe → V)O solid solution.

Generalizations of analogous ionization for *M*–O and M_1–M_2–O systems of balanced mean Group Number are justified in many cases.

It should be noted that the point defects in the metal ion lattice, to which the single-metal NaCl-type oxides are prone, have the natural tendency to provide host sites for second metals. These need not be divalent, and for instance in MgO trivalent ions have been observed, such as Cr^{+++}, Fe^{+++} and Ti^{+++}, associated with a vacancy. A large amount of work has been done on such point defects in NaCl-type oxides, their nature, pairing, distribution, diffusion and the effect on properties, especially as relevant in ceramics and oxidation. Magnesio-wüstite (MgO–FeO) is a case in point where structure and volume changes upon oxidation and reduction have been studied.

The CaF_2-type *dioxides* mainly dominate the actinide and Rare Earth series. They readily form mutual solid solutions (e.g. ThO_2–UO_2), but again the available information is incomplete. Purely ternary isomorphs also occur (see *Table 8.5*). In fact, the so-called 'stabilization' of ZrO_2 and HfO_2 means the formation of the cubic fluorspar structure towards lower temperatures by second-oxide additions, in ceramic practice mainly CaO and MgO. In principle, this stabilization is provided by the Groups to the *left* of Group IV, namely the Sc, Y, Rare Earth and Alkaline Earth oxides. Thus e.g. the zirconia–yttria system[126]—Y adjoins Zr in Group III—proves that already 6 mol. per cent Y_2O_3 suffice to produce the cubic ZrO_2 solid solution fully at all temperatures, extending up to 55 per cent Y_2O_3. The cubic structure of Y_2O_3 is already related to the CaF_2 version of ZrO_2, containing vacant O sites that can absorb oxygen from ZrO_2 and encourage such wide solid solutions. (Indirectly, this also reflects the scavenging effect of yttrium.) CeO_2, on the other hand, and the other Rare Earth dioxides, already of CaF_2 structure, are completely intersoluble with cubic ZrO_2 after first depressing the tetragonal/monoclinic transformation point[126]. Similar effects occur through scandia and other Rare Earth oxide additions, and such replacements can cause considerable thermal expansion changes[122b], among other effects.

However, the CaF_2 structure in ZrO_2 would also be stabilized by the actinide oxides (ThO_2–UO_2) which, of course, already possess this structure (*Figure 8.12*, Graph 4). This would, however, diminish the refractory properties, creating semi-metallic compounds which could become new substances of interest. Furthermore, upon adding the Group V and VI oxides to the CaF_2 types, the transition to the rutile structure can become either incipient or induced. Thus, a system such as ZrO_2–NbO_2, not studied yet (as against ZrO_2–Nb_2O_5 which has), should be of special interest, and similarly the transitions $ThO_2 \rightarrow TaO_2$, $UO_2 \rightarrow WO_2$, etc. For the ternary fluorspar-type oxides summarized in *Table 8.5* the degree of non-stoichiometry and ease of substitution has been expressed[41] by the general formula

$$A_K{}^{z+}B_{4-K}{}^{4+}O_{8-K/2(4-z)} \quad (K \text{ between 1 and 2})$$

where A = Ni, Co, Fe when B = Ti; A = Mn, Cr, V, Mo, W when B = Zr (properly reflecting the adaptation in ionic sizes and shift with Group number); z^+ is the valency of A. In the ZrO_2–UO_2 system, the non-stoichiometry has been established[91] in the range

$$U_{0\cdot9-0\cdot7}\,Zr_{0\cdot1-0\cdot3}O_{2\cdot04-2\cdot17}$$

(Further systematic trends traced[91] include a comparison with ThO_2–UO_2 solid solutions.)

The type of bonding position and the degree of ionization of the oxygen atom and thus, effectively, its size, clearly differ greatly between the above FCC (NaCl- and CaF_2-type) families, and so do the thermal and electrical properties. As the fluoride series also includes such widely varying members as the Rare Earth oxides (CeO_2, etc.), ThO_2, UO_2 and AmO_2, the purely ternary members (cf. *Table 8.5*) can no doubt also comprise much wider composition ranges, so that those listed are unlikely to be complete.

(5) *Spinels*—The double oxides of spinel-type structure (*Table 8.5*) constitute a large group and fundamentally one of the most interesting. They are of a diversity important in practice, as shown by their occurrence and use in ceramics, ferrites, various types of semi-conductor, in minerals and ores, in basic refractories and gemstones. They also occur in mill scale and other corrosion or oxidation products and protectants. The spinels become stable when there is only a moderate difference in the two metal ion sizes, in contrast to the greatly dissimilar sizes required e.g. in columbites, etc. and in di/trivalent combinations. Spinels are principally limited to the 1st Long Period; there are exceptions, however (cf. *Figure 8.12*, Graph 5). The members of the spinel family far exceed the transition-metal block (*Table 8.5*).

The spinels have been introduced already in the binary section but are basically ternary, with $Fe^{++}OFe_2{}^{+++}O_3$ as a singularity. The two types of metal-atom site can in some cases be inverted (*Figure 8.6*), to form pairs of type AB_2O_4 and A_2BO_4. This occurs in FeV_2O_4/Fe_2VO_4—and essentially, so it appears, only where one of the early transition metals (Ti or V) participates. This site interchange is not possible in all spinels. In the 'normal' AB_2O_4 spinel, the A^{++} ions always occupy the 8-fold, the B^{3+} ions always the 16-fold sets of position, with their octahedral environment of 6 oxygen atoms (cf. *Figure 8.5*). However, in the case of the 'inverse' spinels, e.g. $MgFe_2O_4$, there can be a redistribution[20] such that half the Fe^{+++} ions occupy the 8-fold positions, the other half plus the Mg^{++} ions the 16-fold ones. Structurally, the 'normal' spinels may thus be described as AB_2O_4, the partially exchanged ones as $BABO_4$. $ZnAl_2O_4$ is typical of the first, $FeMgFeO_4$ (or $FeTiFeO_4$) of the second. The ratio A:B can also vary in degree, with intermediate occupancies.

Another important feature of the spinels is that of *vacant* metal-atom sites, generally with random vacancy distribution. The cases of γ-Fe_2O_3/Fe_3O_4 and of γ/α alumina are classical examples, but the same can of course occur in two-metal spinels. A reduction to $21\frac{1}{3}$ Fe sites (instead of 24) can occur[19,127] in a defect Fe_2O_3 spinel, which gives an indication of the order of vacancy number that might be expected. The possibility of simultaneous oxygen deficiencies is not excluded.

A special application of the spinels is that they form one of the classes of 'ferrite', sought after for their unique magnetic properties. The general composition of spinel ferrites is AFe_2O_4, where A is any of the divalent atoms Mn, Fe, Co, Ni, Cu, Zn, Cd or Mg or a mixture of these (e.g. Mn/Mg or Ni/Zn). The degree of inversion here becomes important, and this, together with the magnetic properties, can be influenced critically by the manner of manufacture and by heat treatment. The magnetic properties of these spinels and their temperature transitions have been the subject of extensive research in recent years on V_2FeO_4 and on vanadium spinels and their properties[128–9]. Singificant effects here are those of anisotropy, or order–disorder changes, dependence of the properties on crystal growth and boundary energy, as well as that of migration of cation vacancies as a function of heat treatment.

As for the NaCl-type oxide structures (CoO, etc.), Fe-, Co-, Ni-bearing spinels are prone to anisotropic lattice contractions on cooling below the antiferromagnetic ('Neel') temperature. These effects are closely associated with the alignment of magnetic dipoles. Thus $NiCr_2O_4$ becomes tetragonally deformed[3c,d] below 30°C, from cubic symmetry above this, whilst Fe_3O_4 suffers a rhombohedral distortion below −160°C, $CoFe_2O_4$ a tetrahedral one below 180°C. Also in a series of solid solutions such as Fe_3O_4–$FeCr_2O_4$ the transition is sensitive to composition as well as to temperature. A low-temperature tetragonal distortion occurs for $FeFe_{2-x}Cr_xO_4$ (with $x \approx 1$) which, with increasing Cr, becomes orthorhombic; then, at $x = 2$, there is a reversion to the tetragonal form, but now with an inverted axial ratio ($c/a < 1$ instead of > 1 for $x = 1$). Neutron diffraction has been greatly utilized in recent years to determine the precise structure of these ferrites, and its change with temperature, e.g. above and below Curie point. It is thus possible to follow the degree of inversion and to locate the oxygen positions, as well as changes in magnetic structure, e.g. in magnanese ferrite[130]. Changes like that of conductivity frequently accompany those in magnetic properties of ferrites, often associated with anisotropy effects. This temperature sensitivity of anisotropy is of course not unique to the spinels and MO but applies to other oxide structures, such as the columbites, rutiles and perovskites.

It is instructive to refer back to *Figure 8.6* in relation to ionic size and to compare the regions of prevalence and overlap of the spinel structure $AO{\cdot}B_2O_3$ and of their close relatives AO and B_2O_5, i.e. the rocksalt and hematite types. Seemingly quite irregular spinels have been synthesized, such as $7ZnO{\cdot}Sb_2O_5$, though the electronic balance must obviously always be maintained[103a]. *Hexagonal* deviants to the cubic spinels have also been reported.

One further feature of some spinels may be noted: that, at the critical composition, AB_2O_4, they can become *supersaturated* in $B_2^{3+}O_3^{2-}$, with an incipient tendency to precipitate as hematite type. An example is provided by the $MgO-Al_2O_3$ system where a variation in the $MgO:Al_2O_3$ ratio from 1:1 to 1:7 can be accompanied by extensive metastable supersaturation. Similar phenomena are also common to the ferrite, chromite and other spinels. They are linked with precipitation effects of great significance which, in the example referred to, occur in 3 stages: (*a*) pre-precipitation (Guinier–Preston-zone type) lattice stresses; (*b*) a metastable phase of monoclinic distorted but still spinel-type lattice, and (*c*) precipitation of α-Al_2O_3. The significance of this type of mechanism, analogous to age-hardening effects in alloys, is quite a general one amongst oxide structures and would be particularly interesting to investigate for the suboxides when in contact with the metal and the higher oxides.

The tentative Fe–Cr–O diagram[105] shown in *Figure 8.16* highlights a point of principle. Whilst FeO coexists with Fe_3O_4, it ceases to do so beyond a certain minimum Cr content in the spinel because the dissociation pressure of FeO is (below 575°C) lower than that of $FeCr_2O_4$ but higher than that of

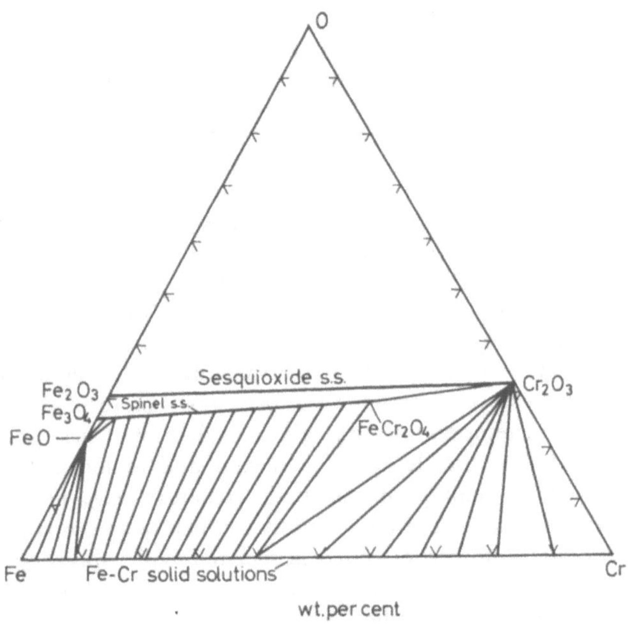

Figure 8.16. Cr–Fe–O *ternary diagram above* 575°C (From Woodhouse and White[105])

By courtesy of the British Ceramic Society

Fe_3O_4. A similar position arises in the Fe–Al–O system (spinel $FeAl_2O_4$), but with quantitative displacements[131]. Starting from Fe–Cr alloys, FeO-bearing oxide scales at low Cr contents give way to spinel ones for higher Cr, whilst, above 50 per cent Cr, both these are superseded in favour of Cr_2O_3. The NaCl-type → spinel → hematite-type transition is thus reflected by the phase transitions occurring in the oxide series based on Fe → Cr alloys.

It may be noted in *Figure 8.16* that the spinel → hematite transition in the oxide scales formed coincides with σ-phase occurrence on the Fe–Cr side.

This is not fortuitous and raises a basic question on a broader basis: do structural changes in oxide films formed upon binary alloys correlate with the incidence of specific intermetallic compounds (σ, Laves, etc.) in the base alloys themselves? Electronically and structurally this seems almost certain, but it would be interesting to assemble more systematic evidence.

(6) *Hematite and derivatives*—The members of the hematite-type (α-Fe_2O_3, Cr_2O_3, α-Al_2O_3, etc.) families are, in the main, binary-based and have been discussed already. Their intersolutions are extensive (e.g. Fe_2O_3–Cr_2O_3, V_2O_3–Ti_2O_3 complete) but with similar qualifications as to size and valency factors as before*. Thus, when the Group numbers separate, the structural distortions of the ilmenite type set in. $FeTiO_3$ and $MnTiO_3$ are typical (see *Table 8.6*). Structurally, these are relatively minor distortions from the original rhombohedral hematite unit cell, the 6-coordination of the cations being nearly maintained. Alternatively, the ilmenites can of course be considered as binary $A^{++}O{\cdot}B^{4+}O_2$ compounds rather than $A_2{}^{+++}O_3$, $B_2{}^{+++}O_3$ (e.g. $FeOTiO_2$—$(Fe,Ti)_2O_3$), so that in reality the state of ionization will be an intermediate and temperature-sensitive one. The whole subject again reverts to that of the ferrite materials, exemplified by the spinels, including the complex hexagonal deviants such as $(Ba,Sr,Pb)^{++}{}_{0\cdot6}(Fe,Cr,Al,Ga)_2{}^{+++}O_3$ and other (Ba,*M*)-modulated Fe_2O_3 where M = Mn,Fe,Co, Ni.

The double hematites and ilmenites are substantially linked to the 1st Long Period, but Nb- (probably Ta-, possibly Zr- and Hf-) bearing members also exist where there is considerable latitude in stoichiometry. The basic structure of Cr_2O_3 is still maintained, with a slight distortion after limited Mo or W addition.

The list of hematites in *Table 8.6* and in *Figure 8.12* is no doubt incomplete, and others very probably can be, or have been, synthesized in connection with ceramic applications. *Figure 8.12* readily suggests such possibilities. Titanium need not therefore be unique to the ilmenites, and extensions to higher Groups are probable; in fact, two ferrimagnetic ilmenites, $MnCoO_3$ and $MnNiO_3$, are known. In contrast to the straight hematite series with *random* cation arrangement (*Table 8*, No. 6.1), we can also have *ordered* hematites. This leads to the niobate/tantalate $A_2B_4O_9$ variants[133] (6.2). $Nb_2Co_4O_9$ and $Nb_2Mn_4O_9$ are antiferromagnetic (no doubt others), with critical temperature dependence. The disorder ↔ order transition clearly reflects itself again amongst the rutile and columbite derivatives upon oxygen addition from 60 to 66·7 at. per cent O (*Table 8.6*, Nos. 7–9), but with modified metal ratios.

* The Fe_2O_3–V_2O_3 system (studied by neutron diffraction and magnetic methods) typifies some of the chief effects. Fe_2O_3 dissolves V_2O_3 almost completely at 1,000°C, and there is probably complete solution at higher temperatures[132]. Up to 40 per cent V_2O_3, the magnetic behaviour is as in hematite, then weak ferrimagnetism sets in, increasing to a maximum at 65 per cent V_2O_3, and from 85 per cent the properties are V_2O_3-like. These changes are accompanied by a steady change in magnetic structure up to 60 per cent V_2O_3 in Fe_2O_3, then followed by a transition from long- to short-range order. Here again, a weak tendency to ilmenite-type configuration is suggested as a partial reason, as occurs more pronouncedly in the Fe_2O_3–$FeTiO_3$ system. The FeO–TiO_2 system contains a further stable compound, Fe_2TiO_5, pseudo-brookite, which varies structurally but is still related to ilmenite.

Starting from the basic hematite, extensive deviants in ABO_3 structures exist; these are typified and revealed in opposite Periodic Table directions from the transition-metal block by (1) Sc_2O_3 which becomes cubic, Y_2O_3, La_2O_3 and the [Rare Earth]$_2O_3$, and (2) Ga_2O_3, In_2O_3, Tl_2O_3; As_2O_3. Stoichiometrically, the *perovskite* family ($CaTiO_3$, etc.) of course belongs here too, and although their structure is distinct, the link with the hematites is still of interest, as part of the broader pattern of metal substitutions.

The hematite → ilmenite → perovskite series of ABO_3 compounds clearly follows increasing atomic size ratios of the metal partner atoms.

The crystal chemistry of the oxide families with trivalent cations has been surveyed by Schneider, Roth and Waring[134]. *Figure 8.17* shows the distribution of stability regions for various ABO_3 structure types, as a function of

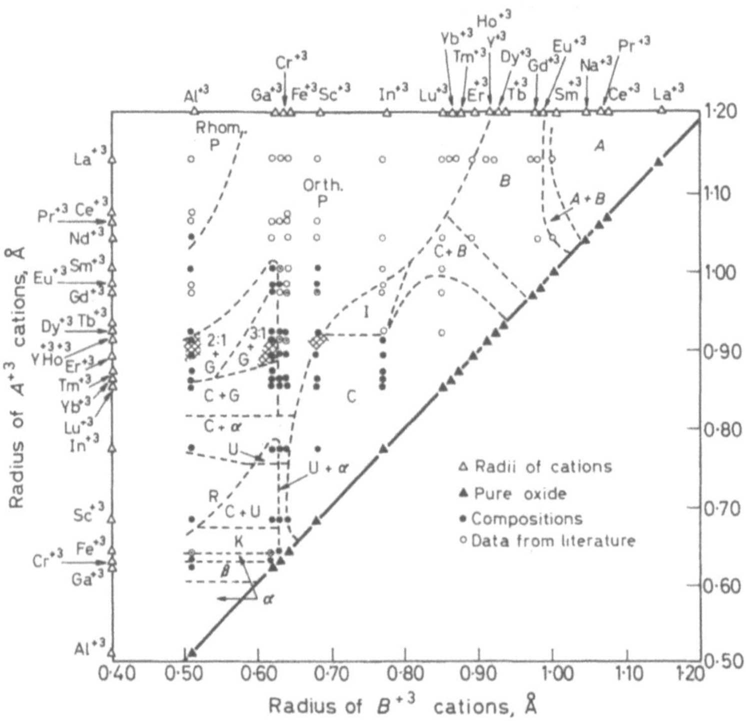

Figure 8.17. Occurrence of phases of ABO_3-type structures with varying A^{+++} and B^{+++} cation radii (From Schneider, Roth and Waring[134])

A—A-type, B—B-type, C—C-type: Rare Earth oxide
G—garnet
I—unknown type
α—corundum
β—beta gallia
K—kappa alumina
U—unknown type similar to kappa alumina
P—perovskite
R—unknown type
2:1—unknown type
3:1—unknown type

By courtesy of the National Bureau of Standards, Washington, D.C.

radii of the two cations (including Fe_2O_3, Cr_2O_3, Al_2O_3, Ga_2O_3, In_2O_3, Sc_2O_3, Y_2O_3); the structures are indicated. The work included the trivalent Rare Earth oxides as a major item, and the *A*, *B* and *C* types of the oxides shown refer to the following series of related polymorphous forms which can be adopted by them[135]:

$A = La_2O_3, Ce_2O_3, Pr_2O_3, Nd_2O_3$

$B = Sm_2O_3, Eu_2O_3, Gd_2O_3$

$C = Tb_2O_3, Dy_2O_3, Ho_3O_3, Er_2O_3, Tm_2O_3, Yb_2O_3, Lu_2O_3$
and their intersolutions.

Although each of these oxides has only the single stable type polymorph shown, the other forms can occur *metastably* at lower temperatures for the alternative groups. This behaviour is very similar to that found[16d] for Nb_2O_5, and as a general principle such metastable $\leftrightarrow$ stable polymorphs with consequent irreversible transitions are a characteristic feature of many of the higher oxides; by addition of further solutes, these transitions can be varied to occur at different temperatures and rates. The intersolutions between these Rare Earth oxides and the conditions favourable to transition to the perovskites have also been described by Schneider and Roth's[135] comprehensive work at the National Bureau of Standards which should be referred to for details (cf. Bibliography).

The *perovskite* family (prototypes: $CaTiO_3$, $BaTiO_3$, $NaNbO_3$) always contains one non-transition metal. The unit cell is basically cubic and very simple (*Figure 8.18*). It is, however, liable to tetragonal, orthorhombic or higher distortions towards lower temperatures which are also a function of composition. These have been investigated in detail[136] (cf. Bibliography), owing to the important dielectric properties which they possess. They too are associated with considerable degrees of non-stoichiometry. Such variations can consist in the progressive vacation of the large atom (body-centre) site to give an ultimate unit cell as shown in *Figure 8.18*, right-hand side. This is that of ReO_3, and intermediate homologous series will occur, constituting the so-called 'tetragonal bronzes' such as the tungsten bronzes Na_xWO_3 ($x < 1$) or the Nb or Ta bronzes which are indeed of semimetallic appearance as well as behaviour.

The perovskite family contains numerous members, and the extent of multiple substitution of ions in the lattice is best illustrated diagrammatically rather than listing it here (*Figure 8.19*, intended[137] to represent only the general trends because not all combinations necessarily form perovskites). Useful details of perovskite properties, structure and tolerances for ion replacement are now available, and many quaternary and higher perovskites can readily still be predicted. The anion replaceability of F, (OH), etc. for oxygen is an interesting feature where readjustment of electron balances amongst the metals is clearly required. As in the spinels, 'inverse' perovskites can occur, with metal ions partly or wholly exchanging positions. For further references on perovskites and associated phases, such as the bronzes, cf. Bibliography.

Figure 8.20 gives one of Megaw's[136] classical results regarding the perovskites, namely the structural variation in $BaTiO_3$ with temperature, showing the transition to low-temperature anisotropy—a research which set in train many of similar type. Structural sequences such as for $NaNbO_3$, monoclinic $\rightarrow$ pseudo-tetragonal $\rightarrow$ tetragonal $\rightarrow$ true cubic, are typical[138].

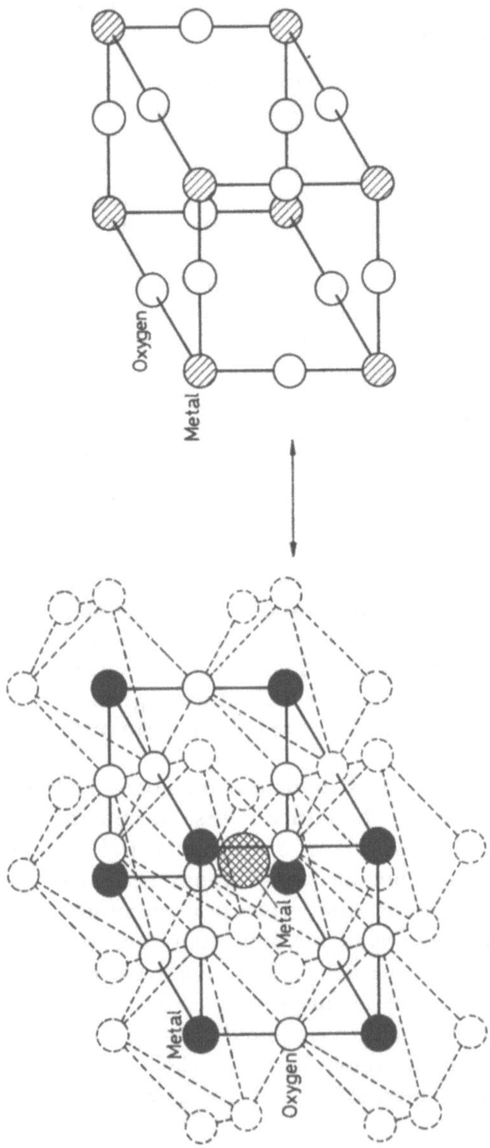

Figure 8.18. The unit cells of perovskite-type and ReO_3*-type niobium bronzes formed through intermediate occupation* (From Goldschmidt[16c])

By courtesy of the Editor, *Metallurgia*

Other types of variation can occur in any one two-metal system containing a perovskite. Thus, for example, in $CaO–TiO_2$ there exist, apart from the plain cubic $CaTiO_3$, two closely related phases $4CaO{\cdot}3TiO_2$ and

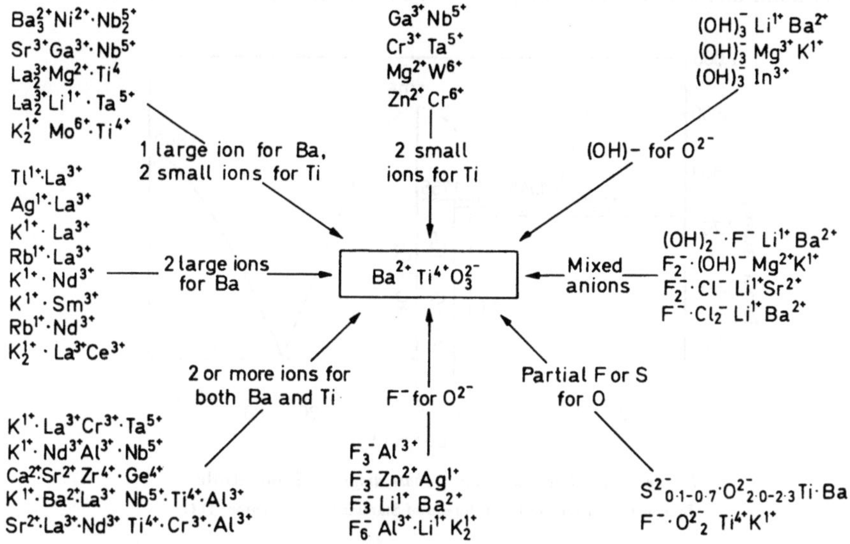

Figure 8.19. Types of multiple ion substitution in the perovskite lattice (From Roy[137])
By courtesy of the American Ceramic Society

$3Ca{\cdot}2TiO_2$. Although distinct, i.e. they melt peritectically as against the congruent melting point of $CaTiO_3$, the phases differ mainly in the degree of pseudo-tetragonal distortion and in the multiplicity of unit cells (phase

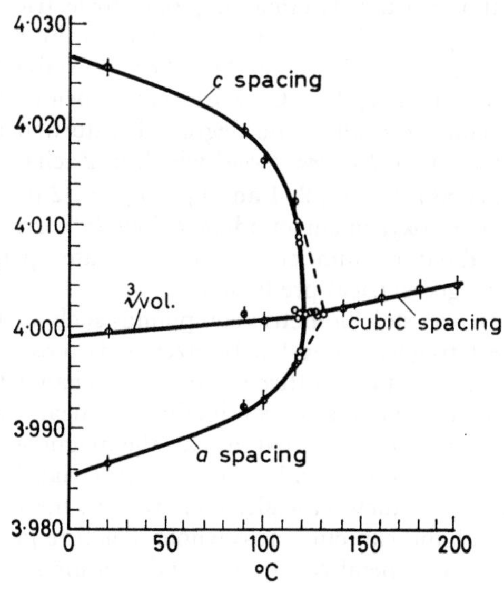

Figure 8.20. Transformations and spacing variations with temperature of the perovskite structure in $BaTiO_3$ (From Megaw[136])

Vertical lines through points = limits of experimental error

By courtesy of the Royal Society

diagram[80,81b]: *Figure 8.21*). Thus, the 4:3 compound has *triple*, the 3:2 compound *double* perovskite layers, each interleaved with CaO layers[81b]. We may regard the types of (*a*) temperature distortion (*Figure 8.18*), (*b*) compositional distortion (the three phases of *Figure 8.21*) and (*c*) defect structure

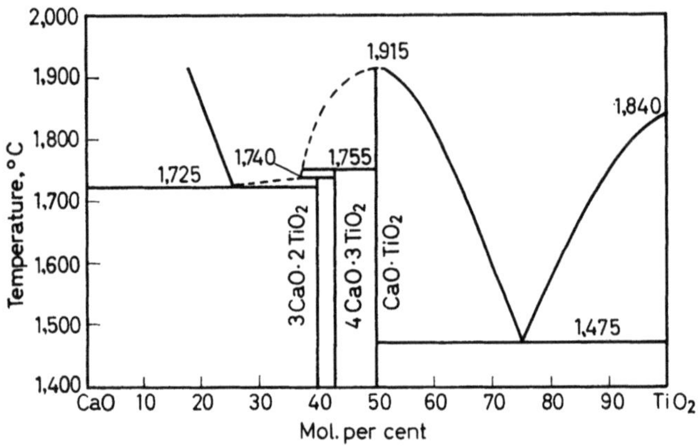

Figure 8.21. The CaO–TiO_2 *phase diagram* (From Roth[81b])

By courtesy of the National Bureau of Standards, Washington, D.C.

sequence in the perovskite → bronzes → ReO_3 series, as three manifestations of the wide structural adaptability of the perovskite lattice. This further expresses itself in (*d*) the multiple ionic substitutions possible (*Figure 8.19*). *Figures 8.19, 20* and *21* thus reflect in different ways the same basic phenomenon. This is perhaps the most pronounced in the perovskite form, but it applies quite generally to the higher ionic oxides, although with some modifications. This is clearly of basic practical consequence in connection with the lattice dynamics e.g. of ferroelectrics and the oxidation mechanism of alloys in which such phases can occur in scale layers.

In general, the perovskites become stable for a radius ratio $r_B/r_X > 0{\cdot}41$ when in $ABX_3(X = \text{O,OH,F, etc.})$; below this and for r_A/r_B nearer unity, ilmenites prevail. The degree of distortion in the perovskites, however, is subject to a '*tolerance factor*' which in Zachariasen's early work[22b] was established as $r_B/r_X > \sqrt{2}-1$ and $r_A + r_X = \sqrt{2}\,(r_B + r_X)$, for octahedral coordination of oxygen surrounding *B* ions (a correction being necessary when the coordination number increases). Later perovskite findings verified these early geometrical predictions.

In non-stoichiometric compounds such as those of the perovskite type and the tetragonal tungsten bronzes, it is possible, by 'manipulating' the ionic replacement and lattice vacancies, to change the electrical conductivity from near-metallic to semi-conducting and near-insulating. This can be critically temperature-dependent, as can the magnetic and other properties. A whole new and fruitful field of ceramic materials has thus been developed. Quite clearly, in these basically ionic phases, the borderline to metallic bonding is broad and can cut across whole structure groups. In ternary compounds of transition metal/*B* Group metal/non-metal combinations, the validity of a kind of 'Hägg's rule' can be shown among the perovskite-type compounds[139], but with a limiting ratio somewhat higher than 0·59; when it exceeds 0·65, coordinations more familiar in the borides arise.

A further ABO_3 deviant class are in the *garnets*, of complex cubic structure with 8 molecules per unit cell and general composition $A_3^{2+}B_2^{3+}C_3^{4+}O_{12}^{2-}$. Originally[140], only silicate members were known (i.e. 'C' = Si, e.g. $Fe_3Al_2Si_3O_{12}$). However, multiple synthetic replacements have since become possible, in defined lattice sites[122c,141–2], by partially substituting divalent and tetravalent ions. The Rare Earths are here prominent, and compounds such as $Y_3Fe_2Fe_3O_{12}$, $Y_3Al_2Al_3O_{12}$ or $Y_3Fe_{5-x}Ga_xO_{12}$ (yttrium iron garnet) have been prepared. The 'Rare Earth–iron garnet' and derivatives have now become important for their magnetic properties, some embodying Ga for Al, Ge for Si, etc. (a good review has been given by Geller[122d]).

Some examples, not exhaustive, of more recent work on hematites and their derivatives are referred to in the Bibliography.

(7)–(9) *Rutiles, columbites and ABO_4 deviants*—The double dioxides of this class form once more a consecutive set of related structures caused by the increasing radius difference of A and B ions in $(A,B)O_2$. All binary oxides which are straight rutile, as in *Tables 8.1* and *8.2*, will form complete solid solutions, but limited by major ion size and electronic differences. Two main types of structural deviants occur, each with more refined subdivisions: the ABO_4 ($CrVO_4$, wolframite, $FeWO_4$, and scheelite, $CaWO_4$) classes and the A_2BO_6 (columbite or tantalite) class [(*Table 8.6*, Nos. (8) and (9)]. Both can be considered to be progressive distortions from the rutile parent. Viewed overall, the two sequences develop, with increasing cation-size differences, as follows:

I		II	
$(A,B)O_2$ rutiles (including trirutiles)	$(A,B)_2O_3$ hematites	$(A,B)_3O_4$ spinels	$(A,B)O$ rocksalt type
↓	↓	↓	↓
wolframites	ilmenites	[no strict analogies: drastic change in composition and structure when radius and valency differences become excessive]	
↓	↓		
columbites	perovskites, garnets, etc.		
↓			
scheelites			
↓			
zircons			

Variations in form occur, depending on relative Group and Period numbers. Nevertheless, there is an overlap, and polymorphic forms, stable at high and low temperatures, appear, such as rutiles or columbites. Such transitions cease on reversion to the metal-richer oxides, as the spinels and MO under II, or quite change in character.

Amongst straight rutile solid solutions we have[143] those of TiO_2–NbO_2, TiO_2–MoO_2, NbO_2–CrO_2, NbO_2–MoO_2. *Figure 8.22* shows the change in lattice dimensions—unit cell volumes and axial ratio—for some mixed rutile series.

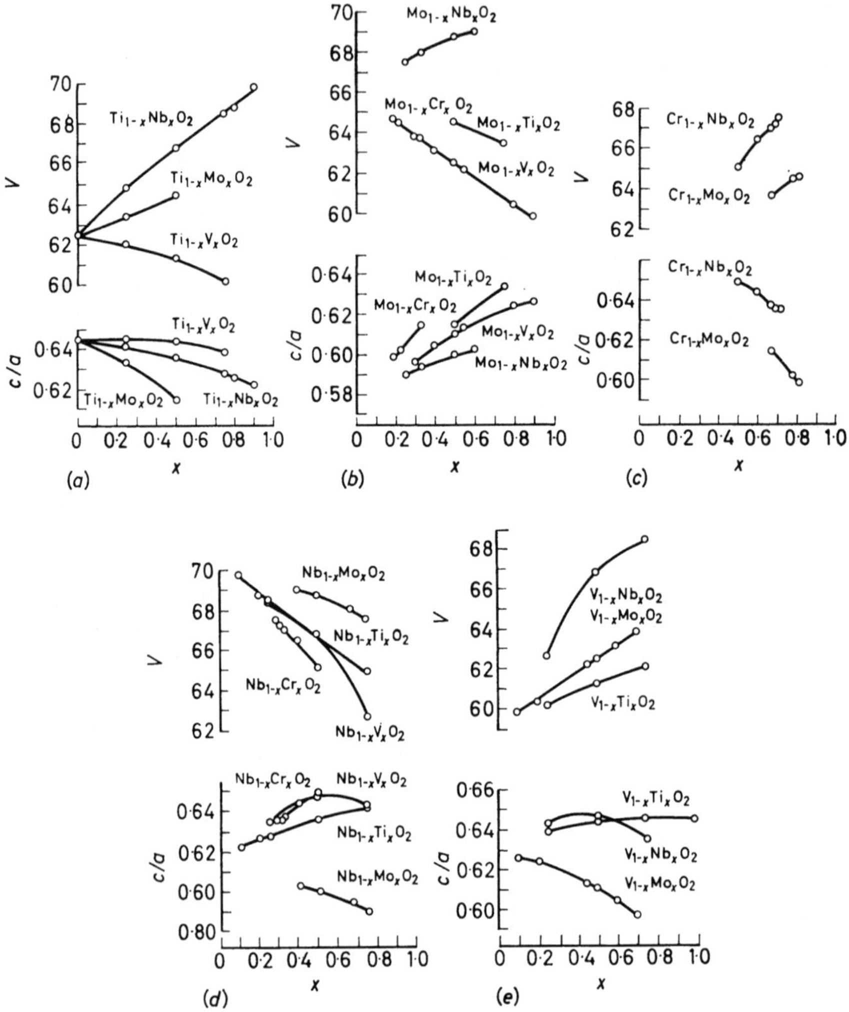

Figure 8.22. Lattice dimensional changes within ternary rutile solid solutions (From Marinder *et al.*[143])
(Unit cell volumes, A^3, and axial ratios, c/a)
By courtesy of Messrs. E. Munksgaard, Copenhagen

Both in the rutile and in their derivative structures, considerable deviations occur from the ideal $(A_xB_y)O_2$ compositions, with defect lattices and $(A + B)/O$ as well as x/y tolerances, depending on radius ratio and electron concentration. Thus some of the work[16c] on systems between Nb_2O_5 and second oxides has shown the prominent stabilization of rutiles and of columbites by the later metals of the 1st Long Period, namely by Ni, Co, and Fe and also by Cr, V, Ti (no doubt also by Mn) for rutile only; by Cu and Zn for columbite only; and by Mg for both. The resulting (schematic) diagrams are shown in *Figure 8.23.* These bring out several points, the most relevant being the dimorphy of rutile/columbite with temperature; this disappears

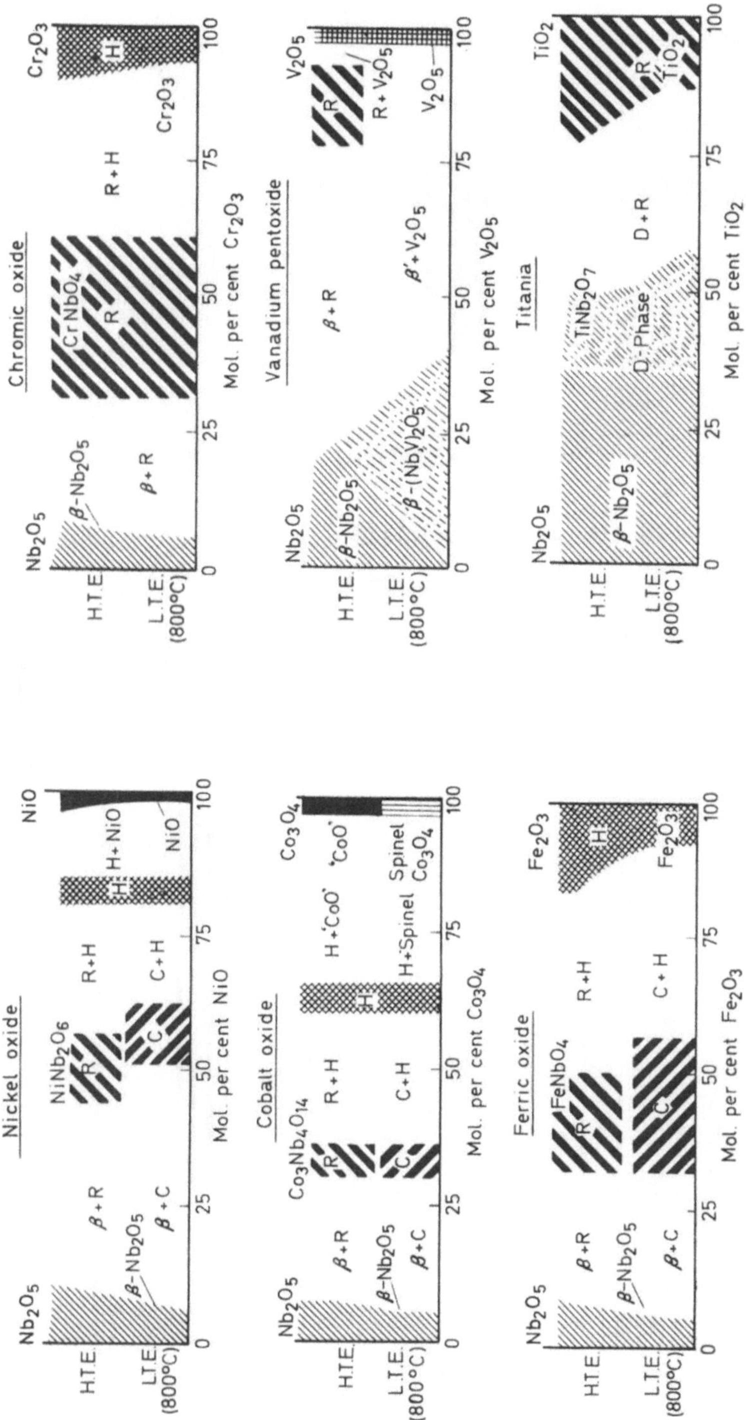

Figure 8.23. Constitutions of systems Nb_2O_5 + *additional oxides, in order of Periodic Table* (From Goldschmidt[16c]).

By courtesy of the Editor, *Metallurgia*

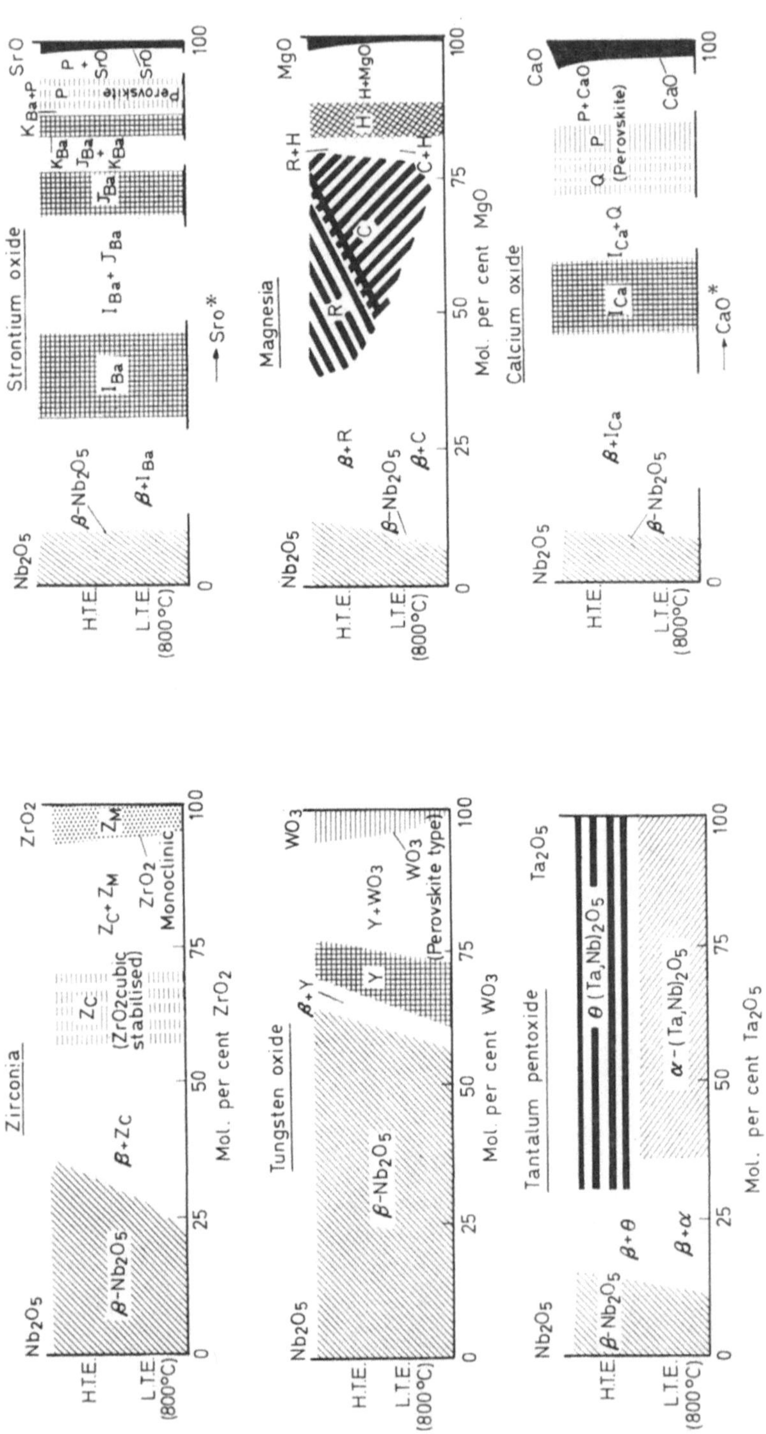
Zirconia
Tungsten oxide
Tantalum pentoxide
Strontium oxide
Magnesia
Calcium oxide
H.T.E.
L.T.E. (800°C)
Mol. per cent ZrO2
Mol. per cent WO3
Mol. per cent Ta2O5
Mol. per cent MgO
β-Nb2O5
(ZrO2 cubic stabilised)
ZrO2 Monoclinic
WO3 (Perovskite type)
θ (Ta,Nb)2O5
α-(Ta,Nb)2O5
(Perovskite)

Figure 8.23b

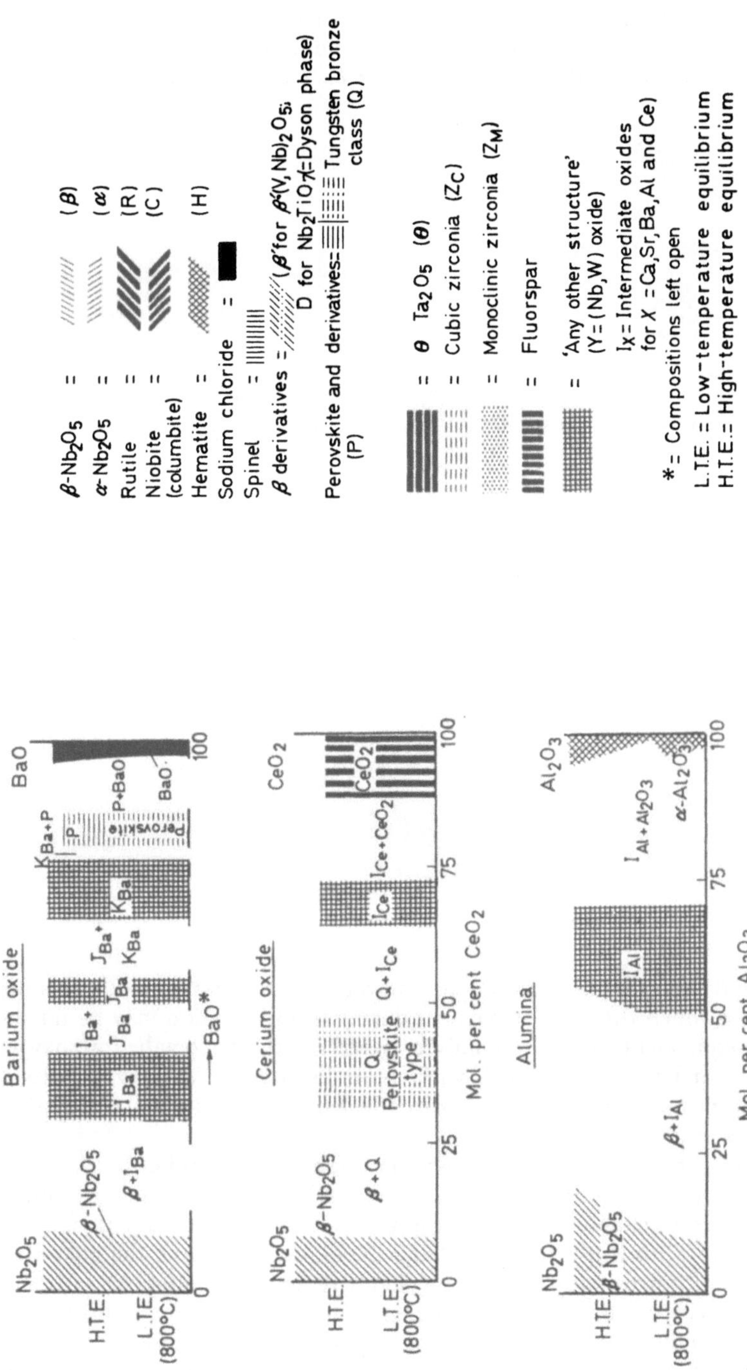

Barium oxide
Nb2O5
BaO
H.T.E.
L.T.E. (800°C)
β-Nb2O5
β+IBa
IBa
IBa+JBa
JBa
JBa+KBa
KBa
KBa+P
P
Perovskite
P+BaO
BaO
→BaO*
0
100
Cerium oxide
Nb2O5
CeO2
β-Nb2O5
β+Q
Q Perovskite type
Q+ICe
ICe
ICe+CeO2
CeO2
0
25
50
75
100
Mol. per cent CeO2
Alumina
Nb2O5
Al2O3
β-Nb2O5
β+IAl
IAl
IAl+Al2O3
α-Al2O3
0
25
50
75
100
Mol. per cent Al2O3
β-Nb2O5 = (β)
α-Nb2O5 = (α)
Rutile = (R)
Niobite (columbite) = (C)
Hematite = (H)
Sodium chloride =
Spinel =
β derivatives = (β' for β'(V,Nb)2O5; D for Nb2TiO7(=Dyson phase)
Perovskite and derivatives (P) = Tungsten bronze class (Q)
= θ Ta2O5 (θ)
= Cubic zirconia (ZC)
= Monoclinic zirconia (ZM)
= Fluorspar
= 'Any other structure' (Y=(Nb,W) oxide)
IX = Intermediate oxides for X = Ca, Sr, Ba, Al and Ce)
* = Compositions left open
L.T.E. = Low-temperature equilibrium
H.T.E. = High-temperature equilibrium

Figure 8.23c

towards Groups VI, V, to favour rutile only. These mixed niobates are not tied to the $(A,B)O_2$ stoichiometry, and the metal/oxygen and metal/metal homogeneity ranges are wide; they also include proper the dioxide series, as from NbO_2 rather than Nb_2O_5. *Figure 8.24* compares the structures of

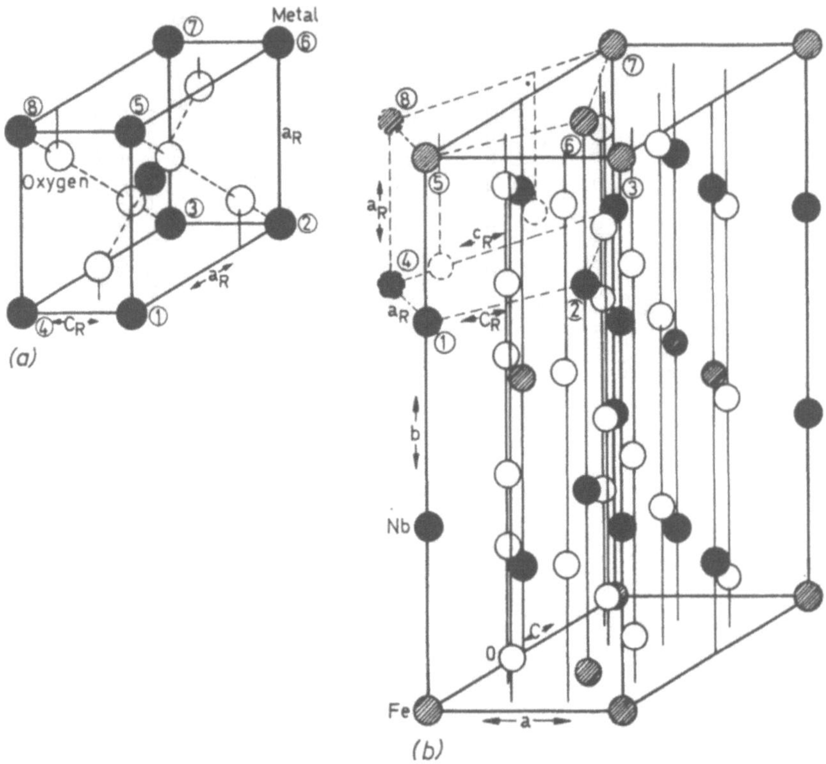

Figure 8.24. Comparison of (a) rutile and (b) columbite structures (From Goldschmidt[16c])
By courtesy of the Editor, *Metallurgia*

rutile and columbite, namely their tetragonal and larger orthorhombic cells, from which the atom movement during transformation may be traced; a rutile-equivalent cell in columbite is indicated, the octahedral oxygen environment being maintained, with some distortion. Varying degrees of order are possible, which are liable to affect the rate and structural path of the transformation.

Further variants are the 'ilmeno-rutiles', tapiolites, trirutiles (of 3 × unit cell) and others, but these may be generically included under rutiles, as they are very closely related. Similar subdivisions occur among the columbites. Thus Roth[81c] reports that $FeNbO_4$ does not have the columbite structure but forms two polymorphs closely resembling columbite, a monoclinic low-temperature variety changing near 1,010°C to an orthorhombic one of columbite with $\frac{1}{3}$ its b axis and a possible order → disorder change. The rutile forming at high temperatures represents[81c] a decrease in Fe; this, however, can be well reconciled with the extensive rutile homogeneity ranges characteristic of these phases (see *Figure 8.23*), which are associated with considerable changes in spacing.

These transformations are of considerable practical importance, as are the implied volume changes, defect lattices and ordering effects which influence adhesion and diffusion rates when present in a surface scale, and the physical properties of rutile ceramics.

Linked with the above systems, TiO_2–Nb_2O_5 and ZrO_2–Nb_2O_5 diagrams[81a], including several new compounds, are shown in *Figure 8.25*.

The *wolframites*, $FeWO_4$, may be considered as an intermediate stage, transitional to rutiles and columbites. They are of monoclinic symmetry and of ordered structure. However, a mineral form, 'ixiolite', has been reported, ABO_4, having the wolframite structure but *disordered*[81c,144] (see also members in *Table 8.6*, No. (8)).

A further evolution in ABO_4 structures for more widely separated Group Numbers, passing outside the transition-metal block, is that to the *scheelite* group ($CaWO_4$, $BaWO_4$, $PbWO_4$, $KReO_4$ and many others, linking on to $MgWO_4$, the wolframites). This is favoured by radius ratios $r_A/r_O = 0{\cdot}65$–$0{\cdot}85$, $r_B/r_O = 0{\cdot}44$–$0{\cdot}52$; a further size decrease of the B ion leads directly to the sulphate structures of $CaSO_4$, etc. The tolerance factor in the scheelites is very large indeed, akin to that in the perovskites for ABO_3.

From here we pass, for smaller B ions of the same valency, to the $CaCrO_4$ structure which is already that of zircon, $ZrSiO_4$, with a lead-over to the silicates. A further typical deviant, to the *columbites*, is La_2MoO_6, containing a tetragonal layer lattice with alternating sheets of La_2O_2 and MoO_4 groupings. Additional subdivisions and deviations are derived from the brookite and anatase forms of TiO_2 itself, from MnO_2, ReO_2, etc. with numerous rutile-like modifications. The dimorphy in germanium dioxide, GeO_2 (rutile and quartz types), provides an interesting development to silica. Either transformation can be promoted or prevented by judiciously chosen two-anion substitutions.

Among the ABO_4 oxides of Nb and Ta have been mentioned the fergusonites (as wolframite variants) which are of special interest[145]. These are Group III and Rare Earth niobates and tantalates, such as $LaNbO_4$ or $YTaO_4$. A new type of trirutile, $A_2^{+++}B^{+}O_6$ (e.g. Cr_2WO_6) has also been reported, against the normal $A^{++}B_2^{5+}O_6$ (e.g. $CoTa_2O_6$), with ion replacements by Al,Fe,Te. Inversions as in spinels take place, with some readjustments.

Recent work[3e] on many synthetic ABO_4 compounds between (*a*) the Rare Earth and Group III oxides and (*b*) Nb_2O_5 or Ta_2O_5 (e.g. $LaNbO_4$, $YNbO_4$, etc.) has detailed the structural and spacing changes, including distortions and transformations at high temperatures.

Many specialist investigations have been made in connection with the minerology of Nb, Ta and related *ores*. Amongst these the columbites are prominent, e.g. in the huebnerite–ferberite series $MnWO_4$–$FeWO_4$ or the columbite–tantalite series $FeNb_2O_6$–$FeTa_2O_6$ (see also Bibliography).

Lastly, we should recall the *fluorspar*-type structure, which is also a dioxide, so that any system such as TiO_2–ThO_2 or stabilized ZrO_2 is bound to show either transitional rutile/fluorspar forms or a straight duplex equilibrium.

Little appears to be known about this, except as noted in *Table 8.5*. However, we may well consider the CaF_2-type compounds, (Zr,Cr)O_2, (Zr,Mo)O_2 and (Zr,W)O_2, as 'stabilized ZrO_2' transitional to the rutiles of CrO_2, etc. In the system TiO_2–ZrO_2, a compound $TiZrO_4$ was found[80] (orthorhombic, with possibly two polymorphs). The diagram is included in *Figure 8.25*, together with TiO_2–NbO_2 for comparison. The 'CaF_2-type' compounds (Ti,Fe)O_2, (Ti,Co)O_2 and (Ti,Ni)O_2 (see *Table 8.6*) have not been confirmed and are suspect, as the Group VIII metal atoms are too small.

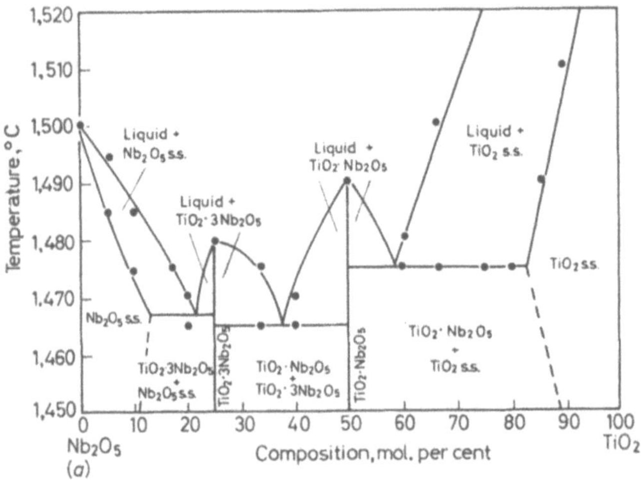

Figure 8.25a. Comparison of the systems TiO_2–Nb_2O_5; ZrO_2–Nb_2O_5; TiO_2–ZrO_2 (From Roth and Coughanour[81a])

(s.s. = solid solution)

By courtesy of the National Bureau of Standards, Washington, D.C.

The compound $CrUO_4$ in the Cr_2O_3–UO_2–O_2 system[104] is of some special interest, as here we have a double oxide based on metals from the same Group. The stoichiometric end members of the pseudo-binary section (CrO_2–UO_2) on which this compound lies are of the rutile and fluorspar structure, and the orthorhombic $CrUO_4$ must evidently be a stage intermediate to these two. In Group IV we had observed a similar case for $TiZrO_4$*. The general question arises whether similar ternary structures tend to interpolate between oxide end members forming the same structure pairs, in this case, the rutile and fluorspar types. The data are insufficient to generalize, and variations with Group number and valency are likely. The phase $TiZrO_4$ has been reported[80] as of orthorhombic symmetry with $a = 4{\cdot}806$, $b = 5{\cdot}032$, $c = 5{\cdot}447$ Å, which could be comparable with the dimensions of $CrUO_4$, assuming half the unit cell. However, a polymorphous change also occurs, and more detailed structural data are needed to test this.

* Unfortunately, the more comparable system TiO_2–ThO_2 does not appear to be known, and a hypothetical compound '$TiThO_4$' could well exist. Equally, the compounds '$VPaO_4$' or '$NbPaO_4$' in Group V are worth searching for.

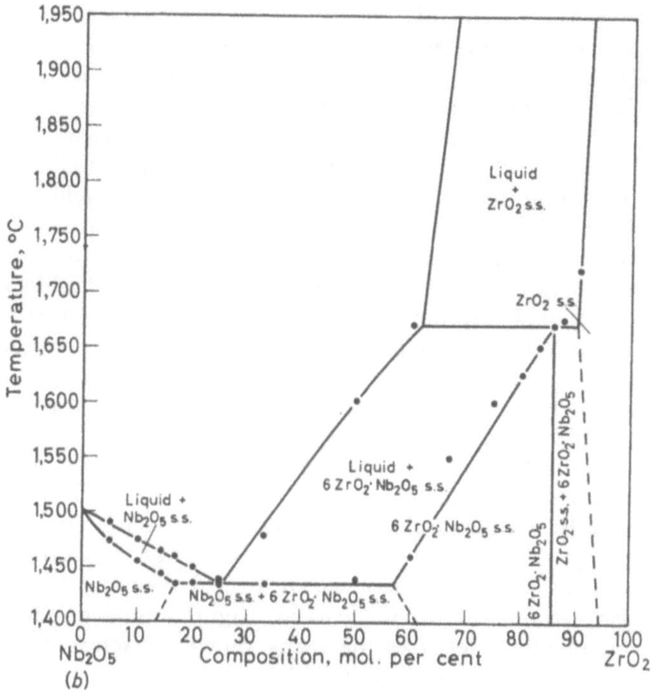

Figure 8.25b

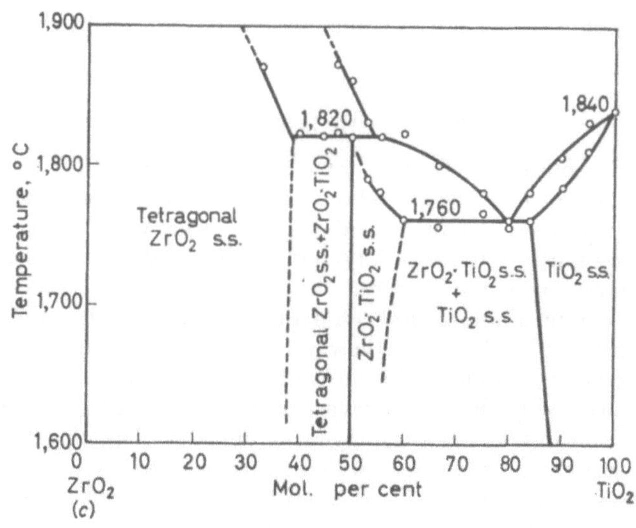

Figure 8.25c

In the V–Y–O system a rhombohedral phase $UO_{2.9}$ $6YO_{1.5}$ is known[146], representing only one of the no doubt numerous further double oxides involving Group III and the Rare Earths. Work on the V–Y oxide system[147] (WO_3–Y_2O_3) is of connected interest.

Double oxides higher than $(A,B)O_2$

The single-metal pentoxides of Group V (V_2O_5, Nb_2O_5, Ta_2O_5), the trioxides of Group VI, the higher Group VII oxides (e.g. ReO_3) and the remaining higher transition-metal oxides have been discussed in Section 8.2. When combined, these systems open into a multiplicity of ionic substitutions and structure changes. A new branch of inorganic chemistry and new types of ceramic materials based on it are rapidly developing, and a complex of inter-solid solution series and of new compounds arises. A further impetus to research on these high-oxygen inter-oxide systems has been given by the development of highly refractory alloys (e.g. Nb,Ta,Mo,W base) and the complex oxide layers formed thereon, protective or otherwise.

As an example, tetragonal phases occur[148], of general formula $M^{I}_2O_5$ $9M^{II}_2O_5$, which appear to be isostructural with Mohanty's $Ta_2O_5{\cdot}2Nb_2O_5$; furthermore, a new V_2O_5–Nb_2O_5 compound is likely to be a prototype to other metal combinations.

The non-stoichiometries, defect lattice effects and homologous series of closely related structures noted amongst the single-metal–O systems now evolve on a greatly extended scale. We should also take the ionic substitutions beyond the transition-metal block (Al^{+++}, Si^{4+}, Mg^{++}, alkalis, alkaline earths, rare earths, etc.) and more widely regard them in their entirety. This has in one direction led to new electrical, magnetic and high-temperature ceramics, and in others leads to the many known refractory materials, to natural silicates and to mineralogy itself (cf. Bibliography).

It is perhaps debatable whether these deviations go beyond our terms of reference, 'interstitial alloys': but, viewing the subject as a whole, the survey would be unbalanced without reference to the complex refractory oxides.

8.5 APPLICATIONS

The practical applications of oxides are so numerous and so diverse that only a brief synoptic sketch can be given here, apart from the existence of authoritative texts, reviews and papers in the special fields which make a fuller account unnecessary. Some applications of specific oxide systems have been referred to in the preceding Sections.

The strictly 'interstitial' oxides, comparable to the carbides and nitrides, are much more restricted in scope, however. The metallic sub-oxides such as occur at interfaces or under reducing conditions are perhaps the only true equivalents, because the higher oxides are set apart by progressive ionization. This does not mean that the metallic or near-metallic character then necessarily becomes lost, and the whole twilight region of semi-conducting oxides, ferroelectrics, magnetic ceramics, etc. can still justly be classed as 'interstitial', although in a modified sense. The boundary between metallic and ionic bonding becomes blurred over wide composition, temperature and pressure ranges and, indeed, from this very fact stem some of the most important applications of oxides.

The concept of 'interstitiality' transgresses the metallic/ionic boundary. It obtains a new significance in oxides when we consider the role of lattice defects of Wagner, Frenkel and Schottky types, and of vacancy movement, implied e.g. in *n*- and *p*-type semiconductors and in oxidation processes. Also the concept of transition as against non-transition metals loses much of its distinctness as atoms become ions. Thus when we pass from metallic to ionic oxides, many transition and non-transition metal ions (e.g. $Fe^{++} \leftrightarrow Mg^{++}$, $Cr^{+++} \leftrightarrow Al^{+++}$) become substituents on similar footing, where the *metals* had been unlike and not replaceable.

Table 8.7 attempts a systematic summary of the main applications of oxides. In basic principle, they can be divided according to their three modes of occurrence: (*a*) in bulk; (*b*) in layers and (*c*) in dispersion. Here (*a*) represents the obvious role of oxides used independently for their mechanical, chemical or special ceramic (thermal, electrical, etc.) properties. A further cross-division is that into 'natural' (rocks) and 'synthetic' oxide types (modern special-purpose ceramics), the first including their role as the 'storehouse' for metals in ores and minerals.

Table 8.7. Applications of oxides

	In volume		*In layers metal contacting*	*In dispersion*	*Main function or use*
	natural	*synthetic*			
1	Most of earth's crust	principal condition strength or strengthenability			mechanical bases; foundation for life
2	*processed*				building materials of all kinds
3	Minerals, ores	principal condition chemical reducibility			'storehouse' for metals
4		refractory materials, acid and basic; principal condition non-reactivity with metals			crucibles, furnace linings, etc.
5		slags; principal condition partial reactivity with metals			metal reducants and protectants
6		lower oxides as reducing means, by forming higher ones	in volume, layer and dispersed form		use of nucleation of metals and low oxides in higher oxides, inverse to oxide dispersions in metals
7		glasses			multiple
8		ceramics: high temperature; principal condition refractoriness, with other attendant special-property requirements; Al_2O_3, SiO_2, ZrO_2, ThO_2, etc.			as solid refractories, linings, coatings, dispersoids, etc.; in 'Pyroceram' towards (7)
9		ceramics for special physical properties		mainly in volume, but in some cases layered or dispersed	ferrites, dielectrics, insulation, semi-conductors, magnetic ceramics; cutting ceramics; luminescent oxides

	In volume natural	synthetic	*In layers metal contacting*	*In dispersion*	*Main function or use*
10			thin films and scale layers on metals		protectants against oxidation or chemical attack (e.g. Cr_2O_3, $FeCr_2O_4$ on stainless steels, $NiCrO_4$ on 'Nichromes', etc.)
11			(synthetic) protective coatings on metals		e.g. at room temperature against corrosion, at high temperatures for oxidation resistance
12			paints, enamels, etc.		protection or decoration
13			oxides as means to anchor compatible phases (e.g. carbides) to metal bases		as transition-layer compounds
14				oxides as dispersion strengtheners	dislocation locking, crystal-growth inhibition, etc.
15			oxides formed during internal oxidation of metals		
16	Metal ceramics (oxide component); belong to all groups: bulk as solid material; layer in interface bondings; dispersion in distribution				high-temperature, nuclear energy and other uses
17			catalyst oxides		to aid multiple reactions (syntheses or decomposition, e.g. N_2O, CO etc.); chemisorption reactions
18	Atomic energy materials (e.g. UO_2, PuO_2, etc.)				utilization of non-stoichiometric effects* in nuclear fuels, etc.
19	Getters and scavengers				utilization of high-activity oxide formation (e.g. of Group III and IV (rare earths)) to purify metals of lower O affinity (e.g. of Groups V and VI)
20	Oxide chemicals				multiple applications in technology, research, biology and medicine

* The utilization of defect lattices in oxides applies to many other fields, e.g. for electrical, thermal, optical and luminescent properties, in diffusion layers, etc.; furthermore, oxides may be rendered more ductile through reduction (e.g. by vacuum or hydrogen treatment).

Classes (*b*) and (*c*) mostly represent the oxides in immediate metal contact: (*b*) in form of scales or thin films, be it protective or cumulative, and including both lower, still metallic, and higher, ionic, oxides; (*c*) in form of dispersed fine particles, such as stable ceramic oxides in a metallic matrix, used for dispersion strengthening.

It is interesting to note the singular way in which classes (*a*), (*b*) and (*c*) fundamentally link up. The refractory properties of ceramic oxides in bulk, e.g. of alumina or spinel components for high-temperature use, recur two-dimensionally in that the same types of oxide phases form as thin films, e.g.

upon heat-resisting alloys. Although the latter generally are also subject to additional requirements, such as compatibility of thermal expansion with the metal substrate, conductivity, etc., the basic ceramic properties play a similar part in both these modes of occurrence. In fine-particle dispersion (*c*), the same refractory properties are again required, and some of the additional conditions of metal/oxide compatibility apply as to class (*b*). The function which is unique to class (*c*) is the micro-physical one of dislocation locking by the fine oxide particles in the metallic mass; in this sense, oxides (e.g. ThO_2 in tungsten) are similar in application to other types of interstitial compound (carbides, nitrides or borides). It might be argued that this extension to non-oxides would also be valid in cases (*a*) and (*b*). However, case (*c*) is special in that the dispersed particles are sealed off from the open (oxygen-bearing) atmosphere, whereas in (*a*) and (*b*) the saturated stable oxides generally equilibrate with it at surfaces, giving the oxides a natural chemical advantage over other interstitial compounds. As pointed out before, even silicides and borides depend, in the final resort, for their protectiveness upon a thin overlay of oxide.

Table 8.7 being self-explanatory, little additional comment should be needed. Several purposes of oxides can be classified under (*a*), (*b*) and (*c*), or under any two. Oxides in *metal-ceramic composites* illustrate a combination of all three classes, if we consider the requirements that (*a*) their use is that in volume form; that (*b*) the metal-oxide grain interface should show good bonding; and, obviously, that (*c*) they are physically dispersed. (In Cermets, however, dispersion strengthening applies only to a limited extent, and fine particle size is not necessarily a condition.)

In high-temperature materials, the oxidation resistance inherent in oxides gives them advantage over metals. Although brittle at room temperature, their mechanical strength can exceed that of metals at high temperatures. Even a measure of ductility can arise and be enhanced under hydrostatic conditions of pressure. Thus modern ceramics have now been developed where, through pressure at temperature, pores and point defect have been eliminated. Most oxides, however, suffer from poor thermal conductivity and, especially, poor thermal shock resistance, though here, too, more recent research has brought great improvements, combining strength, conductivity, thermal shock and oxidation resistance in matched metal-oxide composites. A general aim of much present-day endeavour is to develop plasticity in oxide ceramics, to allow, so to speak, their 'breathing' with a metal substrate or a metal intermixture, and this has had much success already. It is even possible, in special cases, to improve mechanical properties of a base metal by means of a suitably applied oxide layer; for example, steam treatment of sintered iron parts can form an Fe_3O_4 surface layer which, when pure, acts beneficially through its density and adhesiveness to the substrate, possibly helped by a pore-blocking action. Physical sintering of otherwise permeable oxide layers can also be valuable, and much of the function of, and search for, protective high-temperature coatings depends on this, as that for oxidation-resistant (uncoated) high-temperature alloys.

The extensive new field of magnetic oxide ceramics has been referred to earlier. Some of these materials attain their ferrimagnetism only at relatively low temperatures: $TiCo_2O_4$ spinel, for example, has a Curie temperature at 45°C. Generally, by choosing other spinels of appropriate composition and transition point (here magnetic, but also those in other properties), it becomes possible to adjust the ranges of application optimally to each new requirement. Such varying of the isomorphous substitutions in oxides to fit desired applications, i.e. adaptability, is a cardinal feature in modern ceramic developments. This can be by cation exchange (Cr ↔ Fe, etc.) or by vacancy introduction or removal, i.e. by altering the stoichiometry. The same is true for the newer dielectric ceramics, based e.g. on mixed Group IV/V transition-metal oxides (TiO_2, ZrO_2, Nb_2O_5, etc.). These researches have led to wide ranges of application in modern electronics developments, such as in microwave devices in radar and other components, exploiting variable electric or magnetic properties of given oxide groups.

Another such recent application of transition-metal oxides is the use of some NiO-base oxides for *temperature measurements*. These are the so-called 'Thermistors' (Thermally Sensitive Resistors) which utilize the sensitive temperature variation of tiny wire-connected beads of semi-conducting oxides as precise, very small resistance thermometers. Extensive uses have already been found in aerospace and medical research. The temperature range to which application is possible is also sub-zero, viz. about 20°–900°K overall. The basic point is here again the utilization of the borderline position of such an oxide between the metallic and non-metallic electron states. This clearly underlies the critical temperature dependence of physical properties; electrical resistance is but another example, and this dependence (as it were, 'relay fashion') opens the way for many novel applications.

Oxides of rutile (TiO_2) and related structures have also found useful semi-conducting applications in rectifying devices and as high-temperature semi-conductors.

In the *dispersion strengthening* of metals, the finely divided particles of hard, stable and matrix-insoluble oxides can greatly increase the upper limit of service temperatures by improving tensile strength and providing creep resistance very close to the melting point of the matrix. One of the first of such 'oxide-reinforced' metals was 'SAP', i.e. sintered aluminium powder, Al with fine Al_2O_3 dispersions; other early examples were thoriated tungsten (W with ThO_2 or ThC, e.g. in lamp filaments), molybdenum with Group IV metal oxides and lead with PbO_2. More recently, the principle has been fruitfully extended to the refractory alloys of Nb, Ta, Mo and W base, particularly important in raising the attainable temperature ceiling, to Cu-, Ni-, Co- and Fe-base alloys as well as the noble metals. Such alloys dispersion-strengthened by oxides are principally made by powder metallurgy; however, *internal oxidation* is a powerful new approach, particularly as this can achieve finer particles in more uniform dispersion. The fine oxide particles here can occur both between and within the matrix grains; their chief functions are locking of dislocation and inhibition of grain growth and recrystallization.

Oxides as *catalysts* in heterogeneous reactions are more elusive in classification (they probably best fall under groups (*b*) and (*c*), and their role is largely transient. They occur in the form of molecularly thin and unstable adsorbed surface layers, such as oxide-like configurations on noble-metal (e.g. platinum) surfaces or particles. In catalysis, a point of further great importance is the excess oxygen present in oxides, beyond the stoichiometric quantity; this, in effect, is the 'active' oxygen and determines many physico-chemical properties. In catalytic nickel oxide it is this thin excess oxygen which assists in oxidation of CO to CO_2 and of NH_3 to NO. Paravano's[149] work is here outstanding amongst much other. From a systematic viewpoint, this brings out the attractive theme that, whilst the applications of oxides largely centre around their stability and refractoriness in Groups IV–VI, the opposite extreme of instability has its own uses in the later Group VIII metals (Pt and, to some extent, Ni, Co, Fe). The intermediate range of iron, Group VII and numerous mixed-metal oxides reflects itself in the wide range of applications where moderate and controllable stability is needed. Subtle adjustments of lattice defects and electronic structure, in some cases even anisotropically in different directions, are an advantage to induce special magnetic, electric, thermal or diffusional properties. Furthermore, the ascending stability of oxides towards earlier Groups (e.g. Mn $\rightarrow$ Cr $\rightarrow$ V $\rightarrow$ Ti $\rightarrow$ Sc $\rightarrow$ Ca; Re $\rightarrow$ W $\rightarrow$ Ta $\rightarrow$ Hf $\rightarrow$ Rare Earths) clearly reflects the utilization of the earlier Group metals in helping to purify the later ones. Gettering and scavenging by Ti or Zr, and more rigorously by Sc, Y or the Rare Earths, are instances.

The actinide oxides, UO_2, etc., have of course their additional fields of nuclear application, e.g. in reactor fuel elements. Here again the non-stoichiometry $UO_2 \rightarrow UO_{2\cdot3}$ (Willis[14], etc.) has a major influence on performance. The powder metallurgy of UO_2, UO_{2+} and U/UO_2 dispersion-strengthened metal ceramics has been developed to a considerable extent, with the aim of producing strong and efficient nuclear fuels, moderated to a prescribed degree[150–1].

Neutron-*irradiated* oxides, not only those of uranium, but of all the transition metals, prove of particular interest from the viewpoint of the point defects produced. Often entirely new physical properties can ensue, as well as enhanced chemisorption of extraneous gases. We might see in irradiation a means of artificially producing vacancies, either to act as potential hosts for other interstitials or metals in a given parent oxide, or to remain unoccupied. To some extent this could be regarded as a means of attaining chosen electrical, mechanical or thermal properties. Some of the most interesting studies are those by electron microscopy on dislocation movement and radiation-produced interstitial and vacancy loops.

We should also note that in many cases a severe disadvantage of an oxide in one sphere can turn to advantage in another. The low melting point and high volatilization of MoO_3, which limits the high-temperature use of molybdenum to non-oxidizing atmospheres or coated alloys, can become helpful in providing MoO_3 as a lubricant in certain high-temperature manufacturing techniques such as rolling or wire drawing. Similarly, oxides can often be used in fluxes, accompanying halogenides and other compounds.

In the *lower* oxides, applications depend on structures and volume ratio, coherence, orientation and expansion coefficient relative to a metal substrate and the coexisting higher or mixed oxides. Other important factors are ease of sintering, wettability, permeability to gases, etc., i.e. properties which determine the adherence and protection of oxide films on metals in both room and high-temperature service[11,152–5].

A frequent limitation of oxides is their inherent brittleness and, as mentioned, much research is proceeding in giving them a plastic range which, though only slight, can make a critical difference in their adaptability to substrates. Non-stoichiometry has an important influence, and a reducing gas (H_2,CO,N_2O) or high-temperature *vacuum* treatment can be one means to attain this. In hematite-type oxides (Fe_2O_3,Cr_2O_3, etc., even Al_2O_3) an incipient or finely dispersed spinel can be produced, further enhanced by the addition of P-type (NiO, CoO, FeO, MnO, etc.) to N-type semiconductors (e.g. Al_2O_3). In Cr_2O_3, for example, intrinsic oxygen vacancies can be produced ($Cr^{+++} \rightarrow Cr^{++}$; similarly in Al_2O_3 by Cr addition) which are utilized industrially as these can give a significant toughness to brittle oxides. Incipient spinel formation in Al_2O_3 is also able to give some strengthening because it aids the wetting of crystal boundaries and thus improves the bond. Vacuum treatment in MgO can withdraw oxygen to such an extent that metallic Mg nuclei are evolved; the same is possible even in Al_2O_3, with Al nuclei forming under extreme reducing conditions. This possibility of metal precipitation by reduction is important in a variety of oxides in contact with alkali coolants.

Altogether trace impurity control in oxides can critically determine the properties. The conversion of pure NiO, which is an insulator, to a conductor by minor additions of foreign atoms like lithium, mentioned before, is typical. Such trace impurities do not act as though independent 'alloy additions' but as agents for filling or creating vacant sites.

Oxidation of the refractory metals (Nb,Ta,Mo,W) is one of the most severely limiting factors in regard to their high-temperature applications, and much effort has been directed to its suppression and, on the other hand, to measurements of oxidation rates as a function of temperature, time and gas pressure; these are critically related to the oxide structures. For niobium, this depends on the α/β-Nb_2O_5 transformation (discussed earlier), on new oxide phases that can be stabilized by alloy additions and on non-stoichiometry in each[155]. The latter can take the form of deficiency of anions at one temperature, one of cations at another; both would accelerate oxidation. The formation of metastable lower oxides, even if only in thin interfacial layers, such as the metastable 'NbO_x' and 'NbO_y' described[11,58b,61–2], is here equally important.

The kinetics of the oxidation of the refractory metals is well reflected in the phase relationships discussed. Thus in Ti and Nb four parallel reactions have been observed[156]: (1) the solution of O in the metal; (2) nucleation and growth of a suboxide phase on the metal surface, evolving (3) to the higher oxides in their different modifications. The latter process is the severely damaging one, as it gives rise to spalling of the oxide film through differential thermal expansion and interfacial stresses. This, however, can be modulated by alloy addition, and development work is aimed at just this effect[157–8]. Pressure and temperature dependencies are here vital variables, in consonance with stabilization (or otherwise) of different oxide phases, their stages in non-stoichiometry, crystal size and 'crystallinity'. Semi-amorphous oxide films formed at low temperatures generally tend to be more protective than the more crystalline ones for all metals, as electron diffraction work is able to reveal.

Similar and highly complex mechanisms[159] are at work in determining the oxidizability of iron- or nickel-base alloys where Cr and other added alloying elements can have such spectacular improving effects through thin-film formation of stable Cr_2O_3 and spinel types. Stainless steels and the heat-resistant alloys of 'Nichrome', etc. type are familiar examples. The function of chromium lies not in rendering the Group VIII metals 'more noble' but, on the contrary, in the very stability of the oxides it engenders and their structural adaptability to the metal substrate. Analogous cases arise throughout the transition-metal block. The point of principle is that of balanced choice of addition of an earlier Group metal which could improve the protection of the later one, by promoting the suitable oxide structure. Although the literature in this field is extensive, many fundamental problems still await solution.

For the later Group metals, particularly the Pt group, the problem shifts towards the volatility of the oxides. Pd,Pt,Rh,Ir,Ru and Os all show linear rates of weight loss of metal in air at high temperatures, owing to transient formation and vaporization of volatile oxides, combined with surface-adsorbed atomic oxygen[160]. This again is a point of wider principle which is well illustrated for iridium[161]. Iridium has a good inherent oxidation resistance up to 700°C; above this, a thin IrO_2 film forms which decomposes at higher temperatures whilst still solid, and at about 1,100°C the decomposition pressure reaches 1 atm. At the same time, a volatile higher oxide, IrO_3, forms so that Ir metal then loses weight. A similar process occurs for Ru and other Pt-group metals at varying temperatures and, indeed, more concealed and overlaid by non-vaporizing oxides, already in the earlier Group VI metals with their volatile oxides (MoO_3, WO_3, etc.) at high temperatures.

One interesting evolution of this occurs for *silver* (possibly gold?), to the right of the Pt group. Here a 'superoxide' AgO_2 has been reported[162], besides the normal Ag_2O which can vaporize already at moderate temperatures and is also important in the action of Ag as a catalyst.

The subject of oxides merges, as far as applications are concerned, into mineralogy, extraction metallurgy, ore reduction, steel making (e.g. technology of slags, later separate uses of slags; refractory science) and glass technology, to name only a few (cf. Bibliography).

SELECTED LITERATURE

Binary oxides

(Sequence from Group III to VIII in Periodic Table)

(Sc, Y, Rare Earths)–O Gschneidner, K. A. *Rare Earth Alloys*, Princeton (Van Norstrand) 1961; Spedding, F. H. and Daane, A. H. *The Rare Earths*, New York (Wiley) 1961; Brauer, G. and associates in *Progress in the Science and Technology of the Rare Earths I*, 152, Oxford (Pergamon) 1964; Bevan, D. J. M. *J. inorg. nucl. Chem.* 1 (1955) 49; with Kordis, J. 26 (1964) 1509 [MO_2 fluorite → M_2O_3 transitions]

Ti–O[4,24–31] Rutile TiO_2: Ashbee and Smallman[12] [Plastic deformation]; Wachtman *et al.*[37] [Elastic constant]; Knunun, V. Ya. and Tsikin, A. N. *Soviet Phys. solid St.* 4 (1963) 2575; Carnahan, R. D. and Brittain, J. O. *J. appl. Phys.* 34 (1963) 3095 [Point defects; internal friction evidence]; Cardon, F. *Phys. Status Solidi* 3 (1963) 1415; Koikov, A. N. *et al., Fizika tverd. Tela* 4 (1962) 1067; Davidovich, N. M. and Dorin, V. A. *Fizika Metall.* 16

Ti–O[4,24–31] *cont.* (1963) 273 [Electrical properties, electrical aging and regeneration in rutile related to vacancies and interstitial occupancy]; Chester, P. F. and Bradhurst, D. H. *Nature, Lond.* 199 (1963) 1056 [Induced conduction in rutile]

Zr–O[38–40,42] Osthagen, K. and Kofstad, P. *J. electrochem. Soc.* 109 (1962) 204; Gulbransen, E. A. and Andrew, K. F. 96 (1949) 365; Guldner, W. G. and Wooten, L. A. 93 (1948) 223; Debuigne, J. and Lehr, P. *C.r. hebd. Séanc. Acad. Sci., Paris* 25 (1962) 3710, 4168 [Typifying work on reaction kinetics under various pressure/temperature conditions]; Teufer[43] [Tetragonal ZrO_2 structure at high temperatures]; Bailey, J. E. *J. nucl. Mater.* 8 (1963) 259 [Electron microscopy on ZrO_2]; McCullough, J. D. and Trueblood, K. N. *Acta crystallogr.* 12 (1959) 507; Duwez and Odell[126]; *J. Am. Ceram. Soc.* 32 (1949) 180 [ZrO_2 transformations]; Kofstad and Ruzicka[11] [Defects in ZrO_2 and HfO_2]; Aronsson, B. *J. electrochem. Soc.* 108 (1961) 312 [ZrO_2 non-stoichiometry]

Hf–O[45] Nowotny, H. *et al., Radex Rdsch.* (1960) 367 [Constitution]; Geller, S. and Corenzwit, E. *Analyt. Chem.* 25 (1953) 1774 [HfO_2 structure = of ZrO_2]; Curtis, C. E. *et al., J. Am. Ceram. Soc.* 37 (1954) 460 [HfO_2 polymorphy]; Smeltzer, W. W. and Simna, M. T. *Acta metall.* 5 (1957) 328 [Oxidation kinetics]; Adams, J. and Rogers, M. D. *Acta crystallogr.* 12 (1959) 951 [HfO_2 structure]

Th–O Gerds, A. F. and Mallett, M. W. *J. electrochem. Soc.* 101 (1954) 171 [Phase relations]; Zachariasen[22a,b] [ThO_2 structure]; Ackerman, J. R. *et al., J. phys. Chem., Ithaka* 67 (1963) 762 [Thermodynamic study; vaporization behaviour, entropies, etc.]; Brown, A. and Chitty, A. *J. nucl. Energy* 1 (1960) 145 [ThO_2 as liquid metal blanket in atomic energy; basic aspects]

V–O[32,41,47–51,53–4,57] Westman, S. and Nordmark, C. *Acta chem. scand.* 14 (1960) 465; Holtzberg, F. *et al., J. Am. chem. Soc.* 78 (1956) 1536; Heckingbottom and Linnett[55] [Structures of V oxides]; Allen, Kubaschewski and v. Goldbeck[52] [Free energy diagram of V–O system]; Åsbrink, S. *et al., Acta chem. scand.* 13 (1959) 603 [V_3O_5]; Minomura, S. and Drickamer, H. G. *J. appl. Phys.* 34 (1963) 3043 [V–O and Cr–O]

Nb–O, Ta–O, Mo–O *and* **W–O** Speiser, R. and St. Pierre, G. R. in Agard Symposium *The Science and Technology of W, Ta, Mo and Nb and their Alloys*, p. 289, Oxford (Pergamon) 1963

Nb–O[7b,c,58a,b,61–2] Shafer, M. W. and Roy, R. *Z. Kristallogr. Kristallgeom.* 110 (1958) 24; Holtzberg, F. *et al., J. Am. chem. Soc.* 79 (1957) 2039; *J. phys. Chem., Ithaka* 65 (1961) 1192; Frevel, L. K. and Rinn, H. N. *Analyt. Chem.* 27 (1955) 1329; Holser, W. T. *Acta crystallogr.* 9 (1956) 196; Schäfer, H. *et al., Z. anorg. allg. Chem.* 275 (1954) 289; Goldschmidt[16d] [Constitutional; Nb_2O_5 and lower oxides]; Norman, N., Kofstad, P. and Kjollesdal, H. *Trans. metall. Soc. A.I.M.E.* 221 (1961) 285 [Lower oxides, oxidation process]; Kofstad and Anderson[11b] [Nb_2O_5 vacancies; oxidation studies and thermodynamic properties]; Gulbranson, E. A. and Andrew, K. F. *J. electrochem. Soc.* 96 (1949) 365; 105 (1958) 4; 107 (1960) 619; *J. Metals, N.Y.* 188 (1950) 586; Klopp, W. D., Sims, C. T. and Jaffee, R. I. *Trans. Am. Soc. Metals* 57 (1959) 282; Cowgill, M. G. and Stringer, J. *J. Inst. Metals* 91 (1963) 220 [also Ta–O]; Bridges, D. W. and Fassell, W. N. *J. electrochem. Soc.* 103 (1956) 326 [Reaction kinetics of Nb, etc.]; Gebhardt, E. and Rothenbacher, R. *Z. Metallk.* 54 (1963) 443, 623 [O solution and low-oxide formation]; McLintock, C. H. and Stringer, J. *J. less-common Metals* 5 (1963) 278 [Oxidation of Nb]; Douglass, D. L. 5 (1963) 151 [Thermal expansion of Nb_2O_5]; Makkay, R. W. and Fine, M. E. *J. appl. Phys.* 33 (1962) 745; Greener, E. H. *et al., J. electrochem. Soc.* 109 (1962) 600; *J. chem. Phys.* 38 (1963) 133; Janninek, R. F. and Whitmore, D. H. 37 (1962) 3750; 39 (1963) 179 [Electrical conductivity of Nb_2O_5 as defect semiconductor, depending on temperature and pressure; thermoelectric power]; Hicks, W. T. *Trans metall. Soc. A.I.M.E.* 221 (1961) 352 [Oxidation rates of NbO]; Kusenko, F. G. and Geld, P. V. *Izv. vyssh. ucheb. Zaved. Khim. Met.* 3 (1960) 102 [Thermochemical data on Nb oxides]; Valletta, R. *J. chem. Phys.* 37 (1962) 67 ['Doped' Nb_2O_5]

Ta–O Brauer *et al.*,[58b] Norman[61–2]; Kofstad[11c]; *J. Inst. Metals* 90 (1962) 253; 91 (1963) 209, 414 [Ta oxidation; suboxides]; with Krudtaa, O. J. *J. less-common Metals* 5 (1963) 477; with Anderson[11b] [Defects in Ta_2O_5]; Reisman, A. *et al.*, *J. Am. chem. Soc.* 78 (1956) 4514 [Ta_2O_5 structure]; Gerhardt, E. and Seghezzi, H. D. *Z. Metallk.* 50 (1959) 250, 522 [Oxidation; morphology of low oxides to Ta_2O_5]; Schoenberg, N. *Acta chem. scand.* 8 (1954) 240 [Ta oxide structures]

Nb–O and Ta–O equilibria Miller, G. *Niobium and Tantalum*, London (Butterworths) 1961

Pa–O Sellers, P. A. *et al.*, *J. Am. chem. Soc.* 76 (1954) 5935; Zachariasen, W. H. *Acta crystallogr.* 2 (1949) 388; 5 (1952) 19; Roberts, L. E. J. *Q. Rev. chem. Soc.* 15 (1961) 442 [PaO_2, Pa_2O_5 structures]

Cr–O Byström, A. and Wilhelmi, K. A. *Acta chem. scand.* 4 (1950) 1131; Caplan, E. and Butt, A. A. *Trans. Am. Inst. Min. Engrs* 203 (1955) 1052; Glemser, O. *et al.*, *Z. anorg. allg. Chem.* 277 (1954) 113 [Equilibria]; Wilhelmi, K. A. and Jonsson, O. *Acta chem. scand.* 12 (1958) 1532 [CrO_2 = rutile]

Mo–O, W–O Kihlborg[69] [Crystal chemistry of Mo oxides]; Magneli, A. *Acta chem. scand.* 3 (1949) 88; 9 (1955) 1382 [Mo and W oxide structures]; Andersson, G. 7 (1953) 154; Kehl, W. L., Hay, R. G. and Wall, D. *Phys. Rev.* 82 (1951) 774; *J. appl. Phys.* 23 (1952) 212 [WO_3 structure]; Hägg, G. and Magneli, A. *Ark. Kemi Miner. Geol.* 19A (1944) 1; *Rev. pure appl. Chem.* 4 (1954) 235; Andersson, G. and Magneli, A. *Acta chem. scand.* 4 (1950) 793 [MoO_3 structure]; Magneli[7b,c] [X-ray identification of Mo and W oxides]; Schönberg[41c] [Mo_3O, W_3O]; St. Pierre, G. R. *et al.*, *Trans. metall. Soc. A.I.M.E.* 224 (1962) 259 [W–O system, overall survey]; Gerasimov, J. L. *et al.*, *Dokl. Akad. Nauk SSSR* 134 (1960) 1350 [W oxide, free energies]

U–O[22,75] Rundle, R. E. *et al.*, *J. Am. chem. Soc.* 70 (1948) 99; Willis[14] [Structure of UO_2]; Belle, J. *Peaceful Uses of Atomic Energy*, Geneva (United Nations) 1958 [Properties of UO_2; UO_2 as nuclear fuel; references]; Belle, J. (Ed.) *Uranium Dioxide, Properties and Nuclear Applications*, U.S. Atomic Energy Comm., 1961; Thewlis, J. *Acta crystallogr.* 5 (1952) 790; Alberman and Anderson[72] [UO_2 spacing changes in homogeneity range]; Siegel, S. 8 (1955) 617 [U_3O_8 spacing changes in hom. range]; Perio, P. *Bull. Soc. chim. Fr.* (1953) 776 [UO_3 structures of polymorphs]; Chapman, A. T. and Meadows, R. E. *U.S. At. Energy Comm. ORNL* Rep. 3587, 1964 [UO_{2+x} volatility and U-O phase relations]; Marinkovic, S. *J. nucl. Mater.* 8 (1963) 187 [$UO_2 \leftrightarrow U_3O_8$ transformation]; Belbeoch, B. *et al.*, *Acta crystallogr.* 14 (1961) 837 [U_4O_9 structure]; Armstrong, W. N. and Irvine, W. R. *J. nucl. Mater.* 9 (1963) 121 [U oxides]; Blinova, N. I. *et al.*, *Dokl. Akad. Nauk SSSR* 147 (1962) 112 [U_2O_5 magnetic properties, electron structure]; Zhukovskii, V. M. *et al.*, *Soviet Phys. Metals Metallogr.* 14 (1962) 139; *Fizika Metall.* 15 (1963) 210; Tkachenko, E. V. and Viasov, V. G. *ibid.* 239 [U_3O_8 range; electric resistivity and ordering; reduction sequence of U oxides by carbon and ammonia followed structurally]; Westrum, E. F. and Grønvold, F. *Jnl Phys. Chem. Solids* 23 (1962) 39; Grønvold, F. and Haraldsen, H. *Nature, Lond.* 162 (1948) 69 [α- and β-U_3O_7, thermodynamic properties]; Engmann, R. and De Wolff, P. M. *Acta crystallogr.* 16 (1963) 993 [γ-UO_3 structure]; Roberts, L. E. J. *Q. Rev. chem. Soc.* 15 (1961) 442 [Review of actinide oxides]

Mn–O Elefson, B. S. and Taylor, N. W. *J. chem. Phys.* 2 (1934) 58; Jay, A. H. and Andrews, K. W. *J. Iron Steel Inst.* 152 (1946) 15 [MnO structure, x-ray]; Shull, C. G. and Smart, J. S. *Phys. Rev.* 76 (1949) 1256 [Neutron diffraction; ordering at low temps.]; Tombs, N. C. and Rooksby, H. P. *Nature, Lond.* 165 (1950) 442 [X-ray: rhombohedral distortion at low temps.]; Mason, B. *Am. Miner.* 32 (1947) 36 [Mn_3O_4 structure, spinel derivative]; Verwey, E. W. W. and de Boer, J. H. *Recl Trav. chim. Pays-Bas Belg.* 55 (1936) 531 [Mn_2O_3 polymorphs]; Byström, A. M. *Acta chem. scand.* 3 (1949) 163; Brenet, J. *C.r. hebd. Séanc. Acad. Sci., Paris* 230 (1950) 1360 [MnO_2 polymorphs]; Bradt, R. C. and Wiley, J. S. *J. electrochem. Soc.* 109 (1962) 651 [Thermal expansion of MnO_2, anisotropy]

Fe–O Darken, L. S. and Gurry, R. W. *J. Am. chem. Soc.* 68 (1946) 798; Jette, E. R. and Foote, F. *J. chem. Phys.* 1 (1933) 29; Willis, B. T. M. and Rooksby, H. P. *Acta crystallogr.* 6 (1953) 565, 827; Bénard, J. *Bull. Soc. chim. Fr.* (1949) 109; *Acta crystallogr.* 7 (1954) 214; Collogues, R. *ibid.* 213; Aubry, J. and Marion, F. *C.r. hebd. Séanc. Acad. Sci., Paris* 240 (1955) 1770 ['FeO' structure, non-stoichiometric changes; rhombohedral distortion at low temps.]; Greig, J. W. *et al., Am. J. Sci.* 30 (1935) 239; Jay, A. H. and Andrews, K. W. *Nature, Lond.* 154 (1944) 116; Tombs, N. C. and Rooksby, H. P. *Acta crystallogr.* 4 (1961) 474; Rooksby, H. P. and Willis, B. T. M. 6 (1953) 565; Abrahams, S. C. and Calhoun, B. A. *ibid.* 105 [Fe_4O_3 spinel; low-temperature rhombohedral distortion]; Goldschmidt[16b] [FeO and Fe_3O_4 relations]; Hägg, G. *Trans. Am. Inst. Min. Engrs* 105 (1933) 287; Haul, R. and Schoon, T. *Z. phys. Chem.* B44 (1939) 216; Verwey, E. J. W. *Z. Kristallogr. Kristallgeom.* 91 (1935) 65 [γ-Fe_2O_3]; Glemser, O. and Gwinner, E. *Z. anorg. allg. Chem.* 240 (1939) 161 [Further form of Fe_2O_3?]; Jantzen, O. *Naturwissenschaften* 50 (1963) 14 [$Fe_3O_4 \leftrightarrow Fe_2O_3$ transition]; Tannhauser, D. S. *Physics Chem. Solids* 23 (1962) 25 [Electrical conductivity and EMF of iron oxides; correlation with Fe ionization and Fe/O ratio]; Kachi, S. *et al., J. phys. Soc. Japan* 18 (1963) 106 [Transformations in Fe_2O_3]; Izvekov, V. I. *et al., Soviet Phys. Metals Metallogr.* 14 (1962) 195 [Diffusion of Fe in Fe_2O_3]; Blackman, M. and Kaye, G. *J. phys. Soc. Japan* 17 (1961, Suppl.) B 289 [Curie point of Fe_2O_3]; Brown, W. F. and Johnson, C. E. *J. appl. Phys.* 33 (1962) 2572 [Saturation magnetization of γ-Fe_3O_3]; Yamaguchi, S. *Trans. Faraday Soc.* 58 (1962) 1690 [Ferrimagnetism of Fe_3O_4]; Ito, A. *et al., J. phys. Soc. Japan* 18 (1963) 1465 [Low-temperature transitions in Fe_3O_4]; Bénard, J. and Coquelle, O. *C.r. hebd. Séanc. Acad. Sci., Paris* 222 (1946) 884 [Fe oxidation at high temperatures]; Halla, F. and Nowotny, H. *Naturwissenschaften* 45 (1958) 384 [FeO decomposition; kinetics]; Haubenreisser, W. *Phys. Status Solidi* 1 (1961) 619 [Fe_3O_4, theory of electron transport]; Herai, T. and Manenc, J. *C.r. hebd. Séanc. Acad. Sci., Paris* 256 (1963) 684 [FeO decomposition]; Raccah, P. and Vallet, P. 255 (1962) 1919 [Wüstite: two varieties of different O content]; Suzuki, Y. and Azada, E. *J. phys. Soc. Japan* 16 (1961) 1785 [FeO lattice distortion on cooling through Néel temperature]

Co–O Asanti, P. and Kohlmeyer, E. J. *Z. anorg. allg. Chem.* 265 (1951) 90; Tombs, N. C. and Rooksby, H. P. *Nature, Lond.* 165 (1950) 442; Greenwald, S. *Acta crystallogr.* 6 (1953) 396; with Smart, J. S. *Nature, Lond.* 166 (1950) 523 [CoO structures, low-temp. tetragonal form]; Hendricks, S. B. and Albrecht, W. H. *Ber. dt. chem. Ges.* 61 (1928) 2153; Goldschmidt[16c] [Co_3O_4 spinel]; Iida, Y., Paper *Am. Ceram. Soc. Meeting* 1963 [Electrical conductivity of Co oxides]; Eror, N. G. and Wagner, J. B., Paper *Metall. Soc. A.I.M.E. Meeting*, 1963 [CoO, conductivity as function of reduced pressures]; Gulbransen, E. A. and Andrew, K. F. *J. electrochem. Soc.* 98 (1951) 241 [Oxidation mechanism]; Holtermann, C. B. *Annls Chim.* 14 (1940) 121 [Co_3O_4 highest oxide]

Ni–O Cairns, R. W. and Ott, E. *J. Am. chem. Soc.* 55 (1933) 527; Rooksby, H. P. *Nature, Lond.* 152 (1943) 304; *Acta crystallogr.* 1 (1948) 226; Shimomura, Y. *et al., J. phys. Soc. Japan* 11 (1956) 1136 [NiO rhombohedral low-temperature distortions]; Foëx, M. *C.r. hebd. Séanc. Acad. Sci., Paris* 227 (1948) 193; 228 (1949) 1579; *Bull. Soc. chim. Fr.* (1952) 373 [Temperature changes in NiO]; Holtermann, C. B. *Annls Chim.* 14 (1940) 121 [NiO highest oxide]; Finch, G. I. and Sinka, K. P. *Trans. Faraday Soc.* 53 (1957) 623 [A Ni–O spinel?]

Re–O Magneli and Anderson[7f] [ReO_2]; Wilhelm, K. A. *Acta chem. scand.* 8 (1954) 693; Meisel, K. *Z. anorg. allg. Chem.* 207 (1932) 121 [Re_2O_7, ReO_3]; Magneli[7e]; *Acta chem. scand.* 11 (1957) 28 [ReO_2, structure]; Woolf, A. A. *Q. Rev. chem. Soc.* 15 (1961) 372 [Rhenium chemistry including oxides]

Pt group metals–O

Ru–O Goldschmidt, V. M. *Geochemische Verteilungsgetze* VI, 1926 [Early structure work on RuO_2, OsO_2, IrO_2, etc.]

Rh–O Zachariasen, W. H. *Skr. norske VidenskAkad.* no. 4 (1928) [Rh_2O_3]

Pd–O Zachariasen, W. H. *Z. phys. Chem.* 128 (1927) 412; Huggins, M. L. *Chem. Rev.* 10 (1932) 427; Moore, W. J. and Pauling, L. *J. Am. chem. Soc.* 63 (1941) 1392 [Early work on PdO]; Waser, J. *et al., Acta crystallogr.* 6 (1953) 661 [PdO, neutron and x-ray diffraction]

Os–O Zalkin, A. and Templeton, D. H. *Acta crystallogr.* 6 (1953) 106 [OsO_4 structure]

Ir–O Schäfer, H. and Heitland, H. J. *Z. anorg. allg. Chem.* 304 (1960) 249 [IrO_2 stability]; Cordfunke, E. H. P. and Meyer, G. *Recl Trav. chim. Pays-Bas Belg.* 81 (1962) 670 [Dissociation pressures of IrO_2 and thermodynamics of the Ir-O system]

Pt–O Alcock, C. B. and Hooper, G. W. *Proc. R. Soc.* A 254 (1960) 551 [Thermodynamics of the gaseous oxides of Pt Group metals]; Moore, W. J. and Pauling, L. *J. Am. chem. Soc.* 63 (1941) 1392 [PtO structure, contradicted by] Galloni, E. E., with Busch, R. H. *J. chem. Phys.* 20 (1952) 198 [also PtO_2]; with Roffe, A. E. 9 (1941) 895; Waser, J. and McClanahan, E. D. 19 (1951) 413; 20 (1952) 199 [Pt_3O_4 structure]

Np–O, Pu–O Holley, C. E. *et al.* in *Peaceful Uses of Atomic Energy* 6 (1958) 215 [Pu oxides, thermodynamic and phase relations]; Zachariasen[22a]; Mooney and Zachariasen[22c]; Asprey, W. B. *et al., J. Am. chem. Soc.* 77 (1955) 1707 [Pu oxide structures]

Zachariasen[22a]; Makarov[23]; Katz, J. J. and Gruen, D. M. *J. Am. chem. Soc.* 71 (1949) 2106; Asprey, L. B. *et al.* 77 (1955) 1707 [Np oxide structures]

Binary systems exemplifying the TiO – NiO *monoxide series*

Rundle, R. E. *Acta crystallogr.* 1 (1948) 180 [Interstitial oxides of NaCl type; structural theory]; Mott, N. F. *Phil. Mag.* 6 (1961) 287 [Borderline position of monoxides between non-metallic and metallic state]; Hoch, M. *Jnl Phys. Chem. Solids* 24 (1963) 157 [Vacancy interaction in defect lattice for TiO ↔ NiO series]; Yamashita, Y. *J. phys. Soc. Japan* 18 (1963) 1010 [Electron-band theory of TiO and NiO]

Ternary oxides

Spinels: structures (and related)[3a–d,20,130]

Tarte, P. and Preudhomme, J. *Acta crystallogr.* 15 (1963) 227; Rooymans, C. J. M. *J. inorg. nucl. Chem.* 11 (1959) 78; Pines, B. Ya. and Gumen, N. M. *Kristallografiya* 6 (1961) 901 [$CoFe_2O_4$ structure changes during heating in magnetic field]; Yearian, H. J. *et al., J. chem. Phys.* 22 (1954) 1196; *Corrosion* 12 (1956) 561; Bayer[103b] [Survey of spinel, etc. derivatives]

Spinels: properties and applications

Glasser, F. P. and L. S. D. *J. Am. Ceram. Soc.* 46 (1963) 377 [Crystal chemistry of some AB_2O_4 compounds where A = Ba, Sr, Pb, B = Ca, Cr, Fe, Al]; Huber, M. in Kingery, W. D. (Ed.) *Kinetics of High Temperature Processes*, p. 19, New York (Wiley) 1959; Sheftel, I. T. *et al., Soviet Phys. Solid St.* 3 (1962) 1979, 1989 [Electrical properties and structures of semiconductor oxides in MnO–CoO–NiO/O_2 and CuO–NiO/O_2]; Anbo, E. *J. ceram. Ass. Japan* 71 (1963) 109 [Precipitation *from* chromite spinels of M_2O_3 (= Cr_2O_3, Fe_2O_3, Al_2O_3); structure changes]; Wold *et al.*[128] [Structural transitions in $Ni_xFe_{1-x}Cr_2O_4$ type of spinels: ion siting, effect on axial ratio]; Schmalzried, H. *Ber. Bunsenges. Phys. Chem.* 67 (1963) 93 [Ion, electron and vacancy disordering in $CoCr_2O_4$ and $CoAl_2O_4$ spinels; effect on conductivity]; Butler, S. R. and Buessem, W. R. *J. appl. Phys.* 34 (1963) 1754 [Magnetization in $MnFe_2O_4$, dependence during preparation on oxygen pressure and temperature]; Mozzi, R. C. and Paladino, A. E. *J. chem. Phys.* 39 (1963) 435 [Cation and vacancy distribution in $MgFe_2O_4$]; Belov, K. P. and Zaleskii, A. V. *Izv. Akad. Nauk SSSR Fiz. Ser.* 25 (1961) 1434 [Mn–Fe spinel]; Smit, J. and Wijn, H. P. *Ferrite*, Eindhoven (Philips) 1962; Megaw, H. D. *Ferroelectricity in Crystals*, London (Methuen) 1957; Wolf, W. P. and Rodrigue, G. P. *J. appl. Phys.* 29 (1958) 105; Lotgering, F. K. *J. inorg. nucl. Chem.* 9 (1959) 118, etc.; Menyuk, N. *et al., J. appl. Phys.* 53 (1952) 1144 [(Mn, Co) . $(Cr,V)_2O_4$ spinels; magnetic transitions]; Severin, H. *Sprechsaal Keram. Glas Email* 95 (1962) 685; Birks, N. and Rickert, H. *J. Inst. Metals* 91 (1963) 308 [Oxidation of Ni–Cr alloys with spinel formation, etc.]; Foley, R. T. *J. electrochem. Soc.* 109 (1962) 1202 [Oxidation of Fe–Ni alloy]; Dulac, J. *J. Phys. Radium, Paris* 23 (1962) 872 [Germanate spinel]; Paulus, M. *C.r. hebd. Séanc. Acad. Sci., Paris* 255 (1962) 871 [Ferrites]; Sakamoto, N. and Yamagushi, Y. *J. phys. Soc. Japan* 7 (1962, Suppl. B) I 276 [Co–Ti spinels]; Shafer, N. W. *J. phys. Chem., Ithaka* 65 (1961) 2055 [Ni–Fe oxides]; Shirane, G. *et al., Phys. Rev.* 96 (1954) 581 [$NaNbO_3$ and (NaK) NbO_3 transformations]; Zaslavskii, A. I. *et al., Kristallographiya* 7 (1962) 835 [$CuMn_2O_4$ spinel]

ABO_3 perovskites and related structures

Francombe, M. H. and Rooksby, H. P. *Nature, Lond.* 178 (1956) 586; Lewis, B. and White, E. A. D. *J. Electronics* 1 (1956) 646; Rooksby, H. P. and Steward, E. G. *Acta crystallogr.* 6 (1953) 49; Steward, E. G. with Rooksby, H. P. 4 (1951) 53; with Runciman, W. *Nature, Lond.* 200 (1953) 75 [ABO_3 variants of A_3BO_6 type, including alkaline earth (Ca, Ba) uranates, tungstates, niobates and tantalates in various combinations]; Naray-Szabo, S. *Naturwissenschaften* 3 (1943) 202, 466; Magneli, A. *Ark. Kemi Miner. Geol.* 1 (1949) 213; with Blomberg, B. *Acta chem. scand.* 5 (1951) 372 [Alkali tungsten bronzes]; Straumanis, M. E. 14 (1961) 493; Vousden[36]; Megaw[136]; with Wells, M. *Acta crystallogr.* 11 (1958) 858; Wood, E. A. 13 (1960) 682; 15 (1962) 1273 [Review of ABO_3 compounds related to ionic radii and polymorphy]; Geller, S. 9 (1956) 885; 10 (1957) 243; with Wood, E. A. 9 (1956) 563; and Bala, V. B. *ibid.* 1019; Gilleo, M. A. 10 (1957) 161; Rosen *et al.* 9 (1956) 475; Francombe, M. H. 9 (1956) 256; Rooksby, H. P. *Nature, Lond.* 155 (1954) 484; Krylov, E. I. *Dokl. Akad. Nauk SSSR* 104 (1955) 246; with Sharnin, A. A. *Zh. obshch. Khim.* 25 (1955) 1860; Ridgley, D. H. and Ward, R. *J. Am. chem. Soc.* 77 (1955) 6132; Shafer, M. W. and Roy, R. *Z. Kristallogr. Kristallgeom.* 110 (1958) 241

Perovskite derivatives including tungsten bronzes, etc.

Stadelmaier, H. H.[139] [Systematic review of non-stoichiometric ABO_3 phases]; Leonidova, G. C. and Polandov, I. N. *Soviet Phys. solid St.* 4 (1963) 616; with Kudzin, A. Ya. *Soviet Phys. Crystallogr.* 7 (1963) 646 [$BaTiO_3$]; Hardy, B. *Acta crystallogr.* 15 (1962) 179 [$BaMnO_3$]; Newnham, R. E. *et al., Acta crystallogr.* 17 (1964) 240 [$CoTiO_3$ ilmenite structure and magnetic properties]; Keller, G. S. *et al., J. appl. Phys.* 34 (1963) 1033 [$NiTiO_3$ ilmenite, antiferromagnetism at various (low) temperatures]; Blackman and Kaye[85]; Ishikawa, T. *J. phys. Soc. Japan* 17 (1962, Suppl. B1) 239 [Transition $Fe_2O_3 \rightarrow FeTiO_3$: change of Curie print in hematite/ilmenite series and low-temperature magnetic properties]; Leonidova, G. C. and Polandov, I. N. *Soviet Phys. Solid St.* 4 (1963) 1616 [Curie point depression in BaTiO perovskites through high applied pressure]; Kudzin, A. Ya. *Soviet Phys. Crystallogr.* 7 (1963) 646 [Dielectric properties of $BaTiO_3$ perovskite, containing varying small amounts of Co; variation with temperature]; Wood, E. A. *et al., Acta crystallogr.* 15 (1962) 273 [Antiferroelectric/ferroelectric transition in $NaTiO_3$ perovskite; relation to structure distortions]; Hardy, B. *Acta crystallogr.* 15 (1962) 179 [Structure by neutron diffraction of $LiNbO_3$ perovskite, relation to ferroelectric properties; low-temperature polymorph in $BaMnO_3$, perovskite-based, with changes in stacking sequence]; Iwasaki, H. and Ikeda, T. *J. phys. Soc. Japan* 18 (1963) 197 [Structural and ferroelectric changes in Na(Nb, Ta)O_3 perovskite solid solutions]; Lüthi, B. *Jnl Phys. Chem. Solids* 23 (1962) 35 [Yttrium garnets: thermal conductivity as function of (low) temperatures]; Spenser, E. G. *et al., Phys. Rev.* 125 (1962) 1950 [Microwave acoustic losses in yttrium garnets; internal friction]; LeCraw, R. C. *et al., Phys. Rev.* 110 (1958) 1311 [Yttrium–iron garnets, magnetic properties]

Barium ferrites

Filipov, V. S. and Fesenko, E. G. *Soviet Phys. Crystallogr.* 6 (1962) 616 [Structure and spacings of compounds of type Ba_2(Fe, Mn, Y, La, Pt, Na, Sm)–(Ta, Nb)O_6]; Livshits, B. G. *et al., Bull. Acad. Sci. USSR, Phys. Ser.* 25 (1961) 1429 [Phase equilibria in BaO–Fe_2O_3 system; identification and magnetic properties of BaO.Fe_2O_3 and BaO.6Fe_2O_3]; Srikanta, S. *et al., Acta crystallogr.* 15 (1962) 255 [Phase relations and structure in system $BaTiO_3$–$PbTiO_3$–$PbNb_2O_6$]

ABO_4 rutile- and columbite-type phases[3e,94,145]

Bayer[103a,b] [Review of isomorphies and structure variations in two-metal rutiles]; Magneli, A. and Marinder, B. O. *Ark. Kemi Miner. Geol.* 21 (1963) 407; Byström, A. *et al.* 15 (1941) 1; Sturdivant, J. H. *Z. Kristallogr. Kristallgeom.* 75 (1930) 88; Legrand, G. and Delville, J. *C.r. hebd. Séanc. Acad. Sci., Paris* 236 (1953) 944; Rooksby and White[3e] [ABO_4 oxides of rare earths; Nb_2O_5, Ta_2O_5]; Roth, R. S. *J. Am. Ceram. Soc.* 44 (1961) 49 [CdO–Nb_2O_5]; *J. Res. natn. Bur. Stand.* 62 (1959) 27; *Acta crystallogr.*

10 (1957) 437 [PbO–Nb_2O_5; additional columbite-related compounds, e.g. $CoNb_2O_6$, $PbNb_2O_6$]; Roth[81b] [Ca.O–TiO_2 perovskites, etc.]; with Waring, J. L. *J. Res. natn. Bur. Stand.* 65A (1961) 337 [BaO–Nb_2O_5]; and Schneider, S. J. 64A (1960) 309; Schneider and Roth[135] [Mixed rare earth oxides of perovskite, etc. types]; Schneider *et al.*[134] [Mixed oxide systems of $A^{3+}B^{3+}O_3$ type, most involving rare earths]; Roth, R. S. and Wadsley, A. D. *Acta crystallogr.* 19 (1965) 26, 32, 38, 42 [Nb_2O_5–WO_3; structures and intermediate phases]; Coates, R. N. and Kay, H. F. *Phil. Mag.* 3 (1958) 1449 [AB_2O_6 niobate and tantalate ceramics synthesized; dielectric properties]; King, B. W. and Suber, L. *J. Am. Ceram. Soc.* 38 (1955) 306 [V_2O_5 + additional oxides; fusion temperatures, etc.]; Burdese, A. *Annls Chim.* 47 (1957) 804, 824 [VO–FeO–O]

Miscellaneous double oxide systems

Bertaut, F. and Delorme, C. *C.r. hebd. Séanc. Acad. Sci., Paris* 236 (1953) 74 [Fe–Cu oxides]; 238 (1954) 1829 [Co–Cu oxides]; Bertaut, F. *et al.*, *Jnl Phys. Chem. Solids* 21 (1961) 234 [(Nb,Ta)–*M* oxides]; Davey, F. *et al.*, *J. electrochem. Soc.* 98 (1951) 356 [ZrO_2–V_2O_5]; Geach, G. A. and Harper, M. E. *Metallurgia* 47 (1953) 269 [BeO–ThO_2]; Kawakubo, T. *et al.*, *J. phys. Soc. Japan* 15 (1960) 2102 [Ti_2O_3–V_2O_3, etc.]; Muan, A. and Omiya, S. *Am. J. Sci.* 260 (1962) 230 [Fe–Mn oxides]; Rooymans, C. J. M. *J. phys. Soc. Japan* 17 (1962) 722 [Na–Fe_2O_3]; Schönberg, N. *Acta chem. scand.* 8 (1954) 630, 932; *Acta metall.* 3 (1955) 14 [Ternary oxides, Mn–Ta–O, etc.]; Smith, G. W. *Acta crystallogr.* 15 (1962) 1054 [Ni–Mo oxides]; Yakel, H. C. *et al.* 16 (1963) 957 [Rare earth–Mn oxides]; Campbell, W. J. and Parker, J. G. *U.S. Bur. Mines, Rep. Invest.* 5150 (1955) [Columbite/tantalite series (Fe,Mn).$(Nb,Ta)_2O_6$]; Berman, J. and Campbell, W. J. 5300 (1957) [Huebnerite/ferberite series (Fe,Mn) WO_4]

APPLICATIONS; GENERAL ASPECTS

Ceramics

Flörke, O. W. *Ber. dt. keram. Ges.* 40 (1963) 451 [High-melting oxides, crystal chemistry]; Murray, P. *Jnl Br. ceram. Soc.* 1 (1963) 113 [Ceramic oxides]; Popper, P. (Ed.) *Special Ceramics*, 3 vols., New York (Academic Press) 1962–5; Stewart, G. H. (Ed.) *Science of Ceramics*, 3 vols., New York (Academic Press) 1962–7; Hove, J. E. and Riley, W. C. (Ed.) *Modern Ceramics, Some principles and concepts*, New York (Wiley) 1965; Burke, J. E. *Progress in Ceramic Science*, 2 vols., London (Pergamon) 1961; Hague, J. R. *et al.*, *Refractory Ceramics for Aerospace and other High Temperature Applications*, Battelle Mem. Inst. and Am. Ceramic Soc., 1964; Kingery, W. D. (Ed.) *Kinetics of High Temperature Processes*, New York and London (Wiley and Chapman & Hall) 1959 [Oxide ceramics]; McMillan, P. W. *Glass Ceramics*, New York (Academic Press) 1964; Klingsberg, C. (Ed.) *Symposium Proceedings, Ceramics*, Penn. State Univ., 1962, Gordon & Breach; Ryshkewitch, E. *Oxide Ceramics*, New York (Academic Press) 1960; Paper in 5th Plansee Seminar (1964) [Oxide-metal ceramics]; Parker, E. R. *et al.*, *J. Metals, N.Y.* 10 (1958) 351 [Ductile ceramics]; Chesters, J. H. *Steelplant Refractories*, Sheffield (United Steel Co.) 1963; Taylor, H. F. W. (Ed.) *The Chemistry of Cements*, 2 vols., New York (Academic Press) 1964

Magnetic oxides

Standley, K. J. *Oxide magnetic materials*, Oxford Univ. Press, 1962; Goodenough, J. B. *Magnetism and the Chemical Bond*, New York (Wiley) 1963 [Oxide magnets, spinel types, garnets, etc.]

Dielectric materials

Birks, J. B. (Ed.) *Modern Dielectric Materials*, New York (Academic Press) 1960; Birks, J. B. and Hart, J. (Ed.) *Progress in Dielectrics*, London (Heywood) 1963; Jonker, G. H. *Angew. Chem.* 76 (1964) 175 [Mixed oxides for electric applications]; Lucker, K. E. *Soviet Phys. tech. Phys.* 98 (1958) 2428 [Nb and Zr oxides as dielectrics]; Free, J. M. *et al.*, *Br. J. appl. Phys.* 14 (1963) 289 [TiO_2 ceramics as dielectric materials]

Ferroelectrics, etc.

Megaw, H. D. *Ferroelectricity in Crystals*, London (Methuen) 1957; Sachse, H. *Ferroelektrika*, Berlin (Springer) 1956; Jona, F. and Shirane, G. *Ferroelectric Crystals*, Oxford (Pergamon) 1962

Semiconductors

Arikosov, N. Kh. (Ed.), *Semiconductor Materials*, New York (Consultants Bureau) 1964; Biondi, F. J. (Ed.) *Transistor Technology*, New York (Van Norstrand) 1959; Hunter, L. P. (Ed.) *Handbook of Semiconductor Electronics*, New York (McGraw-Hill) 1962; Shockley, W. *Electrons and Holes in semiconductors*, New York (Van Norstrand) 1950

Irradiation effects including oxides

(*a*) Smoluchowski, R.; (*b*) Fan, H. Y. and Lark-Horovita, K. in *The Effect of Radiation on Materials*, New York (Reinhold), 1958; Stobbs, J. J. and Swallow, A. J. *Metall. Rev.* 7 (1962) 951

Oxidation processes

Evans[154]; Kubaschewski, O. and Hopkins, B. E. *Oxidation of Metals and Alloys*, London (Butterworths) 1962; Hauffe, K. *Reaktionen in und an Festen Stoffen*, Berlin (Springer) 1955; Promisel, N. E. (Ed.) *The Science and Technology of W, Ta, Mo, Nb and their Alloys*, Oxford (Pergamon) 1963; Quarrell, A. G. (Ed.) *Nb, Ta, Mo, W and their Alloys*, Amsterdam (Elsevier) 1961; Gray, T. G. *et al.*, *Oxidation of Metals*, New York (Consultants Bureau) 1965; *Oxidation of Metals*, Paris (Gauthier-Villars) 1962; Kofstad, P. *Oxidation of Metals at High Temperatures*, New York (Wiley) 1966

Internal oxides in metals

Grant, N. J. and Preston, O. *Trans. Am. Inst. Min. Engrs* 209 (1957) 349; Grant, N. J. in *The Strengthening of Metals*, (Ed.) Peckner, D., New York (Reinhold) 1964 [Dispersion strengthening by oxides]; Cupp, C. R. *Progress in Metal Physics* 5, London (Pergamon) 1953; Gregory, E. and Smith, C. G. *J. Inst. Metals* 85 (1956) 81 [Internal oxidation]; Ashby, M. F. and Smith, C. G. 91 (1963) 182 [Internal oxidation of Cu alloys]

Thermodynamics

Richardson, F. D. *J. Iron Steel Inst.* 160 (1948) 261; 166 (1950) 137, 213 [Thermodynamics of oxides and deoxidation]; in *The Physical Chemistry of Steel Making*, ed. Elliott, J. F., p. 55, New York and London (Wiley and Chapman & Hall) 1956; Schenck, H. *The Physical Chemistry of Steel Making*, London (B.I.S.R.A.) 1946 [Sections on oxide reactions in slags, refractories, etc.]; Coughlin, J. P. *Bull. U.S. Bur. Mines* 542 (1954) [Heat and free energy of formation of oxides]; Kelley, K. K. 584 (1960) [Heat content, etc.]; Brewer, L. *Chem. Rev.* 52 (1953) 1 [Thermodynamic data]; Tripp, H. P. and King, B. W. *J. Am. Ceram. Soc.* 38 (1955) 12, 432; Schneider[4a] [Melting points of metal oxides]; Kingery, W. D. (Ed.) *Proc. Int. Symp. High-Temp. Technol.* New York (McGraw-Hill) 1960 [Table of oxide combinations, liquidus temperatures, etc.]; Livey, D. T. and Murray, P. *J. less-common Metals* 1 (1959) 145 [High-temperature stability of oxides and sulphides]; Runck, R. J. in *High Temperature Materials*, ed. Campbell, I. E., New York (Wiley) 1956 [On refractory oxides]

Phase constitutions, general

Levin, C. M., Robbins, C. R. and McMurdie, H. F. *Phase Diagrams for Ceramists* Columbus, Amer. Ceram. Soc. 1964; Hansen, M. *Constitution of binary Alloys* and Elliott, R. P., Supplement, New York (McGraw-Hill) 1958, 1965; Pearson, W. B. *A Handbook of Lattice Spacings of Metals and Alloys*, London (Pergamon) 1958, 1966; Schubert, K. *Kristallstrukturen Zweikomponentiger Phasen*, Berlin (Springer) 1964

Structural chemistry

Wells, A. F. *Structural Inorganic Chemistry*, Oxford Univ. Press, 1962; Cotton, F. A. and Wilkinson, G. *Advanced Inorganic Chemistry*, New York (Wiley) 1962

REFERENCES

Lattice defects in oxides

Read, W. T. and Shockley, W. *Phys. Rev.* 87 (1952) 835 [Non-stoichiometry, etc.]; Smoluchovski, R. *Imperfections in nearly perfect Crystals*, New York (Wiley) 1953; Kroger, F. A. *Chemistry of Imperfect Crystals*, Amsterdam (North Holland) 1964; Verwey, E. J. W. *Chem. Weekbl.* 43 (1947) 229; 49 (1948) 705 [Theory of *vacancies*, etc.]; *The Nature of Defects in Crystals* (Part II of an International Symposium, Melbourne (Aust. Acad. Sci.), 1965

Wagner, C. *Z. phys. Chem.* 223 (1933) 181; with Schottky, W. 113 (1930) 163; Schottky, W. 29B (1935) 335; Frenkel, J. *Z. Phys.* 35 (1926) 659; *Phys. Z. SovjUn.* 8 (1935) 185 [Pioneer work on defect structures, including oxides]

Minerals

Bragg, W. L. and Claringbull, G. F. *Crystal Structures of Minerals*, London (Bell) 1965; Hintze, C. *Handbuch der Mineralogie*, Berlin (de Gruyter) 1904 to present; Brauns, R. *Allgemeine und Spezielle Mineralogie*, Berlin (de Gruyter) 1958–9; Dana, E. S. *Dana's Manual of Mineralogy*, New York (Wiley) 1959

REFERENCES

[1] Brewer, L. and Krikorian, O. *Univ. Calif. Radiat. Lab. Rep.* 2888 (1955); *J. electrochem. Soc.* 103 (1956) 38

[2] Kiessling, R. *Metall. Rev.* 2 (1957) 77

[3] Rooksby, H. P. (*a*) *Nature, Lond.* 152 (1943) 304; (*b*) *Acta crystallogr.* 1 (1948) 226; (*c*) Tombs, N. C. and Rooksby, H. P. 4 (1951) 474; (*d*) Rooksby, H. P., with Willis, B. T. M. 6 (1953) 565; (*e*) with White, E. A. D. 16 (1963) 88.

[4] (*a*) cf. Schneider, S. J., 'Compilation of the Melting Points of the Metal Oxides', *Natn. Bur. Stand. Monogr.* 68, 1963; (*b*) Anderson, G. *et al.*, *Acta chem. scand.* 11 (1957) 1614, 1653

[5] Nowotny, H. and Dimakopoulu, E. *Mh. Chem.* 91 (1960) 877

[6] Hägg, G. and Schönberg, N. *Acta crystallogr.* 7 (1954) 351

[7] Anderson, S. and Magneli, A. (*a*) *Naturwissenschaften* 43 (1956) 495; (*b*) Magneli, A. *et al.*, *Acta chem. scand.* 9 (1955) 1402; (*c*) *ibid.* 1382; (*d*) *Acta crystallogr.* 6 (1953) 495; (*e*) 9 (1956) 1038; (*f*) with Anderson, G. *Acta chem. scand.* 9 (1955) 1378

[8] De Vriess, R. C. and Roy, R. *Bull. Am. Ceram. Soc.* 33 (1954) 370

[9] Smith, D. K. and Cline, C. F. *J. Am. Ceram. Soc.* 451 (1962) 249

[10] (*a*) Meyer, H. *Ber. dt. keram. Ges.* 39 (1962) 1187; (*b*) Hinz, J. and Dietzel, A. *ibid.* 534

[11] (*a*) Kofstad, P., with Ruzicka, D. J. *J. electrochem. Soc.* 110 (1963) 181; (*b*) with Anderson, P. B. *Jnl Phys. Chem. Solids* 21 (1962) 280; (*c*) *J. less-common Metals* 5 (1963) 158; *J. electrochem. Soc.* 110 (1963) 491; (*d*) with Espevik, S. 112 (1965) 153

[12] Ashbee, K. H. G. and Smallman, R. E. (*a*) *Proc. R. Soc.* A274 (1963) 165; (*b*) *J. Am. Ceram. Soc.* 46 (1963) 211; (*c*) *Phil. Mag.* 7 (1962) 1933; (*d*) with Williamson, G. K. *Proc. R. Soc.* A276 (1963) 542; (*e*) *Proc. Br. ceram. Soc.* 1 (1964) 201

[13] Hirthe, W. M. and Brittain, J. O. *J. Am. Ceram. Soc.* 45 (1962) 546

[14] Willis, B. T. M. *Proc. R. Soc.* A274 (1963) 122, 134; *Proc. Br. ceram. Soc.* 1 (1964) 9

[15] Preisinger, A. *Naturwissenschaften* 49 (1962) 345

[16] (*a*) Goldschmidt, H. J., with Brand, J. A. *J. less-common Metals* 3 (1961) 44; (*b*) *J. Iron Steel Inst.* 146 (1952) 157; (*c*) *Metallurgia* 62 (1960) 211, 241; (*d*) *J. Inst. Metals* 87 (1959) 235; (*e*) *Research* 10 (1957) 289; (*f*) *J. Iron Steel Inst* 194 (1960) 169; (*g*) *Adv. x-ray Analysis* 5 (1962) 191

[17] Cowley, J. M. *Acta crystallogr.* 6 (1959) 53

[18] Yamagushi, S. *Trans. Faraday Soc.* 58 (1962) 1690

[19] Verwey, E. J. W. *Z. Kristallogr. Kristallgeom.* 91 (1935) 65

[20] Barth, T. F. W. and Posniak, E. *J. Wash. Acad. Sci.* 21 (1931) 225; *Z. Kristallogr. Kristallgeom.* 82 (1932) 235

[21] van Oosterhout, H. A. and Rooymans, C. T. M. *Nature, Lond.* 181 (1958) 44

[22] Zachariasen, W. H. (*a*) *Acta crystallogr.* 1 (1948) 265; 2 (1949) 388; 5 (1952) 664; (*b*) *Skr. norske VidenskAkad.* 4 (1928) 152; (*c*) Mooney, R. C. L. and Zachariasen, W. H. in *The Transuranium Elements*, ed. Seaborg, G. T. *et al.*, II, 1442, New York (McGraw-Hill) 1949

[23] Makaroff, E. S. *Crystal Chemistry of Simple Compounds of U, Th, Pu and Np*, New York (Consultants Bureau) 1959

[24] Bumps, E. S., Kessler, H. D. and Hansen, M. *Trans. Am. Soc. Metals* 45 (1953) 1008

[25] Cadoff, I. and Palty, A. E. *U.S. At. Energy Comm.* NAL–401–14–30, 1953
[26] Jenkins, A. E. and Worner, H. W. *J. Inst. Metals* 80 (1951) 157
[27] Wang, C. C. and Grant, N. J. *J. Metals, N.Y.* 6 (1954) 200
[28] Schofield, T. H. and Bacon, A. E. *J. Inst. Metals* 84 (1955) 47
[29] Jaffee, R. I. *et al.*, *Trans. Am. Inst. Min. metall. Engrs* 188 (1950) 1261
[30] Yao, Y. L. *Trans. metall. Soc. A.I.M.E.* 215 (1959) 851
[31] Bright, N. F. N. *Adv. x-ray Analysis* 4 (1960) 175
[32] Rostoker, W. (*a*) *J. Metals, N.Y.* 4 (1952) 209, 982; (*b*) with Yamamoto, A. S. *Trans. Am. Soc. Metals* 47 (1955) 1002; (*c*) *Trans. Am. Inst. Min. metall. Engrs* 203 (1955) 113
[33] Hoch, M. *J. phys. Soc. Japan* 18 (1963, Suppl. 2) 147
[34] Holmberg, B. *Acta chem. scand.* 16 (1962) 1245
[35] Ehrlich, P. *Z. anorg. allg. Chem.* 259 (1949) 1
[36] Vousden, P. *Acta crystallogr.* 9 (1956) 141
[37] Wachtman, J. B. *et al.*, *J. Res. natn. Bur. Stand.* 66A (1962) 465; 67A (1963) 281
[38] Domagala, R. F. and McPherson, D. J. *J. Metals, N.Y.* 6 (1954) 238
[39] Treco, R. M. *Trans. Am. Soc. Metals* 45 (1953) 872; *Trans. Am. Inst. Min. metall. Engrs* 197 (1953) 344
[40] Dietzel, A. and Tober, H. *Ber. dt. keram. Ges.* 30 (1953) 47, 71
[41] Schönberg, N. (*a*) *Acta chem. scand.* 8 (1954) 1347; (*b*) *ibid.* 221; (*c*) *ibid.* 617; (*d*) *ibid.* 932
[42] Gebhardt, E. *et al.*, *J. nucl. Mater.* 4 (1961) 241, 255
[43] Teufer, G. *Acta crystallogr.* 15 (1962) 1187
[44] Woltern, G. M. *J. Am. Ceram. Soc.* 46 (1963) 418
[45] Rudy, E. and Stecher, P. *J. less-common Metals* 5 (1962) 78
[46] Bain, W. L. *Science, N.Y.* 140 (1963) 1330
[47] Mathewson, C. H. *et al.*, *Trans. Am. Soc. Steel Treat.* 20 (1932) 357
[48] Klemm, W. *et al.*, *Z. anorg. allg. Chem.* 250 (1942) 42; *Optik, Stuttg.* 3 (1948) 75
[49] Seybolt, A. U. and Sumsion, H. T. *Trans. Am. Inst. Min. metall. Engrs* 197 (1953) 292
[50] Kosuge, K. *et al.*, *J. phys. Soc. Japan* 19 (1963) 318
[51] Tucker, C. W. *Acta metall.* 1 (1953) 390
[52] Allen, N. P., Kubaschewski, O. and Goldbeck, O. *J. electrochem. Soc.* 98 (1951) 417
[53] Anderson, G. *Research, Lond.* 6 (1953) 45S; *Acta chem. scand.* 8 (1954) 1599; 10 (1956) 623
[54] Aebi, F. *Helv. chim. Acta* 31 (1948) 8
[55] Heckingbottom, R. and Linnett, J. W. *Nature, Lond.* 194 (1962) 678
[56] Westman, S. *Acta chem. scand.* 15 (1961) 217
[57] Byström, A. *et al.*, *Acta chem. scand.* 4 (1950) 1119
[58] Brauer, G. (*a*) *Z. anorg. allg. Chem.* 248 (1941) 1; *Z. Elektrochem.* 46 (1940) 397; (*b*) Brauer G. *et al.*, *J. less-common Metals* 4 (1962) 533
[59] Argent, B. B. and Phelps, B. *J. Inst. Metals* 88 (1959) 301
[60] Marinder, B. O. *et al.*, *Ark. Kemi Miner. Geol.* 19 (1962) 435; *Acta chem. scand.* 16 (1962) 293
[61] Norman, N. *J. less-common Metals* 4 (1962) 52
[62] Norman, N., Kofstad, P. and Krudtaa, O. J. *J. less-common Metals* 4 (1962) 124
[63] Grube, G. *et al.*, *Z. Elektrochem.* 45 (1939) 881
[64] Wasilewski, R. J. *J. Am. chem. Soc.* 75 (1953) 1001; *Trans. metall. Soc. A.I.M.E.* 221 (1961) 647
[65] Lagergren, S. and Magneli, A. *Acta chem. scand.* 6 (1952) 444
[66] Lux, H. and Proeschl, E. *Z. anorg. allg. Chem.* 259 (1948) 73
[67] Hilty, D. C. *et al.*, *Trans. Am. Inst. Min. metall. Engrs* 203 (1955) 253
[68] Hägg, G. and Schöenberg, N. *Acta crystallogr.* 7 (1954) 351
[69] Kihlborg, L. (*a*) *Adv. Chem. Ser.* 39 (1963) 37; *Acta chem. scand.* 13 (1959) 954; 17 (1963) 487; (*b*) *Ark. Kemi Miner. Geol.* 21 (1963) 473
[70] Sienko, M. J. and Banerjee, B. *J. Am. chem. Soc.* 83 (1961) 4149
[71] Grønvold, F. and Haraldsen, H. *Nature, Lond.* 162 (1948) 69
[72] Alberman, K. B. and Anderson, J. S. *J. chem. Soc.* (1949) 303
[73] Aronson, S. *et al.*, *J. chem. Phys.* 35 (1961) 1382
[74] *Peaceful Uses of Atomic Energy*, Geneva (United Nations) 1958; (*a*) Brabers, M. J., p. 122; (*b*) Holley, C. R. *et al.*, p. 215
[75] Belle, S. and Austern, A. B. in *Kinetics of High Temperature Processes*, Kingery, W. D. (Ed.), New York (Wiley) 1959
[76] Rothwell, E. *J. nucl. Mater.* 6 (1962) 229
[77] Robins, R. G. *J. nucl. Mater.* 7 (1962) 218
[78] Armstrong, W. M. *et al.*, *J. nucl. Mater.* 7 (1962) 133
[79] Brown, F. H. and Duwez, P. *J. Am. Ceram. Soc.* 37 (1953) 129
[80] Coughanour, L. W. *et al.*, *J. Res. natn. Bur. Stand.* 52 (1957) 37
[81] (*a*) Roth, R. S., with Coughanour, L. W. *J. Res. natn. Bur. Stand.* 52 (1954) 37; 55 (1955) 209; (*b*) 61 (1958) 437; (*c*) Personal communication
[82] Dyson, P. *Acta crystallogr.* 10 (1957) 140

[83] Wang, C. C. and Grant, N. J. *J. Metals, N.Y.* 6 (1954) 200
[84] Grieve, J. and White, J. *Jl R. tech. Coll., Glasg.* 4 (1940) 661
[85] Ishikawa, Y. and Syono, Y. *Jnl Phys. Chem. Solids* 24 (1963) 517
[86] Webster, A. H. and Bright, N. F. H. *J. Am. Ceram. Soc.* 44 (1961) 110
[87] Duwez, P. and Loh, E. *J. Am. Ceram. Soc.* 40 (1957) 324
[88] Voronov, N. M. *et al.*, in *Peaceful Uses of Atomic Energy* 6, p. 221, Geneva (United Nations) 1958
[89] Borchardt, H. J. *Inorg. Chem.* 2 (1963) 170
[90] Lamberton, W. A. and Müller, M. H. *J. Am. Ceram. Soc.* 36 (1953) 367
[91] Aronson, S. and Clayton, J. R. *J. chem. Phys.* 35 (1961) 1055
[92] Saller, J. A. *et al.*, *U.S. At. Energy Comm., Battelle Mem. Inst. Rep.* 1023 (1955)
[93] Fischer, W. A. and Hoffman, A. *Arch. EisenhüttWes.* 28 (1957) 743
[94] Brandt, K. *Ark. Kemi Miner. Geol.* 17 (1943) 1
[95] Cox, D. E. *et al.*, *Jnl Phys. Chem. Solids* 23 (1962) 863
[96] Wold, A. *et al.*, *J. appl. Phys.* 33 (1962, Suppl.) 1208
[97] Mathewson *et al.*, *Trans. Am. Soc. Steel Treat.* 20 (1930) 357
[98] Jagar, W. *et al.*, *Arch. EisenhüttWes.* 35 (1963) 291
[99] Holtzberg, F. and Reisman, A. *J. phys. Chem., Ithaka* 65 (1961) 1192
[100] Mohanty, G. P. *et al.*, *Acta crystallogr.* 15 (1962) 1190
[101] Mohanty, G. P. and Ziegel, L. J. *Acta crystallogr.* 15 (1962) 1196
[102] Bertaut, E. K. *et al.*, *Jnl Phys. Chem. Solids* 21 (1961) 234
[103] Bayer, G. (*a*) *J. Am. Ceram. Soc.* 43 (1960) 495; (*b*) *Ber. dt. keram. Ges.* 39 (1962) 535; (*c*) *Naturwissenschaften* 48 (1961) 46
[104] Felton, E. J. *et al.*, *J. inorg. nucl. Chem.* 24 (1962) 839
[105] Woodhouse, D. and White, J. *Trans. Br. ceram. Soc.* 54 (1955) 365
[106] Cox, D. E., Takei, W. J. and Shirane, G. *Jnl Phys. Chem. Solids* 24 (1963) 405
[107] Thomassen, H. *J. Am. chem. Soc.* 62 (1940) 1134
[108] Smith, G. W. *Acta crystallogr.* 15 (1962) 1054
[109] Evans, D. A. and Jack, K. H. *Acta crystallogr.* 10 (1957) 769
[110] Trunov, V. K. *et al.*, *Dokl. Akad. Nauk SSSR* 141 (1961) 1, 114
[111] Chikalla, T. D. *J. Am. Ceram. Soc.* 46 (1963) 323
[112] Hay, R., Howat, D. D. and White, J. *Jl W. Scott. Iron Steel Inst.* 41 (1933) 97
[113] Van Hoek, H. J. and Keith, M. L. *Am. Mineral.* 43 (1958) 80
[114] Mason, B. *Am. Miner.* 29 (1944) 67
[115] Schenck, H. *Introduction to the Physical Chemistry of Steelmaking*, London (B.I.S.R.A.) 1945
[116] Hilty, D. C. and Crafts, W. *Trans. Am. Inst. Min. metall. Engrs* 188 (1950) 425
[117] Aukrust, E. and Muan, A. *J. Am. Ceram. Soc.* 46 (1963) 511
[118] Roiter, B. D. and Paladino, A. E. *J. Am. Ceram. Soc.* 45 (1962) 128
[119] Wriedt, H. A. and Chipman, J. *Trans. Am. Inst. Min. metall. Engrs* 203 (1955) 477
[120] Lisovsky, D. J. *et al.*, *Tsvet. Metallȳ, Mosk.* (1963) 58
[121] Nevitt, M. V. (*a*) in *Electronic Structure and Alloy Chemistry of the Transition Metals*, ed. Beck, P. H., p. 101, New York (Interscience) 1963; (*b*) with Schwartz, L. H. *Trans. metall. Soc. A.I.M.E.* 212 (1958) 700; with Downey, J. W. and Morris, R. A. 218 (1960) 2019
[122] Geller, J. (*a*) *Acta crystallogr.* 9 (1956) 885; 10 (1957) 380; (*b*) with Yavorsky, P. J. *J. Res. natn. Bur. Stand.* 35 (1945) 87; (*c*) with Gilleo, M. A. *Acta crystallogr.* 12 (1959) 243; (*d*) 10 (1957) 243
[123] Mueller, M. H. and Knott, H. W. *Trans. metall. Soc. A.I.M.E.* 227 (1963) 674
[124] Jeitschko, W. *et al.*, *Mh. Chem.* 95 (1964) 1004
[125] Bertaut, F. and Delorme, C. *C.r. hebd. Séanc. Acad. Sci., Paris* 233 (1951) 256
[126] Duwez, P. and Odell, F. *J. Am. Ceram. Soc.* 33 (1950) 274
[127] Dunitz, J. D. and Orgel, L. E. *Jnl Phys. Chem. Solids* 3 (1957) 318
[128] Wold, A. *et al.*, *J. appl. Phys.* 34 (1963) 1085
[129] Rogers, D. B. *et al.*, *Jnl Phys. Chem. Solids* 24 (1963) 347
[130] Nozrik, Yu. Z. and Yamzin, I. I. *Kristallographiya* 6 (1961) 923
[131] Richards, R. G. and White, J. *Trans. Br. ceram. Soc.* 53 (1954) 233, 422
[132] Cox, D. E. *et al.*, *Jnl Phys. Chem. Solids* 24 (1963) 405
[133] Bertaut, E. F. *et al.*, *Jnl Phys. Chem. Solids* 21 (1961) 234
[134] Schneider, S. J. *et al.*, *J. Res. natn. Bur. Stand.* 65A (1961) 345
[135] Schneider, S. J. and Roth, R. S. *J. Res. natn. Bur. Stand.* 64A (1960) 317
[136] Megaw, H. D. *Proc. R. Soc.* 189 (1947) 261
[137] Roy, R. *J. Am. Ceram. Soc.* 37 (1953) 582
[138] Francombe, M. H. *Acta crystallogr.* 9 (1956) 256
[139] Stadelmaier, H. H. *Z. Metallk.* 61 (1961) 758
[140] Menzer, G. *Z. Kristallogr. Kristallgeom.* 63 (1926) 157; 69 (1928) 300
[141] Yoder, H. S. and Keith, M. L. *Am. Miner.* 36 (1957) 519
[142] Bertaut, F. and Forrat, F. *C.r. hebd. Séanc. Acad. Sci., Paris* 242 (1956) 382; 244 (1957) 69

[143] Marinder, B. O. *et al.*, *Acta chem. scand.* 16 (1962) 293
[144] Nickel, E. H. *et al.*, *Can. Mineralogist* 8 (1964) 40
[145] Keller, G. *Z. anorg. allg. Chem.* 318 (1962) 89
[146] Felten, E. J. and Aitken, E. A. *J. inorg. nucl. Chem.* 24 (1962) 35
[147] Borchardt, H. J. *Inorg. Chem.* 2 (1963) 170
[148] Waring, J. L. and Roth, R. S. *Acta crystallogr.* 17 (1964) 455
[149] Paravano, G. *J. Am. chem. Soc.* 75 (1953) 1448
[150] Arbiter, W. and Stern, G. *Plansee Proc.* 1961 (1962) 547
[151] Runfors, U. *Silic. ind.* (1958) 365
[152] Kubaschewski, O. and Hopkins, B. E. *Oxidation of Metals and Alloys*, New York (Academic Press) 1962
[153] Tylecote, R. F. *J. Iron Steel Inst.* 195 (1960) 380; 196 (1960) 135
[154] Evans, U. R. *The Corrosion and Oxidation of Metals*, London (Arnold) 1963
[155] Argent, B. B. and Phelps, B. *J. Inst. Metals* 88 (1959) 301; *J. less-common Metals* 2 (1960) 181
[156] Ong, J. N. and Fassell, W. M. *Corrosion* 18 (1962) 282
[157] Douglass, D. H. *J. less-common Metals* 5 (1963) 151; *Conf. on Radiation Damage in Reactor Materials*, Salzburg 2 (1962) 223
[158] Cox, B. and Johnson, R. *Trans. metall. Soc. A.I.M.E.* 227 (1963) 36
[159] Bénard, J. (Ed.) *Sous la Direction de l'Oxydation des Metaux*, Paris (Gauthier-Villars) 1964
[160] Krier, C. A. and Jaffee, R. J. *J. less-common Metals* 5 (1963) 411
[161] Rhys, D. W. and Price, E. G. *Engelhardt tech. Bull.* 5 (1964) 37
[162] Vol, Yu. T. and Shishakov, N. A. *Izv. Akad. Nauk SSSR, Otd. Khim. Nauk* 11 (1963) 1920

9

HYDRIDES

9.1 INTRODUCTION AND BASIC POINTS

Hydrogen may perhaps be described as the 'will o' the wisp' amongst the interstitial atoms and, at the same time, the most genuinely interstitial atom of all. Its small size is not much larger than the naturally occurring interstices in the host transition-metal lattice, and these can usually accommodate the hydrogen atom or ion to extensive degrees with less strain on the metal lattice than for say C, N or O. Hägg's rule is always fulfilled. When present as a proton, the single donated valency electron will either contribute to the partly vacant *d* orbitals of the metal atom or to the general mass of free electrons characterizing the metallic state. On the one hand, therefore, hydrogen can virtually play the role of the anion in an ionic crystal; this is the case in NaH and the other alkali hydrides, which are quite equivalent, structurally and in the nature of the bond, to NaCl, with rigid stoichiometry and lattice location. On the other hand (and this is characteristic of the transition-metal hydrides), the H atom can be absorbed in the metal lattice to variable extents and at variable interstices, implying wide homogeneity ranges and a high mobility between parent lattice vacancies. As it enters into the pure metal lattice such as that in Groups IV and V, the H atom tends to cause lattice expansion and distortions up to a high compositional limit, sometimes with low-temperature ordering effects, and phase transformation becomes gradual.

The concept of 'stoichiometry' also becomes ill-defined, often meaningless, amongst the transition-metal hydrides. Furthermore, the stability of given hydrides and their saturation contents depend strongly on pressure and temperature. The capacity for supersaturation and supercooling, with accompanying hysteresis effects, is pronounced in some but, as will arise, not all cases, and the manner of bonding (metallic or ionic) can even, for any one composition, be temperature- and pressure-sensitive, as it is in the oxides. A difference, however, is that, whereas the truly interstitial character holds good only for the lower oxides and solid solutions and soon departs from metallic bonding to ionic or covalent oxides, *all* transition-metal hydrides, up to the highest, are interstitial compounds, even though the character of the bond may be variable and sometimes controversial.

This implies that the distinction between 'hydrogen solid solutions' and 'hydrides' becomes blurred. Indeed, in many cases the transition is continuous with varying concentrations, pressures and temperatures. Therefore, much of this Chapter will have to deal with solid solutions and hydrides as one logical entity (some aspects of this have been described already in Chapters 2 and 3). In particular, the important concept arises of the hydrogen atom itself adopting essentially metallic character when built into the lattice, the more so at high pressures. In this sense, hydrogen may be regarded as playing the part of 'the lightest of the alkali metals'—a part, however, from which it can be deflected to the exact opposite, namely to a halide-type ion (as in NaH). The ancestral position of hydrogen as an element, preceding the division into metal and non-metal in the Periodic Table, but containing the seed of each, is thus well reflected in its alternative modes in solid-state behaviour.

It is characteristic of the transition-metal hydrides that they are metallic in properties, with a high degree of lattice mobility (which reaches its climax in palladium); in contrast to other interstitial compounds, they also retain a measure of ductility*. This again is consistent with the hydrogen atom adopting a quasi-metallic character when present in the lattice. The transition of the hydrogen atom from the ionic to the metallic condition becomes effective in Groups III–V, i.e. in the early transition metals where the *d* band of the metal becomes more receptive to the *s* electron of the hydride. Ubbelohde[1] was one of the leading investigators to demonstrate this thermodynamically.

The theoretical possibility of solid hydrogen as an element adopting a metallic form near absolute zero has been considered for many years. For example, it has been calculated that the metallic state could in fact be attained, but only under extreme pressures and producing inordinately high densities. However, it seems possible that just this condition is fulfilled in the hydride: the metallic bonding forces acting upon the hydrogen atom, when *built into* the metal hydride lattice, could effectively exert such pressures and thus render it metallic.

The versatility in state of combination can thus present the hydrogen atom (or proton) in several guises:

(1) as a pseudo-metal, e.g. in transition-metal alloys;

(2) as a pseudo-halogen, e.g. in alkali and alkaline earth hydrides;

(3) as an element forming one covalent bond, as in hydrocarbons;

(4) as an element forming two covalent bonds, e.g. to two oxygen atoms in the formation of organic molecules.

We are not here concerned with (3) and (4) which are mentioned for systematic completeness. Also some covalent features are evident already in the later-Group transition-metal hydrides. The wider aspect of the (covalent) *hydrogen bond* is, of course, a prodigious subject, though outside our scope (but cf. Bibliography). From a systematic viewpoint it is noticeable that here, too, the bonding transition from metallic to saline and to metal-organic hydrides is not sharp but continuous. The hydrogen bond may well be regarded as a unifying link in the whole of chemistry, and the metal hydrides as a singular case.

A classification of hydrides into *metallic*, *ionic* and *covalent* types clearly follows. The latter mainly apply to the *post*-transition metals, namely the Group IIIB–VIIB hydrides, e.g. silanes, germanes, gallanes, etc., frequently gaseous or liquid at room temperature. The ionic hydrides chiefly comprise the *pre*-transition metals: alkalis, alkaline earths, etc.

---

* Despite 'hydrogen embrittlement' (see below) which is physically of a different nature and mostly more moderate than, say, the brittleness inherent in a carbide.

Whilst metallic hydrides will be our principal subject, the ionic class is still relevant, as a degree of ionization characterizing the Group IA and IIA (alkali and alkaline earth) hydrides is retained when passing to the transition-metal hydrides of Groups III–VIIIA. The negative (H^-) ion must clearly possess a proton with two electrons instead of the single valency electron of neutral H. This applies to the alkali metals, to give the very stable halide-like hydrides. However, as the electronegativity of the partner metal increases on passing to the later Groups, the excess electron of hydrogen will tend to be pooled increasingly and captured by the incompleted electron shells of the transition metal; so will also the valency electron, lending to hydrogen varying degrees of 'metallic' or protonic character. The wide variation in properties and stability of the transition-metal hydrides within each binary system with concentration, pressure and temperature, and also with Group progression, can partly be attributed to this variation.

It has been shown[2–4a] that the internuclear distance in ionic metal hydrides approximates to the sum of an H^- radius (1·29 Å, for coordination number 4) and the metal cation radius based on maximum normal oxidation of the metal. In fact, this is found still to apply, with certain corrections, to the transition metals of Groups III, IV and later. It should only be regarded as a formal rationalization, not describing the 'real' mode of bonding, but it is useful and agrees with a delocalized covalent model of Pauling with a single-bond H *atom* radius of 0·3–0·4 Å. These theories imply that the difference between the saline and metallic hydrides is not a sharp one. Really, for the transition-metal hydrides, the interpretations of bonding as ionic, metallic or covalent can all be justified, with due recognition of coordination number and degree of ionization, though with varying emphasis. This situation also re-emphasizes the rather controversial nature of atomic or ionic 'radii', which in the hydrides becomes even more of a sliding concept than is inherent already in the transition metals themselves.

The theoretical radius of a *free* hydrogen anion[5] is about 2·08 Å. On the other hand, the H radius observed in ionic hydrides is much smaller, in LiH e.g. 0·58 Å. This is still larger than the normal covalent radius of H (0·30 Å). Likewise, in a fully saturated hydride such as ZrH_2, the mean distance of interstitial hydrogen atoms is only 2·7 Å, whereas in elemental *liquid* hydrogen it is 3·9 Å—indicative again of the protonic character of the lattice hydrogen.

Support for the concept of hydrogen occurring as *protons* in transition-metal solutions comes from various sources. For instance, in hydrogenated palladium[6] exposed to an electric potential the hydrogen migrates to the cathode, though so slowly as to indicate that the protonic state is only partial, i.e. the single-valency electron is almost but not entirely pooled with the metallic (conduction) electrons.

We may regard the dissolution of a hydrogen atom in a transition metal as an addition of (*a*) a proton and of (*b*) an electron to it, which may, but need not, behave as two distinctive and separate units. They 'modulate the metal' rather than dissolving in it as an entire elemental atom. Therefore, since the sizes of the proton and the electron are themselves insignificant, the 'atomic size' of hydrogen assumes a second-order role where the proton–electron association is retained within the solvent metal. Thus, upon the absorption of molecular (gaseous) hydrogen, the division $H_2 \rightarrow H + H$ is not bound to be a simple dissociation into two atoms but can 'overshoot' to a third, independent state: $\rightarrow 2$ protons $+$ 2 electrons. The latter would enter either by contributing to the only partially filled metal orbitals (in the earlier Groups) or join the electron sea (in the later Groups). Even for a given metal, the degree to which this process extends is variable with pressure and temperature. We see here a fundamental point of different between hydrogen and the previously discussed interstitials (C, N, Si, B, O) which, even in their most electropositive (electron-barest) state, had a substantial size.

This proton/electron separation is, of course, only an extreme model of the behaviour of H in the metallic lattice, and should be qualified by saying that, as soon as the proton enters, it may lose its own electron but exchange it statistically for others. The proton increases the electron density around itself in form of an electron cloud and becomes 'loosely polygamous', in place of its former firm attachment to a single electron[1]. Diffusion of the hydrogen '*atom*', e.g. in palladium, is therefore still meaningful, although its *s*-electron may be transient and borrowed locally during migration. The slow drift of hydrogen observed in saturated Pd when in an electric field to the cathode, however, confirms in part its 'bachelor' or protonic nature.

The system of hydrogen bonding, with increasing electronegativity within the Periodic Table, may be represented as shown.

(1) *Ionic* (saline) →	(2) *Metallic* →	(3) *Covalent bonding*
Alkali — Alkaline earths →	Transition metals →	Zn → Groups to halogens
	Intermediary: Group III Rare earths → actinides	Intermediary: Group IB (Cu, etc.)
Groups I and IIA	Groups IV–VIIIA	

Even outside this family, the transition to other types of hydrogen bond in crystal chemistry is gradual. In the presence of 'second interstitials' like C or O, the formation of CH groups or OH groups eventually provides a bridge from metallic hydrides towards organic bonding, with metallo-organic compounds presenting an intermediate stage. As will be seen in some cases, whilst the presence in the metal hydride lattice of *atomic* hydrogen is assured, intermediate stages occur towards the molecular (H_2) state, localized near surfaces, grain boundaries and (probably) at dislocations. Some of the peculiarities especially of saturated hydrides must also be explained by an 'awareness' of interstitial hydrogen atoms of each other, which becomes operative close to the decomposition temperatures or pressures: HH groups which form a synologue to CH or OH.

A certain degree of covalency of H occurs even in ionic compounds, and we might well regard the transition-metal hydrides in the above classification as transitional between ionic and covalent groups, still revealing some features of each despite the metallic bond.

9.1 INTRODUCTION AND BASIC POINTS

The physical transition from (1) surface (or interface-)adsorbed to (2) occluded and (3) lattice hydrogen may be seen to parallel that from molecular to atomic, in special cases to protonic or anionic hydrogen, with varying intermediate stages of bond type; this can occur even within a single, apparently equilibrated, metallic mass.

An exception to the generally metallic nature of the transition-metal hydrides and to their non-stoichiometric ranges arises for the hydrogen-rich rare earth and actinide hydrides (see later). Also at high *pressures*, a transition can occur to more definite and stoichiometric hydride compounds. Ultra-high pressures may be considered as strengthening in a sense the metallic nature of the H atom, and the compounds arising could well be classed as intermetallic (electron) phases.

Some historical aspects—The first to study the structure of hydrides on any extensive scale was Hägg by *x-ray diffraction*; his observations put it into the general framework of interstitial compounds. It is remarkable in retrospect how many of these comprehensive early results have remained valid. *Figure 9.1* reproduces, as a matter of record, a tabulation given in one of Hägg's early papers[7], and it may be interesting to compare it with the later advances.

Neutron diffraction, perhaps the most powerful of these, helped to locate the hydrogen atom (whether in hydride or solid solution), after x-rays had provided the first structural evidence, of the metal positions only. Shull and his co-workers[8] were the first to 'locate' hydrogen in NaH (and the deuterium atom in NaD*), and much neutron diffraction work, e.g. on TiH_2, ZrH_2, YH_2, etc. has to be referred to. (The study of metallic *hydrides* formed only part of this evolution, in which neutron diffraction also has been able to locate the H atom, and to study the hydrogen bond, wherever it occurred in inorganic or organic chemistry. An example, now classical, is that of potassium dihydrogen phosphate and Rochelle salt, linked with their ferroelectric properties[9].)

Extensive work on the transition-metal hydrides stemmed largely from the 'Manhattan Project' during the War and the special needs of nuclear energy work e.g. on U, Th hydrides, etc. Basic later advances on the thermodynamic and structural sides are due to the work of McQuillan, Wallace, Mallett and the Battelle group, Libowitz and others. One of the general aspects of hydrides brought out for instance by McQuillan's school is that, when *pairs* of early transition-metals are alloyed, hydrogen solubility and magnetic susceptibility vary in relation to electron/atom ratio. This has been shown on Ti–Nb, Nb–Mo and Ti–Mo alloys[10], typical of Group IV/V/VI pairings, where the H solubility decreases with the amount of later-Group metal added. It disappears for a certain E/A ratio (5·6–5·7), in conformity with the observed insolubility of H in Mo; however, the interpretation of E/A, i.e. the precise attribution of free electrons in hydrides, is itself difficult and slightly arbitrary unless formalized values are assumed.

Thermodynamically, the hydride equilibria are best expressed by *pressure/temperature/composition* (in brief, 'P–T–C') *curves*; their detailed trends and

* The investigation of isotopes, namely deuterides, in parallel with the isomorphous hydrides, has become a frequent and useful approach in this field.

System	r	M_4X	M_2X	MX	MX_2
Zr—H	0·29	12a, 4	12b, 4	12a, 4	ThC_2
Ta—H	0·32	—	12b, 4	8a, 4	?
Ti—H	0·32	—	12b, 4	12a, 4	12a, 4
Pd—H	0·34	—	12a, 4	?	?
La, *Ce*, *Pr*, *Nd* —*C*	0·42—0·43	?	?	?	LaC_2
Th—C	0·43	?	?	?	ThC_2
Zr—N	0·45	?	?	12a, 6	?
Sc—N	0·47	?	?	12a, 6	?
U—C	0·48	?	?	?	LaC_2
Zr—C	0·48	?	?	12a, 6	?
Nb—N	0·49	?	?	12a, 6	?
Ti—N	0·49	?	?	12a, 6	?
W—N	0·51	—	12a, 6	?	?
Mo—N	0·52	—	12a, 6	8b, 6	?
V—N	0·53	?	?	12a, 6	?
Nb—C	0·53	?	?	12a, 6	?
Ti—C	0·53	?	?	12a, 6	?
Ta—C	0·53	?	12b 6	12a, 6	?
Mn—N	0·55	12a, 6	12b 6	?	?
W—C	0·55	—	12b. 6	8b, 6	?
Cr—N	0·56	—	12b, 6	12a, 6	?
Mo—C	0·56	—	12b, 6	?	?
Fe—N	0·56	12a, 6	12b, 6	?	?
V—C	0·58	—	12b, 6	12a, 6	?

Figure 9.1. Hägg's Table[7] categorizes interstitial compounds and structures

Carbide and nitride structures with increasing radius ratio (*r*), classed by stoichiometric formulae; hydrides considered similarly.

Notation:

12a — cubic close packing (FCC); co-ordination number (C.N.) = 12
12b = hexagonal close packing. C.N. = 12
8a = BCC; C.N. = 8
8b = simple hexagonal, $c/a = 1$; C.N. = 8

The second numeral (e.g. 12b, *4*) describes the number of nearest neighbours equidistant to each interstitial, i.e. C.N. for each interstitial atom

their variations with the parent metal will be discussed in the following Sections.

Much of the pioneering work on hydride equilibria was due to Sieverts and his school[11] (1920–1940s). Ideally, Sieverts' law states that the solubility of hydrogen in a metal is proportional to the square root of the molecular hydrogen pressure. However, deviations from Sieverts' law are common and, indeed, the discontinuities, gradient changes and plateaux in the primary solution and the 'hydride' ranges of the P–T–C graphs, as well as their merging (e.g. at high temperatures, from duplex regions at lower ones), reflect the whole atomistic character of the phases. Such determinations involving deviations from Sieverts' law, therefore, form a main basis of hydride research. As a first illustration, *Figure 9.2* shows the isotherms and isopleths for the yttrium-hydrogen system[12]. The yttrium metal solid solutions, the YH_2 phase and the (low-temperature) YH_3 phase and their mutual relations clearly arise from the diagrams, which also manifest effects such as the initial

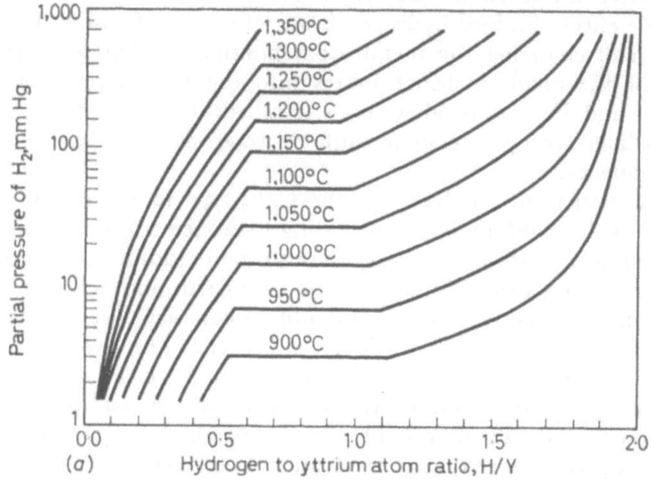

Figure 9.2a. Isotherms and isopleths for the yttrium–hydrogen system, typifying variations: (a) isothermal curves; (b) isoplethic curves (From Lundin and Blackledge[12])

By courtesy of The Electrochemical Society Inc.

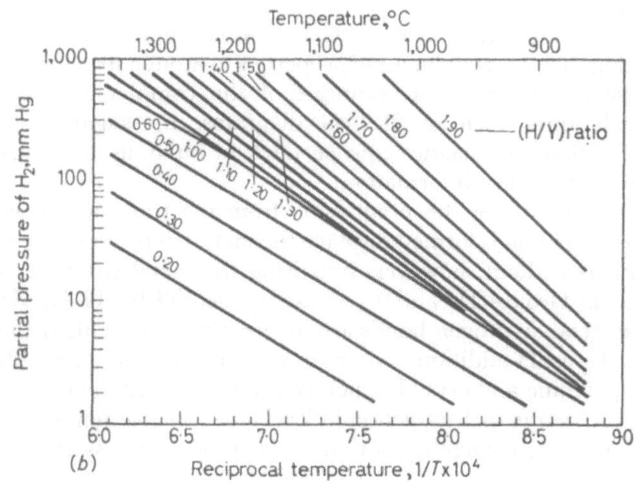

Figure 9.2b

H deficiency of 'YH_2' (substantiated by structure studies) and the gradual approach to YH_2 (i.e. tetradehral site filling) as the temperature decreases towards 900° C.

All hydrides retain relatively simple structures, in which the BCC structure is not essentially transformed but may undergo distortions to lower symmetry (e.g. tetragonal). This evolves, upon hydrogen excess and for the earlier Groups, to the BC cubic → FC cubic or tetragonal change where the FC unit cell can, however, still be alternatively interpreted as a BC one (see also *Table 9.1*).

In the later transition metals where the FCC structure already holds initially, the effect of added hydrogen is one of varying H-atom distribution rather than of changing the metal atom positions. Thus the NaCl-type structure of the alkali hydrides passes into the CaF_2-type structure for Groups III and IV (i.e. octahedral → tetrahedral occupancy). A further deviant in later Groups are some unstable *hexagonal* forms, i.e. changes in stacking sequence.

The transition from the NaCl- to the CaF_2-type lattice can in a given system also be brought about by increasing hydrogen concentration. When the tetrahedral sites are filled nearly to saturation, additional occupation of the octahedral sites can also occur. The rare earth and actinide hydrides are a case in point, and it is by such double occupancy that very hydrogen-rich hydrides can arise, such as MH_3 or higher. The overall pattern of hydride equilibria will, therefore, compositionally clearly reflect many aspects of lattice site and bond behaviour. This is also shown by the structural progression within the Periodic Table amongst neighbouring *M*–H systems on the one hand, paralleling that of H increase in a given *M*–H system on the other. To the extent that H atom addition implies electron donation, this is quite natural—similar trends had been noted amongst carbides, etc.—but any departure from such parallelism with other interstitial compounds is in itself interesting in reflecting back upon the variations in the state of ionization of the H atom.

A cautious synoptic study of hydride systems can thus reveal many useful points, relating their thermodynamic properties to atomic structure. Hydride research provides an excellent example of the complementary value of different techniques such as P–T–C diagrams, investigation of x-ray, neutron diffraction and magnetic properties*, in building up a harmonious but self-checking fund of information.

The fact that, during the progression from the transition metal to the hydride, the *M*–H bonds formed displace the metal atoms from their original lattice positions without, however, impairing the metal structure, becomes most evident in Groups IV and V, as exemplified[4b,14] by Ti–H, Zr–H and Hf–H. The metal–hydrogen bonds are inherently stronger than the metal–metal ones. Upon H addition, two kinds of transition are observed. In the first, energy, volume and crystal structure change discontinuously, as during the FCC → FC tetragonal transition in Zr–H, etc. at low temperatures. In the second kind, the change is continuous, as in the same systems at higher temperature.

Upon hydrogen addition, the HCP structure of α-zirconium forms first the FCC, then the FC tetragonal one. The wide non-stoichiometric ranges clearly correspond to H vacancies in the CaF_2-type lattice which, however,

* One valuable new tool to study metal hydrides is *nuclear magnetic resonance* (here proton resonance). For example, from width of lines and displacement of resonances, it was possible[13] to derive the dispositions and times of duration of H at given lattice sites in Th, U, Ti, Zr, Ta and Pd hydrides, and to calculate activation energies.

is, never quite filled. The nearest neighbours to each M or H atom are the unlike ones. The M–M distances are longer in the hydrides than in the metal, and so are the lattice H–H distances than in those of the H nuclei in the H_2 molecule (but see p. 475). Furthermore, space considerations indicate the presence of the H atom in ionic rather than atomic form. The anisotropy (FCC → FC tetragonal transition) arises through the directional nature of the M–H bond, which suggests[15] also the cause of the embrittlement of metals by hydrogen insertion.

Thermodynamically, the hydrogen solvent metals fall into two classes: (*a*) those dissolving the gas *exothermally* and forming generally stable hydrides; here the solubility decreases with increasing temperature, and Sieverts' law, though still generally obeyed within each isostructural field, is subject to deviations at lower temperatures or higher pressures; deviations occur whenever the structure becomes modulated, and conversely may be taken as a first token for such structural variations; (*b*) those metals dissolving *endothermally*, and to small or limited degrees only, where the solubility increases with increasing temperature, and Sievert's law is generally obeyed.

Class (*a*) comprises the early transition metals of Groups III (with rare earths), IV, VA, the actinides, and one later member, palladium, whereas class (*b*) applies to Groups VI–VIII excluding Pd and the actinides.

The transition along the Periods is, therefore, accompanied by one from exothermic to endothermic solution of hydrogen in the metals, the latter beginning at the chromium group. This does not mean that later hydrides cannot form but that they are basically unstable and their hydrogen can readily be expelled.

In work on metal hydrides, heat capacity and entropy determination have also served as means of detecting or indicating *ordering* effects, which direct observations by neutron diffraction were later able to confirm (see also Bibliography).

A striking feature of the literature is the large number of contradictions between results of different, and sometimes the same, authors, greater for hydrides than for other interstitial compounds. This fact is implicitly due to the fickle nature of hydrides because even slight, and sometimes imponderable, variations in experimental technique can produce the most subtle differences in the stabilities attained. In studies of equilibrium pressures, even the ensuring of equilibrium is often difficult, especially where supersaturation and strong hysteresis effects play major parts. Impurities such as traces of oxygen or nitrogen also have markedly large effects in hydrides, which explains some early erroneous results.

Likewise, the hydrogen atom must clearly have an intensee ffect on the formation, movement and distribution of dislocations in the metal matrix and on the stacking sequence, particularly so if the P–T–C dependence can critically promote either filling of defined lattice vacancies or segregation and precipitation. The H atom can be regarded as an agent of 'steering' dislocations, their distribution and locking, which owing to its bonding nature it does more delicately and critically than other interstitials.

Dislocation theory forms an essential part in explaining *hydrogen embrittlement* in metals, particularly in steels (see also below). The tendency for *molecular* hydrogen to form on diffusion to grain boundaries is an operative

Table 9.1 Single-metal hydrides,

Metal	*Primary solutions for* H	*Hydrides (ranges)*	*Crystal structures*
Alkali (Group I)			
Li	None	LiH	cubic, NaCl type (B–1)
Na	None	NaH	cubic, NaCl type (B–1)
K	None	KH	cubic, NaCl type (B–1)
Rb	None	RbH	cubic, NaCl type (B–1)
Cs	None	CsH	cubic, NaCl type (B–1)
Alkaline earth (Group II)			
Mg	[cf. below, to Sr and Ba]	MgH_2	BC tetrag.
		MgD_2	tetrag., *rutile* type
Ca	to 24 mol. per cent CaH_2 at 890°C	CaH_2 and a low hydride, $Ca_{10}H$	orthorhombic, C–29 type (slightly distorted HCP: related to $PbCl_2$ type)
Sr	Uncertain, but H solubilities and occurrence of unstable lower hydrides likely to be similar to Ca	SrH_2	As CaH_2
Ba	Uncertain, but H solubilities and occurrence of unstable lower hydrides likely to be similar to Ca	BaH_2	As CaH_2
Group III and rare earth			
Sc	Sc–H little known, presumed similar to Y–H	ScH_2	cubic CaF_2 type (C–1)
Y	considerable; H/Y = 0·63 at 1,300°C to 0·55 at 900°C	$YH_{1\cdot95}/YH_{2\cdot00}$	cubic CaF_2 type (C–1)
		YH_3 (low temperature phase only; below 300°C for 1 atm)	HCP (cf. Y metal, HCP with a = 3·6474, c = 5·7306)

from Groups IA–VIIIA

Lattice spacings, Å	*Heat of formation, etc.* kcal/mole	*Physical properties* (ρ = density)	*Remarks*
4·083	−21·7 (melts at 680°C; dissociation pressure 27 mm)	$\rho = 0{\cdot}76/0{\cdot}80$	slight non-stoichiometry known at high temps.; *ionic bonding throughout*; spacing increase follows that of ionic diameter
4·897	−13·5 } decompose before melting	$\rho = 1{\cdot}396$	good electric conductors
5·708	−13·8 } decompose before melting	$\rho = 1{\cdot}43$	photosensitive; very reactive, increasingly from LiH to CsH
6·049	−11·3 } decompose before melting	$\rho = 2{\cdot}59$	strong reducing (NaH and KH) and descaling agents; ignite spontaneously
6·389	−10·1 } decompose before melting (19·9?)	$\rho = 3{\cdot}42$	
$a = 4{\cdot}517$ $b = 3{\cdot}021$; $M = 2$	−18·6		all stable at normal pressures, decomposing only above 1,000°C (in hydrogen atmosphere)
$a = 4{\cdot}5025$ $c = 3{\cdot}0123$; $M = 2$			slight non-stoichiometry at high temps., e.g. $CaH_{1{\cdot}92}$ to CaH_2 (at 800°C) $BaH_{1{\cdot}8}$ to BaH_2 (at 500°C); *ionic bonding throughout*
$a = 6{\cdot}851$ $b = 5{\cdot}948$; $c = 3{\cdot}607$; $M = 4$	−45·1 m.p. > 1,000°C	$\rho = 1{\cdot}9$	eutectic probable in system Ca–CaH_2; peritectic possible, however hydrogen source *in vacuo;* reducing and drying agents; very reactive, increasingly from Mg to Ba
$a = 7{\cdot}358$; $b = 6{\cdot}377$; $c = 3{\cdot}882$	−42·3	$\rho = 3{\cdot}72$	
$a = 7{\cdot}845$ $b = 6{\cdot}801$ $c = 4{\cdot}175$	−40·9	$\rho = 4{\cdot}21$	*rutile* occurrence in MgD_2 gives link to oxides BeH_2 also exists
$a = 4{\cdot}7832$			
$a = 5{\cdot}201$	−44·4 for YH_2 saturation	$\rho = 4{\cdot}293$	
$a = 3{\cdot}674$; $c = 6{\cdot}599$		$\rho = 3{\cdot}958$	

Metal	*Primary solutions for* H	*Hydrides (ranges)*	*Crystal structures*
Group III cont.			
Rare earths	form isomorphous series of stable hydrides (but Gd, Yb and Eu exceptional)		
	solubilities appreciable, e.g. to H/Gd ≈ 0·3–0·4 (atomic)	(RE)H_2 ↔ (RE)H_3 solution series	(RE)H_2 = cubic CaF_2 type, but excess H in octahedral sites towards (RE)H_3
	exceptions: GdH_2 changes from CaF_2 type to HCP structure for GdH_3, with $a = 3{\cdot}73$, $c = 6{\cdot}71$ (similar to PuH_3: *note link of R.E. to actinides*) YbH_2 and EuH_2 are orthorhombic C–29 type, as CaH_2 deuterides: YbD_2 $a = 5{\cdot}871$, $b = 3{\cdot}561$, $c = 6{\cdot}763$ EuD_2 $a = 6{\cdot}21$, $b = 3{\cdot}77$, $c = 7{\cdot}16$		
Group IV			
Ti	H stabilizes β-Ti (eutectoid ≈ 300°C) soln. in β: ≈ 45% at 600° C; in α - Ti: ≈ 8%H at 300°C to 0·1%H at R.T. (for atm. pressure); lattice expansion in α and β	≈ TiH/TiH_2 (55–66·4%H but divergent results)	FCC: defect CaF_2 type (low-temp. tetragonal distortions)
Zr	H stabilizes β-Zr (eutectoid ≈ 550°C); soln. in β ≈ 52 %H, in α ≈ 61%H e.g. β-Zr 51–53 (850–890°C), α-Zr 4·1–7·0 (500–600°C); lattice expansion in α and β	≈ZrH/ZrH_2 (δ) dissociating into 2 phases at lower temperatures, ≈ ZrH–$ZrH_{1\cdot5}$ (δ) and ≈ $ZrH_{1\cdot5}$–ZrH_2 (ε); however, for equilibrium continuous δ ↔ ε transition probable; a further (γ′) phase exists (metastable?)	δ-FCC, defect CaF_2 type; ε BC tetr.; pseudo-cubic high-temp. hydride solution; tetragonal distortion increases with H content; γ′: FC tetr.
Hf	H solubility in β- and α-Hf only small, ≈2%H at R.T. to ≈11% at 900°C	$HfH_{1\cdot5}$ (60·5–63%H) HfH_2 with wide HR, varying from 24–64%H	FC tetr. (pseudo-cubic) to $HfH_{1\cdot53}$ → FCC at $HfH_{1\cdot53/1\cdot70}$; CaF_2-type FC tetr. $HfH_{1\cdot70/1\cdot80}$ (H in tetrahedral holes)
Th	H soln. in Th: 23% at 900°C, 13% at 650°C, 0·002% at R.T.	ThH_2 not attaining stoichiometry; $ThH_{1\cdot78-1\cdot98}$	FC tetr. → cubic (=BC tetr.)
		$ThH_{3\cdot75}$	BCC with M = 4 (16 Th, 60H atoms/cell)
		(Th_4H_{15}) ↔	

Lattice spacings, Å	*Heat of formation, etc.* kcal/mole	*Physical properties* (ρ = density)	*Remarks*
e.g. LaH_2 = 5·667 CeH_2 = 5·581 PrH_2 = 5·517 NdH_2 = 5·470 SmH_2 = 5·376 GdH_2 = 5·303	−49·7 −49·2 −49·7 −50·5 −46·9	ρ = 5·83 for $LaH_{2\cdot76}$ ρ = 5·5 for $CeH_{2\cdot69}$ ρ = 5·56 for $PrH_{2\cdot84}$ (RE)H_2 are conductors, (RE)H_3 are not; certain R.E. hydrides are semiconductors; strong resistance changes within non-stoichiometric range	(RE)H_2/(RE)H_3 form continuous homogeneity range, although two distinct phases, CeH_2 and CeH_3 also reported; CeH_3 can exceed to $CeH_{3\cdot16}$; non-stoichiometric ranges occur at (RE)H_2 ± and (RE)H_3−, even where not continuous R.E. hydride lattices begin to contract from ≈(RE)–$H_{1\cdot8}$ upwards (occupation of octahedral sites before all tetrahedral filled)
a = 4·403–4·445 (for 55 → 66·4%H); $TiH_{1\cdot99}$-cubic, a = 4·454	−22·6 for $TiH_{1\cdot4}$ at 400°–500°C but ranges from −19 to −44 for varying P–T–C; strong hysteresis effect in eutectoid transformation	ρ varying from ≈3·78 at TiH_2 to ≈4·04 at TiH to ≈4·4 at Ti at room temp. ρ = 3·94 for $TiD_{1\cdot97}$; TiH_2 shows anti-ferromagnetic ordering	H atoms in tetrahedral holes but eventual filling also of octahedral holes; two-phase region TiH + TiH_2 disappears above ≈600°C
a = 4·78 a = 9·78 c = 4·449 a = 9·75 c = 4·61	−48·3 (for δ formed from H-rich α-Zr at 400°–550°C), −39·6 for α-Zr at saturation	ρ = 5·7 for $ZrH_{1\cdot5}$; increasingly brittle with H content	reported compounds 'Zr_2H' and 'ZrH' incorrect
a = 4·919, c = 4·363 → 4·708 when cubic; may be regarded as BC tetr. cell with a = 3·452, c = 4·384; M = 2	−33·2 (for hydride formed from H-rich α-Hf)	ρ = 11·48 for $HfH_{1\cdot5}$; brittleness increases with H content, $HfH_{1\cdot70-1\cdot98}$ very brittle	Hf–H similar to Zr–H in many respects, but note differences; FCC → FC tetr. transition occurs between $HfH_{1\cdot80}$ and $HfH_{1\cdot87}$, continuing to $HfH_{1\cdot98}$; supersaturation beyond $HfH_{1\cdot98}$ can occur, H atoms filling octahedral holes
a = 5·80, c = 5·03 but varying a = 4·10, c = 5·03; M = 2 a = 9·11	−36·9 for $ThH_{1\cdot8}$	ρ = 9·20	'ThH_2' approached with high pressures and at low temps.; above ≈1,200°C, Th–ThH_2 phases probably join; miscibility gap at lower temps.

Metal	*Primary solutions for* H	*Hydrides (ranges)*	*Crystal structures*
Group V			
V	≈2% at low temps., but maximum attainable to $VH_{0 \cdot 89}$, with lattice expansion; joining to hydride at higher temps.	V → H, with continuous range from V, above 150°C; 2-phase region below (2–30 %H at R.T.)	BC tetr.; deuteride $VD_{0 \cdot 75}$: ordering of D atoms at very low temps.; probably similar for hydride
Nb	≈5% at low temps., but supersaturation possible; lattice expansion by H (distinction in solubility of atomic and molecular hydrogen); joining to hydride at higher temps.	$NbH_{0 \cdot 7-0 \cdot 9}$ continuous range from Nb above ≈300°C; below Nb + $NbH_{0 \cdot 7}$	FC orth. (may be regarded as deformed BCC FCC (NaCl- or CaF_2-type)); low-temperature ordering
		NbH_2 phase reported	FCC CaF_2-type structure; H atoms random or ordered, depending on heat treatment
Ta	≈4% or 7·5% (conflict) at low temps. but joining to hydride above ≈50°C where >40%; decreasing to 10% at R.T., zero at −15°C; lattice expansion with H	$TaH_{0 \cdot 7}$; continuous range from Ta above ≈50°C; below, Ta + $TaH_{0 \cdot 7}$	FC or BC tetr. (H atoms in octahedral holes); distortion to FC orthorhombic with higher H; towards distorted 'B–3', NaCl type (deformed); low-temp. ordering to Ta_2H
Pa	?	PaH_3	cubic, isostructural with UH_3 (β-W type)
Group VI			
Cr	very low, to $CrH_{0 \cdot 04}$ at R.T.	≈Cr_2H/CrH	hex. CP (B–4, wurtzite type)
		≈CrH/CrH_2 (unstable) *low-temp.* hydride reported up to H/Cr = 1·6; decomp. > −10°C	FCC (B–3, CaF_2 type)
Mo	no solubility	no hydrides	
W	no solubility	no hydrides	
U	solubility of H increasing in order α-U < β U < γ-U < liq. U, but very small: maximum 16 p.p.m. at 1,133°C (γ-U), 30 p.p.m. in liq.	UH_3 4 forms	
		α-(low-temp.) UH_3	BCC; $M = 2$
		ϵ-UH_3	cubic; $M = 8$
		δ-UH_3	cubic; $M = 8$ (complex variants)
		β-UH_3	cubic, β-W type, $M = 8$

Lattice spacings, Å	*Heat of formation, etc.* kcal/mole	*Physical properties* (ρ = density)	*Remarks*
$a = 3{\cdot}02$, $c = 3{\cdot}36$ for $VH_{0{\cdot}94}$ (cf. pure V, $a = 3{\cdot}028$); axial ratio strongly temp.-dependent $a = 6{\cdot}30$ ($\approx 2 \times a_v$)			2-phase region V + 'V_2H' at low temps. disappearing above $\approx$ 150°C
$a = 4{\cdot}84$, $b = 4{\cdot}90$, $c = 3{\cdot}45$, at $NbH_{0{\cdot}89}$ $a = 3{\cdot}42$ (cf. pure Nb, $a = 3{\cdot}301$), expanding to 3·447; on supersaturation (H content increasing from $NbH_{0{\cdot}9}$ to $NbH_{1{\cdot}34}$) lattice spacing *decreases* $a = 4{\cdot}55$	heat of solution of H in Nb increases from 16·1 to 23·3 kcal/mole for H/Nb 0·01 to 0·70; metastable hydrides formed		H more soluble in Nb than in V or Ta metastable high-temp. 'NbH' phase with BCC distorted lattice unconfirmed
$a = 3{\cdot}399$ at $TaH_{0{\cdot}49}$ from 3·303 for Ta; when BC tetr. at $TaH_{0{\cdot}5}$: $a = 3{\cdot}38$, $c = 3{\cdot}41$; when BC orth. at $TaH_{0{\cdot}8}$: $a = 4{\cdot}728$, $b = 4{\cdot}778$, $c = 3{\cdot}428$ (expansion by H greater in Ta than Nb)		embrittlement by H	heat-capacity work indicated *4 forms* of Ta_2H, with transformation points at 33°, 59° and 60°C, attributed to rearrangement of H atoms in ordering and siting; beyond Ta_2H, H atoms enter in tetrahedral and octahedral positions
$a = 6{\cdot}648$			
$a = 2{\cdot}716$, $c = 4{\cdot}422$ at CrH but varying $a = 3{\cdot}865$ (varying)	properties little studied except formation during electrolysis; hardening effects		note BCC → HCP → FCC transformation with added interstitial H; electrolysis can produce HCP hydride (originally thought form of Cr)
			no Mo and W hydrides known, but unstable existence under special conditions still thought possible
$a = 4{\cdot}160$ (for α-UD_3 = 4·150) $a = 6{\cdot}6310$ $a = 6{\cdot}631$ (for β-UD_3 = 6·621)	−30·8; UH_3 dec. at 430°C (for 1 atm); highly exothermic in formation	δ-UH_3 reported ferromagnetic below 173°K; elect. resistivity similar to U metal; ρ = 10·92 for ϵ-UH_3 = 11·1 for α-UH_3	UH_3 strictly stoichiometric in all its forms α-UH_3 forms at low temps. → β at 250°C; electrolytic formation of α also possible; α-UH_3 only forms below R.T. and possibly transitional to β-UH_3

Metal	*Primary solutions for* H	*Hydrides (ranges)*	*Crystal structures*
Group VII			
Mn	low H solubility but varying with polymorph; supersaturation effects pronounced with lattice expansion in α-Mn	no hydrides	
Tc	no data		
Re	no data		
Np	—	$NpH_{3\cdot6-3\cdot8}$ reported (similar to $ThH_{3\cdot6}$?)	
Group VIII			
Fe	very low: in α to ≈ 10^{-3}%H; in γ to ≈4 × 10^{-3}%H; in δ to ≈7 × 10^{-4}%H	no hydrides	
Co	very low	no hydrides	
Ni	very low, but large supersaturation ('surcharging') possible, much in excess of equilibrium	metastable HCP hydride exists (once regarded as form of Ni); also an H-richer FCC Nihydride, $NiH_{0\cdot9-1\cdot0}$, of wider spacing than (Ni–H)solution H-atom positions in NiH could be tetrahedral or octahedral	
[Cu in next Group]	copper hydride, CuH, exists; hexagonal, wurtzite type (B-4)		
Ru Rh	no hydrides known, and probably none exist; only occlusion characteristics studied, further work needed		
Pd	extensive solutions: 5% at R.T. but joining to 'PdH' at high temp.; supercharging possible; lattice expansion	'PdH' with extensive homogeneity range from ≈$Pd_{0\cdot7}H$ upwards, merging to primary Pd → H solid soln. at high temps.; stoichiometry attained at high pressures	FCC, H atoms in octahedral positions, i.e. NaCl type
[Ag, in next Group]	silver forms no hydride (as against Cu and Pd)		
Os Ir Pt Au	no hydrides known, and probably none exist; only occlusion characteristics studied, further work needed; small solubilities only		
Pu	very low solubility of H in Pu polymorphs	$PuH_{1\cdot75}/PuH_{2\cdot74}$	FCC, CaF_2 type, with extra H in solution, occupying octahedral sites
		$PuH_{2\cdot9}/PuH_3$	HCP

Lattice spacings, Å	*Heat of formation, etc.* kcal/mole	*Physical properties* (ρ = density)	*Remarks*
			'surcharging' by H in electrolytic Mn; occlusion rather than lattice solution
$a = 3{\cdot}721$ for $NiH_{0{\cdot}8}$; range 3·71–3·735 (against $a = 3{\cdot}524$ for Ni)			extensive knowledge of solubility isobars in Ni; sensitive to cold work which can increase solubility; FCC (Ni,H) solution and FCC NiH hydride coexist, foreshadowing Pd relations: basic similarity of Ni–H/Pd–H system confirmed
4·027–4·07 (against 3·891–3·84 in Pd solid soln. at R.T.)	−9·79	H permeability in Pd outstanding property	strong hysteresis effects associated with Pd–H solution
$a = 5{\cdot}359$ at $PuH_{2{\cdot}0}$, 5·34 at $PuH_{2{\cdot}5}$; $M = 4$ $a = 3{\cdot}78$; $c = 6{\cdot}76$; $M = 2$	−37·4	x-ray density for $PuH_2 = 10{\cdot}40$, for $PuH_3 = 9{\cdot}61$	structure transitions cubic→hexagonal for $PuH_2 \rightarrow PuH_3$

factor in causing crack formation through volume expansion. However, this tendency is already incipiently present at dislocations. The embrittling effect of hydrogen may be regarded as the dual one of (1) normal dislocation pinning, and (2) the incipient formation of H–H bonds to form molecules with larger expansion, initiating stress and cracking.

The effect of hardening by hydrogen loading at elevated temperatures, and precipitation of 'hydride' (or, rather, H-richer solution) during aging can be pronounced, for instance in the refractory metals (niobium[16]); again, it has to be attributed to initial dislocation locking.

Chromium provides an example for the next stage of this process, where a change occurs in the actual stacking sequence, controlled by hydrogen concentration. This is reflected directly in the BCC → HCP → FCC transformation in the Cr–H system and in intermediate stacking faults linked with dislocations. Technologically this phenomenon underlies important processes such as hard chromium plating.

An interesting feature is that certain *intermetallic compounds* can dissolve hydrogen to an even more marked extent than the pure metals* (see Section 9.3). Examples are CoTh, which was found[17] to develop, upon hydrogen addition, to a phase of the approximate formula '$CoThH_{4.2}$' and NiZr, where similarly an H-rich compound, $NiZrH_3$, exists[2b]. Both derive from the parent compounds, CoTh and NiZr, essentially by lattice solutions. This brings out two fundamental points:

(*a*) Two-metal structures can in principle enhance the hydrogen solubility by providing possible new interstitial sites. (Which sites are taken up in the intermetallic does not seem to be known yet, and this will be structure-variable; it would be interesting to study this systematically for different basic intermetallic compound types.) This possible enhancement is not confined to primary intermetallics but could equally apply to two-metal solid solutions as bases.

(*b*) It seems that 'Early–Late' Group transition-metal combinations in intermetallic compounds are favourable for greatest uptake of H; in the above cases it was a diagonal Group IV–VIII join. It is premature to generalize on this without much fuller exploration of the data, though the fact of combining two metals on the left and the right of the minimum (in Group VI) of hydride-forming tendency may be noted.

Of practical interest is the role of hydrides as sources of supply of *atomic* hydrogen, where the presence of molecular (gaseous) hydrogen would be insufficient for a reaction to take place, even when the equilibrium potential is favourable. This is effectively a catalytic action, the hydride acting as converter to the atomic state. Such transfer has been shown e.g. on palladium[18] where an otherwise sluggish hydrogen absorption at room temperature can be greatly accelerated when the metal contacts pre-prepared hydrides ('trigger hydrides') of some of the earlier transition metals, here of Ti, Zr, Th, Ce and U. (This accelerated transfer takes place if Pd is dusted, e.g. with UH_3 powder.) In effect, the energy for dissociation of the H–H

* This facility of some intermetallic compounds to dissolve hydrogen to copious extents contrasts sharply with the very limited solubilities for other interstitials (C, N, O, B; cf. the preceding Chapters), again reflecting the basic nature of the H atom.

molecule is obviated, and direct feeding by loosely bonded hydrogen in atomic form is possible.

In the hydrogen molecule, H_2, the two atoms are covalently bonded by the sharing of an electron pair. This bond energy is so large that dissociation into atoms will in turn require extreme external energies such as can only be

Table 9.2. Dissociation pressure constants in some hydrides and hydrogen solutions

(A and B in eqn. $\log p_{10} = - A/T + B$; p mm, T °K)

Hydride	*A*	*B*	*Reference*
LiH	8,224	9·926	Cited[2,4a] from various sources
NaH	5,958	10·47	
KH	6,300	11·86	
RbH	4,930	9·51	
CsH	4,410	9·25	
CaH_2	10,870	11·49	
SrH_2	10,400	11·10	
BaH_2	6,450	8·20	
ScH_2	?	?	12
YH_2 (for $YH_{1.9}$)	14,190	13·78	
(YH_2 probably	≈15,000	≈15)	
LaH_2	10,850	10·64	17
CeH_2	10,760	10·63	
PrH_2	10,870	10·53	
NdH_2	11,030	10·48	
GdH_2	10,250	9·72	19
ErH_2	11,780	10·91	
Ti–H	6,671	9·35	20
Zr–H	8,040	9·177	
Hf–H (≈HfH_2)	6,850	8·270	21a
Th–H (ThH_2)	7,500	9·35	22a
(Th_4H_{15})	4,420	9·50	15e,23
V–H	3,175	6·394	(at H/V = 0·05) 24 converted
	3,906	8·474	(at H/V = 0·3)
Nb–H	3,575 ↓	6·170 ↓	(at H/Nb = 0·05) 25
			(increasing)
	4,425	8·765	(at H/Nb = 0·5)
Ta–H	3,480 ↓	6·26 ↓	(at H/Ta = 0·05) 22c
			(increasing)
	4,020	9·26	(at H/Ta = 0·5)
UH_3	4,500	9·28	26
PuH_2	8,165	10·01	27a
PuD_2	7,761	9·71	(note decrease with isotope)
Ni–H solution	761	1·76	28
Ni–D solution	738	1·70	
Ni–T solution	857	1·82	(note increase with isotope)
PdH	1,877	7·48	2,4a

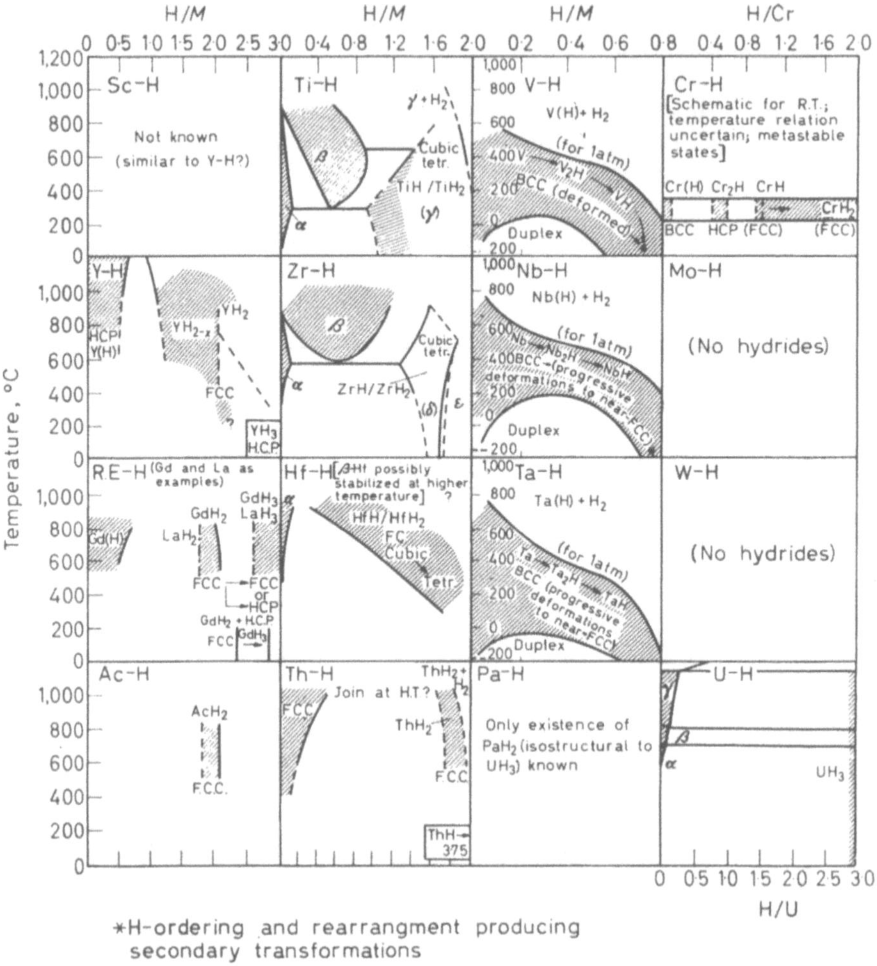

Figure 9.3. Survey of transition-metal–hydrogen binary systems (for atm. pressure)

supplied in an electric arc or discharge. On the other hand, the metal hydride may in this sense be regarded as a convenient 'converter' to atomic hydrogen.

Next we may survey briefly the trend in individual systems, proceeding from Group IA to Group VIII, and deal with each case in some more detail.

Figure 9.3 shows the collated set of metal–hydrogen systems, for atmospheric pressures. A more complete representation would be a full set of P–T–C curves, and a survey of these with Group and Period is presented in *Figure 9.4*. Despite marked secondary variations, the behaviour and property trends of hydrides within any one Group are similar, and the principal differences will be seen to lie *between* rather than within the Groups, except for the special case of the actinides. This Group kinship is more pronounced

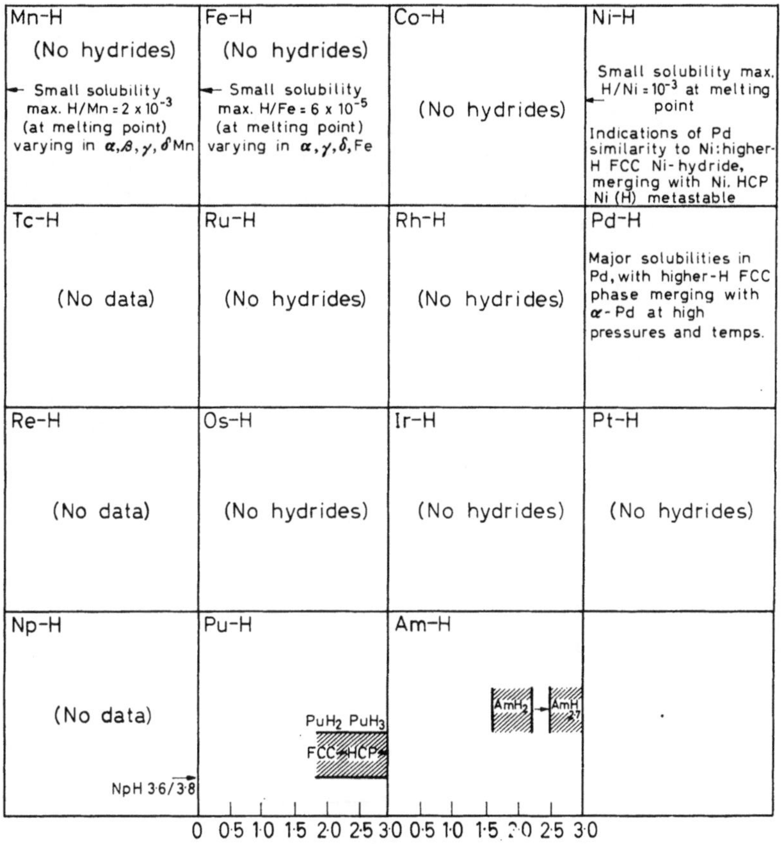

Figure 9.3b

for the hydrides than for any of the previous interstitial compounds (oxides, carbides, etc.) and may well be regarded as a manifestation of the bonding difference of the hydrogen atom.

The most stable transition-metal hydrides are those in Group IIIA and the Rare Earths. In Group IV they are still of considerable but declining stability which rapidly decreases towards Group V. After the minimum in Group VI we have the later but qualified resuscitation in Group VIII, with its second climax at palladium.

Following upon *Table 9.1* which gives a full list of systematized data of hydrides, it is useful also to assemble (*Table 9.2*) the constants A and B of the dissociation pressure equation

$$\log p_{10} = -\frac{A}{T} + B$$

where p = pressure (mm hydrogen), T = temperature (°K). Each isopleth in *Figures 9.2* and *9.4* can be described by this equation. The constants A and B are characteristic of the metal: they may well be taken as a measure of

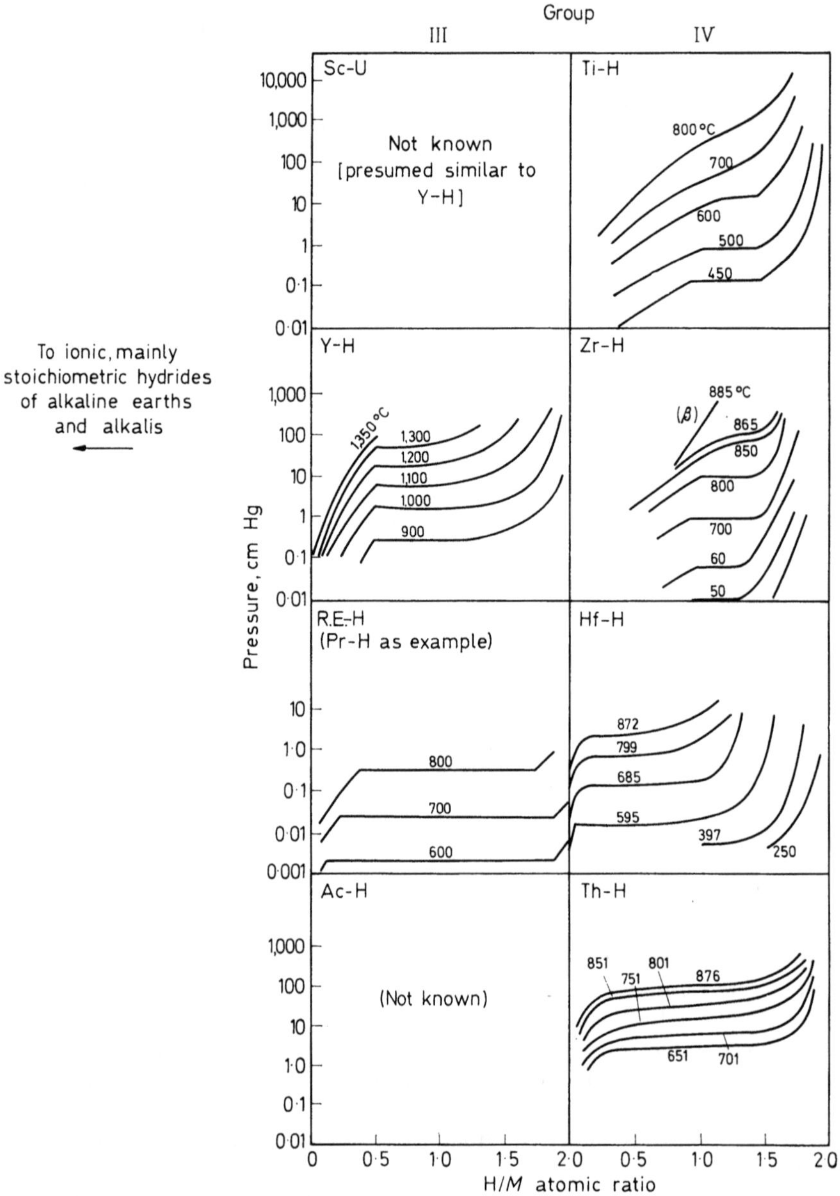

Figure 9.4. Metal–hydrogen binary systems: sequence of pressure–temperature–composition diagrams

stability of the hydride or hydrogen solution. The values given in *Table 9.2* are those from the alkali metals to Group VIII, as far as can be gleaned from the literature.

In *Figure 9.5* a corresponding plot of A and B versus Group number has been attempted. There are still evident uncertainties, and only the relative orders of magnitude should here be noted, whilst some probable

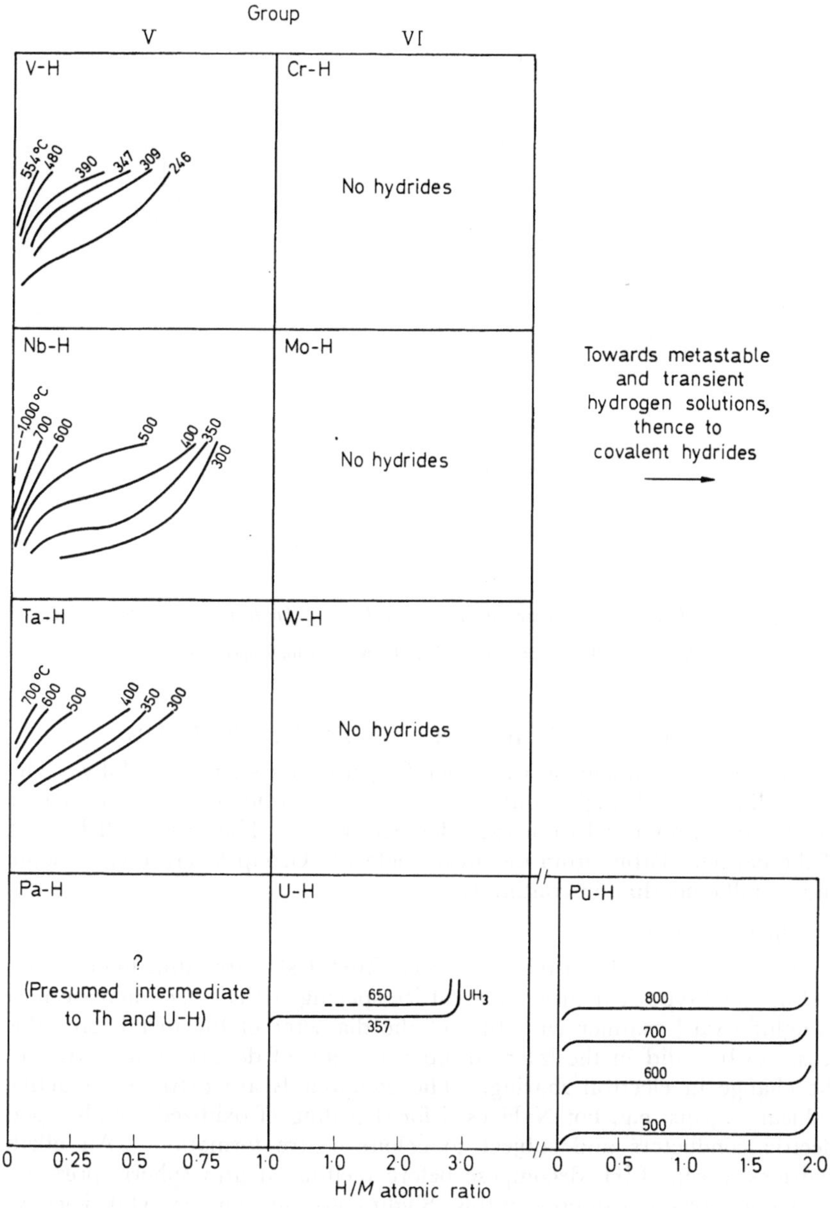

Figure 9.4b

values have been hazarded. The graph, however, brings out several features illustrating points of the following discussion, specially the marked peak in Group III* which decreases later.

---

* The values for scandium hydride do not seem to be available but, it is suggested, could be the highest of all.

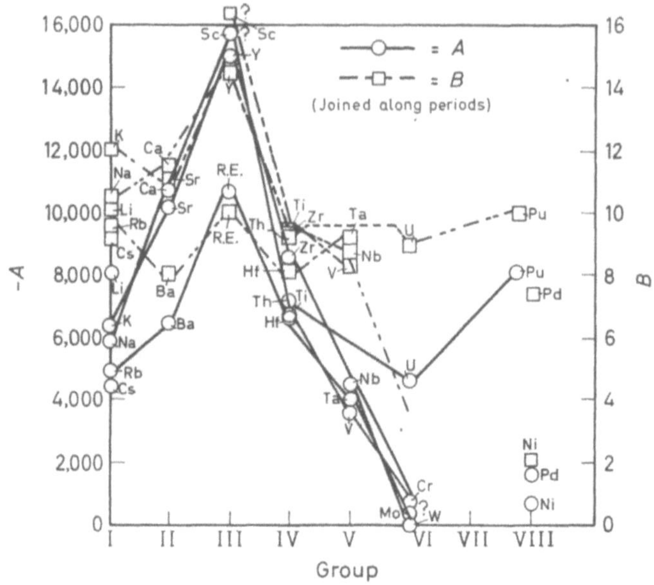

Figure 9.5. Dissociation pressure constants for hydrides as function of Group and Period

Values A and B in eqn. $\log p = \frac{A}{T} + B$ (pressure mm, temperature °K)

9.2 DETAILED BINARY HYDRIDE SYSTEMS

We shall now comment on each metal–hydrogen system more fully. The aim is limited to bringing out the basic points rather than the particular which can be obtained from the references given. The order will be that of the Periodic Table, from the alkali metals to Group VIII, to cover ionic and metallic bonding in continuity.

(1) *Alkali hydrides*

All alkaline hydrides (LiH, NaH, KH, RbH, CsH) are saline in character and of NaCl-type structure. The lattice spacings (*Table 9.1*) increase in a straightforward manner with that of the diameter of the metal ion; this ceases to be valid in the later Groups: a trend of departure indicative of the change in electron sharing. The compounds are extremely reactive reducing agents, e.g. hot NaH used for descaling of oxidized metals, good electric conductors and subject to colour centre formation. All alkali hydrides, except LiH, decompose before melting at atmospheric pressure, but higher pressure can prevent this. Slight non-stoichiometry (H deficiency) is known at elevated temperatures in LiH and NaH.

Even in the alkali hydrides, with a change of temperature the degree of ionization can vary, and such a departure shows itself e.g. in LiH from the thermal vibration amplitudes at low temperatures.

As mentioned already, one basic point about alkali hydrides is their isomorphy and similarity in bonding with the alkali halides. The neutron diffraction work of Shull *et al.*[8] had shown the identity of the H (or deuterium) and the Cl positions in NaCl, NaH and NaD. An $H^- \leftrightarrow Cl^-$ (and other halogen) mode of substitution is implied and has been observed. However, this hydrogen/halogen parallel breaks down for the heavier alkali ions. The CsCl structure is the ordered BCC, not possible in the early hydrides although approached in the later; CsH maintains the FCC lattice.

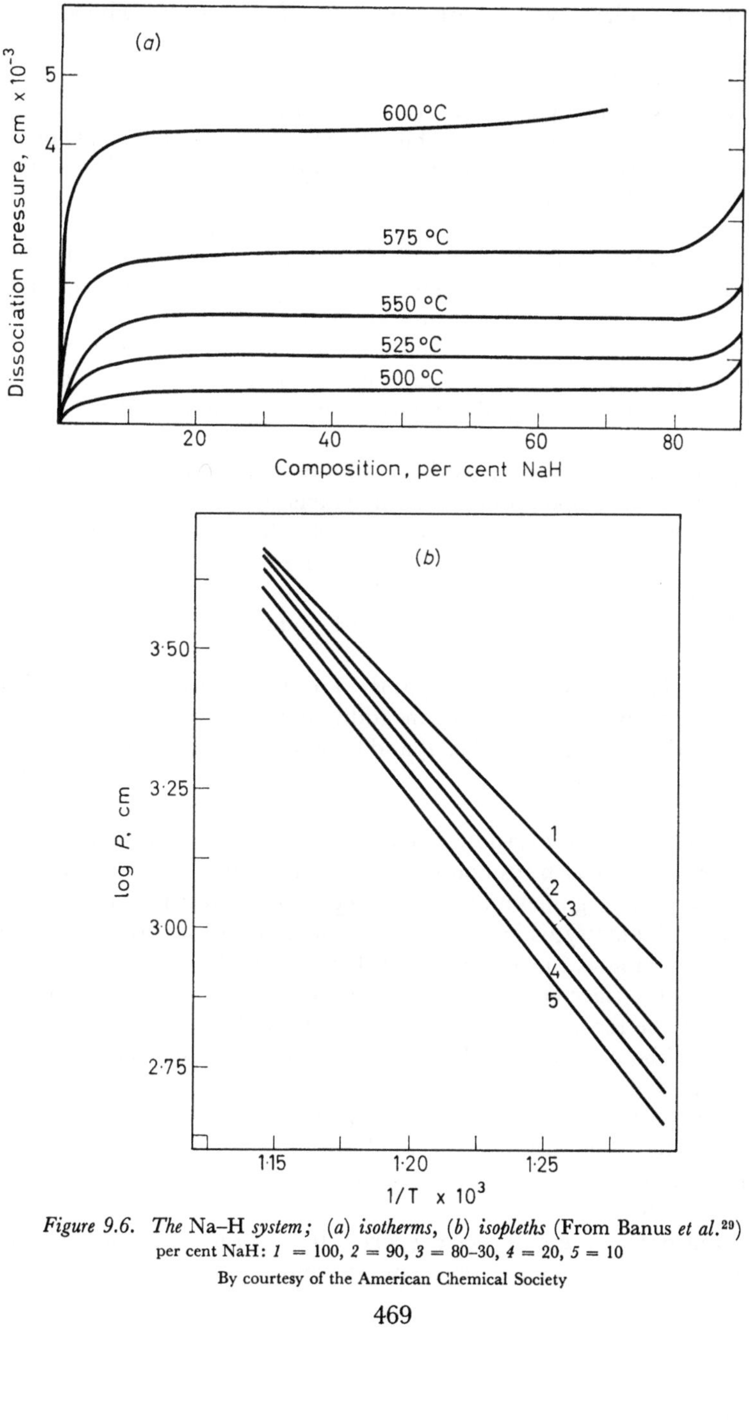

Figure 9.6. The Na–H *system; (a) isotherms, (b) isopleths* (From Banus *et al.*[29])
per cent NaH: *1* = 100, *2* = 90, *3* = 80–30, *4* = 20, *5* = 10
By courtesy of the American Chemical Society

Constitutionally the Na–NaH–H system has been studied by Banus *et al.*[29] at 500–600°C and varying pressures. Their P–T–C graphs are shown in *Figure 9.6*; the dissociation pressure constants obtained are $A = 5{,}070$, $B = 9{\cdot}49$ (cf. *Table 9.2*), the heat of formation for NaH = 11·61 cal/mole.

References on alkali hydrides are numerous; for some, see Bibliography.

(2) *Alkaline earth hydrides*

The alkalines earth hydrides are of lower structural symmetry than the alkali hydrides (see *Table 9.1*). Whilst CaH_2, SrH_2, BaH_2 are all orthorhombic, with steadily increasing spacings, MgH_2, the lightest (in contrast to LiH amongst the alkalis) changes structure. Moreover, the adoption by the deuteride, MgD_2, of *rutile* (TiO_2)-type structure is noteworthy[30]; it would be interesting to search for further hydrides of this structure and to link the TiO_2 and other *oxidic* rutiles with new *hydridic* ones of which, it is felt, MgD_2 may be only a first case. The transition and property change upon, for example, (Mg $\leftrightarrow$ Ti)(H $\leftrightarrow$ O) substitution would be worth studying. We might recall that, in their oxides (CaO, etc.), the alkaline earths maintain the NaCl-type structure, as the hydrides did in the preceding (alkali) Group.

The alkaline earth hydrides, as the alkali ones, are stable at atmospheric pressures; they decompose only at high temperatures but before melting. Owing to their high dissociation pressures (cf. *Table 9.2* and Section 9.4), CaH_2 and BaH_2 are commonly used as sources of free hydrogen under vacuum. They also are reducing or drying agents.

For MgH_2 (which can only be prepared under high H_2 pressures) the dissociation pressure is particularly large.

The Ca–CaH_2 and Ba–BaH_2 diagrams[31] are shown in *Figure 9.7*. It is of special interest that besides CaH_2 an additional *low* hydride exists, stable between 1 and 10 mol. per cent CaH_2 and between 320° and 600°C. Furthermore, there is a considerable solid solubility of H in the high-temperature (BCC) form of calcium metal (β-Ca), namely to a maximum of 24 mol. per cent CaH_2 (= nearly Ca_2H). We might conclude that this foreshadows already the large BCC transition-metal solubilities for hydrogen, as in β-titanium. It clearly signifies a first major departure from strict ionic bonding and from stoichiometry which, though still maintained in CaH_2, gives way to metallic bonding at the low-hydrogen levels. The low-temperature form of calcium, α-Ca (face-centred cubic), however, shows only little H solubility; it may be regarded as structurally the form 'inherited' from NaH to the left, but without room for hydrogen. The new (γ) phase, which is thought by Peterson and Fattore[31a] to form peritectoidally and to decompose eutectoidally, almost plays the part of a third polymorph of the metal ('γ-calcium') but needs a little H in the lattice for stabilization. The crystal structure of γ-(Ca,H) is not known; however, a hexagonal close-packed one would not be surprising.

CaH_2 and BaH_2 have both been reported to undergo a high-temperature phase change, though of an as yet unknown kind.

The crystal structure of CaH_2 and, since SrH_2 and BaH_2 are isomorphous, also of these, has been redetermined[32] and confirmed as of orthorhombic cell (closely related to $PbCl_2$ type as detailed in *Table 9.1*); here the H atom could be successfully located by neutron diffraction.

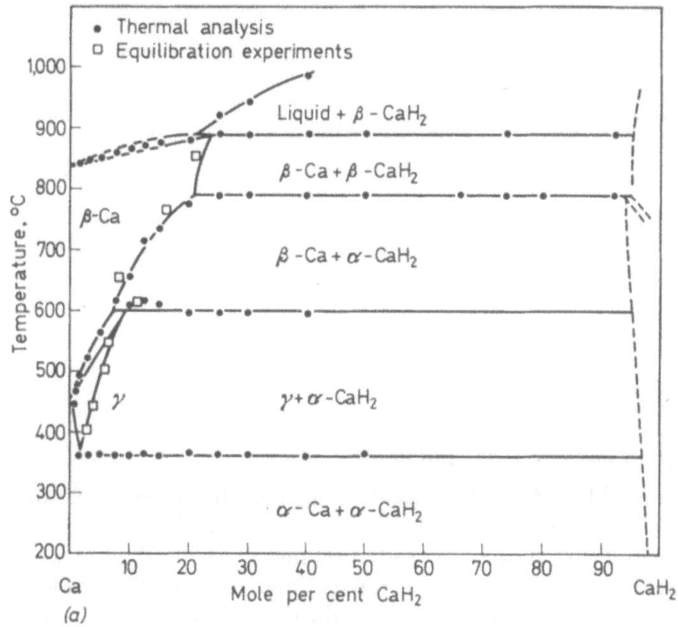

Figure 9.7a

Figure 9.7. The Ca–CaH_2 *and* Ba–BaH_2 *systems.* (From Peterson and Fattore[31a] [(*a*),(*b*)], Peterson and Indig[31b][(*c*)]

(*a*) Ca–H overall; (*b*) Ca end, suggested extent of γ-hydride phase; (*c*) Ba–BaH_2

By courtesy of American Chemical Society

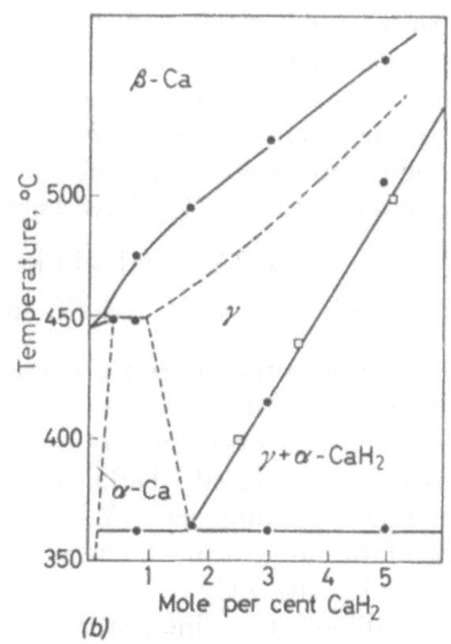

Figure 9.7b

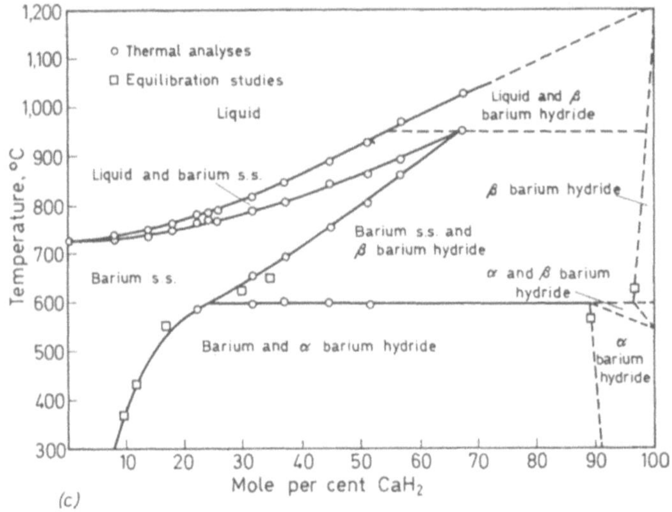

Figure 9.7c

(3) *Group III and rare earth hydrides*

In Group III and the lanthanide elements, the progression towards the metallic solid solutions and the BCC-derivative hydride formation, which will culminate in Ti, Zr and Hf, already becomes marked. This is accompanied by increasing non-stoichiometric ranges (and, indeed, decreasing meaning of 'stoichiometry' in hydrides), and by the development of the metallic bond. There is also a kinship to the *actinides*, as typified e.g. by the development of high-H hydrides (MH_3). At the same time, hydride stability reaches its peak in this Group.

In the 1st Long Period, little seems as yet to be known on the *scandium*–hydrogen system; it may, however, be presumed to be largely similar to *yttrium*–hydrogen which has been quite extensively studied.

In the Y–H system[12] (*Table 9.1*, *Figures 9.2* and *9.3*; cf. Bibliography), the Y solid solution phase and the Y hydride, nominally YH_2, coexist over a wide composition range. We have here the sequence

$$\textit{HCP}\ \text{Y-metal solution} \rightarrow \textit{FCC}\ \text{'YH}_2\text{'} \rightarrow \textit{HCP}\ \text{'YH}_3\text{'}$$
$$\text{(density: at Y} = 4{\cdot}40;\ \text{at YH}_2 = 4{\cdot}293;\ \text{at YH}_3 = 3{\cdot}958)$$

The first transition gives a prognosis of what will occur in zirconium, next to yttrium. The trihydride exists only at low temperatures (< 300°C for 1 atm), and the reversion here to the HCP structure of metallic Y is notable; it may well be asked whether at high pressures and elevated temperatures a continuous solution series Y → 'YH_3' could not be established, to the exclusion of the cubic form which itself would be stable at limited P–T ranges only. Physically, the $YH_2 \rightarrow YH_3$ transformation is accompanied by the metallic mass falling into a fine powder, i.e. by excess H becoming molecular at the grain boundaries with collapse of cohesive forces.

The isotherms (P–T–C curves) and the isopleths in *Figure 9.2* illustrate well the primary solution, the two-phase (plateau) and the (H-deficient) YH_2 regions, where the composition YH_2 is approached as the temperature decreases. We might assume that this means ideal filling of the octahedral sites (in the CaF_2 structure), but not necessarily so, since a limited tetrahedral occupation could still occur, as in NaH, etc. Sieverts' law (solubility $\propto \sqrt{H_2 \text{ pressure}}$) is fairly adhered to in the single-phase fields of this system, but the gradualness of any deviations from it towards the plateau boundaries must be noted, which becomes quite characteristic of these and later hydride systems. The dissociation pressure constants are included in *Table 9.2.*

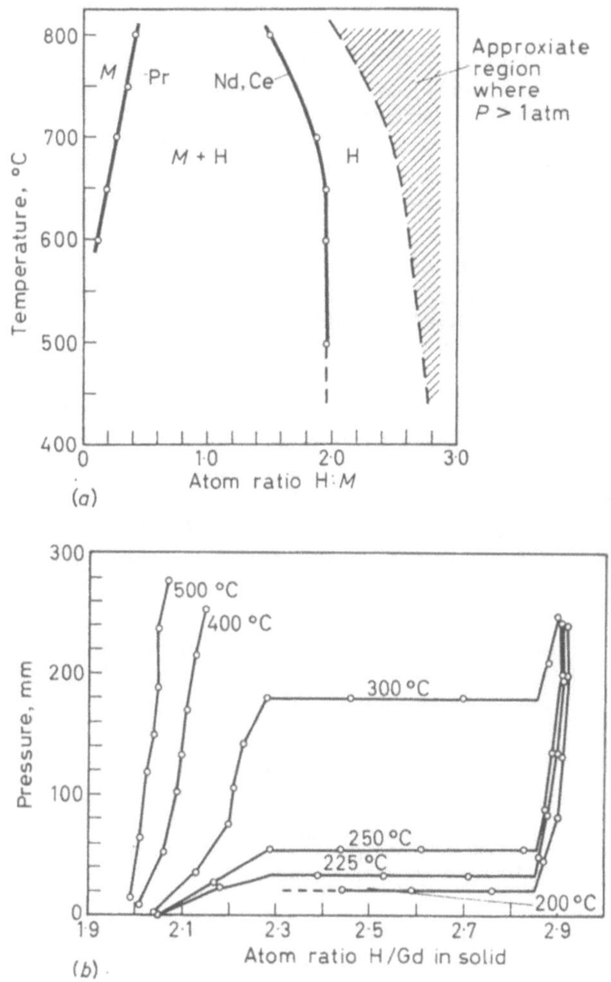

Figure 9.8. *Rare earth–hydrogen systems* (From Mulford and Holley[27c], Sturdy and Mulford[19])

(*a*) Generalized phase diagram for La, Ce, Pr, Nd–H; (*b*) isotherms between GdH_2 and GdH_3

By courtesy of American Chemical Society and Los Alamos Scientific Laboratory

It is also characteristic of yttrium (as of the rare earths) that it can retain large amounts of absorbed hydrogen at relatively high temperatures, but that reduced pressures can liberate the gas with comparative ease; we might see here once more the metal acting as a 'reservoir' for atomic hydrogen which can be tapped at will.

The *rare earth*–hydrogen systems have been investigated comprehensively e.g. by Mulford's[27] and Wallace's[33] schools and by Korst[17] (cf. *Table 9.1* and, for phase diagrams, *Figure 9.8*). Appreciable primary solutions give way to the di/trihydride ranges; these can be regarded essentially as continuous-solution series of FCC structure, with a dissociation tendency at lower temperatures and pressures, conditional, however, to different modes of H-atom entry into the lattice. Thus in cubic (rare earth metal) H_2, the initially defect CaF_2 structure can become supersaturated by H atoms in octahedral sites. Not all rare earths behave alike, however. Anomalies occur for instance for CeH_2/CeH_3 (two phases, and H *overloading* even up to $CeH_{3 \cdot 16}$) and GdH_2/GdH_3 (transformation to HCP). Gadolinium with its two hydride ranges (*Figure 9.8b*) shows already quite close relations to the actinides. Thus the pairs $GdH_2 \leftrightarrow PuH_2$ and $GdH_3 \leftrightarrow PuH_3$ are isomorphous (with the significant avoidance of a 'UH_2' which does not exist, for other reasons).

The cubic dihydrides are all conducting, but the trihydrides are not. The former are ordered at low temperatures and are of essentially metallic bonding; the latter are disordered and show ionic bonding. Nothing could better illustrate the critical borderline nature. All rare earth hydrides are brittle.

P–T–C curves for the La–, Ce–, Pr– and Nd–hydrogen systems by Mulford and Holley[27c] are exemplified by those for Pr (*Figure 9.4*); the heats of formation slightly differ from Korst's[17] (*Table 9.1*). *Figure 9.8a* shows their generalized phase diagram.

The structures of some rare earth hydrides, La, Ce, Pr, Nd and Sm, again determined by Mulford's team[27] using neutron and x-ray diffraction, confirmed the CaF_2 type, with excess H occupying random octahedral sites (spacings, *Table 9.1*). A noteworthy result is that the metal–hydrogen atom distances observed do not conform with the independently known radii of the *M* and H atoms. *Table 9.3* gives the interatomic distances[27d] and the deduced metal (*M*) and H atom radii.

Table 9.3

Hydride	*Distance* *M*–8H, Å	*M radii for* 12 *coordination*	*M radii for* 8 *coordination* Å	H *radii* (*for tetrahedral coordination*)
LaH_2	2·454	1·876	1·823	0·631
CeH_2	2·416	1·825	1·799	0·617
PrH_2	2·389	1·824*	1·780	0·609
NdH_2	2·368	1·820	1·767	0·601
SmH_2	2·328	1·804	1·751	0·577

* Pauling's value; others from measurements at Los Alamos Laboratory

Thus the effective H radius varies with metal surroundings, assuming the metal–H distance to be the sum of the covalent radii of both. Pauling's H radius (for tetrahedral coordination), 0·45 Å, is much smaller than the 0·63–0·58 range derived above which is itself intermediate to the apparent H radii in ZrH_2 (0·55 Å) and ThH_2 (0·67 Å)*. The metal–metal separation is also much larger than in the pure metal, and it can be concluded that *M*–H, not *M*–*M*, bonds mainly contribute to the cohesive energy. Two valency electrons are probably responsible for the *M*–H bond in MH_2, the metallic character being due to additional valency electrons.

The yttrium and rare earth hydrides highlight the point fundamental in hydrides (and, indeed, similarly valid though less pronounced in other interstitial compounds), that the effective H atom (ion) radius is variable and sensitive to environment. Further, we may consider the rare earths with their close mutual electronic similarity as a self-contained 'experimental test ground' of physical principles which apply widely in transition metal/interstitial bonds but which in the later metals may be overlaid by other factors. It may indeed be envisaged that the effective hydrogen 'radius' varies with temperature and pressure as well as with the type of lattice site occupied. The concept of interstitial hydrogen as in a 'metal-like' state (cf. Chapter 2 and above) begins to become operative in Group III, but to varying degrees in which ionic or covalent tendencies compete.

The diminution of the effective H-atom size with increasing H content in a given metal–hydrogen system produces the apparently anomalous result of *contraction* of the lattice-cube edge with additional hydrogen. Such a contraction has been reported[34] in PrH_2, NdH_2 and SmH_2 with added H. In Sm–H and some other systems this culminates in a transformation to close-packed hexagonal (as in YH_3 above). In terms of electron structure, the contraction has been attributed[34] to the fact that these H-rich hydrides are substantially saline in nature, i.e. ionic, and that the additional hydrogen is bonded by the absorption of electrons from the conduction band of the metal. Superimposed on this, contraction may in part also be due to change in *siting* of H atoms from tetrahedral to octahedral.

X-ray work on the *lanthanum*–hydrogen system[35] shows that, by increasing hydrogen pressure, the FCC lanthanum lattice can be progressively *expanded*, but only up to a certain value:

	Spacing, Å
pure La	5·292
at 0·01 mm H_2 pressure	5·650
at 2·2	5·662
at 2·9 with a sudden jump to	5·695;

thence the spacing is pressure-independent. Furthermore, the stable HCP hydride LaH_3, coexisting with the FCC one, disappears under pressure and converts to the latter structure ($a = 5{\cdot}685$ Å). We have here a direct illustration how a metal lattice is receptive to hydrogen up to interstitial capacity by a simple 'pressing into place'. At the same time, an interplay of

* See also earlier remarks on H atom size.

structures occurs, and differences in stacking sequences can be equalized at high pressures. For the later, less stable hydrides, this becomes more difficult, and high temperatures are required.

In contrast to the other rare earth hydrides, those of ytterbium and europium, YbH_2 and EuH_2, behave somewhat differently[17b], owing to a small difference in electronic orbital filling. They have the orthorhombic C–29 structure of the alkaline earth hydrides (CaH_2, SrH_2, BaH_2; for spacings see *Table 9.1*). These hydrides are again more salt-like in behaviour, as CaH_2, etc. This illustrates how even within the limited test family of the rare earths the bonding manner is critically poised between the more alkaline types of the earlier Groups and the more metallic ones of the later Groups. Even the pseudo-cubic rare earth di- $\rightarrow$ trihydrides may be regarded in the same terms as representing a balance between tetrahedral and octahedral occupations characterizing the Group IV and the alkaline hydrides, respectively.

To finalize Group III: *actinium* hydride, AcH_2, again has the FCC structure[36] ($a = 5{\cdot}670$ Å), almost certainly of CaF_2 type. This coexists with a form of Ac metal, also FCC, stable at high temperatures. In the context of the *later* actinides, this agrees with ThH_2, in that of the rare earths, with CeH_2, etc.

(4) *The Group IV hydrides* (Ti–H, Zr–H, Hf–H, Th–H)

In titanium and zirconium, the hydrides reach their first 'peak of interest', both theoretically and in practical application (the second peak occurs later in palladium).

The Ti–H system has been studied by many workers, and *Figures 9.3* and *9.4* give only one of several versions, but probably the most representative. The BCC β-Ti form is stabilized by hydrogen towards a eutectoid decomposition, $\beta \rightarrow \alpha$-Ti + hydride; the hydride possesses an extensive homogeneity range, between approximately TiH and TiH_2, but varying of course with temperature and pressure, as shown. The hydride is FCC, with considerable structural distortions, however. *Table 9.1* summarizes some of the data.

It should be noted that the α-, β- and TiH_{1+x} hydride solid solution series are now almost certainly confirmed as the only phases occurring, and that various earlier Ti hydrides reported have no validity. They possibly occurred through oxygen or nitrogen impurities or, in some cases, were perhaps misinterpreted as independent phases where, in fact, they continuously join on to one of the three basic structures.

The HCP α-Ti lattice is expanded by H: according to Chrétien *et al.*[37] from $a = 2{\cdot}940$, $c = 4{\cdot}680$ Å for pure Ti to $a = 2{\cdot}951$, $c = 4{\cdot}740$ Å at $TiH_{0{\cdot}46}$. There occurs a rapid decrease of solubility with temperature[38], from about 8 per cent H at 300°C to about 0·1 at room temperature (*Figure 9.9*). The lattice expansion within the *FCC hydride* range[37] is shown in *Figure 9.10* (to be compared with the corresponding curves for Zr and Hf, with development to tetragonal distortions).

Increasing the H content within the hydride with temperature is accompanied by an eventual filling of the octahedral holes in the CaF_2 lattice[39]. On the deuteride series $TiD_{1{\cdot}50} \leftrightarrow TiD_{1{\cdot}94}$, neutron diffraction has shown that the CaF_2 filling approaches completion; at lower temperatures and with increased hydrogen, it is accompanied by a cubic → tetragonal distortion[14a].

At $TiH_{1.99}$, the unit cell becomes cubic above 42°C (a = 4·454 Å), with a second-order tetragonal distortion starting at lower temperatures[40]. This consists in a continuous contraction along the c axis and expansion along a, with a minimum c/a of 0·945 at 79°K, indicative of yet a further change at very low temperatures.

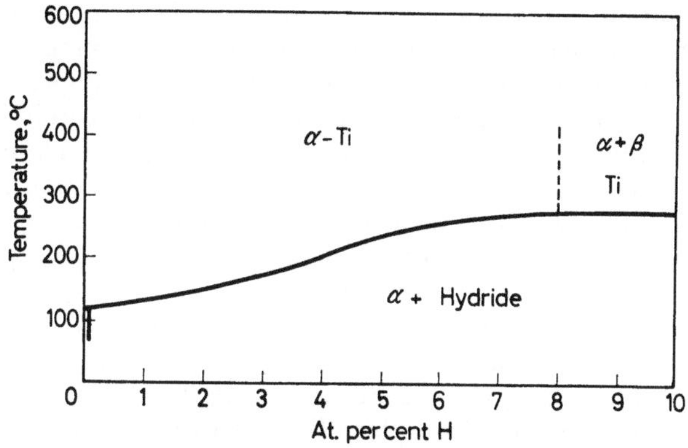

Figure 9.9. Low-temperature solubility limits for H *in high-purity titanium* (From Lenning *et al.*[38])

By courtesy of The Metallurgical Society of AIME

Towards higher temperatures, it is interesting to observe, from high-temperature/high-pressure x-ray work[39], that the FCC unit cell attains a maximum spacing at certain temperatures, which themselves become displaced with varying pressures. The initial expansion is here the normal

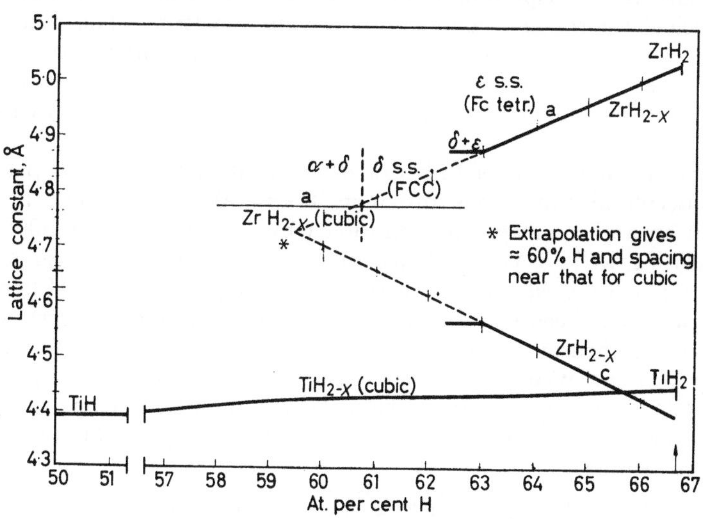

Figure 9.10. Lattice spacing variation in ZrH_{2-x} *and* TiH_{2-x} (After Espagno *et al.*[56c] and Chrétien *et al.*[37])

thermal one, but the higher-temperature contraction is due to a movement of hydrogen from the tetrahedral to the octahedral sites, i.e. a $CaF_2 \rightarrow$ NaCl-type transition. We might recall that this mode of change was quite familiar amongst the oxides (Chapter 8), e.g. $UO_{2-x} \leftrightarrow UO_{1+y}$, etc., and now reappears, in this new version, for the hydrides.

It may well be said that, in the Ti–H system, the above cubic $\rightarrow$ tetragonal transition *within* the hydride-phase region is *preceded*, for lower hydrogen, by a gradual hexagonal $\rightarrow$ cubic transition not dissimilar in kind. The duplex region α-Ti solution + hydride is not one of normal two-phase coexistence but more in the nature of a single-phase field in which the progressive influx of lattice hydrogen promotes a change of stacking sequence ABAB . . . $\rightarrow$ ABCABC . . . over small but growing lattice regions within a crystal. Thermodynamically this is reflected in *Figure 9.4* by the plateau in the P–T–C curves becoming more and more indistinct, more sloping and blending at increasing temperatures with the terminal single-phase solid solution and hydride portions of the curves. It illustrates how the concept of phase boundaries becomes blurred in transition-metal hydride systems (not, of course, only in Ti–H but also in the later ones, especially in Group V). This contrasts with the earlier ionic systems where the delineation was quite sharp.

A related problem has been raised in trying to identify dispersed Ti hydrides in a Ti matrix metallographically[41]; it was found inconsistent that optical anisotropy of the hydride in polarized light should correspond with its presumed FC *cubic* structure. However, x-ray and neutron diffraction data on the Ti deuteride showed that, when dispersed in the Ti *matrix*, the hydride is not cubic but of a more complex tetragonal unit cell*. Its lattice is coherent with the α-Ti lattice and represents a state of non-equilibrium in which the Ti matrix could be supercharged, the TiH_2 subcharged, with hydrogen atoms.

An implication of the Ti–H (and other Group IV) phase relations is the facility of hydride precipitation from the α phase. This occurs upon aging in mostly acicular form and coherently with the α matrix. The hydride is particularly prone to precipitation in zones of strain, whatever the origin. For example, it was found near hardness impressions and at polished surfaces[42–4]. We might liken this to a 'local haemorrhage' of hydrogen atoms.

The antiferromagnetism of Ti hydrides[2e,45] has demonstrated the parallelism of susceptibility and constitution in Ti–H. Wohlfarth[45] had already suggested an antiferromagnetic transition temperature of about 300°K in TiH_2, relating well to Yakel's[40] structural observations mentioned above. His indication of *molecular* hydrogen already playing a part in Ti–H bonding has not, however, been confirmed by later neutron-diffraction work which leaves the tetrahedral or octahedral H-ion atom positions in no doubt. However, a latent 'awareness' of the hydrogen atoms of each other's presence in the FCC lattice seems very probable, which at higher temperatures or lower pressures becomes active and evolves to H–H pairing and to the known gaseous desorption.

* With $a \approx 3{\cdot}12$, $c \approx 4{\cdot}18$ Å, $c/a \approx 1{\cdot}34$, 2 Ti per unit cell (equivalent to FC tetr. with $a \approx 4{\cdot}42$, $c \approx 4{\cdot}18$, $c/a \approx 0{\cdot}946$, 4 Ti per unit cell). This is not the same as the stable tetragonal hydride discussed but probably a related deviant maintained by epitaxial coherency with the HCP matrix lattice.

There is considerable support for the idea of the Ti hydrides and those of Groups III and IV more generally, being largely *ionic*[2a], with an H anion of radius 1·3 Å interstitial to a metal anion lattice and acting as intermediaries in a superexchange mechanism of electrons between the Ti ions; it was suggested that one of the two *s* electrons from the H^- anion is transferred to a Ti cation *d* orbital, leaving a neutral H atom; the other *s* electron, of opposite spin, interacts with another neighbouring Ti ion.

Some of the present theories on the bond nature in Ti and other hydrides of the type still seem difficult to reconcile, although the differences may partly be a matter of formal rather than fundamental interpretation.

In a now classical series of papers[46] McQuillan has studied the thermodynamic character of the solid solution series, deriving the free energy curves, the α/β transformation temperatures and the latent heats. His P–T–C curves and phase diagram are shown in *Figure 9.11.*

The β/α eutectoid reaction exhibits a strong *hysteresis* effect, of about 40°C, independent of the cooling rate[38]. The morphology of hydride precipitates does, however, depend on the cooling rate, giving platelets on slow cooling and fine spheroidal dispersions after quenching. It was also found[38] that hydrogen has little effect on tensile properties of titanium but greatly diminishes the notch toughness (cf. Section 9.4).

Studies of the Ti–H system with regard to hydrogen absorption and desorption[47] provided P–T–C data in general agreement with other work and interesting results on the systems Ti–*deuterium* and Ti–*tritium*; however, no thermodynamic differences between the three isotopes were found here, although on a refined scale some still seem possible, as in the other isotope cases referred to. (See also Bibliography.)

It is of particular interest that, as shown by magnetic resonance[48], the hydrogen atoms possess a preference for the tetrahedral sites in which they are arranged randomly (in the range $TiH_{1\cdot61}/TiH_{1\cdot97}$ from − 196 to + 200°C). It was shown[48] that the diffusional activation energies increase with H content, from 9·4 to 10·2 kcal/g atom (for $TiH_{1\cdot607-1\cdot923}$) and that the diffusion rate is proportional to the number of vacant tetrahedral holes. Therefore, H diffusion appears to proceed via the available vacancies*.

A point of fundamental importance brought out by McQuillan's early work on the Ti–H system[46] is that a phase transformation in a pure metal, as for α-/β-Ti at 882°C, can split into *two* (at 864° and 900°C, shown by thermal arrests) upon trace occlusion of hydrogen. This was initially thought to reflect a third polymorph of titanium, with a structure having a positive heat of solution for H, as against a negative one for α- and β-Ti. This effect is distinctly particle-size-dependent, insofar as it becomes enhanced for fine particles and disappears for coarse-grained samples. This third form will really either be metastable or become inhibited for higher H contents; otherwise it should obviously have a place in the Ti–H diagram and affect the β-eutectoid, which it does not have in the versions proposed

* This work[48] is incidentally of interest also from the viewpoint of technique of investigation, because proton magnetic resonance and the variations derived by it for varying H contents are powerfully adapted to the study of hydride structures; it also gives a correlation with bulk magnetic susceptibilities and suggests from electron-exchange interactions that the titanium–hydrogen atom bond is a combined covalent and ionic one.

so far, including McQuillan's own diagram. Nevertheless, the question is not yet closed, and a modification of equilibria to incorporate this third structure could be envisaged. There is reason to believe that this could be the so-called 'ω phase', a metastable BCC (β) derivative known e.g. in the Ti–Nb and other Ti-rich systems. For *iron* an analogous doubling of the A_3 (α/γ) transformation through small quantities of hydrogen is also known[49], and the effect could in essence be similar[46].

The general theme thus is the stabilization or metastabilization by traces of interstitial hydrogen of third crystalline polymorphs intervening between an otherwise normal phase change. The finding for *calcium* ('γ-Ca' $\approx$ $Ca_{10}H$ phase; see above) is a further case in point, so that we have in Ca, Ti and Fe at least three representatives of the same effect along the First Long Period. It seems possible that further examples will yet arise, e.g. in Mn and Co.

For interstitial *oxygen*, the W_3O type and related phase may be considered in the same sense as a near analogue.

A practical consequence is that small interstitial impurities have a profound effect on the transformation characteristics of titanium; in effect, they

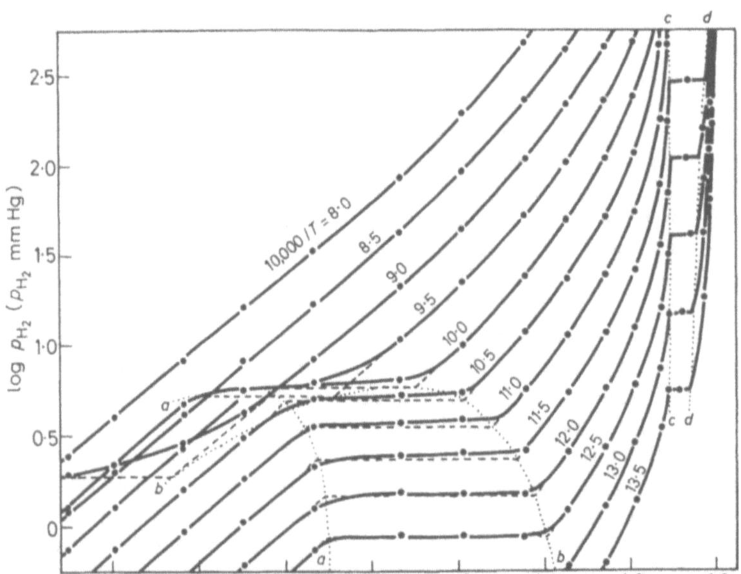

Figure 9.11. The Ti–H *system:* P–T–C *curves and phase diagrams* (From McQuillan[46])

(*a*) Isothermal pressure–concentration curves for the hydrogen–titanium system; (*b*) pressure–temperature curves for range of concentrations; (*c*) hydrogen–titanium constitutional diagram for the condensed phases.

— experimentally determined, – – extrapolated phase boundaries

... probable trends of the system; shaded area: hydrogen pressure > 1 atm

By courtesy of The Royal Society

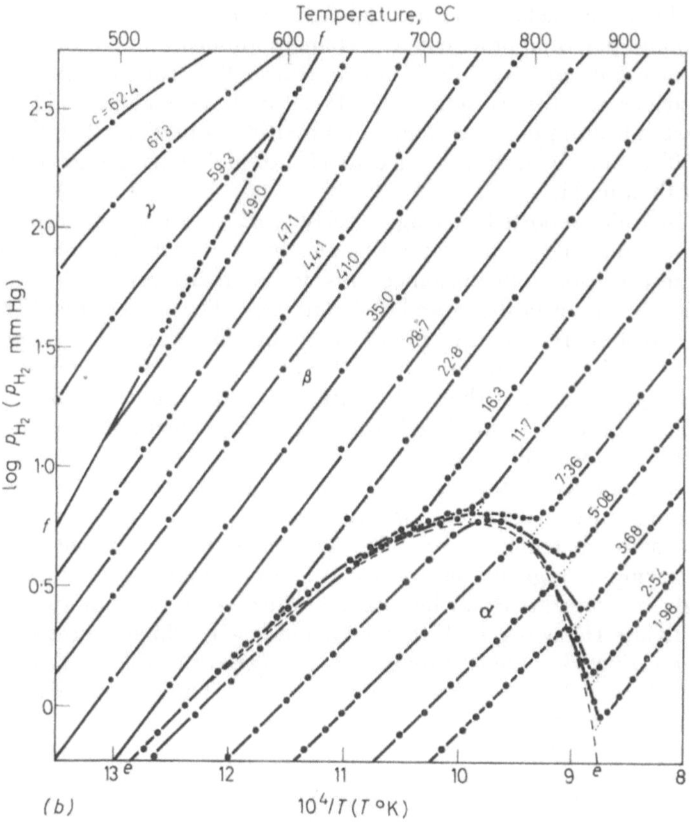

Figure 9.11b

Figure 9.11c

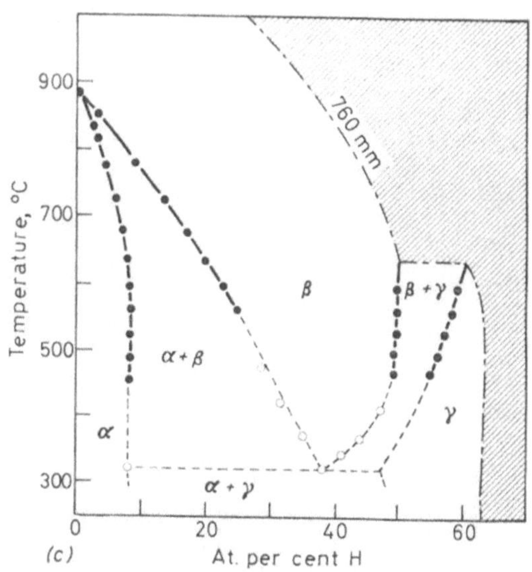

render the 'pure' metal a multi-component system, with the new degrees of freedom and potential new structure changes implied.

Zirconium–hydrogen—The Zr–H system (*Figure 9.3*), compiled by Libowitz[2c] from various sources, is similar to Ti–H both as regards solid solutions —β-Zr stabilization to a eutectoid—and hydride formation. The β eutectoid is at 550°C and about 36 at. per cent H, i.e. higher temperature, lower H content than in Ti.

The hydrides reported in Hägg's early work (a high-temperature Zr_4H, Zr_2H, ZrH and ZrH_2), rather than being superseded, should now be given a new interpretation: for instance, his lower hydrides could have been unstable distortions of the α- and β-Zr lattices. The present two 'ZrH' and 'ZrH_2' hydrides are separate only at lower, but join at higher, temperatures to give a continuous transition from the FCC (δ) to the FC tetragonal (ε) unit cells (for structural data, see *Table 9.1*); the nomenclature of γ-, δ-, ε-Zr–H phases stems from Hägg's original work, sequential to α- and β-Zr, and is still generally adopted.

An additional FC tetragonal (γ′) phase is also known[51]. This may be a high-temperature hydride equilibrating with β-Zr at high, but metastably retained at room temperature. The large FCC hydride range of continuously varying tetragonal distortions at high temperatures has been convincingly demonstrated[52] in high-temperature x-ray work at varying pressures. In the H-poorest hydride we have $c/a > 1$, which decreases linearly with increasing H content through $c/a = 1$ (i.e. FC cubic) at H/Zr ≈ 1·5, to < 1. *Figure 9.12* shows these results in detail. The c/a ratio also increases with temperature[40], as in Ti hydride.

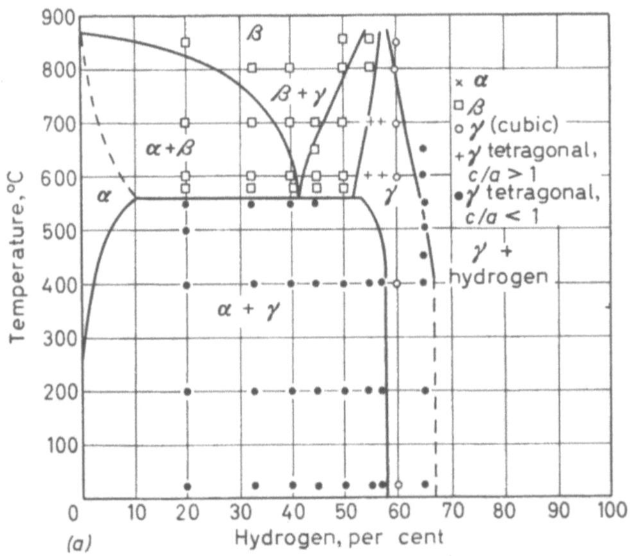

Figure 9.12. The Zr–H *system* (From Vaughan and Bridge[52])

(*a*) Phase diagram; (*b*) change in axial ratio with temperature for 65 at. per cent alloy (≈ ZrH_2); (*c*) change in axial ratio with hydrogen content at room temperature

By courtesy of The Metallurgical Society of AIME

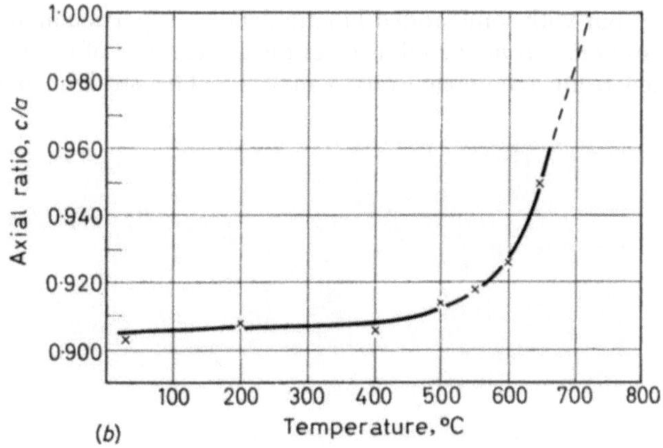

Figure 9.12b

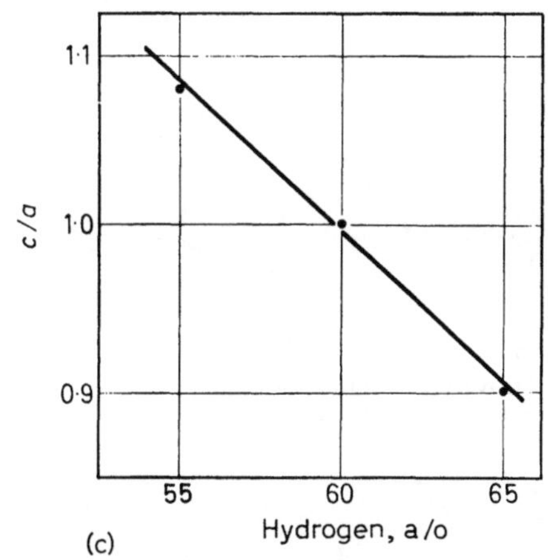

Figure 9.12c

Vaughan and Bridge's[52] basic concept of the Zr–H kinetics differs from that of some other investigators insofar as the validity of at least three independent 'hydrides' in equilibrium previously reported is denied; it supports the idea of continuous structure change in what is really only one extensive hydride region.

Neutron and x-ray diffraction work[15b] confirmed the FC tetragonal Zr lattice in the hydride, with the H atoms in tetrahedral sites, fluorspar-like. The $\delta/\delta + \varepsilon$ region has been defined by x-ray work[50] and by thermoelectric measurements[53].

It is remarkable that the ε phase can be formed by a *martensitic* type of transformation from the δ phase (an effect illustrated by detailed metallography in Whitwham's paper[51]). The ability to form supersaturated H solid solutions is pronounced, and the whole morphology (e.g. grain-boundary segregation of hydrides or even precipitate) is critically dependent on the rate of cooling through the eutectoid range.

An investigation of the Zr–H system, due to Beck[54], provided a diagram, P–T–C curves, structure and metallographic data in good accord with earlier ones but with some added kinetic details, in particular regarding the $\delta \rightarrow \varepsilon$ transition. *Figure 9.13* illustrates the appearance of the ε hydride as precipitated from a homogeneous δ matrix. The straight parallel bands have

Figure 9.13a. Metallographic appearance of ε–Zr hydride (From Beck[55])
a × 200, *b* × 1,500, polarized light; reduced $\frac{4}{10}$ on reproduction
By courtesy of the American Society for Metals

Figure 9.13b

formed by a shear or twinning movement, a coherency only possible between such closely related structures as δ and ε are known to be. A FC tetragonal 'Zr_2H' phase at room temperature, reported by Jack[55], decomposes on annealing at 300°C and seems likely to have been the ε phase.

Neutron diffraction[56c] showed a change in axes and in axial ratio of the δ–ε (Zr–H) series as in *Figure 9.10* (shown together with Ti–H values), and confirmed that, for saturation, stoichiometric ZrH_2 is almost reached. A

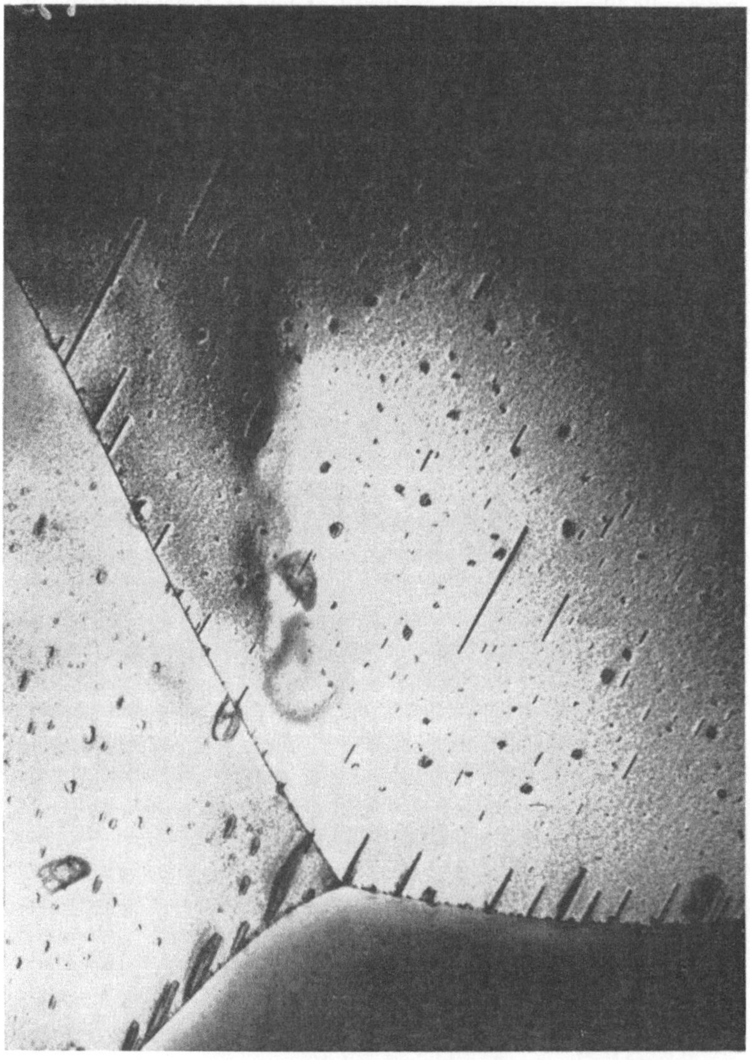

Figure 9.14a. Precipitation of Zr *hydride in zirconium matrix* (From Bailey[57])

Electron micrographs (reduced ⅘ on reproduction); (*a*) preferential precipitation of hydride on grain boundaries (× 11,000); (*b*) needle-like hydride precipitate in Zr quenched in iced brine from 800°C (× 26,000); (*c*) removal of the precipitates shown in (*b*) by electron beam heating (× 26,000)

By courtesy of Pergamon Press Ltd.

linear expansion of Zr metal occurs with hydrogen content absorbed, by 14·95 per cent from pure (α)-Zr to 60·5 per cent H (δ), transgressing the structure change and well illustrating the essential continuity during the transformation sequence. Comparison with Ti–H is of interest. If in Zr–H we extrapolate the *a* and *c* axis curves to lower H, the cubic state is reached at H/Zr $\approx$ 1·5, i.e. at about the ratio at the observed cubic form.

Electron microscopy by thin film[57] on the precipitation of Zr hydride in Zr was able to show the morphology and orientation relation during nucleation. Very fine needle-like precipitates occurred at low H concentrations of the tetragonal hydride (γ') on quenched specimens, representing platelets parallel to the $\{10\bar{1}0\}$ planes of the HCP matrix. *Figure 9.14* shows examples[57] of

Figure 9.14b

their appearance: (*a*) preferential nucleation at grain boundaries (also at emerging dislocations), and (*b*) dense hydride concentrates when quenched from 800°C, which (*c*) can be shrunk and removed by electron-beam heating. The associated dislocations disappear at the same time.

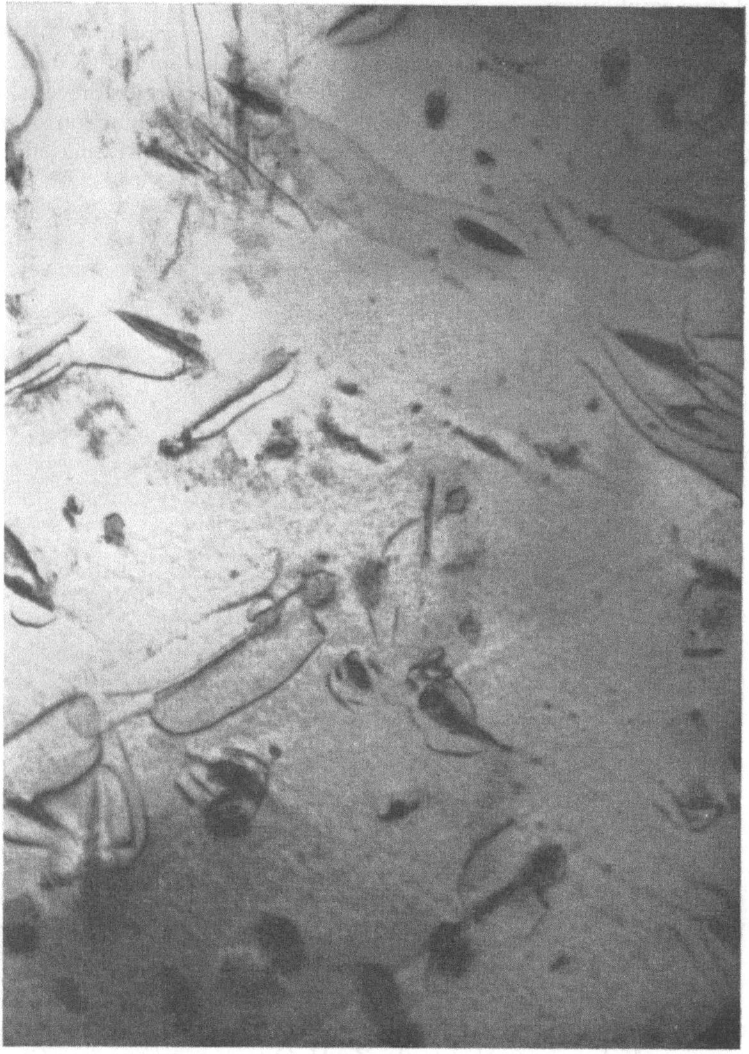

Figure 9.14c

Solubility studies[22b] gave the primary α boundary shown in the diagram, whilst fair agreement in locating the β boundary shown was attained in other work[52,58–60]. However, there remain some considerable conflicts in regard to the actual hydride (δ-phase) boundaries, and Libowitz[2d] proposed a re-location*.

The P–T–C curves and the related phase-boundary determination for high H contents in Zr–H has been extensively studied by Libowitz[2d], whose diagrams *Figures 9.3* and *9.4* essentially represent. Since the $\delta \rightarrow \varepsilon$ change is martensitic, the $\delta \rightarrow \varepsilon$ two-phase region should probably not be considered one of true equilibrium.

The diffusion kinetics and H solubility in Zr at *low* pressures (0·1–4 μ) have shown[22b] further that, at constant pressure, this solubility decreases with increasing temperature and that within each phase region Sieverts' law is well obeyed. Diffusion and 'degassing' coefficients as a function of temperature were also derived in this work, as well as the embrittling effects[61] of the tetragonal hydride in Zr.

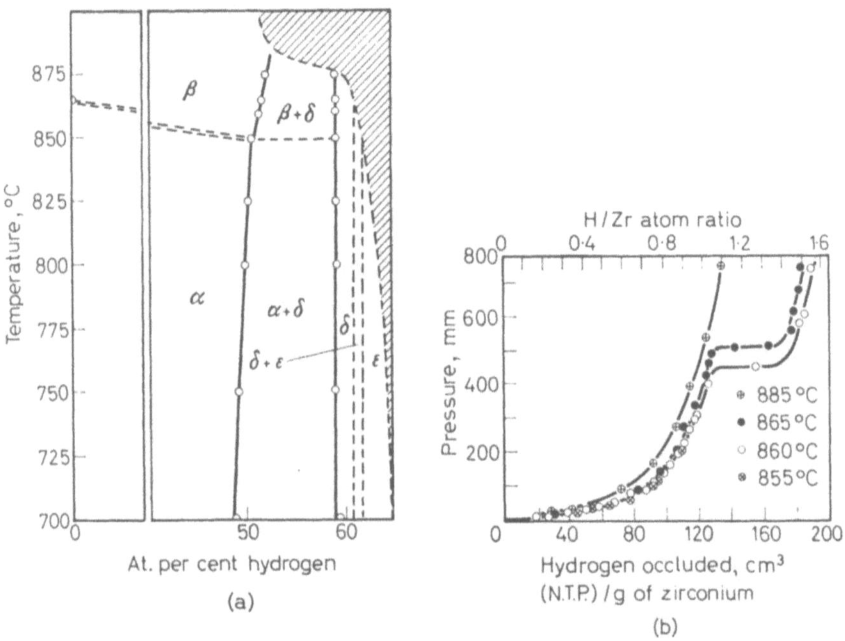

Figure 9.15. Zr–H *system* (From Edwards, Levesque and Cubicciotti[21b])

(*a*) Partial phase diagram for the condensed phases (shaded area where H_2 pressure exceeds 1 atm); (*b*) isothermal pressure–volume curves for 855–865°C range, i.e. near α/β transition point of Zr

By courtesy of American Chemical Society

Extensive solubility and equilibrium studies on Zr–H, due to Edwards and associates[21], have extended to the Zr–H–O ternary system (see Chapter 10). Their P–T–C curves and phase diagram are shown in *Figure 9.15*, amplifying the diagrams of *Figures 9.3* and *9.4*. In contrast to earlier results, lowering of the β- → α-Zr transformation by H seemed here to be only slight.

* This underlines the sensitivity of hydrides to minor experimental conditions, as mentioned before.

An interesting fundamental point arising from this work, related to Section 9.1, is the observation that hydrogen migrates towards the cathode on passing an electric current, which supports the *proton* nature of the H solution in Zr, Th and Pd. They also postulate *metallic* bonding of H to Zr, i.e. the merging of the hydrogen atom's electron in the metallic electron sea, at any rate in the Zr–H solution; in the higher hydrides deviations may occur.

A recent survey on zirconium hydrides[62] has yielded the revised phase diagram shown in *Figure 9.16*. Despite the differences, however, and apart from some quantitative adjustments, the original conclusions are still sound.

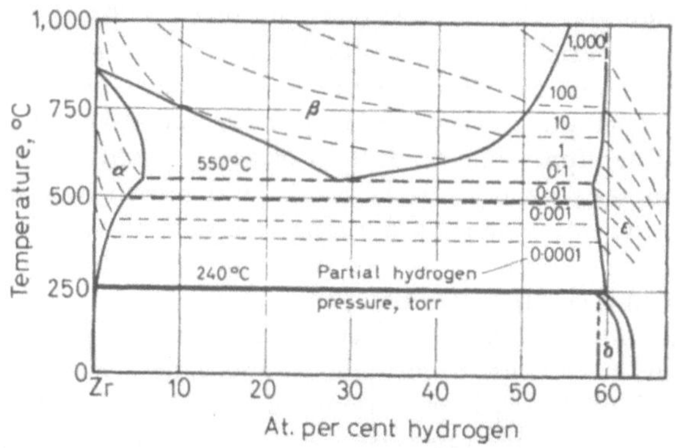

Figure 9.16. Zr–H *diagram, including changes with pressure* (From Motz[62])

By courtesy of the Editor, *Zeitschrift für Metallkunde*

Martin and Rees[63] have given a detailed theoretical discussion of the solubility and P–T–C relationships in the Zr–H system, in terms of a statistical-mechanical model of hydrogen-atom ordering. In the present context, their observation that *oxygen-bearing* zirconium possesses a *lower* hydrogen solubility is significant; it is due to blocking of normal H sites by neighbouring O atoms. This is up to the limit of first oxide formation. This has wider implications on hydride systems generally, again illustrating the competitive nature of interstitial diffusion. However, contrary results have also been obtained[64], of an *increase* of H solubility in α-Zr with oxygen content: whilst hydrogen is occluded or dissolved in the Zr lattice, any oxygen present immediately forms ZrO_2 at surfaces, grain boundaries, etc.

The tensile strength of zirconium hydride, investigated in relation to phase constitution and to nuclear reactor applicability[65], showed definite composition ranges (around 45 and 60 at. per cent H) where high-temperature strength and stability were found to be sufficient for practical service purposes. This incidentally provides an interesting example of directly relating the factors of constitutional stability and of utilization in practice (see also Bibliography).

The *hafnium–hydrogen* system is closely similar to Zr–H, except for a considerably smaller solid solubility. The process of FC cubic → FC tetragonal transformation holds as in zirconium hydrides, though with quantitative differences. The phase diagram of *Figure 9.3* is that of Libowitz[2c]; the P–C–T diagrams (*Figure 9.4*) are those of Edwards and Veleckis[21a], Sidhu, Heaton and Zauberis[14a] and Sidhu and McGuire[14c]. Neutron diffraction was largely instrumental in Sidhu's studies[14]. At least three phases occur, interconnected by continuous transitions: the slightly deformed cubic form coexisting with α-Hf, then extending to $HfH_{1.53}$ and converting to the FCC one around $HfH_{1.63}$, with wide homogeneity range. This is followed by another BC tetragonal phase around $HfH_{1.98}$. (The H atom locations were found here on the equivalent deuterides.)

The effect of *saturated* hydrogen concentration upon bonding for compositions approaching HfH_2 has also been studied[14b]. Here each metal atom has 12 metal neighbours, 8 at a short bond distance (3·289 Å), 4 at a longer one (3·461 Å). However, as the interstitial H atom concentration increases, anisotropy arises in the CaF_2 structure, due to extra metal–gas bonding which strengthens the short and weakens the long metal–metal bonds, resulting in the tetragonal distention. The absorption of new H atoms under sufficient gas pressure imposes such new Hf–H bonds, and this in turn weakens the *M–M* bonds in the 001 plane, by draining electrons away from them and increasing the *a* axis. There is, however, a general strengthening of the short bonds, resulting in a compression along the *c* axis.

It may also help to explain the anisotropy at these high-H saturations by assuming already a tendency to forming H_2 molecules, i.e. (as put earlier) of a 'recognition' of the hydrogen atoms of each other's presence and a polarizing effect. This later progresses to the stage of gas evolution upon reduction of pressure but is already incipient in the parent lattice.

A thermodynamic and structural investigation similar to Edwards's on Zr–H and carried out by the same school[21a] on Hf–H gave P–T–C curves indicating a maximum α-Hf solubility of up to 11 per cent H.

Comparison of the crystallography and dilatometry for Hf–H and Zr–H by Espagno *et al.*[56] has confirmed their essential similarity, but with significant quantitative differences.

In the *thorium–hydrogen* system, there are cross-characteristics of both the Zr–H and Hf–H and of the actinide–H series (for details, see again *Table 9.1* and *Figures 9.3* and *9.4*). The appreciable primary solubility, much higher than in Hf, is natural, as Th metal is itself *FCC* and may thus be regarded as the structural precursor to ThH_2 itself. This hydride again is FC tetragonal (distorted CaF_2 type), as (Ti,Zr,Hf)H_2, and possesses a relatively small H-deficient homogeneity range, $ThH_{1.78}$ to $ThH_{1.98}$. However, the new factor is the occurrence of the higher hydride Th_4H_{15}: this too has a limited homogeneity range. It is of complex BCC structure[30], with 16 Th and (ideally) 60 H atoms per unit cell.

The FC tetragonal structures of ThH_2 (as of ZrH_2), established by neutron diffraction[15b], revealed a much smaller homogeneity range; it brings us back to regarding ThH_2, although still metallic, as again approaching the ionic state, more akin to the Group III and rare earth hydrides; this is indeed to be expected in the actinides.

ThH_2 has an exceptionally open lattice, with *M–M* and *M*–H distances much larger than in ZrH_2. This important feature already foreshadows the position of the later actinides, as in UH_3. Thus, whilst in ZrH_2 the metal–hydrogen bond is still strong, it is weak in ThH_2.

A practical consequence is that, firstly, thorium hydrides are very ready donors of hydrogen, and secondly, they are much more sensitive to oxidation than are, for instance, TiH_2 and ZrH_2. We might say that in the latter the H atom aids in protecting the hydride from oxidation by its still firm hydrogen bond, but this no longer holds for thorium.

The higher hydride Th_4H_{15} (= $ThH_{3.75}$) is the richest of any hydrides known; with its BCC-base structure we may regard it as a further evolution of the open-lattice tendency already present in the FCC ThH_2. Each Th atom is here bonded to 12 H atoms. The lattice is literally 'packed' with H atoms, but loosely so and readily dissociable. The Th–Th distance is so much larger than that in Th metal that the main cohesive energy must be attributed to the Th–H bond. *Two* kinds of H atoms are distinguished in the Th_4H_{15} lattice, of different effective radii according to their position: 0·67 Å of type I, 0·50 Å of type II; both again are larger than the H radius in ThH_2 and far larger than the metallic radius of H (0·27 Å) prevalent in the tightly packed hydrides.—The lattice spacing/composition variations of the FC cubic (→ tetragonal) thorium hydride[17a] also revealed near-stoichiometric Th_4H_{15}.

The isomorphy of Th and ThH_2 makes it possible in principle for a complete solid solution to occur, and indeed this seems to be true[66] at about 1,200°C. Below, a miscibility gap rapidly evolves and widens, so that at room temperature FCC Th dissolves only 0·002 per cent H, equilibrating with nearly stoichiometric ThH_2. We may see in this already a strong resemblance to the Pd–H system, although the hydrogen permeability in Th–H is not as high.

P–C–T curves have been determined in early work[15e,22a,23] (*Figure 9.4*). The system has certain further resemblances to palladium–hydrogen insofar as the 'rift theory' developed for Pd–H (D. P. Smith[67]) can also explain the character of Th–H. According to this, hydrogen is adsorbed at 'rifts' at crystal interfaces, and only here is true equilibrium attained. The external pressure depends on concentration gradients and is not necessarily the same as the equilibrium pressure. Thus the dissociation pressures observed, even within the apparent 'plateau regions', still vary with H content, resembling transient phase 'conditions' rather than equilibria.

Peterson and Rexer[66], in determining the P–T–C variations more specifically for the ThH_2 range and thus also deriving the diffusion coefficients of H in ThH_2, found that Sieverts' law was nearly fulfilled. The composition variation for high pressures in the single-phase hydride region showed that, at low temperatures, this approaches the $ThH_{2.00}$ composition (full tetrahedral occupation). With increasing temperature the lattice vacancies increase.

(5) *The Group V hydrides* (V–H, Nb–H, Ta–H)

The systems of V, Nb and Ta with hydrogen are all characterized by the essential maintenance of the BCC parent structure, but with varying and continuous lattice distortion. The β stabilization which had been observed for Ti, Zr and Hf, but limited by the eutectoid and the formation of new higher hydrides, now evolves to full dominance of BCC variations throughout. As *Figures 9.3* and *9.4* show, at elevated temperatures the structure changes occur within a continuous single-phase region, with a miscibility bay at lower temperatures. Its apex is in the order of room temperature, varying between about + 200°C for Nb and − 50°C for Ta, dependent on pressure.

An excellent investigation covering jointly the vanadium, niobium and tantalum systems with H_2 is that of Veleckis[24]. This, it is felt, supersedes much contradictory previous work, and the diagrams shown are essentially his, with some extra annotations. Although formulae like 'M_2H', 'MH' are used for convenient expression, any ascription of stoichiometry to them here becomes really meaningless, except possibly in the sense of an ordering at very low temperatures (for structural, etc. data see *Table 9.1*).

In *vanadium–hydrogen*, a BC cubic → tetragonal distortion occurs with increasing H. According to x-ray diffraction and magnetic studies[33c], the room temperature V → H solubility is about 2 at. per cent H; the BC tetragonal phase occurs near 'V_2H' (cf. V_2C amongst the carbides), both coexisting in a region ≈ 2–30 per cent H. At raised temperature, this duplex region disappeared around 150°C, with the axial ratio decreasing from 1·1 at room temperature to 1·0 at 205°C—similar to Veleckis's duplex region shown, but with quantitative differences. Susceptibility measurements showed that this property is independent of temperature. However, it decreases with increasing H content, and the indication is that the solute hydrogen is ionized to a proton, with the valency electron joining the metallic 'sea'. *Ordering* of hydrogen atoms occurs at high concentrations, more rapidly so for lower temperatures.

Not much neutron-diffraction work seems to have been done yet on V–H to locate the hydrogen-atom positions, although the results on Ta and Nb below may be presumed to be similar for vanadium. However, vanadium deuteride[68] near 'V_4D_3' has at liquid nitrogen temperatures given an indication of a primitive cubic cell with $a = 6{\cdot}302$ Å (2 × that in V) and of an ordering of H atoms which disappears above 207°K. The actual H-atom locations are not derived; however, internal friction work on V–H indicated[69] that the hydrogen atoms occupied the tetrahedral sites, and a degree of stress-induced ordering at low temperatures.

Thermodynamic work from the same school[70] confirmed that above 165°C a single-phase region prevails for H contents up to at least 32 per cent H, agreeing with the structural results. There is an interesting link-up with the Ta–H system[71] (see below). In both, the cubic → tetragonal transition occurs with decreasing temperature, at around 33 at. per cent H. In tantalum, specific heat and neutron diffraction measurements have shown the tetragonal distortion to be an outcome of the ordering of hydrogen ions in interstitial sites. This is very probably similar for vanadium and niobium, though not yet verified. With increasing H content and/or decreasing temperature the previously randomly arranged H ions begin to order, entailing anisotropy (for reasons to be discussed under Ta–H, and similar to the above cases of Zr–H and Hf–H).

Lastly, it was possible to demonstrate[72] the *mechanical embrittlement* of vanadium by hydrogen for low H and at sub-zero temperatures; the tensile strength, however, is temperature- and composition-dependent, and a fair ductility can be obtained with certain conditions of heat treatment.

Niobium takes hydrogen into limited solid solution at room temperature, with lattice expansion. A hexagonal hydride 'Nb_2H' had first been proposed by Hägg[7], and reported[73a] as rhombohedral. An observation[74] which ought to be revised in modern light, however, is of a FCC hydride 'NbH', probably orthorhombic. Brauer and Hermann[73a], in an extensive study, derived diagrams shown schematically in *Figure 9.17* for Nb–H, Ta–H and the corresponding deuteride systems (at room temperature). The latter are practically identical. In the present light, the phases no doubt are equivalent to those at either end of the miscibility bays in *Figure 9.3*.

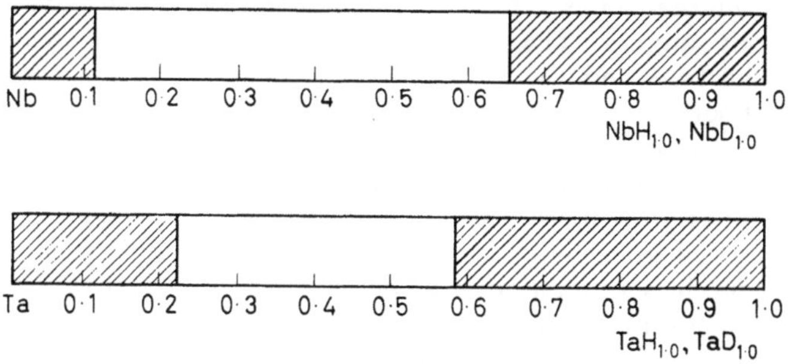

Figure 9.17. The Nb–H (D) *and* Ta–H (D) *diagrams* (From Brauer and Herrmann[73a])

Room temperature, atm. pressure

By courtesy of Messrs. Johann Ambrosius Barth

Brauer and Hermann find the BCC and BC rhombohedral 'NbH' solid solutions to equilibrate. In the latter, the distortion is only slight (near-cubic edge, $a = 3{\cdot}44\ kX$, $\alpha = \beta = 90°$, $\gamma = 89{\cdot}4°$. This phase is sometimes stated as face-centred orthorhombic; but it is the same, with $a = 4{\cdot}83$, $b = 4{\cdot}89$, $c = 3{\cdot}44\ kX$, at $NbH_{0{\cdot}89}$.

It is interesting to note that, for compositions close to the primary Nb → H solid-solution boundary but on the duplex side, the Nb–H solution can be readily retained in a *supersaturated* state, of spacing higher in the single-phase than in the truly two-phase equilibrated condition. This rather conflicts with Veleckis' observation of absence of hysteresis effects (see below). The heat of solution of H in Nb, as throughout Group V, increases with hydrogen content.

A new niobium hydride, NbH_{2+x}, could be prepared chemically from niobium powder and a hydrofluoric acid[73b]. This would have the CaF_2-type (FCC) structure with $a = 4{\cdot}55$ Å, but possible impurities, notably oxygen, are not excluded, and the phase has not yet been confirmed. Nevertheless, not much impurity from Group IV metals may be needed to tip the free-energy balance in favour of the FCC structures of the earlier hydrides or related oxides.

Further structure studies on Nb–H show that a *metastable* 'NbH' hydride is formed[75a] on hydrogenation of Nb at 600°C, of smaller lattice parameter than the equilibrium value. This is attributed to a transient random distribution of the H atoms which, on extended annealing, evolves to ordering in the BCC, more rapidly at higher temperatures. It brings out a basic point of some importance: the time factor in developing interstitial ordering, superimposed on that of H content.

A more detailed high-temperature x-ray study (Wainwright[76]) on Nb–H has demonstrated the orthorhombically distorted BCC structure for high hydrogen contents, the degree of distortion decreasing with temperature (substantially cubic above 100°C). The two-phase region at room temperature is confirmed to be one of coexistence of the BC cubic and BC orthorhombic structures. This high-temperature x-ray work also confirms essentially the closing of the miscibility bay above $\approx$ 250°C to yield the continuous solution series. An interesting side observation concerns *cold-worked* hydride samples. The preferred orientation introduced by cold work into both the low- and the high-hydrogen phases reveals a clear lattice coherency of both. It shows how the hydride phase can evolve from the BCC parent lattice, not by a nucleation-and-growth process, as in a truly separate phase, but by deformation: that is, effectively by the creation of strained lattice regions offering more open sites for hydrogen to occupy. We might thus regard mechanical work in a metal–hydrogen assembly as a further parameter and, in a sense, a novel mode of being equivalent to the traverse of the phase diagram, by activating hydrogen diffusion.

A further x-ray investigation[76b] on Nb–H alloys showed a sequence of fine-structural stages in the distortion for varying H contents, from the BCC to the orthorhombic state, by multiple fine splits of the diffraction-line profiles.

The course of lattice spacings with H/Nb ratio[77–8] is shown in *Figure 9.18*. The magnetic susceptibilities decrease simultaneously, as seen for two temperatures; there is a notable intersection of these curves, indicating a temperature invariancy at a certain H content, $NbH_{0.09}$. A close similarity in susceptibility isotherms exists already in the Nb–H and Pd–H systems, notwithstanding the difference in crystal and electron structure.

The Nb–H equilibria when studied[25] for varying P–T–C conditions, showed a single-phase solid solution throughout, except that below 300°C the P–T–C curves already indicate some initiation of the immiscibility region.

An examination of the P–T–C curves in *Figure 9.4* shows the presence of inflections for isotherms within the seeming 'single-phase regions'. These progressively develop to the two-phase plateaux and cast some doubt on the very meaning of a sharp single phase/duplex boundary in transition-metal–hydrogen systems. This evidences again an early precipitation stage in which a gradual process of segregation of H atoms in separate lattice islands occurs already within the apparent single solid solution. (We recall the similar situation in the Group IV hydrides.)

Albrecht's work[25] also agrees that Sieverts' law holds for low H contents and high temperatures, the Nb lattice expanding with H content. Thus, on studying the Nb–H diffusion kinetics and equilibria in more detail, largely by high-temperature x-ray work, the low-temperature miscibility bay was shown to have a critical point around 140°C, 0·01 mm Hg and H/Nb = 0·3, in fair agreement with Veleckis's results.

A further P–T–C and x-ray study[79] on Nb–H is in excellent general agreement with the findings stated. A new type of transformation below

300°C, certainly related to the miscibility bay, is thought to correspond to a martensitic or to an order–disorder-type reaction. It implies the possibility of *metastable and transitory effects* of many kinds, depending on cooling rates and other heat-treatment history; hardening and similar effects on properties can well be the consequence and merit more study.

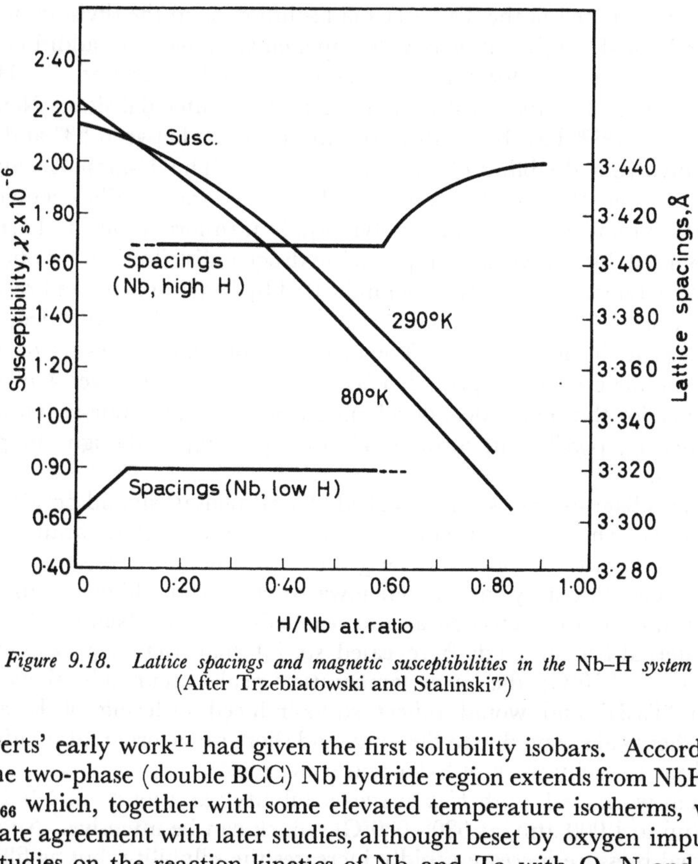

Figure 9.18. Lattice spacings and magnetic susceptibilities in the Nb–H *system* (After Trzebiatowski and Stalinski[77])

Sieverts' early work[11] had given the first solubility isobars. According to this, the two-phase (double BCC) Nb hydride region extends from $NbH_{0.08}$ to $NbH_{0.66}$ which, together with some elevated temperature isotherms, was in moderate agreement with later studies, although beset by oxygen impurities.

In studies on the reaction kinetics of Nb and Ta with O, N and H by means of a vacuum microbalance, Gulbransen and Andrew[50] also showed from the rates of reaction that Sieverts' pressure function is followed essentially and that the nature of hydrogen diffusion is *atomic*. An interesting, though somewhat controversial, observation[80] was of *two* solubilities of H in Nb, the higher for molecular, the other for atomic hydrogen. It may be taken as certain that only the atomic solubility is the true lattice one; the 'molecular' solubility value would then probably correspond to adsorptive occlusion at free surfaces, crystal boundaries and 'rifts'. In this sense, the differentiation of a distinctive metastable *molecular* solubility is, however, a very pertinent one that could be extended to all the present hydride systems. Its value depends, of course, on the degree of subdivision and other physical factors.

The *tantalum-hydrogen* system (*Figures 9.3, 9.4, Table 9.1*) closely parallels Nb–H, and several of the papers quoted[73a,78] have covered both. Many of the thermodynamic data, as for Nb–H, stem from Wallace's school[4c,33–4]; thus vapour pressures were measured by Kofstad[70b] who derived heats of formation and the areas of agreement and deviation from Sieverts' law. At high H contents, such deviations can again be attributed to the progressive but varied filling of interstial sites.

Most transformation effects observed are linked either to the distortion of the BCC unit cell of the Ta $\rightarrow$ H solid solution or to the division into the high- and low-H (miscibility bay) components. However, additionally a disorder $\rightarrow$ order transformation is known, notably around Ta_2H (or Ta_2D), of H (or D) atoms distributed amongst the interstial sites. Neutron-diffraction work[33b] has shown this ordering to occur between 53° and 20°C and to involve a doubling of the unit cell side. The H-atom location[33a] is such that the 16 available tetrahedral sites are statistically occupied by 8D (or H) atoms, but secondary polymorphic transformations in Ta_2D can be accounted for by various types of hydrogen distribution. Four such apparent polymorphs had also been indicated by heat-capacity and residual-entropy work[81] (change points at 306°, 332° and 333·5°K), with decreasing degree of order in this sequence. The final stage of such progressive ordering, that of complete order, is preceded by one of randomness over $\frac{2}{3}$ of the D sites, order for the remainder, developing from full randomness. However, *local* ordering, possibly in form of H-atom pairing, is thought to persist throughout.

In phase-diagram terms, the so-called 'Ta_2H' phase[71] should be identified with the H-richer BC tetragonal phase in the low-temperature duplex region. It should not, however, be given a stoichiometric meaning, as the process is a continuously varying one towards H/Ta ratios higher than Ta_2H, where it appears that tetrahedral and octahedral H sites begin to be filled.

An internal friction study[70c] revealed six internal friction peaks between $-190°$ and $+250°$C; the lower temperature peaks occur only above composition 'Ta_2H' and would reflect stress-induced ordering of H atoms, whilst other peaks seem due to the long- and short-range order, also indicated in the heat capacity work of Saba and colleagues[81].

Russian work dealt with *high-pressure* solubility variations[82] in Nb–H and Ta–H (up to 1,000 atm, at 620°–680°C), which can be intensive. An initial isothermal pressure increase rapidly increases the solubility, but as filling of the lattice to saturation is approached, this slows down to follow a logarithmic law.

Dealing with all three systems, V–H, Nb–H and Ta–H, Veleckis[24] had also shown that the maximum solubility limits of hydrogen in V, Nb and Ta are independent of temperature, at the atomic ratios H/V = H/Nb = 0·89 and H/Ta = 0·71. The observed P–C–T relationship could be expressed by a semi-empirical equation (based on statistical and mechanical considerations and supported by agreement of observed and calculated entropies):

$$\ln p^{\frac{1}{2}} = k + m \ln [r/(s-r)] + 1/T\,(a + br + cr^2 + dr^3 + er^4)$$

(p mm, T °K) where r = atomic ratio of H per metal atom, s = ratio of number of available sites/number of metal atoms (i.e. degree of occupancy);

the remainder are constants for each metal. The values calculated for the critical points (apices) of the duplex regions were

	Atomic ratio	*Critical temperature,* °C	*Critical pressure,* mm hydrogen
For V–H:	H/V = 0·213	47·1	$6{\cdot}87 \times 10^{-4}$
For Nb–H:	H/Nb = 0·373	183·2	$5{\cdot}10 \times 10^{-2}$
For Ta–H:	H/Ta = 0·025	−58	$4{\cdot}56 \times 10^{-11}$

The formula proved to hold quite reliably under all P–T–C conditions, in agreement with experiment. Veleckis[24] concludes further:

(1) Sieverts' law is obeyed by the isotherms at low solution ranges and at high temperatures—proof that hydrogen is dissolved as atomic H; (2) under all conditions, hydrogen is more soluble in Nb than in V or Ta*; (3) V, Nb and Ta are all proved to occlude hydrogen exothermically, with negative heats of solution; (4) the entropy values in all three systems are similar, which tends to show that H is occluded by the same mechanism (Lacher's theory of nearest-neighbour interaction, to be mentioned later for palladium, is here considered and found insufficient); (5) the passivity of metals towards hydrogen at low temperatures can be overcome (and this may hold more broadly) by saturating the sample with H by repeated sorption and desorption, followed by a rapid quench; (6) there is no evidence for thermal hysteresis effects in V–H, Nb–H and Ta–H, in contrast to Group IV and later Groups; Veleckis finds that the equilibrium pressures for a given composition in the solution region are independent of whether they are attained by sorption or desorption.

For the very limited information on the *protactinium-H* system, see *Table 9.1*. Sellers *et al.*[83] were able to prepare PaH_3 and found it isostructural with UH_3.

(6) *The Group VI hydrides* (Cr–H, Mo–H, W–H, U–H)

In Group VI, hydride stability decreases sharply as from the earlier Groups. Only Cr and U appear to be capable of hydride formation, not Mo and W.

In *chromium–hydrogen* a 'hexagonal form of Cr', reported in early work[84], proved to be a metastable hydride with a range ($Cr_2H \leftrightarrow CrH$). A phase of FCC structure beginning at CrH but dissolving hydrogen up to about CrH_2 is now known[85]. Thus the transition [$Cr_2H \rightarrow CrH$] $\rightarrow CrH_2$ corresponds structurally to a transition of wurtzite → fluorspar type, each with a wide H defective homogeneity range.

More specifically the process is one of H atom entry in the wurtzite structure at positions $00\frac{1}{6}$; $\frac{1}{3}, \frac{2}{3}, \frac{2}{3}$, with Cr at 000; $\frac{1}{3}, \frac{2}{3}, \frac{1}{2}$, until filling is complete. Then for any *extra* H atoms to be accommodated, the Cr-base lattice has to change its symmetry from hexagonal to cubic close-packing,

* This interesting maximum in the 2nd Long Period is paralleled by that in miscibility-bay apex. It may also be seen to parallel, in *Group IV*, the maximum hydride stability for zirconium.

probably progressively via stacking faults and local lattice domains where statistical H-atom enrichment had occurred.

The chief means of producing chromium hydrides is by electrolysis. They should be considered as transient but 'persistent', rather than stable, states. In an illuminating paper on the formation of chromium hydrides and their structures, Knödler[86] shows that the former greatly depends on specific electrolytic conditions, hydrides being favoured for instance by pulsating d.c., low bath temperature and high current densities. In fact, variations in deposition temperature would vary the H content and thus the structure; also a high degree of supersaturation of H atoms in BCC parent chromium can be produced, with associated lattice strains (cf. note on x-ray work on 'hard-chromium' plating, p. 69). Excess H then produces the HCP form. The phase diagram in *Figure 9.3* shows Knödler's suggested BCC/HCP/FCC boundaries for room temperature; in supersaturation, the boundary values can be greatly exceeded. — Electrolytic chromium can thus occur in all three forms (BCC, HCP and FCC) characteristic of the pure-metal structures, but here varied by interstitial hydrogen. An α-Mn-type form of hydrogenated chromium has never been confirmed, but the possibility should not, it is felt, be entirely discounted.

Early work[87] showing hydrogen absorption by Cr found a maximum solution of only 6 p.p.m. at 1,200°C. This may well be considered as the *equilibrium* solute value, and all higher contents as the result of supercharging; indeed, it basically brings Cr into line with Mo and W.

In the Mo–H and W–H systems, no hydrides have been reported. Nevertheless, this may be incidental to the fact that molybdenum and tungsten are not produced in electrolytic deposits, as chromium is; if they were, it might quite conceivably be found that a significant (though unstable) H solubility, and Mo or W 'hydrides', could exist, analogous to those of chromium.

The *uranium–hydrogen* system has been fairly extensively investigated. The UH_3 (and no other) hydride is stable, forming in fact highly exothermically. The compound exists in a number of polymorphous forms, at least 3, probably more (see *Table 9.1*). In line with the actinides, here is a return approach to more ionic and stoichiometric conditions.

By neutron diffraction, UH_3 has been found to have the cubic β-W-type structure[15b], with 8 U atoms in the W lattice positions. The 24 H atoms lie in the centre of deformed U tetrahedra. It would be misleading to regard W_3O and UH_3 in any way as 'anti-structures', as the metal/non-metal relation is not reversed. On the contrary, it might be said that the BCC (γ-)U lattice is 'packed' with H atoms, to produce a degree of distortion equivalent to that in BCC by lesser amounts of oxygen. This refers to the elevated-temperature form, β-UH_3, which is stable[27b] at about 250°C. However, when U hydride is prepared at *low* temperatures (R.T. to −80°C), a new ('α'-UH_3) form is found, stable below 100°C; it can also be formed electrolytically[88]. The α form is *BCC*, with the normal two U atoms per unit cell, but loaded with 6 interstitial hydrogens placed at $\pm$ ($\frac{1}{4}$, 0, $\frac{1}{2}$), $\pm$ ($\frac{1}{2}$, $\frac{1}{4}$, 0) and $\pm$ (0, $\frac{1}{2}$, $\frac{1}{4}$), with U at (0,0,0) and ($\frac{1}{2}$, $\frac{1}{2}$, $\frac{1}{2}$). Thus α-UH_3 is isostructural to metallic γ-uranium, and the (γ-U/α-UH_3) $\leftrightarrow$ β-UH_3 relation is the same as in W $\leftrightarrow$ W_3O.

In a wider context, therefore, it may be said that (*a*) the addition of hydrogen stabilizes the high-temperature BCC form of U to low temperature; (*b*) a temperature increase in the BCC hydride thus formed produces the same effect as *oxygen* addition to BCC tungsten. Considering further the polymorphy of pure uranium (γ-U = BCC, β-U = σ type, α-U = its own orthorhombic type), various interesting questions arise: one is whether a σ-type hydride of W and of U could not exist under high-pressures and very low temperature, another, whether this would appear in a mixed oxyhydride.

Furthermore, we may note that, as the β-W-type structure is stabilized

in uranium by *hydrogen*,
in tungsten and molybdenum by *oxygen*,
so it is in molybdenum and chromium by *silicon*.

In each case, this occurs from an initial BCC, and the essential metallic bond and properties are preserved.

Rundle[15b,c] has discussed the UH_3 structure in two classical papers. It is proposed that metal/metal bonds are absent and replaced by U–H–U bridge bonds*, with the H atom still forming a quasi-metallic linkage between two types of uranium atom distinct by their positions; the bridges contain one electron pair for the two bonds. (Rundle's general interstitial 'half-bond' theory has been referred to in Chapter 2.) The resulting UH_3 structure is thus fairly stable, but becomes unstable above 435°C (at 1 atm). UH_3 can be considered as a nearest approach amongst the hydrides to the stable carbides, nitrides and (metallic) oxides[88b]. Neutron diffraction, in the deuterides, has made it possible to locate the H (D) atoms in distorted tetrahedra, equidistant from 4 U atoms. The fact of the existence of U–H–U bridge bonds still holds, but the mode of bonding assumed at first is rather more complex. The hydrogen atom is located in an enormous hole within a U-atom tetrahedron which, if it were not for the assumption of bridge bonds, would imply for it a 'radius' of 0·80 Å, i.e. 3× its normal radius (for the detailed structure, see Rundle's paper[15d]).

In precision-calorimetric determinations of the heats of formation of UH_3 and its two isotopes, the interesting observation of an isotopic dependence was made[89]. The values slightly increase in the sequence hydride → deuteride → tritide:

$$UH_3\ -30{\cdot}352;\quad UD_3\ -31{\cdot}021;\quad UT_3\ -31{\cdot}141\ \text{kcal/mole}$$

The reason is not known, but this tendency, though slight, is significant; it seems to conform more generally with neutron-diffraction indications on other hydrides. Related work on UH_3 and UD_3 kinetics[18b] shows a higher decomposition pressure of the deuteride.

Thermodynamically, too, the P–T–C dependence for uranium is considerably less pronounced than in the other hydrides, consistent with a stable compound. Spedding *et al.*[26], in determining this relationship, found a strong *hysteresis* effect in the hydriding and dehydriding isotherms. This

* In hydrogen bridge bonds, a pair of electrons occupies a vacant orbital on each of the metal atoms adjacent to hydrogen. We might put it that here the hydrogen atom is not strictly 'interstitial' but occupies positions separating the metal atoms. This bridge bond probably also applies to some extent to the actinides generally and to the earlier rare earth hydrides.

can be explained by the mechanical stresses and large volume changes bound up with UH_3 formation and dissociation (see also Bibliography).

The solid solubility of H in U is sharply dependent on the polymorphic forms (α,β,γ-U). *Figure 9.19* shows this for atmospheric pressure; within each form, as function of pressure, Sieverts' law is well obeyed.

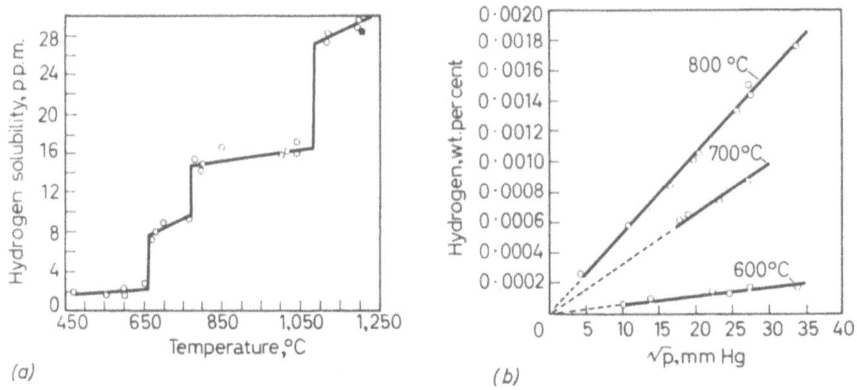

Figure 9.19. The U–H *system* (From Darras and Caillat[88])

(*a*) Solubility of hydrogen in uranium, in equilibrium at atm. pressure; (*b*) solubility as a function of H pressure at 600°, 700° and 800°C

By courtesy of Pergamon Press Ltd.

(7) *The remaining actinide*–H *systems*

The *transuranic metal–hydrogen* systems, Np–H and Pu–H, may appropriately follow at this point, as they are more closely related to U–H and only loosely to Groups VII and VIII where they would strictly fall in the sequence.

Of *neptunium* only little is known, except a high-H hydride[90] of composition range $NpH_{3\cdot6-3\cdot8}$. In hydrogen/metal ratio it is thus similar to the earlier actinide $ThH_{3\cdot6}$; however, data are insufficient to substantiate a relation.

The *plutonium*-hydrogen system has been fairly fully studied in connection with reactor applications[27a,30,91]. The PuH_2 and PuH_3 hydrides are stable: the first of FCC fluorspar type[27d], the second of a HCP unit cell. However, PuH_2 can take a large excess of hydrogen into solution, up to about $PuH_{2\cdot7}$. The mechanism is one of entry into the octahedral sites of the CaF_2 lattice, with lattice contraction, quite as it was in the early Groups discussed (e.g. rare earths, Th, etc.). Then, near $PuH_{2\cdot7}$, i.e. before complete filling of the octahedral sites, the FCC $\rightarrow$ HCP transformation sets in to reach a single-phase state at PuH_3.

It may be noted that the FCC $\rightarrow$ HPC transition is the reverse of the usual one in that it occurs with *increasing* interstitial content, not decreasing as in Cr–H and many previous cases such as the carbides. This shows again the very different nature of these highly saturated hydrides, of such an open lattice, packed with hydrogen, that the direct metal–metal bond has entirely vanished. The metal stacking sequence is now dictated by the *M*–H–*M* bridge bond and thus by the progressive filling of the various sites of H-atom entry. It seems quite conceivable that a tendency towards H-atom pairing (H_2 molecules) here begins to operate already.

In close-packed *hexagonal* PuH_3, the two types of interstitial H sites again appear to be inherited from those in CP cubic PuH_{2+x}; Zachariasen suggests the sites to be

$$\begin{array}{ll} 2\text{Pu at } \pm\ (\tfrac{1}{3}, \tfrac{2}{3}, \tfrac{1}{4}) & \\ 2\text{H at } \pm\ (0, 0, \tfrac{1}{4}) & \text{(type I hydrogen)} \\ 4\text{H at } \pm\ (\tfrac{1}{3}, \tfrac{2}{3}, z) & \multirow{2}{*}{\text{(type II hydrogen)}} \\ \quad\text{and } \pm\ (\tfrac{2}{3}, \tfrac{1}{3}, \tfrac{1}{2} + z) & \\ z = 0{\cdot}607 & \end{array}$$

Viewing the actinide series overall, it is notable that we have a resumption in structural kinship of hydrides from the early to later Groups (Th → Pu), with a departure in between (Pa, U, Np). Furthermore, in the 1st, 2nd and 3rd Long Periods, this is paralleled by a recovery in hydride stability after the trough in Groups VI and VII. Although the mode in which this is reflected along each Period is different, it can be traced to similar basic causes in electron structure.

(8) *The Group VII metal–hydrogen systems* (Mn–H, Tc–H, Re–H)

In the Mn–H system, no hydrides are known, only primary solid solutions, varying in the α-, β-, γ- and δ-Mn forms (cf. *Figure 3.2*). However, instead of hydrides, we have in manganese the capacity of high physical *occlusion* for hydrogen, illustrated by an apparent H solubility in electrolytically produced Mn[92]. This capacity had already increased in chromium along the 1st Long Period and now reaches its height in Group VII. It is manifested by 'surcharging' in electrolytic Mn (as in Cr) and in supersaturation of the α-Mn lattice; when the occluded hydrogen is removed by annealing, it is liberated as free molecular hydrogen, without any hydride stage. In principle this will hold true, with some qualifications, from now on, i.e. in the iron and platinum families of Group VIII.

No data are available on Tc–H and Re–H.

(9) *The Group VIII metal–hydrogen systems* (Fe, Co, Ni and Pt *family*–H)

The series of *iron-group* metals Fe → Co → Ni, when equilibrated with hydrogen, represents a gradual recovery in hydride-type bonding, if we interpret 'hydride' in a sense rather modified to that for the early transition metals. The trend is concomitant with the evolution of the FCC structure of the parent metals which has the inherently higher solubility for interstitials than the BCC, although the BCC, as we have seen, is much more accommodating for hydrogen than for other interstitials (C,N,O, etc.). Factors of site availability and electron sharing become operative which make the potential H solubility in the BCC, FCC and HCP lattices of comparable magnitudes.

In *iron* and *cobalt*, no hydrides are known, only limited primary solubilities (see Chapter 3), less than in manganese. Solubility figures in Fe quoted by Bastien[93] are (cm^3/100 g Fe)

α-Fe (910°C)	3·2
γ-Fe (910°C)	4·65
δ-Fe (1,450°C)	6·5
liq. Fe (1,535°C)	30·2

The hydrogen embrittlement in iron, which becomes so important an effect in steels, is pronounced; it is not, however, connected with hydride precipitating but rather with the differential solutions in liquid, δ-, γ- and α-iron and with hydrogen-atom segregation at dislocations, stacking faults and crystal boundaries. The capacities for occlusion and supersaturation, of type as in Cr and Mn and of the modified and amplified type in Ni and Pd, combine in α- and γ-iron. For the latter, it may well be surmised that the tendencies to be described for nickel are already innate, whilst for α- and δ-iron those described for chromium are still effective. Resistance measurements on hydrogenated Fe wires[94], comparable to Ni, and precision lattice spacings[95] of α-Fe saturated with H serve to illustrate this.

The solubility of hydrogen in pure iron[93,96] (*Figure 9.20*) and the differential solubilities in the α, γ, δ and liquid forms clearly arise. Sieverts' law is well

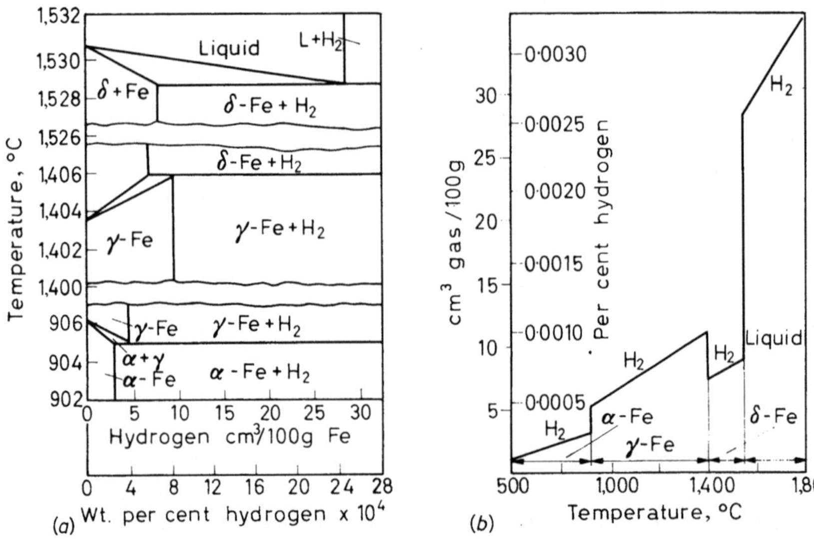

Figure 9.20. Solubilities of hydrogen in iron (From (*a*) Geller and Tak-Ho-Sun[96]; (*b*) Bastien[93b]

By courtesy of Verlag Stahleisen and the Editor, *Öntöde*

obeyed within the range of each form. The solubility of H in α-Fe also showed that quite significant amounts can be retained at low temperatures by *high-pressure charging*[97] around 600°C; this residual hydrogen, which is readily diffusible, is not so much lattice-held but held in 'traps' thought to be dislocations or fine fissures evolving from them (see also p. 522).

Surprisingly little appears known on the Co–H system. However, H solution inhibits the $\alpha \leftrightarrow \beta$ (=HCP $\leftrightarrow$ FCC) hysteresis effect[98] which occurs around 450°C in pure cobalt and stabilizes the β form fully. On outgassing, it reverts to an HCP α phase of lattice expanded compared to pure Co, as is the β lattice.

In *nickel*, whilst again stable hydrides do not form, the 'surcharging' effect is prominent, much in excess of equilibrium. It can be increased by mechanical work: this creates lattice distortions and additional, if transient, sites for hydrogen at dislocations, grain boundaries and stacking faults. This effect is associated with high diffusibility of H in Ni.

Altogether we may see in the nickel–hydrogen relationship already a foreshadowing of the 'palladium effect' (see below), though as yet less developed. Ni, of course, neighbours Pd in the same Group.

There also exists a CP hexagonal 'form of nickel', produced electrolytically. This is created metastably by hydrogen. It is interesting to view this in the light of the cobalt dimorphy and to consider the FCC $\rightarrow$ HCP transformation with lower *temperature* in Co as being produced by the composition parameter of lattice *hydrogen* in nickel. This might seem to conflict with the above findings[98] on Co, but then a *second, FCC* Ni–H phase was reported[99], additional to that of primary Ni–H solid solutions and coexisting with it. This is H-enriched and of larger spacing (to 3·735 Å): a relation quite analogous already to that for palladium. However, the phase is metastable, and the hydrogen desorbs spontaneously, with lattice contraction. But a further such isomorph can form, of dimensions 3·83 Å, which is thought to be very stable and enhanced at the grain boundaries of the nickel matrix. Conceivably this high-hydrogen Ni is the true equivalent to the H-stabilized β-Co.

By neutron diffraction[100], it could be shown that the H atoms in the FCC Ni lattice occupy the octahedral sites, as in the NaCl-type structure. Beginning with excess vacancies, the atomic H/Ni ratio in the *saturated* hydride reached 0·6; other values suggested are higher still (see below). Such an excess over 0·5 in the FCC lattice may be supposed to show that H would also enter some tetrahedral positions. Wollan *et al.*[100] find this 'hydride' to be unstable at room temperature. This contrasts to the analogous *Pd* 'hydride' which is stable.

It was also possible to prepare saturated nickel hydride by the electrolytic charging of nickel strip[101]. The H/Ni ratio attained was 0·8, the FC cube edge being increased from 3·524 Å for pure Ni to 3·721 Å at room temperature, the linear expansion amounting to 5·6 per cent. It is here noted that the hydride formation is discontinuous as from nickel, which would be consistent with a miscibility gap rather of the nature of that in Pd–H. However, the instability is evidenced by the progressive decomposition (H loss) with time at elevated and even at room temperatures, the rate depending on grain size and cold work.

Figure 9.21 shows the appearance of the hydrated nickel surface after decomposition. The massive cracks appearing are in a sense equivalent to 'grain-boundary precipitation' such as, in the case of carbon, of carbides or graphite; only here it is gaseous hydrogen and thus 'vacant space'. The electrical resistance behaviour of H-charged nickel wires showed an initial *increase* of resistance through grain-boundary cracking[102], followed by a *decrease* in time upon degassing, due to the gradual repair of lattice defects. A similar effect was found[94] on hydrogenated *iron* wires.

It is interesting that the solubilities for constant pressures and temperatures decrease with increasing mass of the isotope (see also Bibliography).

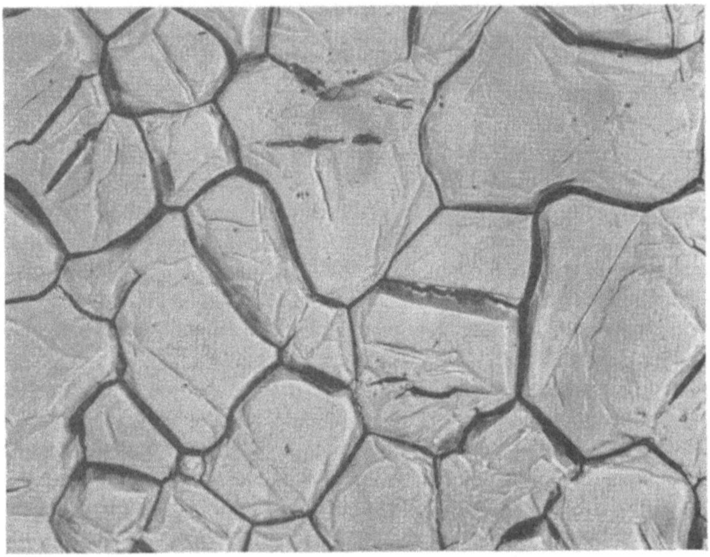

Figure 9.21. Crack formation at surface of nickel, first hydrogenated, then degassed; × 500 (From Boniszewsky and Smith[101])

By courtesy of Pergamon Press Ltd.

The effect of hydrogen charging upon the *plastic deformation* of nickel revealed serrated yielding (the so-called Portevin–Le Chatelier effect[103]), owing to the interaction of H atoms with dislocations. The serrations, i.e. momentary reduction in stress and stepwise yielding, reflect a process of dislocations multiplying; electron microscopy was able to show this, and also how hydrogen atmospheres pin down the dislocations after they are entrapped by tangles, to be released again under sufficient stress.

Coming to the equilibria between hydrogen and the *platinum*-group metals (Ru, Os, Rh, Ir, Pd, Pt), the metal of paramount interest is *palladium*; for the others no hydrides are known. Nevertheless, some of the Pd–H type of effects may well apply to Pt, Rh, etc. on a reduced scale—certainly to their *alloys* with palladium.

As to the Pd–H system (*Table 9.1, Figures 9.3, 9.4*) FCC palladium dissolves up to 5 at. per cent H at room temperature, with a limited lattice expansion from $a_0 = 3{\cdot}891$ Å to 3·894 Å ($\approx$ 0·1 per cent linear). Thence a second FCC phase coexists, with $a_0 = 4{\cdot}027$ Å, which is hydrogen rich, of H/Pd $\approx$ 0·7. This 'hydride' then further expands, on H addition, to $\approx$ 4·07 Å (i.e. a total of 4·6 per cent linear expansion). This, however, is highly temperature- and pressure-dependent, such that at high P and T, the two phases merge to become intersoluble. It is interesting to reflect that the room-temperature capacity of absorption of Pd for H is about 1,000 times its own volume.

Numerous studies have been made on the spacing changes with pressure, temperature and hydrogen content, on the thermodynamic relationships and theories surrounding them; only a brief outline will be given here.

Neutron diffraction[104] has shown that, as in Ni–H, the H atoms in Pd are located in the octahedral vacancies, approaching the NaCl-type structure. An important side observation here is that the total neutron scattering cross-section for hydrogen is abnormally low, indicating by their values that the H atoms are more nearly free protons than for hydrogen compounds of other transition metals. This is quite consistent with their high diffusivity and 'looseness' of bonding in the lattice. Fine palladium powder will absorb[104] hydrogen up to $PdH_{0.7}$ (at room temperature and 1 atm). Whilst thus only 70 per cent of the octahedral sites are randomly occupied at saturation, this increases at higher temperatures and pressures. Many vacancies remain, however, again aiding easy diffusion. Thus at least two causes favour the high diffusion rates and give the Pd–H system its unique position: (*a*) the exceptionally loose hydrogen bond and (*b*) the availability of vacancies. Furthermore, the *catalytic* action of palladium during hydrogenation processes is due to the ready ability of this metal to supply free H atoms from the hydrogen-saturated lattice. This avoids the large activation energy required in the presence of H_2 molecules to dissociate these. Consequently, palladium is a potent 'reservoir' for free atomic hydrogen—for whatever purpose!

The mechanism of the lattice occlusion of hydrogen has also been studied by numerous workers (see Bibliography). From x-ray lattice parameters, it has been possible[105] to effectively trace the phase diagram (miscibility bay and critical points) and to derive the P–T–C curves shown in *Figure 9.22*. The heat of reaction $2Pd + H_2 \leftrightarrow 2PdH$ in the plateau region of *Figure 9.22a* calculated from this is $\Delta H = -9{\cdot}5$ kcal/mole, virtually the same as Nace and Aston's value[106] from calorimetric absorption work. The critical solution temperatures and pressures, i.e. those above which the Pd–H solution becomes single-phase, are derived as 308°C and 19·7 atm, to be compared with those in V–H, Nb–H and Ta–H above.

Isotherms of the high-Pd–H solution have been measured[107] and the heat of formation deduced as −9·786 kcal/mole. The *maximum* solubility of H in Pd is 0·694 occupied sites per Pd atom, a value claimed to be independent of temperature.

Nace and Aston[106] have reported extensive thermodynamic researches on the Pd–H system, providing absorption and desorption isotherms and molar heats, and deducing the free energy and entropy changes when 'Pd_2H' is formed which was suggested as a possible true compound. The entropy of Pd_2H on approaching absolute zero and the part played by the absorbed hydrogen in the build-up of the heat capacity are considered and could tend to support this idea. The presumption is that *if* Pd_2H were a true compound, a completely ordered state would arise and there would be no residual entropy at 0° abs. In fact, this proves not to be so. Some degree of randomness remains and thus Pd_2H is really not a unique state, as is indeed more consistent with the general nature of hydrides evolved. The occurrence of *covalent* bonding of H atoms at low temperatures is suggested as probable, and it seems that the H absorption could be due to some kind of chemical bond Pd–H–Pd developing at low temperature from a looser metallic bond at higher ones.

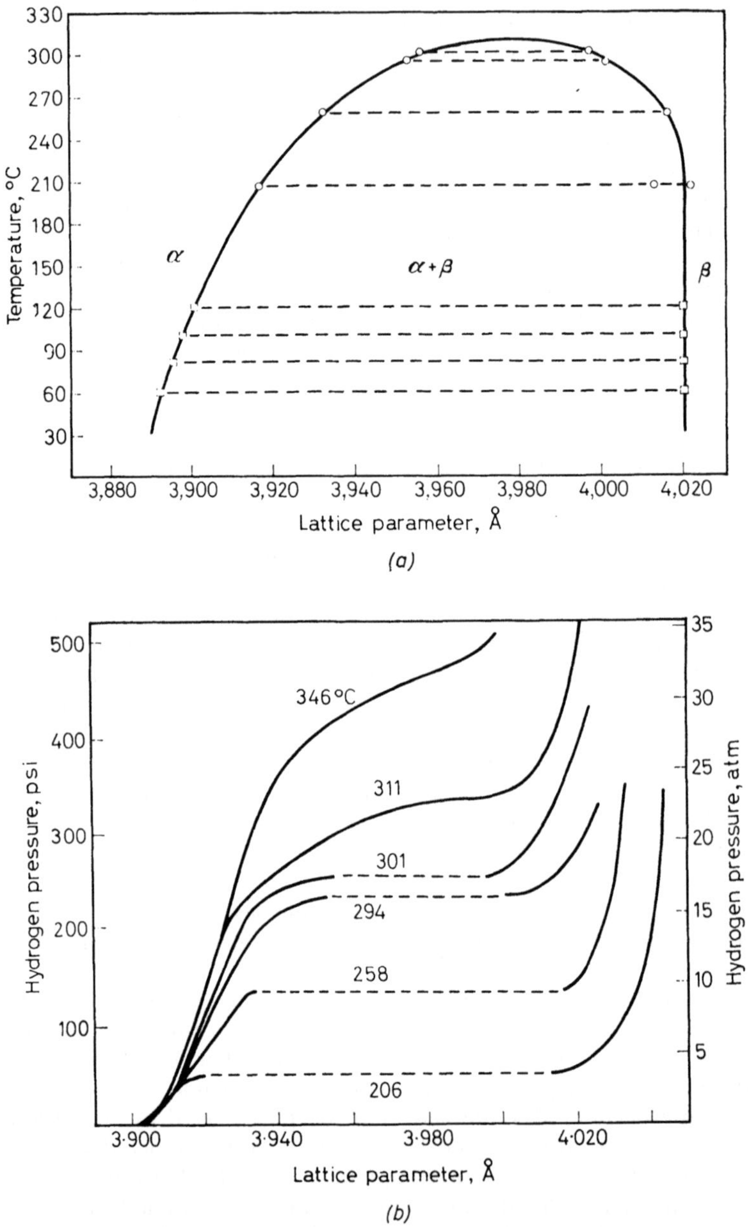

Figure 9.22. The Pd–H *system* (From Maeland and Gibb[105])

(*a*) Lattice parameter/temperature diagram, showing miscibility bay; (*b*) hydrogen evolution isotherms, in terms of lattice parameter/pressure relation for temperatures indicated

By courtesy of American Chemical Society

In a further analysis[106c], the 'deuteride' Pd_2D was compared with the 'hydride' Pd_2H, using heat capacity studies (between 15° and 303°K). A model is derived in which a molecular unit 'PdH_4' is assumed (similar to NH_4 or CH_4 molecules), such that one in every 8Pd atoms has 4H (D) atoms covalently bonded to it; the general formula Pd_2H analyses into $7Pd{\cdot}(PdH_4)$. This bond gradually dissociates towards higher temperatures, by a decrease in the number of H bonds. According to these ideas, diffusion through Pd is regarded as a kind of 'handing-on' (or 'swinging on') process of the hydrogen atom, involving its *rotation* about a Pd atom to which it is bonded, followed by rotation about a neighbouring Pd atom. The theory would also explain the absorption/desorption hysteresis effects and the low-temperature miscibility gap in Pd–H. Although controversial, this work clearly has a more general bearing also on the earlier metal–hydrogen systems such as those of BCC base in Group V (Nb–H, etc.) where the similar miscibility gap may reflect similar causes.

The model has been further developed[4d,108] for the Pd–H systems more generally, beyond 'Pd_2H'. Using heat-capacity dependence on temperature, an anomalous peak at 55°K is found, whatever the H/Pd ratio, linked with a similar peak in resistivity[109]. The theory is that this could be due to a transition from normal octahedral type of H-atom occupancy to a preferred one in certain lattice planes, and to a clustering to the 'PdH_4 molecules'. The liberation of protons from this bonding gives rise to what is called 'warm drifts' underlying H diffusion; by measurement of these proton drifts, the activation energies can be derived.

It is relevant here to note work by Ash and Barrer[110] on diffusion in a transition-metal system with a concentration discontinuity in general, but referring to Pd–H as an example. Hydrogen permeation through Pd membranes showed a break in the concentration gradient which is reflected in the diffusion coefficients and time lags determined.

Related to this, *hysteresis* effects in Pd–H have shown[111] how, during the solution process, the high-H constituent nucleates and grows within the primary low-H solution. Since this involves large volume changes (cf. spacing differences above), strains are set up, and the primary phase becomes plastically deformed as the elastic limit is exceeded. *Absorption* is then linked with growth of the high-H phase whilst, on *desorption*, the hydrogen is mainly supplied by the high-H phase. This gives rise to such hysteresis effects, which can be calculated from known yield stress data and isotherms.

The Pd–H hysteresis effects were also studied by high-temperature x-ray analysis in a H_2-pressured camera[112] and by observing the equilibrium pressures[113], with a palladium sponge to absorb and desorb hydrogen. Such hysteresis loops are shown in *Figure 9.23* which includes, as in permanent magnetic materials, subsidiary (scanning) loops. It is interesting to compare this with early spacing hysteresis[112b], also shown in the figure. A detailed statistical theory had been proposed by Lacher[114]*, and Everitt and Nordon[113]

* Lacher's model assumed that, in Pd, H_2 molecules first dissociate to atoms, then to protons and electrons, the electrons joining the *d*-band vacancies of the Pd atom, the protons occupying definite sites in the Pd lattice (cf. [115]). Mass spectrometry on palladium subjected to H diffusion indeed shows that, under electron bombardment, different degrees of ionization of the H atom can be generated, to H^+, H^{++} or H^{+++}.

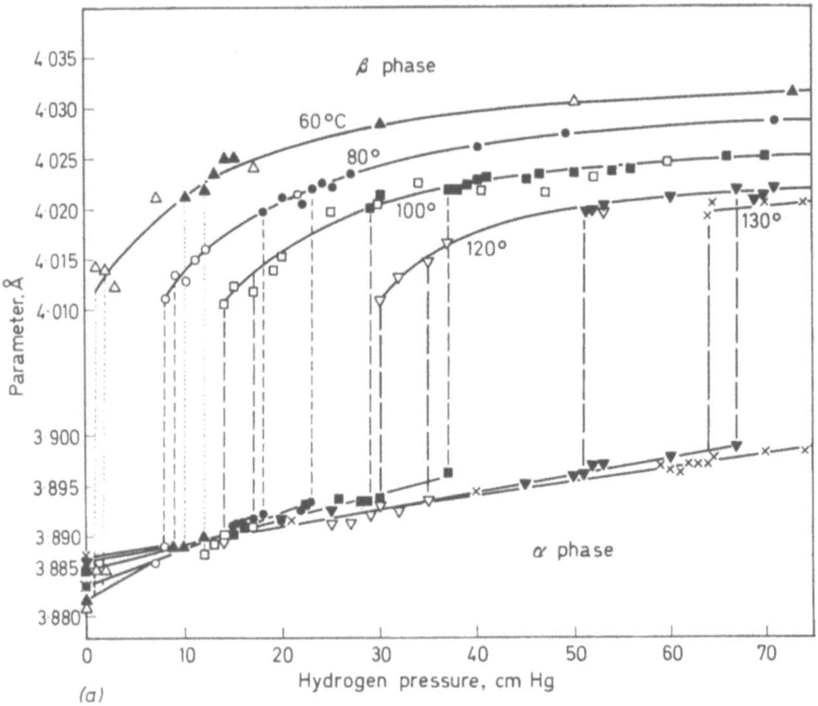

Figure 9.23. Pd–H *hysteresis effects* (From Owen and Williams[112b]; Everitt and Norden[113a])

(*a*) Lattice spacing hysteresis; (*b*) atomic ratio H/Pd as function of hydrogen pressure; hysteresis loops and 'scanning' loops in successive cycles (at 120°C)

By courtesy of The Institute of Physics and The Physical Society, and of The Royal Society

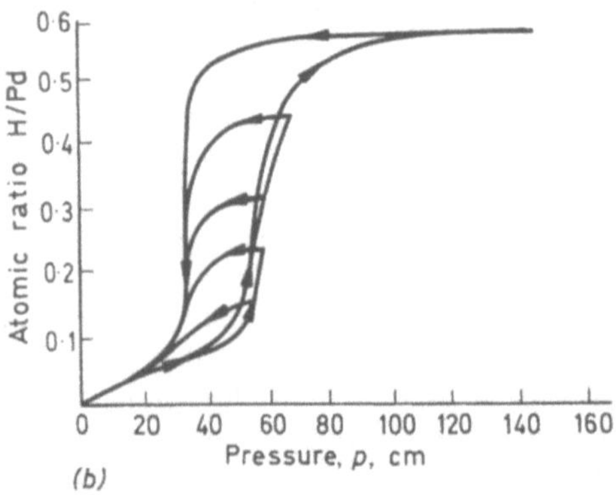

Figure 9.23b

consider this critically (cf. earlier discussion by Smith[67]). The high-H and low-H phases are characterized by different energy parameters of absorption of H atoms in the light of their different manner of interaction in the two phases. Everitt and Norden[113] also suggest a theory to allow for the occurrence of these separate phases during absorption and desorption of hydrogen, with pressure variation.

The hysteresis analogy holds even to the extent of predicting[113b], as in permanent magnets, a *domain* theory of the Pd–H hysteresis, linking this to the internal stresses created by the volume difference between the two isomorphous phases. Here again, the common factor of the development of miscibility bays should be noted, which also underlie permanent-magnet alloy systems (for instance, of the 'Alnico' (Fe–Ni–Al–Co) and similar types).

The Pd–H system, although probably the most spectacular in its effects, ought basically to be viewed in its relationship to all other transition-metal–hydrogen systems. The phenomena which are here the most pronounced and the conditions most favourable for revealing them, are in principle likely to apply, if more concealed, more widely to the noble metals and indeed to earlier transition-metal hydrogen systems. They may actually perhaps culminate in the other metals under different optimal pressure–temperature conditions not yet determined.

On a wider plane and with modifications, the effects could also typify certain metal–*oxygen* equilibria.

Ubbelohde's early work[1] on Pd–H must here be recalled, where he stipulated the *metallic* nature of the hydrogen atom, now supported by much experimental evidence. In this light we should indeed regard the Pd–H system as an 'intermetallic' one of a unique kind.

(10) *Group IB—hydrogen systems*

A brief look at the next higher Group, (Cu, Ag, Au)–hydrogen, may complete the picture.

A *copper hydride*, CuH, is known, according to early work[116] of hexagonal wurtzite, ZnS (B–4) type of structure, later confirmed[117] after some conflicting evidence, using neutron diffraction[118], the Cu atoms in the hexagonal unit cell could be located at (0,0,0) and $(\frac{1}{3},\frac{2}{3},\frac{1}{2})$, with hydrogen at $(0,0,\frac{3}{8})$ and $(\frac{1}{3},\frac{2}{3},\frac{7}{8})$. The cell dimensions are $a = 2{\cdot}89$, $c = 4{\cdot}61$ Å. Thus, as the end of the transition-metal series is reached, we have a reversal towards true stoichiometry as met in the very early Groups (alkalis, etc.).

Copper already well reflects the transition from metallic to covalent bonding with hydrogen, which further develops in the zinc and later Groups (stannanes, etc.; see Section 9.1).

The formula 'CuH' holds precisely, and the compound is stable up to about 60°C when it rapidly decomposes[119]. It has been prepared from the mixed hydride, $LiAlH_4$, and copper iodide by a reaction $LiAlH_4 + CuI \rightarrow LiI + AlI_3 + CuH$. From the perspective of the classification at the beginning of this Chapter, we might express this as: ionic hydride (pre-transition metal) + halide → metallo-covalent hydride (post-transition metal) + ionic halides. It would be of considerable interest to analyse such a process in closer detail and see whether, on slowed reaction, there could not occur as intermediate stages 'trasition-metal-type' hydrides of the kind discussed.

The solubility of hydrogen in copper can be increased[120] by cold work to 0·06 at. per cent H, from that in the strain-free lattice 10,000 times less. Similar effects, particularly with hydrogen, apply more generally, as seen above and in Chapter 3.

In the systems Ag–H and Au–H, no hydrides appear to occur. However, the possibility of unstable hydride type bonding, as in Pd–H, ought, it is felt, not to be finally excluded.

9.3 TERNARY ALLOY HYDRIDES

The development of *two-metal* hydrides has for the most part been of an 'ad hoc' nature, to attain particular industrial properties or to inhibit others*.

One aim, for example, was to achieve maximum atomic hydrogen-bearing capacity in *nuclear reactor* applications, i.e. to obtain the highest H atom concentrations per unit volume, usually referred to as H/M ratio (atomic) or the value

$$N_H = \text{number of hydrogen atoms per cm}^3 \times 10^{-22}$$
$$= \text{density/molecular weight} \times H/M \text{ atom ratio} \times 6{\cdot}023 \times 10^{23}$$

Interest in this case centres around Ti and Zr bases†. Thus, for instance, H *retention* on many Ti- and Zr-base alloys has been studied[121]. It was discovered that the addition of vanadium (14 at. per cent V) to titanium increases the H/M ratio to 1·96 from a base level of 1·75 for Ti alone; molybdenum addition (18 at. per cent Mo) raised the ratio to 2·16. A maximum N_H of 10·1 has been attained for a Ti/69 at. per cent V alloy. Other additions like Zr, Nb, Mo or Mn to titanium also increased its retentive capacity, while Cr decreased it. For Zr-base alloys (optimal H/Zr = 1·92), Nb, V, Cr, Mo and Sn decreased this capacity, whilst nickel first increased H/M to 2·17 (at 2·3 at. per cent Ni), then decreased it to only 1·04 at 33 per cent Ni). (Trzeciak's paper[121] also gives a wider discussion and experimental data on this aspect.)

It clearly makes a vital difference whether the two-metal-base phase is a solid solution or an intermetallic compound. Thus in the Zr–Ni–H system, a definite hydride $ZrNiH_3$ forms[2b], with a maximum H/Zr ratio of 2·93. This is isostructural with the orthorhombic intermetallic compound ZrNi and really represents a high-hydrogen solution in it. Similarly, the intermetallic Zr_2Ni base[4e,122] yielded H loading up to $Zr_2NiH_{4{\cdot}85}$. These compounds, however, apart from mechanical brittleness, have considerably larger dissociation pressures than the Zr hydride. They also show large hysteresis effects on thermal and pressure cycling. Further work on the Zr–Ni–H system[123] suggests that $ZrNiH_{3-x}$ is preceded by a monohydride, ZrNiH, at lower temperatures, with an appreciable heat of formation. On the other hand, only some Zr–Ni compounds absorb hydrogen; others do not, while yet others decompose.

A further example of a ternary hydride is one in the Th–Co–H system[17a], where the compound ThCo can absorb considerable quantities of hydrogen, up to about $ThCoH_{4{\cdot}2}$.

* This implies that, for scarcity of data, a systematic presentation of ternary hydride systems in tabular form similar to that in the Carbide, etc. Chapters is not at present possible.

† Theoretically, Group III and rare earth-base hydrides might be even more attractive but would be impractical.

Two new fundamental themes, therefore, arise.

(1) The use of intermetallic compounds as base structures for hydrides rather than the BCC, HCP or FCC solid solutions of the metals. Just as, by variants of these simpler structures, the binary 'hydrides' are obtained, so ternary hydrides can be derived from known intermetallic structures with H interstition, and not necessarily involving complete or abrupt structure changes. The theme has much scope for extension to σ- or χ-type phases or to η-carbides*.

(2) *Amongst* the intermetallic compounds, a selective capacity for hydrogen solution, clearly depending on structure and available vacancies. Any phase change occurring in thermal treatment can have catastrophic consequences in suddenly evolving (or absorbing) large amounts of hydrogen.

Much attention has been paid to the Zr–U-base hydrides used as fuel moderators in reactors. The stability of the hydride '$ZrU_{0.04}H_{1.4-1.8}$' under practical working conditions has been described[124], as have the creep properties and transformation kinetics[125] of a Zr–U–H alloy of similar type, $ZrU_{0.03}H$. Here the Zr–H phase relations discussed earlier (α/β, hydride eutectoid, etc.) still substantially apply, with metastable states arising which depend on thermal paths in the reactor. (For details of mechanism and rates of transformation of alloys of Zr–U–H, etc. base[125–6], including thermal, tensile, etc. properties, cf. Section 9.4).

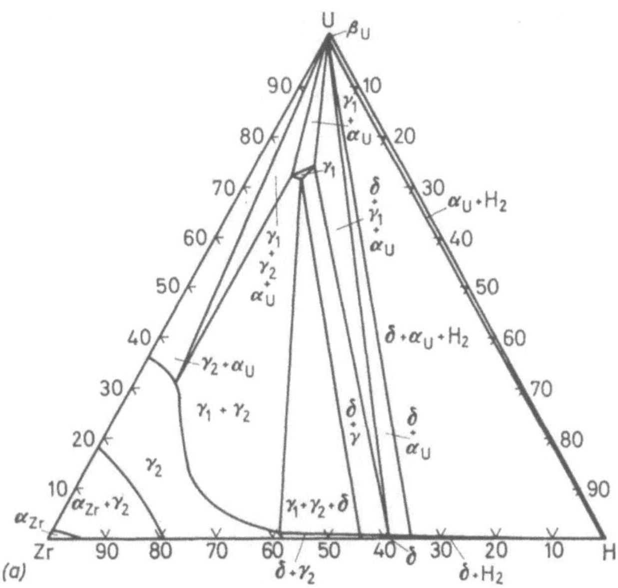

Figure 9.24. The Zr–H–U *ternary system* (From Krause, Bigony and Doig[4f])

(*a*) 627°C Section; (*b*) 675°C Section; (*c*) 750°C Section

By courtesy of American Chemical Society

* Indeed, mixed carbides, nitrides, oxides, silicides and borides may offer scope for the future as bases for hydrogen solutions.

The Zr–U–H system, extensively studied (with many thermodynamic data[4f]), resulted in the phase diagram (3 isothermal sections) shown in *Figure 9.24*, which exemplifies several general points.

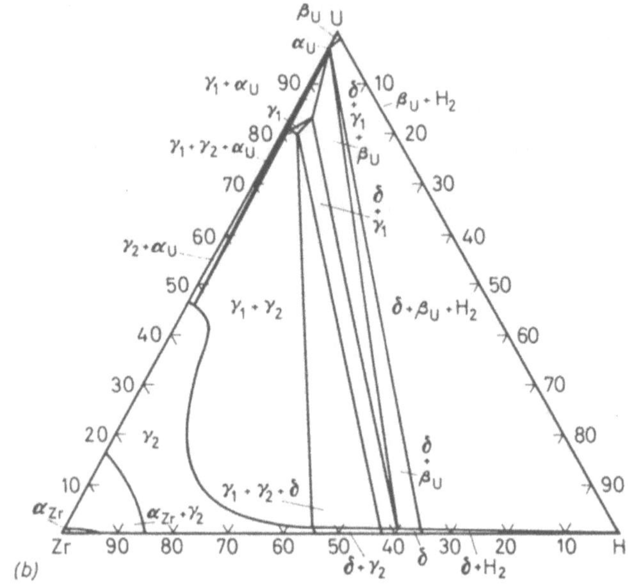

Figure 9.24b

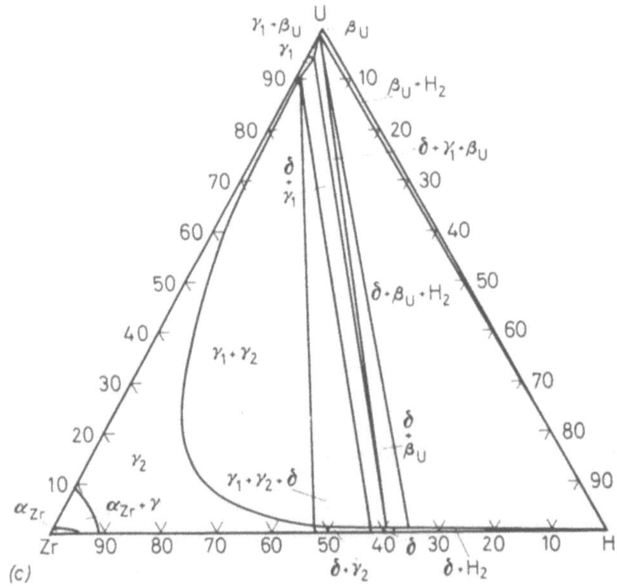

Figure 9.24c

The complete solid solution between the β-(Zr–H) primary *solution* and the (Zr,U) intermetallic *compound* of BCC structure is also of note in demonstrating the preferential H solubility in such isomorphous phases. This phase region extends at higher temperatures to form the complete BCC β-Zr/γ-U solution series; high hydrogen solubility, however, only evolves towards the Zr end. Note also the miscibility bay forming in the isothermal sections*.

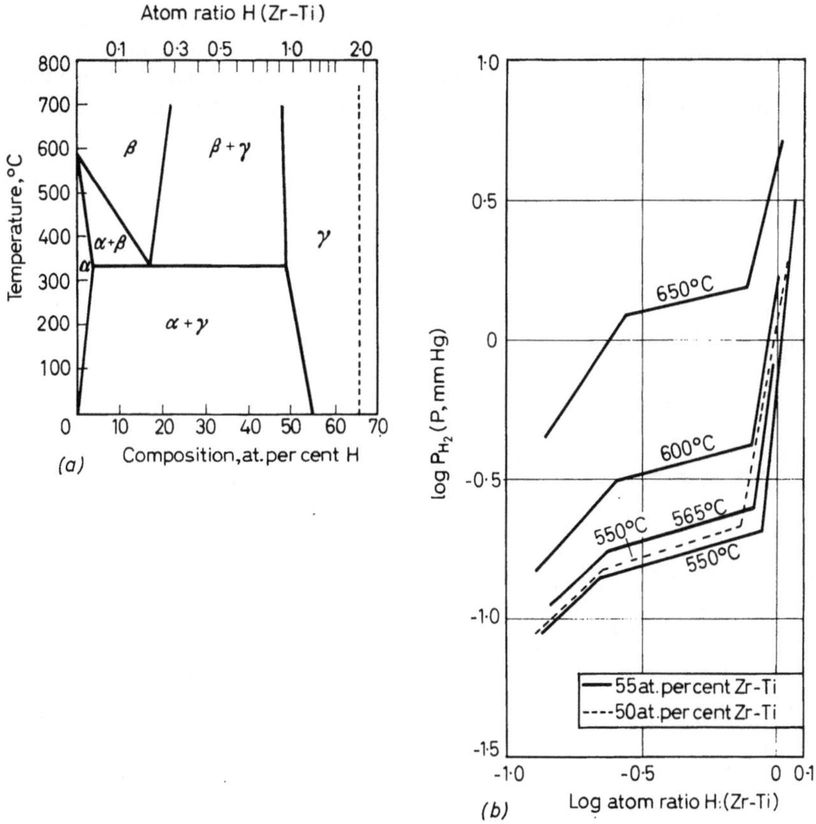

Figure 9.25. The Zr–Ti–H *system* (From Labaton, Garner and Whitehead[127])

(*a*) Phase diagram for a 55 at. per cent Zr–Ti (base) alloy and hydrogen (provisional); (*b*) isothermal curves for the Zr–Ti–H system at the 55 and 50 at. per cent Zr levels

By courtesy of Pergamon Press Ltd.

The Zr–Ti–H system has been investigated[127] for varying pressures and temperatures, and for a Zr/55 atomic Ti alloy; the tentative phase diagram is shown in *Figure 9.25* (to be compared with the binaries, *Figure 9.3*), and

* Bearing in mind that in Zr ↔ U we have a diagonal combination of Group IV ↔ Group VI metal, it seems suggestive to compare the hydrogen-stabilized miscibility bay ($\gamma_1 + \gamma_2$) in *Figure 9.24* to the quite similar 2-BCC miscibility bay in Group V–H systems, e.g. Nb–H (*Figure 9.3*). Although one here evolves in the composition, the other in the temperature plane, both will of course combine in P–T–C space, and the electronic and crystal structures in the central (Zr–U)–H region of this system seem likely to imitate those discussed in the binary Group V–hydrogen systems.

structurally the phases follow the isomorphs predictable from the binaries. The Zr–Ti–H equilibrium diagram has also been studied by Pessall and McQuillan[128], for an isothermal section of 750°C, by constitutional and equilibrium-pressure determination. One interesting observation is that, for a certain Ti/Zr ratio ($\approx$15 at. per cent Zr), the hydride homogeneity range $(Ti,Zr)H_{2-x}$ becomes exceptionally large, from about 30 to 66·6 at. per cent H, and the hydride attains increased stability. This is shown in *Figure 9.26a*; it is known that in the Ti–Zr binary system the $\beta \rightarrow \alpha$ transformation is depressed to lower temperatures for intermediate Ti/Zr ratios[129] (*Figure 9.26b*), and the striking fact is that this is paralleled, upon interstitial H addition, by an early stabilization of the hydride. In more general terms, the

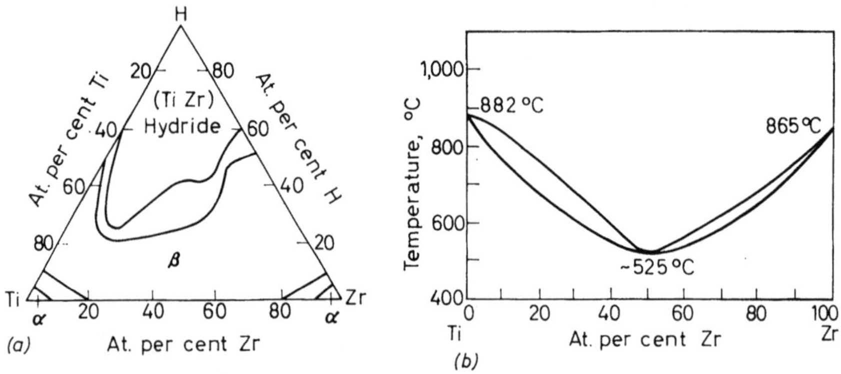

Figure 9.26. (*a*) *The* Ti–Zr–H *ternary section at* 750°C; (*b*) *the* Ti–Zr *binary system* (From Pessall and McQuillan[128] and Hansen[129]

By courtesy of The Metallurgical Society of AIME and Messrs McGraw-Hill

HCP $\rightarrow$ BCC $\rightarrow$ FCC (or FC derivative) sequence is, for certain alloy contents, facilitated by lower interstitial addition than for the *pure* Group IV metals. The minimum in the $\beta \rightarrow \alpha$ change along Ti–Zr is reached at about 550°C, and in Ti–Zr–H space this no doubt links to the β-eutectoid in the Ti–H and Zr–H binaries (*Figure 9.3*). For further researches by McQuillan and associates on ternary hydride systems amongst the early transition metals see Bibliography*.

The amount of work done on hydrogen solubility in *iron-base alloys*, i.e. alloy steels, is considerable, and good reviews on the subject are available. Only a few examples can be given (cf. p. 522, etc.). It was found that in Fe–Ni and Fe–Cr alloys the H solubility increased with Ni, and decreased with Cr contents in the *liquid*[132]. On solidification, as the solubility decreases, the degree of hydrogen precipitation increases with increasing solubility

* Other recent work in this field of ternary hydrides with direct useful applications is on hydrogenated Zr–Nb and on the effect of tin additions (in connection with the 'Zircalloy' alloys[130] containing Nb and Sn); it is found that Nb and Nb–Sn improves hydrogen embrittlement. Studies of the sorption of hydrogen in Ti–Zr and Ti–Mo alloys over wide pressure ranges[131] showed the value of *oxygen* addition to the hydrogen used to prevent its adsorption (supporting observations earlier in this Chapter).

difference between liquid and solid solubilities. The following solubilities have been quoted:

		ml./100 g
γ-Fe	(1,400°C)	12·0
δ-Fe	(1,400°C)	9·0
Ni	(1,400°C)	18·8
liquid Fe	(1,600°C)	28·8
liquid Ni	(1,600°C)	43·5

Nickel as austenite stabilizer simultaneously increases the H solubility in solid or liquid states, consistent with the recovery in hydride-forming tendency towards the later Groups (Ni,Pd). BCC chromium, on the other hand, behaves conversely in this respect, consistent with the trough in hydride stability for the Centre Groups.

The H solubility in liquid Fe alloys containing Nb, Ta and Ti shows pronounced increases in solubility by such Group IV and V metal additions, proportionate to the square root of pressure[133]. Ready supersaturation on solidification can occur, with intensive delayed precipitation, however, at later stages. No data are available for the solid, but this may be presumed to be highly composition-dependent, in line with intermetallic (Laves-)phase formation.

The addition of Si and of Cr to iron (i.e. α-stabilizing solutes) can considerably retard, and even inhibit, hydrogen evolution[96]. These additions also greatly decrease the rates of diffusion of H in the Fe phases, both ferritic and (in 18/8 type steels) austenitic. Much specialized research has proceeded on the effect of hydrogen on alloy steels (see e.g. below, p. 522).

The unique properties of *palladium* have naturally stimulated work on Pd-base alloy–hydrogen systems, but the information is still only fragmentary, and it seems that a much broader approach to the overall systems is required. Most of the studies are limited to neighbouring noble (or semi-noble) metal additions (cf. Bibliography).

The absorption and evolution of hydrogen in Ni–Pd alloys[134] shows clearly how, with increasing nickel (>23 per cent Ni) the two-phase immiscibility of Pd–H tends to disappear; the proportion of Pd^+ ions in the lattice decreases, assuming the presence of a pseudo-ionic compound, as discussed under Pd–H above, in favour of more normal metallic bonding. Addition of copper and other adjacent metals to Pd decreases the H solubility. Earlier work[135] had shown already that the solubility of H in palladium sharply decreases through the addition of silver, with no solubility at all from 70 per cent Ag upwards. At the same time the paramagnetism of Pd disappears.

At the opposite end of the Periodic Table, much effort has been directed to a study of the mixed alkali-base alloys, and hydrides such as $LiAlH_4$, $NaAlH_4$ and others are of interest from the viewpoint of developing powerful reducing agents.

Wiberg and Henle[119] have contributed much to the knowledge of mixed hydrides. In a study of the Ag–Al–H system, a hydride $AgAlH_4$ was found to exist but only below -50°C, decomposing above this temperature. This raises the important point that many ternary hydride phases may really be latent and await discovery, becoming stable only at very low temperatures *or* high pressures; this does not exclude the possible stabilization of occasional compositions of such hydrides towards higher temperatures. Thus, already for *binaries* of Groups IB and IIB (i.e. continuing on the right of the transition-metal block), Wiberg and Henle have discovered hydrides of the following decomposition temperatures at atmospheric pressure:

CuH	+ 60°C	ZnH_2	+ 90°C
AgH	− 50°C	CdH_2	− 20°C
AuH ↓	−155°C	HgH_2 ↓	− 125°C

The bonding here is transitional between homopolar and metallic. Similarly, many *ternary* hydrides may yet be undisclosed, some also no doubt incorporating the transition metals themselves. With the present incomplete picture, only singular cases seem to emerge towards normal temperatures and pressures —peaks probably connected to an 'iceberg' of isomorphous hydride families at low temperatures.

The interest in ternary, and higher, hydride systems is thus clearly a wide one. Its four main centres for ternary metal additions have so far been around (1) titanium/zirconium, (2) iron, (3) palladium and (4) lithium/sodium, with subsidiary centres at Th, U, Cr and the rare earths, each for reasons of particular application. A research interconnection amongst these, so far rather isolated, fields seems desirable, both for systematic considerations and new practical uses.

9.4 APPLICATIONS

The transition-metal hydrides have found numerous practical applications, but only a brief survey is here intended. Mainly adverse aspects such as hydrogen embrittlement, however, are often linked with the favourable ones and physically evolve from them. It is appropriate, therefore, to consider both in conjunction.

The applications can be broadly classified into those utilizing the physical and the chemical properties of the hydrides. Hydrides are often used essentially as a '*holding device*' for hydrogen atoms. This is required, for instance, in nuclear energy for hydride neutron moderators, where the aim is to present the maximum sectional concentration of hydrogen atoms to the neutrons; the actual holding metal is, subject to certain conditions, of secondary importance, as long as it possesses this solvent capacity and retains it at the temperature prevailing in the reactor.

In other applications, the hydrides are used as *supply sources* for free *atomic* hydrogen, as it were, 'a reservoir on tap' for H-atom release wherever this may be needed. This broad concept comprises many facets in application. The hydrogen released may be required to provide a high-purity gas supply as such, or to produce other transition-metal hydrides in the adjoining system; to counteract oxidation in processes such as in welding, brazing or soldering; to produce the metal of highest purity (e.g. Ti from Ti hydride) under conditions where otherwise oxidation might be a hazard. For instance, in powder metallurgy the hydride can serve as primary powder in the high-temperature fabrication of otherwise oxidizable articles or in the joining of metallic layers to ceramics. In this sense, the function of the hydride is essentially that of an '*anti-oxide*', in whatever guise this may be.

Furthermore, there are the important *catalytic* properties of some hydrides, whilst in palladium the uses of the high solubility and permeability for hydrogen are unique.

An obvious gradation in applications occurs from the early to the late Groups of the transition-metal block, and the progression, in this order, from ionic to metallic and covalent bonding is clearly reflected in the areas of application, as is the progression of diminishing hydride stability from Groups III to VIA with its later, partial but modified recovery. It is reasonable to include here the alkali and alkaline earth metals (Groups I and IIA), although pre-transition metals, because their applications overlap and form a fundamentally integral part of the same picture.

To summarize then, we may divide the hydrides into the three functional groups of (1) '*Shelves*' for hydrogen, (2) '*Reservoirs on Tap*' for hydrogen, and (3) *Anti-oxides*, of course with areas of overlap; e.g. catalysis and diffusion in Pd, etc. may be considered under all three aspects. To this we should add: (4) the role of hydrides through their miscellaneous physical properties for which they prove useful in their own right. *Table 9.4* sets out these applications; some amplifications are added here.

Reactors—Hydrogen is in theory the most effective moderator for thermal neutrons in reactors, and the aim is to find materials with as high a volume concentration of H atoms as possible, i.e. high H/M ratio and density; they must also be able to retain them to elevated temperatures, with fair stability, mechanical strength and favourable neutron cross-section of the metal solvent. Zirconium hydride has found most general use, with some advantage in special cases for the Ti, Th, U, Pu, Y and rare earth hydrides*. A review has been given by Mueller and Blackledge[136], discussing their role in different portions of the reactor (moderators, reflectors and shields). Transition-metal hydrides have the clear advantage over other H-bearing materials, e.g. organic, water, of providing this relative refractoriness, though limited by the thermal equilibria discussed above. The hydrogen saturation concentrations for the hydrides do not change much until the actual decomposition pressure becomes too high.

The development of complex alloy hydrides (see also Section 9.3) has the special object of devising hydrides with exceptionally high H solvabilities —some intermetallics are promising here—and of good high-temperature properties.

A point of principle is brought out by the ternary zirconium-based fuel rod alloys such as $ZrU_{0.03}H$ mentioned above. These are suitable for low-temperature, low-power reactors only, owing to the fact of high hydrogen mobility at higher temperatures. This and the complex nature of the Zr–H

* The use of NbH and TaH has attractions through the refractoriness of the parent metals, higher than for Ti and Zr, but is limited by the lower stability of the hydrides.

Table 9.4. Applications of hydrides

Aspect	*Application*	*Examples**
Nuclear energy	hydrides as neutron retarders in reactors (for moderators, reflectors and shields) aim at maximum H concentration consistent with stability at temperature	principally Zr, Ti hydrides; more limited Nb, Ta, U, Th, Pu hydrides; special alloys devised for maximum H-retentive power
As hydrogen source ('reservoir')	to release atomic H required e.g. in preparing other hydrides; as source of pure hydrogen, both on the laboratory and industrial scale; CaH$_2$ as 'portable' source of hydrogen LiH used as prime material for quantity syntheses of new hydrides, e.g. covalent, and in chemistry	CaH$_2$ $\rightarrow$ Zr hydride Zr hydride $\rightarrow$ UH$_3$ UH$_3$ $\rightarrow$ H by vacuum heating; NaH, CaH$_2$, etc. by water addition; Ti hydride heating for very pure supply of hydrogen
Joining (cf. embrittlement)	(*a*) in welding, brazing, soldering metals, to obviate local oxidation; (*b*) metal to non-metal, e.g. brazing metal to an oxide ceramic or two ceramic parts with metal interlayer; for joining corrosion- or oxidation-resistant coatings, e.g. silicides, to metals; for metal–glass or metal ceramic vacuum seals	Ti or Zr hydrides in all these operations Ti interlayer onto other (refractory) metals, e.g. Nb
Gettering	Ti or Zr hydrides used to produce getters deposited on walls of valves before seal-off	in electronic valves and vacuum tubes; vacuum technology generally
Hydrides as starting materials for producing the metals	to produce pure Ti or Zr metal or retain it during manufacturing processes where oxidation would be a hazard; also to produce pure carbides, nitrides, borides or silicides of these metals	(*a*) in powder metallurgy (to protect particle surfaces); (*b*) in bulk manufacture, e.g. forging, rolling, etc. (mainly applies to Group IV metals but could to others); TiB$_2$, NbSi$_2$, etc. using Ti or Nb hydrides
Reducing or drying agent	alkali or alkaline earth hydrides used as strong reducing agents, e.g. for *de-scaling* of ferrous metals, etc.; NaH most common but CsH most powerful as desiccants for organic liquids, e.g. ether, oils; CaH$_2$ for analytical measurement of traces of water in salts or gases; also as reducing agent for producing later-Group metals, e.g. CaH$_2$ for Ti, and their hydrides	
Catalysis	hydrides used as catalysts, (*a*) for producing other hydrides (or metal compounds); (*b*) for hydrogenation in organic chemistry; (*c*) for altering the passivity of metal electrodes	Group IV, V and Pd hydrides; for (*a*) and (*b*) also NaH and other ionic hydrides important; (*c*) known for Ta/Ta hydride electrodes
Diffusion	Pd, for hydrogen permeation into closed systems (e.g. vacuum); for purification of hydrogen; on more limited scale this also applies to some other transition metals and alloys	

Table 9.4 cont.

Aspect	*Application*	*Examples**
Hardening of metal by hydriding	(limited:) surface hardening by hydrogen absorption; aging preceding actual embrittlement; electrolytic hard chromium deposits can be considered as hydrides	Ta surfaces used in acid- or wear-resisting articles (machinable before, hardened by) hydration
Miscellaneous	(*a*) *semiconductors* (*b*) *superconductors*	rare earth hydrides and (possibly) Nb and Ta hydrides have appreciable superconductive transition temperatures
	(*c*) use in *photocells* as sensitive coatings;	NaH, KH, RbH, CsH
	(*d*) foam production, to make sponge metals by hydrogen evolution from hydrides, e.g. ZrH_2;	
	(*e*) production of thin films on surfaces, where plating or other coating techniques not possible (for hydrides decomposing at moderate temperatures)	copper hydride, CuH, to give Cu coating on ceramics

* Potentially the Group III (Sc and Y) and the rare earth hydrides also come under several of these headings, but this is limited by their scarcity. However use is made of their powerful scavenger action for oxygen in small-scale metallurgical quantities.

phase equilibria would cause strong phase segregation and non-uniform hydrogen distribution due to thermal gradients occurring in the plant: typifying one of the most important factors to which applied hydride systems would generally be sensitive, and that must be allowed for.

Hydrogen 'source'—This implies the use of atomic hydrogen that can be released from one hydride, e.g. UH_3, to overcome passivity in preparing another. In laboratory experiments, uranium or titanium hydrides are also convenient sources of supply of very pure gaseous hydrogen, evolved by heating under reduced pressures. One advantage is that the temperature range over which either release or absorption of hydrogen atoms occurs is quite narrow, so that the supply becomes easily controllable by slight changes in heating or cooling. The hydrogen 'tap action' from controlled temperature changes clearly illustrates this point. TiH_2 can often be used as source of exceptionally pure hydrogen where needed, particularly in closely adjoining areas, e.g. in welding (see below).

This is also where the alkali and alkaline earth hydrides come into their own. By adding water to NaH, LiH or CaH_2, large-scale supplies of hydrogen gas can be obtained, used industrially as portable sources of hydrogen. For instance, the yield of 1 lb. of CaH_2 is 16 ft.^{3} of H_2; of NaH, 15 ft.^{3} and of LiH, 45 ft.^{3}, all with potentially high pressures and rates of supply.

The hydrides as sources for hydrogen fall thus into two classes: (*a*) those giving *molecular* H_2 with water, and (*b*) those giving *atomic* hydrogen under vacuum at temperature, and in direct contact. Class (*a*) represents the Early (saline) hydrides: their superior affinity for oxygen and OH groups over that for hydrogen is instrumental in this property. It may be recalled that H_2O is effectually a covalent hydride, so that reactions run:

(*a*) ionic hydride + covalent hydride $\rightarrow H_2$

(*b*) ionic (or metallic) hydride + heat (or reduced pressure) $\rightarrow H$

Lithium hydride finds a special place as a primary material for producing many *new* hydrides in quantity, particularly covalent hydrides and their derivatives, in reactions such as (for silane):

$$4LiH + SiCl_4 \rightarrow SiH_4 + 4LiCl$$

or generally

$$LiH + MX \rightarrow MH + LiX \text{ where } X = \text{halogen}$$

An even more powerful starting hydride was found in mixed hydrides like $LiAlH_4$, mentioned already; these compounds formed the basis for synthetic reactions of many kinds, also in organic chemistry, such as for reductions and hydrogenations, helped by the fact that they are slightly soluble in ether.

Welding—The applications of hydrides in welding, brazing and soldering technology are some of the most important, and the literature is extensive (cf. Bibliography). Here the 'anti-oxide' and 'atomic-hydrogen supply' functions *in situ* are immediately evident. They are used to join not only metal to metal but also metal to non-metal; for instance, a metallic layer to a carbide or an oxide ceramic, or for coating applications. An example is the bonding of silicides, etc. to a metal substrate to be protected by such layers. Corrosion- or oxidation-resistant coatings on Ti, Zr, Nb or Ta can thus be obtained. Again, Zr and Ti hydrides are the most important for these purposes. Furthermore, in joining *non-metals* together, hydrides are used to form effectively a thin metallic or lower oxygen interlayer between oxide masses. Titanium hydride has become prominent by its value in brazing metals to glasses and by facilitating the making of gastight metal–glass seals. By depositing a Ti hydride layer onto a ceramic, a Ti-metal layer is produced onto which other metals can then be soldered. This is particularly used in the electronics industry, for valves, etc.

In these joining applications, however, the favourable property of hydrides to obviate oxidation is often counterbalanced by the drawback of embrittlement, and much research has gone into minimizing this.

Gettering—Ti and Zr hydrides are used as getters in radio valves and vacuum tubes, deposited on the inner walls during evacuation and just before sealing off. Zirconium hydride has here the added advantage over zirconium metal of being less pyrophoric, particularly in the form of a fine powder.

Raw material for metals—Ti and Zr hydrides can serve as starting materials in producing or fabricating the pure Ti or Zr metals in bulk, where heating is an integral part of the process (forged blocks, etc.). In powder metallurgy it can be especially important because the pyrophoric nature of powders or the quiescent oxidation of particles is counteracted by compacting and sintering the hydrides. One might regard this application as that of an 'internal getter'.

Similarly, the hydrides of Groups IV and V are used for producing the borides, carbides, nitrides and silicides of these metals, and in general alloy production. In general terms, the interstitial hydrides serve as intermediaries in providing interstitial compounds other than oxides—essentially to counteract the oxides.—A large technological and patent literature exists in this field.

Reducing or drying—The alkali hydrides are the most powerful reducing agents, used for instance for descaling of metals. NaH, in Na(OH) as solvent, reduces iron oxides on steels, and generally the oxides are converted to free metal, leaving the metal substrate intact. This virtue is also used for the quantitative measurement of oxidizability of alloys. The effect is the stronger at elevated temperatures, and with special precautions these methods are generally employed. KH has similar applications and is even more reactive. (The descaling process has been described e.g. by Gilbert[137a]; see also Hurd[137b].) The strong reducing power of CaH_2 is utilized in producing later-Group metals (Ti, Zr, V, Nb, Ta) and their hydrides.

Calcium hydride, CaH_2, has been used in the laboratory as a drying agent, e.g. of ether and other liquids or, on a larger scale, of oils and, indeed, of hydrogen gas also. In analytical work, CaH_2 is used for trace-water determination in gases or salts.

Catalysts—The catalytic action of transition-metal hydrides such as Ti hydrides is much used for hydrogenation in organic chemistry, and to overcome passivity in electrolytic applications of the metals. Thus, for instance, combined Ta/TaH_2 electrodes have been devised[138] for the solute hydrogen to alter the passivity. Palladium, when in fine subdivision, is a highly active catalyst for hydrogenation reactions.

Pd diffusion—The high absorption capacity and permeation rate for hydrogen of palladium has wide applications, such as supplying the gas into otherwise vacuum-tight enclosures through Pd 'valves' that can, by varying degrees of heating, provide precisely controllable rates of admission. The filter action is also used for the purification of hydrogen gas.

Darling[139] has described thermal and electrolytic palladium alloy *diffusion cells* using a Pd–25 per cent Ag alloy membrane to purify hydrogen to an ultra-high degree; hydrogen can thus be obtained at a rate of 2 l./min, with 1 p.p.m. impurity, but much larger throughputs can now be achieved, to 10 l./min or more.

Although the case of Pd is extreme, it should not in principle be regarded as unique, as the same diffusion characteristics can apply to some other transition metals and their alloys in varying degrees, including, of course, Pd-*base* (e.g. Pd–Ag) and, embryonically, also nickel (see Ni–H/Pd–H comparison). In vacuum technology, much investigation has gone into the question of hydrogen permeation through metals used for vacuum enclosures[140], such as stainless steels, brasses, etc.

Hardening by hydriding—This application, which is effectively equivalent to hardening effects by carburizing or nitriding, is rather restricted for hydrogen, as it is more transient, confined to lower temperatures, and lower hardnesses are attained. Precipitation of hydrides and consequent hardening in the incipient stages is (as explained earlier) also different in nature to that of carbides, etc. However, hardening, e.g. of tantalum through hydrogen absorption is used for making wear- and acid-resistant containers and components. A great advantage is that the parts can be machined when in ductile, pure metallic form, and then hardened as desired by heating in hydrogen at only moderate temperatures.

We might also in a sense consider the hardness attained in electrodeposits, especially of chromium, as due to the presence of lattice hydrogen and strains produced by it.

Miscellaneous properties—Special applications of hydrides are utilizing physical properties such as photosensitivity of alkali hydrides in photocells, superconductivity (of Nb or Ta hydrides) and semiconductivity. Mechanically, hydrogen evolution from hydrides upon decomposition at temperature can be used for foam- and sponge-metal production or, indeed, for deliberate fragmentation of the parent metals where this is desired.

The use of CuH to deposit thin films of metallic copper, e.g. on ceramics, is an example of turning the low decomposition temperatures of the later hydrides to advantage, both as a coating technique and for thin-film production (this links with 'joining' but differs in using the unstable hydrides).

Hydrogen Embrittlement in Metals

Embrittlement through hydrogen absorption and release in transition metals is the principal hazard, referred to already in Section 9.1. Its cause and avoidance have been widely investigated, but much of this work was of an empirical nature to deal with specific industrial problems met in alloy production. However, there were some important groups where the approach was fundamental, such as Andrew's work on steel embrittlement[141] and the causes of hairline cracks.

In general, embrittlement can be traced to atomic hydrogen which had been a lattice solute at high temperatures earlier in heat-treating history, and then segregates, still atomically at point defects and dislocations, molecularly at grain boundaries, voids, pores or surfaces. The consequent volume expansion and stress build-up can be very large, resulting in crack formation.

In iron and steels, this is more specifically associated with the $\gamma \rightarrow \alpha$ transformation, and the solid solubility graph in *Figure 9.20* is illustrative. Since the 'precipitate' is free hydrogen, not as for carbon a solid carbide of atomic volume commensurate with that of the matrix, the disruption can be of quite a different order of severity. Basically, the effect becomes the more pronounced towards the *later* Groups of the Periodic Table because free hydrogen rather than solid 'hydrides' is formed, as in Groups III–V. We can in this sense regard hydride* formation and large solid solubilities as an ameliorating factor. Embrittlement in the early Groups (and in Pd, for the same reason) is therefore less severe but becomes more so at higher H levels where again the tendency would be towards molecular hydrogen formation.

* Under 'hydride' we should here include the hydrogen-richer phases in Ti–H, Nb–H, etc. which really form a continuous solution with the metal at higher, but segregate at lower, temperatures.

In *steels*, the internal pressure build-up can be enormous, not necessarily through trapping of gaseous hydrogen but, where an iron carbide, Fe_3C, is present, it can be through *methane* produced by the reaction $2H_2 + Fe_3C \rightarrow CH_4 + 3Fe$. The theoretical mechanism of the process has been studied by Bilby and Hewitt[142] who show that trace amounts of hydrogen dissolved in iron already suffice to produce great pressures and permanent microcracks devolving from dislocations. As the initial cracks are wedge-shaped, it is quite understandable how the propagation and, as it were, the effect of 'prising open' the wedge become cumulative.

The basic part played by atomic $\rightarrow$ molecular hydrogen transition[141b] should be stressed. The effect e.g. of cold work can become significant, in providing increased space at crystal boundaries and facilitating molecular hydrogen forming. Since the diffusion rates of hydrogen will vary with the constitution of the steel (particularly in view of the differing hydrogen solubilities in the α and γ forms and in intermediate stages of martensite), and thus with steel composition and heat treatment, it is in principle at least possible to combat hydrogen embrittlement by adjusting these factors and, short of rigidly excluding the element, to render it more innocuous. Indeed, in special cases it seems conceivable to turn the presence of hydrogen to advantage by directing it towards hydride (in the same sense as carbide) formation, such as by Group III–V metal additions, and by controlling diffusion rates. This would, however, only apply to moderate- or low-temperature uses.

A possibility, on the same lines, is that additions of *stable* hydride formers from the Early Groups, such as the rare earths, to steels (and other Later Group alloys) would help during hydrogen-introducing processes such as welding, by their *scavenger action*. There is, therefore, much scope for technical improvements in combating embrittlement, if only the hydride stability relations discussed in this Chapter are judiciously applied.

The trapping of molecular hydrogen in *cold-worked* steel, according to Podgurski[143], confirms that gaseous methane is indeed formed in microvoids ($2H_2 + Fe_3C \rightarrow CH_4 + 3Fe$ as described); this methane could be isolated quantitatively. The subject is controversial, however, and free H_2, not CH_4, has been held to be responsible for the embrittlement of carbon steels[144]. Both views can well be reconciled, however, depending on composition, heat treatment and carbide morphology.

Some *internal friction* work on H solutions in iron has also proved fruitful. It was possible to show[145] that atomic solution hydrogen could appear at lattice imperfections in H_2 molecules; the high pressures arising are calculable. In Cr–Ni steels, internal friction peaks linked with stress-induced diffusion of H_2 molecules have been observed[146].

Quite generally, even traces of hydrogen tend to increase the ductile/brittle transition temperature of transition metals. In iron more specifically the hydrogen-induced yield point has been found[147] to be similar in principle to the effect of interstitial C or N, including the effects of initial lattice straining and aging; however, the corresponding effects become operative for hydrogen at lower temperatures than for carbon and nitrogen, and much of this work becomes interesting in sub-zero applications.

The effect of hydrogen on the properties of arc-welded steel has received extensive attention. In *welding*, hydrogen absorption can give rise directly to stress embrittlement (macro and micro), crack-formation and porosity. Supersaturation, which occurs initially in the liquid (cf. *Figure 9.20* for solubility curves), then results in heavy hydrogen evolution on solidification, apart from the lesser, but still important, later evolution during the $\gamma \rightarrow \alpha$ transformation (some references are given in the Bibliography).

The phenomenon of 'fish-eyes' may also be noted: small cavities formed upon slow deformation of tensile specimens at fracture faces through molecular H_2 evolution from initial dislocation[148].

Other hydrogenating processes occur during *acid exposure* and, most potently, during electrolysis, involving hydrogen absorption, hardening and embrittlement during *electroplating*; the electrochemical literature abounds with examples.

A cause of embrittlement can also be, instead of internally precipitated H_2, hydrogen introduced externally at surfaces by high-pressure atmospheres. This acts in what might be regarded as the 'reverse way', namely first by adsorption of the H_2 molecule which then dissociates, leading to absorption of H atoms in the steel. It was demonstrated by Hofmann *et al.*[149] who also proved that small O_2 or N_2 additions to the H_2 can actually prevent such embrittlement, both by blocking interstitial sites and inhibiting the effect of hydrogen (see earlier discussion and, for H embrittlement in metals other than Fe, the Bibliography).

One curious and basically interesting effect may here be noted. It is possible[150] to prevent embrittlement of tantalum by mere surface contact with another metal which has a low hydrogen potential and is cathodic in the same environment (here HCl). The noble metals are suitable for this, particularly platinum, and contact need only be over a small area. To quote: 'contact by riveting, welding, electrodeposition or merely rubbing the noble metals on to the Ta surface in a small spot at one end of a strip, protected the other end, 6 in. away, from hydrogen embrittlement'. The reason for this is not very clear (Pd-like effusion of hydrogen?), but the principle could possibly have extended applications to other metals (e.g. steels) as a means of avoiding embrittlement.

One positive use that can be made of hydrogen embrittlement is for the crushing of metals and alloys which normally are ductile where this is desired. Examples are in powder metallurgy, or for research preparations of strain-free powder specimens e.g. in crystallography, assuming of course that the hydrogen absorbed can later be driven off at permissibly low temperatures. This can even be modified to producing fine-particle flakelets delineated by an array of cracks created by the action of molecular hydrogen[151].

Various mechanical properties of the Ti–H, Zr–H, V–H, Nb–H, etc. systems have been referred to briefly in earlier Sections. A paper on the susceptibility of Zr and its alloys to *delayed* failures by hydrogen embrittlement[152] illustrates well the *time* dependence of the kinetics of hydrogen segregation. This time factor is, of course, already implicit in the phase diagram and thus has an immediate bearing on behaviour in practice when the alloy is held at sub-eutectoid temperatures.

SELECTED LITERATURE

General and basic

The hydrogen bond

Pauling[5]; Latimer, W. N. and Rodebush, W. H. *J. Am. chem. Soc.* 42 (1920) 1419; Sidgwick, M. V. *J. chem. Soc.* 125 (1924) 527

Cochran, W.; Magat, M. *Proc. 3rd Int. Congr. Crystallogr.* Paris, 1954

Pimentel, G. C. and McClellan, A. L. *The Hydrogen Bond*, San Francisco (Freeman) 1960

Libowitz[2c]; Hurd[137b]; Rao, S. *J. scient. ind. Research* 20A (1961) 308; Green, M. L. H. *Angew. Chem.* 72 (1960) 719; Renson, M. *Industrie chim. belge* 24 (1959) 1177 [Reviews of metal hydrides]

Stone, F. G. A. *Hydrogen Compounds of the Group IV Elements*, Englewood Cliffs (Prentice-Hall) 1962

Size of hydrogen atom, and basic theory[5,63]

Lacher[114] [Basic statistics of hydrides, mainly Pd–H but of wider significance]; Anderson, J. *Proc. R. Soc.* A185 (1946) 69 [Defect lattices in hydrides; theory]; Libowitz, G. G.[2h] [Development of vacancy theory]

Libowitz[2c]; Smith[67]; Hurd[137b]; Samsonov and Umanskii[75a] [Thermodynamics of hydrides]; Samsonov[75b] [Classification]; Jeffes, J. H. E. and McKerral, H. *J. Iron Steel Inst.* 202 (1964) 666 [Thermodynamics of non-transition metal hydrides only]

Libowitz[2f] [P–T–C relations to non-stoichiometry and energy of vacancy formation; statistical mechanics of U and Pd hydrides]; Fowler and Guggenheim[115b]; Lacher[114] [Statistics of solid solutions and molecular hydrogen equilibria]; Gibbs[4e] [Various models of V, Nb, Ta hydrides, assuming atomic, protonic or anionic bonding]

Smith[67]; Barrer, R. M. *Diffusion in and through Solids*, Cambridge Univ. Press, 1951; Cupp, C. R. *Prog. Metal Phys.* 4 (1953) 105; Smithells, C. J. *Gases in Metals*, London (Chapman & Hall) 1937 [Hydrogen solubility and diffusion]

Mueller and Blackledge[136] [Metal–hydrogen systems related to nuclear applications]

Binary systems

Alkali metals–H[8,29] Messer, C. E. *et al.*, *J. phys. Chem., Ithaka* 62 (1958) 220; *U.S. At. Energy Comm. Rep.* NYO 3959 (1959) [Li–LiH equilibria]; Ahmed, M. S. *Phil. Mag.* 42 (1951) 997; Cochran, W. *Rev. mod. Phys.* 30 (1958) 47 [LiH structure]; Roos, A. *Chemy Ind.* 85 (1961) 86; Pretzel, F. E. *et al.*, *J. appl. Phys.* 33 (1962) 510 [LiH properties and uses]

Libowitz[2c]; Shull[8]; Banus *et al.*[29]; Hurd[137b]; Schlesinger, A. H. *J. Am. chem. Soc.* 69 (1947) 1199, 2692 [Na–H]

Alkaline earths–H[30–2]

Bergsma and Loopstra[32] [CaH_2 structure]

Peterson and Fattore[31a] [Ca–H]

Peterson and Indig[31b] [Ba–H]

Zachariasen[30] [MgD_2 structure]

(Rare earths, Sc, Y)–H[34–5] Lundin and Blackledge[12] [Y–H, thermodynamics, structure; heat of formation and entropy]; Funston, C. S. *Trans. metall. Soc. A.I.M.E. Spec. Rep.* 7 (1960) 51; Roach, J. D. 221 (1961) 646; Parker, D. S. *G.E.* (*U.S.A.*) *Rep. APEX*-558 (1960) [Y–H, physical and mechanical properties]

Korst and Warf[17b]; Mulford *et al.*[27]; Wallace *et al.*[33d] [Magnetic characteristics of the rare earth hydrides; relation to electron structure]

McGuire, J. C. and Kempter, C. P. *J. chem. Phys.* 33 (1960) 1584 [Sc–H; ScH_2]

Libowitz[2c]; Sturdy and Mulford[19]; Mulford and Holley[27c]; Holley *et al.*[27d]; Pebler and Wallace[34] [Structures; heats of formation, dissociation pressure equations, variation of solubilities]

Goon[39b] [La–H]

Ti–H[37–48] Melkonian, G. A. *Z. phys. Chem.* 17 (1958) 120 [α and β ranges in Ti–H; enhalpies]; Kant, A. *Rep. WAL–TR*-541/1, 1958 [Solubility relations; non-stoichiometry]; Craighead, C. M. and Jaffee, R. I. *J. Metals, N.Y.* 6 (1954) 367 [Ti–H equilibria]; Sofina, V. V. and Pavolvskaya, N. G. *Zh. fiz. Khim.* 34 (1960) 1104 [Heats of formation, P–T–C curves at low

Ti–H *cont.* pressures; comparison of Ti–H/Zr–H systems, concluding increased stability of Zr hydrides]; Sofina, V. V. *et al., Soviet Phys. Crystallogr.* 3 (1958) 544 [Structure]; Stalinski, Coogan and Gutowsky[48] [Diffusion kinetics of H in Ti hydride; energy of H atoms in different lattice positions]; Louthan, M. R. *Trans. metall. Soc. A.I.M.E.* 227 (1963) 1166 [Orientation of hydride particles during precipitation in Ti under applied stress]

Zr–H[2d,21, 50–65] Espagno *et al.*[56b] [Measuring bulk dilatometry of Zr hydrides; thermal expansion related to structure and eutectoid transformation]; Chang, R. *J. nucl. Mater.* 2 (1960) 335 [Crystallographic and internal friction work on transformations in ZrH_{2-x}, related to diffusion and activation energies]; Fitzwilliam, J. *et al., J. chem. Phys.* 9 (1941) 678 [Magnetic susceptibilities of Zr hydrides]; Korst[17b]; Vetrano, J. B. and Atkins, D. F. *Metall. Soc. A.I.M.E. Spec. Rep.* 7 (1960) 57 [Zr–H]; Östberg, G. *J. nucl. Mater.* 5 (1962) 208 [Zircaloy base–H_2]; Hon, J. H. *J. chem. Phys.* 36 (1962) 759 [H diffusion in ZrH_2]; Droge, J. W. *U.S. At. Energy Comm.* BMI–1502 (1961) 364 [Thermal diffusion in β-Zr solution]; Douglas, T. B. *J. Am. chem. Soc.* 80 (1958) 5040 [Thermodynamic data at high temperatures]; Westlake, D. G. *J. nucl. Mater.* 7 (1963) 346 [Enthalpy data]; Sofina, V. V. *et al., Kristallografiya* 3 (1958) 539 [P–T–C curves, etc.]; Raymond, J. W. *U.S. At. Energy Comm. Rep.* NAA–SR 9374, 1964 [Dissociation pressures as f (C, T) of Zr–H δ and ε phases; new diagram proposed]; Singh, K. P. and Parr, J. G. *Trans. Faraday Soc.* 59 (1963) 2256 [Zr–D system: solubility of D in Zr]; Sawatzky, S. *J. nucl. Mater.* 9 (1963) 364 [Heat of Transport of H in Zr and Zr alloys]; La Grange, L. D. *et al., J. phys. Chem., Ithaka* 63 (1959) 2035 [Zr–H phase equilibria; high-temperature x-ray and dissociation pressure]

Hf–H[14,21,56]

Th–H[15b,c, 22–3,30,34,66] Peterson, D. T. and Westlake, D. G. *Trans. Am. Inst. Min. metall. Engrs* 215 (1959) 444 [H solubility in Th]; Peterson and Rexer[66] [Diffusion coefficients of H in Th]; Nottorf[23a]; Rundle *et al.*[23b] [Th–H equilibria; structure]

V–H[24,33d, 68–70,72] Nemnonov, S. A. and Finkelshtein, L. D. *Fizika Metall.* 9 (1960) 530 [Electronic nature of bonding in V hydrides]; Sieverts *et al.*[11] [Early important work on V → H solubilities by equilibrium pressure methods, in Veleckis's view[24] affected by oxygen impurity]; Betsuyaku, H. *et al., J. phys. Soc. Japan* 19 (1964) 1089 [N.m.r. study of V and H atoms with varying C–T]; Trzeciak, M. J. *et al., Battelle Mem. Inst. Rep. B.M.I.* 1112, 1956 [V–H equilibria]

Nb–H[24,33, 73–80,82] Paxton, H. W. *et al., Trans. Am. Inst. Min. metall. Engrs* 215 (1959) 725; Knowles, D. R. *I.G. Rep.* R/C–190, 1957 [Nb–H, Ta–H]; Umanski, Ya. S. *J. phys. Chem. USSR* 14 (1940) 332 [Nb–H]; Rudd, D. W. *et al., J. phys. Chem., Ithaka* 66 (1962) 351 [Permeability of hydrogen diffusing through niobium; temperature relationship]

Ta–H[24–5, 33,71,73,78, 80–2] Pietsch, E. and Lehl, L. *Kolloidzeitschrift* 68 (1934) 226 [Crystallographic work]; Mallett and Koehl[22c] [P–T–C equilibria in Ta–H and thermodynamic constants confirm single-phase → duplex transition]; Kelley, K. K. *J. chem. Phys.* 8 (1940) 316

(Ti, Zr, Nb, Ta)–H Gulbransen and Andrew[50] [Kinetics]

Cr–H[84–7] Trzeciak *et al.*[121] [Hydrides CrH_{1+x} formed unstably by electrodeposition]

Mo, W–H Hill, M. L. *J. Metals, N.Y.* 12 (1960) 725 [Solubility of H in Mo]; Iwase, K. and Fukusima, M. *Nippon Kink. Gakk.* 1 (1937) 202; Martin, E. *Arch. EisenhüttWes.* 3 (1929) 407 [Solubility in W and Mo]

U–H[15b,c,d, 27b,44c,88–9] Besson, J. and Chavallier, J. *C.r. hebd. Séanc. Acad. Sci., Paris* 258 (1964) 5888 [Non-stoichiometric UH_{3-x} formation; properties]; Wicke and Otto[18b]; Flotov and Abraham[89]; Katz, O. M. and Gulbransen, E. A. *J. nucl. Mater.* 5 (1962) 269; *J. Metals, N.Y.* 13 (1961) 677; Mattraw, H. C. *J. phys. Chem., Ithaka* 59 (1955) 93; Karchevsky, A. J. and Buryek, E. M. *Zh. éksp. teor. Fiz.* 42 (1962) 375 [Thermal and physical studies]
Spedding *et al.*[26] [Classical work on UH_3, etc.; preparation and properties]; Darras and Caillat[88] [Effect of hydrogen on uranium; solid solubility of H]

Pu–H[27a,d, 30,91] Mulford and Sturdy[27a] [Thermodynamic data of PuH_2 and PuD_2; P–T–C curves]

Fe–H[49, 93–7] Bastien[93] [Diffusion coefficients and Fe–H solubility]; Schenck, H. and Winsch, H. *Arch. EisenhüttWes.* 32 (1961) 779 [H solubility in liquid Fe]

Co–H[98] Busch, T. and Dodd, R. A. *Trans. Am. Inst. Min. metall. Engrs* 218 (1960) 488 [H solubility in Fe, Co and Ni]

(Co, Ni, Cu)–H Weinstein, M. and Elliott, J. F. *Trans. metall. Soc. A.I.M.E.* 227 (1963) 285 [Solubility of H in *liquid* Co, Ni, and Cu, as function of temperature]; Abeldo, C. R. (North Western Univ. Thesis, 1961) [Chemisorption of H_2 on cobalt, using its effect on saturation magnetization]; Dietz, R. E. and Selwood, P. W. *J. appl. Phys.* 30 (1959) 1015 [Similar for nickel–H_2]

Ni–H[98–103] Reddy, A. K. N. *Acta crystallogr.* 17 (1964) 443 [Critical discussion of HPC and FCC nickel hydride deposits; texture]; Baranowski, B. and Smialowski, M. *Jnl Phys. Chem. Solids* 23 (1962) 429 [Comment on results of Boniszewski and Smith[101]; point to some preceding papers on nickel hydrides, suggesting already the analogy to the Pd–H-type two-phase system]; Szklarska-Smialowska, Z. and Smialowski, M. *J. electrochem. Soc.* 110 (1963) 444 [Electrochemical Ni–H build-up and decomposition]; Cable, J. W. *et al.*, *CERN Symp.*, Grenoble, 1963; '*La diffraction et la diffusion des neutrons*', 1964, 36 [Structure of NiH unstable, FCC by neutrons; H location]; Majchzak, S. and Jarmolowicz, H. *Bull. Acad. pol. Sci.* 12 (1964) 155 [Ni hydride formed on cathodic charging of Ni; orientation by x-rays shows epitaxy]; Bauer, H. J. and Schmidbauer, E. *Naturwissenschaften* 48 (1961) 425; *Z. Phys.* 164 (1961) 367; Andra, W. *Phys. Status Solidi* 1 (1961) 135 [Magnetic properties]; Hawkins[28] [Thermodynamics of solubilities in Ni of hydrogen and its two isotopes, deuterium and tritium; heats of solution; corresponding study of an austenitic stainless steel]

Pd–H[1, 104–15] Nakhutin, I. E. *Zh. fiz. Khim.* 36 (1962) 235 [Solubility as function of pressure]; Bucur[107] [H solubility in β-Pd–H; thermodynamic data]; Rosenhall, G. *Annln Phys.* 18 (1933) 150 [Solubility; lattice-spacing changes]; Lambert, B. and Gates, S. F. *Proc. R. Soc.* A108 (1927) 456 [Hysteresis]; Gillespie, L. J., with Galstaun, L. S. *J. Am. chem. Soc.* 58 (1936) 2565; with Downs, W. R. 61 (1939) 2496 [Presumed stoichiometric 'Pd hydrides' and their decomposition]; Wright, P. *J. phys. Soc., Lond.* A63 (1950) 727; Krüger, F. and Gehm, G. *Annln Phys.* 116 (1933) 174; Linde, J. O. and Borelius, G. 108 (1927) 456 [Electrical resistance versus H_2 pressure; x-ray work]; Maeland and Gibb[105] [Structure by x-rays]; Sieverts *et al.*[11] [P–T–C relations in Pd–H];. Worshan, J. E. *et al.*, *Physics Chem. Solids* 3 (1957) 303 [Neutron diffraction; Pd and H atom positions in Pd–H]

Cu–H[116–20] Eichenauer, W. and Pebler, A. *Z. Metallk.* 48 (1957) 373 [H solubility in Cu]

Ternary systems (M_1–M_2–H)

Trzeciak[121] [Ti- and Zr-based alloys with other transition metals; discussion of solubility variations; experimental details]

Zr–Ni–H[2b, 122a, 123] Peterson, S. W. *et al.*, *CERN Symp.*, Grenoble, 1963; '*La diffraction et la diffusion des neutrons*', 1964, 27 [$NiZrH_3$ and $NiZrD_3$ neutron diffraction; H-atom location]

Zr–U–H[124–6]

Zr–Ti–H[127–8]

Zr–Nb–H[130]

Zr–Ti, Zr–Mo–H[131]

Fe–Ni–H, Fe–Cr–H (liquid)[132]

Fe–(Nb, Ta, Ti)–H[133]

Nb–Mo–H, Nb–Re–H, Mo–Re–H Jones, D. W. and McQuillan, A. D. *Physics Chem. Solids* 23 (1962) 1441 [Nb–Mo, Nb–Re and Mo–Re alloys, their hydrogen affinity and magnetic susceptibility, relation to electron structure and electron/atom ratio]

Ti–Nb–H, Ti–Mo–H, Nb–Mo–H Jones, Pessall and McQuillan[10] [Ti–Nb, Ti–Mo and Nb–Mo alloys with H; solubility falling to zero in Groups IV - VI]; Upadhyaya, G. S. and McQuillan, A. D. *Trans. Am. Inst. Min. metall. Engrs* 224 (1962) 1250 [Ti–Nb hydride structures]

U–Nb–H Katz and Gulbransen (cf. U–H) [Dissociation pressures, P–T–C relations, structures and H loading; solubilities]

Fe–W–H Rezukhina, T. N. *et al.*, *Zh. fiz. Khim.* 25 (1951) 305
Bastien[93] [Effect of hydrogen on alloy steels]

Pd–Ni–H [134a]

Pd–Pt–H Carson, A. W. *et al.*, *Trans. Faraday Soc.* 56 (1960) 363, 371

Pd–Co–H, Pd–Ni–H Vert, Zh. L. and Tverdovsky, I. P. *Trudy̆ gos. Inst. prikl. Khim.* 43 (1960) 169

Pd–Au–H Bernard, J. and Talbot, J. *C.r. hebd. Séanc. Acad. Sci., Paris* 222 (1946) 493; Mosevich, I. A. *et al.*, *Trudy̆ gos. Inst. prikl. Khim.* 42 (1959) 173

Pd–Ag–H[11,135]

Pd–(Cu, Ag, Au)–H[134b]

Pd–Cu–H Karpova, R. A. and Tverdovsky, I. P. *Zh. fiz. Khim.* 33 (1959) 1393

Li–Al–H, etc. Hurd[137b]; Renson (see *General and basic*) [Ionic character, physical and chemical properties of systematically paired ionic hydrides]

Ag–Al–H and related systems[119]

Hydrogen effects

Fe *and steels*[144–9]—Carmicheal, D. C. *et al.*, *Trans. metall. Soc. A.I.M.E.* 218 (1960) 826 and Tetelman, A. S. with Robertson, W. D. *Trans. metall. Soc. A.I.M.E.* 224 (1962) 770; *J. Metals N.Y.* 13 (1961) 690 [Lattice absorption and liberation of hydrogen in α-iron using x-rays; effect of H_2 on Fe and Fe–Si single crystals; embrittlement related to dislocations, stresses and cracks]

Welded steels—McNabb, A. and Foster, P. K. *Trans. metall. Soc. A.I.M.E.* 227 (1963) 618 [Diffusion of H_2 in ferritic steels]; Andrews, W. *Trans. Inst. Weld, Jn.*, 8 (1945) 119; Cotterill, P. *Prog. Mater. Sci.* 9 (1961) 201; Elanigan, A. E. *Weld., Lond.* 26 (1947) 193; Schaeffler, A. L. *et al.* 31 (1952) 283; Cavett, R. H. and Van Hess, H. C. 42 (1963) 316; Van der Willigen[148a]; Hummitzsch[148b] ['Fish-eyes' effect]

Bastien *et al.*[93] [Influence of hydrogen on steel weldings andcastings; H_2 segregation or precipitation during solidification; embrittlement]; Ramachandran, E. E. and Dasarathy, C. *Nature, Lond.* 191 (1961) 998 [Hydrogen impregnation in austenite]

Hydrogen in Steel—B.I.S.R.A. Conference. *Spec. Rep. Iron Steel Inst.* 73 (1963); Smialowski, M. *Hydrogen in Steel*, Oxford (Pergamon) 1962; Meredith, B. E. and White, D. *I.G. Rep.* 145/38, 1959 [Briefer survey]; Veyseeyre, H. *et al.*, *C.r. hebd. Séanc. Acad. Sci., Paris* 254 (1962) 281; Plusquelle *et al.* 255 (1962) 518; Heal, T. *U.K.A.E.A. Rep. I.G.* 145 (RDIC) 35 (1959) [Effect of irradiation]

Hydrogen embrittlement in non-iron metals—Cotterill (see *Welded steels* above); Williams, B. *J. Inst. Metals* 91 (1962) 147 [Titanium]; Clausse, A. *C.r. hebd. Séanc. Acad. Sci., Paris* 255 (1962) 2760 [Tantalum]; Jungkönig *et al.*[130] [Effect of H_2 on mechanical properties of 'Zircalloy'-type alloys]; Mihelich, J. L. and Troiano, A. R. *Nature, Lond.* 209 (1963) 996 [Delayed failure through embrittlement in a hydrogenated Ni alloy]; Moroz, L. S. *et al.*, *Fizika Metall.* 16 (1963) 739 [Effect of H_2 on mechanical properties of nickel and copper]

REFERENCES

1 Ubbelohde, A. R. *Proc. R. Soc.* A159 (1937) 295

2 (*a*) Libowitz, G. G., with Gibb, T. R. P. *J. phys. Chem., Ithaka* 60 (1956) 510; (*b*) Hayes, H. F. 62 (1958) 7; (*c*) *J. nucl. Mater.* 2 (1960) 1; (*d*) 5 (1962) 228; (*e*) *Acta metall.* 6 (1958) 133; (*f*) *J. appl. Phys.* 33 Suppl. 1 (1962) 399; (*g*) *J. chem. Phys.* 27 (1957) 514

3 Gibb, T. R. P. and Schumacher, D. P. *J. phys. Chem., Ithaca* 64 (1960) 1407

REFERENCES

[4] *Non-stiometric Compounds*, ed. Ward, R., Washington (Am. Chem. Soc.) 1963; (*a*) Chapter 7 (Libowitz); (*b*) Chapter 8 (Sidhu *et al.*); (*c*) Chapter 11 (Wallace *et al.*); (*d*) Chapter 10 (Aston and Mitacek); (*e*) Chapter 9 (Gibb); (*f*) Chapter 12 (Krause *et al.*)

[5] (*a*) Pauling, L. *The Nature of the Chemical Bond*, Ithaca (Cornell Univ. Press) 1960; *J. chem. Phys.* 46 (1949) 135; (*b*) Pauling, L. and Ewing, F. J. *J. Am. chem. Soc.* 70 (1948) 1660

[6] Coehn, A. and Specht, W. *Z. Phys.* 62 (1930) 1

[7] Hägg, G. *Z. phys. Chem.* 12 (1931) 33

[8] Shull, C. G. *et al.*, *Phys. Rev.* 73 (1948) 842

[9] Bacon, G. E. *Neutron Diffraction*, Oxford Univ. Press, 1962; *Applications of Neutron Diffraction in Chemistry*, Oxford (Pergamon) 1963

[10] Jones, D. W., Pessall, N. and McQuillan, A. D. *Phil. Mag.* 6 (1961) 455

[11] Sieverts, A. and Danz, W. *Z. phys. Chem.* 34 (1936) 158; 38 (1937) 46, 51; Hagen, H. and Sieverts, A. A165 (1933) 1; Sieverts, A., with Hagen, H. A174 (1935) 247; with Moritz, H. *Z. anorg. allg. Chem.* 247 (1941) 124

[12] Lundin, C. E. and Blackledge, J. P. *J. electrochem. Soc.* 109 (1962) 838

[13] Spalthoff, W. *Z. phys. Chem.* 29 (1961) 258

[14] Sidhu, S. S. *et al.*, (*a*) *Acta crystallogr.* 9 (1956) 607; (*b*) 7 (1954) 447; (*c*) with McGuire, J. C. *J. appl. Phys.* 23 (1957) 1257

[15] Rundle, R. E. (*a*) *Acta crystallogr.* 1 (1948) 180; (*b*) 5 (1952) 22; (*c*) *J. Am. chem. Soc.* 73 (1951) 4172; (*d*) 69 (1947) 1719; (*e*) Rundle, R. E. *et al.*, *US. AECD Rep.* 2120 (1948)

[16] Wilcox, B. A. and Huggins, R. A. in *Niobium, Tantalum, Molybdenum and Tungsten*, (Ed.) Quarrell, A. G., Amsterdam (Elsevier) 1961

[17] Korst, W. L. (*a*) *U.S. At. Energy Comm. Rep.* NAA–SR 6881 (1962); (*b*) SR–6880 (1960); with Warf, J. C. *Acta crystallogr.* 9 (1956) 452

[18] (*a*) Wicke, E. and Otto, K., with Küssner, A. *Acta 2nd int. Conf. Catalyse* (Paris) 1 (1960) 1035; (*b*) *Z. phys. Chem.* 31 (1962) 222

[19] Sturdy, G. E. and Mulford, R. N. R. *J. Am. chem. Soc.* 78 (1956) 1083

[20] McQuillan, A. D., private communication

[21] (*a*) Edwards, R. K. and Veleckis, E. *J. phys. Chem., Ithaka* 66 (1962) 1657; (*b*) Edwards, R. K. *et al.*, *J. Am. chem. Soc.* 77 (1955) 1307

[22] (*a*) Mallett, M. W., with Campbell, I. E. *J. Am. chem. Soc.* 73 (1951) 4850; (*b*) with Albrecht, W. M. *J. electrochem. Soc.* 104 (1957) 142; (*c*) with Koehl, B. G. 109 (1962) 611

[23] Nottorf, R. W. *et al.*, *U.S. AECD Rep.* 2984 (1945)

[24] Veleckis, E. *U.S. Dept. Comm. O.T.S.* AD282, 433 (1960)

[25] Albrecht, W. M., Mallett, M. W. and Goode, W. D. *J. electrochem. Soc.* 106 (1959) 981

[26] Spedding, F. H. *et al.*, *Nucleonics* 4 (1949) 1

[27] (*a*) Mulford, R. N. R. and Sturdy, G. E. *J. Am. chem. Soc.* 77 (1955) 3449; (*b*) Mulford, R. N. R. *et al.* 76 (1954) 297; (*c*) with Holley, C. E. *J. phys. Chem., Ithaka* 59 (1955) 1222; (*d*) Holley, C. E. *et al.*, *ibid.* 1226

[28] Hawkins, N. J. *U.S. At. Energy Comm. Rep.* KAPL–869, 1953

[29] Banus, M. D. *et al.*, *J. Am. chem. Soc.* 77 (1955) 2007

[30] Zachariasen, W. H. *et al.*, (*a*) *Acta crystallogr.* 16 (1963) 352; (*b*) 6 (1953) 393; (*c*) quoted[91]

[31] (*a*) Peterson, D. T., with Fattore, V. G. *J. phys. Chem., Ithaka* 65 (1961) 2062; (*b*) with Indig, M. *J. Am. chem. Soc.* 85 (1960) 5645

[32] Bergama, J. and Loopstra, B. O. *Acta crystallogr.* 15 (1962) 92

[33] (*a*) Wallace, W. E. *J. chem. Phys.* 35 (1961) 2126; (*b*) Zanowick, R. L. and, 36 (1962) 2059; (*c*) Wallace, W. E. *et al.*, *Pure appl. Chem.* 2 (1961) 281

[34] Pebler, A. and Wallace, W. E. *J. phys. Chem., Ithaka* 66 (1962) 148

[35] Dreyfus-Alain, B. and Viallard, R. *C.r. hebd. Séanc. Acad. Sci., Paris* 237 (1953) 806

[36] Farr, J. D. *et al.*, *J. inorg. nucl. Chem.* 18 (1961) 42

[37] Chrétien, A. *et al.*, *C.r. hebd. Séanc. Acad. Sci., Paris* 238 (1954) 1423

[38] Lenning, G. A. *et al.*, *J. Metals, N.Y.* 6 (1954) 367

[39] (*a*) Goon, E. J., with Malgiolio, J. *U.S. AEC Rep.* NYO–7547 (1958); (*b*) *J. phys. Chem., Ithaka* 63 (1959) 2018; (*c*) Goon, E. J. *et al.*, *Rev. scient. Instrum.* 28 (1957) 34

[40] Yakel, H. L. *Acta crystallogr.* 11 (1958) 46

[41] Jaffee, L. D. *J. Metals, N.Y.* 8 (1956) 861

[42] Frost, P. D. *et al.*, *Trans. Am. Soc. Metals* 46 (1954) 231

[43] Phillips, C. W. *Bull. Inst. Metals* 2 (1955) 233

[44] Lamborn, J. I. *J. Aust. Inst. Metals* 5 (1960) 86

[45] Wohlfarth, E. P. *Acta metall.* 4 (1956) 225

[46] McQuillan, A. D. (*a*) *Nature, Lond.* 164 (1949) 24, 537; (*b*) *Proc. R. Soc.* A205 (1950) 309; (*c*) *J. Inst. Metals* 79 (1951) 371

[47] Haag, R. M. and Shipko, F. J. *J. Am. chem. Soc.* 78 (1956) 5155

48 (*a*) Coogan, C. K. and Gutowski, H. S. *J. chem. Phys.* 36 (1962) 110; (*b*) with Stalinski, B. 34 (1961) 1191
49 Esser, H. and Cornelius, H. *Stahl Eisen* 53 (1933) 885
50 Gulbransen, E. A. and Andrew, K. F. (*a*) *J. electrochem. Soc.* 101 (1954) 474; 96 (1949) 364; (*b*) *J. Metals, N.Y.* 7 (1955) 136; (*c*) *Trans. Am. Inst. Min. metall. Engrs* 188 (1959) 586
51 Whitwham, P. *Mém. scient. Revue Métall.* 57 (1960) 1
52 Vaughan, D. A. and Bridge, J. R. *J. Metals, N.Y.* 8 (1956) 528
53 (*a*) Bickel, P. W., quoted by Libowitz[2c]
54 Beck, R. L. *Trans. Am. Soc. Metals* 55 (1962) 542
55 Jack, K. H., 1955 (communication to Edwards[21], cf. Libowitz[2c])
56 Espagno, L. *et al.* (*a*) *C.r. hebd. Séanc. Acad. Sci., Paris* 247 (1958) 83; (*b*) 249 (1959) 1105; (*c*) *Mém. scient. Revue Métall.* 57 (1960) 254; (*d*) 59 (1962) 182; *C.r. hebd. Séanc. Acad. Sci., Paris* 250 (1960) 4352
57 Bailey, J. E. *Acta metall.* 11 (1963) 267; *Int. Conf. Electron Microsc.* 1 (1962), paper K9
58 Ells, C. E. and McQuillan, A. D. *J. Inst. Metals* 85 (1956) 89
59 Hall, M. N. A. *et al.*, *Trans. Faraday Soc.* 41 (1944) 306
60 Douglas, T. B. *J. Am. chem. Soc.* 80 (1958) 5040
61 Schwartz, C. M. and Mallett, M. W. *Trans. Am. Soc. Metals* 46 (1954) 646
62 Motz, J. *Z. Metallk.* 53 (1962) 770
63 Martin, S. L. H. and Rees, A. L. G. *Trans. Faraday Soc.* 50 (1954) 22, 343; Rees, A. L. G. *ibid.* 335
64 Brown, A. and Hardy, D. *J. nucl. Mater.* 4 (1961) 110
65 Beck, R. L. and Mueller, W. M. *Met. Soc. A.I.M.E. Inst. Met. Div. Spec. Rep. Ser.* 10 (1960) 63
66 Peterson, D. T. and Rexer, J. *J. inorg. nucl. Chem.* 24 (1962) 519
67 Smith, D. P. *Hydrogen in Metals*, Chicago University Press, 1948
68 Roberts, B. W. *Phys. Rev.* 100 (1955) 1257
69 Butera, R. A. and Kofstad, P. *J. appl. Phys.* 34 (1963) 2172
70 (*a*) Kofstad, P., with Wallace, W. E. *J. Am. chem. Soc.* 81 (1959) 5019; (*b*) Kofstad, P. *et al.* 81 (1959) 5015; (*c*) with Butera, R. A. *J. appl. Phys.* 34 (1963) 1517
71 Waite, T. R. *et al.*, *J. chem. Phys.* 24 (1956) 634
72 Eustice, A. L. and Carlson, O. N. *Trans. metall. Soc. A.I.M.E.* 221 (1961) 238
73 (*a*) Brauer, G., with Herrmann, R. *Z. anorg. allg. Chem.* 274 (1953) 11; (*b*) with Mueller, H. *Plansee Proc. 3rd Seminar* (1959) 257
74 Aschermann, G. *et al.*, *Phys. Z.* 42 (1941) 349
75 (*a*) Samsonov, G. V., with Antonova, M. M. *Zh. fiz. Khim.* 35 (1961) 900; (*b*) *Russ. J. inorg. Chem.* 8 (1963) 685; (*c*) with Umanskii, Ya. S. *NASA tech. Transl.* TTF–102 (1962), Ch. 2
76 Wainwright, C. *Bull. Inst. Metals* 4 (1958) 68; (*b*) Wainwright, C. *et al.*, *J. less-common Metals* 6 (1964) 362
77 Trzebiatowski, U. and Stalinski, B. *Bull. Acad. pol. Sci.*, Cl. III, 1 (1953) 317
78 Horn, F. H. and Ziegler, W. T. *J. Am. chem. Soc.* 69 (1947) 2702
79 Komjathy, S. *J. less-common Metals* 2 (1960) 466
80 Vukanovic, V. M. *et al.*, *C.r. hebd. Séanc. Acad. Sci., Paris* 241 (1955) 1228
81 Saba, W. G. *et al.*, *J. chem. Phys.* 35 (1961) 2148
82 Perminov, P. S. *Dokl. Akad. Nauk SSSR* 121 (1958) 1041
83 Sellers, P. A. *et al.*, *J. Am. chem. Soc.* 76 (1954) 5935
84 Ollard, E. A. and Bradley, A. J. *Nature, Lond.* 117 (1926) 122
85 Snaveley, C. A. *et al.*, *J. electrochem. Soc.* 92 (1947) 537; with Vaughan, D. A. *J. Am. chem. Soc.* 71 (1949) 313
86 Knödler, A. *et al.*, *Metalloberfläche* 17 (1963) 161
87 Martin, E. *Arch. EisenhüttWes.* 3 (1929) 407
88 Caillat, R. *et al.*, *C.r. hebd. Séanc. Acad. Sci., Paris* 237 (1953) 812; Darras, R. and Caillat, R. *Prog. nucl. Energy* 2 (1957) 19
89 Abraham, B. M. and Flotow, H. E. *J. Am. chem. Soc.* 77 (1955) 1446; *U.S. A.E.C.D. Rep.* 3074 (1951)
90 Fried, S. and Davidson, N. *J. Am. chem. Soc.* 70 (1948) 35
91 Coffinberry, A. S., with Ellinger, F. H. *U.S. Report A/Conf.* 8/P826 (1955); with Waldron, M. G. *Prog. nucl. Energy Ser.* 5 (1956) 354
92 Potter, E. V. and Lukens, H. C. *Trans. Am. Inst. Min. metall. Engrs* 171 (1947) 401
93 Bastien, P. (*a*) *Fonderie* (1961) 131; (*b*) *Öntöde* 12 (1961) 169; (*c*) Bastien, P. *et al.*, *C.r. hebd. Séanc. Acad. Sci., Paris* 249 (1959) 1105; (*d*) with Azon, P. *Rev. Métall., Paris* 49 (1952) 1837
94 Van Oijen, D. J. and Fast, J. D. *Acta metall.* 11 (1963) 211
95 Brainin, T. E. *et al.*, *Izv. Akad. Nauk SSSR, Otdel. tekh. nauk, Metallurgiya Topl.* (1961) 115
96 Geller, W. and Tah-Ho-Sun *Arch. EisenhüttWes.* 21 (1950) 423

97 Hill, M. L. and Johnson, E. W. *Trans. metall. Soc. A.I.M.E.* 221 (1961) 622
98 Michel, A., Drain, J. and Bridelle, R. *C.r. hebd. Séanc. Acad. Sci., Paris* 238 (1954) 107
99 Janko, A. and Michel, P. *C.r. hebd. Séanc. Acad. Sci., Paris* 251 (1960) 1001
100 Wollan, E. O. *et al., J. phys. Chem., Ithaka* 24 (1963) 1141
101 Boniszewski, T. and Smith, G. C. *Phys. Chem. Solids* 21 (1961) 113
102 Van Oijen, D. J. *Phys. Chem. Solids* 23 (1962) 1173
103 Wilcox, B. A. and Smith, G. E. *Acta metall.* 12 (1964) 371
104 Worshan, J. E. *et al., Jnl Phys. Chem. Solids* 3 (1957) 303
105 Maeland, A. J. and Gibb, T. R. P. *J. phys. Chem., Ithaka* 65 (1961) 1270
106 Nace, D. M. and Aston, J. G. *J. Am. chem. Soc.* 79 (1957) (*a*) 3619; (*b*) 3623; (*c*) 3627
107 Bucur, R. *Rev. Math. Phys., Buc.* (1961) 583
108 (*a*) Mitacek, P. and Aston, J. G. *J. Am. chem. Soc.* 85 (1963) 137; (*b*) *Nature, Lond.* 195 (1962) 70
109 Schindler, A. quoted[108b]
110 Ash, R. and Barrer, R. M. *Jnl Phys. Chem. Solids* 16 (1960) 246
111 Scholtus, N. A. and Hall, W. K. *J. chem. Phys.* 39 (1963) 868
112 (*a*) Owen, E. A., with Jones, R. I. *Proc. phys. Soc., Lond.* 49 (1937) 587, 603; (*b*) with Williams, E. J. 56 (1944) 52
113 (*a*) Everitt, D. H., with Norden, P. *Proc. R. Soc.* A259 (1960) 341; (*b*) *Trans. Faraday Soc.* 50 (1954) 1077; 51 (1955) 1551
114 Lacher, J. R. *Proc. R. Soc.* A161 (1937) 525
115 (*a*) Fowler, R. H., with Smithels, C. J. *Proc. R. Soc.* A160 (1937) 37; (*b*) with Guggenheim, E. A. *Statistical Thermodynamics*, Cambridge University Press, 1939; (*c*) Guggenheim, E. A. *Mixtures*, Oxford (Clarendon Press) 1952
116 Mueller, H. and Bradley, A. J. *J.* (*Trans.*) *chem. Soc.* 129 (1926) 1669
117 Warf, J. C. and Feitknecht, W. *Helv. chim. Acta* 33 (1950) 613
118 Goodkoop, J. A. and Anderson, A. T. *Acta crystallogr.* 8 (1955) 118
119 Wiberg, E. and Henle, W. *Z. Naturf.* 7B (1952) 250
120 Carnuth, W. *Z. angew. Phys.* 15 (1963) 291
121 Trzeciak, M. J., quoted by Libowitz[2c]; *U.S. At. Energy Comm. Rep.* BMI 1112, 1956
122 Gibb, T. R. P. and Krutenat, R. C., quoted by Libowitz[2c]
123 Tannenbaum, I. R. *et al., U.S. At. Energy Comm. Rep.* NAA–SR 7132 (1962)
124 Taylor, R. E. *U.S. At. Energy Comm. Rep.* NAA–SR 7736 (1963)
125 Bokros, J. C. *J. nucl. Mater.* 3 (1961) 216, 320
126 Merton, U. *et al., Prog. nucl. Energy* 3 (1961) 3; *J. nucl. Mater.* 10 (1963) 201
127 Labaton, V. Y. *et al., J. inorg. nucl. Chem.* 24 (1962) 1197
128 Pessall, N. and McQuillan, A. D. *Trans. Am. Inst. Min. metall. Engrs* 224 (1962) 536
129 Hansen, M. *Constitution of Binary Alloys*, New York (McGraw-Hill) 1958
130 Jungkönig, W. *et al., Metall* 17 (1963) 1
131 Grushina, V. V. and Rodin, A. M. *Zh. fiz. Khim.* 37 (1963) 559
132 De Kazinszy, F. and Lindberg, O. *Jernkont. Annlr* 144 (1960) 288
133 Kurnoukhov, M. M. and Morozov, A. N. *Izv. Akad. Nauk SSSR, Otdel. tekh. nauk, Metall. Topl.* 12 (1948) 1845
134 (*a*) Tverdovsky, I. P. and Vert, Zh. L. *Dokl. Akad. Nauk SSSR* 88 (1953) 305; (*b*) Tverdovsky, I. P. *et al., Sb. Trud. gos. Inst. prikl. Khim.* 42 (1959) 182
135 Rosenhall, G. *Annln Phys.* 24 (1935) 297
136 Mueller, H. and Blackledge, J. P. *Met. Soc. A.I.M.E. Inst. Met. Div. Spec. Rep. Ser.* 10, 13 (1964) 679
137 (*a*) Gilbert, W. W. *Metal Finish.* 43 (1945) 492, 553; (*b*) Hurd, D. T. *An Introduction to the Chemistry of Hydrides*, New York and London (Wiley and Chapman & Hall) 1952
138 Elze, J. and Oelsner, G. *Metall* 16 (1963) 111
139 Darling, A. S. *Platin. Metals Rev.* 7 (1963) 126
140 Gorman, J. K. and Nardella, W. R. *Vacuum* 12 (1962) 19
141 (*a*) Andrew, J. H. *et al., J. Iron Steel Inst.* 146 (1942) 181; (*b*) with Lee, H. 165 (1950) 382
142 Bilby, B. A. and Hewitt, J. *Acta metall.* 10 (1962) 587
143 Podgurski, H. H. *Trans. metall. Soc. A.I.M.E.* 221 (1961) 389
144 Broomfield, G. H. *Nature, Lond.* 190 (1962) 824
145 (*a*) Fast, J. D., with van Ooijen, D. J. *Philips tech. Rev.* 24 (1963) 221; (*b*) *Schweiss. Schneid.* 9 (1957) 572
146 Chin-Tin, K. and Tin-Sui, K. *Soviet Phys. Metals Metallogr.* 5 (1957) 64
147 Adair, A. M. and Hook, R. E. *Acta metall.* 10 (1962) 741
148 (*a*) van der Willigen, P. C. *Schweiss. Schneid.* 9 (1957) 517; (*b*) Hummitzsch, W. *ibid.* 386
149 Hofmann, W. *et al., Acta metall.* 10 (1962) 688
150 Bishop, C. R. and Stern, M. *J. Metals, N.Y.* 13 (1961) 144
151 Brainin, I. E. and Kharchenko, V. A. *Trudӯ Donetsk. politekh. Inst., Ser. Met.* 56 (1961) 5, 27
152 Weinstein, D. and Holtz, F. C. *Trans. Am. Soc. Metals* 56 (1964) 284

10

MIXED INTERSTITIAL COMPOUNDS

10.1 INTRODUCTION

Following the discussion of the interstitial carbides, nitrides, borides, silicides, oxides and hydrides of the transition metals in the preceding Chapters, we now consider their combinations. The number of interstitial compounds and solution series possible by mutual combinations is clearly enormous. It is only intended here, however, to bring out some points of principle and to give typical examples demonstrating trends. A systematic treatment is handicapped by the fact that information on many mixed systems is still lacking or not accessible. However, the widespread isomorphies shown to exist, for example, between carbides and nitrides, and the associated inter-solubilities (where size factor and electron structure permit) help in the task. Thus pseudo-binary joins MX_m–MX'_n frequently arise between structurally similar phases. These can involve either (*a*) complete solid solutions or (*b*) limited terminal ones with a two-phase eutectic, or (*c*) formation of ternary compounds such as borocarbides, etc. In any one pseudo-binary system, contingencies (*a*), (*b*) or (*c*) could possibly all arise at different temperatures or pressures. Also, *metastable* phases are commonly formed*. Similar stoichiometry of end members, though usual, is not a necessary condition. Indeed, the occurrence of wide non-stoichiometric ranges, present already in some of the single-interstitial compounds, implies that lattice vacancies must enter as an important further variable in the mixed ones. In Chapter 2, when classifying the interstitial species, we had considered lattice vacancies as atom-equivalent units, and in this sense even the single-interstitial compound, say TiC_{1-x}, may be regarded as already a 'double' one, and an *n*-fold compound really $(n + 1)$-fold. The appearance of vacancies introduces a further freedom in stoichiometry by the addition of the new interstitial, sometimes to the extent of a complete solution junction not only between compounds but also between these and a primary interstitial solid solution in one of the pure metals. Furthermore, the known *two-metal* ternary compounds described in the preceding Chapters, such as the carbides, nitrides, oxides and intermetallics of the η-carbide-type of structure, can form mutual solution series (Fe_4W_2C–Fe_4W_2N, etc.) which clearly come under the present heading.

Table 10.1 shows the formal combinations of interstitial pairs which will serve as a guideline in this Chapter, with brief comments. *Figure 10.1* presents this at a glance and includes 'vacancies' (of metal or interstitial) as integral participants. It is important here also to recall briefly the type

* Such transient two-interstitial compounds (better termed 'solid-state conditions') arise as intermediate, non-equilibrated stages during industrial processes. Reduction reactions of oxides by carbon or hydrogen come readily to mind, as do nitriding, carburizing or boridizing processes.

Table 10.1. Two-interstitial combinations in transition-metal compounds

Combination	*Examples*	*Ultimate (metal-free) compounds*
C–N	e.g. Ti(C,N); solid solutions commonly occur	cyanogen, C_2N_2; gaseous*; non-conforming*; C and N_2 would evolve separately
C–B	ternary compounds, e.g. Nowotny type and $M_{23}(C, B)_6$	boron carbide, B_4C; solid*; very stable, hard refractory compound; conforming*
C–Si	ternary compounds, e.g. Nowotny phases	silicon, SiC; solid; very stable, hard refractory compound; conforming
C–O	limited replacements possible, e.g. Ti(C, O); (carbonates non-conforming except possibly for initiation of C–O bonding already in solution, towards CO_3 configurations?)	carbon monoxide and dioxide, CO, CO_2; gaseous; stable; conforming
C–H	limited H solutions in carbides possible (lead to C–H bonds in organic materials)	methane, CH_4, and numerous hydrocarbons; gaseous, liquid and solid; methane conforming
N–B	ternary compounds, e.g. Nowotny phases	boron nitride, BN; solid; very stable refractory compound; conforming
N–Si		silicon nitride, Si_3N_4; solid; very stable refractory compound; conforming
N–O	limited replacements possible, e.g. Ti (N,O); (nitrates non-conforming, except for possible N–O bonding already initiated)	nitrogen oxides, N_2O, NO, N_2O_4; gaseous; non-conforming, unlikely to form
N–H	limited H solutions in nitrides possible; N–H bonds, as in organic materials and NH_3 units, not included	N_3H, NH_3, N_2H_4; hydrazoic acid, unstable liquid; ammonia gaseous; hydrazine liquid; non-conforming, unlikely to form, except possibly NH_3 under special conditions
B–Si	ternary compounds and extensive solid solutions	silicon borides, SiB_3, SiB_6; solid, stable; conforming; SiB_4 recently found, isomorph to CB_4
B–O	limited intersolutions possible; (borates non-conforming, except strong B–O affinity and bond probably; will occur already initially with metal)	boric oxide, B_2O_3; solid; stable; conforming
B–H	little data available	numerous hydrogen borides from B_2H_6 to $B_{10}H_{14}$ (boranes); gaseous, liquid and solid; stable and unstable ones; non-conforming; readily decompose to B and H at elevated temperatures
Si–O	ternary compounds and partial solutions (silicates non-conforming)	silicon monoxide and silica, SiO, SiO_2; solid; SiO_2 very stable, refractory; conforming
Si–H	little data available	SiH_4 and many other silicon hydrides (silanes); gaseous or liquid; stable and unstable ones (comparable to hydrocarbons); could be conforming under some conditions, though improbable
O–H	hydroxyl (O–H) bond strong; hydroxides not here included	water; hydrogen peroxide, H_2O, H_2O_2; liquid; stable; conforming

* 'Conforming' or 'non-conforming' means that the compound does or does not stably equilibrate with the metal/two-interstitial compounds; 'gaseous', 'liquid', 'solid' apply to room temperature and atm. pressure.

of compound formed *between* the two interstitial atoms when without the metal (such as SiC, Si_3N_4, BN, CO, etc.), as this will be the phase to equilibrate and precipitate when saturation of the highest interstitial compounds is exceeded. Some of these intra-interstitial compounds are highly stable and refractory, others are unstable and equilibration with the two separate elements is preferred; this is shown in the third column of *Table 10.1.*

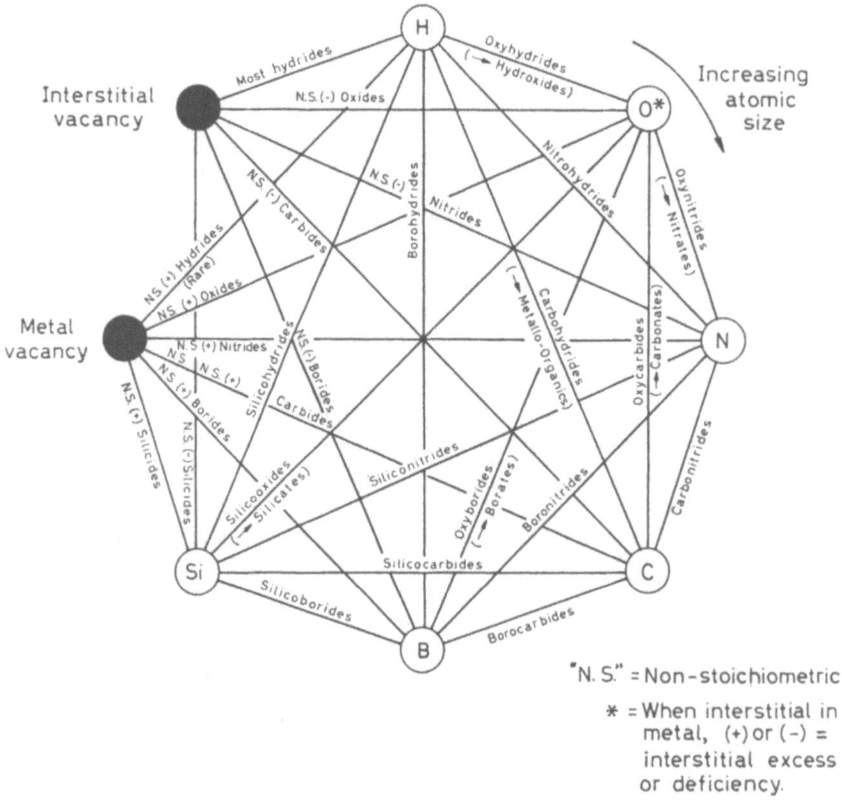

Figure 10.1. Scheme of mixed interstitial compounds

Work on mixed interstitial compounds has been intense in recent years, largely because of their special refractory interests and high-temperature applications. No account can be up to date in this constantly expanding field (but cf. Bibliography). Geach and Jones[1] have examined the interactions, on arc-melted samples, of selected pairs of carbides, borides and silicides. *Table 10.2* summarizes their results. (They also explored some of the melting points and constitutions of the reaction products between TiC, TaC, Mo_2C, $MoSi_2$, $TaSi_2$, Ta_5Si_3 Nb_5Si_3, Mo_2B, TiB_2 and ZrB_2 mutually partnered, in varying proportions.)

Methods of dispersion strengthening and induction of other physical effects in alloy matrices had given practical interest to such heterogeneous systems. In the *ferrous* field, the pairing of interstitials has an extremely important influence; the work of Jack[2] on iron carbo-nitrides, and the effect of boron–carbon and silicon–nitrogen interactions are but some examples.

Table 10.2. Intersolubilities of refractory phases
(From Geach and Jones[1])

	Mo_2B	TiB_2	ZrB_2	Mo_2C	TaC	TiC
$TaSi_2$	*C*	*O*	*O*	*C*	*C*	*O*
Ta_5Si_3	*C*	*C*		*C*	*C*	
Nb_5Si_3	*C*	*C*		*C*	*C*	
$MoSi_2$	*C*	*O*	*O*	*C*	*C*	*O*
TiC	*O*	*O*	*O*	*O*	*M*	
TaC				*O*		
Mo_2C	*O*	*C*	*C*			
ZrB_2		*M*				

M = completely soluble, *O* = partially soluble, *C* = compound formation

In considering the compounds in the order shown in *Table 10.1*, the order of the Periodic Table will be followed within each class with, however, some cross-comparisons; each Section is headed by a brief comparative summary of the non-metal atom pair concerned.

10.2 CARBONITRIDES

	C	N	
Atomic radii, Å	0·77	0·70	*Ratio* (C/N) ≈ 1·1
	(varying with metal surround)		
Group No.	IVB	VB	
Electron structure*	$(1s)^2(2s)^2(2p)^2$	$(1s^2(2p)^2(2p)^3$	

* Here and later, bracketed figures signify the orbital states, indices the number of electrons in this state.

The transition–metal carbides and nitrides are so similar, structurally and in the size of the interstitial atom, that large intersolubilities prevail, almost as extensive as between the carbide/carbide and nitride/nitride pairs (cf. *Figure 4.3*). Ternary compound formation does not arise, except for a tendency in the later (iron) Groups towards metastable structures (carbonitrides of 'Jack' type).

Constitutional surveys of Group IV/V carbonitride systems have been made by Duwez and Odell[3], Kotelnikov[4] (including solutions with borides and silicides), and Bittner *et al.*[5]; their findings are summarized in *Table 10.3.*

As TaN is of HCP structure, in contrast to that of TaC and the other compounds, its incomplete solution is explained. This puts tantalum in a different category from niobium and already in a borderline position with regard to carbide/nitride relations. It seems probable that HfC and HfN as partners are qualitatively similar to ZrC and ZrN. *Figure 10.2* shows the lattice-spacing curves[3,5]; distinct but variable deviations from Vegard's law should be noted. Also included is a recent spacing determination[6] in the system 'VC–VN' for the metal-rich and metal-poor ends of the stoichiometric range. The magnetic susceptibilities have been determined in some of these systems[5].

Table 10.3. Solubilities of Group IV–V carbides and nitrides

	Solid solutions	
complete	*extensive*	*small or none*
TiC → TiN, ZrN, VN, NbN VC → TiN, VN, NbN NbC → TiN, VN, NbN[3,4] ZrC → ZrN[4]	TiC → TaN (to 70 mol. %) ZrC → TaN (to 80 mol. %)[5]	VC → ZrN, consistent with atomic size relations and the '15% difference' rule; for NbC–ZrN (*b*), little or no interreaction was obtained[3,4] even at highest temperatures; TaC → TaN[3,4]
TiC → ZrC, NbN ZrC → ZrN, NbN NbC → ZrN(*a*), NbN TaC → NbN } [5]		

* Results (*a*) and (*b*) are in evident conflict; the complete solution appears to be the more probable reaction.

One point of principle arises here. Although, *in equilibrium*, TiC–TiN and similar series are completely soluble, this may not always materialize in practical alloys. The titanium carbonitrides in alloy steels, e.g. of Fe–Cr–Ni base, are frequently found to be the two-phase, Ti(C,N) and Ti(N,C), with spacings mutually approaching but without solution having been attained. In the metallic mass, such phases can also be observed to separate physically, depending on their thermal origin, which indeed would counter the chance of diffusion[7,8]. Thus, two distinct Ti(C,N) can be extracted from a steel, at different C/N ratios and spacing ranges[9].

In the Ti–C–N system, the Ti-rich region (C and N solubilities in α and β-Ti) are known for different temperatures[10] (*Figure 10.3*; Stone and Margolin's work[10] also included Ti–C–O and Ti–N–O; see below). Although strictly concerned only with the primary solutions, we may here recognize in the large and nitrogen-extended α-solid solutions the precursor, in Group V, of the HCP (Nb_2C type) carbonitride series, to which the above Ti primaries may conceivably join. The α-Ti(C,N solution) $\rightarrow$ Ti $(C,N)_{1-x}$ phase assembly should also be viewed in conjunction as a change in stacking sequence, with all the possible implications of intermediate stacking-fault effects and dislocation locking by interstitial/vacancy interaction.

With regard to applications, the TiC–TiN system is also important in sintered hard metals where the presence of some TiN solution in the titanium carbide is general. Also 'TiC' or 'TiN' inclusions in alloy steels are generally Ti(C,N).

It is instructive in this connection also to show the full Nb–NbC–NbN and Ta–TaC–TaN diagrams[6b,c] with a recent one[6a], of V–VC–VN (*Figure 10.4*). Firstly, they contrast with Ti–TiC–TiN by the restricted primary metal solubilities, and structurally by this 'reappearance' of the HCP phase (α-Ti solid solution) in the form of the hemicarbide series. Secondly, the M_2(C,N)

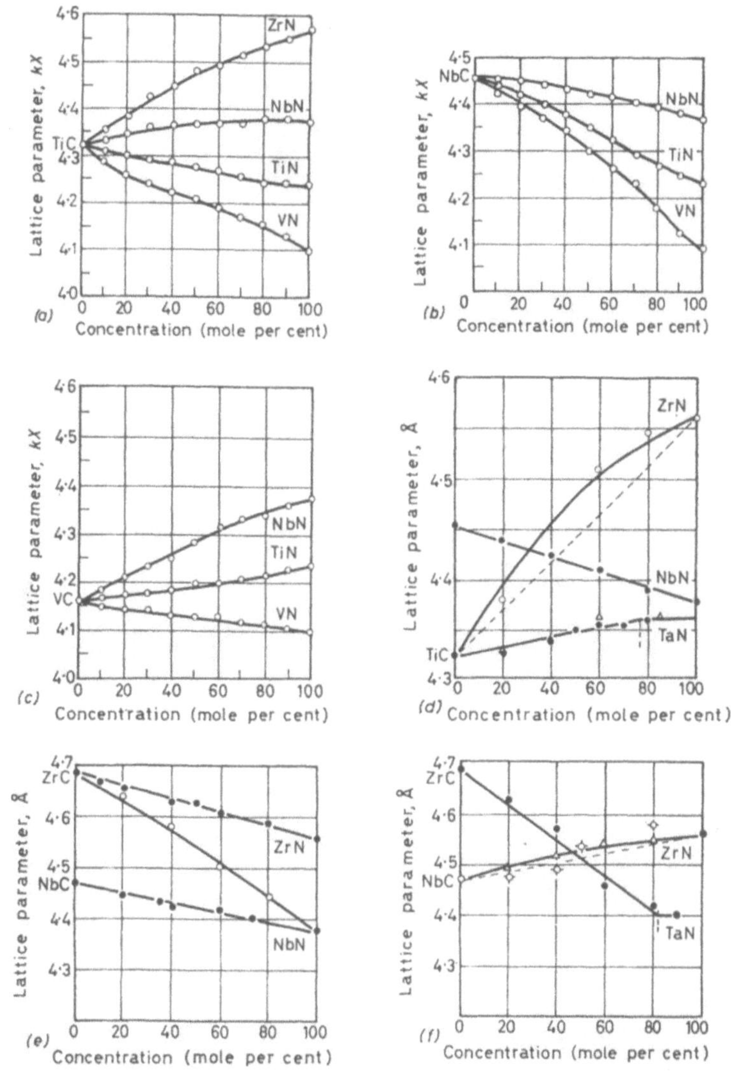

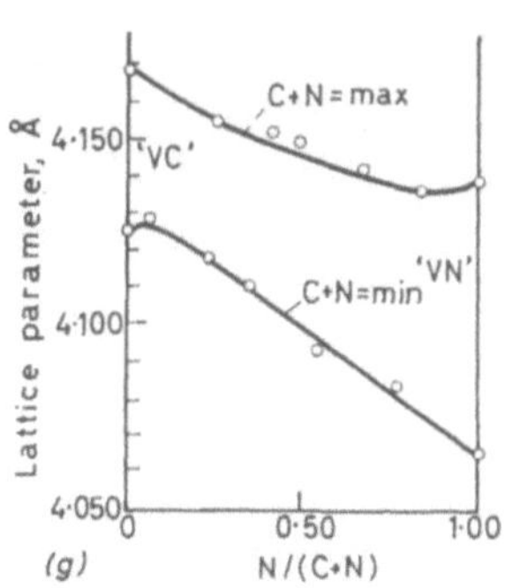

Figure 10.2. Lattice-parameter curves in carbide–nitride systems of Groups IV–V (a–f); (g) 'VC'–'VN' system for non-stoichiometric extremes (From Duwez and Odell[3] (*a–c*); Bittner *et al.*[5] (*d–f*); Brauer and Schnell[6] (*g*))

Spacings *a-c* in *kX*, *d-f* in Å; conversion factor 1:1·002

By courtesy of The Electrochemical Society, Inc., of Springer-Verlag, Vienna and of Elsevier Publ. Co., Amsterdam

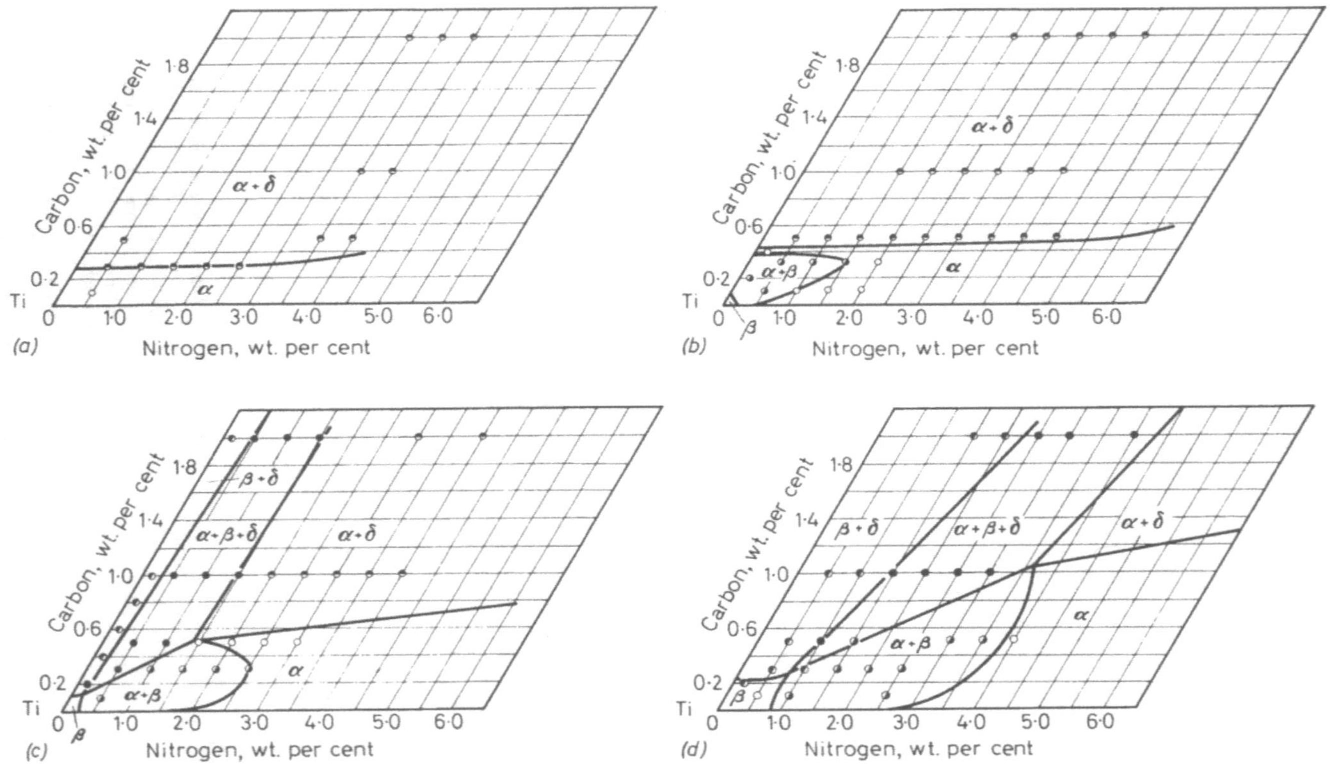

Figures 10.3a–d. Ti–C–N, Ti–C–O *and* Ti–N–O *phase equilibria* (From Stone and Margolin[10])

Isothermal sections in the systems: (*a*)–(*d*) Ti–C–N at (*a*) 800°C, (*b*) 900°C, (*c*) 1,000°C, (*d*) 1,300°C; (*e*)–(*h*) Ti–C–O system, at (*e*) 800°C, (*f*) 900°C, (*g*) 1,000°C, (*h*) 1,300°C; (*i*)–(*k*) Ti–N–O system, at (*i*) 1,000°C, (*k*) 1,400°C

By courtesy of the *Journal of Metals, New York* and the Metallurgical Society of AIME

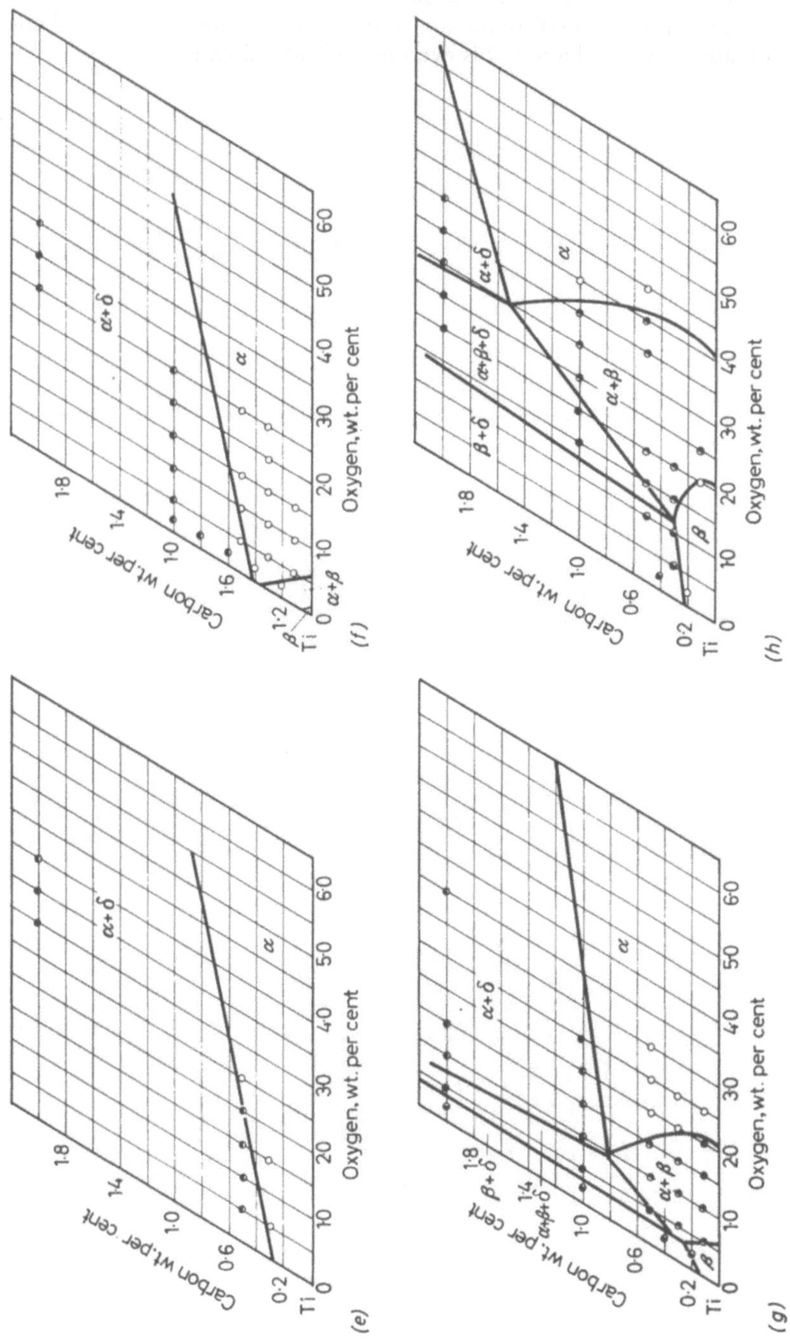

Figures 10.3e–h

series, whilst continuous for niobium, is no longer so for tantalum which evolves further to the complete breakdown of solution in the tantalum *mono*carbide series. A small reduction in the non-stoichiometry of NbN and structure anomaly foreshadow this even in niobium already.

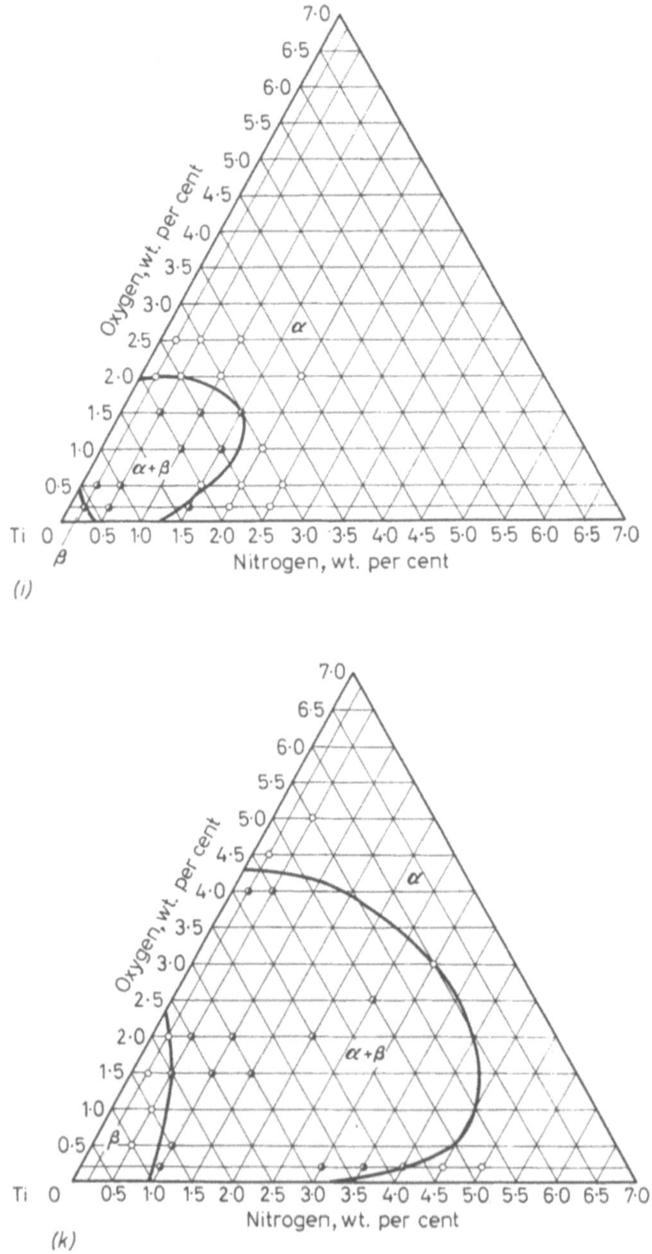

Figures 10.3i–k

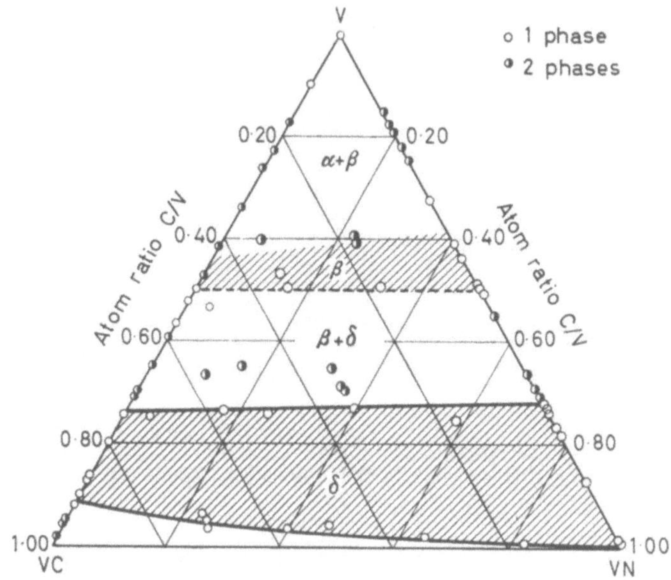

Figure 10.4. Ternary diagrams V–VC–VN, Nb–NbC–NbN *and* Ta–TaC–TaN (From Brauer and Schnell[6a], Brauer and Lesser[6b,c])

(*a*) V–VC–VN, Section for 1,100°–1,400°C range

(*b*) Nb–NbC–NbN } for 1,250°–1,450°C range

(*c*) Ta–TaC–TaN

By courtesy of Elsevier Publ. Co., Amsterdam and the Editor, *Zeitschrift für Metallkunde*

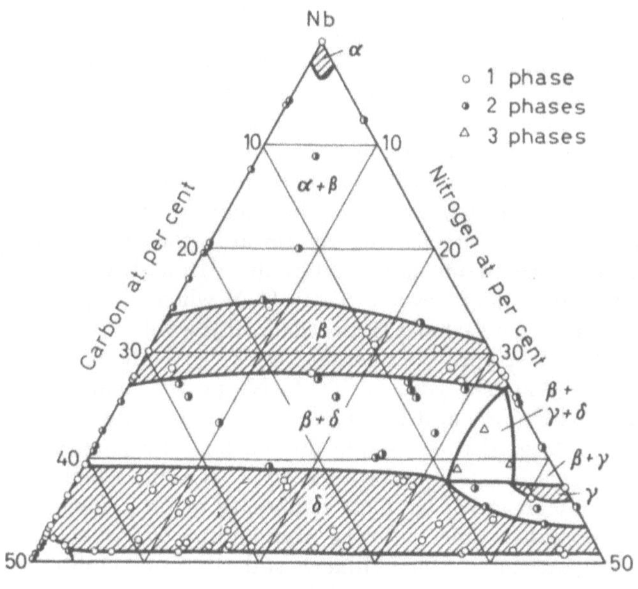

Figure 10.4b

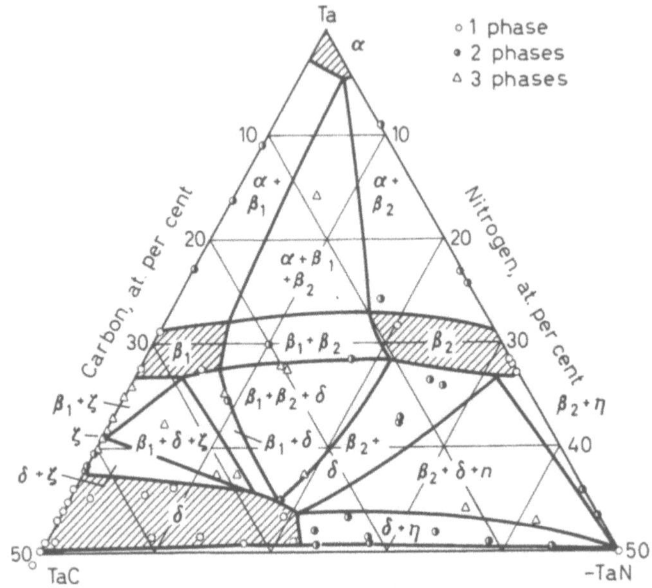

Figure 10.4c

The mutual solubilities of Group IV and V monocarbides and nitrides may then be represented as in *Figure 10.5* which also includes corresponding carboboride and other combinations for comparison.

Little seems to be known of the Group VI metal carbonitrides, except for limited work on kinetic reaction diffusion in the Mo–C–N system[11], indicating a HCP solution series of $Mo_2(C,N)$, and in the W–C–N system[12], incorporating both the W_2C and WC structures with N in solution. The full ternary systems, as that of Cr–C–N, still seem to require much more investigation, except for what might be predicted from known isomorphies, etc. The N solubility in $Cr_{23}C_6$, Cr_7C_3 and Cr_3C_2, likely to be considerable, is not known (the dissimilarity of Cr nitrides, though no doubt having extensive intersolubilities with the carbides, may curtail new ternary structures). Again, a critical position will be occupied by the chromium-base system which probably tends already towards the type of carbonitride transformations accentuated later in Groups VII and VIII; in fact, the Jack-type structures might evolve in Cr–C–N when nitrogen solubility in the primary chromium carbides is exceeded; this would invite investigation.

On the U–C–N system, considerable work by Williams and Sambell[13] and Street and Waters[14] (also on Th–C–N) proved that the continuous solid solutions ThC–ThN and UC–UN exist, with lattice-spacing changes as in *Figure 10.6*. Vegard's law is well obeyed for solid-state diffused alloys, but there are deviations from it for fused UC–UN alloys which could suggest a tendency to segregation at high temperatures or a non-stoichiometric effect. In regard to the diffusion mechanism, *near*-stoichiometric UC possesses strong interatomic forces that resist the diffusion of nitrogen[14], whilst ThC and ThN, which are less metallic, show more rapid interdiffusion.

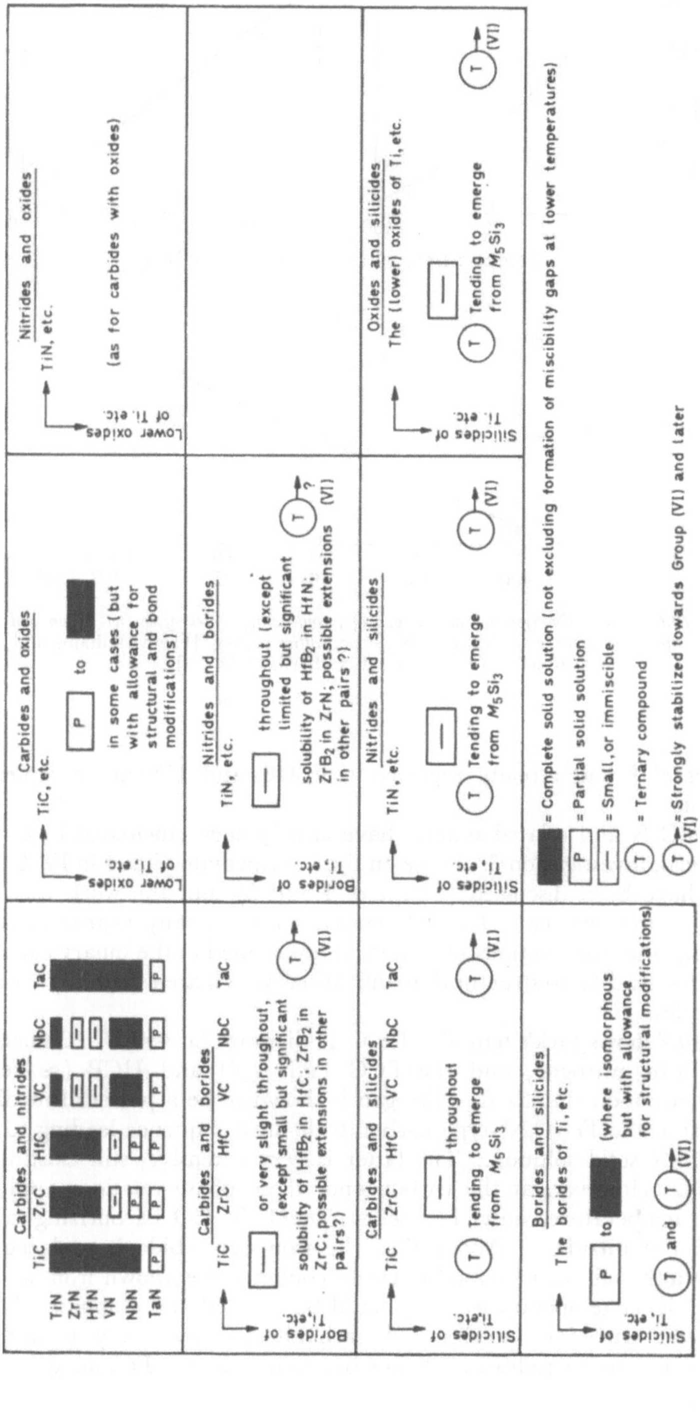

Figure 10.5. Intersolubilities of paired interstitial compounds of Groups IV and VA

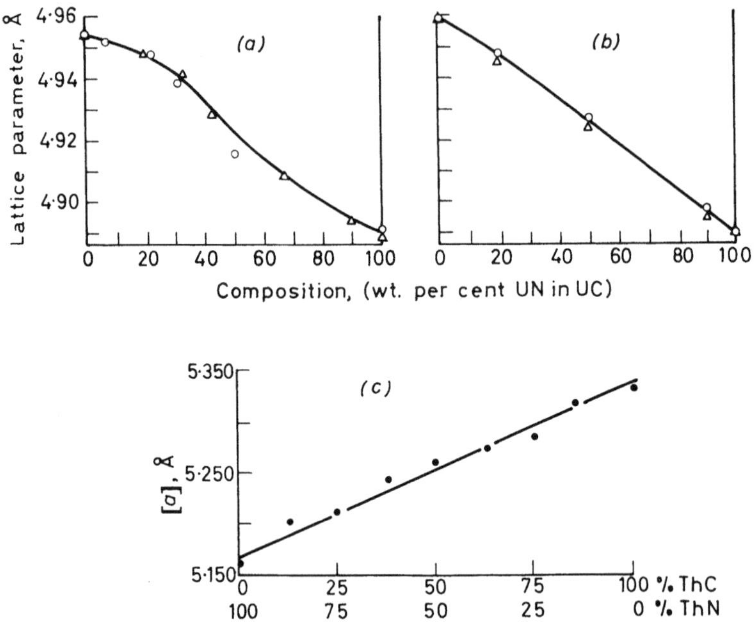

Figure 10.6. UC–UN *system, lattice spacings, contrasting states after* (*a*) *fusion and* (*b*) *diffusion in the solid;* (*c*) ThC–ThN *system, lattice spacings* [From Williams and Sambell[13] (*a, b*), Street and Waters[14] (*c*)]

By courtesy of Elsevier Publ. Co., Amsterdam

Rundle *et al.*[15] had already reported that UC and UN are completely intersoluble.

The Fe–C–N and related systems have largely been elucidated in Jack's, now classical, investigations[2]. Those on the intermediate phases in Fe–C and in Fe–N have been discussed (Chapters 4 and 5), but the work was also extended to the ternary system. Extensive, but critically heat-treatment-dependent, primary solutions of N and C are formed in the binary carbides and nitrides, leading to structural modifications which are essentially ternary carbonitrides.

Figure 10.7 shows Jack's tentative Fe–C–N diagram (at 450°C). Cementite dissolves little nitrogen, and the FCC Fe_4N (γ') and HCP ($\approx Fe_3N$) structures are soon stabilized. Nitrogen is shown to be a powerful medium in stabilizing the Fe_2(C,N)-type carbonitride in a sequence leading to the straight Fe_2N solid solution. The latter is very extensive, still existing at $\approx Fe_8C_3N$. However, at the carbon end, these phase relations apply to moderate temperatures only, i.e. below 500°C (in CO carburizing treatments of iron nitrides). Above this, the iron percarbide is replaced by cementite although, at increased nitrogen contents, the known iron nitrides persist to higher temperatures. (It should be recalled and emphasized that all these Fe carbides, nitrides and carbonitrides are really metastable structures, the true equilibrium being between α- and γ-Fe and graphite/nitrogen gas.)

For a discussion of detailed structures and lattice-spacing changes of the phases observed (ζ, χ, ε, γ' carbonitrides), Jack's original papers[2] should be referred to (especially that of 1948). One significant observation is that the effective atomic radius of C becomes less than that of N in the ε- and ζ-carbonitrides (overall contraction as C replaces N, though anisotropically; in the lower (C,N) phases (cementite/Fe_4N) carbon still appears as the larger atom, as in the earlier transition-metal carbonitrides, e.g. Ti(C,N), and as is normally assumed).

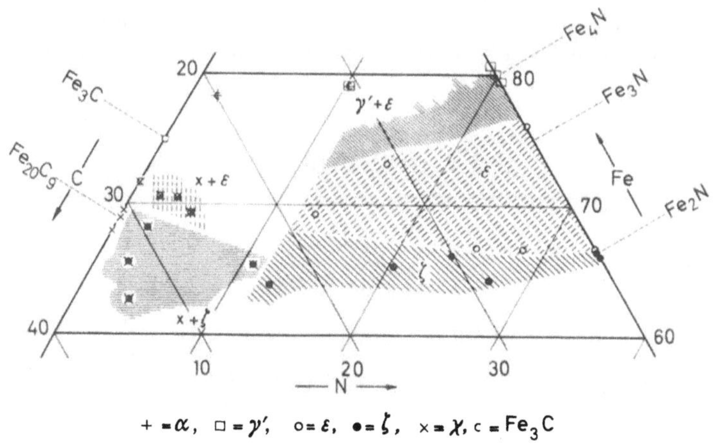

Figure 10.7. The ternary diagram Fe–C–N (20–40 at. per cent C and N range) for 450°C (From Jack[2a])

By courtesy of The Royal Society

These reactions have obvious and far-reaching consequences in steels, such as in carburizing and nitriding reactions for surface hardening. The Fe carbonitrides are also closely associated with the *nitrogen* and *carbon* martensites. The latter form a continuous solution themselves, but upon low-temperature tempering can nucleate the carbonitrides as fine precipitates: a process which critically depends on temperature, time, pre-heat treatment and, of course, on the C/N ratio and the degree of diffusion attained. This can be further varied and controlled in important ways by *alloy additions* to a steel. *Manganese* similarly forms carbonitrides[2b], whilst *cobalt* carbonitride, γ-Co_3 ($C_{0.18}N_{0.87}$) also exists[2d], isomorphous and continuously soluble with the hexagonal γ-Co_3N (Chapter 5) and ε-Fe_3N. Jack[2c] further observed a nickel cyanonitride, ε-Ni_3(C,N), isomorphous with ε-Fe_3N. It seems likely that the simpler hexagonal and cubic structures favoured by the Fe Group nitrides as against carbides will be further stabilized by the addition of earlier transition metals, Cr and, as far as soluble, Mo and V (cf. Bibliography).

These complex carbonitrides are capable of nucleating from the matrix at grain boundaries, sub-grain boundaries and, in incipient form, as (C + N) atom atmospheres around dislocations (Cottrell), with considerable influence in terms of strength of the material. In electron microscope work on thin films[16] of this type of material, distinct 'ridges' have been observed at crystal boundaries, produced when carbonitride particles intersect a surface during electro-thinning, and around them by aggregation of C and N atoms.

We will not here discuss any *non*-transition-metal carbonitrides, except perhaps to note the existence of stable compounds[17] such as Al_5C_3N (hexagonal structure, $a = 3{\cdot}280$, $c = 21{\cdot}55$ kX; $M = 2$) which could come into play in Al-bearing steels.

In general, we should observe that the loss of stability and disappearance of compounds towards the later transition metals, as found for the carbides and nitrides, clearly also applies to the carbonitrides. This is not necessarily so for other mixed interstitial compounds, where the interstitial pair of atoms is less similar, and ternary compounds can in fact be more stable than the binary ones.

10.3 CARBOBORIDES

	B	C	B/C
Atomic radii, Å	0·86	0·77	*Ratio* ≈ 1·1
	(variable, subject to metal surrounds)		
Group No.	IIIB	IVB	
Electron structure	$(1s)^2(2s)^2(2p)^1$	$(1s)^2(2s)^2(2p)^2$	

With the carbon atom always interstitial in transition metals, the boron atom still generally so but borderline, the difference is sufficient to have major effects both on structure and on the properties of alloys incorporating C and B. Ternary carboborides are liable to occur, but more so in the later than the earlier Groups.

Group IV—The Ti–C–B system contains no ternary compound, and the highly stable TiB and TiB_2 equilibrate[18a] directly with TiC (*Figure 10.8a*). There is hardly any intersolubility, and the stable coexistence of the near-pure carbide and borides is a dominant feature. The hard and stable boron carbide, B_4C, enters upon boron excess, as the alternative to graphite; it equilibrates with TiB_2 rather than with the carbide[19]. $TiC + B_4C$ mixtures reacted above 1,920°C to give TiB_2, free carbon and a Ti boride, yet higher in B than TiB_2 (this, however, has not been confirmed). The pseudo-binary section TiB_2–B_4C forms a simple eutectic[20] at 1,900°C 70 mol. per cent B_4C. Geach and Jones[1] and Samsonov[21a] have also studied the pseudo-binary section TiC–TiB_2. It should be noted how strongly the TiC_{1-x} phase and the coexisting TiB and TiB_2 phases contrast in their ranges of non-stoichiometry—a direct reflection of the structural differences between the borides and carbides discussed.

The ternary systems Zr–C–B and Hf–C–B are essentially similar[18b] to Ti–C–B (*Figure 10.8b* and *c*). Again, the MC_{1-x} carbides equilibrate with the MB_2 borides; surprisingly, however, no 'ZrB' analogous to TiB and HfB occurs. One interesting feature is that the cubic carbide is now capable of dissolving limited amounts of boron, increasingly so in the sequence $TiC_{1-x} \rightarrow ZrC_{1-x} \rightarrow HfC_{1-x}$ and with rising x. This shows that the defective

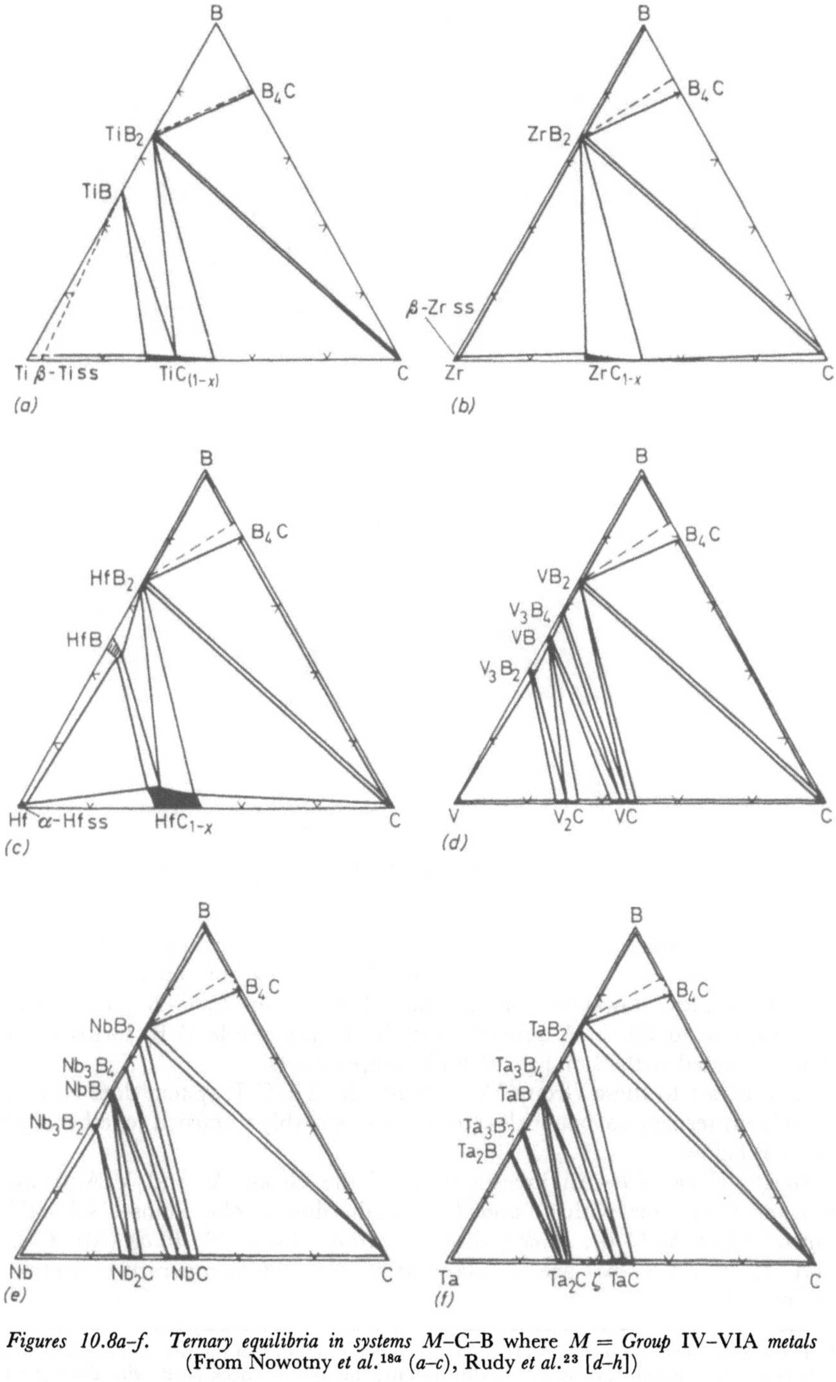

Figures 10.8a–f. Ternary equilibria in systems M–C–B where *M = Group* IV–VIA *metals* (From Nowotny *et al.*[18a] (*a–c*), Rudy *et al.*[23] [*d–h*])

(*a*) Ti–C–B (1,400°C section); (*b*) Zr–C–B (1,400°C); (*c*) Hf–C–B (1,500°C); (*d*) V–C–B (1,450°C); (*e*) Nb–C–B (1,750°C); (*f*) Ta–C–B (1,750°C); (*g*) Mo–C–B (1,800° and 1,300°C); (*h*) W–C–B (1,700°C)

By courtesy of the Editor *Zeitschrift für Metallkunde* and of Springer-Verlag, Vienna

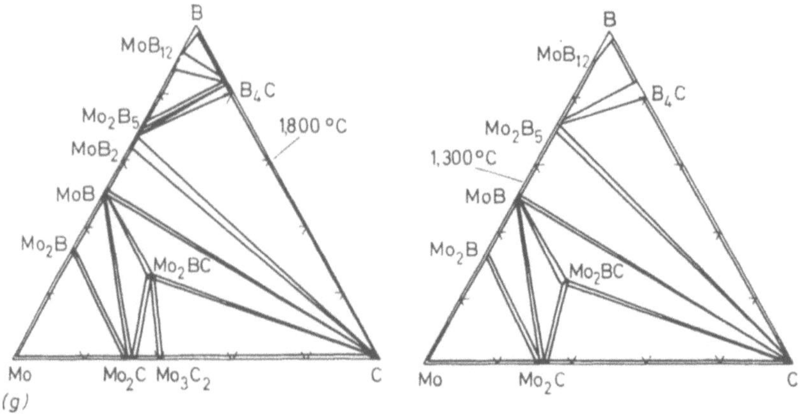

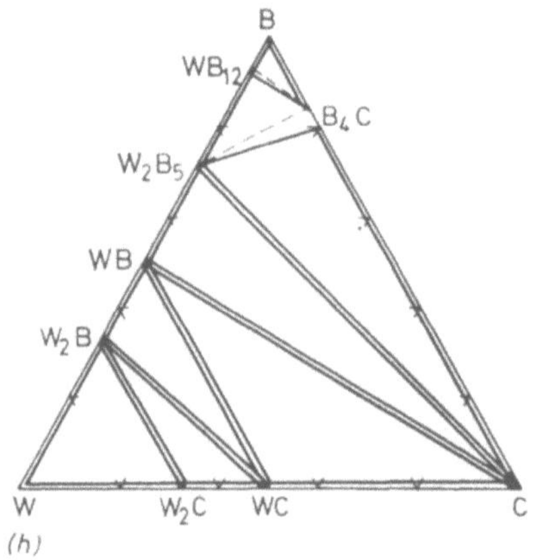

Figures 10.8g,h

carbon-lattice sites in the carbide permit a limited occupation by boron atoms, for HfC up to 6 at. per cent B. No converse solubility of C occurs in the borides, however. Experimental work on Zr–C–B and related systems, due to Glaser[22], proved that the highly stable ZrB_2 forms when ZrC is reacted with B or B_4C at high temperatures.

In contrast to these Group IV systems, the Th–C–B system does contain prominent ternary carboborides (to be more suitably discussed together with U–C–B below).

Groups V and VI—An investigation of the Group V and VIA transition-metal systems with C and B is again due to the Plansee school[23]; *Figures 10.8d* (V–C–B), *10.8e* (Nb–C–B), *10.8f* (Ta–C–B), *10.8g* (Mo–C–B) and *10.8h* (W–C–B) are representative for the temperature sections indicated*.

* The Plansee work on ternaries also included redetermination of the relevant carbides and borides, their structures and lattice dimensions; the results check fairly with those given in previous Chapters; it should be stressed that this work throughout applies to hot-pressed powder samples, and discrepancies in equilibration from the melted samples are not impossible.

No ternary compounds occur in the Group V base systems, and the binary carbides and borides directly come to equilibrium as shown. The close similarity between the V, Nb and Ta bases is evident. There is negligible intersolubility, and the carbide homogeneity ranges again divide their equilibria with the sequence of stoichiometric borides*.

In *Group VIA*, the Mo–C–B system, in contrast, is characterized by the intervention of a ternary compound, Mo_2CB; this is accompanied by a much stricter adherence to stoichiometry, as found in the carbides. The new ternary carbide is very stable, hard and of high melting point ($\approx$ 2,800°C) without decomposition. The structure of Mo_2BC, recently solved, showed an orthorhombic unit cell, with $a = 3{\cdot}086$, $b = 17{\cdot}35$, $c = 3{\cdot}047$ Å; $M = 4$. The detailed structure (*Figure 10.9*) proves to be a compromise between the MoB and the various MoC types. Indeed, this ternary phase is effectively a

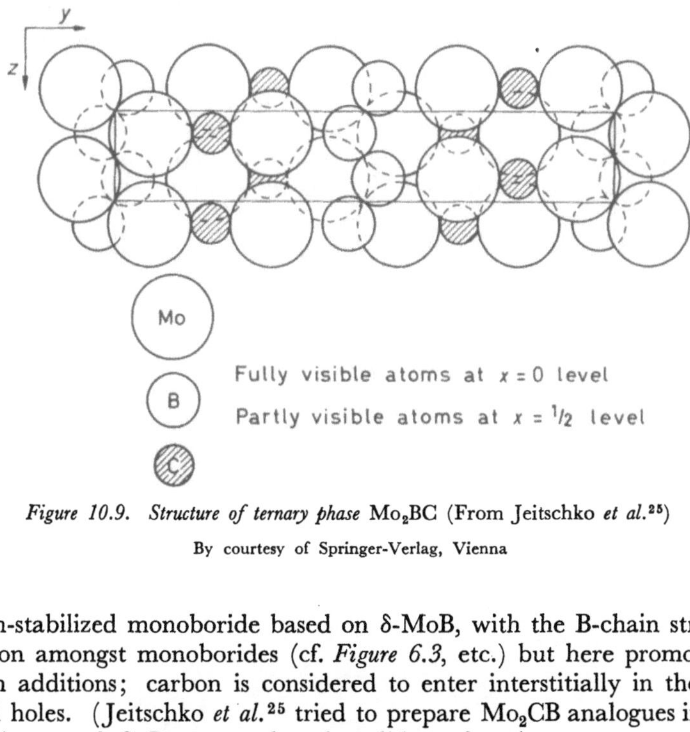

Figure 10.9. Structure of ternary phase Mo_2BC (From Jeitschko *et al.*[25])

By courtesy of Springer-Verlag, Vienna

carbon-stabilized monoboride based on δ-MoB, with the B-chain structure common amongst monoborides (cf. *Figure 6.3*, etc.) but here promoted by carbon additions; carbon is considered to enter interstitially in the octahedral holes. (Jeitschko *et al.*[25] tried to prepare Mo_2CB analogues in other transition metal–C–B systems, but they did not form.)

In the W–C–B system (*Figure 10.8h*), no analogous ternary compound exists. This seems anomalous but may well be linked with the familiar 'MoC' problem, that is, the known structural instability in molybdenum monocarbide (cf. Chapter 4), in contrast to the stability of WC which equilibrates with WB; we might consider the ternary Mo_2CB formation as an endeavour to defer the destabilization of the monocarbide (or,

* Technical developments of refractory materials on Ta borocarbides and Ti boronitrides are due to Forney and Marshall[24]. Generally, materials based on the present systems have received much active attention in applied research.

conversely, as the attempt of boron effectively to produce a stable 'Mo monocarbide' structure).

Viewing the Group IV → V → VIA borocarbide systems as a whole it also emerges that the *di* borides, which are the only ones to equilibrate with carbon in the first two Groups, allow the lower borides to do so in Group VIA, because now the carbides become so much less stable.

Glaser's contributions to the Group IV–VIA metal/C/B systems[22a,b] showed that the diborides generally have greater stability than the carbides of the same metals, and proved it to be complete in contact with carbon. These and other papers[26] represent pioneering experiments on the carbide/boride interreactions, apart from stabilities of the compounds formed.

Little systematic work is so far known on the Cr–C–B system, but the high capacity of the $Cr_{23}C_6$ carbide to dissolve boron (to form a ternary Cr borocarbide) is now well established and technologically important. The boron-atom positions in this structure have been determined by Stadelmaier *et al.*[27] (their paper includes a discussion of electron structure of this and other $M_{23}C_6$ isomorphs); cf. Chapter 6.

Actinides—The U–C–B system[28a] is shown in *Figure 10.10*, together with Th–C–B as the companion actinide system[28b]. Both are characterized by ternary compound formation. For uranium this is of approximate composition UCB, but with considerable homogeneity range and equilibrating

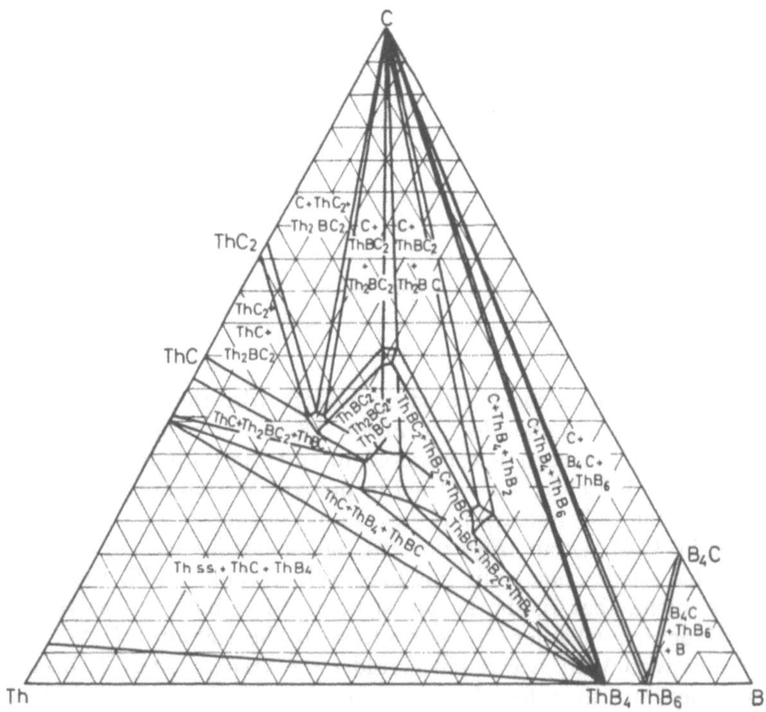

(a)

Figure 10.10. Ternary systems Th–C–B *and* U–C–B (From Toth *et al.*[18])

By courtesy of Springer-Verlag, Vienna

UC, UC_2, carbon, UB_2 and UB_4. On the other hand, pseudo-binary equilibria occur between UC and UB_2, and between UB_4 and graphite. The new ternary phase clearly lies on the UC_2–UB_2 pseudo-binary section; indeed, we may well regard its formation as a structurally necessary transitional stage of boron- and carbon-atom distribution, intermediate to that of the end members. The structure[28a] of UCB (*Figure 10.11*) shows an orthorhombic unit cell with space group D_{2h}^{17}, spacings (Å)

$a = 3{\cdot}591$	$a = 3{\cdot}591$
$b = 11{\cdot}95$	$b = 11{\cdot}96$
$c = 3{\cdot}372$	$c = 3{\cdot}346$
at. per cent: 25 B, 40 C	40 B, 25 C

It is related to both the CrB and $ZrSi_2$ types.

The Th–C–B system is yet more complex. Four ternary carboborides have been found, $ThCB_2$, ThCB, ThC_2B and Th_2C_2B, with further ones possible (for equilibria, see *Figure 10.10*). Of these, $ThCB_2$ is of hexagonal structure related to the C–32 type met in binary borides, and with the spacings (Å)

at high-carbon boundary	at high-boron boundary
$a = 3{\cdot}868$	$a = 3{\cdot}860$
$c = 3{\cdot}810$	$c = 3{\cdot}793$

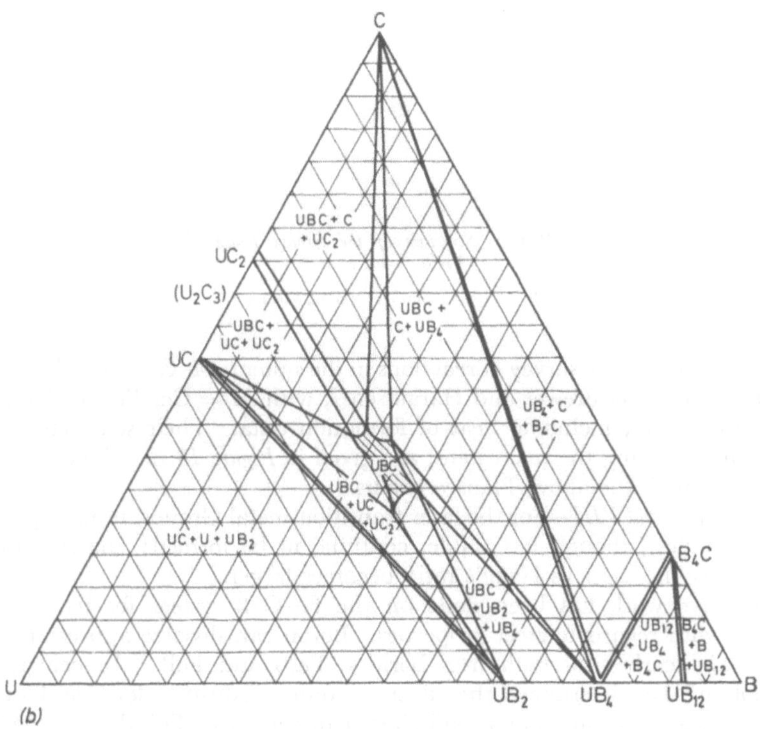

Figure 10.10b

The other structures are not yet solved, but ThCB is closely related to UCB. Otherwise the complete insolubility of C in the borides and of B in the carbides, both for thorium and uranium, is a feature to be noted.

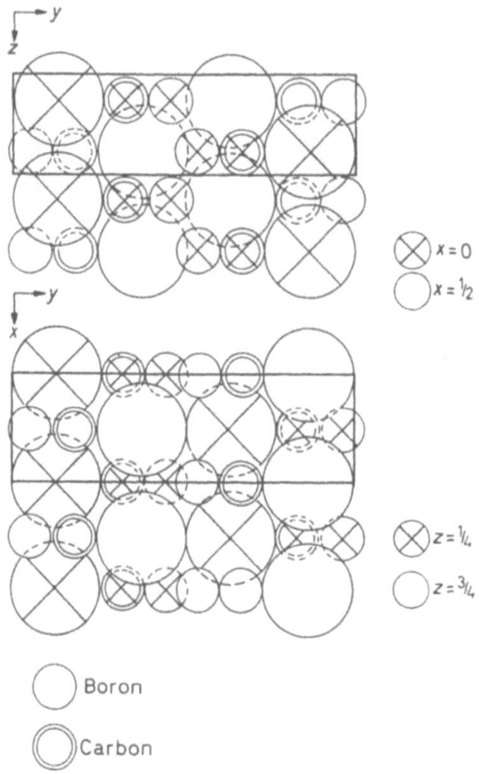

Figure 10.11. Structure of the ternary phase UBC (From Toth *et al.*[28])

By courtesy of Springer-Verlag, Vienna

A thermodynamic survey of reactions in the systems *M*–C–B (and *M*–N–B) has been given by Brewer and Haraldsen[29], with *M* = Ce, Ti, Zr, Th, Nb, Ta, Mo and W, including heat of formation data. Their semi-schematic diagrams, showing the main joins, are given in *Figure 10.12*, which may be compared with the above Plansee results.

Groups VII and VIII—For the later transition-metal Groups, ternary phases (borocarbides), though less stable, continue to be formed; the structural complexity increases as Hägg's ratio becomes larger.

Extensive work has been done on the Fe–C–B system for its obvious interest with regard to steel properties. Boron can dissolve appreciably in Fe_3C, and carbon in both iron borides, Fe_2B and FeB. In cementite, Nicholson[30] has determined the effect of boron substitution for C and found it to cause a significant alteration in lattice dimensions. This does not occur, as might have been expected, by a steady expansion but anisotropically, by

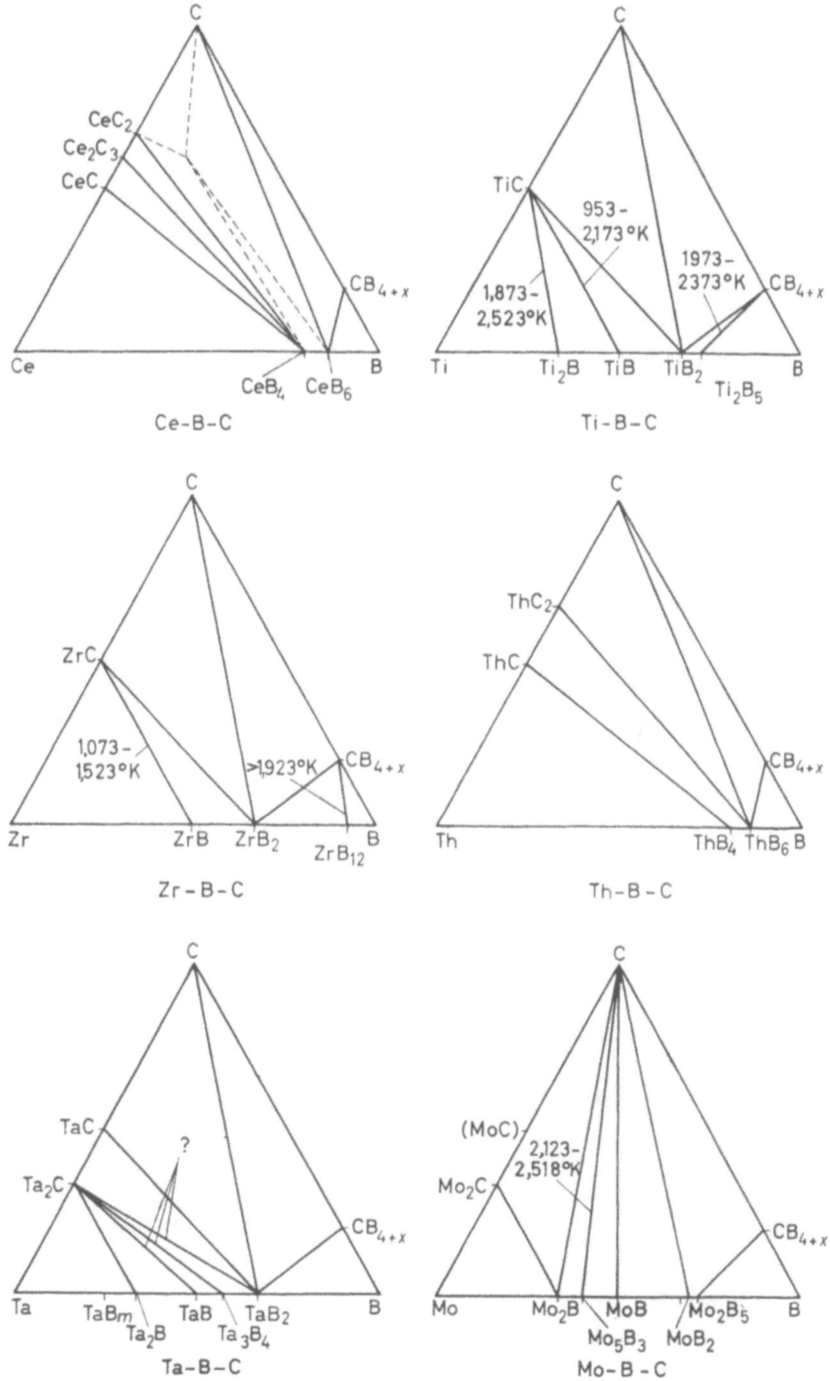

Figure 10.12. Ternary equilibria M–B–C and M–B–N (From Brewer and Haraldsen[29])

By courtesy of The Electrochemical Society, Inc.

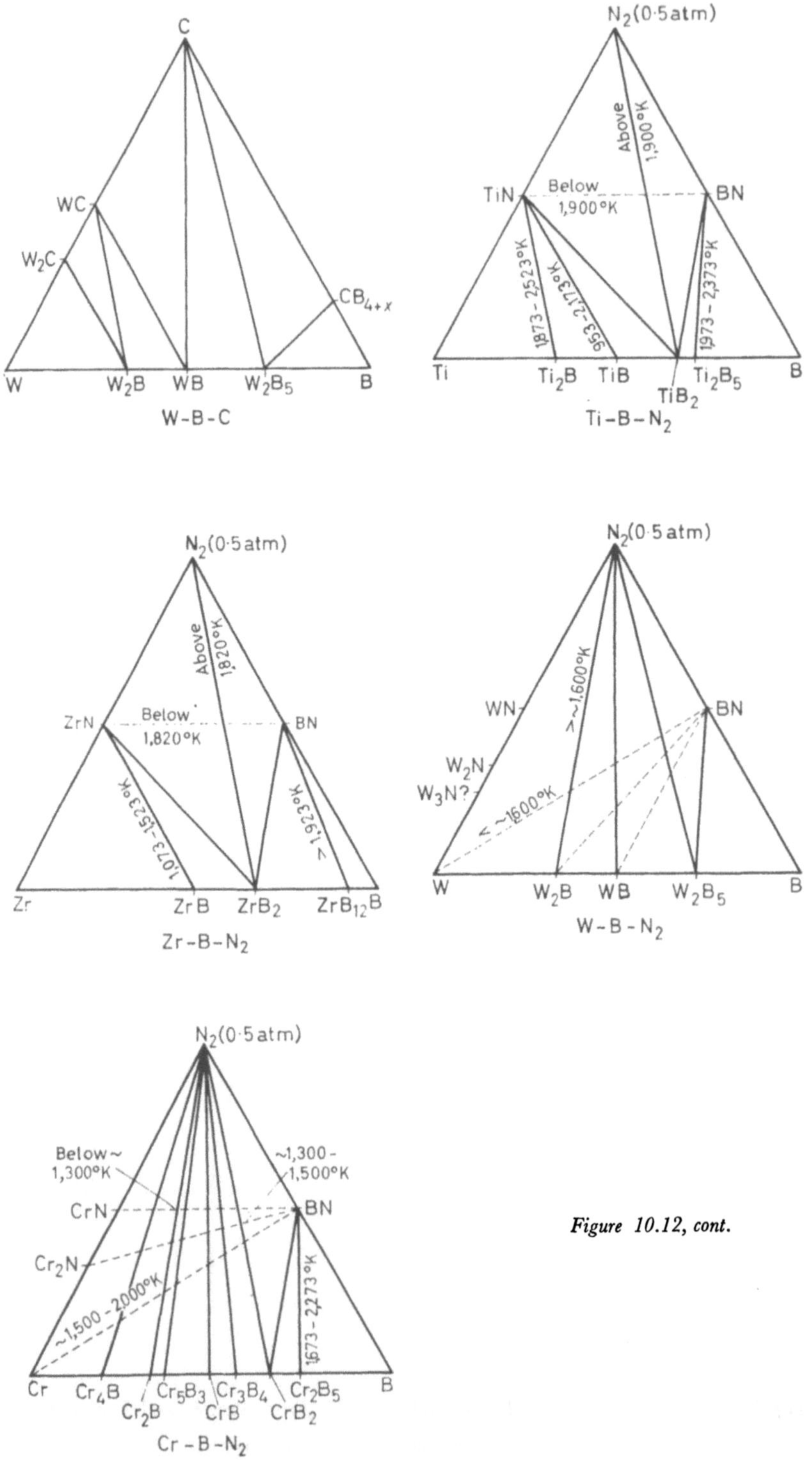

Figure 10.12, cont.

a contraction along the *a* and *c* axes ($a = 4{\cdot}5150$–$4{\cdot}4500$ Å for no boron, $c = 6{\cdot}7480$–$6{\cdot}6685$ Å for maximum boron) and an expansion along the *b* axes ($5{\cdot}0790$–$5{\cdot}4052$ Å) of the othorhombic unit cell. The indication, therefore, is that substitution of B atoms for *both* C and Fe can occur, varying preferentially with crystallographic planes in cementite. This is a recognized feature of boron in iron-based solutions and characteristic of the borderline size of the B atom. However, the detailed structure of this borocementite still invites determination, and neutron diffraction to locate B and C would be attractive.

It is also known[30] that in cementite, Fe_3C, boron can substitute for carbon up to 80 at. per cent B, at 1,000°C, to give $Fe_3C_{0{\cdot}2}B_{0{\cdot}8}$. The saturation magnetic moments and the Curie temperatures likewise increase with boron content. (The electronic aspects of B and N replacement are also discussed in Nicholson's paper[30].)

One of the major features, however, is the occurrence[31] of the *ternary iron borocarbide*, $Fe_{23}(C,B)_6$, isomorphous with $Cr_{23}C_6$. It clearly implies the stabilization of this structure (see *Figure 4.7*) by boron towards the higher Group number, and a manganese isomorph may also be presumed to exist. The general formula can then be expressed as $(Fe,Mn,Cr)_{21}(Mo,W)_2(C,B)_6$. (This includes the further $M_{23}C_6$ analogues with B sub-group metals like Al,Zn, etc. discussed in Chapter 4—Stadelmaier phases with which the above will be soluble.)

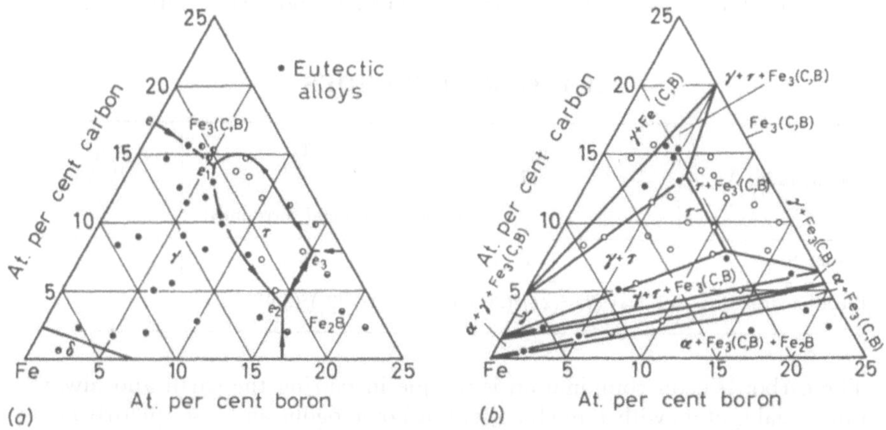

Figure 10.13. The Fe–C–B *system* (*Fe corner*) (From Stadelmaier and Gregg[27b])

(*a*) Melting-point surface (projection); (*b*) 800°C section

By courtesy of the Editor, *Metall*

This iron carboboride has been more fully investigated[27b] and its composition defined as $Fe_{23}C_3B_3$, with a considerable C/B homogeneity range. This is shown in *Figure 10.13,* representing the Fe corner of the Fe–C–B diagram at the melting-point projection and the 800°C section. It shows that the phase, referred to as 'τ', melts congruently. The lattice spacings vary from $a = 10{\cdot}59_4$ to $10{\cdot}62_8$ Å from the C-rich to the B-rich end. The two associated phases are the boron-bearing cementite mentioned, divided into two separate-phase coexistence groups by the τ phase, and the carbon-bearing Fe_2B. However, the borocementite remains metastable like normal Fe_3C, and is found to decompose in time to $Fe_3C_3B_6$ + carbon.

One notable feature is the close link of the ternary phase with the γ-(*austenite*) solid solution. The range of existence of τ extends only over the width of its coexistence with the austenite ($\gamma + \tau$) region, giving way to α/cementite assemblies on either side. This is fully in line with the close structural relation of the $M_{23}C_6$ carbide with austenite, discussed already in Chapter 4, and its mode of precipitating from it (as in stainless steels).

The consequent effect of boron on the mechanical properties of steel (cf. Chapter 6) can be profound and often beneficial, but since it largely depends on the controlled precipitation of borides at crystal boundaries and, incipiently, as B and C atom aggregates around dislocations, the simultaneous presence of traces of carbon and boron can greatly alter it by affecting the nature of the precipitant phases. This can be critical, since the primary solubilities in α- and γ-iron (and the derived alloy steels) are so very small (cf. Chapter 3) and of comparable orders of magnitude.

Little is known about other Group VII or VIII metal/boron/carbon equilibria and phases (e.g. Mn–B–C, Ni–B–C, platinum metal/B/C), and there is clearly wide scope for further research. A particular problem is whether ternary carboborides could still occur much more generally than so far realized, in view of the non-stability of the later-Group carbides but the persisting stability of the borides (cf. e.g. Pt-metal borides, Chapter 6).

10.4 CARBOSILICIDES

Atomic radii, Å	Si 1·17	C 0·77	Si/C *Ratio* ≈ 1·5
	(variable with metal surrounds)		
Group No.	IVB	IVB	
Electron structure	$(1s)^2(2s)^2(2p)^6(3s)^2(3p)^2$	$(1s)^2(2s)^2(2p)^2$	

The carbon–silicon combination is unique in pairing the small and always interstitial C atom with a much larger but homologous and also quadrivalent Si atom. The study of transition-metal carbosilicides has been very active, from the aspect of both ceramic compounds and, of course, of steel technology in the later Groups. (The combination arises as part of the steel-*making* process and as deliberate alloy planning for structural and *properties* control.) Ternary compound formation is frequent but depends on the place in the Periodic Table.

A comprehensive thermodynamic and constitutional study of the refractory silicides in conjunction with carbon (and with nitrogen, see later) is due to Brewer and Krikorian[29b], from the viewpoint of the relative stabilities of silicides and carbides, of heats of formation and reaction kinetics. Some of the ternary M–Si–C diagrams are shown in *Figure 10.14*. They are somewhat

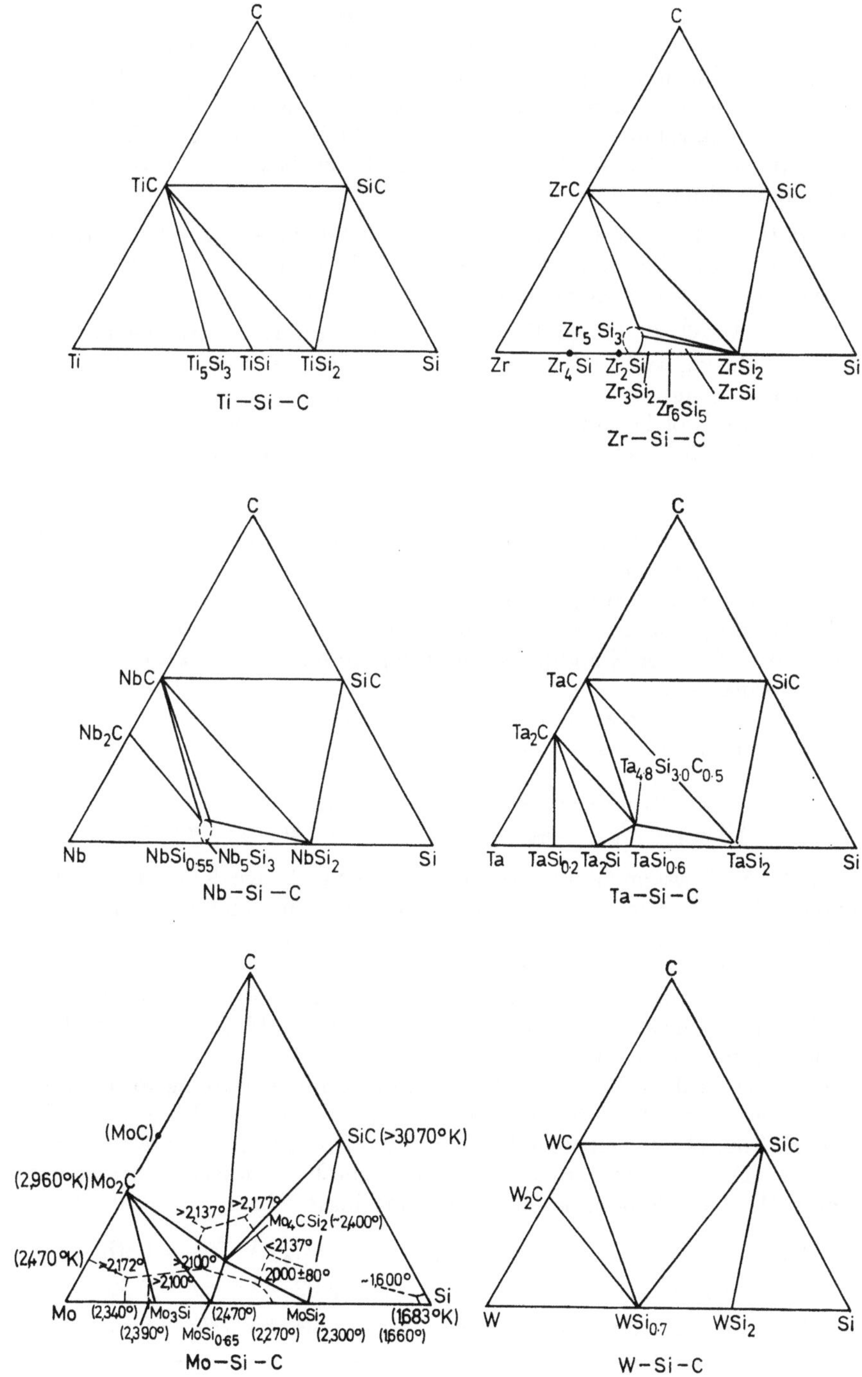

Figure 10.14. Ternary equilibria M–Si–C (From Brewer and Krikorian[29b])

By courtesy of The Electrochemical Society, Inc.

schematic and incomplete; although referring to 'a high temperature', they are not associated with a specific isothermal section and are mainly intended to indicate the basic phase joins and occurrences. The systems are Ti–Si–C, Zr–Si–C, Nb–Si–C, Ta–Si–C, Mo–Si–C and W–Si–C (see also Brewer and Haraldsen's[29a] allied study of *M*–B–Cr above). This work preceded most of that of the Plansee school (see below); the results are generally consistent.

Group IV—TiC and ZrC equilibrate with all the silicides*. Whilst ternary compound formation has not yet evolved, it begins by a considerable solubility for carbon of Zr_5Si_3 and stabilization by it.

In *Group V* (Nb–Si–C and Ta–Si–C) this evolves into a carbon-expanded silicide as shown, and in *Group VI* (Mo–Si–C) into a true and very stable ternary compound, $\approx Mo_4Si_3C$, with an appreciable homogeneity range. The structure is of hexagonal D–8_8 type, with $a = 7{\cdot}285$, $c = 5{\cdot}242$ Å and a calculated density of 6·62. It is significant that the ternary compound formation develops throughout from MSi_2 bases (not from the more stable disilicides) which then equilibrate with the M_5C and MC carbides. The ternary compound clearly lies in the pseudo-binary join $MoSi_2 \leftrightarrow Mo_2C$, whilst in Group V the joins were $NbSi_2$–NbC and $TaSi_2$–TaC. For Ta, a 'ternary' phase, $Ta_{4\cdot8}Si_3C_{0\cdot5}$, arises, of Mn_5Si_3 structure but isomorphous and continuously soluble with Ta_5Si_3. The Mn_5Si_3 structure generally is stabilized by C, O or N additions from the pure silicide which itself is liable to transformations.

To the heats of formation for the pure silicides (*Table 7.1*) may be added, from Brewer's work, two values for *carbosilicides*:

$$\tfrac{1}{3}\,Mo_4Si_3C,\ \Delta H_{298} = -\ 19{\cdot}6 \pm 2$$
$$\tfrac{1}{3}\,Ta_{4\cdot8}\,Si_3C_{0\cdot5},\ \Delta H_{298} = -\ 20 \text{ to } -\ 74{\cdot}7 \text{ kcal/g atom Si}$$

The W–Si–C system, in sharp contrast to Mo–Si–C, reveals no pure ternary compound; this is in good agreement with the later Plansee work where some oxygen or nitrogen was found necessary to its formation.

Throughout, it should be noted that, in the presence of free carbon, no binary silicides can exist (in contrast to carbo*borides*[29a] where binary borides could coexist with carbon). The very stable silicon carbide, SiC, and graphite equilibrate the monocarbide directly.

It will be recalled that the D–8_8 structure, whilst pertaining to the pure silicides in Group IV, required some carbon (or other small interstitial) in the *later* Groups for the lattice to be stabilized (Chapter 7). This now becomes relevant. D–8_8-type ternary silicocarbides have been observed throughout Group VI. The compositions and lattice spacings[18c,32] are

$$Cr_{5-x}\,Si_{3-y}\,C_{x+y} \qquad a = 6{\cdot}97_9, \quad c = 4{\cdot}71_6\,kX;\ c/a = 0{\cdot}67_6$$
$$Mo_{5-x}\,Si_{3-y}\,C_{x+y} \qquad a = 7{\cdot}27, \quad c = 5{\cdot}02\ kX;\ c/a = 0{\cdot}69$$
$$W_{5-x}\,Si_{3-y}(C,N,O)_{x+y}\dagger \qquad a = 7{\cdot}28, \quad c = 4{\cdot}84\ kX;\ c/a = 0{\cdot}67_5$$

* The finite carbide homogeneity ranges MC_{1+x}, not shown in *Figure 10.14*, will divide their joins in sequence from low to high silicides.

† The presence of some N and O is here important for stabilization.

In contrast, the pure 5/3 silicides crystallize as shown in *Table 7.1.* Only very small amounts of carbon are required for transformation to the new structure. As regards W–Si–C, Parthé *et al.*[32b] agree that, for *pure* silicocarbides, the phase does not exist, i.e. Brewer and Krikorian[29b] were proved correct and the nitrogen and oxygen contents were here essential to stabilize a tungsten-bearing D–8_8 structure*. This serves well to illustrate the highly critical nature of phase transformations caused by only trace differences in interstitials contained in these silicides. Similarly, Brewer and Krikorian[29b] found that in Group V the D–8_8 phase can be stabilized from M_5Si_3 only by nitrogen, to give $Ta_5Si_3N_x$ and $Nb_5Si_3N_x$ (possibly also $V_5Si_3N_x$?), in Group IV by oxygen, to give $Zr_5Si_3O_x$ (possibly also $Ti_5Si_3O_x$ and $Hf_5Si_3O_x$?). Considerable non-stoichiometric ranges, i.e. carbon vacancies, are implied.

On *chromium*–Si–C there is little information, but an observation[34] on industrial chrome silicons containing small amounts of carbon seems of interest. For Si contents in the range of 25–42 wt. per cent Si and below 0·04 per cent C, the carbon remained in solution; however, for 0·06 per cent C, a 'carbide' precipitate was found which, with increasing carbon (up to 0·10 per cent C), transforms into a different 'carbide'. Whether or not one or both of these signify a ternary silicocarbide is not clear, but it seems very possible (cf. Mo–Si–C below). On the other hand, for higher Si (42–60 per cent) no such precipitate was formed, the carbon again going into solution. The suggestion then is that CrSi will only dissolve traces of carbon, $CrSi_2$ significant amounts, and that ternary phases are formed in the medium silicon range.

The full Mo–Si–C diagram of Nowotny *et al.*[18c] is shown in *Figure 10.15*, with the 1,600°C isothermal section and a melting-point surface (from Kieffer and Benesovsky[35]). The ternary compound is very stable, melts congruently though only with a shallow maximum at 2,070°C (varying over the homogeneity range), and is surrounded by eutectic troughs. Also it directly equilibrates with carbon, SiC and MoC, as well as with $MoSi_2$ and the Mo_3Si_2 from which it is derived structurally. Technologically this is clearly most important in view of the known refractory applications in which $MoSi_2$, SiC, carbon and their composites are involved.

Groups VII, VIII—Coming to the later-Group metals, Searcy and Finnie[36] have surveyed the phase stabilities in the ternaries of *rhenium* and the *Pt-metals* with silicon and carbon. The general outcome is that only the metal silicides and SiC occur in equilibrium, without ternary compounds. Some x-ray structural and thermodynamic data are also known[36] of these. *Figure 10.16* shows the Re–, Ru–, Os–, Rh–, Ir–, Pd– and Pt–Si–C equilibria obtained. The stability of the silicides is limited by the SiC-forming reaction and varies considerably between metals.

The amount of work done on the Fe–Si–C system is enormous, as it is basic to so many and varied aspects of steel and cast-iron technology. The transformation kinetics in the iron–carbon alloys, complex already, is

* In corroboration, it was also found[33] that, whilst the pure ternary 'W_4Si_3C' compound could not be attained, it can be stabilized by limited additions of Cr or Mo; 4·6 at. per cent Cr and 11·5 at. per cent Mo (substituting W) were required to give '(W, Cr or Mo)$_4$ Si_3C'.

rendered even more so by silicon addition, as the Si atom which electronically might have substituted carbon, say in cementite, is considerably too large for this. Indeed, it substitutes the iron atom in ferrite solution, and the same might have been possible in Fe_3C. However, silicon is known further to destabilize the cementite, already metastable, and to promote graphitization. Furthermore, in contrast to carbon, it acts as a ferrite stabilizer, counteracting carbon by constricting the γ-field and moving it to higher Si so as to join the α- and δ-Fe fields. This is shown in *Figure 10.17* by temperature sections for various Si contents[37].

There has been much controversy regarding the question of existence of a ternary compound in Fe–Si–C. That an iron silicocarbide does exist, had been suspected for many years (in some versions denoted '*X*-constituent'), and substantiated[38]. However, this does not appear to be an equilibrium phase but metastable and appearing only under special types of heat treatment; moreover, there is evidence that it may not even contain silicon. Two Fe silicon carbides have been found[39–40] in quenched silicon cast irons, of

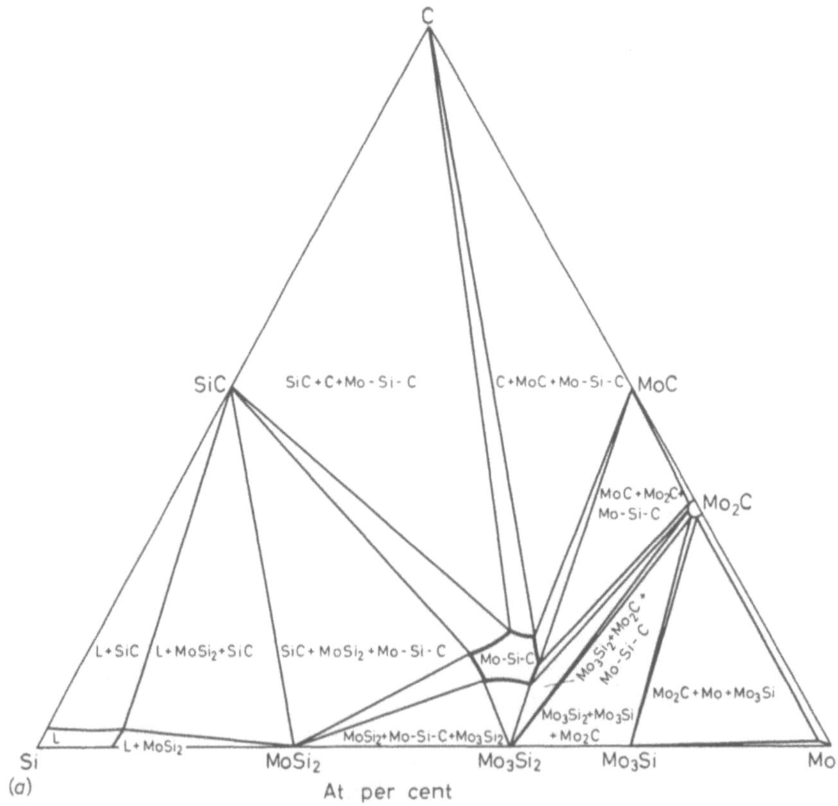

Figure 10.15. The Mo–Si–C *system* (From Nowotny *et al.*[18c] and Kieffer and Benesovsky[35])

(*a*) Section at 1,600°C; (*b*) melting-point diagram with bordering binaries

By courtesy of Iron and Steel Institute and of Springer-Verlag, Vienna

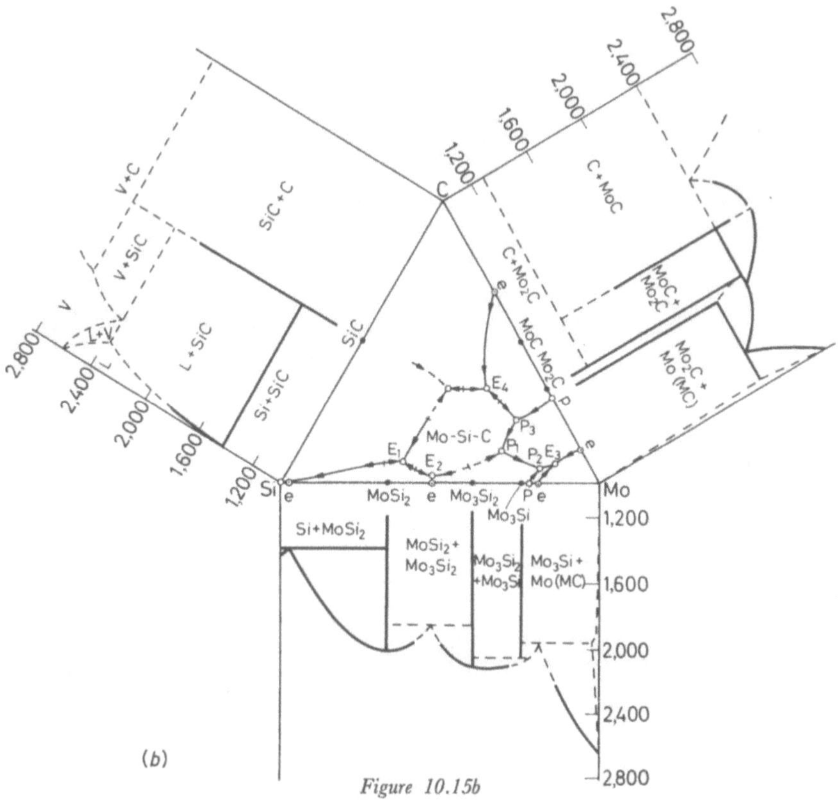

Figure 10.15b

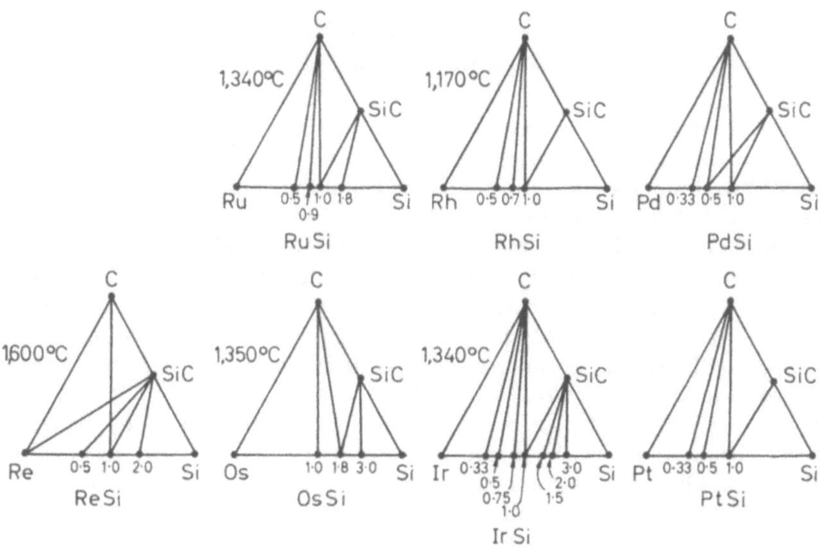

*Figure 10.16. The platinum metal/*Si–C *ternary systems* (From Searcy and Finnie[36])

By courtesy of the American Ceramic Society

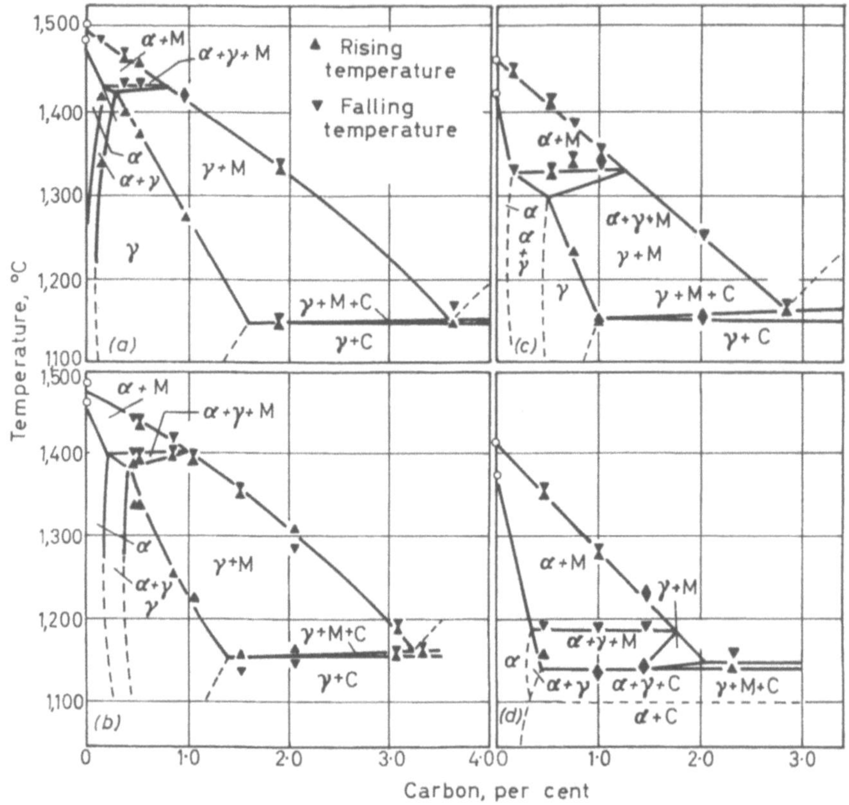

Figure 10.17. The Fe–Si–C system (From Hilliard and Owen[37])

Vertical sections at (*a*) 2·3, (*b*) 3·5, (*c*) 5·2, (*d*) 7·9 per cent Si

By courtesy of The Iron and Steel Institute

different composition. Owen and Street[38] identified the above—different—constituent formed on slow cooling; they discuss x-ray and micro-structural data of all three phases, and conclude from the morphology of the '*X*' constituent that it appears to be a carbide precipitating in conjunction with cementite: less stable than the latter and able to precipitate it. But the phase must owe its existence, however transient, to at least *some* silicon in lattice solution, enriched during the cementite precipitation and variable in amount. Whether Si here replaces Fe or C, effectively to give formulae $(Fe,Si)_3C$ or $Fe_3(C,Si)$, is not established; possibly both. The constituent can also form from austenite.

Figure 11.08 shows the Fe–Si–C diagrams (basal projection) by Scheil[41] and by Kříž and Pobořil[42], in contrast to the newer isothermal sections of Hilliard and Owen[37] (the points c, l, γ and α represent the invariant compositions of carbide, melt, austenite and ferrite, respectively). The older diagrams are here still of interest, as the carbide point (c) falls in the one at cementite, in the other at a finite and appreciable Si content (when extrapolated); this may be looked upon as a first manifestation of a true

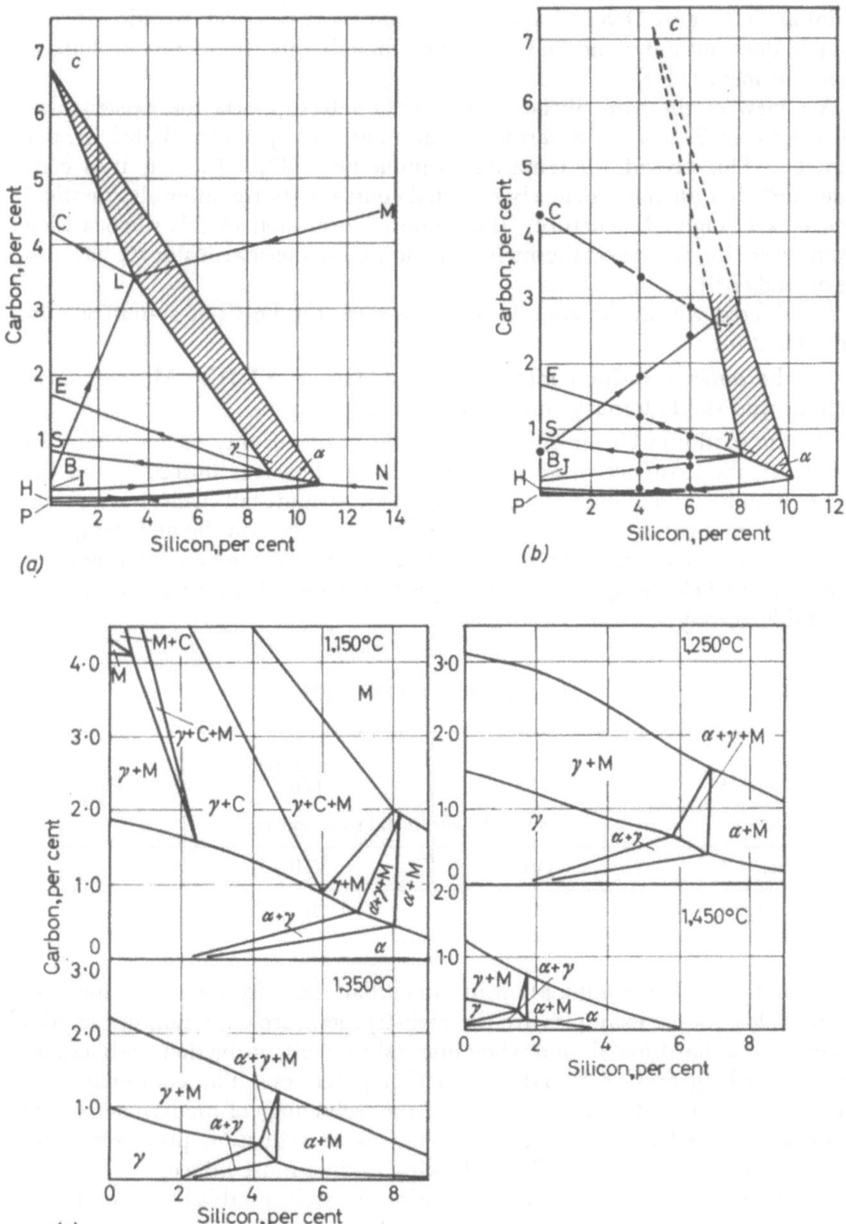

Figure 10.18. Aspects of Fe–Si–C *phase relations* [From Hilliard and Owen[37], Scheil[41] (*a*), Křiž and Pobořil[42] (*b*, *c*)]

(*a*) Basal projection; (*b*) basal projection in metastable diagram; (*c*) sections at 1,150, 1,250, 1,350 and 1,450°C in metastable diagram

By courtesy of The Iron and Steel Institute

carbosilicide. However, Hilliard and Owen[37] firmly place the real carbide point at Fe_3C, and Owen[38b] also shows from x-ray work on carbide extracts that Si does not enter the Fe_3C carbide (though this would not exclude it doing so metastably).

A review of the iron silicocarbide[43a] definitely supports the existence in > 3 per cent Si irons of a carbide containing 9 wt. per cent Si, 6·5 wt. per cent C. This provides a tentative formula $Fe_{10}Si_2C_3$. In fact, it is considered[43b] that all the previously reported constituents are, after all, identical and of a hexagonal structure; the mode of formation of this silicocarbide from austenite, ferrite or the melt, depending on thermal history*, has also been studied[43b].

Considered within the context of this survey, the Fe–Si–C system should be related to:

(*a*) the *earlier* transition metals where in Group VI (Cr, Mo) we have definite and stable ternary compounds†, and

(*b*) the platinum-Group metals which do not form any (see above). In this light, the existence of an *iron* carbosilicide, however unstable, and the more dominant tendency to graphitization would fall well into place. Further complications arise through the effect of silicon on the high-temperature eutectic, peritectic and eutectoid reactions.—An extensive literature exists in this field, largely from the aspect of steel-melting thermodynamics (cf. Bibliography).

10.5 CARBO-OXIDES

Atomic radii, Å	C 0·77 (variable with metal surrounds)	O (unionized) 0·66	C/O *Ratio* ≈ 1·2
Group No.	IVB	VIB	
Electron structure	$(1s)^2(2s)^2(2p)^2$	$(1s)^2(2s)^2(2p)^4$	

The carbon–oxygen combination in interstitial alloys impinges of course on many fields. Some examples are the high-temperature oxidation of carbides when used in hard metals and when integral in steels, or the thermodynamics of steel making and of carbide-production processes. The combination is important in protective coatings and in the technology of modern aerospace materials as well and, conversely, in the oxide reduction processes using carbon, e.g. ore extraction. The following outline must suffice.

In *titanium carbide*, oxygen substitution is well established to give a series Ti(C,O), along with Ti(C,N)‡. In fact, all refractory monocarbides

* Whether or not $Fe_{10}Si_2C_3$ is a true equilibrium or a metastable phase, is still somewhat ambiguous, also in the light of other work[44–6].

† Mn–Si–C is uncertain, but silicocarbides seem likely to form and be more stable than for Fe. The position of Co–Si–C and Ni–Si–C is also in doubt, but silicocarbides, if forming, are likely to be quite unstable.

‡ For the importance of possible Ti(C,O) solution in cemented carbide cutting tools and the need for its avoidance, cf. Chapter 4.

(and hemicarbides) of Groups IV and V are thus capable of dissolving oxygen, initially by C/O substitution but in oxygen excess by higher oxide formation and progressive degrees of ionization of oxygen. Conversely, TiO_2, ZrO_2, etc. can be reduced by carbon, a process in which the rutile or fluorspar structures, after only very limited carbon solution, break down (with CO liberation) to the lower oxide or, eventually, metal structures.

The process is complicated by the possible intermediate formation of the whole homologous series of oxides (Ti_3O_5, Ti_2O_3, TiO, etc.; cf. Chapter 8). The thermodynamic potential of the Ti–O bond greatly increases from TiO to TiC in the solid solution series[47] and, at higher temperatures, the equilibrium pressure of CO during the TiO + C reaction falls sharply and progressively from TiO to TiC.

For titanium, the TiO_2 reduction[21b,47b] to TiC proceeds in the three distinct stages: (*a*) $TiO_2 \rightarrow Ti_2O_3$, (*b*) $Ti_2O_3 \rightarrow TiO$ and (*c*) $TiO \rightarrow TiC$. This means that structurally we must first 'step down the ladder' from rutile to the lower oxide TiO, which already *is* the TiC isomorph and where the $O \rightarrow C$ substitution is easy within the same lattice.

Ashbee and Smallman[48] showed by electron microscopy how TiC single crystals could oxidize to TiO_2 rutile, and how thin films of the latter developed epitaxially on the (100) faces of TiC; they also discussed the orientation relations and dislocation and stacking-fault behaviour, and their bearing on plasticity*.

One of the basic methods of *producing* carbides is, of course, that of reduction of oxides by carbon ($M_aO_b + C \rightarrow aM + bCO$), e.g. $TiO_2 + 3C \rightarrow TiC + 2CO$; $Nb_2O_5 + 7C \rightarrow 2NbC + 5CO$, etc.), often with simultaneous presence of hydrogen or hydrocarbons, however, particularly in the early Groups III and IV.

Stone and Margolin[4], in their work on the Ti–C–N system, also include, Ti–C–O (cf. *Figure 10.3* and Bibliography).

In the V–C–O system, neighbouring Ti–C–O, oxygen addition produces three phases[49], of the composition and spacing ranges

(1) $VC_{0\cdot42-0\cdot50}O_{0-0\cdot1}$ HCP structure $a = 2{\cdot}875$–$2{\cdot}895$, $c = 4{\cdot}545$–$4{\cdot}570$ *kX* (i.e. 'V_2C' base)

(2) $VC_{0\cdot5-0\cdot7}O_{0-0\cdot14}$ FCC structure $a = 4{\cdot}115/4{\cdot}130$ *kX*

(3) $VC_{0\cdot7-0\cdot96}O_{0-0\cdot05}$ FCC structure $a = 4{\cdot}150/4{\cdot}160$ *kX* (i.e. 'VC' base)

These phases must clearly be non-equilibrated members of the V_2C and VC types, with lattice defects partly filled by oxygen. The fact that $VC \leftrightarrow VO$ form one continuous solution series when equilibrated, has also been proved[50] (a linear spacing change occurs). This exemplifies perhaps a basic point, that 'fragmented' phase-composition *bands* can result even in systems which, in equilibrium, form continuous solutions. This applies to many

---

* This might provide a cue for a more general point regarding mixed interstitial systems. The observation of direct growth, during oxidation or nitriding reactions, of certain oxide or nitride structures *upon* the carbides marks these directly as those equilibrating; and any interfacial layer of new structure would indicate a mixed ternary compound. New techniques such as microprobe analysis, if applied with due caution, could thus serve as a useful shortcut to the older types of equilibrium-diagram techniques. The same applies to other interstitial phase pairs and the corresponding surface-diffusion treatments.

processes and industrial materials; the appearance in a steel of two or more such $M(C,N,O_{1-x})$ isomorphs as physically separate inclusions is just one example.

In Nb–C–O, the existence in the same sense of hexagonal and cubic oxycarbides has been demonstrated[51]; their stabilities decrease with increasing temperature and decreasing pressure. A kinetic point of interest during oxycarbide formation is that while O atoms can readily penetrate into the NbC lattice, C atoms do not appreciably enter the NbO lattice; clearly this is linked to the difference in bonding nature (metallic and ionic) in each end member. Large homogeneity ranges are found both in the cubic and hexagonal Nb oxycarbides.

The Ta–C–O system has been studied, with Ta–C–N and Ta–N–O, by Schönberg[52]. By oxidizing TaC and reducing Ta_2O_5 with C, three phases were found. One is a metallic ternary compound of composition $\approx Ta_2$ (C,O) of unknown C/O ratio, having the NaCl-type structure, tetragonally distorted to dimensions $a = 4{\cdot}303$ and $c = 4{\cdot}097$ Å—in fact, isomorphous to Mo_2N and $Nb_2(N,O)$ (see below). The other two phases were undefined. Whether the ternary compound Ta_2 (C,O) forms from TaC by continuous tetragonal distortion through oxygen insertion and, perhaps, a major C-site vacation, or else from hexagonal Ta_2C by stacking-sequence change upon O insertion is uncertain.

Altogether, these metallic interstitial M–C–O systems (with M = Group IV—VI metal) require much further systematic investigation under equilibrium conditions; only isolated kinetic observations such as the above seem to be available at present.

The oxidation kinetics of *chromium carbides* has been studied by Kosolapova and Samsonov[53] who measured the oxidation rates as a function of temperature and time for $Cr_{23}C_6$, Cr_7C_3 and Cr_3C_2. There are, however, little constitutional and equilibrium data—except for a finding of Cr_2O_3 forming on Cr_3C_2 surfaces[54] (see below)—to describe the transition from each to Cr_2O_3. Such research would be of practical interest, perhaps on similar lines as TiC–O tests above. Considerable, and more applied, information is available on carbide/oxide interactions from the aspect of metal ceramics, which we can only mention here except one instance, the reaction at high temperatures[51b] between tungsten carbide and the oxides of other metals (M = Ti, Zr, V, Nb, Ta and Cr). For the Group V and VI oxides, reduction went here so far that *W/M alloys* were obtained low in carbon content (with CO evolution). This is typical of the less stable carbides of Group VI, a process, however, which will be more difficult for the more refractory Group IV and V carbides in their reaction with oxides.

The oxidation mechanism of hard carbide alloys for high-temperature uses has also been extensively studied, e.g. on TiC, WC, Cr_2C_3 and (Ta,Nb)C oxidation[54] in the presence of cobalt. The rutile structure of TiO_2 can form prominently; it takes Co into solution, leading to ilmenites after first forming also CoO or Co_3O_4. However, for higher TiC or WC contents, this deflects to Co titanate or tungstate formation. The rutile structure is also formed for a (Ta,Nb)C–TiC base giving a compound $(Ti,Co,Ta,Nb)O_2$. Thin oxide formation at surfaces of this type has important effects on the 'lubricating' action during machining operations; there is also a direct relation of this to oxidation resistance. On the other hand, complex tungsten oxides form WC (see Chapter 8), Cr_2O_3 forms on Cr_3C_2 surfaces.

Considerable attention has been given to the U–C–O system, in relation to nuclear applications, since both uranium carbides and oxides are important in reactor uses. So are mixtures of the two, in several respects: of progressive oxidation of carbide elements, of integral carbide/oxide composites and of reactions in a CO atmosphere. The high-temperature phase equilibria in the U–C–O system gave reaction kinetic data[55] (pressure/temperature dependence) for carbon monoxide and other gaseous surroundings. UO_2, UC_2 and UC equilibrate directly, allowance to be made for considerable non-stoichiometry and $C \leftrightarrow O$ substitution during the reactions $UO_2 + 4C \leftrightarrow UC_2 + 2CO$ and $UO_2 + 3UC_2 \leftrightarrow 4UC + 2CO$. The respective pressure functions are given as $\log_{10} P_{CO} = -18{,}000/T + 8{\cdot}23$ and $-16{,}000/T + 7{\cdot}26$. There is no ternary compound.

Oxygen can substitute for carbon in the UC lattice up to a spacing of $a = 4{\cdot}950$ Å (compare $a = 4{\cdot}949$ in UC, 4·91 Å in UO); Stoops and Hamme[56] studied the conditions for the stabilization of such a series. It is probable that most of the phases reported in the literature as 'UC' are really U(C,O) or U(C,N,O) solid solutions of varying degrees (including also non-stoichiometry), or that they have become such solutions in the course of service. The oxidation of UC and UC_2 in various atmospheres and temperatures[57] showed linear rates increasing with temperature. It is interesting that in oxygen atmospheres, U_3O_8 resulted, in carbon monoxide, UO_2 plus UC_2; in carbon dioxide UO_2, plus carbon, and in nitrogen non-stoichiometric UN_2 (cf. Section 10.2).

The literature on the Fe–C–O system, including the oxidation of plain carbon steels and of cementite, the reducing effect of carbon or of CO on the iron oxides (e.g. $Fe_3O_4 + CO \leftrightarrow FeO + CO_2$) and that on carbon and oxygen potentials for steels at high temperatures and of C/O interaction during steel processing, is naturally voluminous (cf. Bibliography).

The Fe–C–O diagram[58], as far as steel-solidification reactions are concerned, is shown in *Figure 10.19*. Though semi-schematic, it is instructive in describing the effects of pressure, of (CO_2 or CO) gas composition (and also of increasing Mn content) during solidification of a rimming steel ingot. During this process CO and CO_2 are evolved from the liquid steel, and *Figure 10.19a* essentially gives the liquidus surfaces. (The two binary diagrams Fe–C and Fe–O should be turned upwards to match the ternary.) The main point here is to note the critical changes in these equilibria during melting and freezing, and the almost immediate, but pressure-dependent, destruction of the solubility of carbon in γ- and δ-iron by the presence of oxygen. It results in the evolution of both the gas (CO/CO_2) and a solution of FeO and, possibly, Fe_3C 'molecules' in the liquid (see also Section 4.3).

In the solid we have to distinguish the manifold oxidation and reduction reactions of the iron oxides in the presence of CO or CO_2, such as $Fe_3O_4 + CO \leftrightarrow 3FeO + CO_2$ or $FeO + C \leftrightarrow Fe + CO$, and the carburization and decarburization reactions, such as $3Fe + 2CO \leftrightarrow Fe_3C + CO_2$.

Little seems known on the actual solid solubility of carbon atoms in the FeO, Fe_3O_4 or Fe_2O_3 lattices or, conversely, that of oxygen in Fe_3C. The formation of iron *carbonate* (occurring as siderite ores) at low temperatures should also be recalled. $FeCO_3$ decomposes at 405°C, in 1 atm of CO_2. The carbonates may actually be considered as ionic compounds of the metal oxides with carbon *di*oxide, e.g. $FeO.CO_2$, in the same sense as carbonyls are compounds of the metal with carbon monoxide. The carbonates no longer classify as interstitial compounds, however; they are of course non-metallic and, with their ionically bonded CO_3 groups, structurally more comparable with the silicates, nitrates or borates (also here excluded).

The *metal carbonyls* should be mentioned in this context, being *M*–C–O compounds ($M(CO)_n$) though in a very different and fascinating category. In these, the carbon monoxide molecule is strongly covalently bonded as a

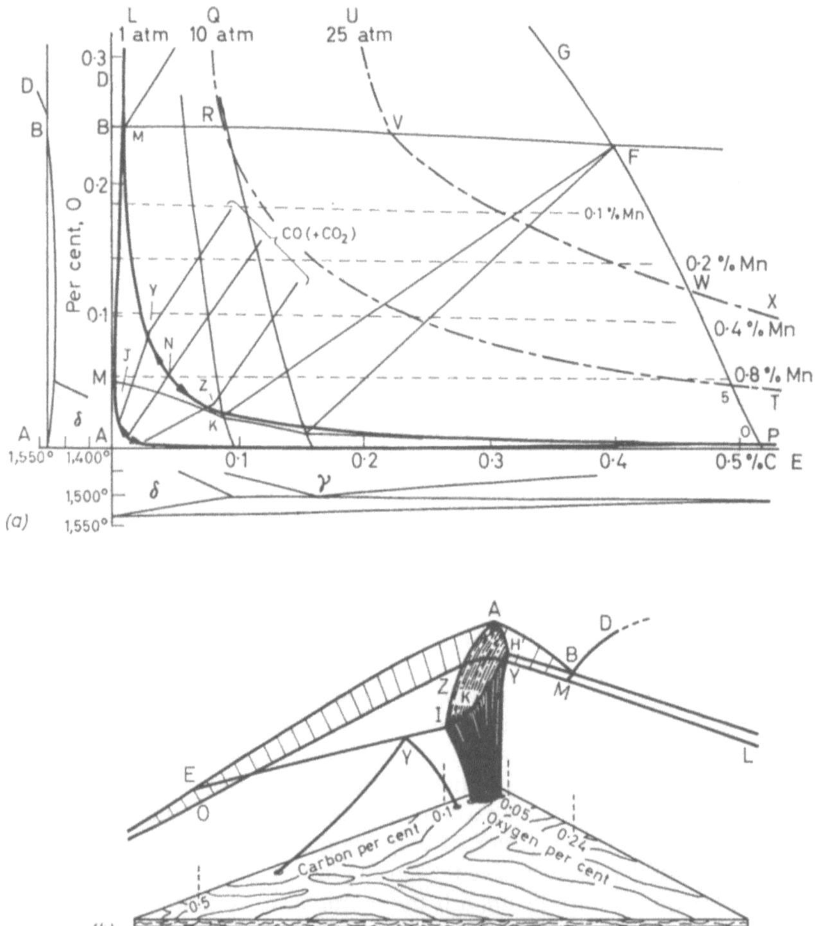

Figure 10.19. Aspects of the Fe–C–O *equilibrium diagram in the solidification region of steel ingots* (From Edwards[58])

(*a*) Tentative iron–carbon–oxygen equilibrium diagram, also indicating the influence of manganese and pressure; projection on composition plane; (*b*) iron–carbon–oxygen system; part of solidus and changes in solid solutions at temperatures below those of complete solidification

By courtesy of the Editor, *Jernkontorets Annaler*

unit by a reaction such as $Fe + 5CO \rightarrow Fe(CO)_5$, and the phases may be viewed as interstitial ones of these CO groups bonded to the metal atoms or, probably better, as the metal interstitial in the CO groups. Historically, the iron, nickel and cobalt carbonyls were the first to be discovered and are probably still the most important. They are the only ones that can be synthesized by direct reaction of CO with the metals; the others require suitable salts or other complexes. However, the carbonyls now known span almost the whole transition-metal block of the Periodic Table, except the early Groups IV and IIIA. *Table 10.4* shows the binary metal carbonyls at present known. It also includes, besides the neutral carbonyls, the carbonyl

Table 10.4. Binary neutral metal carbonyls, their anions and cations at present known (From Abel[59])

Group V	VI	VII	VIIIA		
$V(CO)_6$	$Cr(CO)_6$	$Mn_2(CO)_{10}$	$Fe(CO)_5$ $Fe_2(CO)_9$ $Fe_3(CO)_{12}$	$Co_2(CO)_8$ $Co(CO)_{12}$	$Ni(CO)_4$
$[V(CO)_6]$	$[Cr(CO)_5]^{--}$ $[Cr_2(CO)_{10}]^{--}$ $[Cr_3(CO)_{14}]^{--}$	$[Mn(CO)_5]^-$ $[Mn(CO_6)]^+$	$[Fe(CO)_4]^{--}$ $[Fe_2(CO)_8]^{--}$ $[Fe_3(CO)_{11}]^{--}$ $[Fe_4(CO)_{13}]^{--}$ $[Fe(CO)_6]^{++}$	$[Co(CO)_4]^-$	$[Ni_2(CO)_6]^{--}$ $[Ni_4(CO)_9]^{--}$
	$Mo(CO)_6$	$Tc_2(CO)_{10}$	$Ru(CO)_5$ $Ru_3(CO)_{12}$	$Rh_2(CO)_8$ $Rh_4(CO)_{11}$ $[Rh(CO)_3]_4$	
$[Nb(CO)_6]$	$[Mo(CO)_5]^{--}$ $[Mo_2(CO)_{10}]^{--}$ $[Mo_3(CO)_{14}]^{--}$			?	
	$W(CO)_6$	$Re_2(CO)_{10}$	$Os(CO)_5$ $Os_3(CO)_{12}$	$Ir_2(CO)_8$	
$[Ta(CO)_6]$	$[W(CO)_5]^{--}$ $[W_2(CO)_{10}]^{--}$	$[Re(CO)_5]^-$ $[Re(CO)_6]^+$	$[Os(CO)_6]^{--}$	$[Ir(CO)_3]_4$	

anions and cations, themselves capable of bonding to further ions and molecular groups; indeed, we may regard the carbonyls as a gateway to the whole of metallo-organic chemistry. Many new such compounds found in recent years incorporate carbonyls as integral structural groups. On the other hand, the ready decomposition of the carbonyls at moderate temperatures, for instance in hydrogen, has made them useful as intermediary ('carrier') compounds in the preparation of carbides and for other synthetic processes.

From inspection of *Table 10.4* it might be noted that the area of stability of the carbonyls coincides with that part of the transition-metal block in which the metal carbide stability begins to wane and disappear; it is clearly also pressure-dependent.

The stabilities, reaction kinetics, structures, bond nature, physical properties, general chemistry and technological uses of the carbonyls have been the subject of much investigation (see Bibliography).

10.6 CARBOHYDRIDES (METALLIC)

Atomic radii, Å	C 0·77	H (unionized) 0·28	*Ratio* (C/H) ≈ 2·7
	(subject to variations with metal surrounds)		
Group No.	IVB	(qualified) IA or VIIB	
Electron structure	$(1s)^2(2s)^2(2p)^2$	$(1s)^1$	

The *M*–C–H systems, in the context of metallo-organic chemistry and the carbon–hydrogen bond (see Section 10.1), exceed the present subject by far. Nevertheless, even in solid solution in the primary transition-metal carbides or hydrides, the H and C atoms are bound to 'take notice of each other' with a latent tendency to form C–H bonds which, given suitable thermal conditions, may become active. We may recall, for instance, that in practice one method of preparing carbides from the oxides employs hydrocarbons to speed the oxide reduction, by introducing secondary reactions, $MO + CH_4 \rightleftarrows MC + H_2 + H_2O$* or $M + C_aH_b \rightleftarrows MC_a + \frac{b}{2} H_2$. The reverse reaction obviously also materializes in carbides exposed to these gases at high temperatures.

Although a transient condition, hydrogen occlusion in the carbide lattice at moderate and low temperatures is thus very possible (diminishing with the carbides from Group IV to VI) and aided particularly by the non-stoichiometry in carbides and carbon vacancies. At high temperatures, however, the carbide-held hydrogen will be liberated as H_2 gas, not excluding the possibility that in the less stable carbides the H atom will take some of the carbon along to form methane and leave the lattice yet more sub-stoichiometric. In the later Groups, containing the iron carbides, this becomes particularly critical, in that cementite and the primary carbon-bearing γ-solution can be decomposed by hydrogen, with hair-line crack formation and allied effects referred to already in Section 9.4.

Regarding actual ternary *M*–C–H systems in the metallic solid, the systematic information available is only limited. However, considerable insight has been gained recently by Rexer and Peterson[60]. In the systems Ti–C–H, Zr–C–H, Hf–C–H, V–C–H, Y–C–H and La–C–H, a series of new compounds MC–MH_x occurs, the hydrogen content (x) of the carbide varying over considerable solution ranges. The metal lattice in these phases is hexagonal close-packed, which is to be noted with particular interest from the systematic viewpoint developed especially in Chapter 4. A dominant feature in pure carbides had been the absence in Group IV (Ti, Zr, Hf) of M_2C HCP phases, which developed only in Group V, as V_2C, Nb_2C, Ta_2C. Now hydrogen atoms may be seen to be instrumental in bringing Group IV into line with Group V, i.e. stabilizing the HCP structures as from the

* The deposition of carbides from the vapour phase, such as from halogenites, e.g. $ZrCl_4 + CH_4 + [H_2] \rightarrow ZrC + 4HCl$ (+ C,H, with H_2 as carrier gas), is also a well established process; it is carried out, for ZrC, in the 2,000°C range, for TiC around 1,500°C. The carbonyls provide a similar starting material favoured in the later Groups, e.g. $W(CO)_6 + H_2 \rightarrow W_2C$ + (C,H,O) at around 500°C and reduced pressures.

carbide. Their addition then appears to be essentially equivalent to the effect of electron donation along the Periods. For vanadium, this new phase directly links to V_2C in that the binary HCP carbide simply dissolves hydrogen, with a lattice expansion, whereas for titanium, as for Zr and Hf, we have this as a true ternary phase. A hexagonal solution link will arise beyond doubt in quaternary systems such as Ti–V–C–H. However, recalling that the primary α-Ti, Zr and Hf metals are themselves HCP in their low-temperature forms, this raises the question whether even the new ternary carbohydrides are not really solution-linked with the α forms of base metal itself, although at different temperature levels. If this is the situation, the C–H combination provides stability rather as oxygen does in the Ti–O system to give the pseudo-oxide 'Ti_3O' (cf. Chapter 8). For yttrium and lanthanum as bases, similar considerations apply.

Furthermore, in the Th–C–H system[61] (*Figure 10.20*), two ternary thorium carbohydrides are formed on the ThC–ThH_2 section, namely ThC.ThH_2

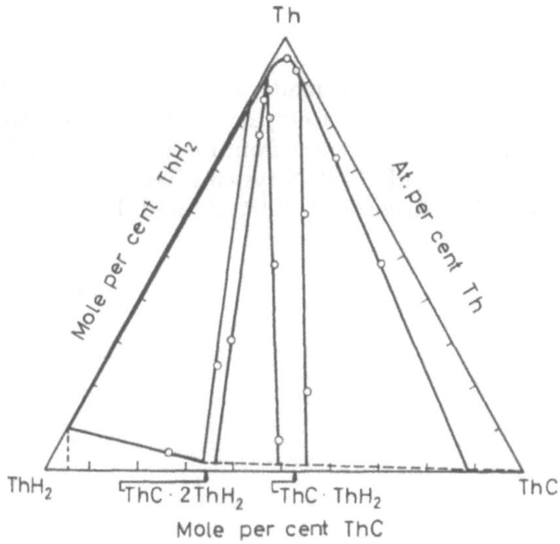

Figure 10.20. The Th–ThC–ThH_2 *system,* 850°C *section* (From Peterson and Rexer[61])

By courtesy of Pergamon Press Ltd.

and ThC.2ThH_2. The crystal structures are, for the first, hexagonal close-packed (a = 3·816, c = 6·302 Å), for the second, however, monoclinic (a = 6·50, b = 3·80, c = 10·91 Å; β = 119°). In other words, hydrogen insertion in the lattice causes a stacking sequence change of the FCC → HCP type which continues to the lower symmetry before the BC tetragonal structure of ThH_2 is attained. The compounds, as ThC and ThH_2, are metallic and good conductors. (Peterson and Rexer's work[61] also provides useful thermodynamic data on these phases, including decomposition isotherms.) It may be seen from *Figure 10.20* that the whole compound series equilibrates with thorium metal. In retrospect, Hägg's early (1931) zirconium hydride, ZrH_2, also reported to be HCP, could well have been a mixed ZrC.ZrH_2 phase analogue to the above (a contention supported by relative unit-cell volumes). The existence of such hexagonal *M*C.*M*H_2 isotypes is thus affirmed throughout Group IV, i.e. from Ti to Th. It seems likely to extend also to Nb and Ta in Group V, though this is unconfirmed.

10.7 BORONITRIDES

Atomic radii, Å	B 0·86	N 0·70	B/N *Ratio* ≈1·2
	(variable subject to metal surround)		
Group No.	IIIB	VB	
Electron structure	$(1s)^2(2s)^2(2p)^1$	$(1s)^2(2s)^2(2p)^3$	

The position for boronitrides is largely similar to that for borocarbides.

Brewer and Haraldsen[29a] have made a survey of the *M*–B–N systems along with *M*–B–C (*Figure 10.12*). Comparison of the joins shows the boride/nitride equilibria and their differences from those of the carbides.

The Ti–B–N, Zr–B–N and Hf–B–N systems have again been studied by the Plansee school (Ti, Nowotny *et al.*[18a], Zr and Hf, Nowotny *et al.*[18d],

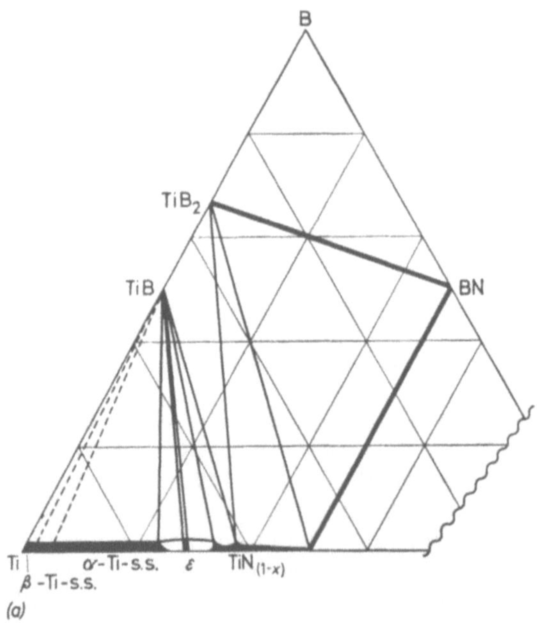

Figure 10.21. The Ti–B–N, Zr–B–N *and* Hf–B–N *systems,* 1,500°C sections (From Nowotny *et al.*[18a,d], Rudy and Benesovsky[23b])

By courtesy of Springer-Verlag, Vienna

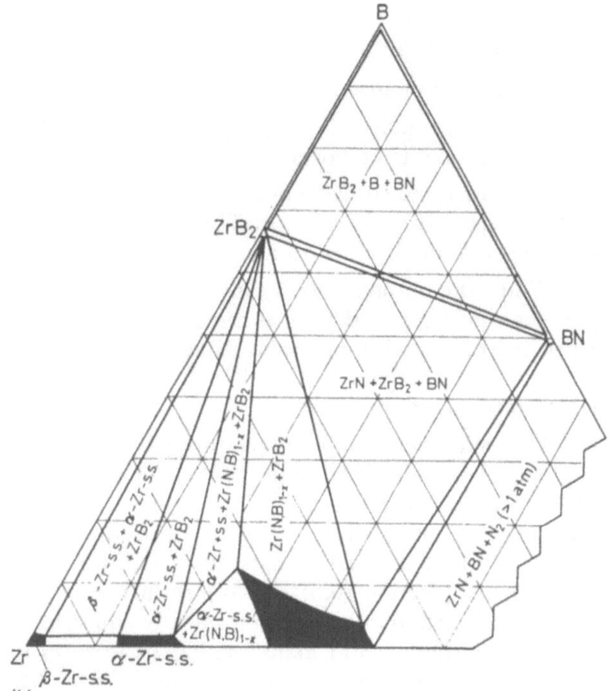

(b)

Figure 10.21b

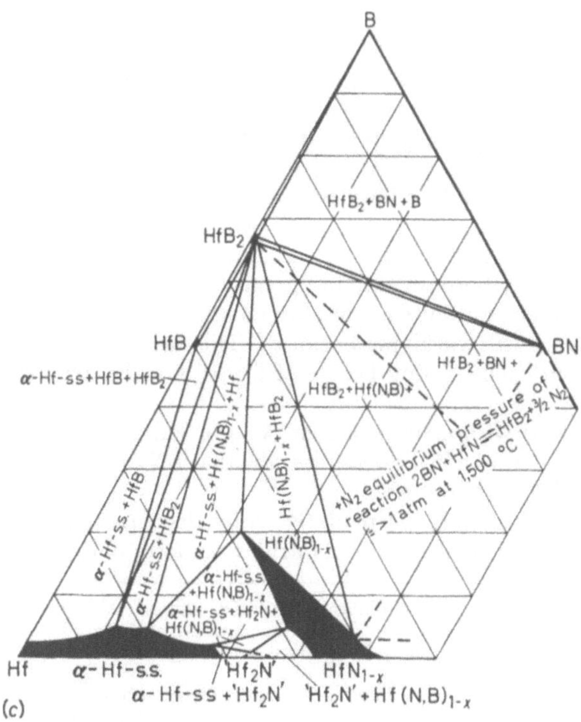

(c)

Figure 10.21c

Rudy and Benesovsky[23b]). The three ternary systems are shown in conjunction in *Figure 10.21* and reveal many instructive features:

(1) No ternary phases occur.

(2) The TiN_{1-x} phase directly equilibrates with TiB and TiB_2, dividing along a line inside its homogeneity range.

(3) The nitrogen-stabilized α-Ti solid solution, which may be considered as a HCP 'pseudo-nitride', also equilibrates the relatively stable TiB and HfB. For Zr, where monoborides do not exist, this occurs with ZrB_2.

(4) Whilst the boron solubility in the Ti nitrides is only small (but rather higher than in the carbides), it is remarkably high in the Zr and Hf mononitrides, increasing with increasing N-atom vacation in the FCC lattice and at the same time expanding it. For hafnium, the α-Hf (N) solubility for boron also becomes appreciable.

(5) The converse solubilities of nitrogen in the Ti, Zr, Hf borides are negligible, in natural conformity with the difference in lattice structure between borides and nitrides.

One point to note is that it is solely boron nitride, BN, which equilibrates with the *M*N and the MB_2 phases, just as, in the case of the metal–boron–carbon systems (Section 10.3), it was free carbon which did so. Considering the B–N combination as being electronically equivalent to C, and BN having the graphite structure and many analogous features in common (e.g. convertibility with pressure to diamond lattice), it might be said that in B–N excess the graphite structure again becomes stabilized from the MB_2 and cubic close-packed phases*.

The TiB_2–TiN system had, in one of Samsonov's studies[21c], shown already that TiN dissolves appreciable TiB_2, but not conversely so. Apart from a report[24] on Ti and Ta boronitride and borocarbide materials, little or no information seems to be accessible on the Group V metal (V, Nb, Ta)–B–N systems.

The thermal stability of the Cr, W and Fe borides in ammonia at various temperatures has been discussed by Kiessling and Liu[63a] who found all the borides to produce the metal nitrides and, significantly again, the very stable boron nitride. There are no ternary boronitrides, but great differences exist between different borides in their stability against ammonia attack, the resistance generally increasing towards the higher borides†. This fact and the types of products obtained from each starting boride of Cr, W and Fe are illustrated in *Table 10.5*.

The interaction of boron nitride with metallic Ti, Zr, V, Nb, Ta, Cr, Mo and W and with their respective carbides[64] showed generally that BN decomposed in favour of the most stable metal borides.

The two iron borides (*Table 10.5*) are particularly easily decomposed, again to boron nitride and the various Jack-type iron nitrides. Indeed, boron

* In contrast, for carbon, boron carbide, B_4C, only equilibrated the MB_2 phases. It is of interest generally in this Chapter to observe systematically the stabilization of the hard inter-interstitial phases (SiC, Si_3N_4, B_4C, BN, SiO_2) from the saturated metal-bearing structures.

† Such nitriding experiments, though kinetically very informative, are really no substitutes for equilibrium determinations based on solid-state reactions, conspicuously lacking here.

Table 10.5. Phases present after nitriding Cr, W *and* Fe *borides in a dry ammonia stream*; 10 *h; specimens quenched from reaction temperature* (After Kiessling and Liu[63a])

Cr–B–N *initial boride*	Cr_2B	Cr_3B_2	CrB	Cr_3B_4	CrB_2
Temperature °C	*Product*				
552	Cr_2B	Cr_3B_2	CrB	Cr_3B_2	CrB_2
800	$CrN+C_2N+BN$*	Cr_3B_2 + (CrN)	CrB	Cr_3B_4	CrB_2
954	$CrN+Cr_2N+BN$	—	—	—	CrB_2
1050	$CrN+Cr_2N+BN$	—	—	—	CrB_2
1180	Cr_2N+BN	Cr_2N+BN	Cr_2N+BN	Cr_2N+BN	Cr_2N+BN

W–B–N, *initial boride*	W_2B	'WB'	WB_2
Temperature °C	*Product*		
700	W_2B	WB	WB_2
800	W_2B	$WB+(W+W_2B+BN)$	—
850	$W_2B+W+(W_2B(\gamma$ form) $+BN$	$W+WB+(\gamma\text{-}W_2B+BN)$	$W+WB_2+(\gamma\text{-}W_2B+BN)$
900	$W+W_2B$ (β form) $+BN$	$W+(WB)+BN$	$W+(WB_2)+BN$
1100	$W+BN$	$W+BN$	$W+BN$

Fe–B–N* *initial boride*	Fe_2B	FeB
Temperature, °C	*Product*	
352	Fe_2B	FeB
400	$\zeta+BN$	$FeB+\zeta+BN$ (?)
448	$\varepsilon+BN$	—
505	$\varepsilon+\gamma'+BN$	—
550	$\gamma'+\varepsilon+BN$	$\gamma'+\varepsilon+BN$
602	$\alpha\text{-}Fe+\gamma'+(\varepsilon)+BN$	$\gamma'+\varepsilon+\alpha\text{-}Fe+BN$
702	$\alpha\text{-}Fe+BN$	—
768	$\alpha\text{-}Fe+BN$	$\alpha\text{-}Fe+BN$

* Phases stated in order of decreasing amount but BN always last; lattice dimensions indicate that no B is in solution in the nitrides, nor N in the borides; Greek letters under Fe–B–N denote the known iron nitrides.

nitride has been identified as a distinct phase in boron steels with low carbon[65] and its presence demonstrated by electron microscopy and diffraction. The effect on the steel of such free BN is one of alleviation of strain aging by removing the nitrogen from solution. The occurrence of free silicon* nitride, Si_3N_4, in silicon- and nitrogen-bearing steels does indeed act similarly (see below).

* This raises a fundamental point concerning double interstitial systems: namely, the effect of *one* interstitial element to bond another in form of a stable (metal-free) compound. By so doing, they remove one another from primary solid solution where one of them could otherwise have had its strain-aging solution strengthening, or other characteristic, effect on the parent lattice. Such an effect can be either deleterious or put to advantage, much depending on circumstances. This applies to the formation not only of solid 'inter-interstitial' compounds (BN, B_4C, Si_3N_4, possibly SiC and SiO_2) but also to gaseous ones (e.g. carbon monoxide or methane nucleation leading to stress and embrittlement effects and crack formation, cf. *Table 10.1*). This may similarly hold true for *ternary* compound formation such as $Fe_{23}C_3B_3$ nucleating out in a steel jointly containing C and B.

Fountain and Chipman[66] have done comprehensive work on the solubility and precipitation of BN in Fe–B alloys, including much thermodynamic and equilibrium information on the Fe–B–N system which entails notable hardenability and strain-aging effects.—A nitrogen analogue to the $M_{23}C_6$-type borocarbide, $Fe_{23}C_3B_3$, does not seem to have been reported, but its existence, or that of some other ternary iron boronitride, seems possible.

10.8 SILICONITRIDES

	Si	N	
Atomic radii, Å	1·17	0·70	*Ratio* Si/N ≈ 1·7
	(subject to variations with metal surround)		
Group No.	IVB	VB	
Electron structure	$(1s)^2(2s)^2(2p)^6(3s)^2(3p)^2$	$(s)^2(2s)^2(2p)^3$	

Brewer and Krikorian's work[29b] on silicide reactions with carbon also included some of those with nitrogen, namely of Ti, Zr, Nb, Ta and Ce with Si and N, and thermodynamic data, particularly on Ta–Si–N. Whilst at lower temperatures, the nitrides of these metals are the more stable, at higher ones the silicides become so. The limited work done on Ti–Si–N and Zr–Si–N seemed, however to indicate a wide solubility of Si in TiN, without a ternary compound forming. In Ta–Si–N, a stable ternary compound $Ta_{5-x}Si_{3-y}N_z$ exists, equilibrating the binary Ta nitrides, silicides and silicon nitride.

The ternary Ta–Si–N diagram is shown[29b] in *Figure 10.22*; this is valid for 1,430°C, but the position radically changes with temperature. At 1,840°C, $TaSi_2$ equilibrates stably with nitrogen *gas*, and Si_3N_4, the ternary compound and all nitrides have decomposed, by a reaction such as $Ta_2N + Si_3N_4 \rightarrow TaSi_2 + N_2$ gas. The Nb–Si–N_2 system is similar, containing a compound $Nb_{5-x}Si_{3-y}N_z$.

It is informative to compare Ta–Si–N in *Figure 10.22* with Ta–Si–C in *Figure 10.14*. The ternary Ta siliconitride gains greatly in stability as against a doubtful silicocarbide. The direct $TaSi_2$–TaC join is broken, in the case of TaN, by the ternary compound, and the resulting SiC and Si_3N_4 equilibria differ quite considerably.

Work on the refractory siliconitride systems by Nowotny, Lux and Kudielka[18e] and Kieffer, Benesovsky and Lux[67a] showed that, if $TiSi_2$ is exposed to a nitrogen atmosphere, TiN forms directly, but that a ternary compound can form for Zr when heated at 1,800°C with silicon nitride which itself decomposes.

Ternary compounds also occur in the Nb–Si–N and Ta–Si–N systems, as shown by Brewer and Krikorian[29b]. They have again the D-8_8 structure, derive from M_5Si_3 and compositionally were defined[32b] as $Nb_{5-x}Si_{3-y}N_{x+y}$ and $Ta_{5-x}Si_{3-y}N_{x+y}$. In contrast, no such structure was obtained from Mo_5Si_3.

There seems little information on Group VI and VII silicon nitrides.

In *iron alloys*, the simultaneous presence of Si and N_2 has been much investigated, largely from the viewpoint of silicon-transformer sheet materials. In high-silicon (4 per cent Si) steels it is known that silicon nitride, Si_3N_4, can form as precipitate for small amounts of nitrogen; it has been identified as the stable constituent and extracted[68] in its α and β forms. In low-Si steels,

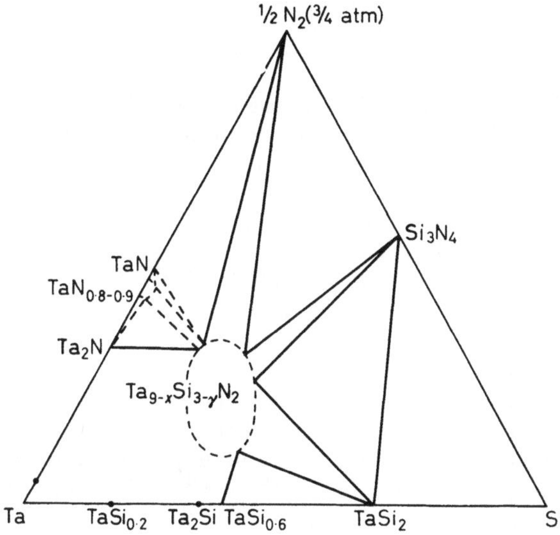

Figure 10.22. *The* Ta–Si–N *system,* 1,327°C section (From Brewer and Krikorian[29b])

By courtesy of The Electrochemical Society, Inc.

Si_3N_4 can occur through nitriding. It might have been surmised that, as in the Fe–Si–C combination silicon is a graphite stabilizer, so it would liberate nitrogen in Fe–Si–N. This is inhibited, however, by stable formation of silicon nitride which would also take nitrogen from active solid solution in ferrite.

A new and different type of Si nitride also forms, after certain treatments, in low-Si steels[69], of composition SiN and isomorphous with hexagonal AlN.

A number of distinct internal friction peaks have been observed in Fe–Si–N alloys[70–1]; some are due to interaction of nitrogen atoms with neighbouring silicon atoms. Although N and Si are here in α-Fe solid solution, this precedes the precipitation of iron nitrides in definite stages[72] controlled by the presence of silicon atoms.

No ternary iron siliconitride has been reported, and although a full Fe–Si–N ternary diagram does not seem to have been established yet, the equilibration of Si_3N_4 with (Fe,Si,N) ferrite must mean the prevention of a join between the binary iron silicides and the iron nitrides, unless it is a metastable join. No doubt, the Fe–Si–N system is of the same complexity as Fe–Si–C.

It has also been shown[73] that silicon reduces the solubility of nitrogen in α-iron and that Si_3N_4 only precipitates at lower temperatures.

10.9 OXYNITRIDES

Atomic sizes, Å	N 0·70	O (unionized) 0·6	*Ratio* N/O ≈ 1·05
	(subject to variations with metal surround)		
Group No.	VB	VIB	
Electron structure	$(1s)^2(2s)^2(2p)^3$	$(1s)^2(2s)^2(2p)^4$	

The oxygen/nitrogen combination as interstitial solutes in the transition metals must apply wherever air exposure occurs at high temperatures, though in general to oxygen preferentially rather than to nitrogen. Much can be predicted from the many close and frequently isomorphous relations between nitrides and oxides shown in *Tables 5.1* and *8.1*. In the early refractory metals, this leads back to the α-Ti(C,O,N)- and Ti_{1-x}(C,O,N)-type solid solutions discussed already. In Stone and Margolin's Ti–N–O phase relations[18] (cf. *Figure 10.3*) the continuous α-Ti(N,O) series is the most striking feature, stabilized to highest temperatures (compare also Ti–N and Ti–O binaries in *Figures 5.1* and *8.1*).

Similarly, the mononitrides and monoxides TiN_{1-x}–TiO_{1-x} form a continuous series of solid solutions. But lattice-spacing and density changes observed[74] have indicated complications and only partial miscibility TiN ↔ TiO; this is thought to be due to the change-over from the more metallic to the ionic bonding and, at the same time, from the moderately to the highly defect lattice from TiN to TiO (16 per cent cation and anion vacancies in TiO, only 2 per cent in TiN at 1,700°C). It seems doubtful, however, whether this ought to be taken as the equilibrated state, and it is also not clear how the complex homologous series of Ti oxides would join to the simpler one of Ti nitrides. This could be a fruitful field of research; the question of O → N replaceability in TiO_2 rutile is one such problem.

In Zr–N–O and Hf–N–O, the HCP α-solution regions for N and O are similarly likely to join; Zr–N–O samples[52b] with (N,O)/Zr ≤ 1 showed a solubility of N and O in α-Zr up to ≈ Zr_2(N,O) *and* a continuous series of solutions* ZrN–ZrO. But open questions are how far nitrogen is soluble in the ZrO_2 polymorphs (and capable of stabilizing them), and likewise in the fluorspar structure of HfO_2.

In a comprehensive study[75] on the nitridation of the refractory oxides TiO_2, Al_2O_3 and especially ZrO_2, it was shown that oxynitrides are readily formed, and their structure, bond nature (division between metallic and ionic bond), metallic behaviour and constitution were defined. In the pseudo-binary system ZrO_2/Zr_3N_4, three Zr oxynitrides were found. The phases are $Zr_7O_{11}N_3(\beta')$, $Zr_7O_8N_4(\beta)$ and $Zr_2ON_2(\gamma)$; stabilities are shown in *Figure 10.23*. They are derived essentially from the fluorspar structure of the high-temperature form of ZrO_2, through a systematic series of vacancy formation. The resulting intermediate oxynitride structure is of the cubic Tl_2O_3 type for γ (determined in detail), with orthorhombic modifications for β and β'.

* If 'ZrO' exists, which seems to be in doubt.

Coming to Group V, there is for *vanadium*[76] a regular V(N,O) phase series of NaCl-type structures, with linear lattice expansion from VN to VO; this need not necessarily be so, e.g. when substantial ionization of the oxygen atom occurs, as in TiN–TiO.

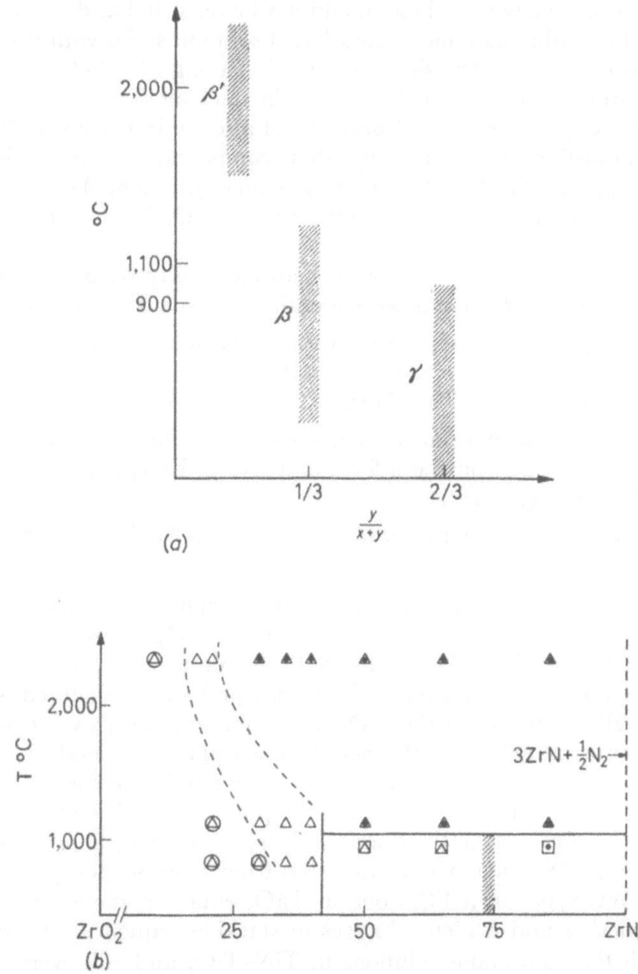

Figure 10.23. Tentative data on Zr–N–O *phase relations* (From Gilles[75b])

(*a*) Phases ZrO_xN_y formed by ammonia attack upon ZrO_2 (β' = phase '$Zr_7O_{11}N_2$'; β = '$Zr_7O_8N_4$'; γ = 'Zr_2ON_2'); (*b*) phase diagram suggested between ZrO_2 and $ZrN_{4/3}$ (assemblies after nitriding ZrO_2 at different temperature levels: ○ = ZrO_2, △ = oxynitrides β or β', □ = oxynitride γ, ● = $3ZrN + \frac{1}{2}N_2$)

By courtesy of the Société Chimique de France

For *niobium*, a tetragonally distorted NaCl-type phase, $Nb(N,O)_{0\cdot78}$, and a cubic one, Nb(N,O), occur[6d]; in this series the O:N ratios can vary widely, from 1:8 to 1:25; this was confirmed[52b], giving for the tetragonal phase the homogeneity range $NbN_{0\cdot6}O_{0\cdot2}$ to $Nb_{0\cdot6}O_{0\cdot3}$ and spacings a = 4·370–4·386,

$c = 4{\cdot}386$–$4{\cdot}325$ Å. This indicates preferential vacancy filling by O atoms along the *a* axis. For the cubic NaCl-type phase, $\approx NbN_{0\cdot90}O_{0\cdot10}$ was found[52b], with spacings ranging from 4·373 to 4·390 Å.

Thus it would appear that the lattice distortion is distinctly associated with increased lattice defection, which itself is linked with an increased one in oxygen. It seems likely that these should not be regarded as discrete ternary oxynitrides but rather as binary NbO-based solution series with progressive lattice distortions; the transition to NbN, having the WC-type simple hexagonal lattice, requires a major change in stacking sequence.

In the Ta–N–O system, it had been found already in early work[77] on the heat of combustion of Ta nitrides that compounds occur of formulae '$TaN_{0\cdot846}O_{0\cdot060}$' and '$TaN_{0\cdot955}O_y$', *y* being a small fraction. However, these may again be presumed to be solid solutions of Ta_2O_5 in non-stoichiometric TaN.

Schönberg[52c] studied the Ta–N–O system more fully by oxidizing tantalum nitrides. Four compounds were identified on the TaN → TaO join:

(1) $TaN_{\sim 0\cdot90}O_{\sim 0\cdot10}$, hexagonal with a superlattice as from parent hexagonal TaN, and doubled *a* axis: $a = 10{\cdot}34$, $c = 5{\cdot}802$ Å (cf. TaN in *Table 5.1*); based on the ε-TaN form

(2) $TaN_{\sim 0\cdot75}O_{\sim 0\cdot25}$, hexagonal, again with TaN-based superlattice and doubled *a* axis but based on the δ form of TaN, $\approx TaN_{0\cdot80-0\cdot90}$; $a = 5{\cdot}988$, $c = 2{\cdot}879$ Å ($a \approx 2a_\delta$, $c \approx c_\delta$)

(3) $TaN_{\sim 0\cdot65}O_{\sim 0\cdot35}$, again hexagonal and δ-TaN-based but with $a = 10{\cdot}34$, $c = 2{\cdot}864$ Å ($a \approx 2\sqrt{3}a_\delta$, $c \approx c_\delta$)

(4) $TaN_{\sim 0\cdot50}O_{\sim 0\cdot50}$, hexagonal superlattice of sub-dimensions $a = 5{\cdot}939$, $c = 2{\cdot}866$Å, neither based on the ε- nor on the δ-TaN forms.

Figure 10.24 shows Schönberg's Ta–N–O diagram for (N,O)/Ta ratio ≤ 1, with these four compounds marked. It may perhaps be doubted whether they should all strictly lie on the TaN–TaO join, especially as non-stoichiometry may enter largely with the δ-TaN-based phases (2) and (3). However, the basic point, and the reason for giving fuller details, is that the transition from the pure mononitride to the pure monoxide can structurally be performed only in a distinct series of steps involving a progression of ordered lattices. We have here a transition from a simple hexagonal lattice in TaN (=WC type) to a FCC one in TaO, entailing these intermediate stages in ordering and evident changes in stacking sequence. It contrasts sharply with the continuous solutions in TiN–TiO, and still even those in Group V, to VN–VO. The break-up starts in NbN–NbO (see above) where, however, we may still see an adherence to cubic-based symmetry, even near the NbN end. The transformation from cubic-based to hexagonal-based still occurs for Nb at much higher N/O ratios than for Ta.

Finally, 'a black ternary phase' of unknown complex structure (and non-metallic) has been reported[52c] when Ta_2O_5 was treated in ammonia.

Linking the Ta–N–O and Zr–N–O systems, the occurrence of a quaternary compound ZrTaNO has been claimed[52c], of a new type of hexagonal close-packed (metal) structure, with $c/a \approx 1$. Modulated ternary structures are developed further in the W–O–N system, neighbouring Ta–N–O in

Group VI. A ternary compound series appears here[63b], based on a metal-defective NaCl-type structure, of composition $W_{0.62}(N_{0.62},O_{0.38})$. In this W(N,O) lattice, only 62 per cent of the *metal* positions are occupied, but the interstitial positions are probably all occupied in the N:O = 62/38 ratio. The spacing is 4·138 Å.

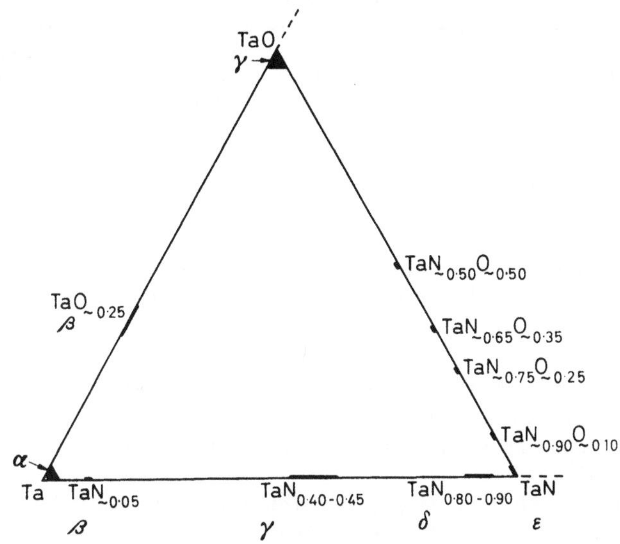

Figure 10.24. The Ta–N–O system, *showing occurrence of oxynitrides for* (N,O)/Ta ≤ 1 (From Schönberg[52c])

By courtesy of Messrs. Munksgaard, Copenhagen

Furthermore, the tungsten mononitride WN, previously presumed pure, is now believed to be also an oxynitride of WC structure, again W-deficient but of ordered W vacancies—in the above phase they were disordered. The formula is $W_{0.75}(N,O)_{1.00}$, the N:O ratio being uncertain but greater than 62/38.

These findings are interesting on two scores: (1) We have a distinctive metal-ordering effect caused by varying interstitial balance; and (2) the cubic NaCl-type structure is stabilized here as a *ternary* phase, not in existence in either the pure nitrides or oxides.

10.10 HYDRONITRIDES

Little or no information on hydride–nitride interreaction and equilibrium systems seems accessible, even though for the action of ammonia on the transition metals much specialist literature exists. In general, we may regard this simply as one effective means of nitriding, such as in

$$Fe + XNH_3 \leftrightarrow FeN_x + 1{\cdot}5\,XH_2$$

where the iron nitrides are the only stable ones. This may, however, not necessarily hold true for the earlier Group metals, and in Groups III and

IV, coexistence at moderate temperatures and higher pressures of the known hydrides and metal–hydrogen solutions with the nitrides seems distinctly possible, including mutual solid solutions and even ternary hydronitrides.

10.11 SILICOBORIDES

	Si	B	*Ratio* Si/B
Atomic radii, Å	1·17	0·86	≈ 1·4
	(subject to variations with metal surround and bonds)		
Group No.	IVB	IIIB	
Electron structure	$(1s)^2(2s)^2(2p)^6(3s)^2(3p)^2$	$(1s)^2(2s)^2(2p)^1$	

The metal–B–Si systems have been the subject of considerable investigation, as were the binary borides and silicides, largely in view of interest in their refractory properties and potentialities as high-temperature oxidation protectants which in some cases were found to be excellent.

The close structural kinship between silicides and borides (cf. *Figures 6.2* and *7.2*) also gives an added basic interest to the mixed compounds; intermediate phases as well as wide mutual solid solution ranges between binaries are likely to arise.

The refractory *M*–B–Si systems of Groups IV–VIA have been comprehensively studied by the Plansee School[18b,35b,78]; Aronsson[79] has summarized the ternary phases occurring, their crystal structures and lattice dimensions (*Table 10.6*). The boron atom is largely instrumental in stabilizing the D–8_8 and related structures ('Nowotny phases'), as was carbon before, and the silicide or boride types reappear here in the pure ternary forms (the later-Group (VII and VIII) data will be referred to below).

Table 10.6. Crystallographic data of ternary M–Si–B phases (From Aronsson[79a])

(Confined to those phases whose homogeneity range probably does not extend to a binary composition)

Approximate composition	*Structure type*	*Lattice parameters,* Å *a*	*c*
Zr_5Si_3 (*X*)	Mn_5Si_3(D–8_8)	$7{\cdot}88_6$	$5{\cdot}55_8$
Hf_5Si_3 (*X*)	Mn_5Si_3(D–8_8)	7·89	5·56
$V_5Si_3B_{\sim 0{\cdot}45}$	Mn_5Si_3(D–8_8)	7·18	4·89
V_5SiB_2	Cr_5B_3(D–8_l)	5·81	10·79
$Nb_5Si_3B_{\sim 2}$	Mn_5Si_3(D–8_8)	7·54	5·25
$Ta_5Si_3B_{\sim 2}$	Mn_5Si_3(D–8_8)	7·47	5·23
Cr_5Si_3B	Mn_5Si_3(D–8_8)	7·06	4·73
Mo_5SiB_2	Cr_5B_3(D–8_l)	6·01	11·03
$W_5(Si,B)_3$	Cr_5B_3(D–8_l)	6·047	10·99
Mn_5SiB_2	Cr_5B_3(D–8_l)	5·61	10·44
Fe_5SiB_2	Cr_5B_3(D–8_l)	5·54	10·32
$Fe_{\sim 4{\cdot}8}Si_2B$	W_5Si_3(D–8_m)	8·82	4·34
$Co_{\sim 4{\cdot}7}Si_2B$	W_5Si_3(D–8_m)	8·62	4·25
Ni_6Si_2B	Fe_2P(C–22)	6·105	2·875

Geach and Jones[1] had already reported such new phases in the alloyed mixtures TiB_2–Nb_5Si_3; TiB_2–Ta_5Si_3; Nb_5Si_3–Mo_2B; Ta_5Si_3–Mo_2B; $TaSi_2$–Mo_2B and $MoSi_2$–Mo_2B. This has been substantiated and detailed in the Plansee work which defines particularly the various Nowotny phases and their equilibria. Nowotny, Lux and Kudielka[18e], dealing more widely with the effect of B, C, N and O additions to silicides, report that, starting from M_5Si_8, the D–8_8 structure is effectively stabilized by carbon, somewhat less so by boron, only weakly by nitrogen and not at all by oxygen. Boron shows an appreciable initial solid solubility in all the M_5Si_3 phases, whether of D–8_8, 'T–1' or 'T–2' (Cr_5B_3-type) structures (α- or β-Nb_5Si_3 types). It is noteworthy that, as from the pure silicides, B additions

(*a*) *expand* the D–8_8 lattice,

(*b*) *contract* the T–1 and T–2 lattices

indicating in (*a*) entry into interstitial Si vacancies, in (*b*) direct Si substitution.

In Group IV, the Ti–B–Si and Hf–B–Si systems are only very incompletely known. The Zr–B–Si diagram (*Figure 10.25*) has been derived[32c] for 1,350°. The outstanding features are the occurrence of the ternary compound, again of D–8_8 structure, and of an appreciable solid solubility of $ZrSi_2$ in the extremely stable zirconium diboride which dominates the system by equilibration with almost all phases, the metal (D–8_8), ZrSi, $ZrSi_2$, Si and B; no doubt also the borosilicides SiB_3, SiB_4 and SiB_6, not here marked.

The silicoboride systems of Ti, Zr, Hf; V, Nb; Cr, Mo and W are shown in *Figure 10.25*, in form of a series of phase equilibria collated from various Plansee sources (cf. Bibliography).

The V–B–Si system—*Figure 10.25* shows a sketch marking only with joins—contains two ternary phases, of compositions V_5SiB_2 (T–2 type) and V_5Si_3, with small boron in solution (D–8_8 type); the first is analogous and isomorphous to the Mo_5SiB_2 phase.

The Nb–B–Si system, on the other hand, studied more fully[25b], shows these two ternary phases of T–2 and D–8_8 type to have moved to close proximity and to have a very considerable homogeneity range. Both phases are descendent from Nb_5Si_3 and equilibrate the Nb borides in a complex sequence as shown. Of special interest are the direct pseudo-binary equilibrations $NbB_2 + NbSi_2$, NbB_2 + free Si and NbB_2 + the silicon borides. The T–2 phase again reaches the composition Nb_5SiB_2 by Si $\rightarrow$ B substitution, the D–8_8 phase Nb_5Si_3B with filling of vacancies*.

The Ta–B–Si system has been studied only to a limited extent, with preliminary data of the Ta_2Si–Ta_2B section[78]; however, the existence of ternary silicoborides similar to those for Nb seems in little doubt.

In Group VI, the Cr–, Mo– and W–B–Si systems form one consistent set (see *Figure 10.25*). All contain the T–1- and T–2-type phases, but for Cr the latter naturally takes a binary root version, as Si solid solution in Cr_5B_3 expanding the lattice; for Mo and W, this develops into the true ternary compounds Mo_5BSi_2 and W_5BSi_2. Moreover, the D–8_8 type ternary phase is only retained for Cr, at about Cr_5Si_3B, with a = 7·05, c = 4·72 kX. These

* The oxidation resistance of Nb–Si–B alloys at high temperature, also studied in this context, shows considerable differences between phase constitutions of the base.

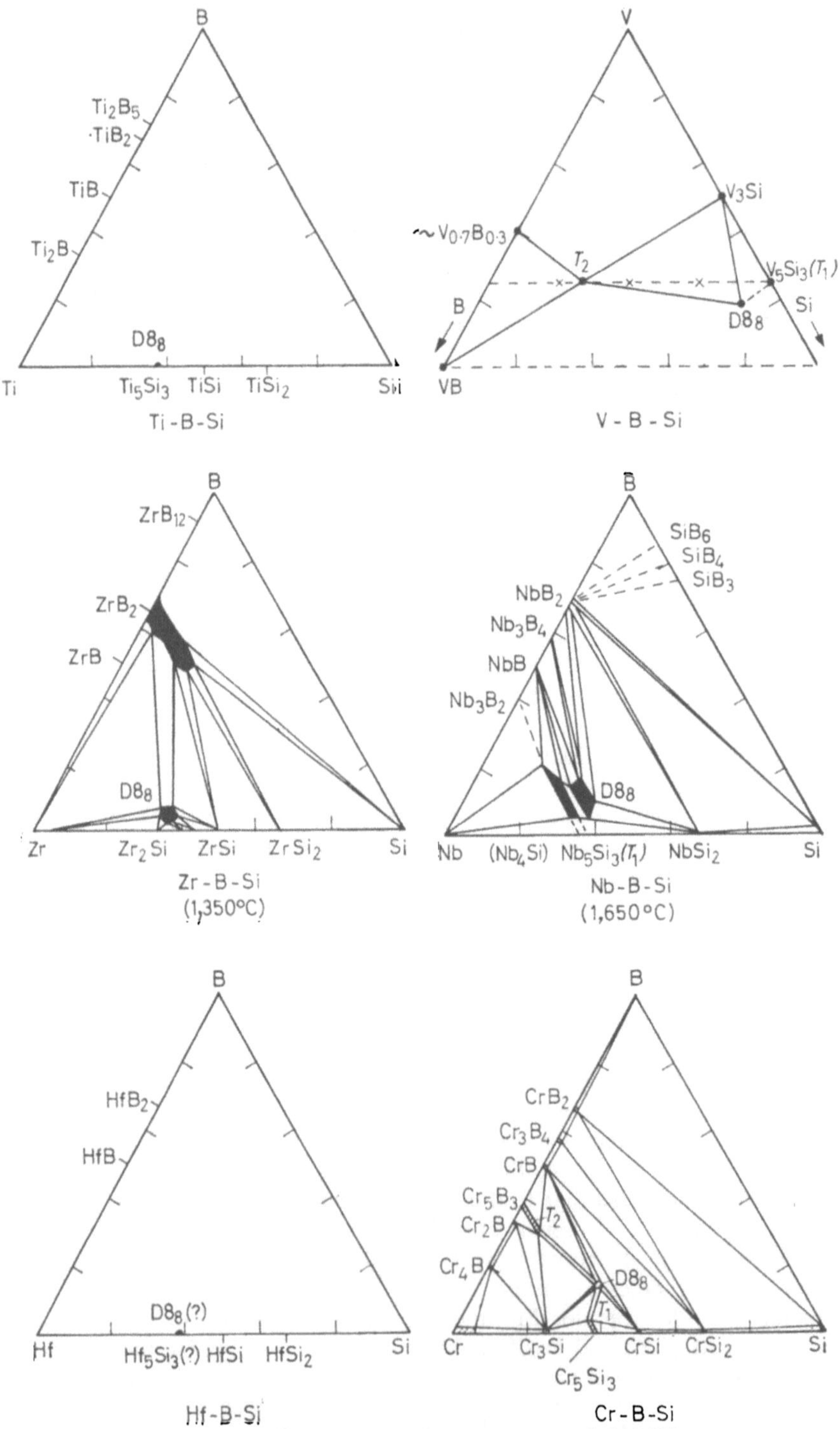
B
Ti2B5
TiB2
TiB
Ti2B
D88
Ti
Ti5Si3
TiSi
TiSi2
Si
Ti-B-Si
V
V3Si
~V0·7B0·3
T2
V5Si3(T1)
B
Si
D88
VB
V-B-Si
B
ZrB12
ZrB2
ZrB
D88
Zr
Zr2Si
ZrSi
ZrSi2
Si
Zr-B-Si
(1,350°C)
B
SiB6
SiB4
SiB3
NbB2
Nb3B4
NbB
Nb3B2
D88
Nb
(Nb4Si)
Nb5Si3(T1)
NbSi2
Si
Nb-B-Si
(1,650°C)
B
HfB2
HfB
D88(?)
Hf
Hf5Si3(?)
HfSi
HfSi2
Si
Hf-B-Si
B
CrB2
Cr3B4
CrB
Cr5B3
T2
Cr2B
Cr4B
D88
T1
Cr
Cr3Si
CrSi
CrSi2
Si
Cr5Si3
Cr-B-Si
(1,300°C)

Figure 10.25

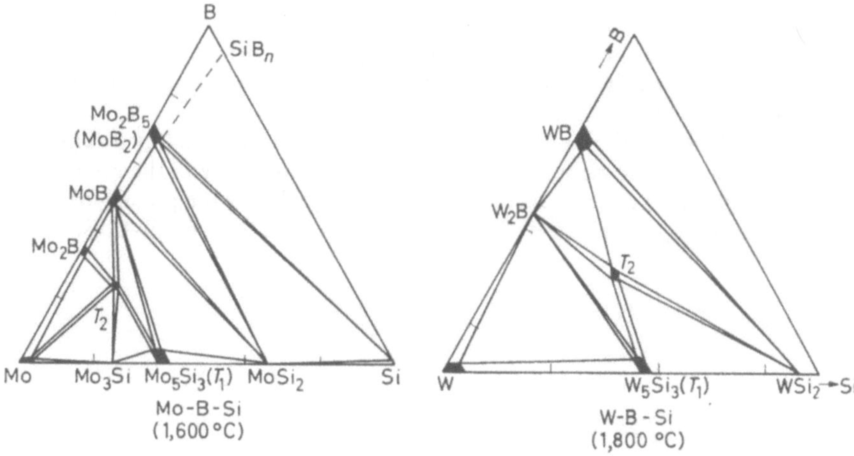

Figure 10.25. M–B–Si systems, Groups IV–VIA (From Kieffer and Benesovsky[32c] (Ti,Zr,Hf); Kudielka *et al.*[78] (V); Nowotny *et al.*[18c,f,g] (Nb,Cr,Mo,W)

By courtesy of the Editor, Planseeberichte für Pulvermetallurgie and of Springer-Verlag, Vienna

ternary phases are very stable, high melting and of greater hardness (1,670 kg/mm^2 at 50g load for Mo_5SiB_2) than any of the binary borides and silicides.

Regarding Mo–Si–B, the reaction kinetics during gas siliconizing of Mo borides has been studied[79b,80], the diffusion zones being identified in terms of the known phases.

The systems $MoSi_2 \rightarrow MoB_2$ and related ones, which are of such interest in view of the high stability and oxidation resistance of $MoSi_2$, are plain two-phase without an intermediate compound. With metal excess, transformation to new ternary phases sets in, however; possible deleterious effects must be allowed for at boundary layers in any practical phase assembly (e.g. protective coatings, cermets, etc.).

Schwarzkopf and Glaser[64] who reacted some of the refractory metal disilicides with elemental boron, showed that this yielded the respective metal *borides*; they also observed that the borides reacted with silicon are more stable than the silicides reacted with boron.

Passing to Groups VII and VIII, the systems[79a] *M*–B–Si (where M = Mn, Fe and Co) again contain ternary compounds, Mn_5SiB_2 and Fe_5SiB_2, of Cr_5B_3(T–2)-type structure (*Table 10.6*), whilst the W_5Si_3 (T–1) type projects itself in phases $Fe_{\sim 4\cdot 6}Si_2B$ and $Co_{\sim 4\cdot 7}Si_2B$. However, two further ternary Fe–B–Si compounds exist[79d], one of cementite (Fe_3C) structure, the other of yet unknown structure but of approximate composition Fe_4SiB.

Probably the most noteworthy discovery is that of the Fe_3C-type borosilicide. This possesses a small homogeneity range with lattice dimensions:

boron-rich: $a = 4{\cdot}46$, $b = 5{\cdot}36$, $c = 6{\cdot}66$ Å; V = 159 Å^3

boron-poor: $a = 4{\cdot}47$, $b = 5{\cdot}30$, $c = 6{\cdot}67$ Å; V = 158 Å^3

This is indeed smaller than the atomic volume reported[30] for $Fe_3C_{0\cdot 2}B_{0\cdot 8}$ (cf. Section 4.3). This can well be explained by assuming either Fe substitution by Si in the cementite lattice, as in ferrite, to give $(Fe,Si)_3B$, or iron vacancies to give $Fe_3(Si,B)$. The equilibria formed by this phase are not yet certain: it should be possible by further x-ray and neutron-diffraction work to decide between these alternatives.

Figure 10.26 shows the Fe–B–Si and Co–B–Si systems derived by Aronsson and co-workers[79]. The complexity for iron contrasts indeed with the relatively simple equilibria for cobalt, the outstanding features being the pseudo-binary sections Co_2B–Co_2Si and CoB–CoSi as well as, effectively, Co_2B–CoSi and

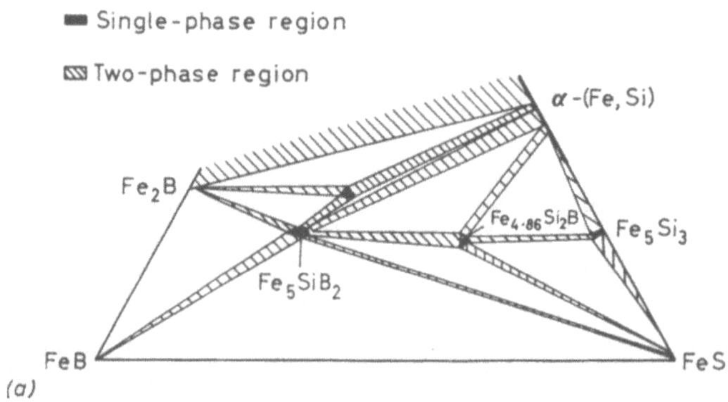

Figure 10.26. The Fe–B–Si *and* Co–B–Si *systems* [from Aronsson and Engström[79d] (*a, b*); Aronsson and Lundgren[79c] (*c*)]

(*a*) The Fe–Si–B system in the composition range 50–70 at. per cent iron at around 1,000°C; the narrow *a*–(Fe,Si,B) single-phase region is only represented by a line; (*b*) outline of Fe–Si–B system; most of the two-phase regions only represented by a line; (*c*) Co–Si–B system at 1,000°C; one-and two-phase regions greatly exaggerated for clarity

By courtesy of Messrs. Munksgaard, Copenhagen

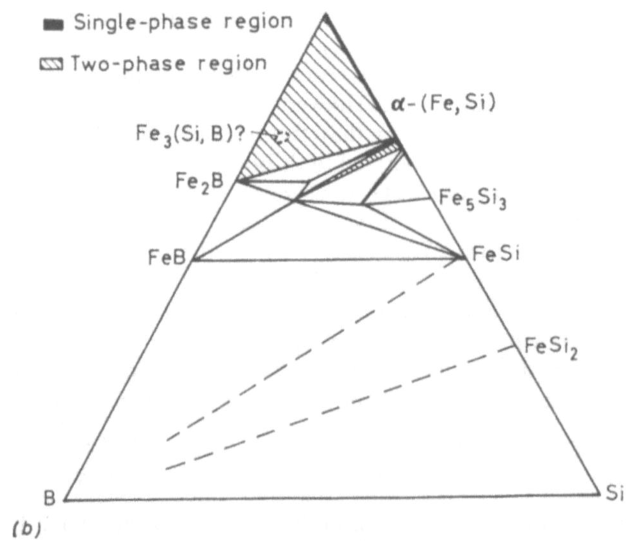

Figure 10.26b

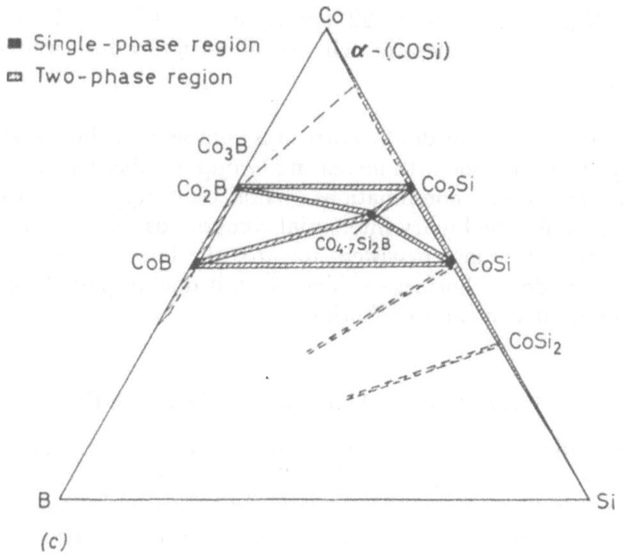

Figure 10.26c

Co_2Si–CoB, with the $Co_{4\cdot7}SiB_2$ phase as intermediate compound. The latter possesses a small homogeneity range, reflected by slight lattice-dimensional variations.

The Fe–Si–B diagram for 1,000°C (*Figure 10.26*) shows the 4 ternary silicoborides

$Fe_{4\cdot86}Si_2B$, isomorphous with $Co_{4\cdot7}Si_2B$ (W_5Si_3 structure)

Fe_5SiB_2 (Cr_5B_3 structure), with a fair homogeneity range implying B ↔ Si replacement to give $Fe_5Si_{1-x}B$; $x \approx 0$–$0{\cdot}1$

$Fe_2Si_{0\cdot4}B_{0\cdot6}$ (structure unknown)

Fe_3(Si,B) (the Fe_3C-type phase)

The binary iron borides and iron silicides are, with one exception, completely insoluble in each other, i.e. no Si–B interchange takes place as in the earlier Groups. For instance, free Fe_2B occurs already in Fe–Si ferrite with very little boron present. The exception is Fe_5Si_3 which can take some boron into solution, as it is again of the D–8_8 structure with its octahedral holes into which boron can fit. The same occurs in Mn_5Si_3.

The Fe–Si–B system is in a unique position in containing among its various compounds one representative of each of the (binary) M_5Si_3 and M_5B_3 types of structure discussed above.

The Mn–Si–B system, only cursorily studied in Aronsson's work, indicated even greater complexity. It includes a compound Mn_5SiB_2 analogous to the above, so that Cr_5B_3 and the Mn_5Si_3-type structures equilibrate here.

Viewing the series of systems Cr–Si–B → Mn–Si–B → Fe–Si–B → Co–Si–B in conjunction, we note the continuity of structure changes.

Finally, in the system Ni–Si–B, not determined yet, a compound Ni_6Si_2B has been reported[81], structurally isomorphous to the phosphides Mn_2P,

Fe_2P and Ni_2P (hexagonal C–22 type): $a = 6{\cdot}105$, $c = 2{\cdot}895$ Å; $V = 99{\cdot}4$ Å^3; $M = 1$ molecule/unit cell (its full structure has been successfully determined).

Looking at the silicoborides together, it is notable how the marked decrease in stability of compounds with increasing Group number which prevailed for the smaller-interstitial combinations, tends to disappear in the *M*–Si–B systems and as, for the binary interstitial compounds, the stability of ternary phases becomes the more persistent towards the later Groups, in the order nitrides → carbides → borides → silicides, so it does in paired combinations, most pronouncedly in the silicoborides.

10.12 OXY- AND HYDROBORIDES

Little systematic information appears to be available regarding the interstitial systems *M*–B–O and *M*–B–H as distinct from the extensive knowledge on the non-metallic compounds, borates and boron–hydrogen complexes.

Insofar as some methods of technological preparation of the borides involve reduction of metal oxides with B_2O_3 by third elements (e.g. Al, Si, Mg, C) or, on the other hand, vapour-phase deposition using hydrogen and halogens (e.g. $ZrCl_4 + BCl_3 + H_2 \rightarrow ZrB_2 + HCl$), it would follow that, at some intermediate stage, lattice oxygen or hydrogen could be retained before final purification of the boride; this, however, does not appear to have been studied from the constitutional aspect.

10.13 SILICO-OXIDES

Atomic radii, Å	Si 1·17	O (unionized) 0·66	*Ratio* Si/O ≈ 1·8
	(subject to variations with metal surround)		
Group No.	IVB	VIB	
Electron structure	$(1s)^2(2s)^2(2p)^6(3s)^2(3p)^2$	$(1s)^2(2s)^2(2p)^4$	

The vast subject of silicates is here clearly excluded, but it may be well to recall the eventual tendency to ionization and to formation of the SiO_4 tetrahedral configurations, which must be latent even in metallic and compound solid solutions of Si and O, readily becoming active under suitable thermal conditions, e.g. at the oxygen-exposed free surfaces of silicides. The lead-over to high-temperature oxidation and protective silicide coating layers is obvious.

In their work on the interreactions of the refractory silicides with C and N, Brewer and Krikorian[29b] have also, in lesser part, considered that with oxygen. *Figure 10.27* shows their provisional diagram for Zr–Si–O (1,680°C). Small additions of oxygen behave similarly to those of the two other interstitials in stabilizing the D–8_8 structure from Zr_5Si_3, but here it is a binary-based solution. Likewise, work[18e] in the [Group V and VI metal]–Si–O_2

systems had shown this type of structural relationship (as discussed already for C and N), but not in form of genuine ternary compounds. The individual results for various base metals are

V: V_5Si_3 + 10 at. per cent O_2 gave the T–1-type phase with built-in oxygen (nothing on Nb)
Ta: Ta hydride + Si + SiO_2 alloys (unspecified) gave T–2 + some D–8_8; for higher oxygen also Ta_2Si* (→ Ta_xO?), presumed to be due to SiO forming
Cr: Cr + Si + SiO_2 alloys gave T–1 [Cr_5Si_3(O)], but with higher oxygen an increase in Cr_3Si, indicative of a Cr_3Si→Cr_3O* solid solution
Mo: Mo + Si + SiO_2 similarly gave T–1 [Mo_5Si_3(O)] and Mo_3Si* which again would probably be oxygen-bearing (Mo_3Si→Mo_3O)
W: W–Si + SiO_2 gave T–1 [W_5Si_3(O)] and free tungsten*.

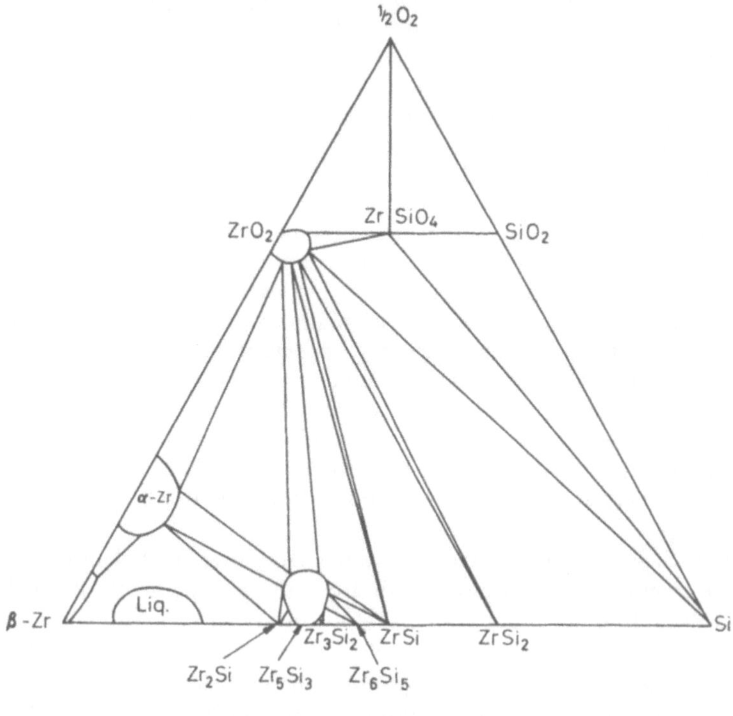

Figure 10.27. The Zr–Si–O *phase equilibria,* 1,680°C (From Brewer and Krikorian[29c])

By courtesy of The Electrochemical Society, Inc.

Whilst in Group IV the metal oxide can coexist with the silicide, this is not so in Groups V and VI; no free metal oxide remains stable in the presence of the stable silicides or of SiO_2.

Referring to Brewer and Krikorian's[29c] Zr–Si–O systems (*Figure 10.27*), we should note the join now of silicon to ZrO_2, precluding $ZrSi_2$ from SiO_2. The reason is that silicon monoxide at the high temperatures in question

* Note the stabilization by oxygen of the lower silicides, particularly of β–W and related types, suggestive of formation of a lower silicide → oxide series, M_3(Si,O) (Ta_2Si → Ta_xO?) and simultaneous reduction SiO_2 → SiO. The tungsten exception is interesting in that it also supports the non-existence of W_3Si as the lower silicide.

already has a considerable partial pressure, so that ZrO_2 and silicon would react

$$ZrO_2 \text{ (solid)} + 4Si \text{ (liquid)} \rightarrow 2SiO \text{ (gaseous)} + ZrSi_2$$

and $ZrSi_2$ can form without preventing Si and ZrO_2 from coexisting. The direct joins from the Zr silicides to ZrO_2 are noteworthy, as well as the equilibration of Zr_3Si_2 with 'Zr_3O' (which is the α-Zr-derived pseudo-oxide; see Chapter 8). A further dominant feature is that the stable, and as a refractory material so important, $ZrSiO_4$ (zircon) equilibrates with free oxygen, in line with ZrO_2 and SiO_2, as well as with silicon. For Zr, the $ZrSi_2 \leftrightarrow SiO_2$ equilibrium is prevented in Group IV, but such an equilibrium $MSi_2 \leftrightarrow SiO_2$ does materialize in Groups V and VI where the pure metal oxides become progressively less refractory and more votatile at high temperatures. Thus $MoSi_2$ is certainly known to coexist with SiO_2, which partially explains its high oxidation-resistant quality (cf. Chapter 7), whilst the less protective nature of the *niobium* silicides ($NbSi_2$, Nb_5Si_3) may in part be connected with their borderline ('labile') equilibrium relation vis-à-vis silica. Few details on *M*–Si–O diagrams for the Group IV–VI refractory metals other than Zr seem to have yet become available, in contrast to the fully explored Fe and Mn Group silicate systems.

During the oxidation of the refractory silicides, especially $MoSi_2$, etc. with their excellent oxidation resistance at high temperatures, these borderline equilibrium relations become obviously important. Thus whilst the M_5Si_3 compounds will dissolve significant amounts of oxygen, the disilicides will not, but immediately shroud themselves in a silica layer*, which could be amorphous, protective and self-healing at high temperatures (see Chapter 7). At the same time, free metal oxide formation (MoO_3, Nb_2O_5, etc.) would be inhibited but become operative upon break-down ($MoSi_2 + O_2 \rightarrow MoO_3 + 2SiO_2$).

The Fe–Si–O equilibria have been thoroughly investigated in connection with the steel-making processes, both from the viewpoint of silicon being used as deoxidizer in liquid iron ($2FeO + Si \rightleftharpoons SiO_2 + Fe$) and of reactions of iron oxides with silica in refractory bricks or linings. The vast subject of slag reactions (cf. Bibliography) impinges on this. The oxidation of silicon steels (in the solid) has also given interest to the system. Gokcen and Chipman[82], for example, have studied the Si/O equilibria in liquid iron and derived the activity coefficients and rates of oxidation.

Ferrous oxide, magnetite and hematite form stable equilibria with silica in its various forms (vitreous, cristobalite, tridymite, quartz), and iron oxides themselves can strongly affect these transformations (e.g. produce tridymite in silica bricks). They also form a stable ternary compound, fayalite ($2FeO.SiO_2$), and a metastable one, grunerite ($FeO.SiO_2$); *Figure 10.28* includes the relevant FeO–SiO_2 diagram. It may be noted that SiO_2 forms a ternary compound only with ferrous, not with ferric oxide. Their dissociation conditions and equilibria are sensitively dependent on the prevalent gas atmosphere and pressure. Not only oxygen is important here but also

---

* Or intermediary silicon monoxide, SiO, which reaches a high vapour pressure at moderately high temperatures already.

CO and CO_2 and their ratios. Basically the [Fe–O–CO–CO_2–SiO_2] quaternary system should really be regarded as one equilibrium complex in logical conjunction, not only because of its practical importance in iron and steelmaking but also because the carbonaceous and the siliceous oxides behave homologously towards iron, though obviously differing in their states. Darken[83] has studied these Fe–Si–O relationships in detail, and *Figure 10.28b–d* shows the three diagrams derived by him, as a function of gas composition and temperature. The various phases, FeO, Fe_3O_4, Fe_2O_3, SiO_2 and 2FeO.SiO_2, enter into the varying equilibria indicated.

The partial system FeO–Fe_2O_3–SiO_2 has been investigated by Muan and Osborn[84a] and by Muan[84b], the equilibria between wüstite, hematite, fayalite and silica being determined in detail, including the melting-point surfaces. One specific item of interest is that the fayalite (2FeOSiO_2) homogeneity area lies slightly inside the ternary field, not strictly on the

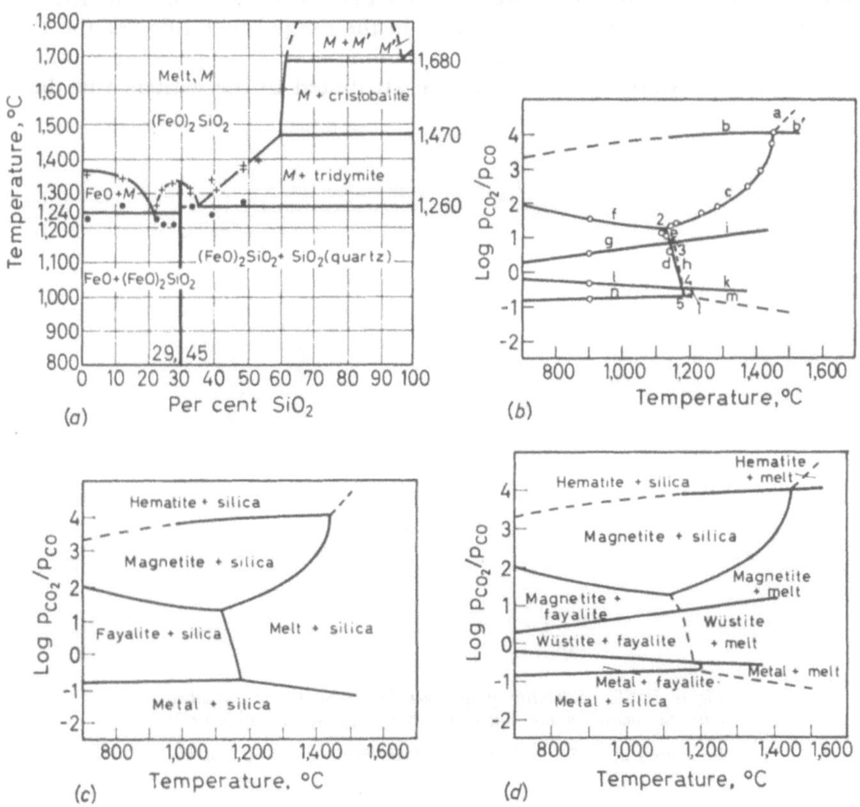

Figure 10.28. The Fe–Si–O *system*: (*a*) FeO–SiO_2 *section*; (*b*)–(*d*): Fe–Si–O *equilibria under varying gas pressure and temperature conditions* [From Schenck[90] (*a*) and Darken[83] (*b–d*)]

(*b*) possible univariant equilibria; the invariant points shown correspond to the following equilibrating phases: *1* Fe_2O_3, Fe_3O_4, SiO_2, melt, gas *2* Fe_3O_4, 2FeO.SiO_2, SiO_2 melt, gas *3* Fe_3O_4, 'FeO', 2FeO.SiO_2, melt, gas *4* 'FeO', 2FeO.SiO_2, metal, melt, gas *5* 2FeO.SiO_2, metal, SiO_2, melt, gas (2FeO.SiO_2 = fayalite); (*c*) case of excess SiO_2 present (as in contact with refractory); (*d*) low Si/Fe ratios (as in a silicon steel when oxidizing)

By courtesy of The American Chemical Society and of the British Iron and Steel Research Association

FeO–SiO_2 pseudo-binary join, which means that some ferric ions are required for its stabilization or, indeed, that there is a Fe-defect lattice, analogous to 'FeO'.

A basic point here is that of silicate stabilization by FeO and by MnO*; this is quite in contrast to the earlier-Group (IV–VI) oxides where the stable, though oxygen-bearing, *silicides* tend to prevail except for the special case of the zircon ($ZrO_2.SiO_2$) structure of Group IV.

The solid solubility of oxygen in the iron silicides is very limited; however, conversely it seems possible that primary Fe_3O_4 is capable of dissolving significant amounts of silicon, considering the newer Si-bearing spinels; this, however, would need verification.

10.14 HYDROSILICIDES

Insufficient information seems available on interstitial *M*–Si–H systems. However, the solubility of hydrogen in molten Fe–Si alloys was determined[85], showing that, for any given temperature in the range 1,400–1,650°C, the H solubility decreases with Si content to a minimum at 50 at. per cent Si, and then increases towards Si. This can be attributed to the existence in the

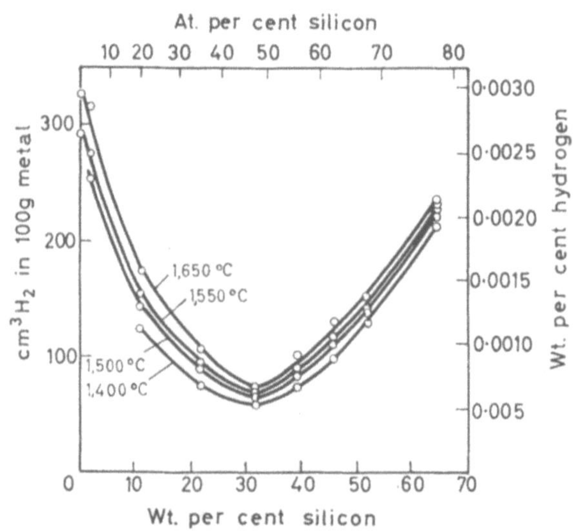

Figure 10.29. Solubility of hydrogen in Fe–Si *alloys: variation with* Si *content in the liquid* (From Liang, Bever and Floe[85])

By courtesy of the Editor, *Transactions of the Metallurgical Society of AIME*

liquid of particles of the undissociated FeSi phase with lower inherent H solubility. The Fe–Si diagram (*Figure 7.2*, maximum melting point at FeSi) could well support this suggestion. The solubility steadily increases with temperature at a given silicon content (*Figure 10.29*).

---

* To give the very stable compounds $2MnO.SiO_2$ (tephroite) and $MnO.SiO_2$ (rhodonite; note its stability, important in slag reactions, in contrast to unstable $FeO.SiO_2$).

10.15 OXYHYDRIDES

	O (unionized)	H	*Ratio* O/H
Atomic radii, Å	0·66	$0{\cdot}28^{+}$	≈ < 2·4
	(subject to variations with metal surround)		
Group No.	VIB	—	
Electron structure	$(1s)^2(2s)^2(2p)^4$	$(1s)^1$	

We have to concentrate on interstitial solid solutions and to exclude the whole extensive field of metal hydroxides (e.g. $Fe(OH_2)$, etc.). Reference to the wide literature available on hydroxide chemistry, wet corrosion, etc. must suffice. Nevertheless, the tendency to form the strong hydroxyl bond will exist latently already even in solid solutions of oxygen in a metal hydride or of H in an oxide, and become active at elevated temperatures (before actual dissociation).

The Ti–O–H system has been studied from the thermodynamic viewpoint[86], and the equilibrium hydrogen pressures in alloys with varying Ti–O ratios have been determined. *Figure 10.30* shows the Ti corner of the ternary

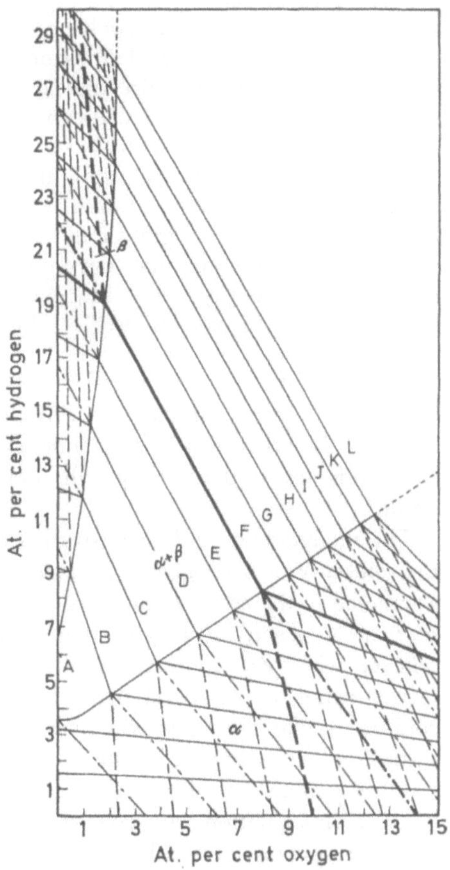

Figure 10.30. The Ti–O–H *system,* 800°C *section* (From Hepworth and Schuhmann[86])

Symbol	A	B	C	D	E	F
$^{\alpha}H$	2·18	3·00	4·00	5·00	6·00	7·00
$^{\alpha}O$	0·000	1·000	2·118	3·424	4·927	6·644
$^{\alpha}Ti$	0·9640	0·9294	0·8935	0·8579	0·8241	0·7911

	G	H	I	J	K	L
$^{\alpha}H$	8·00	9·00	10·00	11·00	12·00	13·00
$^{\alpha}O$	8·596	10·804	13·296	16·051	19·060	22·315
$^{\alpha}Ti$	0·7597	0·7299	0·7019	0·6754	0·6506	0·6274

Activity of hydrogen ———
oxygen - - - - -
titanium – – – –

By courtesy of the Editor, *Transactions of the Metallurgical Society of AIME* system thus derived, for 800°C, including the boundary variations of the α- and β-Ti fields. It is also of special theoretical interest that from the Ti and O atom activities the indication is of an *ordered* structure 'Ti_3O' in HCP Ti–O solid solution, with $\frac{1}{3}$ of the octahedral sites being filled; this could be deduced by extrapolation of the activities in the ternary (H-bearing) alloys.

The solubility of hydrogen in oxygen-bearing α- and β-Ti diminishes with increasing oxygen content[87]. When relating these effects to the creep resistance of the Ti–O–H alloys, it was found that, whilst oxygen improves creep resistance by enhancing dislocation locking, hydrogen reduces it by cutting down the solubility and lowering the transformation temperature. A comparison of the Ti–O and Ti–H diagrams (*Figures 8.1* and *9.1*) makes the contrary effects immediately apparent.

Somewhat similar results have been obtained on the Zr–O–H system by Edwards and Levesque[88] (in extension of their work on Zr–H; cf. Chapter 9). A selection of P–T–C graphs, for three oxygen levels and comparable H content, temperature and pressure scales, is shown in *Figure 10.31* (see also *Figure 9.1* as Zr–H base line); also included is the provisional isothermal section (for 750°C and 1 atm) of their ternary system Zr–ZrO_2–ZrH_2. This proves most instructive, showing (1) how the basic Zr–H solutions discussed can themselves extensively dissolve oxygen, but decreasingly so, for increased hydrogen, up to O:H $\approx$ 1:3 (atomically) in α-(Zr–H) and $\approx$ 1:7 in δ-(Zr–H); and (2) how zirconia equilibrates with the α solution and the ε-'hydride' ZrH_2, which is really the distorted version of the BCC β-Zr–H solution. Oxygen can essentially substitute the interstitial hydrogen, though preferentially in octahedral sites, up to the saturation level, when ZrO_2 forms. Conversely, ZrO_2 cannot dissolve any hydrogen, consistent with its bond nature which is rigidly ionic, as against the metallic bond in the Zr oxyhydrides. Edwards and Levesque[88] conclude that 'each oxygen atom which enters the [hydride] lattice denies more of the interstitial sites to hydrogen than to oxygen itself. This would seem to indicate a form of repulsion through general lattice strain and rule out any special stabilization one might have anticipated to arise from proximate hydrogen. Effects due to the difference in electron affinities between O and H thus seem to be absent'. In other words, no tendency to hydroxyl bonding arises yet in zirconium—as it doubtlessly will, however, in the later transition-metal groups.

Thermodynamic (and lattice-statistical) work[89] on the effect of oxygen on the solubility of hydrogen in Zr confirms that oxygen diminishes it; this is attributed to the fact that lattice sites normally available to H in tetrahedral sites are 'blocked' by neighouring O atoms in octahedral ones. This is relieved when ZrO_2 forms, but the H solubility is still determined by the concentration of Zr metal then remaining. Though here specific to zirconium, similar reasoning will certainly remain valid more generally to the transition metals, except for more immediate oxide formation as the hydride stability diminishes with increasing atomic number.

Little information appears accessible on the lower-than-hydroxide *M*–O–H systems for the later Groups (other than Fe–O–H), apart from their many applied aspects, e.g. of oxide reduction processes and the use of hydrides as primary materials for pure-alloy production where the hazard of oxygen has to be counteracted.

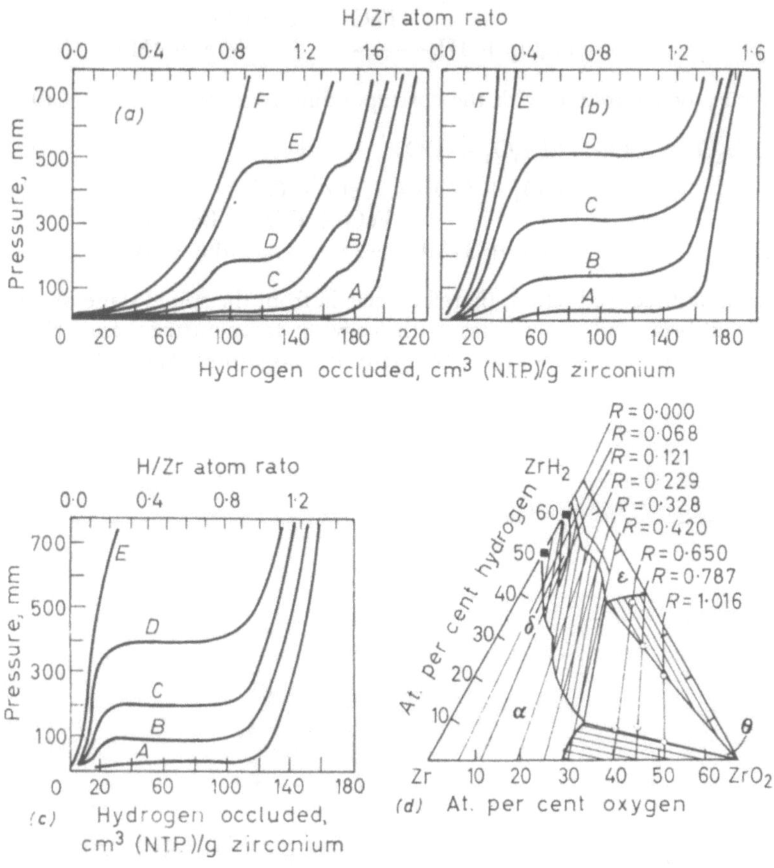

Figure 10.31. *The* Zr–O–H *system*: P–T–C *curves and a ternary section* (From Edwards and Levesque[88])

(*a*)–(*c*) isothermal pressure–volume curves for alloys containing initially (*a*) 6·3, (*b*) 18·6, (*c*) 29·6 at. per cent oxygen, in the temperature range 600°–900° (*a*, *b*) 600°–850°C (*c*); (*a*) R = 0·068; (*b*) R = 0·229; (*c*) R = 0·420; A 600°, B 700°, C 750°, D 800°, E 850°, F 900°C; (*d*) ternary system Zr–O–H; provisional diagram for the condensed phases; isothermal cross-section at 750°; two-phase regions shaded, lines approximating tie lines; three-phase regions solid; extension of boundaries beyond experimental data by inference; lines leading to R values represent experimental composition traverses as hydrogen was successively added to Zr–O solid-solution specimens of constant O/Zr atom ratios, R; dotted line contour of variations in composition at constant hydrogen pressure of 760 mm

By courtesy of The American Chemical Society

A multiplicity of researches (together with Fe–O–C), has been devoted to the Fe–O–H system in a sense throughout history, ever since the reduction of iron oxides to iron became interesting. The relevant reaction stages by H_2

(note again the structural hematite- $\rightarrow$ spinel- $\rightarrow$ NaCl- $\rightarrow$ (FCC,BCC) type series discussed in Chapter 8) are

$$3Fe_2O_3 + H_2 \leftrightarrow 2Fe_3O_4 + H_2O$$
$$Fe_3O_4 + H_2 \leftrightarrow 3FeO + H_2O$$
$$FeO + H_2 \leftrightarrow Fe + H_2O$$
$$Fe_3O_4 + 4H_2 \longleftrightarrow 3Fe + 4H_2O$$

They may be compared with the corresponding reaction stages for CO

$$3Fe_2O_3 + CO \leftrightarrow 2Fe_3O_4 + CO_2$$
$$Fe_3O_4 + CO \leftrightarrow 3FeO + CO_2$$
$$FeO + CO \leftrightarrow Fe + CO_2$$
$$Fe_3O_4 + 4CO \longleftrightarrow 3Fe + 4CO_2$$

The ease of direct reduction, therefore, sets in from the spinel stage onwards. The equilibria and reactivities with varying temperatures and H_2O/H_2 pressures have been worked out by many investigators; *Figure 10.32* shows

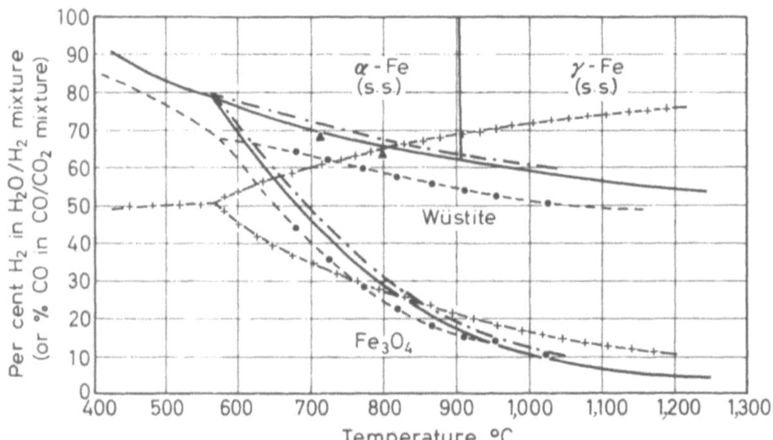

(a)

Figure 10.32. Fe–H–O *equilibria*: (*a*) *phase relations of* H_2–*water vapour mixtures over iron and iron oxides* (From Schenck[90]); (*b*) Fe–H–O *at* 1 *atm.* (From Wiberg[91]); (*c*) Fe–C–O *at* 1 *atm for comparison with b* (From Wiberg[91])

By courtesy of the British Iron and Steel Research Association and of the Editor, *Jernkontorets Annaler*

a summary phase diagram* by Schenck[90] together with one by Wiberg[91]. A related one of Wiberg's on Fe–C–O is also included, as it brings out the similarities and differences. There is no counterpart to the metastable cementite, although we might liken it to ferrous hydroxides existing at low temperatures.

* Schenck comments that the bold lines are probably the more correct ones.

The complex ferrous and ferric hydroxyl equilibria, involving principally FeO(OH)* in several crystallographic forms, and the minerals limonite, goethite, lepidocrocite, turgite, with associated subjects such as wet corrosion and rusting kinetics, can only be hinted at here (cf. e.g. Evans' book[92]).

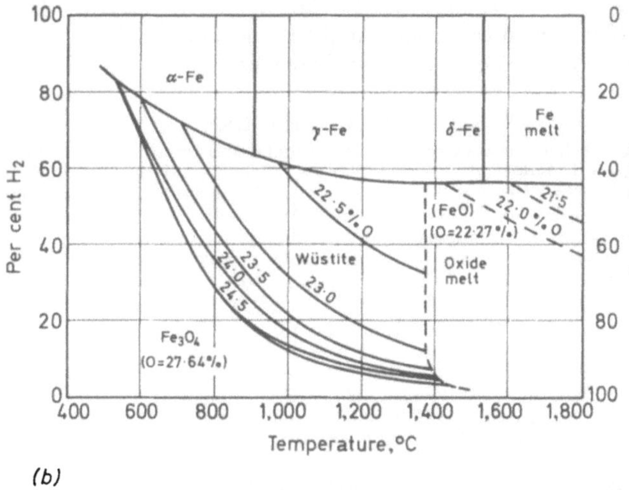

Figure 10.32b

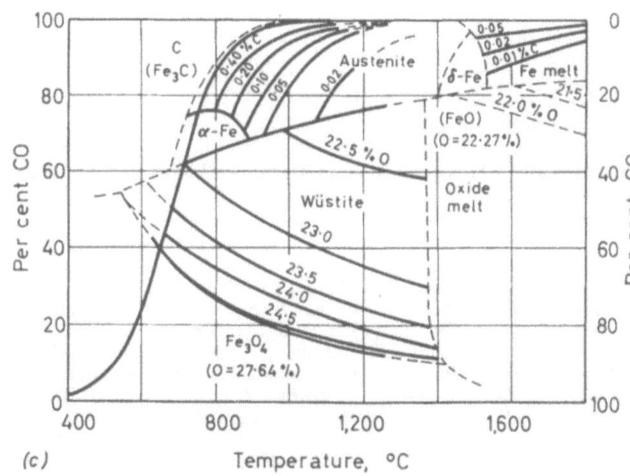

Figure 10.32c

One final point might be made in the context of hydride–oxide equilibria. It had been noted in Chapter 9 that some of the early hydrides and deuterides, MgH_2, MgD_2, can adopt the rutile-type structure of TiO_2. This would,

* Hydroxides reported with more water molecules than corresponds to the $Fe_2O_3.H_2O$ stoichiometry are now believed not to be crystallographically genuine but rather to have this water loosely held by adsorption, possibly in defects and microvoids.

in principle, suggest the wider possibility of forming complete or partial solid solutions between such terminal oxides and hydrides, and of replacement, say in the rutile lattice, of oxygen by hydrogen ions, provided that balanced metal/metal replacement occurs (depending on size and electron factors). The sodium chloride-type structure similarly recurs in both the Group IIA oxides (CaO, etc.) and the Group IA hydrides (NaH, etc.), so that similar balanced substitutions might be possible, and also, metastably, in some of the later-Group oxides and hydrides.

SELECTED LITERATURE

General note—Comprehensive work on mixed hard metal and refractory compounds and the underlying systems is largely due to the Plansee, Vienna, and the associated American Yonkers schools[5,18,22–3,26,28,32,35,67,73]; to Brewer and associates[23] (more on thermodynamics); to Stadelmaier[27] (constitutional), Brauer[6], Aronsson[79], Geach and Jones[1]; in Russia, the school of Samsonov[21,53] is amongst those prominent in this field. Alloy steels and the underlying mixed systems, the principal other branch of mixed interstitial compounds, are studied at numerous centres.

M–C–N

L'vov, S. M., Nyenchenko, V. P. and Samsonov, G. V. *Ukr. fiz. Zh.* 7 (1962) 331 [TiC–TiN electrical properties—resistivity, Hall coefficients, EMF]; Samsonov, G. V. *et al.*, *Dopov. Akad. Nauk. URSR* (1958) 838 [Magnetic susceptibilities]; Vainshstein, E. E. and Zhurakovskii, A. *Dokl. Akad. Nauk S.S.S.R.* 140 (1961) 560 [Hardness, thermal and chemical stability of TiC–TiN, also in relation to electron structure changes]; Beattie, H. J. and Ver Snyder, F. R. *Trans. Am. Soc. Metals* 45 (1953) 397 [Alloy steel constituents bearing Ti, C and N]

Nb and Ta–C–N systems: Brauer, G. in *Niobium, Tantalum, Molybdenum and Tungsten*, Ed. Quarrell, A. G., Amsterdam (Elsevier) 1961; Dulis, E. J. and Smith, G. V. *Trans. Am. Soc. Metals* 44 (1952) 877 [Nb–C–N]; Schönberg[52] [Ta–C–N; solubilities and lattice dimensional changes in $Ta_2(C,N)$ and Ta(C,N) series]

Fe–C–N system: Jack, K. H.[2]; *Australas. Engr.* (1950) 55 [Summary of more technological implications of his findings, surface hardening of steels by nitriding, carburizing or carbonitriding]; Bridelle, R. and Michel, A. *C.r. hebd. Séanc. Acad. Sci., Paris* 239 (1954) 274; *Rév. Metall.* 51 (1954) 278; 52 (1955) 1397; [Co–C–N phases:] Clarke, J. and Jack, K. H. *Chemy Ind.* 46 (1951) 1004; Köster, W. *Arch. EisenhüttWes.* 4 (1930) 145 [Fe–C–N system: early, still valuable]

M–C–B[18–31]

Forney and Marshall[24] [Ti boronitride and Ta borocarbide-based materials systems]

M–C–Si[18c,29b,32–46]

Fe–Si–C systems (basic and applied aspects of iron and steel making): Chipman, J. and Baschwitz, R. *Trans. metall. Soc. A.I.M.E.* 227 (1963) 473; Chipman, J. *J. Iron Steel Inst.* 180 (1955) 99; Smith, R. P. *J. Am. chem. Soc.* 70 (1948) 2734; Rickett, R. L. and Fick, M. C. *Metals Technol.* 13 (1946) 9; Bergkampf, E. S. *Stahl Eisen* 64 (1944) 200; Darken, L. S. *Trans. Am. Inst. Min. metall. Engrs* 140 (1940) 204; Andrew, J. H. *et al.*, *Iron Steel Inst., Spec. Rep.* 25 (1938) 293; Hanemann, H. and Jass, H. *Honda Anniv. Vol.*, Sendai Univ. (1936) 793; Zhukov, A. A. *Izv. Nauk. SSSR, Otdel. tekh. nauk, Metallurgiya Topl.* 6 (1961) 88; Malinochko, Ya. M. *et al.*, *Metallov. term. Obrab. Metall.* 7 (1961) 15; Lux, B. and Tannenberger, H. *Mod. Cast.* 41 (1962) 57; Vashchenko, K. I. and Rudio, A. P. *Izv. vyssh. ucheb. Zaved. Khim. Met.* 3 (1961) 11 [Surface tension]; Mayer, A. and Pfeiffer, I. *Z. angew. Phys.* 15 (1963) 243; Tkachenka, F. K. and Zubarev, V. F. *Russ. Cast. Prod.* 8 (1962) 378; van Eeghem, J. *Métallurgie, Mons* 2 (1961) 115; Payson, P. *Metal Prog.* 78 (1960) 78; Chandhok, V. K. 80 (1962) 82 [Decarburization of steel by silicon]; Feldmann, U. *Arch. EisenhüttWes.* 34 (1963) 109 [Thermodynamic activities in Fe–Si–C system at steel-making temperatures]; H. Schenck's and J. F. Elliott's books on the *Physical Chemistry of Steel Making* (see below) and similar texts on steel making and on cast irons (e.g. papers in *B.C.I.R.A.Jl*) also cover the important Si–C reactions with *manganese*, *sulphur* and other participating elements besides those of iron)

Fe–C–Si, Fe–C–O, Fe–Si–O and allied systems, jointly [basic equilibria]: Bodsworth, C. *Physical Chemistry of Iron and Steel*, London (Longmans) 1963; Ward, R. G. *The Physical Chemistry of Iron and Steel Making*, London (Arnold) 1962; Elliot, J. F. (Ed.) *The Physical Chemistry of Steel Making*, New York and London (Wiley, Chapman & Hall) 1956; Schenck[90] [Older but more detailed and still valuable]

REFERENCES

M–C–O[21b,47–59]

Samsonov and Umanski[21c] [Discussion of Ti–C–O(–H) reaction kinetics; also those of $ZrO_2 \rightarrow ZrC$; $V_2O_5 \rightarrow VC$, $Nb_2O_5 \rightarrow NbC$, $Ta_2O_5 \rightarrow TaC$; $Cr_2O_3 \rightarrow Cr_3C_2$; $MoO_3 \rightarrow Mo_2C$ and $WO_3 \rightarrow WC$]

Fe–C–O system: Cameron, D. J. and Barratt, K. V. *J. Iron Steel Inst.* 201 (1963) 21; Krasheninnikov, M. G. and Filippov, S. I. *Izv. vyssh. uched. Zaved., Chernaya Met.* 9 (1961) 21; Wiberg, M. *Jernkont. Annlr* 124 (1940) 179; Hultgren, A. and Phragmén, G. *J. Iron Steel Inst.* 140 (1939) 1216; Edwards, C. A. *Spec. Rep. Iron Steel Inst.* 27 (1939); Matoba, S. *Honda Anniv. Vol.*, p. 348, Sendai Univ. 1936; Ziegler, N. A. *Trans. Am. Soc. Steel Treat.* 20 (1932) 73; Jänecke, E. *Z. anorg. allg. Chem.* 178 (1929) 73; Schenck[90] [Thermodynamics of Fe–C–O reactions; heats of formation and equilibria involved]

Carbonyls

Reviews: Abel[59]; Guy, R. G. and Shaw, B. L.; Fisher, E. O. and Fritz, H. P. *Adv. inorg. Chem. Radiochem.* 1 (1959) 55; 4 (1962) 77; Chatt, J., Pauson, P. L. and Venanzi, L. H. in *Organometallic Compounds*, Ed. Zeiss, New York (Reinhold) 1961, p. 468; Coates, G. E. *Organometallic Compounds*, London (Methuen) 1960; Cable, J. W. and Sheline, R. K. *Chem. Rev.* 56 (1956) 1; Anderson, S. A. *Q. Rev. chem. Soc.* 1 (1947) 33

Ferguson, I. F. *et al.*, *Nature*, Lond. 202 (1964), 1327 [$Mo(OH_6)$]

M–C–H[60–2]

M–B–N[18a,d,21c,23b,24,29a,63–6]

Brewer and Haraldsen[23b] [Thermodynamic data]; Samsonov and Petrash[21c] [Melting-point variations; hardness; electrical resistance; expansion; oxidation resistance]

M–Si–N[18f,29b,32c,67–73]

Karnaukhov, M. M. and Morozov, A. N. *Izv. Akad. Nauk SSSR* (1947) 733 [Fe–Si–N]

M–O–N[10,52b,c,63b,74–7,9]

Kiessling, R. *Acta crystallogr.* 7 (1954) 635 [W oxynitrides]

M–Si–B[1,18b,f,g,32d,35b,64,78–81]

Kieffer, R. and Benesovsky, F. *Powder Metall.* 1/2 (1958) 145 [(Ti,Zr,Hf)–B–Si]; Nowotny, Kieffer and Benesovsky[18g] [(V,Nb,Ta,Mo,W)–B–Si]; Kudielka, Nowotny and Findeisen[78] [(V,Ta)–B–Si]; Nowotny, Benesovsky, Rudy and Wittmann[18d] [Nb–B–Si]; Nowotny, Piegger, Kieffer and Benesovsky[18b] [Cr–B–Si]; Nowotny, Dimakopoulou and Kudielka[18g] [(Mo,W)–B–Si]; Nowotny[18h] [Summary of Plansee work on mixed B–Si systems]

M–Si–O[18f,29b,c,82–4]

Gokcen, N. A. and Chipman, J. *J. Metals, N.Y.* 4 (1952) 171; Ward's, Bodworth's, Elliott's and Schenck's books on the physical chemistry of steel making [Fe–Si–O slag reactions]; Chesters, J. H. *Steel Plant Refractories*, Sheffield (United Steels) 1963

Levin, M. E., Robbins, C. R. and McMurdie, H. F. *Phase Diagrams for Ceramists*, Columbus, Ohio (American Ceramic Society) 1964 [Numerous *M*–Si–O diagrams, more from the silicates aspect); Matoba, S. and Ban-ya, S. in '*The Physical Chemistry of Steel Making*', (Ed. Elliott), p. 42 [Part of Fe–Si–C–O system in the liquid: C–O equilibrium in molten C-saturated iron; effect of silicon addition up to 5·2 per cent; Fischer, W., *ibid.* p. 117 [Deoxidation equilibria by Si in liquid iron]; Darken[83] [Silica activities; melting points of iron oxides in contact with silica for various (O_2, CO or CO_2) pressures, also in relation to scaling process in the solid]; Darken, L. S. and Gurry, R. W. *Physical Chemistry of Metals*, New York (McGraw-Hill) 1953

M–H–Si[85]

M–O–H[86–92]

REFERENCES

1 Geach, G. A. and Jones, F. O. *Planseeber. PulvMetall.* 3 (1955) 80

2 Jack, K. H. (*a*) *Proc. R. Soc.* A195 (1948) 41; (*b*) 208 (1951) 216; (*c*) *Acta crystallogr.* 3 (1950) 392; (*d*) 5 (1952) 404

3 Duwez, P. and Odell, F. *J. electrochem. Soc.* 97 (1950) 299

4 Kotelnikov, R. B. *Zh. neorg. Khim.* 3 (1958) 841

5 Bittner, H. *et al.*, *Mh. Chem.* 94 (1963) 518

6 (*a*) Brauer, G., with Schnell, W. D. *J. less-common Metals* 7 (1964) 23; (*b*) with Lesser, R. *Z. Metallk.* 50 (1959) 487; (*c*) *ibid.* 512; (*d*) with Jander, J. *Z. anorg. allg. Chem.* 270 (1952) 160

7 Andrews, K. W. *J. Iron Steel Inst.* 163 (1949) 384

8 Goldschmidt, H. J. *J. Iron Steel Inst.* 162 (1948) 345; 163 (1949) 381

[9] Hume-Rothery, W., Raynor, G. V. and Little, A. T. *J. Iron Steel Inst.* 145 (1942/1) 143P
[10] Stone, L. and Margolin, A. *J. Metals, N.Y.* 5 (1953) 1498
[11] Archarov, V. I., Konev, V. N. and Gerasimov, A. F. *Fizika Metall.* 9 (1960) 695
[12] Gerasimov, A. F., Konev, V. N. and Timofeeva, V. A. *Soviet Phys. Metals Metallogr.* 11 (1961) 106
[13] Williams, J. and Sambell, R. I. *J. less-common Metals* 1 (1959) 217
[14] Street, R. S. and Waters, T. N. *J. less-common Metals* 5 (1963) 295
[15] Rundle, R. E. *et al., J. Am. chem. Soc.* 70 (1948) 99
[16] Philips, V. A. *Acta metall.* 11 (1963) 1139; *Trans. Am. Soc. Metals* 56 (1963) 600
[17] Stackelberg, von M. *et al., Z. phys. Chem.* A175 (1935) 127
[18] Nowotny, H., with Benesovsky, F., Brukl, C. and Schob, O. *Mh. Chem.* 92 (1961) 403; (*b*) with Rudy, E. and Benesovsky, F. *ibid.* 393; (*c*) 91 (1960) 963; (*d*) with Parthé, E., Kieffer, R. and Benesovsky, F. 85 (1954) 255; (*e*) with Lux, B. and Kudielka, H. 87 (1956) 447; (*f*) 89 (1958) 611; (*g*) *Planseeber. PulvMetall.* 5 (1957) 86; (*h*) *Rep. OTS PB* 145949 (1960)
[19] Greenhouse, H. M., Accountius, O. E. and Sisler, H. H. *J. Am. chem. Soc.* 73 (1951) 5086
[20] Portnoi, K. I., Samsonov, G. V. and Frolova, K. I. *Zh. prikl. Khim., Leningr.* 33 (1960) 577
[21] Samsonov, G. V. (*a*) *Vop. poroshk. metall. Prochn. Mater.* 7 (1959) 72; (*b*) with Petrash, J. V. *Metallov. term. Obrab. Metall.* 24 (1955) 19; (*c*) with Umanskii, Ya. S. *Hard Compounds of Refractory Metals, NASA tech. Trans.* F–102, Washington, 1962
[22] Glaser, F. W. (*a*) *J. Metals, N.Y.* 4 (1952) 391; (*b*) *Powder Metall. Bull.* 6 (1951) 54
[23] (*a*) Rudy, E. and Benesovsky, F., with Toth, L. E. *Z. Metallk.* 54 (1963) 345; (*b*) *Mh Chem.* 92 (1961) 415
[24] Forney, G. J. and Marshall, J. J. *Rep. OTS PB* 162925 (1961); AD 253668
[25] Jeitschko, W., Nowotny, H. and Benevsosky, F. *Mh. Chem.* 94 (1963) 65
[26] Steinitz, R. *Powder Metall. Bull.* 6 (1951) 54
[27] (*a*) Stadelmaier, H. H., with Draughn, R. A. and Hofer, G. *Z. Metallk.* 54 (1963) 640; (*b*) with Gregg, R. A. *Metall* 17 (1963) 412
[28] Toth, L. E., Nowotny, H., Benesovsky, F. and Rudy, E. (*a*) *Mh. Chem.* 92 (1961) 791; (*b*) *ibid.* 956
[29] (*a*) Brewer, L. with Haraldsen, H. *J. electrochem. Soc.* 102 (1955) 399; (*b*) with Krikorian, O. 103 (1956) 38, 701; (*c*) *U.S. At. Energy Comm. Publ. Univ. Calif. Rad. Lab.* U.C.R.L. 2544 (1954); 3352 (1956).
[30] Nicholson, M. E. *J. Metals, N.Y.* 6 (1957) 185
[31] Carroll, K. G., Darken, L. S., Fisher, R. M. and Zwell, L. *Nature, Lond.* 174 (1954) 978
[32] (*a*) Parthé, E. *et al., Mh. Chem.* 86 (1955) 859; (*b*) *ibid.* 182; (*c*) unpublished work, quoted [35b]
[33] Matkovich, V. I. and Rogers, H. H. *J. electrochem. Soc.* 108 (1961) 261
[34] Kurnakov, N. N. *Izv. Sekt. fiz.-khim., Analiza Inst.* 17 (1949) 209
[35] Kieffer, R. and Benesovsky, F. (*a*) *Spec. Rep. Iron Steel Inst.* 58 (1954) 242; (*b*) *Powder Metall.* 1/2 (1958) 145
[36] Searcy, A. W. and Finnie, L. N. *J. Am. Ceram. Soc.* 45 (1962) 268
[37] Hilliard, J. E. and Owen, W. S. *J. Iron Steel Inst.* 172 (1952) 268
[38] (*a*) Owen, W. S., with Street, R. S. *J. Iron Steel Inst.* 172 (1952) 15; (*b*) 167 (1951) 117
[39] Hurst, J. and Riley, R. V. *J. Iron Steel Inst.* 149 (1944) I 213
[40] Marles, D. *J. Iron Steel Inst.* 158 (1948) 433
[41] Scheil, E. *Stahl Eisen* 50 (1930) 1725; 52 (1932) 129
[42] Kříž, V. A. and Pobořil, F. *J. Iron Steel Inst.* 122 (1930/2) 191
[43] (*a*) Humphreys, J. G. *B.C.I.R.A. Jl* 10 (1962) 551; (*b*) with Owen, W. S. *J. Iron Steel Inst.* 198 (1961) 38
[44] Leslie, W. C. *et al., Acta metall.* 7 (1959) 632
[45] Konoval, G. *et al., Nature, Lond.* 184 (1959) 1862
[46] Harry, E. D. *J. Iron Steel Inst.* 167 (1951) 241
[47] Meerson, G. A. (*a*) *Russ. Metall. Fuels* 3 (1962) 27; (*b*) *Izv. Sekt. fiz-khim., Analiza Inst.* 6 (1943) 179
[48] Ashbee, K. H. C. and Smallman, R. E. *Phil. Mag.* 7 (1962) 1933
[49] Gurevitch, M. A. and Ormont, B. F. *Dokl. Akad. Nauk SSSR* 96 (1954) 1165
[50] Krainer, H. and Konopicky, C. *Berg- u. hüttenm. Mh.* 19 (1947) 166
[51] Shveikin, G. P. *et al.,* (*a*) *Zh. neorg. Khim.* 8 (1963) 689; (*b*) *Poroshkovaya Metall. Akad. Nauk USSR* 2 (1962) 71
[52] Schoenberg, N. *Acta chem. scand.* (*a*) 8 (1954) 1460; (*b*) *ibid.* 620, 627; (*c*) *ibid.* 208
[53] Kosolapova, T. Ya. and Samsonov, G. V. *Zh. fiz. khim.* 35 (1961) 363
[54] Hinnüber, J. *et al., Plansee Proc.* (1955) 27
[55] Piazza, J. R. and Linnot, M. J. *J. chem. Engng Data* 7 (1962/1) 451
[56] Stoops, R. F. and Hamme, J. V. *U.S. At. Energy Comm., Nucl. Sci. Abstr.* 2663 (1961) 5; 2664 (1962) 9
[57] Peakall, K. A. and Antill, J. E. *J. less-common Metals* 4 (1962) 426

REFERENCES

[58] Edwards, C. A. *Jernkont. Annlr* 122 (1938) 377
[59] Abel, E. W. *Q. Rev. chem. Soc.* 17 (1963) 133
[60] Rexer, J. and Peterson, D. T. *U.S. OTS Rep.* IS–875, 1964; *Chem. Abstr.* 62 (1965) 12543*c*
[61] Peterson, D. T. and Rexer, J. *J. inorg. nucl. Chem.* 24 (1962) 519
[62] Hägg, G. *Z. phys. Chem.* B 12 (1931) 33
[63] (*a*) Kiessling, R., with Liu, Y. H. *J. Metals, N.Y.* 3 (1951) 639; (*b*) with Peterson, L. *Acta metall.* 2 (1954) 675
[64] Schwarzkopf, P. and Glaser, F. W. *Z. Metallk.* 44 (1953) 353
[65] Butler, J. F. *Trans. metall. Soc. A.I.M.E.* 224 (1962) 84
[66] Fountain, R. W. and Chipman, J. *Trans. metall. Soc. A.I.M.E.* 223 (1961) 677; 224 (1962) 559
[67] Kieffer, R., Benesovsky, F. and Lux, B. *Planseeber. PulvMetall.* 4 (1956) 30
[68] Sloman, H. A. *J. Iron Steel Inst.* 182 (1956) 307
[69] Arrowsmith, J. M. *J. Iron Steel Inst.* 201 (1963) 699
[70] Leak, D. A. *et al., Acta metall.* 3 (1955) 501
[71] Rawlings, R. and Robinson, P. M. *J. Iron Steel Inst.* 197 (1961) 211
[72] Nancarrow, D. J. F. and Rawlings, R. *Metal Treat. Drop Forg.* 29 (1962) 175
[73] Corney, N. S. and Turkdogan, E. T. *J. Iron Steel Inst.* 180 (1955) 344
[74] Schmitz-Dumont, O. and Steinberg, K. *Naturwissenschaften* 44 (1954) 117
[75] (*a*) Collongues, E., Gilles, J. C. and Lejus, A. M. *Bull. Soc. chim. Fr.* 11/12 (1962) 2113; (*b*) Gilles, J. C. *ibid.* 2118
[76] Epelbaum, V. A. and Breger, A. K. R. *Zh. fiz. Khim.* 20 (1946) 458
[77] Neumann, B., Kröger, C. and Kunz, H. *Z. anorg. allg. Chem.* 218 (1934) 379
[78] Kudielka, H., Nowotny, H. and Findeisen, G. *Mh. Chem.* 88 (1957) 1048
[79] Aronsson, B. *Acta chem. scand.* (*a*) 14 (1960) 1414; (*b*) 12 (1958) 3; (*c*) with Lundgren, G. 13 (1959) 433; (*d*) with Engström, I. 14 (1960) 1403
[80] Konev, V. N., Nesterov, A. F. and Glazkova, P. *Fizika metall.* 16 (1963) 86
[81] Rundqvist, S. and Jellinek, F. *Acta chem. scand.* 13 (1959) 425
[82] Gokcen, H. A. and Chipman, J. *Trans. metall. Soc. A.I.M.E.* 194 (1952) 177
[83] Darken, L. S. *J. Am. chem. Soc.* 70 (1948) 2046
[84] (*a*) Muan, A., with Osborn, E. F. *J. Am. Ceram. Soc.* 39 (1956) 123; (*b*) *J. Metals, N.Y.* 7 (1955) 203
[85] Liang, H., Bever, M. B. and Floe, C. F. *Trans. Am. Inst. Min. metall. Engrs* 167 (1946) 394
[86] Hepworth, M. T. and Schuhmann, J. *Trans. metall. Soc. A.I.M.E.* 224 (1962) 928
[87] Kornilov, I. I. and Yakimova, A. M. *Russ. Metall. Fuels* 3 (1962) 69
[88] Edwards, R. K. and Levesque, P. *J. Am. chem. Soc.* 77 (1954) 1312
[89] Martin, L. H. and Rees, A. L. G. *Trans. Faraday Soc.* 50 (1954) 343
[90] Schenck, H. *Introduction to the Physical Chemistry of Steel Making*, London (B.I.S.R.A.) 1945
[91] Wiberg, M. *Jernkont. Annlr* 125 (1940) 179
[92] Evans, U. R. *The Corrosion and Oxidation of Metals*, London (Arnold) 1960

APPENDIX

'INTERSTITIAL MATTER': SOME WIDER ASPECTS

The concept of 'interstitial units' appears, not only in metallurgy but much more universally in physics and other sciences, such as mineralogy, inorganic and organic chemistry and biology, and can thus be considered from a wider point of view. These structures, cutting as they do across different disciplines, fit into a pattern in nature of which interstitial alloys are only one manifestation.

An effort has been made to analyse 'interstitial matter' in this sense more widely in Table (*A*), both in respect of scale and kind.

In mineralogy, for instance, the silicon–oxygen tetrahedral framework of many mineral structures provides large interstices that can, but need not, be occupied by ions of various types such as interstitial oxygen, as distinct from the O^{--} of the framework or, amongst cations, numerous metals or metalloids. Glasses provide an extreme case, where these ions or ionic groups enter near-randomly and mingle with vacancies, drastically disarranging the central structure. Such groups can also enter in various degrees of order, as in some quartz crystals and many crystalline silicates, the feldspars here providing a notable series. In silicate glasses, the addition of interstitial Na^+, K^+, Pb^+, Ba^{++} among other ions loosens the SiO_2 framework, at the same time entailing some adjustment in number of oxygens to maintain the electronic balance; the same applies to Si^{4+}, replaceable by other ions like Al^{3+}, Ge^{4+} or P^{5+}.

The comparison with interstitial alloys is valid and instructive. The tendency towards crystallinity in the metallic state is much greater because of the stronger and very different bonds involved; but even the 'glassy' state of minerals and inorganics has its metallurgical counterpart, though less extreme in disarray. Examples are the highly distorted alloy lattices arising through interstitial locking during transformation and precipitation processes, as in the martensitic change; through strain by external work or conditions of finest dispersion or of poor crystallinity, as in surface layers (what used to be known as the Beilby layer), in diffusion zones and heterogeneous interfaces and, locally, in regions surrounding dislocations, grain boundaries and similar lattice imperfections acting as sinks for interstitials and vacancies. There is much in common between these local atmospheres in metals and the vitreous state promoted elsewhere by ionic or covalent bonding throughout the mass.

In zeolitic minerals, we have again very open framework structures of aluminosilicates, with holes and channels on a somewhat grosser than atomic scale from which 'interstitial' water can be readily absorbed or expelled or be replaced by other components like carbon dioxide, ammonia, alcohol or mercury; these structures also possess readily exchangeable large cation bases: calcium for example may be replaced by sodium or by artificial additives. The collapse of a zeolitic structure nucleates crystallites of new and more densely packed lattice, and at the same time defects and impurity clusters will aggregate in the remaining regions. This has its obvious counterpart in alloy reactions.

Table A. Interstitial entities in Nature

Scale	*Enclosed entity*	*Examples*
$\lesssim$ 1 Å	electrons and lattice vacancies	defect lattices in metals, salts, minerals or organic substances; *P*- and *N*-type vacancies in semiconductors
~ 1–3 Å	atoms	interstitial metal solid solutions and compounds; ionic interstitials in silicates, minerals and ceramics
~ 5–50 Å	molecules	nucleic acid units and connected substructures as elements of living matter; clathrates, water of crystallization in inorganic crystals; zeolitic structures
~ 10–100 Å	atomic or molecular clusters	interlayer H_2O pockets in clays; interstitial aggregations at dislocations and grain boundaries, and in regions of lattice strain in metals; radiation-produced sub-gas nuclei; cell structure (biological); protein structures: haem groups, etc.
~ 50–1,000 Å (= 0·1 μ)	submicroscopic particles	living cells and their contents in polypeptide chains: stored interstitial oxygen or H_2O groups; nucleating phases and pre-precipitates in alloys and analogous states in salt, mineral or organic structures
~ 0·1–10 μ	fine particle constituents	small inclusions or precipitates in many solids; dispersion-hardening particles in metals
~ 1–100 μ (= 0·1 mm)	secondary constituents of normal micro-types	grosser types of inclusions, carbides, intermetallics in steels, etc.; similar encased minor phases in all solids
~ 0·01–1 mm	near-macro enclosures, textural	cored structures in alloys; cemented carbides; some segregates in alloys; enclaves and occlusions in many types of building material, e.g. cements where reinforcing; mineral deposits; in biological matter
> ~1mm	macro-heterogeneous structures, artificial or natural	hardened zones in alloys; reinforcing cores in many solids; seams in mineral deposits; one biological tissue enclosed in another

Notes

(*a*) Each case can occur in 1-, 2- or 3-dimensional (fibre, layer or volume) form, sometimes bridging the above classes; also in oriented or random directions.

(*b*) Each case can occur in random or superlattice type distribution, with intermediate degrees of order.

(*c*) The vacancies referred to evolve of course down this Table and could be repeated in different terms, to end in voids, cracks, pores or, indeed, 'space' adjoining open surfaces.

Barrer's work on zeolites[1] shows how large atoms or molecular complexes like argon or methane are able to move through the cavities. The zeolites have proved exceptionally important test media for studying the migration of such ions or of neutral atoms and molecules of different sizes, their mutual interference and the limits over which exchange is possible. A close relation to the zeolites, the 'felspathoids', contain, instead of water, salt ion pairs. The transition is continuous, however, in that water and salt ions can replace each other in some of these substances. The zeolites have also become important as molecular 'sieves' (so-called Linde sieves), to separate ions which can or cannot travel through, be it for size or electronic reasons, or because of blockage by obstructing atoms. Analogues amongst interstitial alloys (e.g. the palladium/hydrogen relationship) will be recalled in this connection.

This 'sieve' aspect is quite fundamental in the present broad theme of 'interstitial matter', and we may see in this a new facet of, or even a new criterion for, 'interstitiality': namely the ease or difficulty of migration through the matrix lattice, or '*sievability*'. In its wider sense, porous filters in all their grades from the molecular upwards are continuations of the same phenomenon at the more macroscopic end of the scale.

The interstitial diffusion coefficients can clearly vary between widest extremes. The zeolites lie at the high-diffusion end amongst the silicates: so much so that, as Barrer pointed out, the diffusivity e.g. of Na^+ ions in the solid is not much lower than that of the simple Na^+ ion in aqueous solution. Again, this kind of transition from a very open lattice to the free medium has its clear counterpart amongst the interstitial alloys, e.g. at surfaces and boundary layers towards the free gas or liquid, perhaps most pronounced in the hydrides.

A novel kind of 'interstitial' substance, not dissimilar in kind to the zeolites, is provided by the *clathrate* structures amongst organic substances. They are very open cage structures, in which stranger molecules can be entrapped to fill the holes, though not necessarily all. Such encaged interstitial molecules (often called 'guest' molecules) can be H_2S, HCl, SO_2, CH_3CN or the larger inert gas atoms, A, Kr or Xe. The framework (host) substance consists of quinol or related molecules, the principal condition being that the molecular framework should in itself be especially tightly held together. The clathrates share this with the transition metals which in a comparable manner encage their C, N, O, and other interstitial atoms, in line with the smaller interspaces available. The difference is, of course, in the type of bonding: in the clathrates, the interstitials are only feebly linked through hydrogen bonds or exist in virtually neutral form, as in the inert gas case; therefore, the interstitial mobility or permanency of their entrapment will accordingly differ from that in metals, where the bond through electron exchange becomes so much stronger and more directional. Nevertheless, there are many factors in common and, with due allowances, one may consider the clathrates as organic counterparts to interstitial alloys, even to the extent of non-stoichiometry and wide solution ranges. The zeolites are clearly a counterpart amongst ionic crystals, and again we see a cross-linkage between the disciplines—here organic and inorganic chemistry and metallurgy.

It may be noted that the clathrate host lattices *without* their guest molecules are usually thermodynamically unstable. They cannot exist 'empty' and require the guest units for stabilization. In the same way, many interstitial metal compounds, e.g. the HCP or FCC carbides, *require* the interstitial atoms (carbon, etc.) for metal lattices to be stabilized; this requirement holds in the early transition-metal Groups but becomes less stringent in the later Groups where the FCC and HCP metal lattices can increasingly be adopted without the interstitials (cf. Chapter 2).

In protein structures, new and exciting discoveries seem to provide perfect examples of 'interstitials' of a varied kind. The multimolecular and complex polypeptide chains in mioglobin and haemoglobin enclose their interstitial haem groups to which are attached the 'stored' and, therefore, expendable oxygen or water groups. This storage of interstitial oxygen is one of the central functions of these structures in the living organism. We may then look upon these 'storage structures' as variants of the more general principle that, despite the vastly greater complexity and difference, their essential role is still that of impermanent interstitial holding, as in so many interstitial alloy systems and minerals.

The mioglobin and haemoglobin work provides another aspect of interstitial insertion which might be called the 'marker' aspect: that of deliberate placing of heavy metal atoms into definite holes in the protein structure, where they least disturb the main structure but give experimentally a vital help in its elucidation by x-ray analysis.

We may recall the great recent successes of biochemists in solving the enormously complex structures of the nucleic acids, RNA and DNA*, as basic materials of life, and at the same time detecting the genetic key and reproductive code these long-chain molecules bear within them. This has been much described in the scientific (and more popular) press, and here is not the place to recount this mechanism. However, two features are relevant: (1) that these are obviously 'interstitial' structures *within* themselves which can part and recombine according to a precisely determined plan; and (2) that the nucleic acid molecules are interstitial as an entity within the living cell. So are the protein molecules: the DNA and RNA act as permanent information carriers, the protein as transient instruction givers to the former. (By the way, we may also note that the atom *species* constituting each of these molecules are C, N, O, P and H, i.e. the atoms familiar to us in interstitial alloys. So this is what happens in nature when, under suitable conditions, they are left to themselves!)

In the light of bonding forces, the concept of interstitials as occurring in metallurgy, in mineralogy, inorganic and organic chemistry and in biology, would represent a steady progression (in this order, but with overlaps) from the strong metallic bond through the ionic, covalent and van der Waals bonds to, finally, the virtually non-existing bond which amounts to an 'occlusion'.

* RNA = ribonucleic acid, DNA = deoxyribonucleic acid; DNA is the principal constituent of chromosomes, and the sequence in which the submolecular bases within the long helical molecule are strung together is now known to carry the individual reproductive code. RNA, in contrast, is dispersed mostly through the cytoplasm of the cell; the sequence of the base group along the molecule in RNA is itself determined by that in the DNA which is in turn directed by the sequence of amino acids in the protein molecules.

The concept of dislocations, though at first mainly associated with metals, is another aspect that cuts across the disciplines, and so is its attribute of attracting interstitials and vacancies in their wider sense. The equivalent occurs in ionic crystals, minerals, organic and biological substances, and—going up in scale from the atomic—reappears in its modified forms in materials of each of these sciences more macroscopically: in the extreme, from 'faults' in geological formations to certain faulting phenomena in living tissues, each attracting interstitial matter ('Cottrell atmospheres') of its own kind.

Looking at the concept of interstitiality from the technological approach of recent composite materials of high strength or other special properties, we might well regard much of this new class of materials as yet another aspect of it. Fibre or laminate composites represent a macroscopic and deliberate, man-made imitation of what happens naturally on a fine-structural scale, e.g. in heat-resisting superalloys containing precipitates of second phases nucleated at dislocations, along parent lattice planes and at grain boundaries. In principle, the idea of reinforcement by such fibre or laminate is quite similar. Though generally the dispersion is on a grosser scale, this is not even necessarily so, and in some of the newer materials the size ranges of the synthetic and the natural metallurgical (dispersed) second phases are blurred over a wide band from about $\approx$ 50 to 10,000 Å or more. One might indeed view the whole field of composite materials as a chain of grand design in which, at the small-scale end, the bond of interstitial atoms in the metal lattice forms a logical and integral part. Many such new materials come readily to mind, such as plastics reinforced by fibreglass or long-chain polymers, alloys reinforced by ceramics or a second metal such as aluminium containing a net of finest steel wires or glass fibres, 'S.A.P.'-type materials or refractory metals internally strengthened by second-metal bands. The recent active researches in single-crystal 'whisker' growth largely gain their practical impetus from such applications. These subdivided phases are only in part used to form inherently strong interstitial bonds to the surrounding matrix. A second function common to these types of interstitial units, whether fibrous, lamellar or globular, is that of reinforcement by providing impediments* to stop the propagation of *cracks* within the matrix material. Matching of properties, e.g. in lattice dimensions and expansion coefficients, between these disparate materials forms of course an essential part of the engineering design features of composites, whether synthetic or internally evolved. The extreme slenderness of the built-in fibres is in itself indeed an important attribute from the viewpoint of strength per unit area in both cases. In alloys, it might well be said that the intricate network of martensite needles, which plays an important part in the new development of high-strength steels, is a direct synologue in purpose and function to the deliberate filamental composites.

In a translated sense, Nature has already blazed the trail in composite materials, and the strength of biological structures such as bones, muscles,

* Another form of such impediment to crack propagation, rather than a particle, can be *another crack* at right angles, or a void upon which the initial crack impinges—in interstitial terms, a 'vacancy'.

wood and other plant structure, is in principle well comparable to that of modern high-strength materials. Again, they are composites of heterogeneous (tough but weaker) + (strong though possibly brittle) units which together are many times stronger than each part. The existence of anisotropy is also directly comparable amongst the disciplines.

REFERENCE

[1] Barrer, R. M. *Proc. Br. ceram. Soc.* 1 (1964) 145

INDEX

(M=metal)

MIX
Papier aus verantwortungsvollen Quellen
Paper from responsible sources
FSC® C105338

If you have any concerns about our products,
you can contact us on
ProductSafety@springernature.com

In case Publisher is established outside the EU,
the EU authorized representative is:
Springer Nature Customer Service Center GmbH
Europaplatz 3, 69115 Heidelberg, Germany

Printed by Libri Plureos GmbH
in Hamburg, Germany